Hervé Chamley

Clay
Sedimentology

With 243 Figures and 65 Tables

Springer-Verlag Berlin Heidelberg New York
London Paris Tokyo HongKong

Prof. Dr. Hervé Chamley

University of Lille I
F-59655 Villeneuve d'Ascq

University of Paris VI
F-75252 Paris cedex 05

ISBN 3-540-50889-9   Springer-Verlag Berlin Heidelberg New York
ISBN 3-387-50889-9   Springer-Verlag New York Berlin Heidelberg

Library of Congress Cataloging-in-Publication Data
Chamley, Hervé. Clay sedimentology/Hervé Chamley. p. cm. Bibliography: p. Includes index.
ISBN 0-387-50889-9   1. Clay minerals. I. Title. QE389.625.C47   1989   549′.6–dc20

© Springer-Verlag Berlin Heidelberg 1989
Printed in Germany

The use of general descriptive names, registered names, trademarks, etc. in this publication does not imply, even in the absence of a specific statement, that such names are exempt from the relevant protective laws and regulations and therefore free for general use.

Product Liability: The publisher can give no guarantee for information about drug dosage and application thereof contained in this book. In every individual case the respective user must check its accuracy by consulting other pharmaceutical literature.

Typesetting, Printing and Binding: Brühlsche Universitätsdruckerei, Giessen
2132/3145-543210   Printed on acid-free paper

To Marie, my wife,
and to Camille, Noëlle and Vincent, my children,
who shared much more than expected the efforts
linked to the redaction of this book

Preface

1. Purpose

Clay is the dust of the Earth. Clay sedimentology deals with the composition, properties and significance of the argillaceous fraction of recent and ancient sediments. Clay minerals represent the most ubiquitous components of all sediments, from desert or beach sands and sandstones to deep-sea oozes and muds. Clays constitute a dominant percentage of some of the most common sedimentary rocks such as mudstones, shales, and marls, from Proterozoic to Cenozoic age.

Surprisingly, textbooks devoted to clay minerals in natural environments are rare. The most known work is "Géologie des Argiles" translated as "Geology of Clays" by G. Millot (1964, 1970). Millot's book constitutes an invaluable reference that is still largely used and quoted. One may envisage various explanations to the scarcity of geologic books about clays: the relatively small number of clay mineral families, that can hardly be thought of as reflecting the very diverse natural environments, the difficulty for geologists to easily memorize the complex clay structures and to handle devices which usually belong to physicists and chemists, or the fact that clay just looks like a dirty, earthy, and unattractive stuff. Whatever the reason, here is a book that tries to present some information on "Clay Sedimentology" to the patient reader.

One of the main points of interest in writing a textbook is to summarize data that are published in various research papers on a given subject, and therefore to tend toward a relatively objective and dispassionate document. As far as clays in sediments are concerned, it is in addition particularly exciting to put together information arising from European, North-American, and other laboratories, whose respective ways of reasoning sometimes differ.

2. Bibliographic Data

The literature on sedimentary clays is immense. The development during the few past decades of new geochemical and microscopic devices, applied problems on the argillaceous fractions from soils and rocks, and mineralogical sections in numerous geological departments, has determined a staggering increase of publications. For example, Volume 50 of "Bibliography and Index of Geology", compiled for the year 1986 by the American Geological Institute Compilation, reports a total of 1110 references under the topic "Clay Mineralogy", covering areal and experimental studies, mineral data and theoretical studies. Under the same title 117 references are quoted in the issue of June 1987. It is, of course, very difficult and tiresome to read and synthetize such a huge bibliography, which itself is necessarily incomplete. In addition, to summarize such an extensive literature would result in an indigestible catalog. For this reason I have tried to present selected examples of the available data, by combining classical and most recent studies. G. Millot's book (1964, 1970) is considered as a basic document, to which the reader is often referred.

One should add that the different topics covering the sedimentology of clay sometimes correlate to a very uneven number of references. Bibliographic data on well-debated subjects like clays in soils and in estuaries, or on applied problems such as the clay evolution in sandstones are countless; by contrast, accurate references on less-common subjects such as the early to middle diagenetic evolution, or the transition from detrital supply to authigenesis are fairly rare. "Clay Sedimentology" attempts to give a basic idea on most topics dealing with sedimentary clays, and to present the uncertainties due to the frontiers of knowledge.

3. Contents

"Clay Sedimentology" is divided into six parts, each of them containing two to five chapters. The successive topics considered are the following:

- Part I. Clay minerals and weathering, covering a short recall on the main natural minerals, and a summary of clay formation during land- alteration processes.
- Part II. Clay sedimentation on land, from deserts, glaciers, and rivers, to lakes of various chemical types.
- Part III. Origin and behavior of clay minerals and associated minerals in transitional environments such as estuaries and deltas, influence of sorting in sea water and of aeolian supply, and importance of the recent terrigenous input within the sea.
- Part IV. Genesis of clay in the marine environment, before appreciable burial of sediment has taken place: alkaline-evaporative deposits, ferriferous clay granules, organic sediments, metalliferous deep-sea clays and volcano-hydrothermal deposits.
- Part V. Post-sedimentary processes intervening during early and late diagenesis, or controlled by tectonic, lithologic or hydrothermal factors.
- Part VI. Use of clay stratigraphic data for the reconstruction of past climate, marine circulation, tectonics, and other paleogeographical characters, supplemented by an overview on the transition from the paleoenvironmental to diagenetic expression of clay associations considered from a geodynamical viewpoint.

It is sometimes difficult to make a choice between different methodological or scientific approaches. For instance, I have chosen to generally follow a classical distinction between single clay minerals and random mixed-layers, rather than to apply a statistical classification based on continuously transitional terms between smectite and illite (calculation of the mixed- layering rate); in fact, the former classification seems often more convenient for designating clay associations devoid of burial modifications, is more widely used and quoted, and often allows better characterization and understanding of complex natural assemblages. The term diagenesis is employed in its most frequent European meaning, and designates processes occurring after the sediment is appreciably buried and protected from exchanges with the open-sea water (Part Four). As some very different environments, such as terrigenous deposits and deeply buried sediments, may contain the same clay minerals (e.g., illite and chlorite), and symmetrically as a given environment may include very distinct clay minerals (e.g., metalliferous clays with either terrigenous, hydrogenous, volcanogenic, or hydrothermal species), it is sometimes necessary to choose a type of presentation, and to make some cross-references or repetitions. In a didactic aim, I have deliberately chosen to put in distinct chapters some data on allochthonous and autochthonous processes, although they may jointly occur in the same sediments; the last chapter provides some indications on combined processes, and the conclusion of each chapter tries to integrate a summary of results as well as general, somewhat critical considerations. Finally, some topics are only briefly considered, since they belong to disciplines adjacent to clay sedimentology, or are extensively covered by high-quality books; this is the case with the structure and classification of clay minerals, the weathering processes and associated clay formation, and the diagenesis due to depth of burial. I apologize for the disagreements induced by these drawbacks in the presentation and contents. "Clay Sedimentology" is a textbook written by a geologist for geologists or for people interested in clay, and could severely disappoint some mineralogists, crystallographers, and geochemists.

4. A Few Lessons

Some general considerations arise from the present state of research on the sedimentology of clays. First, the understanding of the significance of clay-mineral successions greatly benefits from data on the geological and sedimentological context, on the detailed chemical and morphological nature of individual particles, and on logging and modeling experiments.

Much progress is expected from the systematic analysis of the clay fraction from the diverse lithologic types that constitute a sedimentary pile. To concentrate the investigations only on the clay-rich facies greatly diminishes the interest of clay stratigraphic investigations.

A great deal of new information results from investigations on the materials sampled during the Deep Sea Drilling Project and the Ocean Drilling Program. The clay sedimentology of submarine series is frequently better known than that of exposed or land-borehole sections. A continuous effort of research has to be pursued on exposed series from Proterozoic to Cenozoic ages, as well as on transects extending from continental to deep- sea sedimentary facies.

Some common mistakes or improper conclusions should be fought patiently and replaced by up-to-date information, knowing humbly that today's evidence can change tomorrow into crude errors due to technical or scientific improvements. For instance, smectite is not systematically of volcanic origin in marine sediments and may have various sources, including geologic and pedogenic continental sources. Palygorskite and sepiolite do not systematically form in sediments where they are identified, and can be reworked by wind or water from distant soils or chemical sediments. Kaolinite does not exclusively derive from laterite-type soils, and may issue from various other weathering profiles or ancient sediments. The formation of iron-rich granules of the glaucony type does not necessitate a hot climate. The deep-sea red clays do not automatically comprise only hydrogenous or volcano-hydrothermal minerals and may include noticeable land-derived components. The diagenesis linked to depth of burial does not systematically show intermediate mixed-layered terms, and appears to depend noticeably on high temperatures of fairly short duration rather than on moderate pressure over a long time. Numerous transitional situations occur between exclusively detrital series and solely authigenic series, as far as the sedimentary clay fraction is concerned. The time is over to passionately discuss and fight about the either allochthonous or autochthonous character of clay fractions in sediments. It is time to better understand the boundary situations and the kinetic control by pedologic, sedimentary, and diagenetic environments.

Some genetic processes responsible for the formation of clay minerals in past sediments obviously do not exist any longer today; this is the case of the sedimentary precipitation of most fibrous clays and probably of some smectite-rich deposits. By contrast, it is likely that some present-day kinds of clay genesis were very restricted or even lacking during some ancient periods; this is probably the case with illite- and chlorite-rich sediments that abundantly form climatically in peri-glacial regions, and were hardly deposited during warm Mesozoic times. One should therefore be cautious in applying some present-day data to ancient sediments.

Finally clay-stratigraphy data appear to represent a useful tool in understanding the control of successive internal and surficial geodynamical factors on the large-scale history of a sedimentary basin subject to plate-tectonic constraints.

5. Acknowledgements

I have benefited by fruitful scientific discussions with numerous colleagues during my search on clay sedimentology the last 20 years, as well as during the writing of this book. I would especially like to mention, among others, Pierre E. Biscaye, Jean Blanc and Laure Blanc-Vernet, Chantal Bonnot-Courtois, Anne Bouquillon, Jean-Paul Cadet, Norbert Clauer, Pierre Debrabant, Jean-François

Deconinck, Alain Desprairies, Liselotte Diester-Haass, Gilbert Dunoyer de Segonzac, Claude Froget, Michel Hoffert, Thierry Holtzapffel, Glenn A. Jones, James P. Kennett, Claude Latouche, Margaret Leinen, Jacques Lucas, Frédéric Mélières, Georges Millot, André Monaco, Gilles S. Odin, Hélène Paquet, Léo Pastouret, Christian Robert, Michel Steinberg, François Thiébault, Norbert Trauth, Colette Vergnaud-Grazzini. Efficient and high-level technical support was provided at Lille University for the achievement of the book, notably by Martyne Bocquet, Jean Carpentier, Jean-Claude Deremaux, Françoise Dujardin and Philippe Récourt. "Clay Sedimentology" is dedicated to Professor Georges Millot, who was my first teacher in geology, and an exceptional specialist in the geology of clays.

Clay families are approximately as numerous as the notes of the musical scale. The mixed-layers and other transitional minerals may be compared to accidentals in music. There is a huge diversity of clay mineral assemblages in natural environments. Knowing the great variety of musical works of art that man has produced in a few thousand years, it is easy to understand the extraordinary variety and the numerous sequences of argillaceous environments that nature has built in the sedimentary record over a few billion years.

Villeneuve d'Ascq, April 1989   Hervé Chamley

Table of Contents

_____ PART I   Clay Minerals and Weathering

_____ **Chapter 1   Clay Minerals** _____

1.1   Introduction   3
1.2   Clay Structure   3
1.3   Main Clay Minerals   6
        1.3.1   Main Divisions   6
        1.3.2   Kaolin-Serpentine Group   7
        1.3.3   Pyrophyllite-Talc Group   8
        1.3.4   Mica Group   8
        1.3.5   Brittle Mica Group   12
        1.3.6   Vermiculite Group   13
        1.3.7   Smectite Group   13
        1.3.8   Chlorite Group   14
        1.3.9   Palygorskite-Sepiolite Group   15
        1.3.10 Mixed Layers   16
1.4   Main Associated Non-Clay Species   18
1.5   Conclusion   19

_____ **Chapter 2   Clay Formation through Weathering** _____

2.1   Main Weathering Processes   21
        2.1.1   Introduction   21
        2.1.2   Hydrolysis   22
2.2   Main Soils and Clay Content   26
        2.2.1   Zonal Soils   26
                Parallel Change of Temperature and Humidity . . . 26
                Antagonistic Change of Temperature and Humidity . . . 32
                General Distribution of Zonal Soil Clays 37
        2.2.2   Azonal Soils   37
                Halomorphic Soils   37
                Hydromorphic Soils   38
                Volcanic Environments   38
                Epeirogenic Environments 41
        2.2.3   Paleosols and Clay Content   41
                Quaternary   42
                Tertiary   43

Mesozoic 46
Paleozoic, Precambrian 48
2.3   Conclusion 49

_____ PART II   Clay Sedimentation on Land

_____ **Chapter 3   Deserts, Glaciers, Rivers** _____
3.1   Deserts 53
       3.1.1 Desert Dusts 53
       3.1.2 Aeolian Sediments 56
3.2   Glaciers 58
       3.2.1 Origin of Glacial Clays 58
       3.2.2 Paleoenvironmental Applications 59
3.3   Rivers 61
       3.3.1 Origin of River Clays 61
       3.3.2 River suspended minerals 64
       3.3.3 Environmental Applications 66
3.4   Conclusion 72

_____ **Chapter 4   Lacustrine Clay Sedimentation** _____
4.1   Freshwater Lakes 75
       4.1.1 Detrital Supply 75
       4.1.2 Authigenic Clays 79
              Biosiliceous Environment 79
              Iron-Clay Granules 79
              Volcanic Environment 79
4.2   Saline Lakes 80
       4.2.1 General Data 80
              Introduction 80
              Saline Minerals 81
              Volcanic Environment 83
              Detrital Clays in Saline Nonvolcanic Lakes . . 84
       4.2.2 Authigenic Clay in Recent Saline Lakes 86
              Ferriferous and Magnesian Environment 86
              Volcanic Environment 89
              Biosiliceous Environment 92
       4.2.3 Fibrous Clays in Lacustrine Environment . . . 92
4.3   Conclusion 93

_____ PART III   From Land to Sea

_____ **Chapter 5   Estuaries and Deltas** _____
5.1   First Studies on Estuarine Clays 97
5.2   Estuarine Clays and Continental Sources 101
5.3   Mechanisms of Clay Distribution 103

          5.3.1 Differential Settling and Flocculation 103
          5.3.2 Mixing of Water Masses 107
          5.3.3 Marine Transgression 111
          5.3.4 Cation Exchange and Chemical Modification 114
5.4   Conclusion 115

Chapter 6   Clay Sorting and Settling in the Ocean
6.1   Introduction 117
6.2   Differential Settling 118
          6.2.1 Recent Sediments 118
          6.2.2 Ancient Sediments 122
          6.2.3 Mechanisms 124
6.3   Particle Aggregation and Advection 127
6.4   Conclusion 130

Chapter 7   Aeolian Input
7.1   Wind Transport over the Ocean 133
          7.1.1 Evidence of Dust Supply 133
          7.1.2 Characterization of Aeolian Dust 134
          7.1.3 Importance of Aeolian Input 136
7.2   Identification of Aeolian Minerals in the Oceans 139
          7.2.1 Mineral Composition of Aeolian Dust 139
          7.2.2 Comparison of Aeolian Dust and Surface Sediments . . . 142
7.3   Distribution of Aeolian Sediments 146
7.4   Aeolian Influence in Past Marine Sedimentation 152
          7.4.1 Quaternary 152
          7.4.2 Pre-Quaternary 155
7.5   Conclusion 159

Chapter 8   Terrigenous Supply in the Ocean
8.1   Climatic Control 163
          8.1.1 Distribution of Kaolinite and Chlorite 163
          8.1.2 Other Clay Minerals 167
8.2   Petrographic Control 171
          8.2.1 Polar Regions 171
          8.2.2 Arid Regions 173
          8.2.3 Recognition of Volcanic Sources 174
          8.2.4 Mixed Geologic and Pedologic Sources 176
8.3   Hydrodynamic Control 180
          8.3.1 Offshore Currents 181
          8.3.2 Resedimentation Processes 183
8.4   Combination of Terrigenous Supply and Transportation Agents . 185
8.5   Conclusion 190

_____ PART IV   Clay Genesis in the Sea

_____ **Chapter 9   Alkaline, Evaporative Environment** _____

9.1   General Features 195
9.2   Evaporative Clay Sedimentation, Paleogene in France 197
        9.2.1   Basin of Paris 197
        9.2.2   Southeastern France 200
        9.2.3   Geochemical Implications 204
9.3   Other Marine Environments 205
        9.3.1   Early Miocene in Southeastern United States 205
        9.3.2   Uppermost Miocene, Mediterranean Sea 208
        9.3.3   Illitization by Wetting and Drying 209
9.4   Conclusion 211

_____ **Chapter 10   Ferriferous Clay Granules and Facies** _____

10.1   General Features 213
10.2   Nature and Distribution of Recent Green Clay Granules 216
        10.2.1   Habits 216
        10.2.2   Zonal and Bathymetric Distribution 218
10.3   Genesis of Green Clay Granules 222
10.4   Greensands, Ironstones 227
10.5   Celadonite Facies 230
10.6   Conclusion 232

_____ **Chapter 11   Organic Environment** _____

11.1   Influence of Organic Activity on Clay 235
11.2   Late Cenozoic Sapropels, Mediterranean Sea 238
        11.2.1   Sapropel Characters and Environment 238
        11.2.2   Clay Evolution 242
11.3   Cretaceous Black Shales, Atlantic Domain 251
        11.3.1   Black Shale Environment 251
        11.3.2   Clay Sedimentation 255
11.4   Conclusion 257

_____ **Chapter 12   Metalliferous Clay in Deep Sea** _____

12.1   Deep-sea Clay Environment 259
        12.1.1   Introduction 259
        12.1.2   Metalliferous Nodules 261
        12.1.3   Allochthonous Versus Autochthonous Origin 262
12.2   Smectite Formation 265
        12.2.1   Preliminary Data 265
        12.2.2   Domes Area, Northeast Pacific 266
        12.2.3   Central South-Equatorial Pacific 267
        12.2.4   Nazca Plate 270
                 Geochemistry of Surface Sediments 270
                 Clay Mineralogy of Surface Sediments 272
                 Mineralogy and Geochemistry, Northeast Nazca basin . . 273

              Mineralogy and Geochemistry, Bauer Deep 274
12.3   Fibrous Clay in Deep-Sea Sediments 276
12.4   Transitional Environments 279
       12.4.1   Pacific Ocean 279
       12.4.2   Indian Ocean 282
12.5   Ancient Deep-Sea Metalliferous Clays 284
12.6   Conclusion 287

_____ Chapter 13   Hydrothermal Environment _____
13.1   Submarine Alteration of Volcanic Rocks 291
       13.1.1   Low-Temperature Alteration 291
                Basaltic Glass 292
                Crystalline Basalts 293
                Silicic Glass 296
       13.1.2   High-Temperature Alteration 297
13.2   Basalt Intrusion in Sediment 299
13.3   Hydrothermal Deposits 302
       13.3.1   Main Precipitates 302
       13.3.2   Pacific Ocean 304
                East Pacific Ridge 304
                Galapagos Rift Mounds 305
                Southwest Pacific 308
       13.3.3   Mid-Atlantic Ridge 310
       13.3.4   Red Sea 312
       13.3.5   Old Hydrothermal Deposits 316
13.4   Palygorskite and Sepiolite Deposits 317
13.5   Sedimentary Impact of Volcano-Hydrothermalism 320
       13.5.1   Lateral Extension 320
       13.5.2   Vertical Extension 323
       13.5.3   Authigenic Formation and reworking 325
13.6   Conclusion 325

_____ PART V   Clay Diagenesis

_____ Chapter 14   Early Processes _____
14.1   Introduction 333
14.2   Smectite and Early Diagenesis 336
       14.2.1   General Features 336
       14.2.2   Lathed Smectites, Clay Mineral Overgrowths 343
                Atlantic Ocean 345
                Indian Ocean 349
14.3   Specific Environment 350
       Early Effects of Burial 350
       Chlorite, Palygorskite Formation 351
       Diagenesis in Slightly Buried Volcaniclastic Sediments 352
14.4   Conclusion 357

Chapter 15   Depth of Burial

15.1   Basic Evolution 359
    15.1.1   Introduction 359
    15.1.2   An Example: Gulf of Mexico Coast Sediments 361
    15.1.3   Other Cases 364
15.2   Mechanisms of Evolution 370
    15.2.1   Open Versus Closed Systems 370
    15.2.2   Transformation Versus Dissolution – Precipitation . . . 372
    Illitization 372
    Chloritization, Kaolinization 379
15.3   Modification of Physical Properties 380
    15.3.1   Compaction 380
    15.3.2   Application to Oil Exploration 382
15.4   Conclusion 386

Chapter 16   Tectonic, Lithologic and Hydrothermal Constraints

16.1   Diagenesis through Tectonics 391
    16.1.1   Lateral Effects 391
    16.1.2   Tectonic Overburden 394
16.2   Chemically-Restricted Environment 395
    Deeply Buried Smectite 395
    Limestone – Marl Alternations 397
16.3   Highly-Porous Rocks 398
    16.3.1   Introduction 398
    16.3.2   Diagenetic Clay Minerals in Sandstones 399
    16.3.3   Diagenetic Processes and History 402
    16.3.4   Clay Diagenesis and Reservoir Properties 408
16.4   Volcanic Environment 409
    16.4.1   Igneous Rock Accumulations 409
    16.4.2   Bentonites, Tonsteins 411
16.5   Hydrothermal Environment 414
    Introduction 414
    Vein-Filling and Wall-Rock Minerals 415
    Vein-Filling Clay Sequences 416
    Massive Alteration 418
16.6   Conclusion 420

PART VI   Clay Stratigraphy and Paleoenvironment

Chapter 17   Paleoclimate Expression

17.1   Bases and Conditions 425
    17.1.1   Fundamentals 425
    17.1.2   Conditions of Application 429
17.2   Late Quaternary 432
    17.2.1   Direct Climatic Expression 432
    17.2.2   Indirect Effects 436

17.3   Cenozoic 441
         17.3.1 Atlantic Ocean 441
         17.3.2 Other Oceans 445
         17.3.3 Mediterranean Sea 448
                 Pliocene, Pleistocene 448
                 Messinian 450
17.4   Pre-Cenozoic 452
         Middle-Late Mesozoic 452
         Limestone-Marl Alternations 453
         Pre-Jurassic 454
17.5   Conclusion 455

_____   **Chapter 18   Paleocirculation and Tectonics** _____

18.1   Identification of Sources 459
18.2   Paleocurrents and Resedimentation 462
         18.2.1 Introduction 462
         18.2.2 Mediterranean Sea 464
         18.2.3 Southern Ocean 467
18.3   Tectonic Control 470
         18.3.1 An Example: The Pliocene of South Sicily . . 470
         18.3.2 Applications 473
                 Mediterranean Range, Late Cenozoic 473
                 New-Zealand Region, Cenozoic 474
                 North Atlantic Domain, Late Mesozoic 476
                 Other Regions 477
         18.3.3 General Expression of Tectonic Activity by Clay . . . 478
18.4   Conclusion 483

_____   **Chapter 19   Paleoenvironmental Reconstruction** _____

19.1   Chronological Evolution of Hatteras and Cape Verde basins . . . 487
         19.1.1 Introduction 487
         19.1.2 Major Similarities 490
         19.1.3 Major Differences 493
19.2   Paleogeographic Evolution of Atlantic Regions 496
         19.2.1 North Atlantic Basins at Albian Time 496
         19.2.2 North Atlantic Ocean, Cretaceous Times 500
         19.2.3 South Atlantic Ocean 502
         19.2.4 Adjacent Regions 505
19.3   Specific Paleoenvironmental Successions 507
         19.3.1 Intraplate Volcanic Environment, Late Cretaceous of the
                 Mariana Basin 507
         19.3.2 Tectono-Eustatic Environment, Late Miocene of Sicily . . 513
         19.3.3 Cretaceous-Tertiary Transition 518
                 Large-Scale Studies 518
                 Cretaceous-Tertiary Boundary Layer 521
19.4   Conclusion 524

_____ **Chapter 20   Clay and Geodynamics** _____

20.1   Diverse Significance of Sedimentary Clay Assemblages 527
20.2   Cenozoic Clay Sedimentation in the North Pacific Ocean 530
20.3   Late Mesozoic Clay Sedimentation in the Eastern Atlantic and
        Western Tethyan Domains 533
        20.3.1 Comparison between Cape Verde and Senegal Basins . . 533
        20.3.2 Generalization to Atlantic and Tethyan Domains 543
20.4   Late Tertiary Clay Sedimentation in the Tyrrhenian Domain,
        Western Mediterranean Basin 545
        20.4.1 Western Tyrrhenian Sea 545
        20.4.2 Comparison with the Senegal Basin 550
        20.4.3 Regional Comparisons 553
                Clay Sedimentation on the Sardinian Margin 553
                Comparison with Sicily 555
                Significance of Chloritic Minerals in the Tyrrhenian Sea . 558
20.5   Conclusion 560

_____ **References** 563

_____ **Subject Index** 621

Part I

Clay Minerals
and Weathering

Chapter 1

Clay Minerals

1.1 Introduction

Clay minerals consist of *hydrous layer silicates* that constitute a large part of the family of phyllosilicates. Clay minerals are particularly abundant in clayey oozes, claystones, mud, mudstones, shales, and argillites, a group of fine-grained rocks called physil rocks by Weaver (1980). Clay minerals also occur in virtually all other types of soft and hard sedimentary rocks, including coarse silicoclastites and saline evaporites. This explains the increased interest on the part of sedimentologists in studying these minerals.

We begin with a short summary of the structure, composition, and classification of clay minerals. This will allow the inclusion of some basic data, and the definition of the technical terms used in the following chapters.

Detailed information on clay mineralogy and chemistry is provided notably by Grim (1968), Weaver and Pollard (1973), Dixon and Weed (1977), Van Olphen and Fripiat (1979), Brindley and Brown (1980), Caillère et al. (1982), Velde (1985), whose data are extensively used in this chapter.

1.2 Clay Structure

Clay crystals fundamentally consist of silicon, aluminum or magnesium, oxygen, and hydroxyl (OH), with various associate cations according to the species. These ions and OH groups are organized into two-dimensional structures of two types, called *sheets*. (1) *The tetrahedral sheets* have a general composition T_2O_5 (T = tetrahedral cation; mainly Si, with varying Al or Fe^{3+} content). Silicon is located at the center of the tetrahedra, oxygen anions form the four corners. The individual tetrahedra are connected with adjacent tetrahedra by sharing three corners (the three basal oxygens), constituting a hexagonal mesh arrangement (Fig. 1.1A). The fourth tetrahedral corner points in a direction normal to the sheet. Its oxygen (the apical oxygen) forms part of the *octahedral sheet*. (2).*The octahedral sheets* comprise medium-sized cations at their center (usually Al, Mg, Fe^{2+}, or Fe^{3+}), and oxygens at the eight corners. The individual octahedra are linked laterally with the neighboring octahedra (Fig. 1.1B), and vertically with the tetrahedra,by sharing oxygens. The smallest structural unit of the octahedral sheet contains three octahedra. If all three octahedra have octahedral cations at

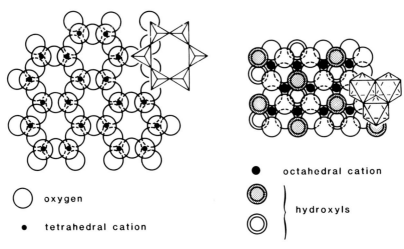

oxygen

tetrahedral cation

octahedral cation

hydroxyls

Fig. 1.1. Structure of hydrous layer silicates. (After Bailey 1980a). **A** Plan view of a hexagonal tetrahedral sheet. **B** Plan view of an octahedral sheet with inner hydroxyl groups of 2:1 layer shown as *vertical or horizontal hatched circles*

their center (bivalent ions like Mg^{2+}, Fe^{2+}), the sheet is called *trioctahedral*. If two octahedra only are occupied and one octahedron is vacant (trivalent ions like Al^{3+}, Fe^{3+}), the sheet is called *dioctahedral*.

The plane of junction between the tetrahedral and octohedral sheets comprises the apical oxygens shared by the tetrahedra and the octahedra, plus unshared hydroxyls (Fig. 1.2). The OH groups are located at the center of each tetrahedral sixfold ring (hexagonal arrangement), at the same level as the apical oxygens. The structure resulting from the assemblage of tetrahedral and octahedral sheets is called a *layer*. Two main types of layers are recognized. (1) The *1:1 layer* or T.O. layer consists of the assemblage of one tetrahedral sheet with one octahedral sheet. Such a layer, typical of the kaolinite group (Fig. 1.2A), comprises an unshared plane of anions in the octahedral sheet consisting of OH groups. (2) The *2:1 layer* or T.O.T. layer links two tetrahedral sheets in an external position with one octahedral sheet. The relative disposition of both tetrahedral sheets is inverted, so that all apical oxygens point toward the octahedral sheet and can be shared. Most clay species, such as for instance smectite (Fig. 1.2B), belong to the 2:1 layer type.

The space located between two successive 1:1 layers represents an *interlayer*. The interlayers are devoid of any chemical elements if the layers are electrostatically neutral (i.e., all structural cations are compensated by oxygens or hydroxyls). Many clay minerals present an excess of layer negative charge, which is neutralized by various interlayer materials: cations (mainly K, Na, Mg, Ca), hydrated cations, or hydroxide octahedral groups. The hydroxide interlayer groups often join laterally to form an additional octahedral sheet (usually referred to as a "brucitic sheet"), giving way to a 2:1 or T.O.T.O. assemblage. Such a structure characterizes the family of chlorites.

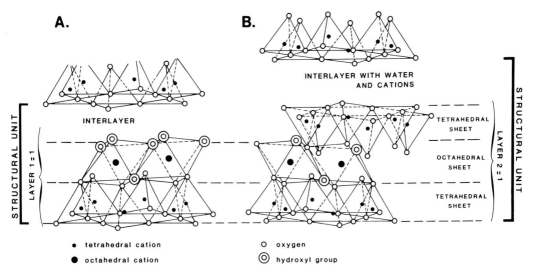

Fig. 1.2. Tri-dimensional sketch of the structure of **A** kaolinite, **B** smectite. (Grim 1968)

The sum of a layer plus an interlayer is called a *structure unit* (Fig. 1.2). It corresponds to a specific chemical *formula unit*. Its thickness is between 7 and 18 Å, depending on the type of layer and the interlayer content of each clay species. The basal spacing between two successive units is 7 Å for the 1:1 type, 10 Å for the 2:1 type, and 14 Å for the 2:1:1 type. The piling-up of many structure units constitutes a *clay particle*, the size of which is usually less than a few micrometers (μm). The clay particles represent the most common components of the *clay fraction* of sedimentary rocks, in which they occur as individuals or as aggregates of various sizes and shapes. In most geological studies, the clay fraction of a given rock is conventionally considered as the fraction less than two micrometers. This fraction may include many other components beside clay minerals, like aluminum or iron oxides, carbonates, phosphates, free silica, various aluminosilicates and organic matter.

The structural order of clay minerals is diminished by both crystallographical and chemical constraints, as compared to most other mineral groups. (1) The lateral dimensions of tetrahedral and octahedral sheets are not exactly the same. The two types of sheets do not fit precisely when linked together through sharing oxygens. This misfit determines strained planes of junction, and an alteration of the resultant size, shape and structure of the crystals (see Bailey 1980a, Caillère et al. 1982). (2) Numerous clay minerals present a *partial substitution* of tetrahedral Si^{4+} by Al^{3+}, or of octahedral Al^{3+} or Fe^{3+} by Fe^{2+} or Mg^{2+}. Such a substitution of tetravalent cations by trivalent ones, or of trivalent cations by bivalent ones, determines a deficit of positive charge and an excess of negative charge within the layers. This disequilibrium of the layer charge can be neutralized by the presence of a positive charge in the interlayers (cations, water, hydroxides). Such substitutions and additions of cations determine an increase of the crystal disorder and instability: structural misfits and deformations due to the large size of some univalent and bivalent cations, possibility of ion exchanges due to the changing chemical composition of the interstitial fluids, close interactions between inorganic and organic components of sediments. The ionic substitutions in clay minerals are primarily responsible for some of their characteristic properties, such as adsorption-desorption, expandability-retraction, and formation of organo-mineral complexes.

1.3 Main Clay Minerals

1.3.1 Main Divisions

The layer type and charge allow the identification of *eight groups of layer silicate minerals;* these include the common clay minerals, the brittle micas, and the so-called fibrous clays (palygorskite, sepiolite). A further division into subgroups and species is based on the octahedral sheet type, the chemical composition, and

Table 1.1. General classification of phyllosilicates related to clay minerals. (After Bailey 1980a)

Layer type	Group	Subgroup	Examples of species
1:1	Kaolin-serpentine ($X \sim 0$)	Serpentines	Chrysotile, antigorite, lizardite, amesite, berthierine
		Kaolins	Kaolinite, dickite, nacrite
2:1	Pyrophyllite-talc ($X \sim 0$)	Talcs	Talc
		Pyrophyllites	Pyrophyllite
	Smectite ($X \sim 0.2$–0.6)	Montmorillonites (dioc.)	Montmorillonite, beidellite, nontronite
		Saponites (trioc.)	Saponite, hectorite, sauconite, stevensite
	Vermiculite ($X \sim 0.6$–0.9)	Dioctahedral vermiculites	Dioctahedral vermiculite
		Trioctahedral vermiculites	Trioctahedral vermiculite
	Mica ($X \sim 1.0$)	Dioctahedral micas	Muscovite, paragonite, illite, glauconite
		Trioctahedral micas	Phlogopite, biotite, lepidolite, (illite)
	Brittle mica ($X \sim 2.0$)	Dioctahedral brittle micas	Margarite
		Trioctahedral brittle micas	Clintonite
(2:1:1)	Chlorite (X variable)	Dioctahedral chlorites	Donbassite
		Trioctahedral chlorites	Chlorite s.s., clinochlore, chamosite, nimite
		Di, trioctahedral chlorites	Cookeite, sudoite
2:1	Palygorskite-sepiolite (= fibrous clays)	Palygorskites	Palygorskite
Inverted ribbons	(X variable)	Sepiolites	Sepiolite, xylotile

X = layer charge per formula unit. X refers to an $O_{10}(OH)_2$ formula unit for smectite, vermiculite, mica and brittle mica.

the geometry of layer and interlayer superposition. A summary of the main groups and subgroups is proposed by Bailey (1980 a and b), according to the recommendations of the nomenclature committee of the AIPEA (Association Internationale Pour l'Etude des Argiles. Table 1.1; Fig. 1.3). The mixed- layered or interstratified clay minerals, formed by the real or apparent superposition of two or more single structure units, should be added to this list. Note that some non-crystalline hydrated silicates, such as allophane and imogolite, present compositions similar to those of clay minerals. Allophane is a hydrous aluminosilicate clay (Al_2O_3, $2SiO_2$, nH_2O) with a short-range order and predominance of Si–O–Al bonds. Imogolite (SiO_2, $2.5H_2O$) consists of paracrystalline assemblies of a one-dimensional structure unit, forming long smooth and curved threads.

In the following summary some symbols will be used:

R^{1+} univalent cations
R^{2+} bi(di)valent cations
R^{3+} trivalent cations
x,y,z rate of substitution in tetrahedral or octahedral sheets
▦   vacancy of positive charge

1.3.2 Kaolin – Serpentine Group

The 1:1. (T.O.) clay group comprises both dioctahedral and trioctahedral minerals (Table 1.2). The main species are the following:

Kaolinite $Al_2 SiO_2 O_5 (OH)_4$.

Kaolinite is a triclinic and well-ordered dioctahedral clay mineral, whose crystals often show distinct hexagonal outlines. Kaolinite mainly forms in surficial environments through pedogenetic processes, but may also develop during early diagenesis. Dickite (monoclinic) and nacrite (monoclinic almost orthorhombic) are very close, but they usually correspond to specific hydrothermal environments and display large crystals.
Halloysite refers to a hydrated kaolinite, commonly with a frequent tubular or glomerular shape. The progressive loss of water leads to intermediate minerals (metahalloysite). Halloysite is essentially a weathering mineral.The fireclays represent disorganized and irregularly superimposed layers of kaolinite.
Serpentines are trioctahedral minerals marked by significant differences in the chemical composition and in the rate of substitution. Antigorite, chrysotile, greenalite, and amesite show no or few substitutions, while cronstedtite and berthierine present notable tetrahedral and octahedral substitutions.

Examples:

Chrysotile $Mg_3 Si_2 O_5 (OH)_4$.
Berthierine $(Fe^{2+}, Mn^{2+}, Mg)_{3-x} (Fe^{3+}, Al)_x (Si_{2-x} Al_x) O_5 (OH)_4$.

Table 1.2. Classification of major 1:1 clay minerals. (After Caillère et al. 1982)

```
                          1:1 Layer type (T.O.)
                    ┌──────────────┴──────────────┐
              Dioctahedral                      Trioctahedral
                    │                   ┌───────────┴───────────┐
              Te = 4 Si              Te = 4 Si              Te < 4 Si
              Oc = 12/12            Oc = 12/12             Oc > 12/12
              ┌─────┴─────┐              │                      │
          Stable      Variable        Stable                 Stable
          spacing      spacing        spacing                spacing

      Kaolinite (Al)  Halloysite (Al)
      Nacrite (Al)                  Antigorite (Mg)      Cronstedtite (Fe²⁺, Fe³⁺)
      Dickite (Al)                  Chrysotile (Mg)      Berthierine (Al, Fe²⁺)
                                    Lizardite (Mg, Al)
      Fireclay (Al)              Greenalite (Fe²⁺, Mg, Mn)   Amesite (Al, Mg)
    (disorganized kaolinite)
```

Kaolinite (Al)   Halloysite (Al)
Nacrite (Al)
Dickite (Al)   Antigorite (Mg)   Cronstedtite (Fe^{2+}, Fe^{3+})
                               Chrysotile (Mg)   Berthierine (Al, Fe^{2+})
                               Lizardite (Mg, Al)
Fireclay (Al)   Greenalite (Fe^{2+}, Mg, Mn)   Amesite (Al, Mg)
(disorganized kaolinite)

- dioctahedral: 2/3 of octahedra occupied by trivalent cations.
- trioctahedral: 3/3 of octahedra occupied by bivalent cations.
- Te = number of Si cations in tetrahedra.
- Oc = octahedral charge.

1.3.3 Pyrophyllite – Talc Group

This group comprises 2:1 (T.O.T.) layer minerals, devoid of cationic substitutions or charge deficit ($x = 0$). The minerals therefore present a high structural stability. Pyrophyllite represents the dioctahedral type, talc the trioctahedral (Tables 1.3, 1.4). Both minerals result from hydrothermal or diagenetic processes.

Pyrophyllite $Al_2 Si_4 O_{10} (OH)_2$.

Talc $Mg_3 Si_4 O_{10} (OH)_2$.

1.3.4 Mica Group

The mica structure consists of 2:1 layers marked by noticeable substitutions, and by strong negative charges neutralized by large univalent interlayer cations. The layer charge ($x \approx 1$) results: (1) from substitution of trivalent cations (R^{3+}) for Si^{4+} in tetrahedral positions, (2) from substitution of univalent (R^{1+}) or bivalent (R^{2+}) for R^{2+} or R^{3+} in octahedral positions, or (3) from vacancies in octahedral positions (Bailey 1980a). Most of interlayer cations are univalent and associated with no or only few molecules of water. This broad group is characterized by a stable spacing on X-ray diagrams, and well-shaped but irregular edges on electron micrographs. It includes both dioctahedral and trioctahedral

Table 1.3. Classification of major 2:1 dioctahedral clay minerals. (After Caillère et al. 1982)

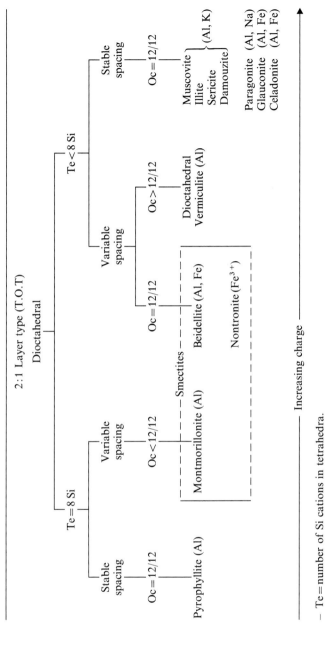

– Te = number of Si cations in tetrahedra.
– Oc = octahedral charge.

Table 1.4. Classification of major 2:1 trioctahedral clay minerals. (After Caillère et al. 1982)

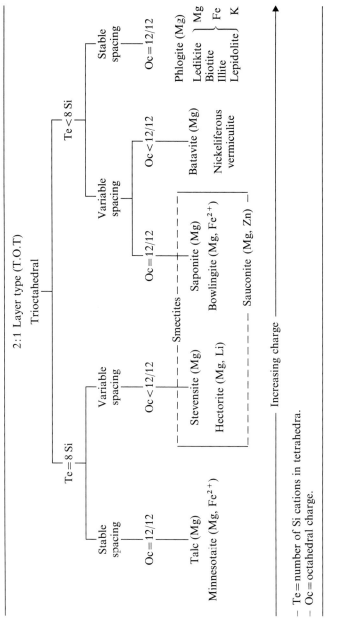

– Te = number of Si cations in tetrahedra.
– Oc = octahedral charge.

minerals (Tables 1.3, 1.4). Most of the species are encountered in clayey, silty and sandy fractions of sedimentary rocks.

Dioctahedral micas (mostly trivalent cations in octahedra)
General formula (Bailey 1980a):

$$(K, Na, H_2O)(Al, Fe^{3+})_{2-x}(Mg, Fe^{2+})_x(Si_{3+x}Al_{1-x})O_{10}(OH)_2$$

Interlayer   Octahedra   Tetrahedra Anion group

Examples:

Muscovite (stable polytype)
$KAl_2(Si_3Al)O_{10}(OH)_2$

Phengite
$K(Al_{1.5}R_{0.5}^{2+})(Si_{3.5}Al_{0.5})O_{10}(OH)_2$

Glauconite
$R_{0.78}^{1+}(Fe_{1.01}^{3+}Al_{0.45}Mg_{0.39}Fe_{0.20}^{2+})_{2.05}Si_{3.65}Al_{0.35})O_{10}(OH)_2$

Celadonite
$R_{0.83}^{1+}(Fe_{0.72}^{3+}Al_{0.49}Mg_{0.63}Fe_{0.21}^{2+})_{2.05}(Si_{3.81}Al_{0.19})O_{10}(OH)_2$.

Muscovite and phengite chiefly originate from metamorphic and silica-rich igneous rocks. Glauconite forms a large part of the green granules present on the upper part of continental margins. Celadonite is an iron-rich mica-like glauconite, with less significant tetrahedral substitutions; it constitutes an alteration product of volcanic rocks.

Trioctahedral micas (mostly bivalent cations in octahedra)
General formula (Bailey 1980a):

$$K_{x-y+2z}\quad[(Mg, Fe^{2+})_{3-y-z}\quad R_y^{3+}\Box_z](Si_{4-x}R_x^{3+})O_{10}(OH, F)_2$$

Interlayer   Octahedra   Tetrahedra   Anionic group

Examples:

Phlogopite
$(K_{0.93}Na_{0.04}Ca_{0.03})(Mg_{2.77}Fe_{0.10}^{2+}Ti_{0.11})(Si_{2.88}Al_{1.12})O_{10}(OH)_{1.49}F_{0.51}$.

Biotite
$(K_{0.78}Na_{0.16}Ba_{0.02})(Mg_{1.68}Fe_{0.71}^{2+}Fe_{0.19}^{3+}Ti_{0.34}Al_{0.19}Mn_{0.01})$
$(Si_{2.86}Al_{1.14})O_{11.12}(OH)_{0.71}F_{0.17}$.

Biotite, a very common mineral, and phlogopite, mainly issue from ferro-magnesian crystalline rocks.

Illite

Illite is the name geologists usually give to the micaceous minerals present in the clay-sized fraction of sedimentary rocks. Statistically illite is predominantly of a dioctahedral type close to muscovite, because muscovite is a very common mineral and one of the more stable micas. Nevertheless illite generally contains more Si, Mg, H_2O and less K than ideal muscovite. In fact the term illite can designate all types of small micas, and sometimes corresponds to a mixture of micaceous minerals of different origins. The dominant type of sedimentary illites can often be determined by using detailed roentgenographic, electron-microscope and microchemical techniques.

1.3.5 Brittle Micas

Brittle micas differ from true micas by a higher layer charge (-2 per formula unit instead of -1), and by bivalent compensation cations in interlayers instead of univalent ones. They result from the erosion and alteration of crystalline schists. They are rarely incorporated in the clay-sized fraction of sedimentary rocks, because of their resistance to fragmentation and low abundance at the surface of the Earth. Brittle micas comprise both dioctahedral and trioctahedral types.

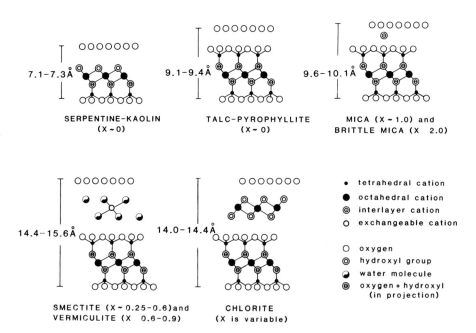

Fig. 1.3. Sketch of major layered clay mineral groups, from a (010) view. (Bailey 1980a). X layer charge per formula unit

Examples:

Margarite: $Ca\,Al_2\,(Si_2Al_2)\,O_{10}\,(OH)_2$
Clintonite: $Ca\,(Mg_2Al)\,(SiAl_3)\,O_{10}\,(OH)_2$.

1.3.6 Vermiculite Group

Vermiculites are T.O.T. minerals with water associated with cations in the inter-layer, determining a basal spacing close to 14 Å (Fig. 1.3). The water molecules are strongly linked to the layer structure (layer charge > 0.6), which considerably limits the possibilities of expandability. Most macroscopic vermiculites are of a trioctahedral type and result from alteration of mica or chlorite.

Example:

$Mg_3\,(Si_3Al)\,O_{10}\,(OH)_2\,Mg_{0.5}\,(H_2O_4)_4$.

Clay-sized vermiculites can be either trioctahedral or dioctahedral, and result from pedogenesis or diagenesis.

1.3.7 Smectite Group

Smectites constitute a large group of T.O.T. minerals, marked by a low layer charge ($x < 0.6$). This characteristic determines a weakness of the linkage between the different layers of a given particle, and allows considerable swelling phenomena as well as possibilities of rotation (turbostatic disposition). The inter-layers are occupied by cations (mainly Na, K, or Ca) and by one to several layers of water (basal spacing from 10 to 18 Å). The shape of smectites varies according to the conditions of genesis: flakes, curls, laths of different sizes, etc. Smectites chiefly belong to the clay-sized fractions of rocks. Smectite minerals may form in different chemical environments, most of them being surficial soils or sediments. This is the reason why they present various chemical compositions, in both dioc-tahedral and trioctahedral subgroups (Tables 1.3, 1.4). Octahedral substitutions are very frequent, tetrahedral substitutions occur commonly. Transition terms exist between some end-member species.

Examples of structural formulas (after Millot 1964):

Dioctahedral smectites (Al, Al-Fe or Fe smectites)

Montmorillonite
$Na_{0.33}(Al_{1.67}Mg_{0.33})Si_4O_{10}(OH)_2,nH_2O$

Beidellite
$Na_{0.36}(Al_{1.46}Fe^{3+}_{0.50}Mg_{0.04})(Si_{3.64}Al_{0.36})O_{10}(OH)_2, nH_2O$

Nontronite
$K_{0.33}(Fe^{3+}_{1.61}Al_{0.39})(Si_{3.67}Al_{0.33})O_{10}(OH)_2, nH_2O$

Trioctahedral smectites (Mg smectites)

Stevensite
$Na_{0.16}(Mg_{2.92}\square_{0.08})Si_4O_{10}(OH)_2, nH_2O$

Saponite
$Na_{0.33}Mg_3(Si_{3.67}Al_{0.33})O_{10}(OH)_2, nH_2O$

Hectorite
$Na_{0.33}(Mg_{2.67}Li_{0.33})Si_4O_{10}(OH)_2, nH_2O.$

1.3.8 Chlorite Group

Chlorites are considered either as a 2:1 layer group with a hydroxide interlayer, or as a 2:1:1 layer group (see Bailey 1980a, Caillère et al. 1982). Their typical structure shows a regular alternation of negatively charged trioctahedral micaceous layers and of positively charged octahedral sheets. The basal spacing of the structure unit is close to 14 Å; it is invariable for true chlorites (Fig. 1.3), because T.O.T. layers and O interlayers are strongly bound. The tetrahedral substitution of Al for Si induces a variable charge deficit, which is partly compensated by an excess of charge in the interlayer octahedral sheet ("brucitic sheet"), and perhaps in the layer octahedra. Most chlorites are trioctahedral, with magnesium as the main cation in both layer and interlayer octahedra. A few chlorites, however, are typically dioctahedral (aluminous, i.e., dombassite), and some species present a combination of dioctahedral layers and trioctahedral hydroxide interlayers (cookeite, sudoite). Chlorites generally display flat particles of various sizes (from <1 μm to >1 mm), with well-defined outlines and sometimes polygonal edges. They chiefly originate from crystalline igneous or metamorphic rocks, or from the alteration of some volcanic rocks.

Examples of simplified formulas (cf Bailey 1980a, Caillère et al. 1982):

Interlayer octahedra	Layer octahedra	Tetrahedra	Anion group
Chlorite s.s.			
$(Mg_{3-y}Al_1Fe^{3+}_y)$	Mg_3	$(Si_{4-x}Al)$	$O_{10}(OH)_8$
Dombassite			
$Al_{2+1/3}$	Al_2	(Si_3Al)	$O_{10}(OH)_8$
Sudoite			
$(Mg_{2.3}Al_{0.7})$	Al_2	$(Si_{3.3}Al_{0.7})$	$O_{10}(OH)_8.$

Iron-rich chlorites, or chamosites, occur commonly in iron-ore deposits. Their X-ray diffraction patterns differ from those of ferriferous kaolinite (berthierine) by the presence of a small (001) reflection (14 Å) and from those of true chlorite by the increasing height of this reflection after heating.

Swelling chlorites or pseudo-chlorites expand like smectites when immersed in water or ethylene glycol, but resist heating like chlorites. They appear to represent an alternation of smectitic layers and octahedral brucitic sheets. They occur, for instance, in some early Mesozoic iron ores and marls.

1.3.9 Palygorskite – Sepiolite Group

Often called fibrous clays or pseudo-layered clays, palygorskite and sepiolite consist of bundles of well-defined and elongated fibers or laths. Their structure comprises continuous tetrahedral sheets forming a hexagonal-mesh arrangement (Si, O), alternating with discontinuous octahedral sheets (Mg, O, OH) extending in a single direction and forming ribbons. The ribbon structure is responsible for the fibrous shape of the clay particles. The ribbons are alternately arranged above and below the continuous hexagonal sheet of tetrahedra, the transverse section resembling a hollow brick structure. The empty spaces are filled with zeolitic water (Fig. 1.4). Palygorskite and sepiolite are both magnesium-rich clay minerals, differing from each other by the width of the inverted ribbons (broader for sepiolite) and by specific chemical characters. Palygorskite is less magnesian and presents more structural diversity than sepiolite. The substitutions lead to in-

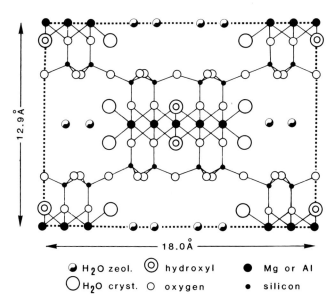

Fig. 1.4. Palygorskite structure, projection (100). (Caillère and Hénin 1961; Bailey 1980a)

☽ H_2O zeol.   ◎ hydroxyl   ● Mg or Al
◯ H_2O cryst.   ○ oxygen   • silicon

12.9 Å

18.0 Å

corporating Al in tetrahedra, and Al or Fe in octahedra. Xylotile is a sepiolite in which Mg is partly replaced by Fe^{3+}. All fibrous clays form under surficial conditions in soils or sediments.

Palygorskite
$$(Mg_{5-y-z}R_y^{3+}\square_z)(Si_{8-x}R_x^{3+})O_{20}(OH_2)_4 \cdot R_{(x-y+2z)/2}^{2+}(H_2O)_4$$

Sepiolite
$$(Mg_{8-y-z}R_y^{3+}\square_z)(Si_{12-x}R_x^{3+})O_{30}(OH)_4 \cdot R_{(x-y+2z)/2}^{2+}(H_2O)_8 \cdot$$
$$R^{3+} = Al, Fe^{3+}. \qquad R^+: Mg.$$

1.3.10 Mixed-layers

Mixed-layered or interstratified clays refer to remarkable phyllosilicate structures, characterized by a vertical stacking sequence of two or more types of single layers. The layers involved can be of 2:1, 2:1:1 and even 1:1 types. The possibility of mixed-layering or interstratification results from the rather weak chemical and structural linkage existing between the successive layers in a given clay particle. As the basal surfaces of all different types of clay layers present similar sheets of oxygen and hydroxyl arranged in a hexagonal mesh, layers with different internal arrangements can stack together and still keep mobile interfaces. Such structural features are almost unique in the mineralogical range and are responsible for the common occurrence of mixed-layered clay minerals (see Weaver 1956, Reynolds 1980). Most mixed-layers can be considered as clay species intermediate between two single clay minerals. They mainly form through weathering or middle-late diagenesis, but also characterize some hydrothermal and sedimentary environments.

The mixed-layered minerals present a roentgenographic behavior that is intermediate between those of the individual minerals involved. Three main types of interstratification are recognized:

1. A periodic alternation of layers of two types A and B refers to regular or ordered mixed-layers:

AB AB AB ...   or   AAB AAB AAB ...

These minerals show well-defined reflections on X-ray diagrams and tend to be given specific names: hydrobiotite for illite-vermiculite, corrensite for chlorite-smectite, etc (Table 1.5). Sometimes they are simply labeled by linking the name of both their components (e.g., regular illite-smectite, regular kaolinite-smectite), or by using a symbolic expression (Lucas 1962):

Examples: $10\text{-}14_S$: regular illite-smectite
$\qquad\qquad 14_C\text{-}14_V$: regular chlorite-vermiculite

Table 1.5. Examples of mixed-layer clay minerals. (After Reynolds 1980)

Layer types	Specific name
A. Regular or subregular alternation of two layer types	
Muscovite-montmorillonite	Rectorite (allevardite)
Illite-smectite	Bravaisite
Glauconite-smectite	
Chlorite-smectite	Corrensite
Dioctahedral chlorite-smectite	Tosudite
Mica-vermiculite	Hydrobiotite
Talc-saponite	Aliettite
Kaolinite-smectite	
B. Irregular alternation of two layer types	
Illite-smectite	
Glauconite-smectite	
Mica-vermiculite	
Mica-chlorite	
Mica-dioctahedral chlorite	
Smectite-chlorite	
Chlorite-vermiculite	
Kaolinite-smectite	
C. Three-component systems	
Illite-chlorite-smectite	
Illite-smectite-vermiculite	

2. A random alternation of each type of layer corresponds to irregular and randomly mixed-layered clays:

A A B A B B B B A A A B A A B A.

This group often corresponds to poorly defined X-ray diffraction patterns, with dome-, plateau- or wedge like shapes, and locations at intermediate positions between those of the single minerals involved . The random mixed-layers are very common and develop peculiarly in soil profiles. They include both two- and three-component systems, identified by using reference charts and curves (e.g., Lucas et al. 1959, Millot 1964, Cradwick and Wilson 1978). They are labeled by the types of layer responsible for the interstratification (Table 1.5) or through specific symbols, the most abundant layer being listed first:

Examples: $(10\text{-}14_V)$: irregular illite-vermiculite
$(14_S\text{-}14_V)$: irregular smectite-vermiculite

3. Partially ordered structures appear to exist, especially in some soils (Mac Evan 1949). Intermediate between both former types, they are little known and not specifically labeled.

Two new approaches have been developed to our knowledge of mixed-layered minerals:

1. Statistical studies indicate that some smectitic minerals include illite-like layers in their structure, in the same way that illites often include expandable layers (see Reynolds 1980, Šrodoń and Eberl 1984). The percentage of illitic layers in the smectite network may change, especially according to the diagenetic conditions. Detailed calculations permit the identification of the illite-smectite mixed-layering in smectitic minerals, and its evolution in relation with thermodynamic changes and time (e.g., Šrodoń 1980, 1981, 1984a, Reynolds 1983).

2. Transmission electron-microscope experiments on diagenetic series suggest that some clay minerals, identified as illite-smectite mixed-layers by X-ray diffraction, consist in fact of mixtures of independent illite and smectite particles (Nadeau et al. 1984a, 1985). The X-ray diffraction pattern is attributed to an interparticle diffraction effect, determined by the very small size (about 10 to 50 Å) of the clay minerals evolving during burial diagenesis (see also Chap. 15).

1.4 Main Associated Non-Clay Species

Many minerals can be associated with clay minerals in sediments and soils. Most of them occur in both clay and nonclay granulometric fractions. Some of them are difficult to identify on X-ray diffractograms based on oriented preparations of clay minerals. In addition, some amorphous components exist in variable amounts beside the crystallized species. They include silico-aluminous (e.g., allophane, imogolite), silico-ferric (e.g., hisingerite) and siliceous (opal) minerals. We briefly quote here the most frequent minerals associated with clays in sedimentary rocks, because they are often useful in paleoenvironmental or diagenetic interpretations. Detailed data and references are provided by Brown (1980).

Iron oxides and hydroxides:
    hematite αFe_2O_3
    maghemite γFe_2O_3
    geothite $\alpha Fe_2O_3, H_2O$
    lepidocrocite $\gamma Fe_2O_3, H_2O$.

limonite: rock mainly composed of geothite, sometimes associated with other oxides and clay minerals.

Aluminum hydroxides:
    gibbsite $\gamma Al(OH)_3$
    boehmite $\gamma AlOOH$.

Silicon oxides: quartz, tridymite, cristobalite: SiO_2 .

Manganese oxides: todorokite

$(Ca_{0.08}K_{0.08}Na_{0.05})\,(Mn^{2+}_{1.17}Mg_{0.17}Mn^{4+}_{5.06})O_{12}\cdot 36H_2O$.

Zeolites:

clinoptilolite, heulandite

$(Na, K)_x(Ca, Mg)_y(Al, Fe^{3+})_{x+2y}Si_{36-[x+2y]}O_{72}\cdot 19$ to $26H_2O$

phillipsite

$(Na, K)_x(Ca, Mg)_y(Al, Fe^{3+})_{x+2y}Si_{16-[x+2y]}O_{32}\cdot \sim 12H_2O$

analcite

$Na_{17}Al_{17}Si_{31}O_{96}\cdot 16H_2O$ to $Na_{14}Al_{14}Si_{34}O_{96}\cdot 16H_2O$

Carbonates:

calcite, aragonite	$CaCO_3$
dolomite	$CO_3(Mg, Ca)$
siderite	$FeCO_3$
rhodochrosite	$MnCO_3$.

Phosphates:

carbonate apatite

ex: $Ca_{10-p-q}Na_pMg_q(PO_4)_{6-x}(CO_3)_xF_{2+y}$.

Sulfides: pyrite, marcasite: FeS_2 .

Sulfates:

anhydrite	$CaSO_4$
gypsum	$CaSO_4\ 2H_2O$
baryte	$BaSO_4$
alunite	$KAl_3(SO_4)_2(OH)_6$
jarosite	$KFe_3(SO_4)_2(OH)_6$
natrojarosite	$NaFe_3(SO_4)_2(OH)_6$
melanterite	$FeSO_4\ 7H_2O$

Various feldspars, amphiboles and other nonclay silicates.

1.5 Conclusion

1. Sedimentologists studying the clay minerals in present and past series need more and more accurate information on their nature. Knowing the different clay groups of a given formation is frequently no longer sufficient. It is often necessary to determine the clay species, their chemical composition, and micromorphology. This is the reason why *the methods and nomenclature used should be clearly defined*, and the terms employed in a proper sense. For instance, the reasoning and interpretation about mixed-layer clays may differ completely if these minerals are identified by X-ray reflections located between those of single clay minerals (e.g. Lucas 1962, Holtzapffel 1985), or by statistical calculations in-

cluding the position of the single minerals themselves (i.e., Reynolds 1983, Środoń 1984a, Środoń and Eberl 1984).

2. The precise characterization of clays is especially difficult, because these minerals are often very small, poorly or diversely crystallized, and associated with subamorphous compounds. They frequently include transitional terms and form complex associations. The precise identification of natural clay assemblages often requires special efforts. Reliable quantitative estimations are still mostly based on a good practice of diffractograms rather than on computerized X-ray signal processing. There are still no preparation, identification, or quantification methods universally acknowledged (see Pierce and Siegel 1969, Brindley 1980, Brindley and Brown 1980, Stucki and Banwart 1980, Mann and Fischer 1982, Thorez 1983). This is the reason why the methodological approach employed should be precisely explained in each case, verifiable calculations such as peak height ratios added to the quantitative estimations based on surface percentages, and excessively precise values avoided.

3. Despite these difficulties, our knowledge concerning the identification and interpretation of sedimentary clay is progressing very fast. An increased number of sophisticated devices usually employed by chemists and physicists are used by sedimentologists or even installed in sedimentology laboratories. The following examples of techniques and general references illustrate some modern ways of better knowing the nature and composition of clay minerals:

X-ray diffraction: Carroll 1970, Thorez 1975, Eberhart 1976, J.C.P.D.S. 1978–1979, Brindley and Brown 1980, Caillère et al. 1982, Desprairies 1983, Holtzapffel 1985.

Differential thermal and gravimetric analysis: Mackenzie 1970, Wilson 1987.

Electron microscopy: Beutelspacher and Van Der Marel 1968, Borst and Keller 1969, Gard 1971, Sudo et al. 1981.

Infrared spectroscopy: Van Der Marel and Beutelspacher 1976, Wilson 1987.

Geochemistry: Pinta 1971, Weaver and Pollard 1973, Velde 1977 and 1985, Newman 1987.

Microprobe analysis, microgeochemistry: Stucki and Banwart 1980, Dritz 1981, Debrabant et al. 1985, Rautureau and Steinberg 1985.

Chapter 2

Clay Formation Through Weathering

2.1 Main Weathering Processes

2.1.1 Introduction

Sedimentary clays may have various origins. One of the most expected origins consists in the erosion, transportation, and deposition of geological and pedological formations that are exposed at the surface of land masses. It is necessary to summarize the conditions of clay genesis in such formations in order to appreciate to what extent the sediments may incorporate the materials issued from their erosion.

The surficial processes resulting from the interaction between the rocks (lithosphere), the air (atmosphere), the water (hydrosphere) and the organisms (biosphere) are referred to as *weathering*. Weathering includes all mechanisms responsible for the rock fragmentation, the production of dissolved ions, and the development of pedological formations at the surface of the Earth. Weathering tends to reach a stage of equilibrium if environmental conditions remain stable. The intensity of weathering is mainly controlled by lithology, climate, and morphology. Rocks are all the more sensitive to weathering since they are softer, more porous, heterogeneous or fractured. Climate and morphologic factors determine the dominantly physical or chemical character of weathering.

Physical weathering leads to rock fragmentation and disintegration. It intervenes mainly in sloped areas subjected to large variations of temperature or humidity: mountains, coastal cliffs, deserts. It acts largely by crystallization of water or salts (halite, gypsum) within the rock cracks, because of frost-thaw alternation or strong evaporation. The crystallization determines an augmentation of the volume, which induces rock fragmentation. Physical weathering also includes the wetting-drying cycles on argillaceous sediments, giving way to the formation of clay chips easily reworked by the wind. Animal burrowing and growing of plant roots belong with other biological actions to the same type of process.

Chemical weathering is by far the most widespread phenomenon. It intervenes alone or in addition to physical weathering. Determined by the action of natural waters, chemical weathering proceeds by the following fundamental reaction:

Parent rock + ion-depleted water → weathering complex + ion-enriched water
PRIMARY + ATTACK   → SECONDARY   + LEACHING
MINERALS   SOLUTION   MINERALS   SOLUTION

The weathering complex tends to evolve with time, and gives a *soil* that constitutes the natural transition between the parent rock and the atmosphere. Clay-sized fraction and clay minerals form the major components of most weathering complexes and soils.

Chemical weathering comprises four different types, according to the composition of the attack solution (Table 2.1). (1) *Acidolysis* develops in organic and acid environments, where organo-mineral complexes from easily. (2) *Salinolysis* characterize evaporitic saline environments (Na, K, etc.) under normal conditions of pH. (3) *Alcalinolysis* corresponds to basic conditions and to water charged with Ca or Mg. (4) *Hydrolysis* consists in the attack of rocks by water little ionized, under medium pH conditions. Hydrolysis is by far the most developed and best-known process. Extended data on weathering are given by Millot (1964, 1970), Loughnan (1969), Carroll (1970), Duchaufour et al. (1977), Drever (1985), Colman and Dethier (1986).

Table 2.1. Main chemical weathering processes. (After Pédro 1979)

	pH > 5	5 < pH < 9.6	pH > 9.6
Attack solution depleted in saline elements (Na, K, Ca, Mg)	Acidolysis water rich in dissolved organic acids	Hydrolysis pure water or CO_2-rich water	
Attack solution concentred in saline elements		Salinolysis water charged with salts of strong acids (chlorides, sulphates)	Alcalinolysis water charged with salts of weak acids (carbonates, bicarbonates)

2.1.2 Hydrolysis

Hydrolysis basically represents a chemical reaction between a salt and water, to form an acid and a base. The widely distributed aluminosilicates (e.g., feldspars, pyroxenes, amphiboles, micas) constitute salts of weak acids, which react with water to form dissolved silicic acid and various bases, as well as secondary minerals like clays.

Example:

orthose   + water   → kaolinite   + solution

$$2(Si_3Al)O_8K + 11H_2O \rightarrow Al_2Si_2O_5(OH)_2 + 4Si(OH)_4 + 2(K, OH).$$

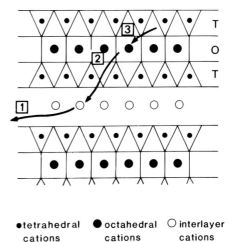

Fig. 2.1. Schematic representation of ionic subtraction through hydrolysis of a mica, under temperate-wet conditions. (After Millot 1967). *1* Leaching of interlayered ions. *2* Migration of octahedral ions into the interlayered sheet. *3* Migration of tetrahedral ions into the octahedral sheet

● tetrahedral   ● octahedral   ○ interlayer
cations   cations   cations

Hydrolysis, which etymologically means destruction by water, consists practically of the progressive *subtraction of ions* from the different minerals of the parent rocks. The subtraction statistically concerns first the more mobile ions, like Na, K, Ca, Mg, and Sr. The transition elements tend to be evacuated later (Mn, Ni, Cu, Co, Fe), but before Si. Al is the less mobile element through the hydrolytic processes. The subtraction starts in the parts of a given mineral that are more exposed to leaching (external surfaces, cleavage planes, cracks). For instance, the hydrolysis of micas under temperate-wet conditions shows the following subtraction steps, as an average (Jackson 1965, Millot 1967): (1) leaching of interlayered cations (K, Na); (2) migration toward the interlayers of some octahedral cations (Mg, Fe), in order to balance the vacant charges. These cations are in turn progressively evacuated into the leaching solution; (3) migration toward the octahedral layers, and finally toward the interlayers, of some tetrahedral cations (Si, Al) (Fig. 2.1).

Hydrolytic processes are favored by the following conditions:

1. Abundance of soluble minerals. For instance the saline minerals are evidently more fragile than carbonates, which are less resistant than silicates. A similar gradation exists among the common silicates sumitted to hydrolysis:

$$\left.\begin{array}{l}\text{peridots} - \text{pyroxenes} - \text{amphiboles} - \text{biotite} \\ \text{Ca feldspars} - - - - - - - \text{Na feldspars}\end{array}\right\} \text{K feldspars} - \text{muscovite} - \text{quartz} .$$

2. Small size of the particles of a given mineral, allowing high specific surface and numerous places of ion subtraction.
3. Presence of organic acids provided by bacterial activity, and taking part in mineral destruction.

4. Good drainage conditions, permitting continuous leaching and good evacuation of ions.

5. High humidity and temperature, determining an acceleration of the reactions. The effects of humidity prevail over those of temperature.

An increase of hydrolysis intensity and ionic subtraction in weathering complexes and soils gives way to secondary minerals that are more and more depleted in cations, especially the more mobile ones. Correspondingly, the leaching solutions receive more and more dissolved elements, that are evacuated downstream by drainage. Table 2.2 summarizes the evolution of potassic feldspars submitted to increased conditions of hydrolysis.

Secondary minerals usually comprise clay species and often associate oxides, resulting from the degradation of primary minerals (= negative transformation, cf. Millot 1964, 1970), or from a neoformation using ions released in leaching solutions. Depending on the hydrolysis intensity and ion evacuation, silicon is available within the soil to form new minerals with two, one, or no tetrahedral sheets. The formation of secondary minerals marked by two tetrahedral sheets (dominantly two Si-sheets) and one octahedral sheet (= dominantly Al-sheet) has been called *bisialitization* by Pédro (1979, 1981). It corresponds to 2:1 layer

Table 2.2. Increased hydrolysis steps of a potassic feldspar (from top to bottom)

Primary mineral	Attack solution	Secondary mineral	Leaching solution $(L = SiO_2/K_2O)$
Orthose $2.3(Si_3Al)O_8K$	$+8.4H_2O$	Al Beidellite (smectite) $K_{0.3}Al_2(Si_{3.7}Al_{0.3})O_{10}(OH)_2$	$(L = 3.2)$ $+3.2Si(OH)_4+2(K,OH)$
Orthose $2(Si_3Al)O_8K$	$+11H_2O$	Kaolinite $Al_2Si_2O_5(OH)_4$	$(L = 4)$ $+4Si(OH)_4+2(K,OH)$
Orthose $(Si_3Al)O_8K$	$+16H_2O$	Gibbsite $Al(OH)_3$	$(L = 6)$ $+(K,OH)$

Table 2.3. Geochemical and mineralogical characterization of hydrolytic weathering processes. (After Pédro 1979, 1981)

Hydrolysis Process	Bisialitization	Monosialitization	Alitization
Number of tetrahedral sheets	2	1	0
Silica removal	Incomplete	Incomplete	Complete
Bases removal (Ca, Mg, …)	Incomplete	Complete	Complete
Main secondary minerals	2:1 layer clays	1:1 layer clays	Al hydroxides
Example	Beidellite	Kaolinite	Gibbsite
Interlayer cations	NA, K, Ca	–	–

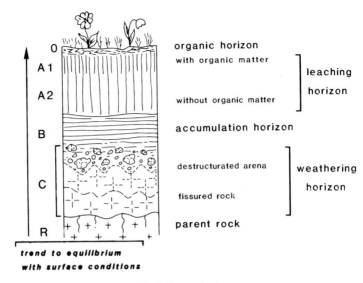

Fig. 2.2. General profile of a soil resulting front hydrolytic weathering

clays. Secondary minerals with only one tetrahedral (= Si) sheet proceed from *monosialitization* (1:1 layer clays), and those devoid of silicon from *alitization* (= Al oxides). Table 2.3 shows the main characteristics of Pédro's classification.

Hydrolysis, sometimes associated with or relieved by other weathering processes, participates strongly in the development of soils. Evolved soils comprise different superimposed *horizons* above the parent rock R (Fig. 2.2): (1) The weathering horizon (C) where the rock fragmentation and the first hydrolysis stages occur; (2) the accumulation horizon (B) where clay usually concentrates, and sometimes iron, silica, carbonates or Al oxides occur; (3) the leaching horizon (A) with organic matter in the upper level; (4) the surficial organic layer (O) where living and dead vegetation concentrate. The thickness and horizon differentiation of soils varies considerably according to lithological, morphological, and climatological constraints.

As the soils progress through weathering by developing roughly horizontal layers above the diversely folded and fractured parent rocks, the weathering front usually corresponds to a morphological unconformity. This *basal pedologic discordance* (= D1; Fig. 2.3) is overlaid with the successive horizons of the soil. If the climate is marked by a strong and short humid season and by rather sparse vegetation, as in West Africa, the leaching horizon is submitted to intense erosion. In this case the clay- and silt-sized particles are actively removed, while coarser materials concentrate in a lag horizon. This phenomenon determines an *upper pedologic discordance* (= D2; Millot 1982). Both weathering and leaching fronts tend to progress downwards in the course of time, at the expense of parent rocks. The progression is faster on morphological heights and slower in depressions. The general result of weathering, combined with erosion, is a widespread planation of continental relief (Millot 1980).

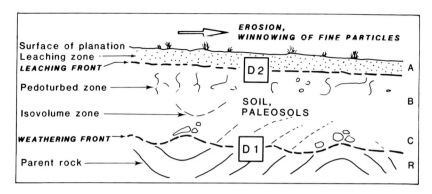

Fig. 2.3. The "double discordance" in soil development and morphological evolution, West Africa. (After Millot 1982). *D1* discordance of the weathering front above geological formations; *D2* discordance of the leaching front above pedological formations

2.2 Main Soils and Clay Content

2.2.1 Zonal Soils

Most types of soils present a global distribution that parallels the latitudinal or altitudinal zonation. Their distribution is mainly controlled by annual temperature and precipitation (Fig. 2.4). They are labeled zonal soils, in contrast with the more local soils developed under specific geochemical or geological conditions, that are referred to as azonal. Both zonal and azonal soils are described in great detail in pedology textbooks (e.g., Millar et al. 1958, Duchaufour et al. 1977). Their clay mineralogy is presented in several review papers or books (e.g., Millot 1964, 1970, Loughnan 1969, Dixon and Weed 1977, Pédro 1984). Extended mineralogical data exist for some countries, such as Japan (Sudo and Shimoda 1978) and Great Britain (Loveland 1984, Wilson et al. 1984).

Parallel Change of Temperature and Humidity

Very Cold Climate
Glacial and peri-glacial climate is characterized by the absence of free water or the presence of ice. The shortness of the summer season, the presence of a permanent frozen zone (permafrost) impeding the drainage, and the scarceness of organic matter prevent most hydrolytic processes. The lack of long-lasting running water amounts to a dry climate, even if the annual precipitation of snow is abundant. The corresponding soils consist of very primitive weathering complexes, nearly devoid of humus, forming thin coatings at the rock surface. This type of soil is typical of the arctic, peri-arctic and southernmost regions, as well as the high-altitude areas. They are labeled *lithosols* or tundra sols. Young soils developed on erosional surfaces (regosols), on sand dunes and alluvial deposits,

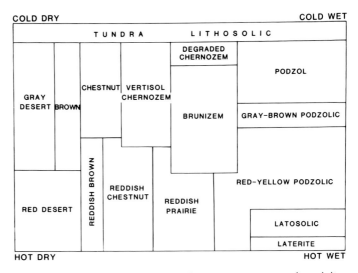

Fig. 2.4. Diagramatic distribution of major soil groups according to temperature and precipitation. (After Millar et al. 1958, in Loughnan 1969)

or on high mountain siliceous and calcareous rocks (protorankers, protorendzinas) present similar characteristics.

Because of the absence of significant hydrolysis, the production of clay depends essentially on physical weathering. The clay fraction of lithosols therefore consists principally of *fragments of the minerals contained in the parent rocks.* For instance, Kelly and Zumberge (1961) point to the absence of noticeable chemical weathering on Antarctic diorite. Weathering complexes and young soils studied in Central Norway contain essentially fragmented micas and chlorites issued from parent rocks (pers. anal.). The thin soils of the southern Pampa plains (Argentina) contain the same illite, associated with rare kaolinite and smectite, as the underlying sandy sediments (Bonorino 1966). The lithosols covering the crystalline rocks in the highest parts of the Alps and Pyrenees (Europe) consist mainly of micaceous and chloritic clays, that locally start to be transformed into vermiculitic mixed-layers (Robert et al. 1980). Most high-latitude soils are characterized by illite and chlorite, because of the abundance of magmatic or metamorphic substrates, whose physical weathering leads to mica exfoliation, feldspar sericitization and silicate chloritization. The weathering of high-latitude noncrystalline rocks gives way to the production of various clay minerals, such as smectite issued from various altered volcanic rocks, or kaolinite, illite and mixed-layers of peri-arctic Mesozoic sediments.

A few examples of chemical weathering effects are reported in the McMurdo Sound region, Antarctica, where irregular mixed-layers, vermiculite, and even smectite are identified in soils developed above igneous rocks and moraines (e.g., Claridge 1965, Ugolini 1977, Ugolini and Jackson 1982). Some of the soils occur on moraines apparently not older than 10,000 years. Nevertheless it is not easy to distinguish clay minerals newly formed from those present in heterogeneous parent rocks like moraines or alluvium, and to ascertain the weathering rate. Most Antarctic soils are much older than Holocene, and result from successive interglacial weathering periods. The present-day hydrolysis processes mainly consist of some metal release (Fe, Mn) and oxide precipitation.

Temperate-Humid Climate
Middle latitude regions are mostly characterized by mild temperatures and by precipitation averaging 50-100 cm/year. Such conditions allow significant chemical weathering. The leaching of silicates determines the removal of some bases (Ca, Mg, etc.) and of little Fe and Si. The resulting soils comprise mainly the *brown and chestnut soils*, that tend to become reddish in the warmer areas (e.g., red mediterranean soils). These soils constitute a rather thick blanket above the parent rocks (10 to 150 cm), and include all fundamental horizons (Fig. 2.2). The accumulation horizon is essentially made of clay that results from both physical and chemical weathering.

As temperate climate allows only partial desalcalinization and desilicification (Table 2.3), the pedologic formation of clay consists mostly of limited ionic subtractions, constituting negative transformations (= degradations). The primary phyllosilicates (micas, chlorites) are first exfoliated by hydrolysis, giving way to "open" illite and chlorite. Successive steps develop if hydrolysis goes on or becomes more active. A classical evolution consists of the successive formation of *irregular* vermiculitic *mixed-layers*, clayey *vermiculite*, smectitic mixed-layers, *degraded smectite*, and finally amorphous compounds (allophane) (cf. Millot 1964, 1970). Such a mineralogic series tends to develop from the base to the top of a given soil profile, and from cool-temperate to warm-temperate regions. Chlorite is usually less resistant to hydrolysis than illite. Noticeable amounts of primary illite and chlorite are preserved in most soils, indicating that parent materials significantly influence the clay mineralogy under temperate conditions. Three main evolutive phases can be recognized:

$$\left.\begin{array}{l} \text{Chlorite} \rightarrow \text{"open" chlorite} \rightarrow \text{chlorite-vermiculite} \\ \quad \text{Mica} \rightarrow \text{"open" illite} \quad \rightarrow \text{illite-vermiculite} \end{array}\right\} \text{vermiculite} \qquad (1)$$

$$\text{Vermiculite} \rightarrow \text{vermiculite-smectite} \rightarrow \text{degraded smectite} \rightarrow \text{allophane} \qquad (2)$$

$$\left.\begin{array}{l} \text{Chlorite-vermiculite} \rightarrow \text{chlorite-smectite} \\ \text{Illite-vermiculite} \quad \rightarrow \text{illite-smectite} \end{array}\right\} \text{degraded smectite} \qquad (3)$$
$$\downarrow$$
$$\text{allophane}.$$

Many examples of such processes are provided in the literature. Camez (in Millot 1964, 1970) describes the development of chlorite-vermiculite, illite-vermiculite and illite-montmorillonite in various rendzinas and brown soils of Northeastern France. Lamouroux et al. (1967) notice the parent-rock control on the pedogenesis of sedimentary rocks of Lebanon, and the accessory formation of irregular mixed-layers and degraded smectites in mediterranean-type soils. Revel and Margulis (1972) also indicate the detrital character of clay minerals in many red soils developed on peri-mediterranean limestones.

Tardy and Gac (1968), Hetier et al. (1969) and Meilhac and Tardy (1970) demonstrate the moderate evolution of most arenas and soils covering the crystalline rocks of Vosges (NE France): detrital illite issues from mica fragmentation and feldspar retrodiagenesis, and is associated with secondary Al-vermiculite resembling chlorite-vermiculite, and sometimes with degraded smectite. Plagioclases constitute the main source of clay minerals issued from granite weathering. Reynolds (1971) indicates the formation of mixed-layer vermiculite-phlogopite, vermiculite and smectite at the expense of igneous and metamorphic rocks in the alpine zone of the Northern Cascades, Washington, USA. He emphasizes the importance of chemical weathering, that appears to develop at a unit area rate much higher than the estimated rate of clay erosion.

Bottner and Paquet (1972) and Vaudour (1979) show the importance of vermiculitization in calcareous soils (rendzinas) from French and Spanish mountain areas. Kha Nguyen and Paquet (1975) observe in clay-rich brownish soils (pelosols) the stopping of weathering processes from mixed-layer stage. Moberg (1975, 1976, in Petersen and Rasmussen 1980) describes in Danish soils a dominance of randomly interstratified minerals consisting of illitic, vermiculitic, and smectitic components. Churchman (1980) describes the following transformations in some New Zealand soils, under normal pH conditions:

Chlorite → swelling chlorite → chlorite-vermiculite
Illite → illite-vermiculite → illite-smectite → smectite
Illite → vermiculite → degraded smectite.

To summarize, the weathering processes developing under temperate climate basically correspond to the formation of altered 2:1 clay minerals, dominated by irregular mixed-layers and poorly to medium-crystallized smectite. In a general way, vermiculite mixed-layers dominate in cool-temperate areas (middle high latitudes, middle altitude mountains), while degraded smectites are more abundant in warm-temperate and humid areas. These processes are sometimes referred to as pseudo-bisiallitization, because 2:1 minerals result from negative transformations rather than from neoformations. *The importance of mixed-layer minerals* in the temperate continental range is emphasized by Russian estimations on illite-vermiculite abundance in surficial horizons of the world soils (Fig. 2.5. Gradusov 1974). Transitional minerals like mixed-layers testify to the moderate and incomplete character of hydrolysis.

As the different mineral species of a given rock do not present identical weathering vulnerability and speed, different secondary minerals may develop at the same time in a given hydrolytic context. For instance, a granite submitted to temperate weathering conditions shows the following evolution (Tardy 1969):

Quartz   → little dissolved quartz
Muscovite   → exfoliated illite, illitic mixed-layers, vermiculite
Plagioclase → kaolinite.

This is the reason why kaolinite and even gibbsite can be encountered in some soils of humid-temperate areas, like recent arenas developed on granites in

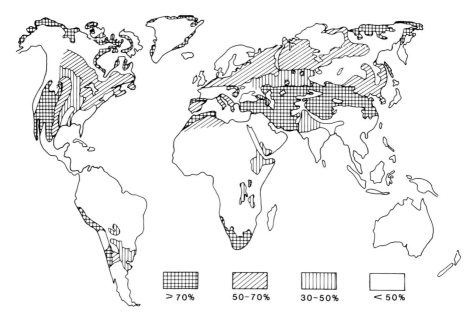

Fig. 2.5. Average percentages of illite-vermiculite minerals in the surficial horizons (A1 and/or A2) of the world soils. (After Gradusov 1974)

the Massif Central, France (Dejou et al. 1968, Maurel 1968) and on gabbros of Quebec, Canada (Clément et al. 1979). The abundance of these secondary siallitic and allitic minerals usually remains low in recent pedologic environments, and does not contradict the widespread importance of transitional 2:1 minerals.

Hot-Wet Climate

Intertropical environments characterized by high rainfall, active drainage, and oxidizing conditions lead to the formation of whitish-yellowish to red soils conveniently grouped under the term *laterite*. Laterites include widespread soils developed in both tropical-humid zones (precipitation: 130 to 300 cm/y) and equatorial zones (200 to 800 cm/y), and issued from leaching by temporary or permanent groundwater. Laterites may reach a thickness of 10 to 30 meters, and even more. Latosols constitute reddish brown soils formed on iron-rich parent rocks (ferruginous sediments, dolerites) under warm-humid conditions, and are close to laterites.

Intense leaching in hot environment prevents the accumulation of organic matter, and determines the flushing out of most of the mobile constituents of parent rocks. Aluminum and titanium remain preferentially in the weathering complex, together with variable amounts of silicon and iron. The combination of these residual elements allows the formation of new minerals, depleted in or devoid of silica. Strong hydrolysis leads to the neoformation of *kaolinite*, whose silica-aluminum 1:1 layer structure corresponds to the chemical process of monosialitization (Table 2.3). Goethite is often associated with kaolinite. Very strong hydrolysis leads to the neoformation of *gibbsite*, more rarely of boehmite, two aluminum oxides that represent pedogenic end-members and characterize alitization processes.

The relative abundance of kaolinite and gibbsite in a given lateritic soil mainly depends on the presence of a permanent groundwater level. A fluctuating or temporary level favors the flushing out of silica, and therefore the formation of gibbsite. However, even in gibbsite-rich horizons, kaolinite is present in noticeable amounts. As a consequence, kaolinite constitutes the most typical pedogenic mineral issued from weathering under hot-wet climate. Extended alitization processes were responsible for the formation of *bauxite* (Al ore) at different geologic periods (e.g., late Jurassic, late Cretaceous of Southern Europe). Bauxites consist of in situ or reworked paleopedologic formations. Bauxitization represents the result of evolved pedogenic mechanisms developed at the expense of various rocks and soils under long-lasting warm and humid conditions, in rather stable areas submitted to very intense leaching (see Millot 1964, 1970, Valeton 1972).

Because of the intensity of hydrolysis and the exportation of most cations, the petrological nature of parent rocks does not influence the mineralogical composition of laterites in a dominant manner. Kaolinite, associated or not with gibbsite and goethite, develops on nearly all types of igneous, metamorphic, and sedimentary rocks under hot-wet climate. Some residual minerals are nevertheless preserved or reworked in some lateritic profiles. This is especially the case with quartz and micas. Micas constitute detrital illitic materials issued either from the fragmentation of granites or from feldspar sericitization (regressive diagenesis; e.g., Lelong and Millot 1966). Some other clay minerals may develop temporarily from the first weathering stages in the lower horizons of lateritic profiles. For instance, irregular mixed-layers, vermiculite, and smectite form transiently in the basal zone (C horizon) of African and South-American soils, especially those characterized by permanent groundwater (in Millot 1964, 1970, Volkoff and Melfi 1978).

Amorphous to diversely crystallized iron oxides are specific formations commonly associated with laterites and other tropical soils. The term laterite was first used to describe ferruginous crusts of south-central India, which emphasizes the close relationships existing between both types of pedological formation. Lateritic soils are sometimes called fersialitic (abundant kaolinite) or ferralitic (abundant gibbsite) because of their high iron content. *Iron crusts,* or *ferricretes,* develop mostly under wet tropical climate marked by a fairly long dry season. Different stages of development are recognized by Nahon (1986): (1) precipitation and crystallization of iron oxyhydroxides from ions leached through parent rock hydrolysis; (2) formation of iron oxide nodules growing by successive generations; (3) formation of pseudoconglomeratic and pisolitic iron crusts. The epigenic accumulation of iron corresponds to the progressive removal of detrital quartz and of kaolinite, in the upper levels of the pedologic profile (Fig. 2.6).

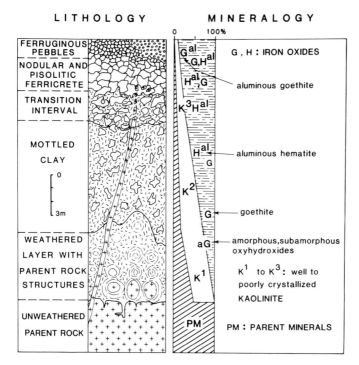

Fig. 2.6. Lithology and mineralogy of a ferricrete profile. (Nahon 1986). The *size of the symbols* corresponds to the relative abundance of each iron compound

Antagonistic Change of Temperature and Humidity

Cool-Wet Climate

Humus tends to accumulate in weathering profiles developed in well-drained forested areas submitted to cool and humid climate. Such conditions determine an acid environment (pH < 5) and the formation of abundant organo-mineral complexes. The weathering process dominantly corresponds to *acidolyse* (Table 2.1) and leads to the formation of blackish-brownish soils called *podzols*. A mature podzol (thickness usually < 1 m) contains different types of organic matter located at various levels: more or less decayed plant debris in surficial horizon (O), black raw humus in leaching horizon (A1), compact and evolved dark-brown humus precipitated in the upper part of accumulation horizon (B1). The rest of the profile is largely composed of sandy and permeable materials.

In acidolytic environments, all cations tend to be removed from the minerals and to form soluble salts (Pédro et al. in Bonneau and Souchier 1977). This is the reason why parent-rock minerals are strongly destroyed during *podzolization*. Aluminum is exported, together with other cations in the leaching solutions, which differs from most of other weathering processes. Silicon constitutes a much less soluble cation, and therefore tends to concentrate within the soil, comparatively to other elements, especially aluminum. As a result, many podzols are

Table 2.4. Clay mineralogy (%) in a podzol profile developed on Vosges sandstones, Alsace, NE France. (After Millot (1964)

Soil horizon	Chloritic minerals	Illite	Illite-vermiculite mixed-layers	Vermiculite
A_o		40	20	40
A_1	10	70	10	10
A_2		70	10	20
B	10	90		
Niederbronn sandstone		100 (mica)		

characterized by the relative abundance of *secondary silica* precipitated as opal or cristobalite (e.g., Swindale and Jackson 1960).

The clay minerals of podzols are essentially *degraded forms of parent-rock minerals*. Strongly altered illite, poorly defined vermiculite and vermiculitic or smectitic mixed-layers are commonly identified (e.g., Bryant and Dixon 1964, Millot 1964, 1970; Table 2.4). If parent rocks contain smectite, this mineral can be preserved at the base of profiles or in little evolved podzols. Kaolinite, gibbsite, and quartz better resist acidolyse, and are often present in upper soil horizons. Very mature podzols may contain exclusively secondary silica, amorphous organo-mineral complexes, and remains of strongly degraded micas.

Podzolization is not only restricted to cool and humid areas characteristic of middle-high latitudes and altitudes. Very good drainage and siliceous substrate favor the development of acidolytic weathering under temperate and even warm climate. Some porous sandstones combine both these conditions, which explains the presence of gray-brown to red-yellow podzolic soils in various regions. This is especially the case in numerous old dune or river deposits (e.g., Quaternary sands of Landes, France; Tertiary sands of the Paris Basin, Robin 1979; late Quaternary fluvial sediments of South Kalimantan, Indonesia; etc).

Hot-Dry Climate
The absence or rarity of rainfall in hot deserts prevents chemical weathering. Hydrolytic processes are very low, and mostly restricted to the removal and precipitation of iron and manganese oxides. Due to the weak leaching, bases are sparsely exported and allow local crystallization of halite, gypsum or lime. Most clay minerals proceed from physical weathering and reflect the *composition of the parent rocks*. The scarceness of liquid water in hot-dry and very cold continental regions determines the production of similar clays and associate minerals. Large deserts like the Sahara are developed in old and stable cratonic areas made of crystalline rocks. This explains the frequent abundance of illitic and chloritic minerals in sediments issued from desert erosion.

Temperate to Warm-Subarid
Temperate to warm areas submitted to a strong alternation of wet and dry seasons are characterized by discontinuous weathering processes. Hydrolysis intervenes actively during the wet season (50-130 cm/y), especially if the tempera-

Table 2.5. Clay mineralogy (%) of a vertisol (black tirs) developed on a clayey calcareous alluvium of Rharb, Morocco. (After Paquet 1970)

Depth below soil surface (cm)	Chlorite	Illite	Irregular mixed layers	Smectite	Kaolinite
0– 10				90	10
15– 30				90	10
50– 70				90	10
90–110	Traces	Traces	80		20
120–130	10	10	60		20
140–150	30	30	20		20
180–190	30	30	20		20

ture is high. On the other hand, the ions released from minerals tend to be stored and to concentrate in the soil profile during the dry season, particularly if this season is longer than the wet one. The resulting soils are labeled *vertisols*, tchernozems, or receive more local denominations (black earths, tirs, regurs,etc.). They occur commonly in various mid- to low-latitude regions, like the Ukraine (= original tchernozems), the Middle East, South India, East Australia, Dakota to Texas, North and West Africa. Tchernozems may reach a thickness of several meters. They are mainly composed of a thick, black, granular A horizon, and of a light brownish, more or less calcareous B horizon. The structure of the soil is often prismatic and unstable; the pH tends to increase downwards, from slightly acid to alkaline values.

The chemical elements released by hydrolysis in the weathering complex comprise various bases, silicon, iron and even aluminum, associated with organic matter in A horizon. During the dry season, these elements combine and form new minerals. The excess of Ca gives way to the development of calcite, whatever the parent rock is like. This is the reason why vertisols belong to the group of calcimorphic soils.

Clay mineralogy is usually characterized by the neoformation of *well-crystallized smectites*, corresponding to a true bisialitization. Depending on the intensity of hydrolysis and therefore on the climatic zone, smectites are abundant to very abundant. They commonly reach 60 to 90 % of the clay minerals in the upper soil horizons, in temperate-warm to dry tropical zones, regardless of the nature of the parent rocks (e.g. Redmont and Whiteside 1967 in Loughnan 1969; Paquet 1970. Table 2.5). Most vertisol smectites are rich in iron and belong to the di-octahedral group *Al, Fe beidellite-Fe beidellite-nontronite* (Trauth et al. 1967, Paquet 1970). Aluminous beidellites from vertisols developed on Togo gneiss and amphibolite (Africa) are reported as neoformation or transformation products by Kounetsron et al. (1977).

In temperate arid regions the neoformation of smectite in vertisols often occurs simultaneously with negative transformations (e.g., open illite, irregular mixed-layers) and the preservation of some parent-rock minerals. In arid tropical areas smectite neoformation often becomes exclusive. The pedogenic smectitization can largely extend over the humid intertropical zone, if the drainage conditions are poor and prevent the normal evacuation of ions. This is the case of some flat coastal zones like the Niger delta, where ions issued from both upward and local weathering concentrate in spite of high rainfall. Of course many, transition situations exist between the typical calcimorphic vertisols and the soils developed in adjacent climatic zones: brunizems and prairie soils characterized by a more organic environment, brown soils and laterites marked by more regular precipitation.

The contrasted rainfall seasonality in arid regions determines sometimes the individualization of *inorganic crusts* through pedogenic processes. This is the case of iron crusts or *ferricretes* associated with laterites, in which iron oxides progressively replace the kaolinitic clays (see 2.2.1). Silcretes and calcretes proceed from similar phenomena. *Silcretes* represent accumulations of silica slabs in various arid to fairly humid and hot environments (e.g., Mauritania, Sudan, South Africa, Australia; see Millot 1964, 1970, Langford- Smith 1978, Thiry 1981, Summerfield 1983). Usually formed by quartz in surficial conditions and by disorganized calcedony and opal in soil profiles, silcretes are mostly fossil formations, whose clay content is very low and often little known. Silcrete may be associated with strong lateritic to little evolved soils (e.g., Meyer 1987).

Calcretes, also called caliche if developed on calcareous rocks, deserve special attention since they represent the lateral equivalent of vertisols and are associated with the genesis of Mg-clays: *Mg-smectite, palygorskite* and even *sepiolite* (e.g., Goudie 1983, Singer and Galan 1984, Singer 1984). They correspond to weathering conditions marked by the combined occurrence of hydrolysis and *alcalinolysis,* processes. Calcretes consitute accumulations of calcite slabs, several decimeters to meters thick. Some of them undoubtedly formed during the middle-late Quaternary, and probably continue to develop today. Calcareous crusts are particularly well known in the subarid belt of the northern hemisphere.

Paquet et al. (1969) explain the presence of palygorskite (attapulgite) in some Morocco crusts by the trapping and organization in the plains of ions issued from upstream soils. In South Nevada, calcretes can contain either dominant palygorskite or dominant sepiolite in the clay fraction (Gardner 1972, Hay and Wiggins 1980). In their review on calcic sols, pedogenic calcretes, and other surficial carbonates of semi-arid region of Southwestern USA, Bachman and Machette (1977) indicate the widespread occurrence of palygorskite compared to sepiolite. Sometimes fibrous clays exist already in the parent rocks, and their pedogenic origin is then questionable (Frye et al. 1974, Shadfan and Dixon 1984). Nevertheless, Millot et al. (1977) demonstrate that palygorskite can undoubtedly form in surface meteoric conditions. Fibrous clays and associate minerals often appear to constitute a transitional phase between the parent-rock silicates and the pedologic calcite accumulations. Palygorskite and Mg-smectite can develop on various crystalline substrates, like granites and green schists, providing that hydrolysis and evaporation conditions are strong enough. In the Adana region, Turkey, Kapur et al. (1987) describe the weathering of Pliocene clays into calcite and palygorskite, the latter mineral being secondarily leached and transformed into smectite; this polygenic evolution ended in a clay coating and a blanketing with acicular rhomboedral calcite.

To summarize, calcretes constitute contracted and little differentiated soils, formed slowly under subarid and warm climate. The clay fraction of calcretes is

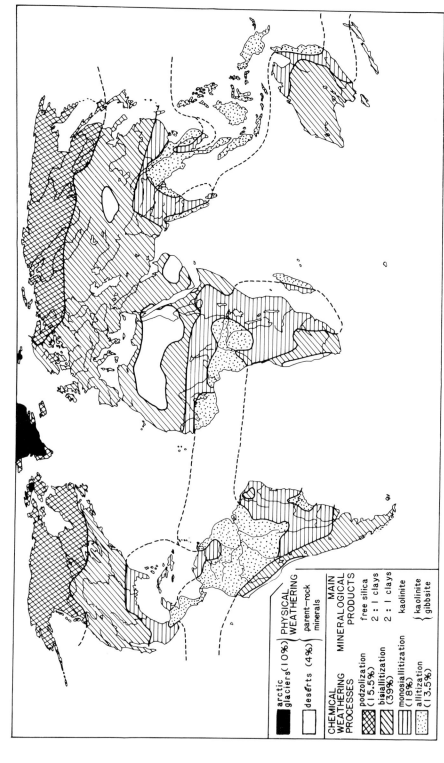

Fig. 2.7. Potential weathering processes at the surface of the Earth, and major resulting soil mineralogical belts (After Pédro 1968). % = surface controlled by a given process

not abundant, and often includes palygorskite or other magnesian minerals. Mg-clays are particularly frequent at the base of the calcitic slabs, and tend to be progressively replaced by a calcitic epigeny.

General Distribution of Zonal Soil Clays

A few types of zonal weathering processes and of soils present a broad distribution on the world land masses.

1. Both very cold and very dry climates determine an essentially physical weathering, and the production of rock-derived clay minerals dominated by illite and chlorite.

2. Cool and humid conditions are responsible for the development of podzols, marked by free silica and highly degraded rock-derived clays.

3. Both temperate-humid and warm-subarid climate favor the genesis of 2:1 layer phyllosilicates. Under temperate conditions the minerals consist of altered clays such as irregular mixed-layers, vermiculite, and degraded smectite (pseudo-bisialitization). Under higher temperature and more contrasted seasonal humidity, they include mainly newly formed ferriferous smectites (true bisialitization).

4. Hot and wet climate induces active hydrolysis and ion leaching. Neoformations occur that depend on the amounts of silica, iron and aluminum still available in the soils. Kaolinite, often associated with goethite, is typical of low latitude soils (monosialitization), and is partly replaced by gibbsite under very aggressive conditions. Figure 2.7 represents the theoretical distribution of the few dominant weathering processes at the surface of the Earth (Pédro 1968). One should notice the widespread importance of bisialitization (2/5 of the surface), that nevertheless corresponds to different clay varieties according to regional pedoclimatic characteristics.

2.2.2 Azonal Soils

Halomorphic Soils

Salts may accumulate through evaporation in depressions of temperate to hot regions, and participate in the formation of halomorphic soils. These soils are characterized by the excess of sodium and sometimes of potassium salts. They include two types: (1) White alkali soils or *solontchaks* form under very evaporative conditions, are fairly thin, and consist mainly of chloride. (2) Black alkali soils or *solonetz* correspond to more humid conditions, comprise both an A horizon with appreciable organic matter and a B horizon with a prismatic structure, and are rich in sodium carbonates. Salts mainly issue from hydrolysis of surrounding slopes, often associated with cyclic desiccation of preexisting lakes.

In most cases, the clay mineralogy of halomorphic soils reproduces that of surrounding parent rocks or underlying soils and sediments. Thus weathering processes related to salt environments, referred to *as salinolyse*, are often of little influence on clay production and evolution. Nevertheless, the upper zone of some evolved solonetz may undergo hydrolytic changes, especially under very hot and evaporative conditions associated with fairly long humid seasons. Bolishev and Kapust-kina (1964) observe in Russian solonetz the development of amorphous materials in A1 horizon, suggesting a degradation of alumino-silicates. Paquet et al. (1966) describe the degradation of vertisolic Fe-smectites in the upper part of solonetz from Chad, by leaching in a sodico-potassic environment (solodization). A few examples of authigenic clay formation are reported in intertropical soil to lake environments, like the borders of lake Chad. These phenomena will be considered together with saline lake deposits (Chap. 4).

Hydromorphic Soils
If the water table is very close or even rises to the Earth's surface, soils are labeled hydromorphic. This is the case of morphological depressions of humid regions located at various latitudes and altitudes: peri-lacustrine areas, alluvial plains, over-deepened glacial basins, peat bogs. Hydromorphic soils may also occur in high terrains where the presence of an impermeable substrate impedes vertical drainage. They usually show abundant organic matter at the surface, and an underlying bluish gray mottled horizon characterized by iron reduction and clay accumulation. This horizon, and by extension the soil, constitutes a *gley*.

Hydrolysis can hardly develop in drowned soils, because of the low mobility of groundwater and of the impermeability of the clayey horizon. Despite its abundance, water is not able to determine an active ion subtraction from parent-rock minerals. As the released ions are not evacuated by running water, they accumulate in the soil and determine both a chemical saturation and the stopping of hydrolysis. As a result, clay minerals of most hydromorphic soils simply reflect the *composition of surrounding parent rocks and soils*, from which they derive. Transitional situations exist toward hydrolytic, acidolytic or salinolytic soils if drainage or evaporation increases (pseudogleys, well- drained peat soils, temporary saline soils).

Volcanic Environments
Weathering tends to progress on volcanic rocks faster than on most other rocks. The abundance of amorphous materials facilitates hydrolysis; the frequency of porous rocks, like pyroclastites, favors leaching and drainage. Soils growing on volcanic rocks rapidly become fairly thick, are grayish and often grouped under the term *andosols*.

The rapidity of the volcanic rock weathering often induces an azonal distribution of soil clays, some minerals usually formed at low latitude or altitude being able to develop under less hydrolyzing conditions. This is especially the case of *smectites*, that form easily on basaltic materials as Fe-Mg species, and on rhyolitic materials rather as Al species. Fairly well to very well crystallized smectites form abundantly on volcanic rocks under temperate-warm to cool climate. Abundant *kaolinitic minerals* proceed from similar alteration in hot to temperate climatic zones. Four steps of pedogenic clay formation can be schematically identified, according to the climate, the drainage, the local petrography, and the duration of weathering.

1. The first weathering step consists of the formation of amorphous to sub-amorphous clays, like *allophane* and *imogolite*. Frequently described in Japanese

andosols (e.g., Yoshinaga and Aomine 1962, Henmi and Wada 1976; see also the syntheses of Wada 1977, and Sudo and Shimoda 1978), imogolite also occurs in surficial weathering crusts on basalts of the Massif Central, France (Torrent et al. 1982). Non crystalline clays appear to be preferentially preserved in young andosols and to present a relatively short life time (Wada and Aomine 1973, Wada 1977).

2. The second stage corresponds to the widespread development of *Fe- to Al-smectites* (nontronite, montmorillonite, etc.), under temperate conditions. Abundant volcanogenic smectites are commonly described in Europe from Mediterranean regions to Iceland, where they may occupy the largest part of soil profiles. For instance, Lamouroux et al. (1967) analyze young soils formed above fresh basalts of Lebanon and exposed to a temperate-warm and subarid climate; exclusive and very well-crystallized smectite forms as soon as the weathering starts, and is present up to the soil surface, where little kaolinite appears. Hétier et al. (1977) also describe the formation of smectite during the first weathering steps of basalts in the Massif Central, France. In organic-rich andosols, organo-mineral complexes may form and prevent the swelling-collapse effects usually shown by successively glycolated and heated smectite; smectite therefore behaves like swelling chlorite or pseudochlorite (Moinereau 1977). Corrensite may form a transitional stage of volcanic rock weathering toward smectite, as reported in ophiolites in North Italy by Brigatti and Poppi (1984).

Hot and humid climate appears too aggressive to allow the permanence of smectites in volcanogenic soils. A compilation by Carroll and Hathaway (1963, in Carroll 1970) emphasizes the decrease of smectite abundance related to the increase of weathering and soil development (Fig. 2.8). Quantin et al. (1975) observe the development of iron smectites in the New Hebrides only within andosols poorly drained and submitted to subarid conditions. Similarly, Kantor and Schwertmann (1974) show that smectites are restricted to the lowest and flattest parts of red-black soils covering basic igneous rocks in Kenya. These Kenyan smectites sometimes present specific morphologies, like curled particles (Van der Gaast et al. 1986).

3. The third step is characterized by the extension of *halloysite*, a hydrated form of kaolinite. Halloysite represents a fairly stable mineral in intertropical soils submitted to high rainfall with a short but significant dry season. This is the case with young soils formed on basalts in Cameroon (Sieffermann et al. 1968) and in the New Hebrides (Quantin et al. 1975). Sometimes, as indicated by Herbillon et al. (1981) in Burundi, halloysite or metahalloysite constitutes in fact kaolinite-smectite mixed-layers (= $7-14_{Sm}$), a transitional phase of smectite alteration into kaolinite. In Central Spain, Martin de Vidales et al. (1987) describe the weathering of olivine basalts and nephelinites into halloysite, allophane, beidellite and a kaolinite-beidellite mixed-layer, which indicates the complexity of alteration processes under semi-arid conditions.

In temperate regions, halloysite can constitute significant amounts of some weathering crusts (e.g., Great Britain, Carroll and Hathaway 1963, in Caroll 1970), of soils on basalts (Massif Central, France; Hétier et al. 1977, Torrent et al. 1982), or of direct alteration products of alkaline trachytic pumices (Central Italy; Quantin et al. 1987). It also forms easily by weathering of airfall tephras, as

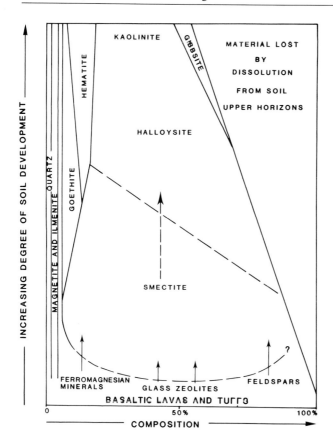

Fig. 2.8. Mineralogical relationships and successions of soils developed on basalts in various localities of Asia, Hawaï and Europe. (After Carroll 1970)

in New Zealand (Lowe 1986). The age and stability field of halloysite is still little known in moderately hydrolyzed environments.

4. Extended kaolinization, often associated with the formation of gibbsite, characterizes the last step of andosol development. *Kaolinite* and *gibbsite* form easily in humid intertropical regions, where the influence of parent rocks is only accessory to climate and leaching influence. Kaolinite replaces halloysite and is pecularly abundant where there is active drainage and no marked dry season (e.g., Sieffermann et al. 1968, Chamley 1969, Kantor and Schwertmann 1974, Quantin et al. 1975). Gibbsite forms in larger amounts if the volcanic eruptions are older and the weathering complexes more evolved, as reported by Rutherford and Watanabe (1966) in Papua-New Guinea.

Under temperate climate, pedogenic kaolinite and gibbsite are never very abundant, but develop more extensively and rapidly at the expense of volcanic rocks than of others (e.g., Hétier et al. 1977).

The four weathering steps described above should not be considered as a rule, and admit some exceptions. In particular, the *airfall tephras* display various situations, because of their large diversity of age, sources, chemical composition, grain size, permeability and reworking processes. For instance, the tephras expelled in 1980 from Mt St Helens (Washington, USA) contained no clay weathering products five months after the eruption, but saponite and zeolites formed hydrothermally prior to the volcanic event (Pevear et al. 1982). The Holocene air-fall tephras in the alpine environment of the Southern Canadian Rocky Mountains show no clear relationships between weathering conditions and products (presence of chlorite, hydrobiotite, vermiculite, smectite), because of the polygenic origin of phyllosilicates derived from both alteration and contamination (King 1986). In Northern Kivu (Zaire, Africa) cinerites submitted to tropical humid conditions contain all the more halloysite and kaolinite since the samples are located farther from the Nyamuragira volcano and therefore the grain size is finer (Gastuche and De Kimpe 1961). An exhaustive study by Lowe (1986) on New Zealand tephras shows that allophane and imogolite possibly represent reaction end points, and that halloysite and gibbsite can form directly from dissolution, whatever the weathering duration. The main control factors consist of the availability of silica and alumina, leaching conditions, organic matter content and pH, and lithological pecularities.

Epeirogenic Environments

In very young mountain regions, the formation of evolved soils does not occur easily, because vegetation hardly develops and erosion is very active on sloped relief. This is the case in alpine environments where lithosols, young rendzinas and rankers often represent the only widespread weathering complexes, and where soil minerals closely resemble those of parent rocks (e.g., French Alps, Chamley and Portier 1974).

If such a strong erosion, due to very sloped landscapes and to epeirogenic activity, occurs in regions submitted to strongly hydrolyzing climate, azonal soils form. These soils appear little evolved because active erosion continuously removes the surficial horizons. Soils do not reach a stage of equilibrium with local climatic conditions. Their clay assemblages falsely suggest the occurrence of a dry or cool climate rather than a humid and warm one. Such tectonic instability explains why many South Japanese soils blanketing crystalline rocks are depleted in neoformed smectite and kaolinite (cf. Sudo and Shimoda 1978), compared to other areas located in the same subtropical humid belt. The abundance of *illite, chlorite, and irregular mixed-layers* in these soils expresses the *strong erosion of tectonically rejuvenated relief*, and not a slightly hydrolytic climate. Similar observations have applied to other instable areas, like various Mediterranean regions (e.g., Peloponnesus, Sicily) since the late Cenozoic. Comparable phenomena may have occurred extensively during the major tectonic periods of geologic history.

2.2.3 Paleosols and Clay Content

Ancient soils are frequently described in geologic series, and can provide useful information on past climate, topography, organic activity, as well as on successive environmental or diagenetic changes (e.g., Wright 1986, Meyer 1987). Paleosols are all the better preserved and recognized since they are younger and harder, and the search for their occurrence in Paleozoic or older formations often represents a difficult task. This is the reason why terms used for designating past

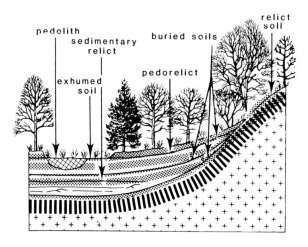

Fig. 2.9. Common terms used in the study of fossil soils. (Retallack 1983)

soils should be defined precisely (Fig. 2.9). Paleosols are usually characterized by tubular root structures, burrows, or even soil horizons. Buried soils represent nearly complete and unmodified soil profiles covered by a sediment influx like alluvium. A relict soil corresponds to an old soil profile still exposed, while a pedorelict designates soil features formed in a horizon or soil different from the one in which they appear. Exhumed soils are buried soils subsequently exposed through erosion. Pedoliths result from the reworking and deposition of old soil components. Clay assemblages of paleosols sometimes contribute in revealing the past climate and other environmental conditions (e.g., Singer 1980, Meyer 1981). Let us consider a few examples about different geologic periods.

Quaternary

Various studies have been devoted in temperate regions to Quaternary paleosols, that mainly developed on *river terraces* during interglacial stages. Clay minerals of interglacial soils usually present an increase of irregular mixed-layers, vermiculite, degraded smectite, and sometimes kaolinite, compared to illite- and chlorite-rich glacial stage deposits (e.g., Redondo 1973). Such differences between paleosols and surrounding sediments obviously do not occur in peri-desert environments, often devoid of important climatic variations over the course of the time (e.g., Coudé-Gaussen et al. 1982, Coudé-Gaussen and Rognon 1986).

In temperate zones, clay assemblages commonly express an *increase in weathering effects in old paleosols*, compared to young ones (e.g., Alimen and Caillère 1964a and b, Icole 1973, Bornand 1978, Mary and Grenèche 1986). For instance the Villafranchian (late Pliocene-early Pleistocene) sediments of the peri-Mediterranean range are often blanketed by rubefied soils rich in kaolinite (Vaudour 1983), contrary to what occurs on middle to late Pleistocene sediments (e.g., Redondo 1973). Soils formed on alluvial and aeolian deposits of Touraine, France, show a decrease in fragile primary minerals and a correlative increase in secondary minerals, from late Pleistocene to late Pliocene (Macaire 1986; Fig. 2.10). The older the soils, the more weathered they appear. Such a change is

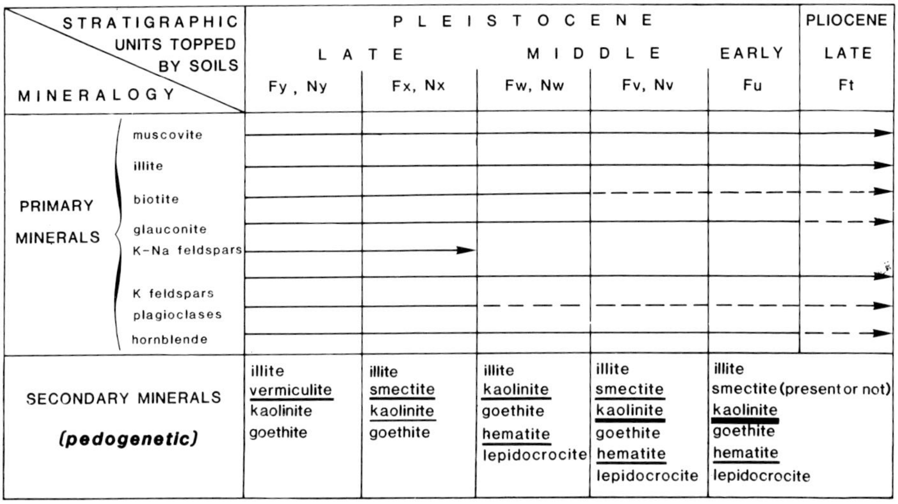

STRATIGRAPHIC UNITS TOPPED BY SOILS / MINERALOGY	PLEISTOCENE					PLIOCENE
	LATE		MIDDLE		EARLY	LATE
	Fy, Ny	Fx, Nx	Fw, Nw	Fv, Nv	Fu	Ft
PRIMARY MINERALS — muscovite						→
illite						→
biotite			— — —	— — —	— — —	— →
glauconite						— →
K–Na feldspars		→				→
K feldspars			— — —	— — —	— — —	→
plagioclases						→
hornblende						— →
SECONDARY MINERALS (pedogenetic)	illite, vermiculite, kaolinite, goethite	illite, smectite, kaolinite, goethite	illite, kaolinite, goethite, hematite, lepidocrocite	illite, smectite, kaolinite, goethite, hematite, lepidocrocite	illite, smectite (present or not), kaolinite, goethite, hematite, lepidocrocite	

Fig. 2.10. Mineralogical evolution in B horizons of Plio-Pleistocene paleosols in Touraine, France. (After Macaire 1986). *F* aeolian deposits, *N* river deposits

partly due to the severe hydrolytic conditions in the early Pleistocene and late Pliocene. But the major cause is the duration and repetition of weathering , a given soil being possibly submitted to successive hydrolysis stages during the Quaternary. This is especially frequent in alluvial terraces, often characterized by long periods of exposure and high permeability. *The cumulative effects of weathering* therefore diminish the potential use of clay associations to define past climate. In addition many Quaternary soils are truncated by erosion, or modified by the post-depositional processes commonly occurring in permeable terraces (Barrière 1971, Singer 1980). In spite of this, the combined study of soil lithology, structure, and mineralogy strongly contributes to establishing *stratigraphic correlations* in Quaternary series (Bourdier 1961, Bornand 1978, Fedoroff 1986).

Climatic implications sometimes arise from lithological and mineralogical studies carried out on nontemperate paleosols. Gile (1967) investigates post-middle Pleistocene nodular calcareous soils in an ancient basin floor in New Mexico, USA. Both palygorskite and sepiolite, often accompanied by smectite, are identified. Especially well represented at the base of calcareous crusts, the fibrous clays indicate a subarid climate, different from the present desert climate. The increase of sepiolite toward the center of the basin is attributed to an increase in ionic concentration and evaporation in the soils. From a general point of view, nearly all recent clay formation processes (see 2.2) are recognized in Quaternary soils, because the climate was roughly similar to the present one, and because many soils started to form during the Quaternary.

Tertiary

Fairly abundant studies carried out on Tertiary weathering complexes indicate various conditions of clay genesis, among which *lateritization* is the most commonly quoted. Thick and extensive kaolinite blankets occur on various

Cretaceous sediments or older substrates, and correspond to Paleogene hydrolyzing climate. Such blankets are reported in various regions, such as around the Massif Armoricain, the Massif Central and the Ardenne, France (Estéoule-Choux 1967a, Voisin 1981, Dubreuilh et al. 1984), in Senegal and the Ivory Coast (Tessier and Triat 1973). The flint clays covering the Cretaceous chalk in the Paris Basin also proceed from Tertiary hydrolytic processes, in addition to the residues of chalk dissolution (Thiry and Trauth 1976a). All these formations represent the products of cumulative weathering, more or less reworked and accumulated in depressions. The broad latitudinal extension of early Tertiary lateritic facies emphasizes the *widespread character of warm and humid climate.*

Some studies allow precising past lateritization processes and recognizing pedologic clay successions. For instance, Desprairies (1963) describes in the Oligocene basin of Brioude (Central France) the progressive transition of a gneissic parent rock to a kaolinitic soil topped by a ferruginous crust. Thiry et al. (1977) show the successive formation of smectite, smectite-kaolinite mixed-layers and kaolinite, from the base to the top of early Eocene pedologic profiles developed on chalks in the Paris Basin (Fig. 2.11), under increasing hydrolytic conditions. Nilsen and Kerr (1978) study a well-preserved lateritic paleosol drilled at about 1270 meters below sea level, on a basaltic plateau located on the

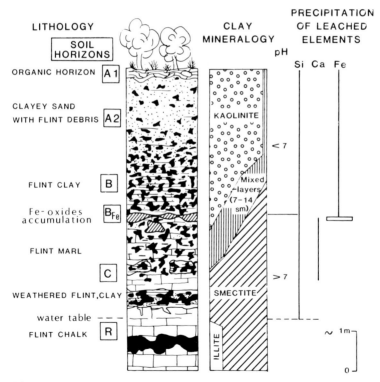

Fig. 2.11. Schematic weathering profile developed during the Eocene above Paris Basin chalk. (After Thiry et al. 1977)

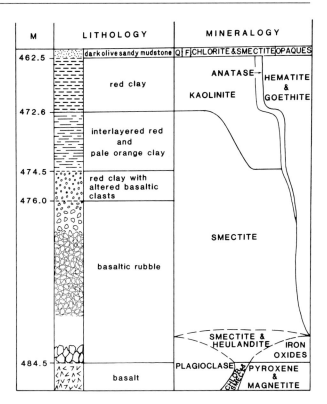

Fig. 2.12. Petrology and
mineralogy of early
Paleogene lateritic
paleosol, Site 336 DSDP,
Ireland-Faeroe ridge. (Nilsen and Kerr 1978)

NE flank of the Iceland-Faeroe Ridge, in the North Atlantic (Site 366 of the
Deep Sea Drilling Project). The paleosol is overlain by about 470 meters of
Eocene-Oligocene marine mudstone and Plio-Pleistocene glacio-marine sediment. Several units are recognized within the paleosol, that show the successive
replacement of basalt feldspars, pyroxenes and magnetite by montmorillonite
and then by kaolinite, hematite and goethite (Fig. 2.12). An upward decrease of
Si, Fe^{2+}, Mg, Ca, Na, and K is recorded, balanced by an increase of Al, Fe^{3+},
and Ti. All these characters demonstrate that the Iceland- Faeroe Ridge was located above sea level before the middle-late Eocene, and experienced a hot and
humid climate. The subsidence and submersion of the ridge, from the middle
Eocene, was responsible for the cessation of weathering and preservation of the
soil profile by the blanketing of subsequent marine sediments.

Evidence of arid climate and contrasted seasonal humidity arises from some investigations on Tertiary pedologic remains. Larqué and Weber (1984) identify in
the Paleogene of Le Puy basin (Massif Central, France) a Stampian kaolinization
of smectite-rich continental sediments, followed by smectitization in a calcimorphic soil. This pedogenic sequence corresponds to a climate successively warm-
humid and warm-subarid. Buurman (1980) investigates Paleocene soils in the
Reading Beds of the Isle of Wight (UK); illite, kaolinite, and smectite clay
minerals express both detrital sources and subarid weathering conditions. Bown

and Kraus (1981) report the existence of lower Eocene alluvial paleosols in the Northwest Wyoming (USA), whose characteristics indicate a temperate-warm to subtropical climate with alternating dry and humid seasons. Laurain and Meyer (1979) observe the formation of calcrete above weathered chalk in the North-East of the Paris Basin, and the concomitant transformation of kaolinite to kaolinite-smectite mixed-layers and smectite; this change is attributed to a climatic aridification. A similar evolution is recorded in Provence, France, at the expense of late Cretaceous bauxite (Chamley et al. 1976). Leguey et al. (1987) describe middle-late Miocene pedogenic processes in the Madrid Basin, Spain, at the expense of alluvial fan and fluvio-lacustrine deposits: smectite formed mainly during flood stages, while sepiolite developed rather under more arid conditions through smectitic clay weathering by Mg-rich leaching water. Many cases of carbonate, silica or iron oxide accumulations exist in Tertiary series, which indicates widespread hot climate and changing humidity (e.g., Gardner 1972, Meyer 1981, Thiry 1981, Retallack 1983, Summerfield 1983). Some paleogeographic reconstructions emphasize the diversity of Tertiary pedogenic processes associated with various sedimentation events, especially in peri-marine areas surrounded by sloped relief and submitted to sea-level fluctuations, like the Paris Basin in the early Paleogene (e.g., Thiry 1981).

Mesozoic

The most common weathering processes described in the Mesozoic series refer to *kaolinization*, and indicate the existence of warm, humid climate. Meyer (1976) identifies the pedologic development of kaolinite above the Wealdian clays (early *Cretaceous*) of the Eastern Paris Basin. The late Cretaceous of Rhodanian Provence (Southeast France) experienced successive weathering phases, developing extensively at the expense of various quartzitic sandstones, sandstones and limestones, and leading to the formation of kaolinite. For instance, the sedimentation on the left side of the Rhone Valley (Uchaux Massif) was interrupted at both middle Cenomanian and early Santonian, and the corresponding marine deposits were weathered into red or white kaolinitic sandstones frequently topped by ferricrete (Parron and Triat 1978). On the right side of the same valley (Gard) three weathering phases are recorded at middle Cenomanian, late Turonian and early Santonian (Parron and Triat 1977). Comparisons made between the different outcrops and areas allow stratigraphic correlations and paleogeographic reconstructions. The kaolinization is preceded, accompanied, and followed by various other weathering effects, from an early desilicification and decarbonation to a final ferruginization (Triat 1982; Fig. 2.13).

Jurassic series also reveal some examples of lateritization. For instance, Abed (1979) describes Toarcian red beds from Central Arabia that contain abundant kaolinite, hematite, and goethite, and are depleted in illite and feldspar. These deposits differ from the common Jurassic illite-rich redbeds, and probably consist of paleosol products reworked into the Toarcian sea. Goldbery (1979) interprets the laterite composition of the upper part of the early Jurassic fluviatile sediments of the Northeastern Negev (Israel) as the result of a strong weathering event. A diagenetic imprint, precisely identified by kaolinite crystallinity, subsequently modified the pedogenic signal (Valeton et al. 1983).

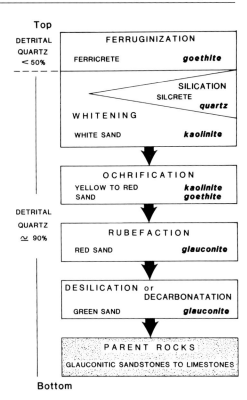

Top

DETRITAL
QUARTZ
< 50%

FERRUGINIZATION

FERRICRETE   *goethite*

SILICATION
SILCRETE
quartz

WHITENING

WHITE SAND   *kaolinite*

OCHRIFICATION
YELLOW TO RED   *kaolinite*
SAND   *goethite*

DETRITAL
QUARTZ
≃ 90%

RUBEFACTION

RED SAND   *glauconite*

DESILICATION or
DECARBONATATION

GREEN SAND   *glauconite*

PARENT ROCKS
GLAUCONITIC SANDSTONES TO LIMESTONES

Bottom

Fig. 2.13. Lithological and mineralogical evolution through surficial weathering of Cretaceous rocks of Rhodanian Provence, SE France. (After Triat 1982). *Arrows* indicate the downward progression of weathering fronts

Laterite formation extended widely in the middle-late Mesozoic, especially during the Cretaceous (Van Houten 1982; Fig. 2.14A). This indicates the development of hydrolyzing climate from low to fairly high latitudes (0-60°). One of the main causes is certainly the existence of low latitude seaways between the Tethys, Atlantic, and Pacific Oceans, allowing active interlatitudinal circulation among continental blocks. The situation was somewhat different in the late Permian and early Triassic, when laterite mainly formed at high latitudes, under moist, warm climate (Fig. 2.14B). At that time the connection of Gondwana and Laurasia prevented interlatitudinal oceanic circulation and determined the preponderance of warm, subarid climate, and therefore of calcrete formation, in low latitudes. The Triassic low-latitude dryness was followed by more humid climate in the early Jurassic, as suggested by the widespread occurrence of laterite and oolitic ironstone in Eurasia and Northeastern Gondwana (Van Houten 1982).

It should be noticed that *very few examples of true Mesozoic paleosols are reported* in the literature. Most descriptions and interpretations available are based either on hard weathering remains like kaolinitic sandstones, ironstones, calcrete and silcrete, or on continental or marine sediments issued from soil reworking (see Parts Two and Five). There are especially very few reports on the existence of old downstream hydromorphic and other *soft soils*. This lack of information results from the vulnerability of soft soils to erosion and from the specific effects of sea-level variations on the reworking of lowland soils, but also

A. CRETACEOUS

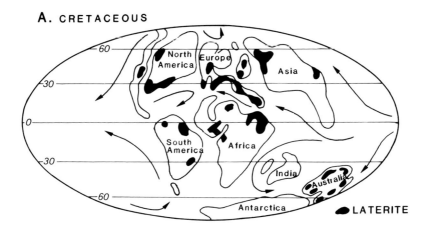

B. LATE PERMIAN-EARLY TRIASSIC

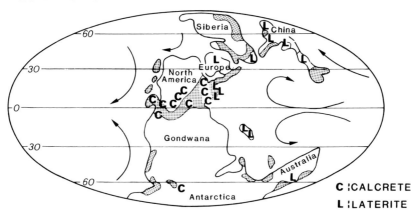

Fig. 2.14. Continental distribution of: **A** Cretaceous laterite; **B** Late Permian and early Triassic laterite and calcrete. (Smith and Bridden 1977, and Scotex et al. 1979, in Van Houten 1982). *Arrows* = estimated major oceanic surface currents

from the little attention often paid by geologists to paleopedologic manifestations.

Paleozoic, Precambrian

Laterite-ironstone, calcrete and silcrete are quoted in some early Phanerozoic and Proterozoic series. For instance in the middle and late Paleozoic, lateritic conditions were favored on dispersed Laurasian blocks of middle and high latitudes, but occurred on assembled blocks only in low latitudes (Van Houten 1982). Warm and humid to arid climate are deduced from the presence of various iron-rich deposits and calcrete in Proterozoic series of Canada, South Africa and India as old as 2300 million years (e.g. Windley 1977, Button and Tyler 1979, Chandler 1980). Silcretes are reported in Precambrian series as old as 1800 million years in the Northwest Canadian shield, and are attributed to arid climate (e.g. Ross and Chiarenzelli 1985). The occurrence of aluminum-rich weathering crusts of 3500 million years in Canada (Serdyuchenko 1968) suggests the existence of still older weathering processes in atmospheric conditions not very different from the subsequent ones.

Despite these data, *few clay mineral analyses* are available on Paleozoic and Precambrian pedologic formations. Wright (1982) identifies illite-smectite mixed-layers in early Carboniferous calcrete of South Wales (UK), and concludes a diagenetic modification of smectitic paleosols developed under a seasonally arid climate. Diagenesis appears to have been very low locally, which permitted the preservation of various pedogenic illite-smectite and kaolinite-smectite mixed-layers (Robinson and Wright 1987). Conrad and Michaud (1987) identify thick post-hercynian kaolinitic paleosols developed on Devonian claystones from the Central Sahara, Algeria. Jiranek and Jirankova (1987) describe the massive kaolinization of arkosic molasse in the Plzen basin, Bohemian Massif (Czechoslovakia), during the late Carboniferous.

The scarcity of mineralogical information about very old paleosols results from several causes, in particular the possibility of truncation and cumulative effects as already expected for Quaternary soils (e.g. Power 1969), and also the increased chances of diagenetic modifications due to the succession of geological cycles. Special efforts should be made to better identify the nature and character of Proterozoic and early Phanerozoic soils (e.g., McDonald 1980, Meyer 1987). The careful study of pedogenic clay assemblages would probably provide some useful paleoenvironmental informations, since meteoric conditions seem to have been not very different from now as far back as 3800 million years (Schopf 1983, Cairns-Smith and Hartman 1986). Some Precambrian sediments and early Paleozoic show no extensive mineralogical changes related to metamorphism or burial diagenesis (e.g., Chamley et al. 1978a, 1980a), suggesting similar characters for interbedded paleosol clays.

2.3 Conclusion

1. *The main types of recent weathering complexes and soils present a latitudinal and altitudinal distribution chiefly controlled by hydrolysis.* The clay assemblages developed through weathering reflect the corresponding pedoclimatic conditions. Schematically, a very cold or hot-dry climate leads to rock-derived minerals, often composed of dominant illite and chlorite. Cool-humid conditions favor the intense degradation of primary minerals and the preservation of free silica and organic matter. Temperate climate gives way to incompletely altered minerals, mostly exfoliated illite and chlorite, irregular mixed-layers, vermiculite, and degraded smectite, with changing proportions according to local petrography, precipitation, drainage, and temperature. Subarid areas favor the formation of well-crystallized Fe-smectite and sometimes fibrous clays (mostly palygorskite), particularly developed in warm and poorly drained regions submitted to long dry seasons. Hot and wet climate is characterized by the neoformation of kaolinite and goethite, and of gibbsite under very intense leaching. *This basic zonation can*

be diversely modified or hidden due to local conditions. In particular, volcanic substrates usually allow an acceleration of weathering processes. Tectonically instable areas are characterized by strong physical weathering and erosion, which prevents the soil development and the production of clay in equilibrium with climate.

2. *Soils undergo active physical and chemical changes in the course of time.* Vertical changes, marked by ions and particle migration and morphological planation (Millot 1982), occur together with lateral changes. For instance, smectitic vertisols developed in poorly drained downward areas of hot-wet regions may migrate upwards and replace formerly well-drained kaolinitic soils through a progressive ionic saturation of weathering profiles (Bocquier et al. 1970). Such dynamic effects may have occurred on a large scale in past geologic times because of sea-level variations,the regressions inducing increased soil leaching, and the transgressions increased chemical saturations.

3. If the clays removed from soil erosion experience no or only few mineralogical and chemical changes during transportation, deposition and burying, one may expect to use them as useful markers of past environments, in the sedimentary series dominantly fed by continental input. Such an objective implies first a *good knowledge* of sedimentary clays. For instance, sedimentary smectites may derive from temperate-humid or from arid weathering processes (2.2.1; see also Tardy et al. 1970), or from volcanic or older sedimentary rocks. The precise characterization of smectite mineralogy, micromorphology and geochemistry often helps to identify their origin. Second, the knowledge of weathering processes is progressing quickly by modern investigations, including detailed petrology (e.g., Nahon and Noack 1983), micromorphology and microgeochemistry (e.g. Drever 1985, Ahn and Peacor 1987a, Tazaki and Fyfe 1987), pedochronology and modelization (e.g., Fritz 1985, Colman and Dethier 1986). The clay sedimentologist needs to know about the prominent aspects of these progresses, in order to better understand the properties of the materials susceptible to further incorporation into basin deposits.

4. *Paleosols* are commonly described in Quaternary and in some Tertiary series. Their interest in reconstructing past climate (e.g. Singer 1980) is often diminished by the effects of cumulative weathering and diagenetic changes, and because of erosion through physical weathering, sea-level changes, or tectonics. The study of paleosols may provide other useful information, especially on stratigraphic correlations, paleogeography, and global tectonics. Pre-Cenozoic paleosols are rather rarely identified and studied in detail. They chiefly include hard pedogenic formations like ferruginized laterite, calcrete, and silcrete. Old soft soils rarely appear preserved, which contrasts with old soft and clay-rich sediments that have occurred extensively since the early Precambrian. As the most suitable ways to form clay at the surface of the Earth lie in soils, and as a large part of modern sedimentary clays issue from soil erosion, it is likely that *most old weathering blankets were reworked toward the basins in the course of geologic time.* In spite of this, special attention should be paid to search remains of paleosols, whose study provides irreplaceable information on past weathering processes, pedoclimatic environments, and dynamical relations between the lithosphere, atmosphere and biosphere.

Part II

Clay Sedimentation
on Land

Chapter 3

Deserts , Glaciers , Rivers

3.1 Deserts

Aeolian environments, together with glacial and fluvial systems, represent the major paths of particle transportation from the weathering complexes and soils into the lacustrine and marine basins. The small, flat and light clay minerals are transported more easily and farther than other materials derived from weathering and erosion, especially through the action of wind and running water. For this reason, clay-sized minerals do not usually constitute dominant components of desert and river deposits, since they chiefly accumulate downstream, in lakes and seas. By contrast the diverse ways of glacial erosion and transportation determine a high diversity in lithology and clay content, from the frontal moraines almost devoid of fine fraction to the heterogeneous, often clay-rich subglacial tills.

3.1.1 Desert Dusts

Aeolian dusts collected with fine-mesh nets in the deserts or at their close periphery usually contain abundant quartz associated with various clay and non-clay minerals (see Pye 1987; Table 3.1). *The mineralogical spectrum is often broad and reflects the source composition.* For instance Coudé-Gaussen (1982) describes the mineralogical composition of a dust sample in Tanezrouft (South Sahara) during a storm in December 1980. The bulk material contains abundant quartz, followed by K- and Ca/Na-feldspars, micas, amphiboles, pyroxenes, calcite, dolomite, gypsum, ilmenite, apatite, olivine, clay aggregates and iron- coated grains. The clay fraction (< 2 μm) is rich in smectite (75%), chlorite, illite and kaolinite (5% each), small quartz and goethite.

Paquet et al. (1984) study the mineralogical composition of about thirty aeolian dusts sampled in May 1974 at 2 meters height, along a 5000 km north-south transect across the Sahara. They note the abundance of clay fraction (> 25%) and identify a wide range of minerals: illite, chlorite, illite-smectite and chlorite-smectite irregular mixed-layers, smectite, kaolinite, palygorskite, quartz, feldspars, calcite, gypsum, probably magadiite (hydrated Na-silicate), and traces of dense minerals. The minerals are distributed in four main groups, indicating

Table 3.1. Mineral composition of continental windblown dusts (studies done between 1966 and 1985; ref. in Pye 1987). X: dominant constituent. x: important constituent

	Quartz	Feldspars	Mica/illite	Chlorite	Kaolinite	Smectite/mixed layer	Palygorskite	Calcite	Dolomite	Gypsum	Halite	Hematite	Amphibole	Talc
Kuwait	X	X	x					X	x	X	x			
N. Nigeria	X	x	x		x									
Arctic	X	x	x	x	x	x		x						
S. Israel	X	x	x	x	x	x		x	x	x				
N. Nigeria	X				X	X								
N. Nigeria	X	x	x	x	x	x							x	
N. Nigeria	X	x	x		x	x								
NW. England	X	x	X		x		x	X	x					
Ireland	X		x					x				x	x	
Scotland	X	x	X	x			x	x						
N. England	X	x	X		x	x		x	x					
Netherlands	X	x	x	x	x		x	x	x				x	
Equatorial Atlantic	X	x	X	x	x	x		x					x	
Barbados	X	x	X	x	x			x						
E. Equatorial Pacific	X	x	X	x	x	x								
N. Atlantic	X	x	X	x	x									
E. Equatorial Atlantic	X	x	X	x	x	x		x	x				x	
E. South Atlantic	X	x	X	x									x	x
E. Atlantic	X	x	X	x	x			x	x				x	x
Turkmenistan	X	X	x					x						
New Mexico	X		x		x			x			x			
Beijing, China	X	x	X	x				x					x	

the importance of local sources and of the erosion of surficial and soft sediments of soils Fig. 3.1):

1. Group 1, from the Mediterranean coasts to the South of the Saharian Atlas: abundance of illite, chlorite, expandable mixed-layers (70-75% of clay fraction), and of calcite (40-50% of total dust). These minerals, associated with little kaolinite, palygorskite and quartz, result mainly from aeolian erosion of sabkha sediments and high plain and mountain soils.

2. Group 2, located in the Mesozoic and Cenozoic plateaux of the Northern Sahara : relative abundance of palygorskite (20-25% of clay fraction) and quartz (25-40% of bulk material), probable presence of magadiite. Calcite is still fairly abundant (25-40%), while the illite group diminishes (illite, chlorite, mixed-layers). This assemblage proceeds from the reworking of local soft sediments, late Cenozoic soils, and calcareous crusts.

3. Group 3, developed in the region of Cambro-Ordovician Tassili sandstones and North Hoggar Precambrian series. The relative abundance of kaolinite (25-30% of clay), the abundance of the illite group (60-70%) and the presence of gyp-

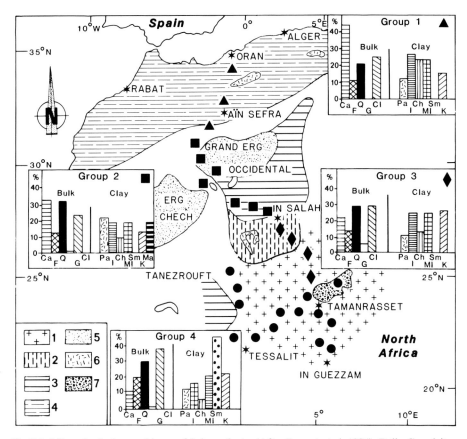

Fig. 3.1. Mineralogical assemblages of Sahara dusts. (After Paquet et al. 1984). Bulk: Ca calcite. F feldspar. Q quartz. G gypsum. Cl clay minerals. Clay: Pa palygorskite. I illite. Ch chlorite. ml mixed-layers. Sm smectite. K kaolinite. *1* Crystalline basement. *2* Paleozoic Tassili sandstones. *3* Mesozoic limestones, Neogene hamadas. *4* Atlas chains, Algeria high plains. *5* Dune fields. *6* Sabkhas. *7* Volcanic massifs

sum correspond to the composition of local parent rocks and of Quaternary surficial blankets.

4. Group 4, characteristic of South Hoggar and Tanezrouft: abundance of smectite (50% of clay) associated with various clay and nonclay minerals. Abundant smectite corresponds to the regional extension of Quaternary soils developed under warm climate and strong seasonal humidity. Toward the South, aeolian dusts become enriched in kaolinite, because of the more hydrolyzing climate determining lateritic soils and paleosols (e.g., Tobias and Megie 1980/1981).

The influence of local sources diminishes with increasing distance of transportation, and also with the increased importance of storms that homogenize the products of different sources and cause severe sorting. The most widespread and resistant minerals, like mica-illite and quartz, tend to be more abundant and more widely distributed. For instance, Glaccuum and Prospero (1980) study the products of Sahara storms deposited in Sal Island (Cape Verde archipelago). The bulk material contains mica-illite (56.7%), quartz (16.5), kaolinite and gypsum (5-7% each), plagioclase (4.9), calcite (4.4), chlorite (4.1) and microcline (2). The clay minerals comprise illite (47%), kaolinite (26.5), chlorite (14), and smectite (12.5) (see also Chap. 7). The quartz abundance tends to decrease compared to illite with increasing distance, because of the average large size and more massive shape of the former mineral. In the French Pyrenees, Bücher and Lucas (1975) and Bücher et al. (1983) observe the preponderance of illite and chlorite in three Saharian dust deposits. Illite, associated with little kaolinite, is also the most noticeable windblown clay species incorporated in a Quaternary ice core of dome C, Antarctica (Gaudichet et al. 1986), where continental microparticles are more abundant than volcanic ones (Briat et al. 1982). Barrios et al. (1987) show the local importance of aeolian dust supply in Spanish soils, and the difficulty in identifying the sources due to the presence of ubiquituous wind-derived minerals like quartz, mica-illite, kaolinite and smectite.

Despite this average dominance of resistant minerals in widely transported aeolian dusts, some fragile minerals that are classically supposed to be rapidly destroyed or sorted can be found at a very great distance from their sources. Bücher et al. (1983) report the occurrence of one smectite-rich dust deposit in the Pyrenees. Bain and Tait (1977) identify abundant palygorskite with illite and chlorite in Great Britain (1968) and Scotland (1971), corresponding to dust expulsions off Morocco. Robert et al. (1984) identify up to 25% of Africa- derived palygorskite, associated with kaolinite, in the clay fraction of sediments from high-altitude lakes of Corsica. SEM examinations show that palygorskite grains are often shaped by wind as felt-like figures, which allows good resistance to further transportation (Coudé-Gaussen and Blanc 1985; Fig. 7.4).

3.1.2 Aeolian Sediments

The mineralogy of aeolian deposits usually reflects the mineralogy of geologic and pedologic sources, the most surficial, erodable, and widely outcropping formations being the best represented. Muller-Feuga (1952, in Millot 1964, 1970) shows that feich-feich, a silty clay deposited in east-Saharian depressions, contains more kaolinite in the vicinity of early Mesozoic Nubia limestones, and more illite close to Cretaceous formations. Swineford and Frye (1955) notice the abundance of probable volcanogenic montmorillonite in Quaternary loess from Kansas, and of glacio-detrital illite, chlorite, and mixed-layers in those from Northern Europe. The loess that accumulated in South Tunisia during the late Pleistocene contains various clay minerals, including palygorskite, inherited from the hinterland (Coudé-Gaussen et al. 1984). Pye (1987) indicates that typical loess contains 50-70% quartz, 5-30% feldspar, 5-10% mica, 0-30% carbonate and 10-15% clay minerals, but may contain various other minerals (Table 3.2) whose proportions vary according to rock and soil sources.

The high permeability of most aeolian deposits favors *secondary leaching and chemical weathering*, that can determine modifications of the original detrital

Table 3.2. Bulk mineralogical composition of various loess (minerals > 5%). (After Pye 1987)

	Quartz	Feldspar	Calcite	Dolomite	Illite	Kaolinite	Smectite	Chlorite	Gibbsite	Biotite
Vicksburg, Mississippi	X	X	X	X	X	X	X			
Nurek, Tajikistan	X	X	X	X	X	X	X	X		
Mingtepe, Uzbekistan	X	X	X	X	X	X	X	X		
Karhlich, West Germany	X	X	X		X	X				
Timaru, New Zealand	X	X	X		X					
Pegwell Bay, Kent, UK	X	X	X	X	X	X				
Tyszowce, Poland	X	X	X	X	X	X				
Wallertheim, West Germany	X	X	X	X	X	X				
Richardson, Alaska	X	X			X			X		X
Meng Xian, China	X	X	X		X	X		X		
Jiuzhoutai, China	X	X	X		X			X		

suites. This is probably one of the reasons why few clay mineral studies have been performed on ancient aeolian deposits. Such secondary changes rarely occur under arid climate, when low precipitation and presence of impervious calcareous crusts prevent significant hydrolysis (e.g., loess of South Tunisia, Coudé-Gaussen et al. 1984). Loess submitted to temperate-humid conditions shows a partial decarbonation and alters into lehm. The clay fraction presents a decrease of chlorite relatively to illite (Alsace, France; Camez 1960, in Millot 1964, 1970), or even the development of mixed-layers, vermiculite and degraded smectite (Wisconsin, USA; Glenn et al. 1960; see also Pye 1987). Humid-tropical conditions lead to strong leaching and formation of authigenic kaolinite, gibbsite, and iron oxides, as for instance in Northeast Australia (Pye 1983a).

Moderate weathering does not destroy all primary mineralogic characters of aeolian deposits, which sometimes permits the recovery of some information on

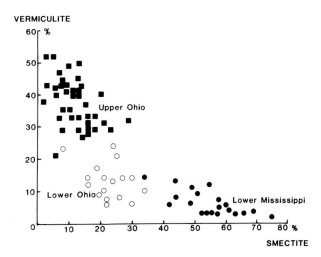

Fig. 3.2. Vermiculite and smectite composition of silty loess, Ohio river basin, USA. (After Ruhe 1983)

sources or wind direction. For instance the late Quaternary loess deposited in the Ohio river basin shows a climatic alteration of vermiculite into smectite. This weathering was not severe, and clay suites represent good indicators of three different sources: Upper Ohio basin, Lower Ohio basin and Lower Mississippi valley (Ruhe 1983; Fig. 3.2). Chamley et al. (1987) study weathering complexes developed in Fuerteventura, Canary Islands, and distinguish the local and distant influences on the Quaternary sedimentation and pedogenesis. The weathering of local volcanic rocks produces abundant Mg- smectite, long bundles or balls of fibrous clays, or corrensite, while weathering of aeolian silt and sand issued from West Africa is characterized by the preservation of detrital illite, kaolinite, and palygorskite aggregates. Various transitional stages between both autochthonous and allochthonous sources may be recognized by close mineralogical and geochemical investigations.

3.2 Glaciers

3.2.1 Origin of Glacial Clays

Glacier ice often incorporates and preserves colluvial and aeolian particles, the study of which provides useful information on past climate, volcanic activity, or other environmental aspects (e.g., Briat et al. 1982, Delmas et al. 1985, Gaudichet et al. 1986). Ice also erodes the various bedrocks successively covered by glacier progression, and therefore incorporates or displaces huge amounts of rock fragments.

The nature of clay suites transported by glaciers and then deposited as glacial, glacio-fluvial, glacio-lacustrine or glacio-marine sediments depends on the petrographic nature of substrates. *Glacier-derived clays are basically detrital.* For instance, New England and East Canada tills and varved clays are rich in illite, chlorite, and vermiculite, minerals abundantly present in local Paleozoic bedrocks and superimposed weathering complexes (Allens and Johns 1960). Alpine and Pyrenean moraines and glacial lakes contain dominant illite and subordinate chlorite, directly issued from surrounding crystalline rocks (Lafond et al. 1961). The sediments deposited in a permanently ice-covered lake in the Northwest Territories, Arctic Canada, are characterized by illite, kaolinite, chlorite-smectite, quartz, and feldspars, inherited from local Cretaceous and Tertiary sediments (Coakley and Rust 1968). Ballivy et al. (1971) compare glacial marine and lacustrine clay of Northwestern Quebec, and find no significant difference in the late Pleistocene clay mineralogy (dominant illite and chlorite, accessory kaolinite, and vermiculite). Huff (1974) investigates the glacial filling of Trièves lake in the subalpine range of Southeastern France, and concludes that mica-illite (65-85% of clay fraction), iron-rich chlorite (10-30%) and rare mixed-layered material are essentially detrital. Ugolini et al. (1981) study five cores selected from the Antarctic Dry Valley Drilling Project. They identify complex clay suites, including mica-illite, chlorite, kaolinite, vermiculite, montmorillonite, and various mixed-layers that result from the reworking of local rocks and little evolved weathering complexes.

The question of in situ chemical weathering in glacial deposits was first discussed by Collini (1956) and Rosenqvist (1961), in order to explain the huge quantity of illitic and chloritic Pleistocene clays in Scandinavia. Rosenqvist (1961) suggested that degraded micas and smectites produced by weathering of the crystalline shield and of superimposed sedimentary blankets were transformed into well-crystallized illite during glacial transportation. In fact, such a complex mechanism is hardly probable, since we know that weathering under glacial conditions is very weak, and that transportation by running water does not allow identifiable mineralogical changes, especially if temperature is low (see 3.3). Moreover Rosenqvist's hypothesis does not clearly explain the abundance of well-crystallized chlorite that, together with mica-illite, naturally constitutes the dominant component of rock flour produced by glacier grinding. Abundant illitic and chloritic clays in Scandinavia probably result, as in North America, from *massive, long physical weathering through the action of glaciers*, the progression of which on gently sloping landscapes favored fine grinding and accumulation of small-sized material.

Such an explanation does not contradict the possibility of reworking alteration products, and of incorporating both bedrock and soil materials in glacier deposits. Feininger (1971) indicates that pre-Pleistocene pedologic blankets could constitute a dominant source of numerous glacially eroded materials. Tiercelin and Chamley (1975) show that clay suites incorporated in Pleistocene glacial deposits of Southeast France proceed from the erosion of both bedrocks and paleosols, and provide information on pedogenic processes formerly developed in upstream zones; these suites differ from the clay assemblages identified in the paleosols covering the successive glacial deposits that give indications of local interglacial climate. Ugolini et al. (1981) observe that Plio-Pleistocene sediments of McMurdo Sound, Antarctica, contain clay minerals predominantly derived from bedrocks and accessorily from minimally evolved weathering complexes. Note that permeable glacier deposits, like frontal or lateral moraines, may experience some in-situ weathering in a manner similar to aeolian sands (cf. 3.1). Such a secondary evolution is suggested by Stewart and Mickelson (1976) for two tills in Wisconsin, USA. Nevertheless, very few examples of severe weathering in glacial environments are reported in the literature. One of them concerns recent fluvio-glacial deposits in East Greenland, where the presence of fairly well-crystallized vermiculite and smectite, beside micas, quartz, and feldspars, is attributed to the K release from particles issued from local gneiss and charnockite (Petersen and Rasmussen 1980).

3.2.2 *Paleoenvironmental Applications*

The essentially detrital character of glacial clay associations, combined with their frequent low vulnerability to post-depositional alteration effects, allows various applications. In the Subalpine Range of the French Alps, Tiercelin (1977) investigates two Pleistocene fluvio-glacial basins, and proposes stratigraphic correlations based on glacial and interglacial clay assemblages, from the Günz stage to the post-Würm. To the North of the same domain, Huff (1974) compares the clay mineralogy of Trièves lake sediments and of surrounding bedrocks. Mica-illite and chlorite, that occupy almost the whole clay fraction, correspond to the clay composition of surrounding late Carboniferous micaceous sandstone and black silty shale, and of early to middle Jurassic argillaceous limestone. These minerals also form the phyllosilicate fraction of the Hercynian basement. W.D. Huff notices the absence in glacial deposits of allevardite-like mixed-layered clay, a mineral characteristic of local Bathonian shales and Oxfordo-Callovian marls called "terres noires". The middle-late Jurassic shales and marls present large outcrops that are chiefly located in the lower part of the Pleistocene glacial basin. The lack of incorporation of allevarditic minerals in the Trièves lake is therefore interpreted as the result of a low surface-erosion rate during Würm time. The crystalline and Paleozoic sedimentary rocks, rich in illite and chlorite and devoid of allevardite, crop out in the higher elevation cirque areas and were easily eroded during periods of glacial maximum. In contrast, the middle-late Jurassic "terres noires" were covered by ice at that time, and were largely protected from mechanical weathering.

The late Wisconsinian tills deposited in southeastern Michigan present interlobate moraines determined by the junction of the Saginaw and Huron-Erie glacial lobes. Till and glaciolacustrine materials supplied by both glacial lobes show similar lithology, color, texture, and macrofabric, which prevents the identification of different sources, glacier progression steps, and local stratigraphy. On the other hand, clay mineralogy shows systematic differences in the relative abundance of illite compared to chlorite, kaolinite, and vermiculite (Rieck et al. 1979). The 7/10 Å peak height ratios of the Kalamazoo moraines formed by Saginaw lobe progression toward the South are systematically higher than those of the Mississinewa moraines built by the northwest displacement of the Huron-Erie lobe (Fig. 3.3). The precise mapping of clay characteristics allows the limit between both influences in the interlobate moraines to be traced, and the local history of glaciation and deglaciation to be followed.

Similar attempts at paleogeographic reconstructions can be made on very old glacial deposits devoid of important diagenetic, tectonic and metamorphic modifications. This is the case of late Precambrian to Devonian glacial formations deposited in the cratonic basin of Taoudeni, Mauritania, rich in impervious clayey rocks (Chamley et al. 1978a, 1980a). The clay suites cover a large diversity: illite, chlorite, vermiculite, smectite, kaolinite, various irregular mixed-layered minerals, and rare pyrophyllite and subregular mixed-layers. The clay stratigraphy usually does not reflect the lithological changes, except in some coarse sandstones where a strong increase of kaolinite indicates active pore diagenesis (see Chap. 16.3). Mineralogic variations chiefly vary according to sources, glacier progression, and climate. During the late Precambrian, the glacial clay assemblages resemble those of bedrocks located to the North, differ from those of southern bedrocks, and suggest a NNE-SSW progression of the inlandsis. The interglacial marine deposits are characterized by higher mineral diversity, more abundant mixed-layers and more poorly crystallized illite, sug-

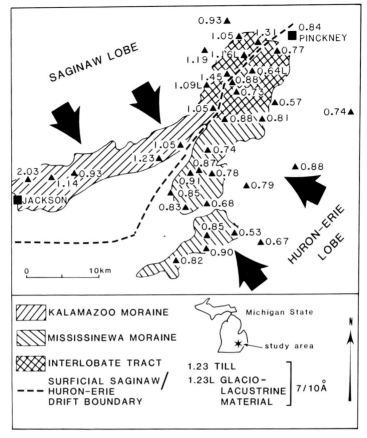

Fig. 3.3. Geographical distribution of 7/10 Å ratios of late Pleistocene glacial sediments in southeastern Michigan. (After Rieck et al. 1979)

gesting the existence of normal chemical weathering conditions allowing moderate hydrolysis. The interglacial conditions of the late Precambrian appear similar to those of modern temperate-humid climate . Glacial associations of late Precambrian and uppermost Ordovician indicate more local sources than interglacial ones, which corresponds to differences in marine circulation and particle dispersion.

3.3 Rivers

3.3.1 Origin of River Clays

Mineralogical investigations of river sediments began around the mid-20th century. These *first studies* suggested a good *correspondence between the clay*

composition of fluvial deposits and of the formations submitted to erosion. For instance, Millot (1953) noticed that Niger river sediments in Sudan contain
abundant kaolinite, a mineral typical of surrounding lateritic soils. When going
down the Durance river, in the Southern French Alps, illite and chlorite issued
from upstream schists are enriched in kaolinite provided by downstream Jurassic
marls. Claridge (1960) observed that the Waipaoa river catchment, in New
Zealand, contains abundant smectite reworked from various bentonitic mudstones and argilites, that represent the most easily erodible rocks of the
watershed. Lafond (1961) identified in the Vilaine river, West France, illite,
chlorite, kaolinite, and pyrophyllite, all minerals belonging to local crystalline
rocks and Cenozoic soils. Müller (1961) found that early Tertiary limestones
were responsible for the supply of fibrous clays in Quaternary alluvium of
Hadramant, South Arabia, and that palygorskite (attapulgite) appeared to be
transported downstream farther than sepiolite. Rawi and Sys (1967) also
identified detrital palygorskite in the flood plain of the Tigris river, that is
depleted in smectitic minerals compared to the Euphrates. Packham et al. (1961)
reported that clay assemblages from nine English river sediments reflected the
nature and composition of the local argillaceous formations.

Recent studies confirm the essentially *detrital character of fluvial clay suites,*
and allow precising the origins. Here are a few examples provided by literature.
The clay suites transported by the three major rivers entering the Northeastern
Gulf of Mexico result from both parent rocks and regional soils (Griffin 1962).
The Mississippi river located westward carries mainly smectitic sediments supposedly reworked from unaltered formations, while the more eastern
Apalachicola river receives weathering products largely issued from kaolinitic
soils; the Mobile river is located in an intermediate situation and therefore
provides an intermediate clay mineral suite to the sea (Fig. 3.4). The importance
of the Mississippi suspended load compared to those of both other rivers
determines a smectite-rich average composition of the inferred marine sediments.
Climatic differences also explain the mineral changes recorded in the Columbia
river basin, Northwestern USA, and British Columbia, Canada: the abundance

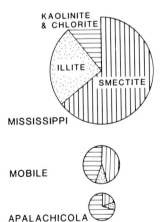

KAOLINITE
& CHLORITE

ILLITE

SMECTITE

MISSISSIPPI

MOBILE

APALACHICOLA

Fig. 3.4. Clay mineral composition of river sediments
entering the Northeastern Gulf of Mexico. (After Griffin
1962). The *area of the circles* is proportional to the load
contributed by each river

of illite in the upper sub-basin, the northwestern half of the middle basin and small parts of the lower sub-basin results from cool subhumid climate acting on sloped plutonic and metamorphic rocks; the rest of the Columbia basin comprises mainly basaltic plateaux undergoing an arid to semi-arid climate, which favors the formation of pedogenic smectite and the downstream increase of the mineral in river sediments (Knebel et al. 1968). On the other hand, in the Santa Ana river basin, South California, McMurtry and Fan (1975) point to the accessory influence of climate on the sediment composition, compared to the direct supply from old sedimentary deposits: all major minerals (smectite, mica-illite, feldspars, quartz) and minor species (kaolinite, chlorite, calcite, amphibole, gypsum) are present in the bedrock, where they vary highly and influence the clay composition of the different tributaries.

The studies performed on European rivers also indicate the detrital character of clay associations. For instance, Italian rivers entering the Adriatic and Tyrrhenian Seas contain abundant illite, chlorite, and mixed-layers when they largely drain the metamorphic zones of the Southern Alps and of Apenninnes, and abundant smectite when they mainly run through Cretaceous and Cenozoic sedimentary formations (Quakernaat 1968, Tomadin 1969, Tomadin et al. 1985). The sediments of the Tech, Têt, and Agly rivers, that drain the Roussillon region, Southern France, comprise different proportions of illite, chlorite, smectite, and kaolinite according to the relative importance in each basin of crystalline rocks and of Mesozoic or Neogene sedimentary formations (Monaco 1971). Chamley and Picard (1970) study the sediment mineralogy at the mouth of 25 rivers of Provence, Southeast France. The river clays that comprise among other minerals various mixed-layers and poorly crystallized species reflect a mixture of rock and

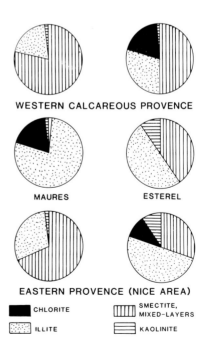

Fig. 3.5. Average clay assemblages in river sediments of Provence, Southeast France. (After Chamley 1971)

soil sources. Specific clay and sand assemblages characterize the successive geological units, from West to East: smectite in calcareous Provence; illite, chlorite, micas, epidote, staurotide, apatite in Maures; illite, smectite, apatite and augite in Estérel; and illite or smectite in the Nice region (Fig. 3.5). In a general way, clay minerals represent particularly good markers of sources in sedimentary areas, while dense sandy minerals are more reliable in metamorphic (Maures) and volcanic (Estérel) areas.

3.3.2 River suspended minerals

The first investigations on river suspensions confirmed the similarity of clay composition in river basin formations with river sediments: parallelism of mineral associations in the Cheyenne river basin, Wyoming, USA: kaolinite, illite, vermiculite, quartz, feldspars (Rolfe and Hadley 1963); presence of detrital illite, chlorite, kaolinite, mixed-layers but also of fibrous clays (sepiolite ?) in the upper Rhone valley, Switzerland (Vernet 1969); confirmation of the occurrence of more abundant smectitic minerals in the Euphrates compared to the Tigris river, Mesopotamia plain (Berry et al. 1970; see Rawi and Sys 1967).

Strong *changes in weather conditions* may determine significant mineralogical modifications in the suspended load. Weaver (1967a) notices an enrichment of illite in the suspended matter of the Arkansas river (USA) during storm periods. Similar changes occur in surface water of the lower part of the Rhone river, France (Chamley 1971): the first annual flood following spring thaw provokes a strong increase of large, fresh mica-illite, that accumulated in the alpine range during winter physical weathering, and was rapidly reworked by spring running water. The subsequent flood-low water alternations show an opposition in the abundance of calcite, quartz, and feldspars, and of illite and chlorite, the flat, small clay minerals being favored during low energy periods. In flood suspended materials from several Eastern Kansas rivers, USA, Schneider and Angino (1980) observe the ubiquitous presence of illite, interlayered smectite, kaolinite, quartz and K-feldspars, that form the mineralogical background issued from surrounding rocks and soils. Vermiculite results from the drainage of Pleistocene glacial material, while mixed-layers are mainly reworked from soils covering Permian rocks. The flood-stage suspended materials contain lower chemical concentrations of trace metals (Mn, Fe, Zn, Co, Cu, Ni) than average discharge materials, which is attributed to a dilution by leached soil minerals strongly eroded via sheet wash.

The mineral distribution in suspended matter depends mainly on the *average size* of the different species submitted to hydrodynamical forces, as demonstrated in the Amazon and the Mississippi (Fig. 3.6). Nonclay minerals occupy mainly large-sized fractions, while increased amounts of chlorite, illite, kaolinite, and smectite are successively recorded in smaller fractions. The particles' shape, buoyancy, and electrochemical characteristics intervene in addition, which explains, for instance, why small kaolinites usually settle much more rapidly than small smectites.

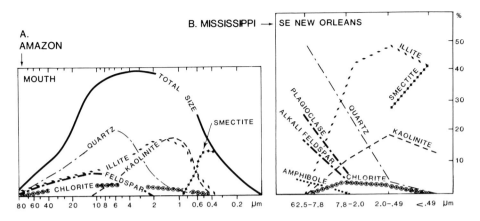

Fig. 3.6. Size distribution of various suspended minerals transported by **A** Amazon and **B** Mississippi rivers. (After Gibbs 1977 a, and Johnson and Kelley 1984)

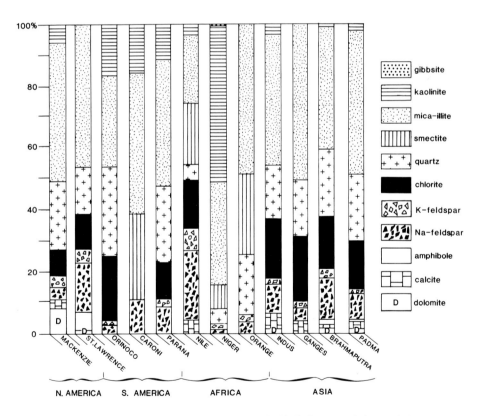

Fig. 3.7. Average abundance of major minerals identified in bulk suspended materials of 12 large rivers. (After Konta 1985 a)

Attempts have been made to *summarize the mineralogical composition of lutite suspensions* in some of the largest rivers in the world (Emeis 1985, Konta 1985 a, b. Fig. 3.7). Clay minerals constitute by far the major component of all river suspensions, and chiefly reflect the composition of soils in the drainage basins. This overall result will be taken into account when discussing the cause of widespread argillaceous sediments in the marine geologic record, and the importance of soil contribution to the genesis of marine deposits (see Chap. 17, 19). Mica-illite represents the principal and omnipresent phyllosilicate identified by the semi-quantitative methods used by the authors, except in the Niger river marked by the preponderance of kaolinite and the rarity of quartz. Illite is particularly abundant in high-latitude rivers (Mackenzie, St Lawrence), as well as in regions draining young mountain chains (Indus, Ganges, Brahmaputra) or in very arid regions (Orange). Kaolinite characterizes large parts of tropical rivers, where chlorite is often rare or absent (Niger, Caroni, Orinoco, Parana; see also Gibbs 1977 a , b for the Amazon). Smectite is identified in the bulk suspensions of tropical and subtropical rivers, like the Caroni, Nile, Niger and Orange. Quartz and feldspars occur commonly, while calcite and dolomite are chiefly recognized in rivers draining either arid or mountainous regions.

3.3.3 Environmental Applications

Many studies are devoted to the environmental significance of mineral changes along river basins. In the Amazon, Gibbs (1967) identifies mainly illite and chlorite in the high relief, physically weathered, andean part of the basin. These minerals add to kaolinite and smectite in the flat, downstream portion of the basin, where chemical weathering becomes very active. Therefore Amazon sediments entering the Atlantic Ocean proceed from both rock and soil erosion. Neiheisel and Weaver (1967) show that river deposits in the Southeastern United States are characterized by specific minerals in the geomorphological sectors successively identified from upstream areas. Kaolinite and hornblende dominate in the piedmont province, smectite increases in the coastal plain, and illite on the continental shelf. The kaolinite/smectite ratio serves to estimate the relative importance of detrital supply from piedmont and from plain in a given deposit, while illite abundance expresses the off-shore contribution to estuaries and coastal areas.

Latouche (1971) compares the clay mineralogy of geologic formations and recent river alluvium in the Garonne basin, Southwest France (Fig. 3.8). The upstream river zone receives very little weathered chlorite- and illite-rich clay suites issued from Pyrenean crystalline schists and plutons. The central zone around the city of Toulouse city is dominated by smectites derived from local Tertiary molasses. When going downstream, many tributaries, especially on the left side of the Garonne, supply medium-crystallized illitic clay suites characteristic of more or less weathered Quaternary silty clays. The lower part of the river course includes a mixture of the various upstream associations. An estimation of the average clay composition of geologic and pedologic formations shows a good

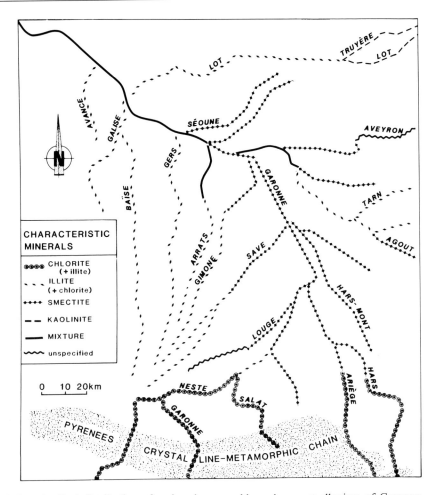

Fig. 3.8. Longitudinal distribution of major clay assemblages in recent alluvium of Garonne river, Southwest France. (After Latouche 1971)

agreement with the composition of lower Garonne river sediments (Table 3.3). The flat morphology of downstream portions of river basins such as the Garonne often determines an active sedimentation and precludes the reworking of local soil products toward the ocean. This is well demonstrated in the Adour basin located South of Garonne, where vermiculitic minerals issued from weathering remain uneroded and do not feed the sedimentation on the adjacent Atlantic continental shelf (Snoussi 1986).

A study by Potter et al. (1975, 1980) on the Mississippi river basin (3,220,900 km²) leads the authors to minimize the contribution of soils, and therefore the importance of climatic control on river sedimentation, in temperate-cool areas. A smectite-dominant association characterizes the western part of the basin from the Rocky Mountains to Indiana, while the eastern part is marked by an illite-chlorite-kaolinite suite (Fig. 3.9). The smectite-rich sediments are related to the

Table 3.3. Average clay composition (%) of geological formations (estimated) and recent downstream river sediments in the Garonne-Basin, Southwest France. (After Latouche 1971)

	Chlorite vermiculite	Illite	Smectite mixed layers	Kaolinite
Geological formations	14	38	33	15
River deposits	13	37	32	18

clay mineralogy of Cretaceous, Tertiary, and Pleistocene substrates rather than to weathering, despite the subarid climate possibly favoring smectitic vertisols (see 2.2.1). The eastern facies are principally attributed to the erosion of middle and early Paleozoic sediments that contain abundant illite and chlorite and are often devoid of smectite; kaolinite and mixed-layers could partly express the climatic imprint. *The importance of direct rock supply under temperate conditions* is also stressed by Chamley (1971), Latouche (1971), and Snoussi (1986) for different French river sediments, whose clay associations nevertheless show reliable modifications determined by local weathering and climate.

The longitudinal topography of river basins strongly controls the upstream-downstream distribution of sedimentary clay suites, which provides different types of information. Let us compare for instance two basins of Provence, Southeast France, both draining various sedimentary substrates but each located

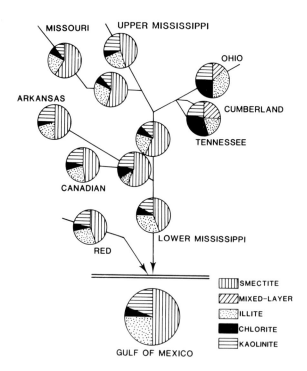

Fig. 3.9. Clay mineralogy of Mississippi river and its principal tributaries. (After Potter et al. 1975)

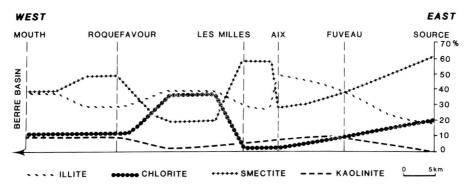

Fig. 3.10. Clay mineral variations along the gently sloping Arc river basin, Western Lower Provence, Southeast France. (After Chamley et al. 1971)

in different morphological situations. The Arc river runs through fairly flat landscape in the western calcareous Provence (Fig. 3.10). *The clay associations change strongly and irregularly* from the source area downstream, and reflect the geologic formations successively eroded and drained: chlorite and illite from eastern Maures and Jurassic limestones (source), smectite from early Paleogene fluvio-lacustrine deposits (Aix), kaolinite from uppermost Cretaceous lignites (Les Milles), smectite from late Cretaceous massifs of South Provence (Roquefavour), and finally illite and chlorite from alpine Plio-Pleistocene deposits of the Durance river (mouth area). The absence of an important East-West slope prevents the distant transportation of clay suites, the establishment of mineralogical gradient, and expresses *successive relay* due to local petrographic changes or tributary influence.

By contrast, the upper course of the Verdon river, in Eastern Provence, is characterized by strong relief, which determines at a distance of 50 km the constitution of *cumulative clay assemblages* (Fig. 3.11). A step by step mineralogic diversification occurs from North to South downwards, and is facilitated by the corresponding succession of different geologic and pedologic formations: illite and chlorite issued from black flysch and hardly evolved mountain soils, illitic mixed-layers provided by low alteration of Aptian black marls, chloritic mixed-layers and little kaolinite eroded from Paleogene limestones, marls, and Annot sandstones, smectite and kaolinite supplied by late Cretaceous limestones and marl-limestone alternations. When compared to the location of the main tributaries and to the surface of the main geologic formations, the *progressive diversification* of mineral assemblages provides indications on the importance of erosion and alluviation processes in the different parts of the Verdon basin. Despite their low outcropping surface (7%), Aptian black marls contribute to the river sedimentation in a proportion similar to widely outcropping late Cretaceous calcareous formations (40%). Useful conclusions may be proposed from such studies about the erodibility of a given river basin and the consequences for environmental protection.

Relatively few studies have been published on clay successions in ancient river deposits, probably because of the difficulty in finding extended fine-grained

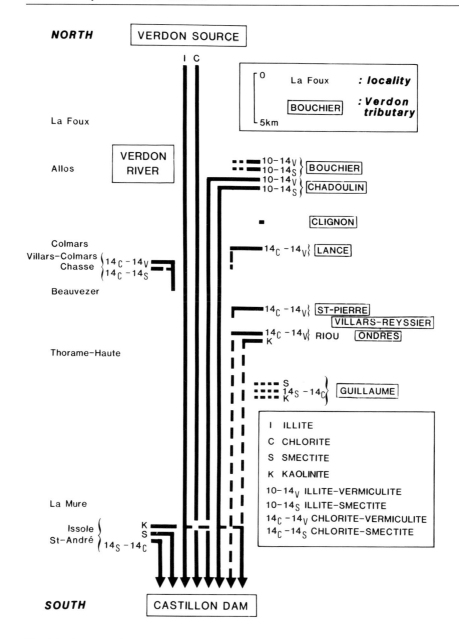

Fig. 3.11. Downwards mineralogical diversification in the strongly sloping upper Verdon river basin, Eastern Upper Provence, Southeast France. (After Chamley et al. 1973)

fluvial sequences devoid of significant diagenetic changes. Some information nevertheless exists about *past climate, detrital sources and diagenesis.* Alimen and Caillère (1964a,b) propose a paleoclimatic explanation for the clay suites identified in the successive Quaternary terraces of the North Pyrenees, France; the early Pleistocene sediments contain kaolinite and sometimes gibbsite, which appear to have been reworked from hydrolyzing soils, while the Würmian and Holocene deposits seem to correspond to a more temperate climate testified by chlorite and mixed-layers. Similar conclusions arise from the study of the super-imposed Pleistocene terraces from the middle part of the Rhone-Isère basin (SE France); fluvial argillaceous deposits, devoid of noticeable post- sedimentary modifications, show in older deposits an increase of irregular mixed-layers and kaolinite, and a concomitant decrease of detrital smectite and chlorite as well as of illite crystallinity (Martin 1963, Bornand and Chamley 1974).

The sources of mineral assemblages can be found in ancient river sediments, especially if morphology and alluviation patterns did not significantly change in the course of the time, or if minerals of specific origin can be identified. Lough-nan (1971) attributes the kaolinite claystone of the Pennsylvanian from Sidney basin, Australia, to the reworking of laterite soils in river and alluvial fan environments. Macaire et al. (1977) determine the presence of clinoptilolite in four Pleistocene terraces of the Creuse river, in Western Central France; this supposedly fragile zeolite is in fact reworked from Cenomanian-Turonian limestones cropping out locally in the basin.

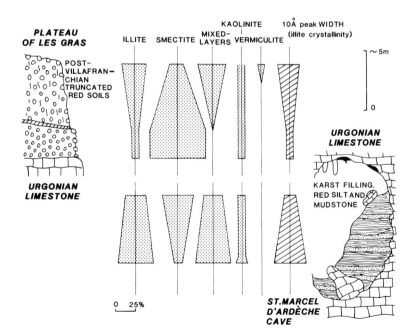

Fig. 3.12. Schematic comparison of clay mineral succession in Plio-Pleistocene red paleosols and in Pleistocene cave filling red beds, Saint-Marcel d'Ardèche, Southeast France. (After Blanc and Chamley 1975)

The much-debated question of the origin of *red beds*, often referred to ancient fluvial deposits (e.g., Turner 1980), can also be approached through mineralogic information. Red beds may include all types of clay minerals and associated species, from kaolinite and gibbsite to illite, smectite, chlorite, mixed- layers, and even palygorskite (Pye 1983b). Diversified clay assemblages often indicate detrital sources (e.g., references in Rao and Shihari 1980; April 1981). This is the case with Pleistocene red beds deposited in the karstic environment of St. Marcel d'Ardèche, France: the cave filling reproduces in reverse order the mineralogic succession identified in local paleosols, that are characterized by upward increasing hydrolysis (development of mixed-layers, poorly crystallized illite, etc). This probably results from the progressive erosion of paleosols, whose deeper and deeper horizon materials were successively deposited by running water into the karstic network (Fig. 3.12). In other cases, red bed clay associations include only a few species, regular mixed-layers or very well crystallized illite and chlorite. The diagenetic imprint may then add to the detrital one, as in late Triassic / early Jurassic fluvio-lacustrine sediments of Connecticut, USA (April 1981).

3.4 Conclusion

1. *Clay suites incorporated in aeolian, glacial, and fluvial deposits mostly reflect the detrital sources* and can be used to reconstruct environmental conditions. Investigations performed on ancient sediments are still limited, notably because many continental deposits are little developed, or easily subject to weathering or diagenesis. Deserts and rivers often constitute only transit environments, since the high transport energy of wind and running water hardly permits the abundant deposition of small, light clay particles, except in specific environments like downstream alluvial plains. In spite of these limitations, studies should be developed in order to better understand past continental environments and post-depositional effects.

2. *Deserts deposits are peculiarly subject to secondary changes* since they are mostly sandy, well-sorted and highly permeable. Conditions of paleoenvironmental reconstitutions are all the better since sediments are fine-grained (aeolian dusts, fine loess), kept under arid climate, and rapidly buried. The preciseness of aeolian source recognition increases when deposits result from normal dust inflow rather than from storms, the latter determining grain size and mineral sorting, as well as an obliteration of different sources due to long transportation and homogeneization. Increased attention should be paid to the nature and significance of aeolian minerals transported overseas and incorporated in thick high-latitude and high-altitude continental ice caps.

3. *Glaciers correspond to the continental nonlacustrine environments most suitable for preserving original environmental messages* borne by clay and associate minerals. Glacial deposits are frequently thick, widely distributed, heterometric, and fairly rich in clay. This explains the relatively abundant studies performed on past tillites, including late Precambrian and Paleozoic inlandsis

products. Information on past sources and climate emerges from studies on clay suites, that also sometimes constitute indirect stratigraphic markers.

4. *Opportunities to study clay suites exist broadly in Quaternary river deposits, but only seldom in older deposits* that are often referred to permeable sandstones or to little extended series. Ancient fine-grained river sediments should be especially investigated, since mineralogic data combined with lithology may provide useful information on the contribution of rocks versus soils, the climate, the importance of alluviation from different tributaries and geologic formations of a given drainage basin, and the morphologic and tectonic changes in the course of the time.

Chapter 4

Lacustrine Clay Sedimentation

4.1 Freshwater Lakes

4.1.1 Detrital Supply

Like desert, glacier and river deposits, lake sediments may include most of the minerals issued from igneous, metamorphic and sedimentary rocks. This is especially the case of freshwater sediments, that commonly include numerous nonclay and clay silicates, various carbonates, Fe-Mn oxides, phosphates, sulfides, and fluorides (see Jones and Bowser's review, 1978). Illite and smectite constitute the dominant species, and are as frequently encountered as quartz. Chlorite, kaolinite, mixed-layers and palygorskite are variously present, depending on the lake location. *A general correspondence exists between the mineral composition of most freshwater lakes and the average clay mineralogy of rocks and soils in the surrounding drainage basins.* Let us consider a few selected examples.

In North America, the clay fraction of sediments from Northeastern Lake Michigan contains about 50% illite, 30% various irregular mixed-layers and 20% chlorite, a composition that reproduces the clay suite of surrounding late Pleistocene tills, and that remains nearly constant with distance from shore, water, and sediment depth (Moore 1961). In the southern part of Lake Michigan similar affinities exist between glacial till deposits and postglacial lacustrine sediments (25-50% illite, 30-40% chlorite, 20-30% expandable layer clays and vermiculite); illite tends to increase in older sediments, contrary to chlorite (Gross et al. 1972). Other data on the Great Lakes, summarized by Jones and Bowser (1978), lead to the same conclusions. In Minnesota the silt- and clay-sized fraction of 46 lakes depends mainly on the lithology of various crystalline and sedimentary bedrocks, and shows an increased crystallinity from Northeast to Southwest (Dean and Gorham 1976). Lakes Pontchartrain and Maurepas, Louisiana, contain dominant smectite associated with illite and kaolinite, an assemblage provided by the Mississippi river drainage system (Griffin 1962). Smectite and illite abundance tends to increase relatively to kaolinite, when the salinity increases through marine water input from the Gulf of Mexico; this is interpreted by Brooks and Ferrell (1970) as the result of differential settling, kaolinite being deposited faster than other species in brackish environments.

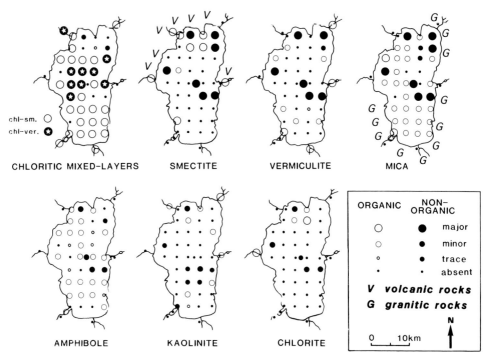

Fig. 4.1. Mineral content of surface sample clay fractions, Lake Tahoe, California-Nevada. (After Court et al. 1972)

Lake Tahoe, located in the Sierra Navada on the California-Nevada border, contains two types of surficial sediments (Court et al. 1972. Fig. 4.1): (1) Organic-rich oozes dominate and comprise flat and well-stratified beds, abundant diatoms and pollen, and a clay fraction largely composed of chlorite-vermiculite and chlorite-smectite mixed-layered minerals (referred to as chloritic intergrades). (2) Clastic gravel, sand and clay occur more locally and include a large variety of textures; diatoms and pollen are rare or absent; the clay fraction is characterized by vermiculite, mica, and smectite. All clay minerals are considered as detrital, even in the organic-rich deposits where no in-situ degradations are reported (see Chap. 11). Smectite is derived from volcanic rock weathering, as is indicated by the areal distribution of sand and gravel constituents, and by the close correspondence existing between the location of volcanic rock outcrops and of lacustrine smectite (North and West). Mica-illite, vermiculite, and chloritic minerals represent characteristic products of physical and chemical weathering of granites that crop out mainly in the southern and eastern parts of the drainage basin. The clastic, nonorganic materials represent the still-exposed sediments deposited during the last glaciation, under a cold regime favoring physical weathering (abundance of granitic mica and volcanogenic smectite), rapid erosion, and slumping into the deep parts of the lake. By contrast, organic oozes resulted from the postglacial erosion of more chemically weathered formations

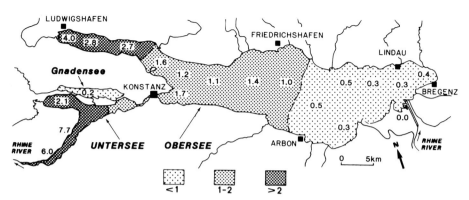

Fig. 4.2. Smectite/illite-mica ratio in surficial sediments of Lake Constance. (After Müller and Quakernaat 1969)

(abundance of mixed-layers), and were deposited widely in Lake Tahoe at a slow rate and without noticeable disturbance (bedded sediments).

Numerous European examples also indicate the close relation between lacustrine clays and detrital sources. For instance Lake Mauvoisin, located at 2000 meters altitude in the Swiss Alps, contains mainly illite and chlorite from local crystalline rocks, together with quartz and feldspars; some kaolinite and very little fibrous clay referred to sepiolite are also identified, and were probably derived from old weathered materials and fracture filling (Lombard and Vernet 1969). Lake Constance, located at the junction between Switzerland, the Federal Republic of Germany, and Austria, comprises dominant mica-illite with chlorite, smectite, and a few mixed-layers in the recent clay fraction of sediments (Müller and Quakernaat 1969). The areal distribution of minerals shows a decrease of illite and chlorite abundance from the eastern mouth of the Rhine river, that drains the crystalline Northern Alps, to the West of the lake developed in Tertiary molasse and Quaternary glacial deposits. This variation is clearly expressed by the smectite/illite ratio (Fig. 4.2). Various studies indicate the illite and chlorite abundance in upper Rhine rocks and river suspended materials, and the presence of smectite, illite, chlorite, and some kaolinite in Cenozoic deposits surrounding the lake. The mineralogic variation observed therefore results from the progressive westward dilution of minerals from an upstream crystalline origin by those derived from downstream sedimentary formations. The very low smectite content of the Gnadensee, a small part of the lower basin of Lake Constance (Untersee, Fig. 4.2), corresponds to glacial clays usually underlying the postglacial sediments and locally cropping out on the bottom of the lake.

Lake Mjøsa, the largest in Norway, is dominated by illite, chlorite, quartz, and feldspars in the sedimentary clay fraction, with very little smectite and mixed-layers (Englund et al. 1976). This association, together with the chemistry of lacustrine water, reflects the major influence of local source rocks and the weakness of chemical weathering (see 2.2.1). Several high-altitude lakes in Corsica contain three mineralogic groups in surface sediments (Robert et al. 1984): (1) Chlorite, illite, irregular mixed-layers, vermiculite, and stilpnomelane (a Fe-aluminosilicate) are derived directly from local little-weathered plutonic and metamorphic rocks (2) Poorly to medium- crystallized smectite, and a fraction of kaolinite, result from weathering under a temperate-humid climate, whose influence is locally enhanced by the presence of vulnerable volcanic rocks (see 2.2.1, 2.2.2). (3) Palygorskite and some kaolinite result from high-altitude aeolian supply from Africa; they are trapped during rainfall periods into lacustrine depressions where slow sedimentation prevents excessive dilution.

To summarize, most recent freshwater lacustrine clays simply result from detrital input from more or less local distances, and diversely express the geologic sources, the weathering conditions and the dispersion effects. Ionic solutions are usually too much diluted and detrital supply is too high to allow noticeable crystallization, especially of complex silicate minerals such as clays.

Similar results arise from investigations on various ancient lacustrine deposits, which can sometimes lead to paleogeographic and paleoclimatic deductions. For instance, Aprahamian et al. (1970) identify in the Isère valley, French Alps, lacustrine sediments successively marked by illite and chlorite, and then by irregular mixed-layers and kaolinite; this change is interpreted as the result of a climatic warming around Eemian time (Riss-Würm interglacial stage). In the same area Huff (1974) attributes the exclusive supply of illite and chlorite, in the glacial Lake Trièves, to the large size of the ice blanket precluding alluviation from lower parts of the drainage basin (see 3.2.2). Lineback et al. (1979) compare glacial and postglacial sediments in Lakes Superior and Michigan, from lithologic and mineralogic aspects. The upward increase of vermiculite, smectitic minerals, and 10-14 Å mixed-layers, opposed to the disappearance of calcite and dolomite (Table 4.1), corresponds in Lake Superior to more hydrolyzing conditions when moving from the late Pleistocene to the Holocene. Some parallel trends exist in both lakes, and this facilitates stratigraphic correlations when combined with lithologic data. Older examples are quoted from the Oligocene of France (Aquitaine, Velay), where illitic lacustrine sediments are partly attributed to sources (Pyrenees, Massif Central) and to sorting during transportation (in Millot 1964, p. 202).

Table 4.1. Some mineralogical data (%) of <2 μm fraction of late Pleistocene to Holocene Lake Superior sediments. (After Lineback et al. 1979). N. B. chlorite varies from 5 to 6%, kaolinite from 2 to 6%, without noticeable trend

Lithology	Expandable minerals	Vermic- ulite	(10–14 Å) Mixed layers	Illite	Cal- cite	Dolo- mite	Quartz	Feld- spars
Holocene								
Surface sand	9	8	26	27	0	0	9	13
Brown silty clay	12	7	22	27	0	0	9	13
Gray clay	10	5	21	26	0	0	12	16
Gray varved clay	8	2	15	26	9	4	9	17
Red varved clay	6	2	10	22	11	7	11	19
Red till	6	3	17	25	8	5	12	14
Pleistocene								

4.1.2 Authigenic Clays

Biosiliceous Environment
Freshwater diatoms represent conspicuous components of lacustrine sediments, especially in lakes characterized by low detrital supply and weak carbonate precipitation (e.g., Lakes Baikal and Tanganyika, and some high-altitude lakes), or in volcanic environments (e.g., Massif Central, France). Diatom frustules suffer noticeable dissolution in the water column and surficial sediments of the American Great Lakes (e.g. Lake Superior, Parker and Edgington 1976). By contrast these algal remains are often extensively preserved in East-African lacustrine sediments; this results from larger inputs of dissolved silica by rivers and springs, from the different nature of organic matter or from more continuous planktonic productivity (see Jones and Bowser 1978). A case of clay mineral formation associated with diatom-rich sediments has been reported in Lake Malawi, Southeast Africa (Müller and Förstner 1973). *Nontronite*, a Fe-rich smectite, occurs in relatively shallow and oxic sediments, sometimes associated with limonite and vivianite (Fe phosphate). Amorphous hydrous ferric oxides are more or less admixed with opaline silica to display an oolitic texture. Nontronite is supposed to form in situ by reaction of silica-enriched hydrothermal water and ferrous iron resulting from sediment and vent interaction. This uncommon example appears chiefly controlled by hydrothermalism, but nontronite is also described in Lake Washington, USA, which is devoid of known hydrothermal activity (Shapiro et al. 1971, in Jones and Bowser 1978, p. 210). Further investigations should be made to explain the origin of nontronite in lacustrine environments (see also 4.2.2), whose high organic productivity often determines oxygen-depleted water favoring silica and iron mobility.

Iron-Clay Granules
Glauconitic minerals are not restricted to marine environments. Parry and Reeves (1968a) have reported the occurrence of a glauconitic mica in pluvial Lake Mound, Texas (USA). The mineral constitutes pellets and more or less diffuse streaks in clastic sand and lacustrine dolomite, together with smectite, illite, and mixed-layers. As there are no detrital glauconitic minerals in the vicinity, glauconitic mica is attributed to an in situ fixation of iron and potassium in smectitic layers, within the lake sediments. The Mound Lake glauconitic mica contains more sodium and less potassium than common green clay granules, and refers to little organized glauconite. Note that glauconitic illite, called ferrous illite or aluminous glauconite, was formerly described in Oligocene lacustrine sediments of the North Aquitaine basin, France, by Jung (1954) and others (see Millot 1964, 1970); the mineral does not form granules but massive green clay beds. This question is further discussed together with the problem of shallow-water marine sediments (Chap. 10).

Volcanic Environment
Lacustrine sediments of volcanic regions often contain *smectite,* that is closely related to the evolution of igneous subamorphous material. The precise origin of smectite is nevertheless frequently difficult to assess, because subaerial and sub-

aqueous alteration of volcanic material a priori leads to fairly similar clay minerals. A close comparison should be made between the mineralogy, crystallinity, geochemistry, and micromorphology of both weathering and lake minerals. The studies performed on freshwater, nonsaline, and nonalkaline lacustrine sediments provide various data and interpretations. Many lakes of volcanic areas contain simply detrital assemblages, especially if the volcanic input is accessory. This is the case of the largest parts of the deep sediments of Lake Tanganyika (East Africa), that are characterized by allochthonous illite and chlorite (Degens et al. 1971); only shallow sill deposits located between the main basins of the lake lack active turbiditic sedimentation and contain smectite and traces of sepiolite that could be authigenic. Lake Nicaragua, a large lake developed in a volcanic terrane in Guatemala, contains volcanic glass, quartz, plagioclase, chlorite, illite, and mixed-layered smectite (Swain 1966); expandable minerals are very poorly crystallized, mixed with organic matter, semi-amorphous materials, and glass, and are difficult to identify precisely. Nelson (1967) studies the Crater Lake, Oregon (USA), whose sediments are exclusively derived from surrounding lavas, ash, pumice, scoria, breccia and tuff. Plagioclase, hypersthene, augite, and hornblende are associated with smectite in both clay and sand fractions; smectite undoubtedly issues from alteration of volcanic glass, but the possibility of subaerial formation of the mineral should be envisaged parallel to the subaqueous one. In the East African Naivasha Lake, Richardson and Richardson (1972) suggest from roentgenographic observations that crystalline minerals have been altered or even destroyed in a volcanic alkaline environment. Such an interpretation is hardly consistent with the persistence of feldspars in the same sediments, and could be due to poor X-ray responses because of abundant organic matter and subamorphous minerals (diatom frustules, opal, glass).

4.2 Saline Lakes

4.2.1 General Data

Introduction
Saline lakes correspond to relatively unusual situations, most lakes of the world being well flushed and not chemically concentrated. The solute load may accumulate beyond the potable range because of an excess of evaporation over inflow, or because of a saline inflow, or both. Saline lakes are favored in endorheic areas of block faulting, and rift and playa zones of warm regions. Saline lakes are therefore common in some parts of the world such as Eastern Africa and Western USA, but in size and total number are much lower than freshwater lakes (see Eugster and Hardie's review, 1978). They range from small temporary ponds as in Saline Valley, California, to deep, permanent, and large basins like the Dead Sea. Saline lakes present a large variety of chemical compositions and concentrations, which determine a wide range of crystallization products. The concentra-

tion may range from about 5000 ppm total dissolved solids, representing the upper limit of biological tolerance, to 400,000 ppm (Eugster and Hardie 1978). The brines can be highly saline as in the Great Salt Lake, Utah (USA), alkaline like those of Lake Magadi, Kenya, or bitter as in Basque Lakes, British Columbia (Canada). The brine regulation greatly depends on local conditions of inflow chemical composition, of rainfall, evaporation, infiltration, and seepage (e.g., Carmouze and Pédro 1977). Most lake brines are dominated by a single cation, *Na*. Ca- or Mg-rich brines occur rarely and K-rich ones almost never. A few saline lakes contain fairly abundant elements usually present only in traces, like Br in the Dead Sea (5000 – 7000 ppm), Sr in the Great Salt Lake (2000 ppm) and PO_4 in Lake Searles (900 ppm) (Eugster and Hardie 1978).

Saline Minerals
Saline lakes devoid of significant volcanic influence are fundamentally characterized by the precipitation of saline minerals, with few or no other species (Eugster and Hardie 1978, Eugster and Kelts 1983). Saline minerals comprise a large range of associations, controlled by the dominant cations and anions present in the brines (Table 4.2). The central zones of highly saline basins are protected from direct freshwater inflow and subject to higher evaporation, which favors the crystallization of the more soluble minerals. The geochemical center of an evaporating lake also receives the heaviest solute load determined by dissolution of surficial efflorescent crusts; this increases the gradient from dilute inflow on the lake borders to the concentrated central brine. Minerals that formed earlier from more dilute brines are therefore fed by more concentrated brines, then dis-

Table 4.2. Major saline minerals of the different brine types. (After Eugster and Hardie 1978)

Brine type	Main minerals	
Ca-Mg-Na-(K)-Cl	Bishofite	$MgCl_2, 6H_2O$
	Carnallite	$KCl, MgCl_2, 6H_2O$
	Halite	$NaCl$
	Sylvite	KCl
Na-(Ca)-SO_4-Cl	Glauberite	$CaSO_4, Na_2SO_4$
	Gypsum	$CaSO_4, 2H_2O$
	Mirabilite	$Na_2SO_4, 10H_2O$
	Thenardite	Na_2SO_4
	Halite	
Mg-Na-(Ca)-SO_4-Cl	Epsomite	$MgSO_4, 7H_2O$
	Kieserite	$MgSO_4, H_2O$
	Bishofite, glauberite, gypsum, halite, mirabilite	
Na-CO_3-Cl	Nahcolite	$NaHCO_3$
	Natron	$Na_2CO_3, 10H_2O$
	Trona	$NaHCO_3, Na_2CO_3, 2H_2O$
	Halite	
Na-CO_3-SO_4-Cl	Burkeite	$Na_2CO_3, 2Na_2SO_4$
	Halite, mirabilite, nahcolite, natron, thenardite	

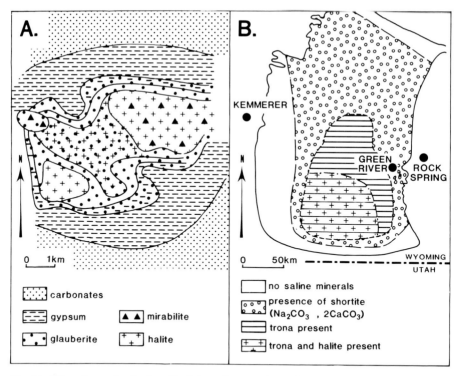

Fig. 4.3. Mineral zonation in the saline lake sediments of **A** Saline Valley playa California. (After Hardie 1968). **B** Eocene Green River formation, Wilkins Peak member, Colorado, Utah and Wyoming. (After Culbertson 1971)

solve and reprecipitate as more saline authigenic species. Some saline minerals such as gaylussite (Na_2CO_3, $CaCO_3$, $5H_2O$), pirssonite (Na_2CO_3, $CaCO_3$, $2H_2O$), glauberite ($CaSO_4$, Na_2SO_4) and anhydrite ($CaSO_4$), result totally or partly from diagenetic reactions. As a general result, the saline mineral precipitation often presents a *concentric zonation*, the more soluble species accumulating in the geochemical center of the lake. Many examples are recorded from the literature, such as the Saline Valley, California, that presently displays a centripetal formation of various sulfates and halite, or the Wilkins Peak Member of the Eocene Green River Formation (USA), marked by successive carbonates and chlorides (Fig. 4.3).

Note that various clay minerals, such as abundant illite, fairly abundant smectite and kaolinite, and rare chlorite, talc, and sepiolite, are reported as authigenic species in the lacustrine Green River Formation (Milton 1971). This assertion appears somewhat questionable, according to the diversity of clay suites and to the dominance of usually detrital minerals. The distribution of clays and associate minerals should be investigated in more detail in this huge and complex formation, in order to better understand the respective influence of detrital sources, authigenesis and diagenesis (see also Eugster and Hardie 1978). Data provided by Dyni (1976) and Tettenhorst and Moore (1978) indicate the formation of stevensite oolithes and of probable secondary illite in some specific environments characterized by geochemical gradients; this indicates the variety of the genetic mechanisms intervening in the Green River Formation.

Volcanic Environment

Saline lakes in volcanic environment basically induce the formation of authigenic zeolites. These silicates correspond to alkaline conditions and result from the reaction of volcanic glass, mostly tuff, with interstitial brines. The rate of glass dissolution increases with both salinity and alkalinity (Hay 1966). Many economically significant zeolite deposits result from the alteration of silicic tuff in saline alkaline lakes (see Pacquet 1969, Surdam and Sheppard 1978, Fisher and Schmincke 1984), although some analcimolites appear not to be related to volcanic influences (see discussion in Millot 1964, 1970).

The mechanism by which zeolites form from volcanic glass is still poorly understood (see discussion and references in Eugster and Hardie 1978, Fisher and Schmincke 1984). A direct precipitation of erionite ($Na_{0.5}$ $K_{0.5}$ $Si_{3.5}$ O_9, $3H_2O$) from trachytic glass is suggested in Lake Magadi, Kenya, by the simple addition of H_2O (Surdam and Eugster 1976). Alumino-silicate gels may first precipitate from the dissolved glass products, and then progressively crystallize into zeolites (e.g., Mariner and Surdam 1970). A third way consists of the secondary precipitation of zeolites from true solutions resulting from dissolution of glass (e.g., Hay 1966). Finally, early diagenetic reactions may determine the transformation of one type of zeolite into another, as for instance erionite into analcite in the Magadi region (Surdam and Eugster 1976). *Potassium feldspars* commonly form by replacement of analcite in the most saline sectors; they probably result from an early diagenetic reaction, since authigenic feldspars occur in many Pleistocene deposits but are lacking in active saline lakes.

A zonal arrangement of authigenic minerals exists within tuff layers of saline alkaline lakes, reflecting the centripetal geochemical gradient. Unaltered glass, zeolites, and K-feldspars successively occur from peripheral to central areas. Such a zonation is well exhibited by Pleistocene sediments of Lake Tecopa near Shoshone, Southeastern California (Sheppard and Gude 1968; Fig. 4.4). The type of alkalic zeolite depends on the nature of the glass and on the chemical composition of the brine. Analcite constitutes an optional step between erionite-type zeolites and K-feldspars.

Bedded cherts represent another authigenic product of saline alkaline lakes. These siliceous beds, known in both North-American and East-African Quaternary lakes (ref. in Eugster and Hardie 1978), are derived from transformation of *magadiite*, a sodium silicate precursor ($NaSi_7O_{13}$ $(OH)_3$, $3H_2O$; Eugster 1969). Magadiite has been reported to form together with other sodic silicates like kanemite ($NaHSi_2O_4$, $(OH)_2$), kenyaite ($NaSi_{11}$ $O_{20.5}$ $(OH)_4$) and makatite ($NaHSi_2O_4$, $(OH)_2$, H_2O) (Maglione 1970, Sheppard et al. 1970).

Clay minerals of saline alkaline lakes from volcanic areas are often less abundant compared to zeolites, other silicates, and subamorphous materials. For instance, Lake Magadi sediments comprise only traces of detrital illite, reworked from the drainage basin with anorthoclase (Surdam and Eugster 1976). Smectite is the more frequently reported clay mineral and is often attributed to an in-situ

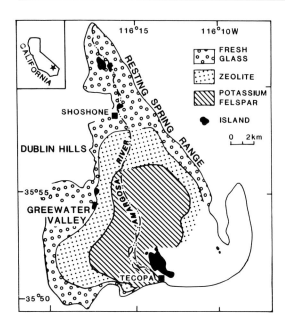

Fig. 4.4. Mineral zonation of Pleistocene Lake Tecopa deposits. (After Sheppard and Gude 1968, in Fisher and Schmicke 1984). Zeolites consist of phillipsite, erionite and clinoptilolite

formation. In fact, *smectite may result from both subaerial and subaqueous alteration* as in freshwater lakes (4.1.2), and may even be unstable in alkaline environments. For instance Jones and Weir (1983) show that smectitic minerals identified in Lake Abert sediments, Oregon (USA), are detrital and reworked from weathered volcanic rocks of the Abert rim; high-charge dioctahedral montmorillonite and smectite-chlorite mixed-layers are not stable in the lake but are partly transformed into illitic and stevensite-like minerals (see 4.2.2). This in- situ change remains quantitatively weak, most of the clay sedimentation being dominated by the detrital supply of chemically weathered volcanic material. Further mineralogic and geochemical investigations are needed on both hydrolysis and halmyrolysis (subaqueous alteration) processes, in other comparable environments.

Detrital Clays in Saline Nonvolcanic Lakes
Clay associations of many modern saline lakes simply reflect the mineral suites resulting from river discharge and rain wash, like the clay associations of most freshwater lakes. For instance the Great Salt Lake, Utah (USA), contains detrital kaolinite and illite, and some smectite whose origin could be either allochthonous, autochthonous, or mixed (Eardley 1938, Grim et al. 1960a). The playa sediments of the Mojave desert, California, show clay suites similar to those of surrounding formations (Droste 1959, 1961); there are no mineralogic changes linked to the depositional environment; smectite is derived mainly from volcanic ash weathering and is reworked within the lakes like other minerals.

Table 4.3. Mineralogical composition of the clay fraction from sediments and soils of Lake Kinneret and its watershed. (After Singer et al. 1972)

Mineral	Lake sediments	Basaltic soils	Brown soils on Neogene sediments	Red soils on limestone	Neogene sediments
Smectite	69	50–60	60–65	45–50	45–50
Kaolinite	18	25–40	25–30	30–35	25–30
Palygorskite	9	–	5–10	–	5–10
Illite	–	–	0– 5	5–10	0– 5
Quartz	4	0–10	0– 5	10–15	–
Free oxides	–	0–10	0– 5	5–10	–

Recent sediments of Lake Kinneret (Tiberias), Israel, contain a mixture of smectite, kaolinite and palygorskite, in addition to carbonates and quartz (Singer et al. 1972). The clay composition of lake deposits reproduces that of soils and sediments within the drainage basin (Table 4.3). The association of a high kaolinite concentration (10-30% of clay fraction) with the Jordan river delta indicates a preferential settling of this mineral in lacustrine zones marked by increased salinity and currents. Palygorskite is mainly concentrated along the eastern shore, where Neogene continental sediments containing this mineral crop out . Smectite is the more abundant species, chiefly derived from various volcanic rocks and blanketing weathering complexes. Only carbonates partly result from an authigenic formation.

The sediments of Lake Turkana, Kenya, consist mostly of laminated terrigenous mud, whose clay fraction is abundant (average 71%) and composed of smectite (average 40%), kaolinite (16%), illite (9%), quartz (28%), and feldspar (7%). Four sedimentologic and mineralogic zones are identified, and are well characterized by the smectite/kaolinite ratio (Yuretich 1979; Fig. 4.5). They express different modalities of detrital supply: (1) The north region of the Omo delta comprises iron-rich silty kaolinitic mud mainly derived from Tertiary volcanics of the Ethiopan plateau and rapidly deposited. (2) The fine-grained sediments of the northern basin contain fairly abundant smectite, that results from slower deposition of the Omo river detrital load. Some metamorphic sand grains (quartz, plagioclase, blue green amphiboles), also present in the southern part of the lake, probably represent an aeolian contribution from South and East. (3) The central delta region is characterized by silty smectitic mud rich in Na_2O, K_2O, with fairly abundant illite and quartz, and little kaolinite. This assemblage results from weathering and erosion of both metamorphic and volcanic rocks cropping out in the Southwest Turkwel-Kerio river system. (4) The south basin comprises argillaceous calcitic silt somewhat depleted in smectite compared to kaolinite, that corresponds to both terrigenous alluviation and lacustrine ostracod and diatom contribution. Essentially detrital clays are also reported by H. Paquet (in Tiercelin, Vincens et al. 1987) in sediment cores from lakes Baringo and Bogoria, Gregory rift, Kenya.

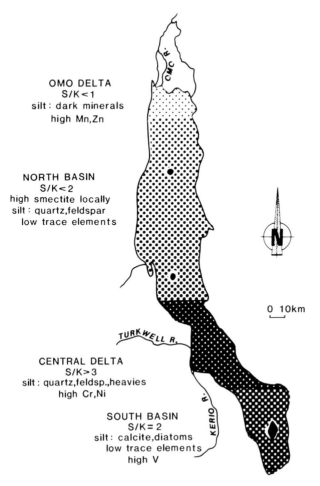

OMO DELTA
S/K < 1
silt : dark minerals
high Mn,Zn

NORTH BASIN
S/K < 2
high smectite locally
silt : quartz,feldspar
low trace elements

0 10km

TURKWELL R.

CENTRAL DELTA
S/K > 3
silt : quartz,feldsp.,heavies
high Cr,Ni

SOUTH BASIN
S/K = 2
silt : calcite,diatoms
low trace elements
high V

Fig. 4.5. Mineralogical provinces of Lake Turkana, and related lithological and geochemical characteristics. (After Yuretich 1979). *S/K* smectite/kaolinite peak height ratio

4.2.2 Authigenic Clay in Recent Saline Lakes

Ferriferous and Magnesian Environment

A few examples of well-studied saline lakes display unquestionable in-situ formation of clay minerals in different geochemical environments. *Lake Chad*, a large closed depression (21,000 km^2) located in Central Africa and mainly fed by the river Chari, is one of them. Maglione (1974) studies the interdune depressions situated to the North of the lake, where the groundwater level varies according to the level of the main lake, and evaporation is very active. He identifies the formation of dominant trona, halite, and nahcolite in the natronière of Liwa (location on Fig. 4.6), and the precipitation of thenardite, halite, and other Na-salts in small lakes of Napal island. Within the unconsolidated muds lying below the sur-

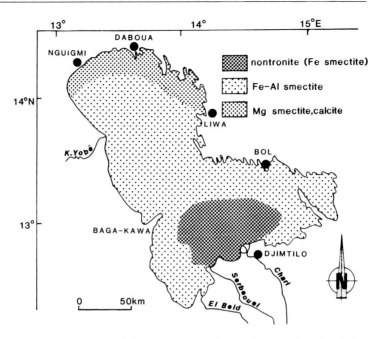

Fig. 4.6. Distribution of smectites in surficial sediments of Lake Chad, Central Africa. (After Carmouze et al. 1977 a)

ficial saline sediments, Mg-calcite, gaylussite, magadiite, and nahcolite form by capillary evaporation. Cheverry (1974) and Tardy et al. (1974) report the occurrence of authigenic magnesian smectite in the same mud, marked by high silica concentration and pH rise. All these neoformations occur at the frontier between sedimentation and pedogenesis, since strong evaporation, periodic groundwater rise, and desiccation determine alternating subaqueous and subaerial environments.

The waters and sediments of Lake Chad itself have been extensively studied in the theses of Carmouze (1976) and Gac (1979). The lake water, 3 to 4 meters deep, has an average salinity of 350 mg/l, and mainly contains Na, K, Ca, Mg, HCO_3, and $Si(OH)_4$. The mineralogy of both Chari suspended matter and lake surficial sediments consists of abundant kaolinite, with fairly abundant illite, quartz, and feldspar, that express the detrital contribution of Precambrian bedrocks and well-drained soils outcropping in the river drainage basin (Table 4.4). Smectites represent another noticeable compound, characterized by increased content in lacustrine deposits. Carmouze et al. (1977 a) think that the mineralogic composition of river suspensions and of lake sediments can be directly compared, and that the smectite supply could not have been significantly higher in the past. They therefore estimate that about 35% of lacustrine smectite is detrital and 65% authigenic, and that large proportions of dissolved SiO_2 (47.5%), Al_2O_3 (80%), Fe_2O_3 (30%) and Mg (28.8%) are extracted through chemical clay sedimentation.

Table 4.4. Major minerals of suspended matter in Chari river (on an annual basis) and of Chad Lake sediments. (After Carmouze et al. 1972 b)

	Kaolinite	Illite	Quartz	Feldspars	Smectite	Subamorphous materials		
						SiO_2	Fe_2O_3	Al_2O_3
Chari	38.5	11.5	9.5	6.0	10.5	2.3	2.8	1.3
Chad	26.5	8.0	11.10	4.5	21.0	4.9	1.4	0.2

Table 4.5. Structural formulae of smectites in Chad Lake area. (After Carmouze et al. 1977a and b)

Chari river	$(Si_{3.47}Al_{0.53})O_{10}(Al_{0.33}Fe^{3+}_{1.17}Ti_{0.29}Mg_{0.21})(OH)_2(Mg_{0.06}Ca_{0.14}Na_{0.02})$
Delta, SE lake	$(Si_{3.83}Al_{0.06}Fe^{3+}_{0.11})O_{10}(Fe^{3+}_{1.76}Mg_{0.20})(OH)_2(Ca_{0.22}Na_{0.04})$
Middle lake	$Si_4O_{10}(Al_{1.32}Fe^{3+}_{0.43}Ti_{0.02}Mg_{0.18})(OH)_2(Ca_{0.14}Na_{0.01})$
North lake	$Si_4O_{10}(Al_{0.24}Fe^{3+}_{0.12}Ti_{0.02}Mg_{1.99})(OH)_2(Ca_{0.19}Na_{0.07})$

Four types of smectite are recognized, from the Chari river suspensions in the Southeast to the northernmost lake deposits (Carmouze et al. 1977a,b; Fig. 4.6, Table 4.5). (1) The Chari carries *Fe-Al smectite* rich in iron, that is mainly derived from erosion of vertisols broadly developed in the downstream, arid zone of the river basin. (2) The subaqueous deltaic sediments of Chari, in the southeastern part of Lake Chad, comprise a surficial granular layer (5-15 cm thick) characterized by oolitic to peloidal particles (Lemoalle and Dupont 1973). The grains mainly consist of goethite and *nontronite*. This Fe-rich smectite is attributed to an in situ precipitation of iron with silica, followed by a migration of silica inside the pellets; silica combines with Fe^{3+} in a hydroxy complex that finally crystallizes into a ferric smectite (Pédro et al. 1978). (3) The major part of the lake sediments contains an *Al-Fe smectite*, whose composition is similar to Al-Fe beidellite formed in vertisols (Trauth et al. 1967, Paquet 1970; see also Fig. 4.7), and that is enriched in Al and depleted in Fe as compared to Chari river smectite. Carmouze et al. (1977a) interpret this smectite as a neoformation product. One could also envisage a reworking from peri-lacustrine vertisols combined to a river supply. The experimental evaporation of Chari water leads to Mg-smectite-like minerals and not to Al-Fe smectite (Gac et al. 1978). The abundance of detrital compounds in Lake Chad (Table 4.4), the presence of abundant smectite-rich soils in the drainage area (Gac 1979), the possibility of differential settling processes in increased saline and calm water, and the widespread occurrence of detrital smectite in saline and freshwater lakes suggest that an allochthonous origin could be as probable as an autochthonous one. (4) The northernmost part of Lake Chad contains *stevensite* (Mg-smectite) mixed with Al-Fe smectite. The autochthonous formation of stevensite is documented by the evaporation of Chari river water, that successively leads to the precipitation of Mg-smectite-like minerals, calcite, and amorphous silica (Gac et al. 1978). If suspended material is removed before

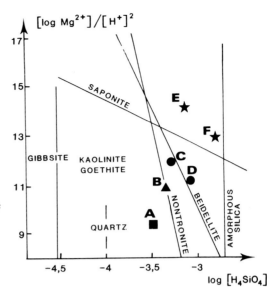

Fig. 4.7. Equilibrium diagram of various clays and oxides, and location of results relative to Chad Lake water. (After Tardy 1981). *A* Chari river, downstream zone; *B* SE delta: non-tronite zone; *C* ephemeral ponds above vertisols, and D, lake: Al-Fe smectite zone; *E* evaporative salt pans, and F, lake northernmost part: Mg smectite zone

evaporation, only amorphous magnesian silicate forms, following Mg-calcite and amorphous silica; this suggests the possible precipitation of magnesian clay on mineral precursors. Note that Mg-rich soils and evaporative sediments, cropping out on the northern border of Lake Chad, also contain stevensite (Tardy et al. 1974), have a composition similar to that of overlying water (Fig. 4.7), and could perhaps account for the partial supply of stevensite to main lake sediments during strong flood periods.

To summarize, surficial sediments of Lake Chad offer a succession of mineralogic zones, characterized by different types of smectites, from the slightly acid fluvial environment to the alkaline northernmost part of the lake (Fig. 4.8). Water tends to concentrate through evaporation from the South to the North, and successive minerals form accordingly, mainly nontronite, stevensite, Na silicates and salts, and hydrous silica (Tardy 1981). These authigenic processes are superimposed on the dominant detrital supply and produce some minerals that are also formed in surrounding soils; this complicates the precise identification of the different possible origins. An unquestionably and exclusively authigenic genesis applies to nontronite peloidal granules in the uppermost sedimentary layer of the deltaic area. Stevensite appears to form mainly in lake sediments, while Al-Fe beidellite could be derived to a significant extent from soil formations.

Volcanic Environment
Lake Mobutu Sese Seko (formerly Lake Albert) lies in the northern part of the Western Rift Valley, Africa. The dark gray to black fine-grained mud deposited during the last 28,000 years contains smectite, illite, irregular mixed-layer illite-smectite (10-14$_S$) and kaolinite (Stoffers and Singer 1979, Singer and Stoffers 1980). The following mineralogic sequence is recorded from the bottom to the

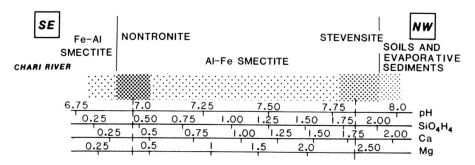

Fig. 4.8. Geochemical characteristics of interstitial waters of Chad Lake sediments, in the different smectite zones. (After Carmouze et al. 1977 a). Concentration in mol 1^{-1}

top of cores sampled in the lake (Fig. 4.9):

$$S(I, K) \rightarrow 10\text{-}14_S (S, I, K) \rightarrow I (10\text{-}14_S, K) \rightarrow 10\text{-}14_S (S, I, K) \rightarrow S(I, K)$$
( ) = minor amounts   S = smectite   I = illite   K = kaolinite

Smectite, illite and kaolinite occurring at between 28,000 and 25,000 years B.P., and from 12,500 to the present time, are considered as detrital and supplied by the Victoria Nile and Semlike rivers from upstream bedrocks and soils. The diatom assemblage suggests normal chemical conditions and a high lacustrine level, while fluctuating manganese values indicate temporary stratification of lake water. From 25,000 to 16,000 years and again briefly after 12,500 years B.P., the occurrence of illite-smectite mixed-layers corresponds to the absence or scarcity of diatoms; this is attributed to lower lake levels and a fairly concentrated en-

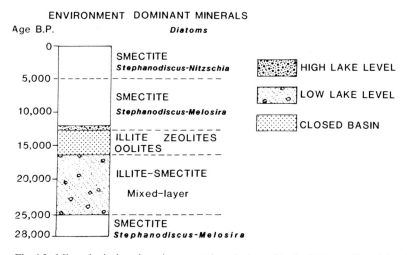

Fig. 4.9. Mineralogical and environmental evolution of Lake Mobutu Sese Seko for the last 28,000 years. (After Stoffers and Singer 1979)

vironment. Between 16,000 and 12,500 years B.P. the increase of illitic minerals coincides with the presence of protodolomitic oolites and of potassium-rich chabazite and phillipsite. *Illite is considered to be the result of smectite transformation in a closed basin submitted to both saline and alkaline conditions, during a stage of very low level due to a dry climate.* Smectite is thought to be illitized by the incorporation of potassium released by evolution of the K/Na-zeolites to Na-zeolites. The mixed-layers represent an intermediate stage in the mineralogic evolution proposed. The possibility of reworking of detrital illite by wind or water during the period of low lacustrine level is not envisaged (see Środoń and Eberl 1984, p. 523). The very small size of illite particles in 16,000–12,500 year-old deposits (Stoffers and Singer 1979) could indicate either a chemical growth or more distant sources.

Another example of illitization is reported in *Lake Abert*, Oregon (USA), located in a volcanoclastic environment (Jones and Weir 1983). High-charge dioctahedral montmorillonite, poorly crystallized smectite, and smectite-chlorite mixed-layers form through subaerial weathering, and are reworked into the lake by runoff. Lacustrine clays and coarser rock debris show the disappearance of pure well-crystallized montmorillonite and the appearance of *poorly crystallized illitic minerals and stevensite-like minerals* (see 4.2.1). The illitic mineral is attributed to the extraction of potassium from the lake and its fixation within smectite layers, possibly during a period of fresher water. A comparable explanation is tentatively presented by H. Paquet (in Tiercelin, Vincens et al. 1987) for a core sampled in Lake Bogoria, Kenya. Środoń and Eberl (1984, p. 523) suggest that illitic material could also have formed during the transportation phase by wetting and drying cycles, because no significant difference exists between the lake sediments and those deposited in the adjacent playa. The trioctahedral stevensite-like clay could form in a similar way to illite, by fixation of magnesium in highly saline water (salinity of 30 to 90 g/kg). Both authigenic minerals are scarce, the clay sedimentation being permanently dominated by detrital input (Jones and Weir 1983).

Hay and Guldman (1987), following an earlier description of smectite and chlorite destruction in Quaternary volcaniclastic sediments of Lake Searles, California (Hay and Moiola 1963), confirm that detrital clays have reacted extensively in sediments with diagenesis. The study of a 693 m-long drilled core (3.18 my) in this playa-lake complex points to the existence of two types of transformation. The upper 291 meters of silicic ash layers, in contact with highly saline, alkaline pore water, were initially altered to phillipsite and merlinoite (zeolites), and then to K-feldspar and searlesite ($NaBSi_2O_6$, H_2O). This first type of transformation required more than 45,000 years and was largely completed in 140,000 years. The lower part of silicic ash was in contact with moderately saline, slightly alkaline pore fluid. Tephra layers altered mainly into smectite and accessorily into clinoptilolite, opal and analcime, the intensity of transformation being inversely proportional to the grain size. Alteration occurred in an almost closed hydrologic system. Note that an illitic clay, attributed to the early transformation of altered tuff beds under the influence of increased salinity and alkalinity, is reported by Fishman et al. (1987) in an upper Jurassic playa-lake complex of the Eastern Colorado plateau.

Biosiliceous Environment

The Plio-Pleistocene continental basin of the *Bolivian Altiplano* is characterized by a strong acid volcanism, especially in the southern area where volcanoes delineate small and closed lakes. This high-altitude region (4000 to 4500 m) is submitted to cold-dry winters, and to wet and fairly warm summers (rainfall 1 to 3 m, evaporation 1 to 1.5 m). The small intravolcanic lakes of the Lipez region contain diatom mud with calcite, clay and various salts according to the location (Badaut et al. 1979, 1983). Electron-microscope observations show the rapid dissolution of *diatom frustules*, that may be partly or totally replaced by tiny sheets. X-ray diffraction on selected areas, dark- field observation, microdiffraction and elemental microanalysis indicate the occurrence of poorly crystallized *Mg-smectite*, whose 14-15 Å reflections do not collapse at 10 Å after heating. This clay is attributed to an autochthonous formation at about 5°C, a temperature noticeably lower than that used to synthetize Mg-clays (20°C; e.g., Siffert 1962, Decarreau 1980). The essential conditions for clay growth from diatom frustules in such an environment appear to be a chemical saturation with respect to amorphous silica, and a pH above 8.2, rather than high Mg concentration or salinity (Badaut et al. 1983). If silica concentration is high and pH low, the frustules remain sound and unaltered.

4.2.3 Fibrous Clays in Lacustrine Environment

Early investigations suggested that palygorskite and sepiolite could form in present-day saline lakes, like those in the Western USA or in Lake Eyre, Australia (Grim 1953, Bonython 1956). Parry and Reeves (1968 a, b) report the presence of sepiolite associated with illite, kaolinite and smectite, in two closed basins of the Southern High Plains, Texas, that were filled with water during the Pleistocene pluvial stages. Sepiolite is considered as the result of the transformation of smectite within the lake, in response to very high salinity due to an arid climatic episode apparently younger than 20,500 years B.P. Such descriptions are very rare in the literature (e.g., Hassouba and Shaw 1980), and *the present-day lacustrine formation of fibrous clays remains highly questionable*. First, modern studies on still active saline and alkaline lakes have not as yet drawn conclusions on any in situ genesis of palygorskite or sepiolite. Second we know that fibrous clay may form in calcrete and caliche under arid conditions, and are both resistant enough to be eroded and transported into sedimentary basins, especially the lacustrine depressions located in the vicinity and submitted to an arid climate (see 3.1 and 4.2.1).

Numerous descriptions of authigenic fibrous clays in ancient lakes are reported in the literature, especially concerning the warm periods of the Early Paleogene (Paleocene, Eocene). Invaluable work has been done on this subject by G. Millot and his colleagues (see review in Millot 1964, 1970), who show that fibrous clays are often arranged with Mg-smectites in vertical and lateral sequences. The lacustrine environment is calcareous, marked by the frequent occurrence of siliceous beds and sulfate accumulation, and referred to "chemical alkaline sedimentation". Recent studies provide detailed information on the transition from typically detrital to exclusively authigenic environments (e.g., Trauth 1977), on paleogeographic implications of clay and rock successions (e.g., Brell et al. 1985, Jones et al. 1986), on the different ways of possible sedimentary formation of fibrous clays (e.g., Post 1978, Galan and Castillo 1984, Leguey et al. 1985), on the role of juvenile water and geochemical variations on the crystallization of Mg-clays (e.g., Khoury et al. 1982, Hay and Stoessel 1984, Hay et al. 1986), etc. The most recent periods where lacustrine clays unquestionably formed in large amounts appear to be (1) the uppermost Miocene of Southeast France where crystallization occurred in small, isolated, strongly evaporating basins (Chamley et al. 1980 b), (2) the Pliocene of the Amargosa desert, Western USA, where Ca-Mg bicarbonate spring water favored Mg-smectite and sepiolite precipitation (Hay et al. 1986), and (3) a few Pleistocene lakes in Kenya and California located in a volcano-hydrothermal environment (Hay and Stoessel 1984, Starkey and Blackmon 1984).

Many studies indicate the *accessory role of salinity* in the genesis of palygorskite, sepiolite, and associate Mg-smectite. The main restriction factors consist of the availability of dissolved silica, the presence in suitable proportion of ions necessary for clay building, the existence of a warm climate, with a fairly long dry season, strong evaporation, and temporarily small detrital supply. These *conditions may be realized in both lacustrine and peri- marine environments*. In fact, fibrous clays and associate minerals and sediments formed equally within lakes and epicontinental seas in past geologic times, the latter being more widely distributed (e.g., Millot 1964, 1970, Trauth 1977, Weaver and Beck 1977, Singer and Galan 1984). We will consider these environments together in more detail in a later section devoted to marine sedimentation (Chap. 9).

4.3 Conclusion

1. *Most freshwater and saline lakes contain clay suites that result exclusively from the erosion of drainage basins.* This explains why most studies on lake minerals do not include geochemical investigations, which usually do not provide relevant additional information. Clay suites therefore constitute reliable tools for identifying the detrital sources, the relative role of bedrocks and soils in alluviation, the sedimentary processes, and past environmental changes.

2. *Saline lakes and evaporative processes do not automatically favor clay authigenesis.* High salinities first determine the formation of various *salts*, while high alkalinity in a volcanic environment rather induces the genesis of *zeolites*. The formation of complex clay silicates supposes the availability in suitable proportion of dissolved Si, Al, Fe, Mg, etc, which is rarely realized in modern lacustrine environments. Smectite does not constitute the systematic alteration product of volcanic glass under continental subaqueous conditions, and is even unstable and transformed into illitic minerals in some Plio- Pleistocene lakes.

3. *Clay genesis is reported in a few recent lakes and corresponds to specific conditions*, namely to highly ferriferous, magnesian or siliceous environments: iron-

rich granules including chamosite, glauconitic mica or even nontronite; probable stevensite in the northernmost and most evaporative part of Lake Chad, and possible Al-Fe smectite in the central part of the same lake; illitic and smectitic mixed-layers in some volcanic environments; disorganized Mg-smectite growing at the expense of diatom frustules in some lakes of the Bolivian Altiplano. The in situ clay formation sometimes remains questionable, since several minerals may form in both subaerial and subaqueous environments (e.g., Icole et al. 1987). Detailed studies performed on lacustrine water and sediments should be associated more systematically with similar investigations on the surrounding drainage basin from which minerals may be reworked.

4. *The lacustrine formation of fibrous clays and associate Mg-smectite* does not obviously occur in Quaternary time, but *was extensive in several geologic periods* marked by a hot climate, strong hydrolysis and evaporation, and high carbonate and silica accumulation. The conditions of genesis were similar in lakes and in some peri-continental seas (see Chap. 9), were not greatly dependent on salinity, and present some parallel characteristics with the formation of clay in calcretes under arid climate.

Part III

From Land to Sea

Chapter 5

Estuaries and Deltas

5.1 First Studies on Estuarine Clays

Sedimentary particles transported from the continent to the sea experience changes in salinity, pH, and other physicochemical characters. Clay minerals, that form the main constituents of suspended matter entering the sea, pass from the river environment depleted in dissolved ions to the chemically more concentrated marine environment. The surface negative charge, determined by layer discontinuities and sheet substitutions (Chap. 1), is more or less balanced by hydrated cations, and is subject to increasing ionic strength in passing from freshwater to sea water (e.g., Aston 1978, Dyer 1986). Some surface charge disturbances and even sheet reorganizations may therefore be expected, especially at the expense of clay species formerly degraded through continental weathering. This is the reason why clay mineral studies were performed on deltaic and estuarine sediments as early as the mid-twentieth century. In-situ investigations were encouraged by the experimental results of Carroll and Starkey (1958), who measured the exchangeable cations of pure minerals (montmorillonite, illite, kaolinite, vermiculite, illite-smectite) immersed in sea water for 10 days. These authors found that Mg from sea water moved into the exchange positions in preference to Ca and Na, and that kaolinite adjusted very rapidly compared to smectite and mixed-layers; their data also suggested the release of appreciable amounts of SiO_2, Al_2O_3, and Fe_2O_3 from the clay minerals to the sea water.

The first studies performed on deltaic and estuarine clays often concluded the occurrence of some in situ chemical adjustments or even strong mineralogic changes. For instance, Griffin and Ingram (1955) observe an increase of chlorite and illite relatively to smectite and kaolinite, from freshwater to saline water in the estuary of Neuse river, North Carolina (USA); they envisage a chemical evolution of detrital clays or associate amorphous material within the saline environment. Johns and Grim (1958) and Milne and Early (1958) describe an augmentation of illite abundance when fluvial waters enter the Northern Gulf of Mexico, in the Mississippi delta, and between the Mississippi and Mobile rivers; this change appears moderate, occurs only locally, and is attributed to a transformation of smectite to illite in some sectors characterized by a.low sedimentation rate allowing significant chemical reactions. Note that Taggart and Kaiser (1960) do not find any appreciable difference, in the same area, between fluvial and marine clay assemblages supplied by the Mississippi and Mobile rivers. In the estuary of the Rappahannock river, Virginia, Nelson (1960) describes the downstream augmentation of illite crystallinity, the diminution of smectite abundance and the appearance of chlorite; he envisages in situ mineralogic changes, but does not exclude the possibility of lateral detrital sources or peculiar conditions of deposition. Grim and Loughnan (1962) consider that clay minerals passing from freshwater into the marine harbor of Sydney, Australia, experience a degradation of vermiculitic and chloritic minerals and a correlative development of illite. In the Rhone river, France, illitic and chloritic clays include a few mixed-layered minerals that tend to slightly shift closer to illite reflections when passing into the marine environment (Chamley 1964); this could be determined by a chemical adjustment, but the possibility of hydrodynamic effects and the absence of similar evolution in the adjacent Camargue salt marshes (Chamley 1968) incite to caution. Jeans (1971) thinks that the large amounts of dissolved Al, Fe, and Si issued from continental weathering and transported by rivers may precipitate in estuaries and deltas and form hydroxide gels; these gels could allow the neoformation of iron minerals and of various silicates, including clay minerals. This hypothesis is supported by calculations on the amounts of dissolved Al, Fe, and Si, and by experiments on the precipitation of Al-Si gels. Nevertheless C.V. Johns estimates such phenomena as of little importance today, in comparison with some geologic periods marked by lower particular and higher dissolved input.

Let us consider in more detail three examples that have led to *controversial interpretations*. The Guadalupe delta and San Antonio bay, in the vicinity of Rockport, Texas, represent a typical river-sea transition in the Northwest Gulf of Mexico. This area was first studied by Grim and Johns (1954), who identified abundant smectite associated with illite and chlorite in the Guadalupe river. The delta and bay sediments display a progressive but strong relative increase of illite and chlorite abundance, which was attributed to the transformation of smectitic minerals entering the saline environment. Morton (1972) investigated the same area from 80 samples located on a transect from the Guadalupe delta to the continental shelf, and conducted a statistical study by measuring linear regression and correlation coefficients. He did not identify any chlorite in the delta sediments but kaolinite, occurring together with smectite and illite. R.D. Morton found no significant difference in smectite, illite and kaolinite abundance between the different fresh, brackish, and marine water environments. The lack of chlorite, and the clay similarity in both river and bay environments suggest that marine mineral assemblages depend exclusively on detrital sources, and do not result from any identifiable chemical modifications. The latter interpretation is supported by the fact that smectite-rich assemblages exist in both river and open sea sediments (Fig. 5.1), which was first attributed to in- situ transformations (Shepard and Moore 1954) and later to a differential deposition process (Postma 1967; see 5.3.1).

The Chesapeake bay constitutes the largest estuary in the middle Atlantic USA region. Its sediment clay mineralogy was first studied by Powers (1954, 1957, 1959), especially in the James and Patuxent river dependencies. Fluvial deposits contain variously crystallized illite with minor amounts of kaolinite and weathered chlorite. Estuarine sediments contain significant amounts of mixed-layers, fairly abundant chlorite, some well-crystallized illite and little vermiculite.

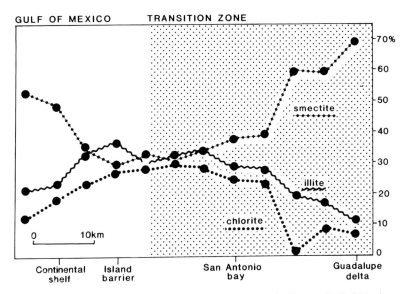

GULF OF MEXICO   TRANSITION ZONE

Fig. 5.1. Clay mineral distribution in sediments of Guadalupe delta and adjacent Gulf of Mexico, Texas. (After Shepard and Moore 1954). The *shaded area* corresponds to the zone investigated by Grim and Johns (1954), on a nearby transect that gave very similar results

The thermal stability of the chlorite increases as the salinity in the estuary increases. M.C. Powers observes a good correlation in the abundance of Mg and K contained within the clay minerals, among the exchangeable ions, and in the interstitial waters. Additional microscope observations and preliminary experiments also suggest possible in situ changes. M.C. Powers concludes an important chemical transformation of weathered illite into chlorite, as the clays are transported from the freshwater to a more saline environment. Vermiculitic and mixed-layer clays constitute a possible step in the progressive development of secondary chlorite. The preferential formation of chlorite is attributed to a preferential adsorption of Mg compared to K, the latter element being partly fixed by illite to regenerate better crystallized micaceous minerals. Powers (1957) estimates that chlorite formation especially occurs in surficial marine environments that are marked by high Mg/K ratios (values close to 5), while illite formation is more frequent in buried sediments (from about 100 m) characterized by low ratios. Similar conclusions arise from the study of Atchafalaya (West of Mississippi, Gulf of Mexico) and San Diego (East Pacific) regions, where chlorite is supposed to form from the evolution of smectitic and partly of illitic minerals (Powers 1957).

Owens et al. (1974) also study the Chesapeake bay, considering the clay mineral variations in a geological context. In opposition to the chemical variations identified by M.C. Powers, that may vary with clay mineralogy and water composition in the absence of real in situ changes, they first notice the immaturity of clay assemblages, that are composed of various minerals, often poorly crystallized and little compatible with significant geochemical reorganization.

This immaturity occurs in both fresh and saline sediments. Illite is the most abundant clay species in the upper part of the bay, and is accompanied by smectite, kaolinite, and chlorite. J.P. Owens and colleagues estimate that chlorite is essentially detrital because of its widespread distribution in Piedmont rocks, particularly cropping out along the Susquehanna river west of the bay. Heavy minerals associated with clay indicate the same terrigenous sources. Furthermore, if degraded illite was presently converted to chlorite in waters of increasing salinity, the same mechanism should probably have acted in the local marine formations of the Miocene age (Chesapeake group), virtually devoid of chlorite (Owens et al. 1974). Finally the authors show that the suspended matter from the slightly brackish Chester river near its confluence with Chesapeake bay contains an average of 11 % chlorite in the clay fraction, a value very close to that of recent and late Quaternary sediments deposited in the bay (13 %). Owens et al. (1974) think that the clay minerals of Chesapeake bay mainly express, as in the Delaware and Hudson estuaries, the detrital supply of minerals removed from continental rocks and moderately weathered soils under temperate climate (see also 5.3.2).

Some conflicting interpretations also arise from studies of the Gironde, the estuary of the Garonne river, Southwest France. Latouche (1971, 1972) observes a progressive decrease of the smectite/illite ratio from the freshwater sediments to the Atlantic shelf deposits, with a local augmentation on the estuarine interchannel shoals. Discussing this variation, C. Latouche thinks that a chemical change probably occurs, that is favored by the long residence time of clay particles within the estuary, before their expulsion toward the ocean and their burying. Smectite is therefore thought to partly transform into illite. On the contrary, Mélières and Martin (1969) and Martin (1971) estimate that Gironde clays are exclusively detrital. Differential settling processes are supposed to extensively control the mineral distribution, smectite being kept in suspension during tidal cycles and directly evacuated toward the ocean (see 5.3.1). Such an interpretation is supported by the general independence existing between mineral suspended particles and associate chemical elements in the Gironde estuary (Jouanneau 1982).

To summarize, most of the studies first performed on estuarine clay assemblages favored geochemical interpretations, the transition from freshwater to saline water environments being considered as the first step of the diagenetic evolution (see also Millot 1964, 1970). Subsequent investigations often led to more cautious approaches, suggesting that the geochemical break occurring between land and sea was not strong enough to permit significant crystallochemical changes of clay minerals. In fact, one may be somewhat suspicious about the diversity of mineralogic transformations envisaged by the various authors, the land-sea transitions being supposed to determine the evolution either of illitic or smectitic minerals into chlorite, vermiculite and chlorite into illite, smectite into chlorite and illite, and smectite into illite only. Such a diversity implies various geochemical mechanisms not easily compatible with the relative homogeneity of salinity and pH changes at the river-sea borders. In addition, detailed investigations have shown that only very little change can affect minerals like illite and chlorite when passing from freshwater to sea water environment. For instance, Hoffman

(1979), in an X-ray diffraction and K-Ar study of the Mississippi river and adjacent Gulf Coast sediments, shows that only 0.2-0.3% K_2O may enter the illite network, and that the clay mineralogy remains unchanged until a burying temperature of at least 50°C.

5.2 Estuarine Clays and Continental Sources

Various data indicate the existence of a general correspondence between the composition of estuarine and deltaic clay assemblages and the mineral composition of rocks and soils in the hinterland. For instance, the *suspended matter* of the Eastern North Sea estuaries, from Denmark (Varde Å) to the Netherlands (Rhine-Meuse), contain fairly similar mineral suites (mainly illite with lesser amounts of chlorite, smectite, kaolinite, illite-smectite), that express the relative homogeneity of Northwestern European substrate petrography, soil composition, and climate characters (Rudert and Müller 1981). The *sediments* of nine representative British estuaries contain a mixture of illite, kaolinite, smectite, and chlorite, whose composition varies according to the location and therefore the geologic context (Biddle and Miles 1972). The Thames is relatively rich in smectite, the Clyde and Forth in kaolinite, and the Severn, Mersey, and Kyle of Durness in illite. The Mersey and Kyle of Durness comprise the highest percentages of chlorite. The clay composition at different points of large estuaries like the Thames and Mersey varies very little in sediments and liquid muds, which results from active mixing under macrotidal conditions. Similar results arise from the study of the Canche estuary, Northern France, where high tidal ranges determine intense mixing; preferential settling of smectite occurs locally only, in left side sectors marked by lower hydrodynamic forces (Despeyroux and Chamley 1986). In the Adour estuary, Southwest France, Snoussi (1986) shows that clay assemblages represent the sum of minerals provided by the main course and the different tributaries, and therefore strictly express the composition of drainage basin rocks and soils. M. Snoussi proposes similar conclusions about Oued Sebou, Morocco, where the supply from smectitic vertisols developed in downstream flat areas and from hard palygorskite-bearing calcareous crusts is rejected in favor of that from upstream actively eroded illitic and chloritic rocks.

Estuaries of the USA have been studied extensively by Folger (1972) and Hathaway (1972). Most bottom sediments accumulating in the estuarine zone consist of terrigenous detritus, biogenic debris, and pollutants. Inorganic constituents mainly comprise clay minerals, quartz, and feldspars. Their distribution, and that of associate heavy minerals, depends on the composition of continental rocks and soils. Especially clay minerals, that commonly constitute over 90% of the fine fraction, display quantitative variations related to regional sources and weathering processes. From New England to Chesapeake bay, on the Atlantic coast, illite and chlorite dominate, and characterize crystalline and old sedimentary rocks as well as little to moderately evolved soils; these minerals were mostly supplied within the estuaries from the adjacent shelf during the

Holocene transgression, after being transported from land to the ocean margin during the last glacial regression (Folger 1972).

Kaolinite, mainly derived from soils developed on the crystalline Piedmont region, is characteristic of the east coast estuaries South of Chesapeake bay, and often predominates over illite, smectite, and mixed-layers. Piedmont-derived kaolinite is frequently associated with hornblende, which helps to distinguish upstream sources from sources controlled by lateral tributaries or by marine currents. The kaolinite association may be diversely diluted by detrital smectite, illite, chlorite, vermiculite, palygorskite, sepiolite, dolomite, phosphorite or sillimanite, depending on local geology and hydrodynamic conditions (Neiheisel and Weaver 1967). For instance, soil-derived kaolinite reaches 80% of the clay fraction in Albemarle Sound, while Coasawhatchie river in the upper estuary includes up to 40% of sepiolite and palygorskite eroded from local Miocene formations. All these variations point to the importance of terrigenous sources. In the Gulf of Mexico eastern kaolinite-rich assemblages (e.g., Apalachicola river) are replaced toward the West by smectitic suites (e.g., Mississippi, Guadalupe), which expresses the westward change of geologic substrates and pedogenic conditions (Folger 1972).

On the Pacific coast smectite also dominates over illite, kaolinite and chlorite, because of the abundance of volcanic rocks, and of sedimentary rocks and soils rich in smectite.

Past sediments from a given area mostly display strong similarities between continental and marine clay suites, which indicates the absence of significant changes in the transition zone. For instance, Ballivy et al. (1971) observe identical clay associations in glacial lacustrine and superimposed postglacial marine muds of Northwest Québec, Canada (illite, chlorite, vermiculite, kaolinite). Faugères and Robert (1976) find no difference in the clay composition of early Pleistocene continental deposits and subsequent middle Pleistocene marine transgressive sediments, from the study of two wells drilled in the Thermaic bay, northernmost Eastern Mediterranean.

A detailed investigation performed by Środoń (1984b) on late Miocene sediments of the Carpathian foredeep, Poland, leads to similar conclusions. Badenian and Sarmatian claystones deposited at a fairly low sedimentation rate (< 8 cm/1000 years) in brackish to hypersaline environment. The clay assemblage of claystones and gypsum beds is identical and comprises randomly interstratified illite- smectite, kaolinite, illite, and sometimes minor chlorite. No correlation exists between the clay composition and the sedimentary environment or the depth of burial (0 to 200 m), although the most precise identification techniques have been applied to study the mixed-layering. The lack of correlation between the salinity and illite-smectite ratio in the mixed-layer clay, as well as the low illite content, indicates that salinity increase does not determine any illitization process. A calculation shows that K availability in pore water and sea water is too low to increase the illite layers in a significant and measurable manner. J. Środoń concludes that smectite and smectite-illite are stable in marine sediments until high temperatures of the deep diagenesis range are reached. This conclusion does not apply for high K-Mg saline and alkaline evaporative basins (see 4.2.3 and Chap. 9). The only change recorded consists of a weak augmentation of quartz and kaolinite in coarser sediments relatively to expanding clays (Środoń 1984b), which probably results from physical sorting (see 5.3). Note that Schultz (1978) also reports negligible differences between freshwater and adjacent marine facies, in the smectite-rich series of the late Cretaceous Pierre Shale formation, Great Plains, USA.

5.3 Mechanisms of Clay Distribution

5.3.1 Differential Settling and Flocculation

Early experiments showed that mixing at river-sea boundary with variable water salinities and current velocities can give rise to *selective transportation and deposition of clay minerals*. Whitehouse and McCarter (1958) observed a preferential settling of kaolinite and illite compared to smectite in low-salinity water, where clay floccule formation often occurs preferentially by surficial adsorption of cations (Powers 1959). In consequence, estuarine conditions favor the deposition of illite and kaolinite and disfavor that of smectite, a mineral rather kept in suspension and passing directly into the marine environment. Further investigations by Whitehouse et al. (1960) showed that the flocculated clay suites settle out as a solid-rich fluid, and that *the differential settling depends on various parameters* (water density, temperature, turbulence, pH, organic matter content, dissolved ions, etc; see Aston 1978). The average sedimentation rate at 26°C in motionless and slightly saline (18‰) water is 1.3 m/day for smectite, 11.8 m/day for kaolinite and 15.8 m/day for illite. The presence of organic matter may considerably modify such values. Smectite settles faster in presence of hydrocarbons and more slowly in presence of proteins, while illite remains insensitive (Whitehouse et al. 1960). Despite the wide range of factors possibly intervening in flocculation mechanisms, *smectite statistically appears to settle less rapidly than other clay minerals*, notably because of its frequent small size, flaky shape, and low density due to close bonding with water.

Various examples of differential settling and flocculation of clay minerals are reported from estuarine environments, and are used to demonstrate the weakness of in situ chemical changes (e.g., Weaver 1959, Manickam et al. 1985). In the lower Loire valley, Western France, Barbaroux and Gallenne (1973) and Gallenne (1974) observe a relative concentration of smectite compared to illite in the dense body of suspended matter that moves alternately upstream and downstream according to tidal cycles (Fig. 5.2). This is interpreted as the result of an early deposition of illite and kaolinite, smectite remaining preferentially in suspension during flood and ebb tide. The high and low tide phases, marked by less intense hydrodynamic conditions, allow the settling of smectite, that accumulates in surficial sediments as a mud cream, and can be easily reworked during subsequent flood or ebb phase. Local variations occur and result mainly from lateral supply, topographic changes, grain-size sorting and particle aggregation. Chemical changes affect the clay assemblages very little or not at all, in con-

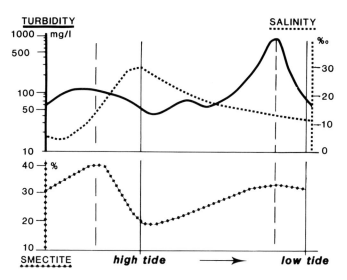

Fig. 5.2. Turbidity and relative abundance of smectite in surficial water of Loire estuary at Le Pointeau, France, during a tidal cycle in summer 1972. (After Gallenne 1974). The maximum of smectite abundance in suspended matter before the high tide is attributed to a preferential settling of illite in sediment due to increased salinity

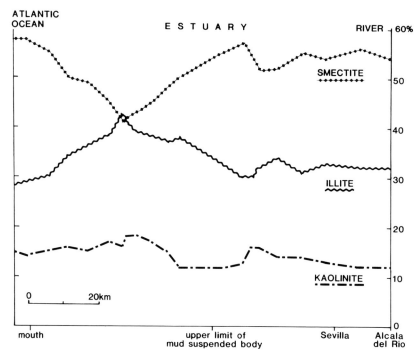

Fig. 5.3. Clay mineralogy of sediments in Guadalquivir estuary region, SW Spain. (After Mélières 1973)

trast to carbonates, phosphates, and amorphous silica that may precipitate locally (Manickam et al. 1985).

Similar results arise from investigations done by Mélières (1973) on the
Guadalquivir river, Southwestern Spain, where smectite abundance decreases in
estuarine sediments by comparison with upstream fluvial and downstream
marine sediments (Fig. 5.3). Preferential settling of illite in the estuary contrasts
with smectite buoyancy. River clay assemblages reform in the marine environments because of the removal of estuarine sediments by tidal currents and
through periodical river floods.

The low tidal range in the Mediterranean precludes the formation of
estuaries, and therefore disfavors settling processes controlled by alternating
salinity changes. At the mouth of some French rivers entering the Western
Mediterranean, a strong local increase of smectite in *sediments* is reported by
Monaco (1971) and Aloisi and Monaco (1975). Smectite issues from hinterland
Pliocene formations during the heavy rainfall periods, and appears to flocculate
in abundance when dense and turbulent freshwater enters the relatively calm
marine environment. Similar interpretations arise from Jacobs and Ewing's
studies (1969) off the Mississippi delta. In the Nile delta, Weir et al. (1975) observe little change in surficial sediments from shore to marine environments
(illite-smectite, kaolinite, illite, chlorite), but envisage chemical changes with the
depth of burial; kaolinite abundance slightly increases when the clay content increases, which could denote discrete sorting.

The minerals contained in *suspended matter* off the Rhone delta reflect the
more or less progressive dilution of land-derived materials in marine waters, according to local climatic conditions. For instance, heavy rainfall in the river basin
determines a high suspended load, a strong and distant penetration of freshwater
and terrigenous minerals into the Mediterranean and, because of the absence of
wind, the formation of a sharp density front toward the East and the South; the
dilution of Rhone suspensions occurs more progressively toward the West, because of the permanent action of the westward ligurian-balearic marine current
(Fig. 5.4A). On the contrary, a drop in river level coupled with strong NW-SE
wind (= mistral) induces a rapid and widespread mixing of poorly charged freshwater with marine water (Fig. 5.4B). Clay minerals and associate species, studied
by X-ray diffraction after filtration, represent useful markers of salinity and
hydrology variations in such instable boundary environments.

Sakamoto (1972) reports the modalities of differential settling processes in
the lower part of the Kunebetsu river and Hakodate bay, North Japan, from experimental and in situ investigations. Floccules mainly form at the freshwater-
saline water boundary, and are larger for smectite (diameter 40 to 500 μm) than
for illite and kaolinite (60 μm). Floccules tend to grow gradually by successive
particle adherence. They include many pores initially occupied by fresh or slightly saline water, which determines a low density and impedes fast sedimentation.
Despite their small size, illite and kaolinite floccules are denser than smectite or
even quartz floccules, both latter aggregates tending to be transported farther
and to settle preferentially in open marine environment. The tide alternation and
river discharge determine the reworking of illite- and kaolinite-rich estuarine
sediments, leading to the periodic reconstitution of river clay assemblages in the

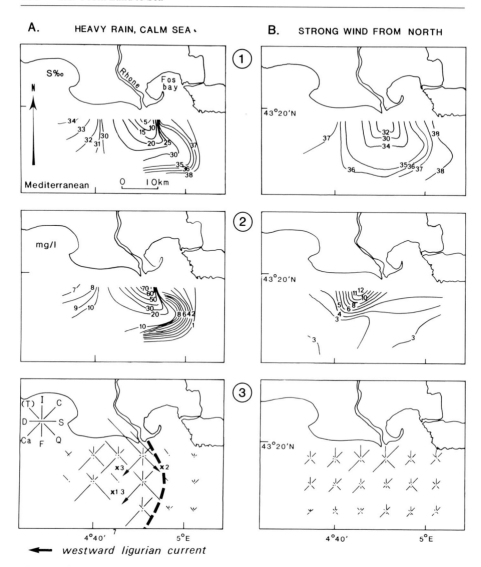

Fig. 5.4. Mineralogical composition of suspended matter off the Rhone delta, Western Mediterranean. (After Blanc et al. 1969; Chamley 1971). **A** Febr. 28–29, 1968, high river level (3.5 m in Beaucaire). **B** March 14–15, 1968, low river level (1.4 m in Beaucaire). *1*. Salinity (‰). *2*. Suspended matter (mg/l). *3*. Mineral abundance (arbitrary scale based on XRD peak height). *I* illite; *C* chlorite; *S* smectite and expanding mixed-layers; *Q* quartz; *F* feldspars; *Ca* calcite; *D* dolomite; (*T*) turbidity

sea. In the Pamlico river estuary, North Carolina, USA, Edzwald and O'Melia (1975) also interpret the downstream clay variations by particle aggregation and flocculation. Experiments and in-situ measurements suggest the early aggregation and sedimentation of kaolinite, and probably of smectite, relatively to illite.

Few data exist about differential settling processes in *past environments*. Stanley and Liyanage (1986) observe that kaolinite abundance in Quaternary sediments of the Nile delta is higher than in the distal offshore Nile cone, which is related to a selective entrapment of this mineral relatively to smectite. Lonnie (1982) compares nonmarine, marine, and transitional clays from the Pleistocene and late Cretaceous sediments of the south shore of Long Island, New York. The augmentation of illite in marine sediments, as well as local variations in kaolinite and mixed-layer abundance, is attributed to differential settling and changes in salinity. Some stratigraphic correlations can be deduced from mineralogic data. Smoot (1960) identifies a clay mineralogic zonation in Mississippian deltaic formations of Illinois, USA. Kaolinite-rich sandstones progressively pass to interstratified-rich fine sandstones and then to illitic shales. T.W. Smoot considers this zonation the result of the preferential settling of large kaolinite particles in proximal turbulent areas, the small mixed-layers being transported farther offshore and subsequently transformed into illite by diagenesis (see also Millot 1964, 1970).

5.3.2 Mixing of Water Masses

An increasing number of papers published in the past decades suggests that *estuarine mineral assemblages result simply from the mixing of river and marine sources, sometimes modified by the addition of local alluviation of colluviation materials.* One of the first studies of this kind was performed by Siegel et al. (1968) in the estuary of the Rio de la Plata, South America. The estuary is divided into a southwest zone (Argentina side) characterized by abundant smectite (average 45%), subordinate illite and kaolinite (27% each), and associate molybdenum, and a northeast zone (Uruguay coast) marked by equal amounts of smectite, illite and kaolinite (33%), and associate manganese. The first zone is attributed to a physical segregation of clay minerals carried by the Rio de la Plata from the Argentine Pampa and Patagonia (rich in volcanic and crystalline rocks). The northeast zone probably results from the influx of sediments entering the estuary from the adjacent Atlantic Ocean and eroded from the coastal Brazilian-Uruguayan shield. The mixing of both sources appears to determine the mineralogic zonation and gradient recorded in the estuary. A similar mixing of river and sea sources is reported in various areas, like the Seine estuary in Western France (Germaneau 1969),the San Francisco bay system in California (Knebel et al. 1977), the coastal mangroves of Casamance, Senegal (Marius and Lucas 1982), or the Garolim bay in Korea (Song et al. 1983).

One of the most striking investigations concerns the James river estuary, Virginia, that constitutes a part of Chesapeake bay and formerly gave rise to controversial interpretations. Powers (1954, 1957, 1959) attributed clay mineral variations to intense geochemical evolution and in situ formation of chlorite, while Owens et al. (1974) showed that chlorite was detrital and that terrigenous input could explain the whole clay suites (see 5.1). Nichols (1972) already proposed that illite, smectite, kyanite, and sillimanite are indicative of river-borne sediments in the James river, and that kaolinite, chlorite, and staurolite mainly derive from marine supply. This question is reconsidered in details by

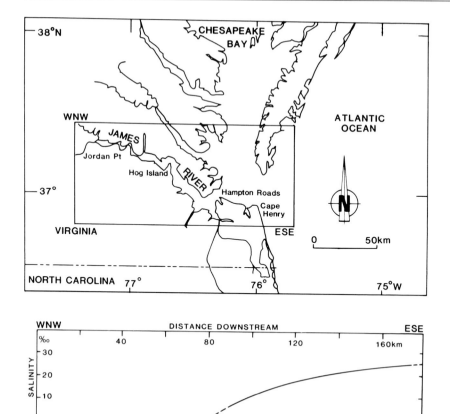

Fig. 5.5. Location map of James river estuary, and average autumn salinity. (After Feuillet and Fleischer 1980). *Dashed lines* extrapolated

Feuillet and Fleischer (1980). The study area is 135 km in length, and extends over the western part of Chesapeake bay from Jordan point to Cape Henry (Fig. 5.5). The James river has a partially mixed estuary, characterized by a two-layer flow with a net downstream flow in the surface layer and a net upstream flow in the bottom layer. The boundary between both flows deepens landwards and represents a surface of no net motion (Fig. 5.6A). There is a point at which the net upstream (landward) flow diminishes to zero; it is positioned close to Hog island (Fig. 5.5) and can move seawards when river runoff increases.

Two clay mineral suites are identified by Feuillet and Fleischer (1980). One is a mixture of kaolinite, illite, and dioctahedral vermiculite, that characterizes the James river, and largely issues from Piedmont formations (e.g., Neiheisel and Weaver 1967). The other comprises illite, chlorite, and smectite, and corresponds to the westward Chesapeake bay entrance. The latter suite reproduces that of the adjacent continental margin of the Atlantic ocean and derives mainly from northern sources by longshore currents (Hathaway 1972, Pevear 1972). Mixed-layer clays occur in both suites, but are more abundant in the James river sedi-

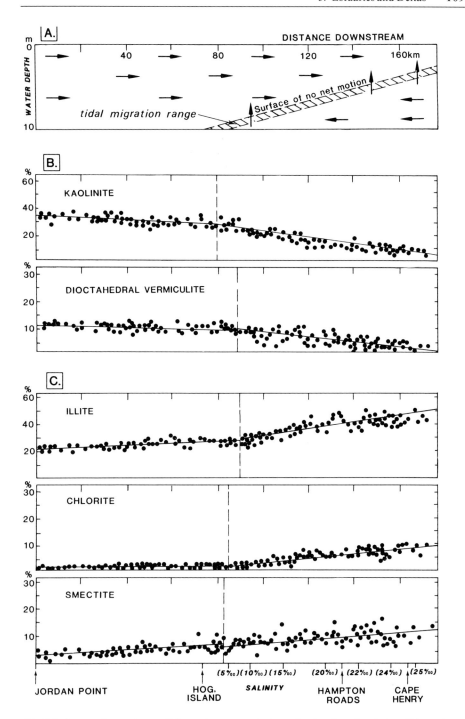

Fig. 5.6. James river estuary. Schematic representation of estuarine circulation (**A**), and percentage distribution of river (**B**) and sea (**C**) supplied minerals. (After Feuillet and Fleischer 1980)

Table 5.1. James river estuary, Virginia. Clay mineral gradients based on least square fits. (After Feuillet and Fleischer 1980). Arrows indicate the direction of clay mineral decrease

Mineral	Upstream section	Downstream section	Downstream location of gradient break (km)
Kaolinite	−0.08	−0.14	80
Dioctahedral vermiculite	−0.02	−0.12	89
Chlorite	0.008	0.09	88
Illite	0.04	0.16	92
Smectite	0.07	0.08	(86) Minor change

ments. The clay mineral proportions show a sharp change in the vicinity of Hog Island: kaolinite and dioctahedral vermiculite suddenly decrease seawards (Fig. 5.6B), while illite, chlorite and smectite increase (Fig. 5.6C). The gradient change is located by an iterative least-squares fit based on segments of ten data points, followed by the same iteration at one-point intervals for the segment displaying the largest gradient change (Table 5.1). Kaolinite, dioctahedral vermiculite, illite, and chlorite all show noticeable gradient changes 80-92 km below the upper part of the estuary (Jordan point), which corresponds to the landward limit domain of the two-layer estuarine circulation. The estuary is thus considered to be the result of the mixing of the river suite with the marine suite transported by net upstream bottom flow. Gradients expected from flocculation settling are not recorded, since no kaolinite maximum occurs near the upper zone of saline water (Hog Island), where such a process should preferentially develop (Whitehouse et al. 1960, Gibbs 1983). Measurable diagenetic effects cannot be recognized by Feuillet and Fleischer (1980), which corresponds to Owens et al.'s conclusions (1974). *Mutual dilution by estuarine mixing is therefore considered the essential factor governing the clay distribution* within the James river estuary.

Examples of source mixing arise from investigations done on other estuaries from the Atlantic coast of USA. Byong-Kwon and Ingram (1975) report that North Carolina sounds and estuaries present a seaward decrease of kaolinite and mixed-layers, and a landward decrease of illite, chlorite and smectite, which is similar to the mineralogic gradient observed in the James river, Virginia. The authors think that this evolution results from the mixing of particles supplied from the continental shelf and from the land, but also from differential settling and perhaps early diagenesis. In Georgia, Windom et al. (1971) compare the sand and clay mineralogy of sediments and suspended matter from three estuaries (Altamaha, Ogeechee, Satilla). Illite and smectite proceed from an offshore supply through flood tidal currents. This suite is mixed with kaolinite, vermiculite, mixed-layers, and talc inherited from Piedmont and transported by the rivers, and also with smectite issued from coastal plain and paralic environments. Conditions of transportation and mixing processes are considered as essential, and give a misleading appearance of chemical transformations of minerals.

Pinet and Morgan (1979) examine two estuaries in Georgia located North of the Florida border. All clay minerals that include minor amounts of palygorskite and sepiolite appear to be detrital, and are largely supplied by the rivers. Local colluvial supply also contributes to the constitution of clay suites, while the landward input from marine sources is thought to have been overestimated by previous authors. Scheinfeld and Adams (1980) think that mineralogic transformations due to the ingestion of sediment by organisms should be taken in account in these Georgia estuaries, in order to explain the mineralogic variations observed. Especially amphipods like *Ampelisca abdita* could determine a degradation of smectite and a decrease of kaolinite crystallinity because of acid intestinal pH and bacteria action. Pinet and Morgan (1980) estimate that such *biological degradations* are likely. However they occur fairly uniformly, independently of the general clay zonation, and are quantitatively much less important than the control by different sources and mineral mixing.

5.3.3 Marine Transgression

The Holocene transgression, like other eustatic ingressions, tended to determine the reworking and landward displacement of sediments accumulated on the continental shelf during the earlier low-level stages. Clayey sediments were abundantly removed in regions where fine sedimentation developed during glacial stages, which occurred particularly in mud flats located off estuaries and deltas. This is the reason why many present estuaries are largely filled by sea-derived sediments. Deltas do not show similar phenomena because the high terrigenous input from river basins determines a seaward progradation and a dilution of sea-derived materials.

Estuarine clay assemblages often reflect the Holocene transgression, whose effects are frequently enhanced today by the landward supply of marine particles provided by active tidal currents in opposition to moderate river input. Postma (1967), Meade (1969), Hathaway (1972) and Van Neuwenhuise et al. (1978), among others, show that the clay suites of various estuaries from the Northern Gulf of Mexico and the Northwestern Atlantic resemble those of adjacent marine embayment, or result from longshore currents whose suspended load enters the river mouths. Müller and Förstner (1975) and Salomon et al. (1975) show, from heavy metal and isotope studies done on clay minerals, that an active landward transport takes place in the Rhine-Meuse and Elbe estuaries (North Sea). The very low content of illite-smectite mixed-layers in the estuarine suspended materials from the Eastern North Sea (from Denmark to the Netherlands) is attributed to the removal of shelf glacial sediments (Rudert and Müller 1981); these Würmian deposits incorporated less degraded minerals derived from continental soils than Holocene deposits, because of lower weathering processes on land (see Chap. 2).

McMurtry and Fan (1975) demonstrate the close resemblance between the clay mineralogy of downstream sediments from the Santa Ana river, South California, and of marine sediments deposited on the adjacent shelf in the Northeastern Pacific Ocean (Fig. 5.7). The seaward decrease of smectite could be determined either by a marine ingression in the lower course of the river, or a

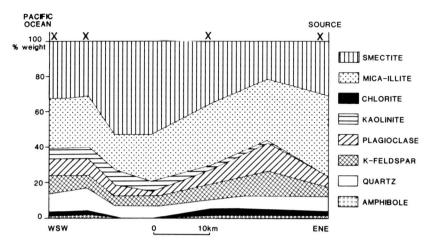

Fig. 5.7. Mineral abundance of Santa Ana river and adjacent Pacific shelf sediments, South California. (After McMurtry and Fan 1975). *X* sectors of the river course where smectite and illite relative percentages are similar

geochemical evolution into illite in K-rich sea water. As the composition of the downstream clay suite has a smectite/illite ratio similar to that of several parts of the upper course, the latter hypothesis is hardly probable. This is confirmed by the parallel increase of mica-illite and of typical detrital minerals like kaolinite, quartz, K-feldspars, and plagioclases, in both estuarine and marine sediments. Smectite-rich downstream sediments therefore result from landward detrital supply linked to recent marine ingression.

Few examples are reported on the effects of marine transgression on *past* estuarine or deltaic clay assemblages. An exhaustive study performed by Brown et al. (1977) on the Desmoinesian (middle Pennsylvanian) of Iowa-Missouri, USA, shows a close relationship between clay suites and the chronologic and geographic evolution of a deltaic system. The successive stages of delta progradation, agradation, and destruction, and the subsequent Carboniferous marine transgression are particularly well expressed in mineral successions investigated in interdeltaic areas (Fig. 5.8). The *progradation* of river sediments first determined a freshwater influx of predominantly illitic clays issued from distant upstream sources, mainly located to the Northeast in Canada. Clays were deposited *seaward* as prodelta mudstones and also transported by longshore currents to adjacent marine embayments. During the following stage of delta *agradation*, the interdeltaic areas were frequently exposed or submitted to acidic marsh conditions: illite weathered into mixed-layer clay with abundant expandable layers. At the same time surficial weathering and pore water circulation provoked the hydrolysis of fluvial channel sand and the formation of kaolinite and sometimes of vermiculite. A seaward increase of illite supplied by channels, and a decrease of kaolinite and mixed-layers formed in interdeltaic zones are therefore recorded at a given time slice (Fig. 5.9). At the end of these progradation-aggradation stages, when river influx decreased and delta *destruc-*

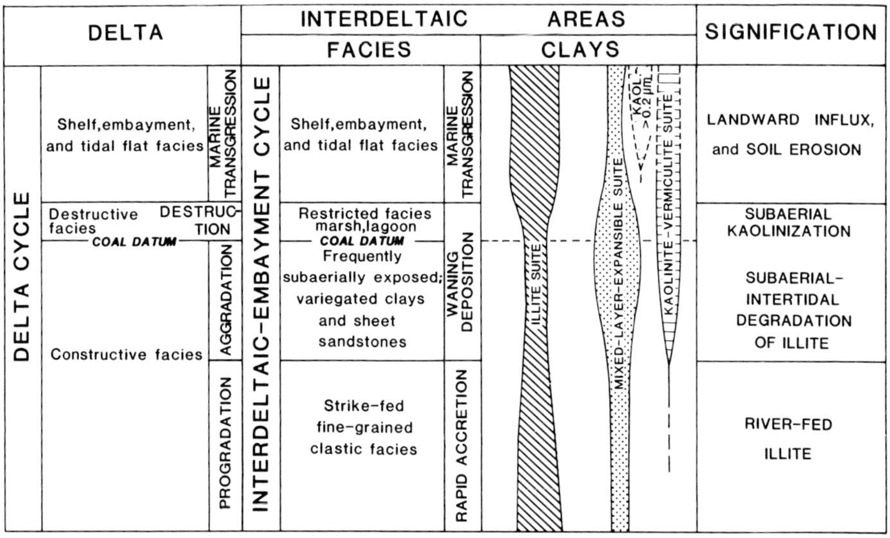

DELTA			INTERDELTAIC AREAS				SIGNIFICATION
			FACIES		CLAYS		
DELTA CYCLE — Shelf, embayment, and tidal flat facies (MARINE TRANSGRESSION)			**INTERDELTAIC-EMBAYMENT CYCLE** — Shelf, embayment, and tidal flat facies (MARINE TRANSGRESSION)		ILLITE SUITE / MIXED-LAYER-EXPANSIBLE SUITE / KAOLINITE-VERMICULITE SUITE / KAOL >0.2 µm		LANDWARD INFLUX, and SOIL EROSION
Destructive facies (DESTRUCTION) — COAL DATUM			Restricted facies marsh, lagoon — COAL DATUM				SUBAERIAL KAOLINIZATION
Constructive facies (AGGRADATION)			Frequently subaerially exposed; variegated clays and sheet sandstones (WANING DEPOSITION)				SUBAERIAL-INTERTIDAL DEGRADATION OF ILLITE
(PROGRADATION)			Strike-fed fine-grained clastic facies (RAPID ACCRETION)				RIVER-FED ILLITE

Fig. 5.8. Idealized delta cycle and clay stratigraphy, Desmoinesian series, middle Pennsylvanian of northern Missouri and southern Iowa (USA). (After Brown et al. 1977)

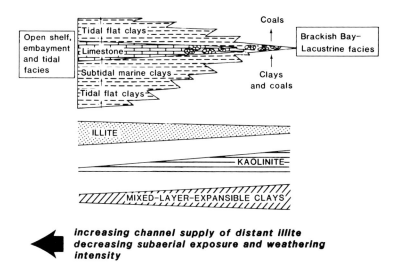

Open shelf, embayment and tidal facies — Tidal flat clays — Limestone — Subtidal marine clays — Tidal flat clays

Coals — Clays and coals — Brackish Bay–Lacustrine facies

ILLITE

KAOLINITE

MIXED-LAYER-EXPANSIBLE CLAYS

**increasing channel supply of distant illite
decreasing subaerial exposure and weathering
intensity**

Fig. 5.9. Lateral relationships of clay associations within a marine to non-marine pinchout, Desmoinesian deltaic system of northern Missouri and southern Iowa. (After Brown et al. 1977)

tion started, both river-derived illite and interdelta kaolinite and mixed-layers were reworked and supplied downstream in marine embayments (Fig. 5.8). The *marine transgression*, that constituted the last phase of the deltaic cycle, provoked the removal of shallow marine sediments and the *landward* input of clays in the lower course of rivers. Illite, irregular mixed-layers, kaolinite, and vermiculite were therefore provided in formerly land-sea transitional environments and in coastal plains, similarly to what happened during the late Quaternary in numerous estuaries. A specific population of detrital kaolinite (particles > 0.2 µm) added to this reworked marine suite, and resulted from the erosion of fluvial plain soils.

5.3.4 Cation Exchange and Chemical Modification

Estuary and delta clays appear to experience very few chemical changes at the freshwater-saline water boundary. *The only widespread phenomenon recorded consists of an exchange of cations, that occurs at the surface or within the interlayers of clay particles*, is reversible, and does not affect the structural characters of crystals. Usually the river-supplied Ca adsorbed on clay is replaced by Na, K, and Mg at the contact of saline water. Such exchange processes are reported from various estuaries and deltas of the USA and Europe (e.g., Pinsak and Murray 1960, Harriss 1967, Monaco 1970, Russell 1970, Drever 1971a, Jouanneau 1982). Russell (1970) demonstrates that ion exchange does not modify the mineralogic nature, relative abundance, or crystallinity of smectite, kaolinite/halloysite, illite, and subamorphous compounds supplied by Rio Ameca from Western Mexico into the Pacific Ocean. The number of equivalents of cations is the same in river clay particles and in marine ones. There is no rapid reaction between clay and sea water to remove cation alkalinity. All dissolved K issued from land can be removed from Rio Ameca clay, but only about one fourth of the Na and Mg. When applied to a world average river, these results show that between 11 and 47% of the dissolved K carried by rivers is removed within a few weeks. Such an exchange, which is not important for Mg and probably not for Na, allows the balance in cation alkalinity that tends to increase by river dissolved influx. Russell (1970) thinks that amorphous materials could theoretically react with some of the dissolved ions, but that all changes identified can be explained without any reaction involving these materials.

Note that some alteration of the chemical composition of river-suspended matter by *biological activity* may intervene significantly, in addition to inorganic exchanges, as reported by Sholkovitz and Price (1980) for the Amazon estuary. In this river, noticeable amounts of Si, P, Ca, Mg, Ti and Mn, and probably of Fe and K, are incorporated into the skeletal and organic constituents of marine phytoplankton, mainly consisting of diatoms. Note also that a part of the land-derived metals like Fe, Ca, K, and Ti may successively precipitate in the cavities of intertidal quartz grains, through water evaporation at ebb phase (e.g., Northern France; Mocek and Vandorpe 1984).

Very few cases of *clay chemical modifications* arise from modern investigations at the transition between land and sea environments. They refer exclusively to intertropical areas submitted to strong evaporation. Gouleau et al. (1982) report the presence of aluminum hydroxides in different environments associated with mangroves in Sine-Saloum and Casamance, Senegal. These uncommon microcrystals form subaerially through clay acidolysis, at the surface of hyperacid sediments subject to long desiccation and evaporation; the Al-hydroxides are subsequently reworked by the wind in both freshwater and saline water sediments located in the vicinity.

Baltzer (1975) observes that Pleistocene and Holocene mangrove sediments in New Caledonia, Southwest Pacific, often contain smectitic clays and fairly abundant quartz, even in regions where rivers drain almost exclusively kaolinitic soils and ultramafic rocks. The same apparent lack of correlation between continental and coastal minerals exists in some parts of Kalimantan, Indonesia (G. Sieffermann pers. commun.). Baltzer (1975) studies the delta of the Dumbea river in New Caledonia, especially the mangrove environment developed in delta marginal depressions. Fluvial water supplies large amounts of dissolved silica and some detrital smectite (Baltzer 1971). Mangrove sediments are enriched in biogenic silica and characterized by the presence of a lens of brackish water, that slowly migrates landward in the direction of an adjacent hypersaline marsh. The strong evaporation of this water determines a concentration of the silica released by diatom frustules, and appears to allow the precipitation of quartz under acid, organic and reducing conditions. Less reducing sediments located at the boundary between the swamp and the hypersaline marsh contain abundant clay-sized nontronite; this Fe-smectite is thought to precipitate in situ from the dissolved ions (especially Si and Fe) accumulated in interstitial water. No information is available on the extension of such processes, the specificity and stability of minerals newly formed, and their possible dispersion in the marine environment.

5.4 Conclusion

1. *First investigations performed on modern estuary and delta clays led to preferring evolutive interpretations*, since the land-sea boundary was expected to constitute a major step of the geochemical cycle. Local increase in illite or chlorite observed in some estuaries was considered as the result of a prediagenetic transformation of smectite and mixed-layer minerals. We now know that such an interpretation is inaccurate in most cases, and that *the geochemical break at the freshwater-saline water transition is not strong enough to allow significant in-situ clay degradation or formation*. Formation of illitic or chloritic minerals is never confirmed by modern studies. Most of the geochemical changes recorded simply express mineralogic variations determined by physical sorting or change in detrital sources. The chemical imprint on clay minerals is usually limited to ion exchanges in surface and interlayer positions, and only very few cases of possible smectite formation are reported from restricted intertropical evaporative environments. Such conclusions were proposed by Pinsak and Murray as early as 1960, but convincing arguments were provided later through the evidence of similar river and sea clay assemblages, and because of progress in the knowledge of settling and mixing processes. Such a lesson applies to Quaternary estuaries and deltas, and appears to be valid for some deltaic environments as old as the Carboniferous; but the question remains open for various geologic periods, especially those characterized by very strong chemical weathering on land and high dissolved load entering past seas.

2. *Source mixing appears to be the best understood mechanism* for explaining the distribution of clay and associate mineral suites in estuaries. Precise examples allow the identification of the respective influence of river input, lateral tributaries or colluviation, and marine influx. Marine transgression or tidal influence tends to determine the estuary filling by sea-supplied sediments, and represents an end-member case marked by one source only. A symmetrical situation characterizes some deltas, where abundant river-fed materials simply disperse in the marine environment according to meteorologic conditions.

3. *Flocculation and differential settling processes* are still imperfectly understood, because of the diversity and complexity of factors involved (physicochemistry of water, organic matter, mutual behavior of minerals), and because of the difficulty in applying experiments to natural situations. When clearly identified, differential settling appears to favor the *preservation in suspension of smectite minerals*, that pass directly in marine environment without settling in estuaries. Other single minerals settle preferentially in estuarine mud, and can be reworked toward the open sea through strong river discharge and high tide periods.

4. One may expect that the main mechanisms of clay distribution recognized at the land-sea boundary are favored by specific environmental conditions. A strong river influx opposed to low tide range and low-charged sea water allows the progradation of deltaic sediments, and the *simple dispersion* of land-supplied minerals in the marine environment (e.g., Rhone, France); some interchannel marshes and ponds of intertropical deltas could permit local in-situ chemical modifications of some minerals, especially in strongly evaporative conditions. Estuaries characterized by a fairly important river influx opposed to a medium tidal range probably combine suitable conditions for *differential settling* (e.g., Guadalquivir, Spain). Estuaries where river and marine particle influx face each other in comparable proportions, under middle to high tidal range, probably displays preferred *source mixing* (e.g., East American bays and sounds). Estuaries draining flat land, rather depleted in solid mineral charge, and entering marine basins characterized by high tidal range and strong longshore currents (e.g., North Sea, English Channel) tend to be filled by sea-derived sand and clay. They reproduce the evolution that probably occurred during most of the past *transgression* stages.

Chapter 6

Clay Sorting and Settling in the Ocean

6.1 Introduction

It is well known that shore and shallow marine sediments often contain few fine-grained particles, even if clay minerals of natural or industrial origin are rarely totally absent from these environmeents (e.g., Griffin 1963, Stone and Siegel 1969). Except on muddy tidal flats and in some coastal depressions, clay particles are usually swept away, which was enhanced in late Quaternary periods by hydrodynamic variations linked to sea-level changes. *The scarcity of the clay fraction in shallow water sediments*, due to winnowing through the action of waves, tide, and currents, *contrasts with the huge accumulation of argillaceous deposits in the rest of the oceanic range*. A general grain size sorting therefore exists in marine basins between shore and offshore sediments, and one may wonder whether such a phenomenon affects the distribution of individual clay species. Early investigations already suggested the existence of clay sorting processes. Van Andel and Postma (1954) distinguished in the Gulf of Paria, Eastern South America, the delta platform characterized by dominant illite with associate kaolinite and smectite, and the other parts of the gulf somewhat enriched in smectite relatively to illite. Differential flocculation was considered as the most probable explanation among the four possible reasons discussed. Gorbunova (1962) observed in Indian Ocean sediments a decrease in kaolinite abundance with increased distance from India, and envisaged changes in conditions of marine transportation. Parham (1966) proposed from a bibliographic compilation the existence of a general clay zonation from land to deep ocean (kaolinite, illite, chlorite, palygorskite, sepiolite), in relation with increased distance of transportation; smectite and expandible mixed-layers were the only minerals considered as only slightly dependent on the transport conditions.

The occurrence of grain sorting was subsequently suggested by the complexity of the distribution of particulate matter in the ocean. For instance Brewer et al. (1976) and Biscaye and Eittreim (1977) showed that suspended particulate matter in Atlantic sea water ranges from 5 to 300 µg/kg and mainly depends on regional features. High concentrations usually characterize surface and bottom water masses, the latter being frequently enriched in inorganic particles and forming a nepheloid layer. A clear-water zone depleted in suspended matter exists at varying mid-depths in the water column, but reflects a geographic distribution similar to that of overlying and underlying more charged water masses: the particule concentration increases toward the continent borders, especially in upwelling areas, and decreases beneath central gyre areas. The nepheloid layer largely results from resuspension and displacement of bottom sediments. The corresponding particulate loads range from about 2.10^6 tons in the equatorial Guyana basin to about 50.10^6 tons in the North American basin (Biscaye and Eittreim 1977). The total suspended load in the Western Atlantic Ocean is almost ten times higher than to the East (111.10^6 vs 13.10^6 tons). The northward flux of suspended particles carried by the Antarctic Bottom Water (AABW) decreases from 8.10^6 to 1.10^6 tons/year between the southern and northern parts of the Brazil basin, in the Southwest Atlantic. All these bathymetric and geographic variations of oceanic fluxes imply sorting processes. Let us consider what happens on clay transportation and distribution in the oceanic environment, and if particle sorting is effectively associated or not with any mineralogic sorting.

6.2 Differential Settling

6.2.1 Recent Sediments

The Niger delta, in the equatorial Atlantic Ocean, displays a convincing example of clay segregation in the marine environment (Porrenga 1966). The existence of a unique and important terrigenous source formed by the wide drainage basin of the Niger river, and the strong hydrolyzing climate that leads to fairly simple soil-derived mineral suites, provide a good opportunity to test sedimentologic processes controlled by the distance to the coast. D.H. Porrenga studies 234 samples from the recent subaerial and submarine parts of the Niger delta, at distances to the shore reaching 120 km. Kaolinite and smectite form the largest part of the clay fraction, that also comprises rare illite and traces of halloysite. Smectite abundance, estimated by the (001) peak-area value, increases from about 30% in downstream channels and coastal sediments to about 50% in offshore deposits. The sediments of the main river stream contain 30 to 60% smectite. This indicates that nearshore smectite does not settle proportionally to its abundance in river suspensions. A zonal distribution appears, that roughly parallels the coastal line and indicates the regular offshore increase of smectite abundance (Fig. 6.1). Coastal sediments depleted in smectite are often silty to sandy, while smectite-rich offshore deposits generally contain an abundant clay fraction. Porrenga (1966) considers this clay zonation the result of the preferential settling of smectite in fairly deep and calm marine environment. Referring to Whitehouse et al.'s experiments (1960; see 5.3.1) and to specific settling experiments on Niger clays, D.H. Porrenga interprets the clay segregation in marine deposits as determined by the late flocculation of smectite relatively to kaolinite. The eastward increase of smectite abundance is attributed either to a preferential supply by the Guinea current (equatorial counter current) or a local erosion of deeply weathered volcanic rocks cropping out east of Calabar (Fig. 6.1). The less-marked mineral zonation in the easternmost part of the studied area could result from a sediment mixing related to an increased tidal range (Porrenga 1966).

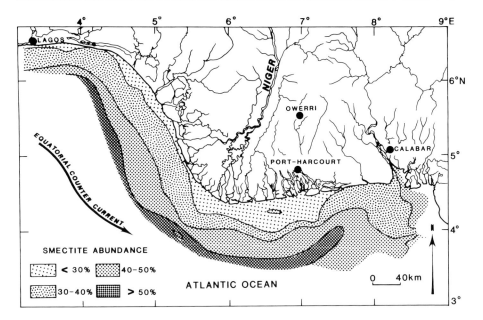

Fig. 6.1. Smectite peak-area percentage in the less than 2 μm fraction of surface sediments of the Niger delta. (After Porrenga 1966)

Other examples of clay differential settling arise from investigations on Atlantic margins. For instance Chamley et al. (1977a) observe an increase of relative abundance of smectite and palygorskite from 1000 m to 3100 m water depth, in sediments located off Northwest Africa coast between 20° and 27° N, which is attributed to a late deposition of these small-sized and low-flocculable minerals. Johnson and Elkins (1979) investigate the mineralogy and geochemistry of recent sediments in the northern North Sea. Smectite abundance, expressed by the smectite/illite X-ray peak ratio, increases together with Fe/Al, Al/K, and Al/si ratios in fine-grained sediments located in offshore depressions, in opposition to what occurs in coarser deposits. Smectite is considered as settling preferentially in areas marked by decreased current and by associate grain sorting.

In the Gulf of Mexico, Pinsak and Murray (1960) and Griffin (1962) find no regular increase in the smectite content in sediments at increasing distance from the shore. Similarly, Scafe and Kunze (1971) recognize no differential settling of clay minerals in gravity cores sampled from the nearshore to the deep-sea areas of the gulf. Doyle and Sparks (1980) investigate the northeastern part of the Gulf of Mexico, and indicate the existence of both *large- and small-scale clay variations, that tend to obliterate differential settling processes.* The presence of various rivers providing smectite-rich suites to the West and kaolinite-rich suites to the East favors lateral differentiation related to different sources. The variable influence of noticeable longshore currents, that strongly depend on meteorologic and seasonal conditions, also interferes with the slow and progressive settling phenomena.

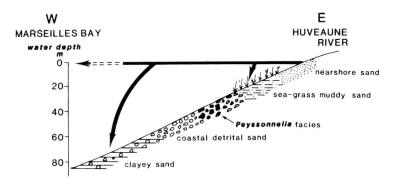

Fig. 6.2. Preferential settling of smectite minerals by biological trapping and decantation in the bay of Marseilles, SE France. (After Chamley 1971)

In the Western Mediterranean Sea, preferential settling of smectite and associate expandable mixed-layers appears to represent a common phenomenon. Chamley (1971), Monaco (1971), and Roux and Vernier (1977) report the frequent increase of expandable minerals at increased distance from the shore of the Gulf of Lion, with local modalities due to either massive flocculation off river mouths or artificial dumping. In Marseilles bay, smectitic minerals tend to settle preferentially in both deep muddy sediments and nearshore seagrass environments, which determines a zonation parallel to the coast, displaying alternating smectite-depleted and smectite-enriched deposits (Fig. 6.2). Seagrass leaves determine a diminution of wave movements in shallow water, which allows the relatively higher deposition of small-sized clay minerals in a way similar to what occurs by decantation in deeper areas. The preferential trapping of clay in seagrass sediments was previously quoted by Giresse (1967) in *Zostera marina* fields located off the South Cotentin peninsula, Western France, where kaolinite appears to settle preferentially. Differential settling is also reported in the Nice and Saint-Tropez regions, Southern France, where the seaward mineralogic zonation is somewhat disturbed by slumping or by longshore currents (Sage and Chamley 1977, Orsolini and Chamley 1980). In the Adriatic Sea, mechanical sorting and flocculation account for the distribution of clay suites derived from the Po and other coastal rivers (Veniale et al. 1972, Tomadin and Borghini 1987).

The ability of smectitic minerals to settle preferentially in calm sea water allows *identification of current activity or quiescence in scarsely dynamic bottom environments.* This is the case on the eastern shelf of the Gulf of Lion, Northwestern Mediterranean, where smectite and associated irregular mixed-layers characterize two groups of areas (Fig. 6.3): (1) Areas enriched in expandable clay, indicating very low currents and preferential decantation; they are located off the Rhone river mouths (A) and on the upper slope (B), and also constitute two strips parallel to the coast on the internal and external platform (C). (2) Areas depleted in expandable minerals, expressing noticeable bottom currents. This second group comprises various sectors: close vicinity of river mouths where the freshwater inflow is still active (D); knee-line at the shelf-slope transition (water depth of about 100 m), where currents perpendicular to the shore develop and

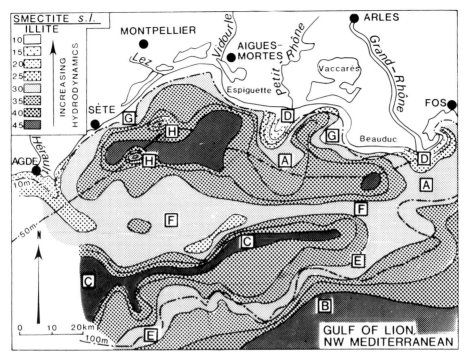

Fig. 6.3. Distribution of expandable clay (smectite, illite-smectite, chlorite-smectite) compared to illite in surficial sediments of eastern shelf of the Gulf of Lion (NW Mediterranean). (After Arnoux and Chamley 1974). *A* to *H* explanations in the text

determine the presence of a large water-filtering benthic community (Ophiurids; E); central zone of the shelf, parallel to the coast and corresponding to the passage of the main branch of the westward Ligurian-Balearic current, enhanced by Rhone inflow (F); nearshore strip characterized by coastal hydrodynamics and eastward North Mediterranean counter current (G); eddy zones (H).

The Northeast Pacific margin does not appear to present numerous examples of differential settling, from the literature available. Baker (1973) investigates the suspended matter in the bottom waters of the Washington continental shelf and slope, USA, and uses the smectite/chlorite ratio to follow the terrigenous input from the Columbia river. The seaward decrease of the ratio could proceed from either differential settling or resuspension. Karlin (1980) discusses the clay mineral distribution off the Oregon coast in a general context. He finds no clear patterns which might be readily explained by selective sorting. This probably results from the diversity of clay suites issued from different rivers, mixing in the marine environment and interfering with sorting, similarly to what occurs in the northern part of the Gulf of Mexico. *In contrast, the Northwest Pacific* reveals several cases of preferential clay settling, especially in Japan Sea, China Sea and West of Japan (Oinuma and Kobayashi 1966, Aoki et al. 1975; in Sudo and Shimoda 1978). A similar explanation is proposed for the Ganges deep-sea fan in the northern *Indian Ocean*, where increased amounts of smectite in hemipelagites are reported from proximal to distal sectors (Bouquillon and Chamley 1986). Southeast of Madagascar, on the eastern border of the Mozambique Channel, the relative abundance of smectite increases in both offshore mud and coastal mangrove mudbanks, relatively to sandy shore and reef facies, where kaolinite amount increases (Chamley et al. 1966). This variation is also referred to preferential deposition or trapping of smectite in calm environments.

6.2.2 Ancient Sediments

Most examples of clay differential settling concern *Cretaceous* series, although some cases are reported from younger and older series (e.g., Japan: Aoyagi 1967, 1968, 1969, in Sudo and Shimoda 1978. Paleogene in the Apennines, Italy: Accarie 1987). In the *Atlantic Ocean* Debrabant et al. (1979) observe slightly increased amounts of smectite and palygorskite at increasing distance from the Armorican basin, in lower late Cretaceous sediments at DSDP Sites 402, 401, and 400 (Northern Bay of Biscay, Eastern Atlantic). This gradient is balanced by a symmetrical decrease of detrital illite, chlorite, kaolinite, and feldspar. The mineralogic variation is attributed to the better buoyancy of smectite and palygorskite, that were probably transported farther in the ocean. Similar trends exist in the late Jurassic and late Cretaceous sediments of Northern France, from onshore to offshore sedimentary environments (Decommer and Chamley 1981). Gillot et al. (1984) report the correspondence between pelagic microfacies and smectite abundance, opposed to the correlation of platform facies and kaolinite content, in the Barremian of DSDP Site 549 (Leg 80), also located in the Northern Bay of Biscay. Similarly, Holmes (1986) shows the dependence of clay suites on sorting processes in early Cretaceous sediments drilled off Cape Hatteras, in the Western North Atlantic (Site 603 DSDP). Kaolinite tends to accumulate in continental sediments, illite in transitional to marine sediments, and smectite in deeper deposits, which is attributed to differential settling. Increasing amounts of illite and kaolinite in deep-sea sediments indicate a general increase in direct terrigenous supply. Some levels especially enriched in kaolinite appear to express the phases of maximum deep-sea fan development, when continental material was quickly supplied within the basin, without noticeable sorting by continental and shelf processes.

Several *exposed series of the Tethyan domain* also suggest differential settling during the Cretaceous, especially in areas characterized by sharp transition between platform and deep-sea environments. The Barremian-Bedoulian sediments in Provence, Southeast France, contain smectite, kaolinite, illite, chlorite, and irregular mixed-layers (Chamley and Masse 1975). Kaolinite is abundant in platform calcarenites, smectite in pelagic micritic or argillaceous limestones. The absence of identifiable diagenetic change, and the close relationship existing between clay mineralogy and depositional environment, suggest that mineral suites depend mainly on hydrodynamic factors. The early Neocomian deposits of Western Switzerland (Neuchâtel and Vaud counties) have been studied from shore to upper basin facies, across the carbonate platform developed during the

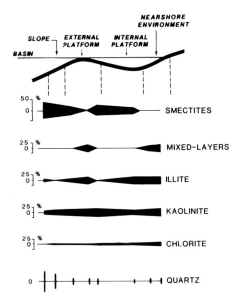

Fig. 6.4. Distribution of clay minerals and quartz on the Berriasian-Valanginian carbonate platform of subalpine range, Western Switzerland. (After Adatte and Rumley 1984)

Berriasian-Valanginian (Adatte and Rumley 1984). Smectite abundance varies in opposition to that of mica-illite, kaolinite, mixed-layers and partly of chlorite (Fig. 6.4). Smectite abundance is maximum in depressed sectors of the internal platform and within the basin, while illite and kaolinite reach highest percentages on the external platform barrier and near the shore. Such relationships between lithology and clay mineralogy indicate the importance of transportation processes but preclude reliable stratigraphic correlations based on clay associations. Detrital quartz abundance increases in upper basin facies, which could result from offshore marine currents (Adatte and Rumley 1984), or perhaps from aeolian supply.

Similar results and interpretations arise from investigations performed on Cretaceous series from the French subalpine range (Darsac 1983, Viéban 1983, Deconinck et al. 1985). Mineralogic sorting appears to be favored by the proximity of shallow carbonate platforms and of fairly deep basins. This is the case with the Bauges platform enriched in kaolinite (e.g., Urgonian limestones) and of the adjacent Vocontian trough characterized by abundant smectite (e.g., Barremian-Aptian hemipelagic marls and limestones). The basinward progradation of carbonate platforms determines a time-lag in the clay mineral changes recorded in the successive series studies from shallow water to deep-sea areas (Deconinck et al. 1985). Three megasequences are identified at Tithonian-Berriasian, Barremian-Bedoulian, and Cenomanian times, each of them being characterized by the progressive progradation of platforms and the trapping of kaolinite, which appears to have induced a relative accumulation of smectite in the adjacent basin. Smectite abundance may vary as much as from 10% to more than 70% of the clay fraction, from platform turbulent environments to basin environments marked by more progressive clay settling. Such a range largely exceeds the variations of clay abundance attributed to differential settling in recent sediments (6.1, 6.3). One cannot exclude the possibility of some in-situ clay formation in certain past oceanic environments, responsible for an additional increase in smectite content, and not identified by the methods used so far (see Part Four). It should nevertheless be mentioned that the sharp transition from carbonate platforms to deep pelagic basins corresponds to specific morphologic features very seldom encountered in modern environments, and whose antagonistic hydrodynamic conditions favored strong segregation among land-derived minerals. This question would probably be further documented through clay mineral investigations carried out in modern environments at the passage from shallow carbonate zones to adjacent basins, especially in warm-humid regions where active soil formation and erosion lead to clay assemblages comparable to those of Cretaceous time.

6.2.3 Mechanisms

Few studies have been performed to explain the detailed mechanisms responsible for the differential settling of clay minerals in marine sediments. Selective *flocculation* is the more frequently quoted, which probably results from the large impact of noteworthy experiments done by workers like Whitehouse and McCarter (1958) and Whitehouse et al. (1960; see 5.3). Precise arguments on this mechanism in natural environment are, however, rarely provided. Most evidence of flocculation processes concerns the freswater-sea water transition under low-salinity conditions (e.g., Whitehouse and McCarter 1958, Sakamoto 1972), rather than the open marine environments.

A striking study on the possible factors controlling the clay mineral segregation in marine water was presented by Gibbs (1977b) for the *Western Equatorial Atlantic, off the Amazon river*. The study area extends from the mouth of the Amazon river along the continental shelf northwestward for about 1400 km (Fig. 6.5). The Amazon mouth represents a point source supplying the material transported northwestward by currents, with other river sources of very little importance. Results are averaged in seven zones from the river mouth to the open sea (Fig. 6.6A). Smectite abundance increases from 27 to 40% of the clay fraction (<2 μm), while illite decreases from 28 to 18% and kaolinite from 36 to 32%. These trends parallel the shore along the shelf; similar trends, although of lesser magnitude and somewhat more complex, occur across the shelf. Other size fractions express comparable gradients.

Gibbs (1977b) successively envisages three possible causes that could account for the clay mineral variations identified off the mouth of the Amazon. *A chemical transformation* appears highly unlikely because alteration in sea water is supposed to form K-rich illite (Whitehouse and McCarter 1958; see also 5.1) and therefore to determine a decrease in relative abundance of smectite, the opposite of what is actually observed. In addition the time span for mineralogical changes envisaged by Whitehouse and McCarter (1958) is far too slow (>5 years) to be observed in a river-ocean system. As the buried sediments sampled by coring in the study area show no noticeable clay variations relatively to surface sediments,

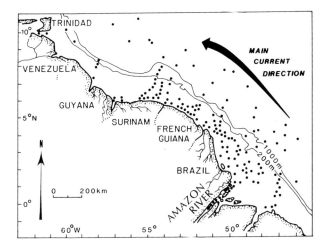

Fig. 6.5. Sample position on the continental shelf off the Amazon river. (After Gibbs 1977 b)

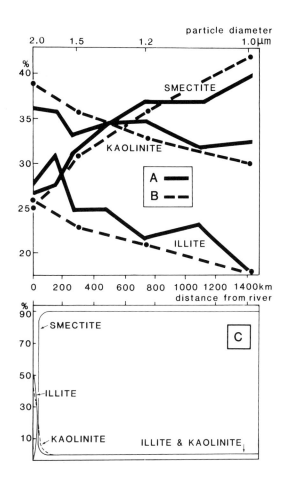

Fig. 6.6. Clay mineral composition of sediments on the continental shelf off the Amazon river. (After Gibbs 1977 b). Comparison of data obtained from samples analysis (*A*), from predicted data based on the grain-size of clay minerals (*B*), and from predicted data based on flocculation processes (*C*)

we can conclude that chemical changes do not account for the mineralogic modifications recorded on the South American continental shelf.

A second explanation consists in *the differential flocculation* of clay species in saline water, following Whitehouse et al.'s statement (1960). Illite and kaolinite settle faster than smectite because of specific electrochemical surface properties, which could justify the relative increase of smectite abundance at increasing distance from the river mouth. But illite and kaolinite flocculate with an augmentation in chlorinity of as little as 2 parts per thousand, which determines a rapid settling already in little saline water (see also Gibbs 1983). Smectite settles more slowly but should have flocculated at a few tens of kilometers off the mouth of the Amazon, where salinity values reach those of normal sea water. Gibbs (1977b) calculates the predicted clay mineralogy of the bottom sediments using data of Whitehouse et al. (1960) on settling velocity of smectite, illite and kaolinite, at different salinities measured off the mouth of the Amazon (Fig. 6.6C). Results indicate an early deposition of all illite and kaolinite close to the river mouth with almost pure smectite settling farther out, which differs completely from real data (Fig. 6.6A). Of course direct application of laboratory experiments by Whitehouse et al. (1960) cannot be made, since natural conditions involve mineral mixtures and turbulent conditions instead of pure mineral and motionless suspensions. But the discrepancy observed between theoretical and real data shows that flocculation obviously could not account for the clay differentiation recorded. This could result either from floccules dissociation by marine turbulence, or from the presence of organic and metallic coatings on clay particles, blanketing specific electrostatic properties (Gibbs 1977b).

The third mechanism proposed by R.J. Gibbs is based on the fact that smectite, illite and kaolinite supplied by the Amazon river have different size distributions (see 3.3.2; Fig. 3.6A). Smectite shows small particle sizes in Amazon suspended load (from 0.1 to 0.9 µm, average 0.4 µm) relatively to illite (from 0.4 to more than 80 µm, mean comprised between 2 and 4 µm); kaolinite presents intermediate values (0.4 to 10 µm, mean 1–2 µm). A simple *physical size segregation* would therefore produce a corresponding clay mineral segregation during deposition, and the statistically smaller smectite particles would be transported farther than coarser illite and kaolinite particles. To test this mechanism Gibbs (1977b) measured the area under the size distribution curve for each component clay mineral for the material coming from the Amazon river (see also Gibbs 1967; Fig. 3.6A), and was able to predict an average clay composition for different reference sizes (2.0, 1.5, 1.2, 1.0 µm; Fig. 6.6B). There is close agreement between the actual clay composition of the bottom sediments (Fig. 6.6A) and the predicted composition. R.J. Gibbs also segregated a sample of Amazon discharged material by size fraction by decantation in a beaker, in order to attain as closely as possible the various size distributions of bottom natural sediments. This last experiment showed good agreement between the resulting clay mineral composition of the beaker simulation experiment and that of marine sediment.

Physical sorting by size therefore appears to represent the most likely mechanism responsible for clay mineral differentiation in western Equatorial Atlantic. This does not mean that such a mechanism accounts for all mineral segregations recorded off river mouths in the marine environment. Accurate studies similar to that of Gibbs (1977b), accompanied by grain-size observations, should be performed in other regions, marked by different clay suites, associate oxides and organic matter, and dispersion patterns. One should notice that kaolinite particles are often much smaller than expected when observed under the electron-microscope, contrary to some smectites, that frequently display large flakes. The close bonding of smectite and adsorbed water, as well as the fleecy shape of most smectite particles compared to the usual sharp outlines of illites, chlorites, and kaolinite, also favor the preferred buoyancy of the former mineral.

6.3 Particle Aggregation and Advection

The possibility of *trapping suspended minerals by planktonic organic matter* is suggested by several studies, as, for instance, on the Western Africa margin (Emery and Honjo 1979). The formation of organomineral aggregates in the plankton is supposed to allow an acceleration of settling processes, individual lithogenic particles being gathered into larger and heavier assemblages. This question has been especially investigated by S. Honjo and colleagues, who deployed sediment traps at several depths in the Panama basin (Honjo 1982, Honjo et al. 1982 a,b). A first experiment consisted of measuring the total mass flux and the main constituents of this flux (organic carbon, carbonates, lithogenic particles), at three depths for an entire year (Fig. 6.7). Two spikes of organic flux were simultaneously recorded at all three depths, and correspond to the high productivity period in February-March (regional upwelling, abundant organic matter), and to a unusual bloom of a coccolithophorid (*Umbellicosphaera sibogae* ; abundance of both carbonate and organic matter) in June-July. The flux of lithogenic particles parallels the surface production but surprisingly increases with increased depth, while the planktonic carbonate flux decreases together with the total flux. These data suggest that suspended minerals are convected with the rapid sinking of organic matter in the water column, but that they do not exclusively derive from surficial water masses and planktonic fecal pellets.

Additional studies in the same area show that the increasing lithogenic flux observed at increased depth is mainly due to smectite, whose abundance rises considerably relatively to chlorite and kaolinite down to a certain level (Table 6.1; Fig. 6.8A). Smectite-rich sediments are reported on the adjacent continental slope of Central America (Heath et al. 1974), which suggests that this mineral is supplied laterally at mid-water depths, not through surface current or bottom nepheloid layers. This hypothesis is strongly supported by the fact that smectite vertical flux at the Panama basin station increases when an easterly current issued from the slope area prevails, and decreases when the current reverses. Honjo et al. (1982a,b) envisage the physical scavenging and agglutination of smectite particles by amorphous organic matter issued from plankton and settled toward the bottom. Such scavenging would resemble flocculation of adhesive clay particles by differential collision, reported in some estuaries (Krone 1978). This packaging would occur at mid-depths and determine the preferential settling of smectite issued from lateral sources. The role of surface filter feeders would be

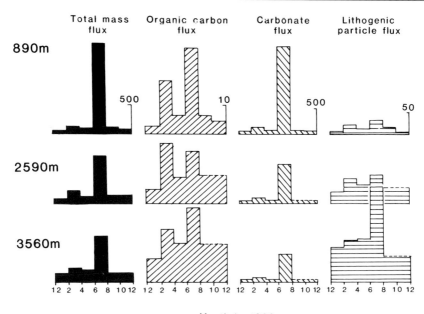

Fig. 6.7. Fluxes of suspended matter constituents at 890 m, 2590 m and 3560 m in the Panama basin (5°22′N, 85°35′W) between december 1979 and december 1980. (After Honjo 1982). Total dry mass flux and organic carbon by combustion-infrared detection, carbonate flux by acetic acid leaching, and lithogenic particle flux by combustion residue and X-ray diffraction. Values given in mg m^{-2} day^{-1}

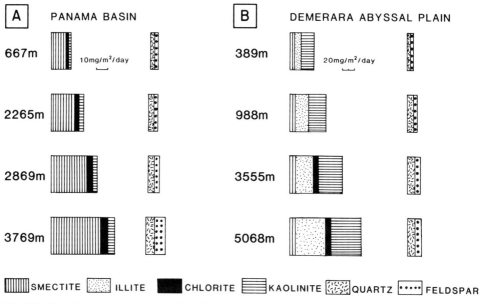

Fig. 6.8. Flux of mineral constituents in sediment traps from different water depths, in Panama basin (equatorial Pacific) and Demerara abyssal plain (Western equatorial Atlantic), from X-ray diffraction assessments. (After Honjo et al. 1982a)

Table 6.1. Distribution of mass flux, lithogenic flux and smectite flux at different depths in Panama basin (5°21'N, 81°53'W), summer-autumn 1979. (After Honjo et al. 1982 b). Fluxes are given in mg m^{-2} day^{-1}

Depth (m)	Mass flux	Total lithogenic flux	Smectite flux
667	114.1	11	10
1268	104.5	14	12
2265	127.5	24	19
2869	158.0	32	27
3769	179.3	50	42
3791	179.6	50	42

accessory and responsible only for the fairly constant vertical supply of chlorite and kaolinite at different depths (Fig. 6.8A).

To summarize, smectite aggregates of the Panama basin are considered as forming marine snow issued from the organic agglutination of particles advected by easterly deep current, while kaolinite and chlorite aggregates mainly issue from subvertical settling of fecal pellets formed through surface planktonic activity. Note that the smectitic material identified refers to beidellite, a mineral thought to be neoformed on the continental slope of Central America (Honjo 1982, Honjo et al. 1982b). This type of question will be discussed later (Chap. 12), but one should notice that beidellite commonly forms in intertropical areas through continental weathering (Chap. 2), that abundant detrital smectite is supplied in the Eastern Pacific Ocean by South American rivers (e.g., Scheidegger and Krissek 1982), that the Panama basin is largely fed with terrigenous clay (Heath et al. 1974), and that smectite augmentation in sediment traps correlates to increased amounts of mainly detrital quartz and feldspars (Fig. 6.8A).

Honjo et al. (1982b) also investigate the Demerara and Söhm abyssal plains, in the Western Atlantic Ocean. The sediment traps moored in the Demerara region, Equatorial Atlantic, display an increased flux of illite, chlorite, and kaolinite at increased depths, smectite flux appearing to be rather constant (Fig. 6.8B). Advective supply of clay minerals issued from the Amazon river could be responsible for this gradient, the physical scavenging of illite and associate species by mucus and fiber aggregates acting in the same way as for smectite in the Panama basin. Note the low amounts of smectite, and the absence of quotation to any preferential settling of smectite in this area, located close to the region investigated by Gibbs (1977b; see 6.2.3). The Söhm abyssal plain, off the eastern US coast, displays a sudden increase of lithogenic flux (illite, chlorite, kaolinite, smectite) at 5202 m trap, what appears to be related to a well-developed nepheloid layer (Biscaye and Eittreim 1977, Honjo et al. 1982b).

In other areas, the formation of fecal pellets and the direct sinking of organomineral aggregates formed in surface water appear to dominate over advective input. This has been reported by McCave (1975), Honjo and Roman (1978), and Scheidegger and Krissek (1982) in different parts of the world ocean. The latter authors, for instance, investigate depositional processes off Peru and

Northern Chile, in the Eastern Pacific Ocean. Coarse silt particles settle out of the water column rapidly over the shelf and slope. Finer components, that include various clay minerals, quartz and feldspars, are incorporated into fecal pellets by zooplankton and nekton. The proportions of quartz and feldspars ingested by planktonic organisms are similar to those found on the underlying sea floor, which suggests the absence of other noticeable mechanism of sorting or deposition. Deuser et al. (1981, 1983) show that the total flux of aluminosilicates in the Atlantic Ocean closely corresponds to the annual cycle of primary productivity, and that the removal of particulate aluminum, expressing the clay flux, is ultimately linked to the rapid downward transit of organic matter from plankton.

As we can see, controversies still exist about the modes of settling of clay-sized particles, and about the processes acting to transfer them to the deep sea. Preferential settling appears to be restricted to continental margins and not to occur in deep-sea basins. Particle aggregation, that induces rapid clay sinking, seems to be much more widespread than formerly thought, and to cover different processes including surface scavenging and deep-water agglutination. In addition, *the transportation and sorting over long distances of single particles or small aggregates* should not be minimized, because of the very common occurrence of such phenomena (see 6.1). For instance fine-grained particles are shown to be advected in the Atlantic for thousands of kilometers in association with various water masses (e.g., Antarctic Bottom Water, Labrador Sea Water; Brewer et al. 1976). Lambert et al. (1984) report lateral transport of fine-grained Al-, Mn- or Fe-rich particulates resulting from land or from the Mid-Atlantic ridge to distances of 1000 km. The advection of aluminosilicates to distances up to 2000 km occurs in deep water associated with the Mediterranean Water outflow (see also 8.4 and 18.2). We do not know the type of clay sorting determined by such long distance transportation processes, if there is any.

6.4 Conclusion

1. *Clay mineral segregation by differential settling represents a common phenomenon on certain continental margins* characterized by a simple influx from river to ocean (e.g., Niger or Amazon). When different rivers carry distinct clay suites within the sea, source mixing processes predominate and interfer with differential settling, especially if longshore currents favor lateral flux exchanges (e.g., Northern Gulf of Mexico). Clay sorting usually determines *the farther transportation of smectite and fibrous clays* relatively to most other clay species. This is the reason why river suites depleted in smectite hardly express clay segregation when entering the sea. Mixed-layer minerals sometimes play the same role as smectite, comparatively to illite, kaolinite, and chlorite. The distance over which clay sorting acts does not seem to exceed a few hundred kilometers (e.g., South American shelf northwestward of the Amazon river), and often appears to represent a few tens of kilometers only. Past sediments also reflect differential settling processes, that seem particularly important at periods like the Cretaceous, when large carbonate platforms passed in a short distance to fairly deep

basins. The clay segregation in Cretaceous environments appears to reach much higher values than in modern environments. As early diagenesis does not seem, from available data, to account for such a strong differentiation, one may expect that clay sorting was reinforced both by a hydrolyzing climate providing abundant kaolinite and smectite to the sea (strongly different buoyancy properties), and by sudden variations of turbulence between platform and basin environments. These phenomena should be further investigated from studies on comparable modern environments and from laboratory experiments. Note that clay sorting processes on continental shelves and slopes disfavor the identification of paleoenvironmental signals of low amplitude, like those determined by short-period climatic changes.

2. *The mechanisms responsible for clay changes recorded on continental margins appear to be dominated by grain size sorting*, as documented by Amazon-derived lithogenic particles. In-situ chemical variations cannot account for such phenomena, since they should result in the authigenic formation of illite rather than of smectite, and are not supported by any clear arguments. Differential flocculation also appears to be unimportant, since most flocculation processes occur under low salinities at the freshwater-sea water transition (5.3.1). The mechanism of grain-sized sorting can nevertheless not be generalized, because of the diversity of factors intervening theoretically, in regard to the small number of detailed studies performed so far: nature and size of clay minerals, associate oxides and organic matter, presence or not of coatings, water turbulence, secondary dissociation of floccules, etc. (e.g., McCave 1984). The particulate size of different clay species may vary greatly, depending on the nature of sources and genetic processes. It should be noticed that the high buoyancy of smectite particles does not always correspond to small sizes, and can also result from low density due to high water adsorbtion and from flaky to fleecy shape. In addition, little knowledge is available on the relationships existing between individual clay mineral sorting and clay aggregation in the ocean.

3. *The aggregation of clay and other lithogenic particles by marine amorphous organic matter appears to represent a widespread phenomenon, responsible for the rapid vertical sinking of land-derived particles.* Clay is mainly incorporated in fecal pellets and other mucous matter within the surface water masses where high planktonic productivity develops seasonally. Clay may also be entrapped and agglutinated as marine snow by organic gelatinous complexes in deeper water masses, especially in regions where swift currents seasonally provide laterally advected particles (e.g., Eastern Panama basin). The influence on sedimentation of direct and rapid subvertical sinking of clay aggregates should nevertheless not be overestimated, since horizontal transport and resuspension of individual particles and aggregates occur widely in the ocean, under the action of surficial, deep and bottom currents. *Long-distance advection of fine-grained particles* in the ocean acts concurrently with particle aggregation and direct sinking, and also with clay mineral sorting. The relative importance of each type of mineral flux is still little known, as well as its time variability and real effects on the sediments definitely immobilized on the ocean floor.

Chapter 7

Aeolian Input

7.1 Wind Transport over the Ocean

7.1.1 Evidence of Dust Supply

Numerous observations demonstrate the importance of aeolian supply over the ocean (see reviews by Péwé 1981, Prospero 1981a, Pye 1987). Darwin (1846) already noticed the correspondence existing between the northeast winds blowing from Africa and the presence of abundant fine dust fall in the tropical Eastern Atlantic. He suggested that the repeated occurrence of dust input into the ocean determined a significant deposition of wind-derived materials in the course of time. Radczewski (1939) reported rather common dust falls as far as one thousand miles from the African coastline. The dusts consisted mainly of quartz, clay, calcite, iron oxides, feldspars, and micas, associated with minor quantities of various dense minerals (hornblende, tourmaline, garnet, epidote, titanite, rutile, zircon). O.E. Radczewski identified quartz grains coated with red-brown hematite, proved to be characteristic of a desert origin. This desert-quartz (= Wüstenquartz) was found in both glacial and postglacial sediments off Africa.

These first observations were followed by abundant studies, all indicating the reality of aeolian supply within the ocean. Let us give a few examples of the data successively reported. Kolbe (1957) identifies freshwater diatoms in Atlantic deep-sea sediments, and concludes the likelihood of an aeolian origin. Goldberg (1961) shows that quartz abundance in Pacific surface sediments displays a latitudinal dependence, with maximum concentrations around 30°N and, less distinctly, around 35°N: this zonation parallels the latitudinal distribution of arid areas on exposed land masses, which suggests a desert origin. Quartz particles exhibit well-sorted chips and shards (main size 1-20 μm), whose abundance in the mid-latitudes increases in deposits furthest from land. This fact, combined with the presence of identical quartz grains in atmospheric fallouts and in the intestines of filter-feeding pelagic organisms living in surface waters, indicates the importance of atmospheric transportation.

Delany et al. (1967) collect wind-borne dust on Barbados, in the Western Atlantic. They identify fungi, freshwater diatoms, and magnetic particles, whose seasonal variations correspond with the seasonal shift in wind patterns off the African coast, in the Eastern Atlantic. This suggests that deep-sea sediments of the tropical Western Atlantic could significantly depend on African aeolian supply. Windom (1969) studies the composition of dust incorporated in permanent snowfields of the Antarctic and Greenland, where aeolian supply represents the only possible origin. Distant sources appear to be responsible for a particle accumulation of 0.1 to 1.0 mm/10^3 years. Calculations suggest that adjacent marine sediments of the Pacific and Atlantic oceans could receive as much as 25 to 75% of their detrital phases from atmospheric fallouts.

Folger (1970) measures the concentration of mineral grains, phytoliths, freshwater diatoms, fungus spores and opaque spherules in air and surface water samples from transects crossing the tropical and temperate North Atlantic. The highest abundance of land-derived particles occurs between 19 and 25°N in the mid and western ocean, where phytoliths and freshwater diatoms are particularly abundant. Between 40 and 45°N most particles are present in the western ocean, and comprise abundant fungus spores. D.W. Folger concludes that most of clay- and silt-sized particles are carried to the tropical Atlantic by the trade winds, and to the mid-latitudes by westerly winds. Aston et al. (1973) collect aeolian dusts at the surface of Eastern Atlantic, Indian, and westernmost Pacific Oceans. Average dust loadings, issued from the lower atmosphere, decrease in the following order: North Atlantic (7.7 µg/m3 of air), Northern Indian Ocean (1.2 µg), South Atlantic (0.78 µg), Southern Indian Ocean (0.68 µg), China Sea (0.21 µg). S.R. Aston and his colleagues relate this distribution to the location of main desert areas, which is also supported by the dominant clay minerals. Kaolinite in the Atlantic northeast trades chiefly expresses the erosion of Northwest African soils and rocks, while illite in the northeast monsoons of the Northern Indian Ocean mainly issues from the Rajasthan desert.

7.1.2 Characterization of Aeolian Dust

Various methods are used to identify and to quantify aeolian transport over the world ocean. The concentration of airborne dust can be measured directly by weight or indirectly by turbidimetry. For instance, Duce et al. (1980) observe a particle concentration of 2.3 µg/m^3 of air in aerosols trapped in April 1979 above Enewetak Atoll, central Tropical Pacific. This spring dust, peculiarly abundant relatively to the concentrations measured subsequently (decrease to 0.02 µg/m3 over the next 5 months), probably derives from China and settles at a deposition rate of about 0.3 mm/10^3 years. R.A. Duce and colleagues estimate that aeolian input significantly contributes to the deep-sea sedimentation in the North Pacific. Péwé (1981) indicates that the dust load at most oceanic sampling sites ranges from 0.02 to 1.0 µg/m^3, although concentrations may locally exceed 100 µg/m^3, and even 13,000 µg/m^3 near the sources. Prospero (1981b) reports the results of the Global Atlantic Tropical Experiment (GATE) held in the summer of 1974. Aerosol concentrations, measured in a network of eight ships and four land stations, fluctuate by factors 10 to 100 from day to day in the Eastern Atlantic, as the outbreaks pass over the stations. The vertically integrated aerosol concentration, measured by the atmospheric turbidity using Volz sun photometers, 'shows

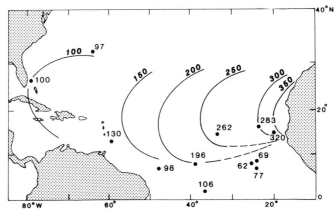

Fig. 7.1. Volz turbidity at 500 nm wavelength (× 1000) during Global Atlantic Transport Experiment (GATE), above the ocean surface. (After Prospero 1981 b)

that values off the coast of Africa are similar to the maximum mean monthly turbidity for a heavily industrialized urban area. The mean Voltz turbidity at a 500 nanometer wavelength regularly decreases by a factor of 3 toward the West across the Atlantic, because of aerosol removal and plume broadening in transit (Fig. 7.1).

The grain size of wind-blown dust falls into two major groups (Péwé 1981). The first group corresponds to dust that is carried a few kilometers to less than 100 km, and is usually restricted to continental areas and very adjacent marine zones. The grain size commonly ranges between 5 and 50 µm. Most dust devils, dust storms, and loess-type deposits belong to this group. The second group corresponds to fine-grained and tropospherically sorted dust that moves as an aerosol, i.e., that remains suspended in the air until brought down by rainfall onto ocean and land surfaces. Dust particles of this group range between 2 and 10 µm, and most of them are smaller than 5 µm (in Péwé 1981). For instance Prospero (1981b) reports from numerous observations that the size distribution of aerosol particles at Sal Island, Eastern Atlantic, nearly constantly ranges between 0.3 and 10 µm, with a peak at 3.6 µm diameter. This group is mostly typical of high-altitude dust transported over long distance and broadly deposited in oceanic or continental areas.

Geochemical analysis of aerosols often provides a good estimation of the abundance of dust constituents. Many authors consider that aluminum, silicon, iron and scandium represent suitable reference elements for continental crust components and land weathering products. Al is most often used because the analytic procedure is simpler than that for Si, and because of its relative abundance in aerosols. For instance Uematsu et al. (1983) measure the concentration of atmospheric aluminum in Asia-derived dust, regularly collected during a twelve month period at seven surface stations in the North Pacific. The highest concentrations of Al are found from February to June, at mid-latitude stations, which corresponds exactly to the highest seasonal activity of the westerlies and to the dust storms issuing from arid regions in Asia. Prospero (1981a)

Table 7.1. Concentration of some land-derived metals in the lower atmosphere. (After Prospero 1981 a). Values in 10^{-9} g/m^3 air

	Al Mean	Al Range	Fe Mean	Fe Range	Sc Mean	Sc Range
Bermuda	140	3 –3000	90	4 –1900	0.02	0.002 –0.4
Tropical East Atlantic	50	12 – 130	40	380 – 110	0.014	0.003 –0.04
Tropical North Atlantic	1600	550 –5700	960	250 –3700	0.29	0.10 –0.87
Hawaii	4	0.5– 50	9	1.0– 5.0	–	–
South Pole	0.6	0.2– 1.4	0.5	0.3– 1.0	0.00012	0.00006–0.0003
Urban regions	1600	340 –3800	1700	380 –4800	0.39	0.11 –1.3

Table 7.2. Mineral aerosol concentrations in the lower atmosphere, based on concentrations of Al[C(Al)], or of Fe[C(Fe)]. (After Prospero 1981 a). Values in g/m^3 air

	C(Al)	C(Fe)	Range
Northern Norway, coast	0.56	1.32	0.08 – 2.63
Bermuda	1.96	2.37	0.04 – 50.00
Tropical East Atlantic	0.70	1.05	0.17 – 2.90
Gulf of Guinea	–	3.16	2.11 – 4.47
Hawaii	0.008	0.24	0.03 – 1.32
South Pole	22.45	0.013	0.003– 0.026
Urban regions		44.74	4.77 –126.34

shows from numerous bibliographic data that the range of Al, Fe, and Sc means for oceanic regions varies strongly according to the location and to the weather (e.g. table 7.1). The element data can be converted to equivalent mineral aerosol concentrations by using a conversion factor derived from the average composition of continental crust and soils (e.g., Table 7.2). The variations recorded above the sea surface commonly are of a factor of 60, excluding areas with very high and very low concentrations (e.g., urban areas, South Pole). The North Atlantic displays the highest concentrations of metals, the region of Hawaii the lowest. However the range is almost as great in the North Atlantic itself, because of varying emission rates from continental sources. Uematsu et al. (1983) estimate from aerosol chemistry that 6 to 12.10^6 tons of dust are transported annually from Asia to the central North Pacific, and that larger amounts probably settle in the western North Pacific, closer to continental sources.

7.1.3 Importance of Aeolian Input

The major sources of dust emissions are situated in the subtropical desert belt extending from West Africa to Central Asia, and in semi-arid regions where soils

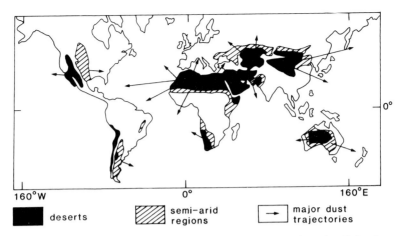

160°W   0°   160°E

■ deserts   ▨ semi-arid regions   [→] major dust trajectories

Fig. 7.2. Distribution of areas with high dust emissions, and major dust trajectories. (After Pye 1987)

are seasonally exposed to strong winds (Fig. 7.2). The greatest concentrations of aerosol particles occur over marine areas located off arid regions and deserts. Most surface materials are removed from North Africa, the Arabian Peninsula and India. This is the reason why the mean mineral aerosol concentrations over the tropical North Atlantic, the North Indian Ocean and the Mediterranean are at least an order of magnitude larger than those over the Pacific Ocean, and over the northern and southern parts of the Atlantic Ocean (Péwé 1981, Prospero 1981b).

The Sahara is probably the world's most important source of dust (Coudé-Gaussen 1982, Middleton et al. 1986). As Sahara dunes chiefly consist of medium- to coarse-grained sand, the abundant fine dust thrown out of Africa over the ocean probably issues essentially from paleosols and little-consolidated formations located on the southern and northern borders of the desert. The budget of aeolian material supplied from the Sahara region over the Atlantic Ocean is known in more detail than in other regions of the world ocean. Intensive insolation in the Sahara creates strong surface winds and large-scale convection, which lift dust particles as high as approximately 6 km. These dense dry air masses move westwards in the Harmattan (Sahara wind) and can reach the Caribbean after one week (Schütz et al. 1981). During this time, the Saharan air masses flow above the southwestward trade wind inversion, that is located at about 1.5 km altitude and also carries dust particles over the ocean. The dust layer extends from about 15 to 25°N. The large particles (> 1 μm) fall out rapidly from both Harmattan and trade wind, while the fine particles are subject to turbulent diffusion of the boundary air masses and remain in suspension. Atmospheric turbidity measurements show that a southward shift of about 8° latitude occurs from summer to winter in the location of the Saharan dust plume. This is due to the seasonal migration of the Intertropical Convergence Zone (ITCZ), that represents an atmospheric barrier to the exchanges between northern and southern hemispheres (in Prospero 1981b, Pye 1987).

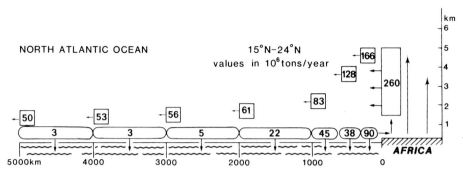

Fig. 7.3. Modelization of the annual mass budget of dust transported over the Atlantic in the northeast trade wind zone. (After Schütz et al. 1981)

Schütz et al. (1981) have developed a model of Saharan dust transport, assuming the size range of aeolian particles is between 0.1 and 20 μm. Although some much larger particles (up to 50 μm) may be transported westward from Africa over distances larger than 100 km (e.g. Coudé-Gaussen et al. 1987), the model provides a good idea of transportation and deposition features in the Atlantic region. Schütz et al. (1981) propose calculations based on a transport channel of 1000 km width (15 to 24°N) and 3.5 km height (1.5 to 5.0 km altitude, above the trade wind zone), the Harmattan being considered to move westward at a constant speed of 6 m/second. The model indicates that approximately 260.10^6 tons leave the Saharan region each year (Fig. 7.3). About 90.10^6 tons return toward the desert areas, under the action of the westerly winds that dominate the near-surface flow within about 300 km of the West African coast. About 170.10^6 tons/year remain as suspended matter and are carried toward the West. Because of the rapid fall of large particles, approximately two-thirds of the total Saharan dust mass settle within the first 1000 km from the coast. Beyond 1000 km, only about 80.10^6 tons/year remain airborne, and are then only slowly removed from the plume by vertical mixing. L. Schütz and colleagues suggest that roughly 50.10^6 tons/year of dust thrown out of West Africa are transported over the Carribean Sea. As a general result, the sedimentation rate due to aeolian supply would be about 10 cm/10^3 year at 300 km from Sahara, 1 cm/10^3 year at 1000 km, and 0.2 cm/10^3 year at 3000 km (Schütz et al. 1981). These values do not differ strongly from actual measurements on deep-sea cores. They appear somewhat overestimated, since marine biogenic particles and continental detritus provided by rivers, mixed with aeolian particles, notably participate in the Atlantic sedimentation (e.g., Kennett 1982). Whatever the exact values, the model of Schütz et al. (1981) provides very useful information about the quantitative distribution of wind-supplied materials in the ocean.

Let us now consider the distribution of minerals in the aeolian dust carried over the ocean, and the impact of airborne particles on marine sedimentation. As aeolian quartz represents a particularly useful and well-studied marker of aeolian supply in the ocean, we will consider its distribution together with that of clay minerals and associate species.

7.2 Identification of Aeolian Minerals in the Oceans

7.2.1 Mineral Composition of Aeolian Dust

Various clay and nonclay minerals are identified in aeolian dusts sampled above or close to the ocean surface (3.1. Table 3.1). The problem of their identification and quantification is closely related to the *sampling techniques.* Most available data have been obtained from aerosol particles collected by means of nylon meshes (1 m³ surface) made of monofilament fiber and suspended normally to the wind. The *sampling efficiency* strongly depends on the particle size, and dramatically decreases for particles smaller than a few micrometers in diameter (Prospero 1981a). This is the reason why very small minerals like clays and especially smectites may often be underestimated (e.g., Glaccum and Prospero 1980). The quality of results increases if aeolian particles are abundant and tend to plug the meshes. But the amount of sampled materials is usually only in the order of milligrams to tens of milligrams, which adds difficulty to the mounting techniques for X-ray diffraction analyses, as well as to mineral determination and quantitative estimation. These analytical problems shoud be kept in mind while discussing the results, that are all the better since they are founded on a comparative basis.

Most available data concern the dusts carried over the *Atlantic Ocean* (Table 7.3). *Illite* constitutes the dominant mineral of Atlantic aerosols (25 to 70% of total minerals) because of its wide distribution in continental rocks and its platelike morphology (e.g., Delany et al. 1967, Emery et al. 1974, Prospero 1981a). *Kaolinite* represents the second abundant clay species, and issues mainly from recent and past soils developed under strong hydrolytic conditions (Chap. 2). For this reason kaolinite abundance increases in the aerosols sampled at low latitudes. *Chlorite*, mainly derived from fresh crystalline rocks, is often not abundant, but tends to vary in opposition to kaolinite (Behairy et al. 1975, Prospero 1981a). *Smectite* abundance increases in subtropical-tropical zones, which is attributed to the alteration of both continental and submarine rocks (Behairy et al. 1975, Chester et al. 1972). Additional clay minerals include *palygorskite*, mostly recognized on electron-micrographs as broken fiber bundles or feltlike aggregates (e.g., Coudé-Gaussen et al. 1987; Fig. 7.4). *Talc* occurs in some dust of the Western Atlantic, especially at mid-latitudes, and seems to derive largely from industrial and agricultural activity (Windom et al. 1967, Poppe et al. 1983). *Quartz* is the major nonclay mineral in aerosols, and tends to decrease westward relatively to illite, because of its preferential settling (Glaccum and Prospero 1980). Note that Atlantic aerosols show little quantitative variations from available data, which results partly from the homogenization of materials from different sources with increased distance of transportation (e.g., Glaccum and Prospero 1980). The Saharan outbreaks appear to be better identified by the total abundance of atmospheric dust than by specific minerals (Prospero 1981a), as are the North Eastern America dust storms thrown out over the Atlantic by westerlies (e.g., Windom and Chamberlain 1978).

Mineralogic data on *Pacific* aerosols are still not abundant, especially when considering the huge oceanic surface exposed to the action of the winds

Table 7.3. Dominant clay and associate minerals of some oceanic aerosols (%; after Prospero 1981a). (N.B. Italic characters = clay concentrations normalized to 100% for clay minerals alone; quartz as weight percent of total sample)

	Chlorite	Illite	Smectite	Kaolinite	Quartz	Mica	Feldspar	Calcite
Atlantic Ocean								
NE, off Dakar	3	40	5	6	16	5	10	10
NE, Sal Island	4	51	3	7	19	2	5	8
Gulf of Guinea	–	26	< 4	45	15	5	0	5
NW, Barbados	4	61	< 4	8	13	2	4	4
NW, off Miami	4	63	< 4	6	14	1	5	7
NW, westerlies	2	68	< 4	2	15	3	4	–
S, 20°–40°S	*25*	*53*	*2*	*20*				
Pacific Ocean								
NW, off Japan	7	52	< 4	4	20	4	10	< 1
NE, tropical	7	54	< 4	6	13	4	10	< 1
NE, equatorial	2	33	10	3	6	<3	25	< 1
SE, equatorial	7	56	< 4	6	17	2	6	–
Indian Ocean								
N, equatorial	*19*	*66*	*6*	*9*	*6*			
Bay of Bengal	*27*	*48*	*14*	*11*	*13*			
SW, Madagascar	*40*	*42*	*4*	*14*	*6*			
SW, South Africa	*7*	*75*	*12*	*6*	*6*			
Mediterranean Sea								
South East	*< 8*	*40*	*< 5*	*50*	*20*			
North East	*< 8*	*46*	*25*	*22*	*9*			

(Table 7.3). The available data, partly questionable because of the low dust concentration over many of the regions studied, indicate both the dominance of illite and quartz, and a fairly large areal variability in mineralogy (Prospero 1981a). The dust from the equatorial Eastern Pacific contains more smectite, pyroxene, and plagioclase in southern regions than in northern ones, probably because of the influence of westward winds from Ecuador, Peru and northern Chile. The western North Pacific aerosols issued from Eastern Asia contain a higher concentration of chlorite, plagioclase, and microcline than Saharan aerosols. The South Pacific area appears to depend on aeolian supply from Australia, as suggested by the abundance of wind-supplied illite, kaolinite, chlorite and quartz in New Zealand snowfields (Windom 1969). Some investigations indicate the importance of *seasonal variations* in mineral composition of dust. For instance, Gaudichet and Buat-Ménard (1982) study aerosols sampled at 20 m altitude above Enewetak Atoll in April and July 1979. Spring dust contains fairly abundant illite (47%), smectite (8%), vermiculite (6%), and plagioclase (5%), attributed to a direct supply from Asian deserts. Summer dust is enriched in quartz (20% instead of 7% in April), kaolinite (12% instead of 2%), talc s.l. (7% instead of 0), Fe (22%) and Cr (4%) particles, which appears to be of American origin characterized by both soil erosion and industrial pollution.

Illite again represents the dominant mineral identified in the *Indian Ocean* aerosols. The few available data (Table 7.3) suggest the existence of noticeable

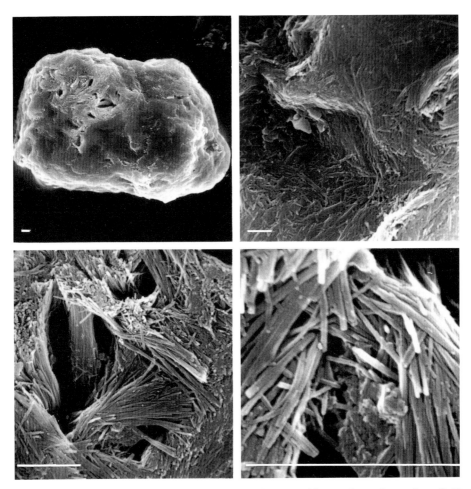

Fig. 7.4. SEM micrographs of an aeolian aggregate of palygorskite, marine sediment from Gulf of Gabès, South Mediterranean. (Courtesy G. Coudé-Gaussen). Bar = 5 μm

mineralogical variations according to region (e.g., Goldberg and Griffin 1970), especially as far as chlorite, smectite, and kaolinite are concerned. In the *Mediterranean Sea*, Chester et al. (1977) show that southeastern aerosols contain fairly abundant kaolinite and quartz issued from Africa. Dust transported from Israel contains particularly abundant smectite (Yaalon and Ganor 1973), while northeastern aerosols are enriched in European illite and smectite (Chester et al. 1977). Tomadin et al. (1984) study the mineral constituents of aeolian aerosols sampled above the northern Ionian Sea and the Adriatic Sea. The southern sources are characterized by increased amounts of kaolinite and palygorskite, both minerals of African origin and also identified in high- altitude lakes of Corsica (Robert et al. 1984). The European dust contains better crystallized illite, and noticeable amounts of chlorite and serpentine. Measurements performed on Corsican "red rain" products lead Loÿe-Pilot et al. (1986) to envisage that

African dust supplied in the Western Mediterranean could be of the same order of magnitude as sediments supplied by the Rhone river (about 4.10^6 tons/year).

7.2.2 Comparison of Aeolian Dust and Surface Sediments

The most reliable and convincing information on the importance of aeolian input within the oceans arises from the direct comparison of minerals contained in wind-borne dust and in underlying marine sediments. In the Eastern *Atlantic Ocean*, Chester et al. (1972) and Behairy et al. (1975) observe that the distribution of kaolinite, illite, and chlorite in the dusts collected between 50°N and 35°S closely parallels that in the underlying deep-sea sediments. The clay mineralogy of aeolian dust mainly reflects the weathering characteristics of the adjacent land masses, which indicates its detrital origin: kaolinite abundance increases at low latitudes marked by lateritic soils on land, while illite and chlorite are more abundant toward desert and middle latitudes where hydrolysis decreases. Smectite, whose relative abundance increases North of the Equator, is tentatively attributed to a partial authigenic origin (Chester et al. 1972); but one should notice that these high values correspond to continental areas where smectitic vertisols increase northward (Chap. 2). In a later paper, Chester (1982) shows that the average distribution of smectite in Northeastern Atlantic is the same in the clay fraction of atmospheric particulates, surface sea-water seston and surface sediments (Table 7.4). In South Atlantic, the smectite abundance increases southward in sediments more than in dust. This difference could indicate the existence of some authigenic processes (Chester et al. 1972, Behairy et al. 1975), insofar as aeolian sources are exclusive. In fact, abundant smectite occurs in Weddell Sea sediments, in the southernmost part of the Atlantic Ocean; this mineral is carried northwards by the Antarctic Bottom Water (e.g., Barker, Kennett et al. 1987). The distribution of quartz along the African margin increases strongly in northeast trade dust and slightly in southeast trade dust, which corresponds to the abundance of the mineral in underlying sediments, and confirms its detrital origin.

Johnson (1979) compares the dominant mineral composition (total clay minerals, quartz, plagioclase, dolomite) of deep-sea surface sediments from the North Atlantic Ocean, and of two reference dusts. One dust (= A) is collected on ships near the shore, the second (= B) in the Central Atlantic within the atmo-

Table 7.4. Average percentage distribution of major clay minerals in dust, sea water, and sediments of the eastern North Atlantic (5°–35°N). (After Chester 1982)

Material	Illite	Smectite	Kaolinite, Chlorite
Atmosphere	45	14	41
Sea-water surface	40	15	45
Surface sediment	55	16	30

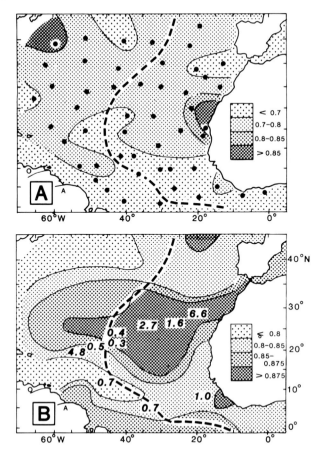

Fig. 7.5. Distribution of mineral similarity coefficient between North Atlantic sediments and: **A** gravitationally deposited dust; **B** airborne aeolian dust. (After Johnson 1979). *Dotted line* Mid-Atlantic ridge axis. *A* Amazon; *O* Orinoco. **A** *Circles* location of stations. **B** *Italic numbers* sediment deposition rate (mm/10³ a) on a carbonate-free basis. (Griffin et al. 1968)

sphere. Reference dust A, that represents the dense material rapidly deposited close to Saharan sources, contains 27% clay minerals, 30% quartz, 10% plagioclase, and 0.7% dolomite. Reference dust B, characteristic of fine aerosols transported over long distances across the Atlantic, contains 46% clay minerals, 12.5% quartz, 5% plagioclase, and 0.8% dolomite. L.R. Johnson calculates similarity coefficients for each dominant mineral quantified in surface deposits and both reference dusts, on the following basis:

$$S = 1 - (X_i - X_j)/r$$

S: similarity coefficicent
X_i: mineral abundance in sediment
X_j: mineral abundance in dust
r: range of the analyses for a given mineral.

The comparison (Fig. 7.5A and B) leads to three main results:

– The sedimentary area of greatest similarity to aeolian dust corresponds to the northeast trade wind belt ($5°$-$35°$N).

– The gravitationally deposited dust A correlates only with mineral distribution in sediments located close to the African coast, in the Senegal - Cape Verde islands region.

– The airborne aerosol B corresponds closely with the wide distribution of minerals in the Eastern Atlantic, and to a lesser extent in the western basin. The sedimentation rates decrease parallel to similarity coefficients with increasing distance from West Africa (Fig. 7.5B), what indicates the large impact of aerosol supply in the eastern North Atlantic. Johnson (1979) estimates the sedimentation rate of aeolian dust in the central North Atlantic is close to 0.7-1.4 mm/10^3 year, which is about half the total rate of noncalcareous particles in the same area (about 2 mm/10^3 yr, Griffin et al. 1968), and confirms the significant aeolian influence.

In the Pacific range, first evidence of a direct correlation between dust and deposit compositions was provided by Rex et al. (1969) from grain size and isotopic measurements on quartz particles. R.W. Rex and colleagues showed the close similarity in size and oxygen-isotope composition between quartz grains collected from Pacific atmospheric dust, from Hawaiian soils, and from surface sediments of the central Eastern Pacific (e.g., Table 7.5). This composition differs strongly from that of hydrothermal quartz from Hawaii, and of Cretaceous cherts from the same area. Rex et al. (1969) concluded that quartz grains from Hawaiian soils are virtually all of aeolian origin, which suggests that the same origin could apply for quartz from most North Pacific sediments. As the abundance of mica-illite from Hawaiian soils displays a covariant relation with quartz, a similar origin is envisaged for the former mineral. This is confirmed by datations and isotopic measurements on Hawaiian micas (Dymond et al. 1974). K-Ar and Rb-Sr apparent ages are approximately 100 times greater than the age of volcanism on Oahu island, Hawaii (230 to 390 m.y. vs. 3.5 to 8.5. m.y.). $^{87}Sr/^{86}Sr$ ratios for mica-bearing soils reach values as high as 0.7273, which greatly contrasts with the nonradiogenic strontium in Hawaiian lavas (0.703 to 0.707).

The similarity between the mineralogy of aeolian dust and that of deep North Pacific sediments themselves was first indicated by Ferguson et al. (1970) and later discussed by several authors. Blank et al. (1985) provide particularly convincing arguments by comparing the mineralogy of the clay fraction (<2 μm) from western North Pacific aerosols and surface sediments (Table 7.6). A very close similarity occurs between both types of materials in the abundance of chlorite, illite, smectite, kaolinite, quartz, and plagioclase, which indicates the essentially terrigenous origin of Northwest Pacific deep-sea red clays. Note that few clay analyses have been performed in Pacific sea water, but the increase of Al concentration in surface masses located under the easterlies zone suggests a strong aeolian contribution from Asia (Orians and Bruland 1986). Smectite abundance appears to be slightly higher in sediments than in aerosols (3% instead of 1%, according to the quantitative method used by Blank et al. 1985), which could indicate discrete autochthonous formation. However the precision

Table 7.5. Oxygen isotopic composition of quartz from different North Pacific origins. (After Rex et al. 1969)

Origin	$\delta^{18}O$, Mean
Atmospheric dust	16
Hawaiian soils	18
Pelagic sediments	18
Hydrothermal deposit (Hawaii)	6
Cretaceaous chert	32

Table 7.6. Mean mineralogical composition of the clay fraction ($<2\,\mu m$) of North Pacific aerosols and surface sediments. (After Blank et al. 1985). Weight percentages are normalized to maximum percentage diffracting by X-ray (80.4%)

	Chlorite	Illite	Smectite	Kaolinite	Quartz	Plagioclase
Aerosols (6 samples)	2.7	39.5	1.1	15.5	10.5	11.1
Sediments (12 samples)	3.0	38.7	3.1	16.4	9.5	9.7

Table 7.7. Clay mineralogy of soil-sized particulates ($<2\,\mu m$) in northern Arabian Sea aerosols and surface sediments. (After Chester et al. 1985). Clay minerals are normalized to 100%

	Chlorite	Illite	Smectite	Kaolinite
Aerosols (7 samples)	30	55	5	10
Sediments	20–40	50	10	10

of the measurements is low because of the small percentages involved and possible undersampling of aeolian smectite by the mesh (see 7.2.1). In addition, we should notice that aerosols, which contain more feldspars than sediments (11.2% instead of 9.7%), have been collected much closer to Asia than sediments; differential deposition could have occurred during transportation, favoring the settling of slight smectite relatively to denser feldspar grains, at increased distance from the land.

Comparisons between dust and sediment mineralogy in the *Indian Ocean* are presented for the northern part of the Arabian Sea (Chester et al. 1985). The aeolian input during the northeast monsoon is among the highest found over marine regions (Al concentrations between 323 and 20,300 ng/m^3 of air). The clay mineralogy of the soil-sized particulates from dust is characterized by relatively abundant chlorite (30%; Table 7.7), which parallels the composition of underlying surface sediments but differs from that of other sediments in the Arabian Sea. R. Chester and colleagues conclude that most of chlorite (and illite) does not issue from the Indus river by runoff, but from aeolian sources located in

Table 7.8. Clay mineralogy (< 2 μm) of Greenland dust and Arctic Ocean surface sediments. (After Windom 1969). Clay minerals normalized to 100%

	Chlorite	Illite	Smectite	Kaolinite
Aerosols	20	53	–	27
(10 samples)				
Sediments				
83°N–93°W	25	54	–	21
83°N–89°W	24	52	3	21

the surrounding arid regions like Iran-Makran, the Arabian Peninsula, or Somalia. Note that Chester et al. (1985) do not quote the presence of palygorskite in Northern Arabian Sea sediments, a mineral abundantly recognized in both aeolian dust and marine sediments of the adjacent Gulf of Oman (Stoffers and Ross 1979, Khalaf et al. 1982). *In the Arctic range*, tentative comparisons arise from Windom's investigations (1969). The clay mineralogy of both Greenland snowfields and Arctic marine sediments is characterized by abundant illite (> 50%) with fairly abundant chlorite and kaolinite (20-25% each; Table 7.8). This suggests the possibility of similar sources, and indicates the probable aeolian influence on Arctic marine sedimentation.

7.3 Distribution of Aeolian Sediments

Let us consider the impact of aeolian supply in the different oceans, from selected information provided by sedimentary quartz, clays, and associate minerals. Note that beside the mineralogical markers, other sediment characteristics may provide useful indications about aeolian influence. this is the case of the grain shape and morphology (e.g., quartz, exoscopy: Krinsley et al. 1973; clay aggregates: Coudé-Gaussen and Blanc 1985, Bryant et al. 1986; Fig. 7.4), of isotopic measurements (e.g., Biscaye et al. 1974, Grousset et al. 1987), and of the sediment reflectance (Balsam et al. 1987) or magnetism (Bloemendal 1987).

In the Atlantic Ocean, a large area collecting abundant Saharan dust extends *westward of the Northwest African coast*, at about 20° N. This is particularly well expressed by the distribution of quartz, the abundance of which decreases regularly from the eastern margin to the mid-oceanic ridge (Fig. 7.6A). The diminution in quartz abundance (Johnson 1979) correlates with a decrease of the mean grain size, that varies from 23.5 μm near to the coast (300 km) to 10.5 μm at 2000 km offshore; correspondingly, the size frequency distributions of quartz change at 1000 km off the coast from positively skewed (coarse tails) inshore to negatively skewed (fine tails) offshore (Dauphin 1983). Deep-sea quartz grains mainly issue from Harmattan dust, that blows from the East and is characterized by red stained grains associated with kaolinite and smectite. The impact of trade winds, blowing from the Northeast and marked by unstained quartz, chlorite, il-

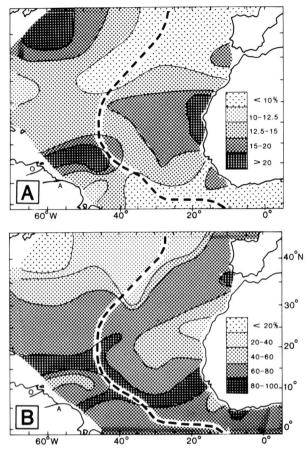

Fig. 7.6. Distribution of quartz (**A**) and total clay minerals (**B**) in North Atlantic sediments. (After Johnson 1979)

lite, and carbonate grains, increases toward the African coast (Lange 1982, Sarnthein et al. 1982a). The total clay minerals display a distribution opposite to that of quartz, because small, light, and platelike clay particles are transported by wind over greater distances and also proceed from marine transportation (Fig. 7.7B). The Saharan influence is also expressed by the distribution of dolomite, whose aeolian origin was formerly suggested by Delany et al. (1967), as well as by that of plagioclase (Johnson 1979. Fig. 7.7A, B). The dolomite- and plagioclase-rich province lies slightly further North than the quartz-rich and clay-poor sediment area, probably because of petrographic differences in the source areas. Notice that origins other than aeolian arise from Johnson's maps (1979). A secondary area rich in quartz and plagioclase and poor in clay minerals extends out from the Orinoco river mouth, while a large tongue of clay-rich material develops out from the Amazon river region. This strongly indicates the existence of other detrital influences acting concurrently with aeolian fallouts.

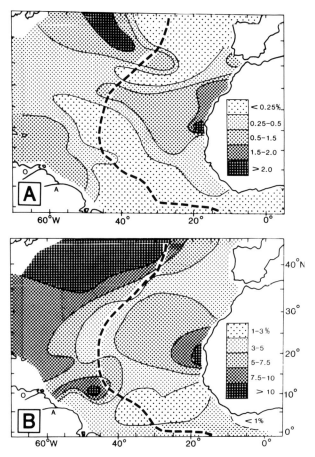

Fig. 7.7. Distribution of dolomite (**A**) and plagioclase (**B**) in North Atlantic sediments. (After Johnson 1979)

In the Northwest Atlantic, the dust storms issued from North America mix with river and coastal discharge to supply quartz, plagioclase, and dolomite in surface sediments (Fig. 7.6A; 7.7A and B).In these areas the river input appears to largely dominate over the wind supply (Johnson 1979). Wind-blown material participates more significantly in the sedimentation around the Mid- Atlantic ridge (Grousset et al. 1983, Grousset and Chesselet 1986): in this region the North American influence, expressed by chloritic clay suites and specific neodymium isotopic composition, is in competition with Icelandic influences marked by smectite, depleted $^{87}Sr/^{86}Sr$ ratios, and heavy rare earth elements (e.g., europium).

In the North Pacific, Rex and Goldberg (1958) first identified a zone of relatively quartz-rich surface sediments extending from Southern Japan across the Hawaiian Islands, almost to the west coast of North America (Fig. 7.10). Further studies confirmed these results, and showed that illite distribution parallels that

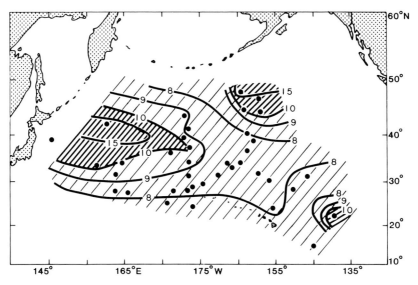

Fig. 7.8. Isopleths of mean quartz grain size (μm) of surface sediments from the North Pacific Ocean. (After Dauphin 1983)

of quartz (Griffin et al. 1968; see Chap. 8, Fig. 8.4). The average mean mass diameter of the silt-sized quartz decreases toward the East, from about 10 μm at 150°E to 7 μm at 140°W (Dauphin 1983; Fig. 7.8). The quartz-rich band extends between 25° and 40°N and reflects the deposition of aeolian material carried by the westerlies from Asia, which is confirmed by the trajectories of Asian dust plumes observed by satellite (Chung 1986). The main source areas of dust in North Pacific sediments are the deserts of Asia and the loess regions of North China (see Pye 1987). The deposition rate of mineral aerosols in the central North Pacific is estimated at 20.10^{12} g/year, which could correspond to 75-95% of the total sediment mass (Heath and Pisias 1979, Leinen and Heath 1981). Aeolian detrital minerals comprise various species associated with illite and quartz, including palygorskite fibers (Lenôtre et al. 1985). Note that another quartz-rich lobe extends in the North Pacific, southeast from Baja California, and reflects deposition of material carried by the northeast trade winds (Heat et al. 1973; Fig. 7.9).

In the South Pacific Ocean, the land-derived runoff is very low, which favors the sedimentary expression of wind influence and of authigenic processes. The Southeast Pacific displays a tongue of quartz-rich sediments northwestward from the coast of Peru and Colombia (Molina-Cruz 1977, Dauphin 1983; Fig. 7.10). As wind and surface current patterns coincide in this region, silty quartz and clay minerals can be transported to the western margin of South America by either agent. The influence of wind appears to become dominant to the seaward side of the Peru-Chile trench, where silt/clay ratios are higher than expected from water-transport processes (Scheidegger and Krissek 1982). Another zone of sediments rich in quartz, illite, and kaolinite extends southeastward from Australia toward New-Zealand (Griffin et al. 1968, Windom 1975; Fig. 7.9, 7.10, 8.4), which corre-

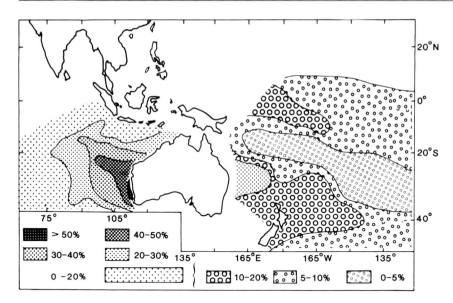

Fig. 7.9. Kaolinite distribution in surface sediments East and West of Australia. (After Windom 1975; Pye 1987)

lates to the trajectory of dust plumes formed in desert zones of Central Australia (in Pye 1987).

In the South Indian Ocean another area of kaolinite-rich sediments extends broadly on the western border of Australia (Griffin et al. 1968, Kolla and Biscaye 1973; Fig. 7.8). Just as what occurs on the Pacific side of Australia, kaolinite is mainly wind-supplied and associated with quartz (Kolla and Biscaye 1977). Easterlies blow offshore from the arid areas of Western Australia and carry the mineral dust above the Southeastern Indian Ocean. The question of aeolian influence in the *North Indian Ocean* is much more complex (in Prospero 1981a, Pye 1987), because of competition with river discharge and coastal runoff. Goldberg and Griffin (1970) think that fairly abundant aeolian dust transported from South Asia falls in the Bay of Bengal; but the high sediment load supplied by the Ganges and other rivers largely masks this aeolian contribution (Kolla and Biscaye 1973, 1977). *In the Northwestern Indian Ocean* quartz, illite and palygorskite issue from winds blowing out of Arabia, Iran and Northeast Africa (Goldberg and Griffin 1970, Kolla et al. 1976, Kolla and Biscaye 1977); but the extension of wind-supplied sediments, and the relative influence of the Indus river and West India coastal rivers remains somewhat unclear.

It is very difficult to estimate the *global impact of wind influence* on present-day marine sedimentation. Clay minerals display too diversified assemblages and origins, and present transport capabilities too high to allow precise identification and quantification of aeolian sources on a world basis. Such an attempt can be envisaged for quartz, which represents a simple mineral species, rapidly deposited and present in most terrigenous, biogenic, and chemical sediments. The distribution of quartz clearly reflects the main aeolian sources : Sahara, East Asia, Baja

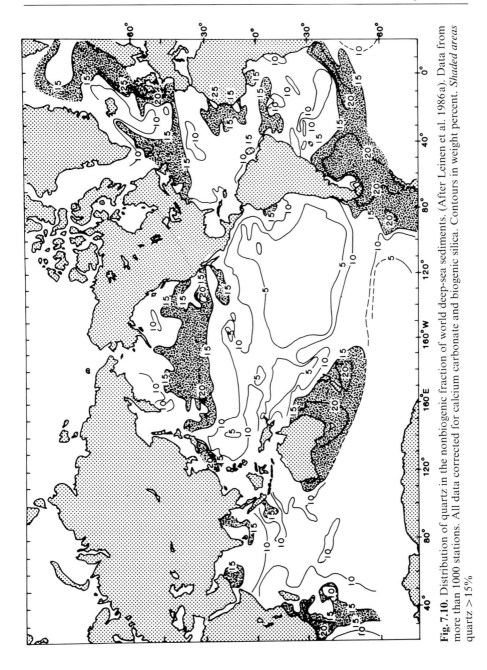

Fig. 7.10. Distribution of quartz in the nonbiogenic fraction of world deep-sea sediments. (After Leinen et al. 1986a). Data from more than 1000 stations. All data corrected for calcium carbonate and biogenic silica. Contours in weight percent. *Shaded areas quartz* >15%

California, Peru-Chile, Australia, Arabian region (Fig. 7.10; Leinen et al. 1986a). Most quartz grains identified in deep-sea sediments deposited far from land appear to be of aeolian origin. Despite the average low dust loads, the wind systems are pecularly well expressed in the Pacific sedimentation, probably because of the accessory influence of river input and of the relatively weak North-South current system. In most parts of the Atlantic and Indian Oceans, the distribution of aeolian quartz is perturbed or obliterated by fluvial input, ice-rafted supply, resedimentation processes, and bottom current activity. As a result, quartz itself must be considered cautiously when used as an indicator of global wind impact within the ocean (Leinen et al. 1986a).

7.4 Aeolian Influence in Past Marine Sedimentation

7.4.1 Quaternary

Most investigations on Quaternary sediments are focused on the relations between the aeolian activity and glacial / interglacial alternations as well as sea-level changes (see Prospero 1981a, Pye 1987). Fairly abundant data refer to the behavior of quartz, especially during the transition from late Pleistocene to Holocene. By contrast, few data exist on the correspondence between the clay aeolian flux and the sea-level and climate changes.

The Eastern Atlantic Ocean off the Saharan region provides the most abundant information. Nearly all available data point to a significant *augmentation of continental aridity and inferred wind velocity during the last glacial stage*, relatively to the postglacial period. This is suggested, for instance, by the increased size of quartz grains and increased amounts of phytoliths and freshwater diatoms in Würmian sediments around the Cape Verde islands (Parkin and Shackleton 1973, Parmenter and Folger 1974), by the increase at 6 – 8°N of quartz relative to illite, which is considered as constantly supplied from Africa (Bowles 1975), and by the extension of the quartz-rich zone off Sahara at 18,000 B.P. (Kolla et al. 1979, Zimmerman 1982; Fig. 7.11). Dauphin (1983) estimates from quartz grain-size data that the wind "vigor", representing the combined effects of wind velocity and dust thickness, increased during glacial times by a factor of 1.7 relatively to interglacials. During glacial times, the sea-level drop favored the accumulation of aeolian dunes on the coastal and shelf areas, which determined subsequent reworking and incorporation of aeolian- material turbidites in deep-sea sediments (Sarnthein and Diester-Haass 1977). The increased abundance of indifferentiated terrigenous material appears to be a less reliable criterion, since it may express an augmentation of either wind activity (e.g., Hays and Peruzza 1972) or river discharge (Diester-Haass 1976).

The question of a southward migration of the desert zone during the last glacial period is still being debated (see discussion in Pye 1987). For instance, Diester-Haass (1976) suggests such a migration, and a correlative southward shift of the more humid Mediterranean zone, from data obtained on quartz, carbonate, and fine terrigenous matter contents from cores located between 15° and

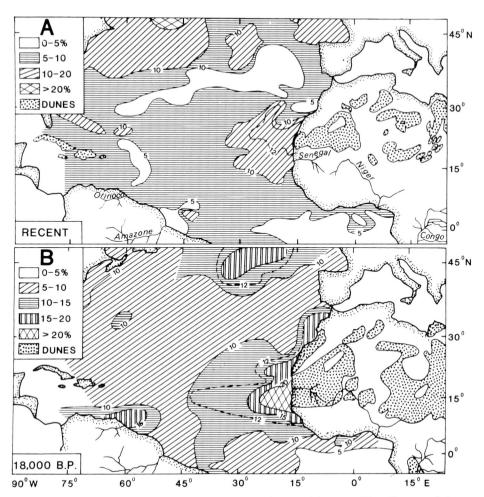

Fig. 7.11. Distribution of quartz in Holocene (**A**) and Late Pleistocene (**B**) sediments of the Central Atlantic Ocean. (After Kolla et al. 1979)

27° latitude N: this deduction arises mainly from the presence of unstained desert quartz derived from Northwest Africa in Holocene sediments North of 20°, and in late Pleistocene sediments South of 20°; by contrast, late Pleistocene sediments of the northern area are depleted in desert quartz. Additional arguments for such a shift may be provided by the presence in late Pleistocene deposits of decreased amounts of chlorite, a rock-derived mineral sensitive to weathering and possibly indicative of more humid conditions (Chamley et al. 1977a), as well as by the southward migration of the quartz-rich and freshwater diatom-rich zone (Kolla et al. 1979, Zimmerman 1982, Pokras and Mix 1985). Sarnthein and Koopman (1980) and Sarnthein et al. (1981, 1982b) show from detailed grain-sized data that the center of dust deposition was located between 15° and 20°N at the glacial maximum, a position very similar to the present-day situation. M. Sarnthein and colleagues provide convincing arguments based on the distribution of terrigenous silt fraction and on meteorological considerations. They conclude that the latitudinal extension of the desert belt and the position of the Intertropical Convergence Zone were fairly constant during both warm and cold stages. Sarnthein et al. (1981) consider that late Quaternary climatic changes in Western Africa do not result from southward shift of latitudinal belts but from the varying strength of tropical disturbances and meridional trade wind components; this could be reponsible for changes in the sources of aeolian clay minerals during glacial and interglacial stages. Recent investigation on freshwater diatoms (Pokras and Mix 1985, 1987) suggest a still more complex control, the position of tropical climatic belts being much more sensitive to changes due to Earth's precessional cycle and inferred seasonal solar radiation forcing than to changes from glacial to interglacial conditions.

The Pacific Ocean provides relatively little information on the variations of aeolian activity during the middle-late Quaternary, chiefly because the sedimentation rates are generally very low and preclude high resolution studies. *In the North Pacific*, Heath (1969) observes a positive correlation between illite and quartz, both minerals increasing in relation with the development of Quaternary glaciation. Dauphin (1983) deduces from grain-sized investigations on quartz that during glacial periods the eastward transport path of Asian dust moved southward, accordingly to the southward shift of the polar front. The opposite trend occurred at least twice during isotopic stage 5, which corresponded to interglacial conditions. Janecek and Rea (1985) isolate by chemical treatment a so-called "aeolian fraction" from two cores, located respectively under the prevailing westerlies in the Northeastern Pacific (Hess Rise) and under the westward trade winds in the equatorial Western Pacific. The "aeolian grain size" is considered as a direct indicator of wind intensity, and indicates periodic fluctuations that appear to be controlled by Earth's orbital parameters of precession, obliquity, and eccentricity. Aeolian accumulation rates are generally higher during interglacial times, when Pacific carbonate accumulation is low, which suggests that the dust source areas in Eastern Asia and Central America were more humid and more covered by vegetation during glacial times. T.R. Janecek and D.K. Rea infer from the coarser median size of dust deposited during glacial periods in the Eastern Pacific that trade winds were more vigorous at those times. Nevertheless the aeolian mass accumulation rates are twice as low in this area as on the Hess Rise in the Northeast Pacific, which indicates the dominance of easterlies over the trade winds during the last 700,000 years.

Indications of increased aridity and aeolian supply during glacial periods arise from studies in other parts of the world ocean and adjacent land-masses. *In the Southwest Pacific*, Thiede (1979) observes a higher input of quartz-rich dust from Australia at the late glacial maximum (18,000 yr B.P.) relatively to the present time. *In the Indian Ocean* the quartz accumulation is generally greater in glacial than in interglacial stage sediments (Kolla and Biscaye 1977), the location and relative importance of aeolian sources appearing to remain unchanged (Arabian Peninsula, India, Northeastern and South Africa, West Australia).

Giant ice cores in *Greenland* (Camp Century) and *Antarctica* (Byrd Station) display a strong augmentation during the latest glacial stage (Thompson 1977).

The Greenland site contains much more abundant particles, predominantly composed of soil-derived minerals, while the Antarctic site comprises fairly abundant volcanic debris. Electron-microscope observations of Briat et al. (1982) show that Antarctic ice-core dust is nevertheless of continental more than volcanic origin, and comprises increased amounts of quartz compared to mica-illite in the last glacial ice. Isotopic and geochemical measurements on ice age aerosols confirm the strong atmospheric circulation conditions during the last glacial maximum (Petit et al. 1981). The fast deposition of aeolian dust in the Antarctic region at the end of the last glacial stage could nevertheless have been favored by higher relative humidity at the ocean surface, as suggested by the deuterium excess recorded in the Dome C ice core (Jouzel et al. 1982).

7.4.2 Pre-Quaternary

Atlantic Ocean. Several attempts (e.g., Diester-Haass 1979, Sarnthein et al. 1982b) have been made to identify the aeolian input into the Eastern Atlantic during the *late Cenozoic*, and to place the influence of past winds among other transportation agents. Sarnthein et al. (1982b) offer a review of the atmospheric and oceanic circulation patterns off Northwest Africa for the past 25 million years. They identify eleven successive stages marked by specific characteristics of surface and bottom circulation, meridional and eastern winds, temperature and humidity changes, hemipelagic and turbiditic sedimentation, upwelling and productivity, and sea-level changes. The northward Neogene drift of Africa determined a progressive southward shift of the position of wind systems and correlated climatic belts. This is especially expressed by the Post- Eocene evolution of clay suites, investigated by Stein (1985) on Site 366 DSDP located on the Sierra Leone rise. Prior to 25 MY B.P., the terrigenous input was mainly restricted to aeolian supply from South Africa by southeastern trade winds, as deduced from abundant illite, very rare kaolinite, and very fine grain size. Near the Oligocene/Miocene boundary, Site 366 drifted across the Equator and was submitted to the northeastern trade winds, identified by R. Stein from increased amounts of kaolinite and coarser grain size. Specific maxima of mass accumulation rates and coarsening of terrigenous material suggest increased meridional atmospheric circulation and more arid climate in South Sahara and Sahel between 6 and 5 MY (late Miocene) and in the last 2.5 MY. South of the Canary Islands, arid climate appears characterized by relatively high contents of chlorite, palygorskite and quartz, and humid climate by substantial amounts of irregular mixed-layers and sandy terrigenous particles (Diester-Haass and Chamley 1978, Chamley and Diester-Haass 1979). From these criteria arid and humid periods are considered to alternate in the late Miocene, the former type being predominant. The Pliocene was generally rather arid and without significant river discharge, except temporarily, as revealed by chlorite and palygorskite diminutions. The Pleistocene sand and clay record indicates numerous strong changes between humid and arid periods, and high transport energy, which agrees with Sarnthein et al.'s deductions (1982b).

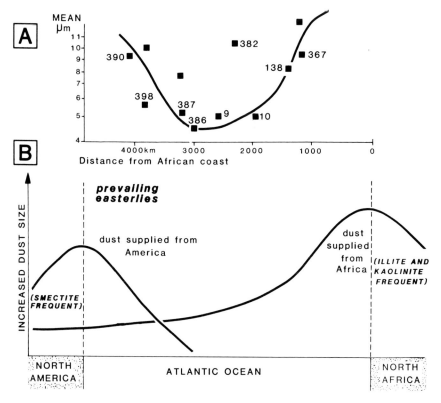

Fig. 7.12. Mean size and skewness for quartz from Late Campanian-Early Maastrichtian North Atlantic sediments (**A**), and tentative model for Cretaceous-Cenozoic dust supply (**B**). (After Lever and McCave 1983)

The Cretaceous and Cenozoic aeolian contribution to the marine sedimentation in the North Atlantic is tentatively considered by Lever and McCave (1983), who investigate eight time slices between the Berriasian-Valanginian (135-125 MY B.P.) and the late Miocene (6.5-5.5 MY B.P.), from about ten DSDP sites. Quartz and clay minerals are considered as essentially detrital and wind-blown. The slight decrease in the size of quartz silt toward the Central Atlantic, and the distinct clay mineralogy from one side to the other, are interpreted as the result of aeolian supply from both borders of the ocean, the eastern input being twice that from West (Fig. 7.12). A. Lever and I.N. McCave observe that the zone of coarsest, least-sorted and most positively skewed aeolian quartz, attributed to the maximum aeolian input, did not change significantly in the course of time; they conclude that the zone of maximum dust input has probably remained in paleolatitudes 20-30°N since the early Cretaceous. Despite the fact that Lever and McCave (1983) restrict their reasoning to latitudinal sources and exclusive aeolian transport, which is surprising for the young and narrow Atlantic Ocean (see Chap. 19), they introduce fruitful concepts. The existence in a large part of North Africa of huge evaporitic accumulations at several periods of the

Mesozoic, from late Triassic to late Cretaceous (Busson 1972, 1984), indicates the frequency of very arid conditions and the likelihood of active aeolian processes.

Pacific Ocean. Detailed investigations on quartz abundance and size, and mass accumulation rates are provided by Leinen and Heath (1981), Janecek and Rea (1983), Leinen (1985) and Rea et al. (1985), on two nearly continuous sections covering the entire *Cenozoic* in the *North Pacific*. DSDP Site 576, located 1500 km Southeast of Japan, East of the Shatsky Rise, indicates an aeolian mass accumulation rate twice as high as that recorded in core GPC3 situated 500 km North of Hawaii. This points to the permanence of main Asian sources and to the broad-scale uniformity of aeolian processes in the North Pacific over the last 70 MY. Both sections display a noticeable increase in aeolian supply in the late Tertiary relatively to earlier times. The low aeolian accumulation rates and fine grain size of early Tertiary dust agree with other evidence on rather warm and humid climatic conditions (e.g., Kennett 1982). About 40-50 MY ago, the Site GPC3 passed through the zone of influence of the northeast trade winds, at a latitude which is close to that of the present day. The wind intensity began to increase slowly about 35 MY ago, in the early Oligocene. Site 576 entered the influence of the westerlies about 15 MY ago, and since that time the rate of sediment deposition has increased. A strong increase in grain size and aeolian accumulation rate occurred about 2.5-3 MY ago, in relation with the onset of Northern Hemisphere glaciation, and has determined high values until the present.

In the *Equatorial and Tropical South Pacific*, the quartz contents are very small and vary only slightly from late Cretaceous to late Cenozoic, the total sedimentation rate being close to 0.2 mm/10^3 yr (Rea and Bloomstine 1986, Schramm and Leinen 1987). Nevertheless the median grain size of wind-supplied minerals increases in sediments younger than 10.5 MY located East of Tahiti, which is interpreted as the result of a significant augmentation in the intensity of atmospheric circulation (Rea and Bloomstine 1986). In the central South Pacific, at site 595 DSDP, the accumulation rate of aeolian quartz is higher in the Cretaceous and Paleocene than in the later Cenozoic. Paleogeographic reconstructions suggest that this decrease results from the migration of the site out of a position where it received aeolian material issued from Australia (Schramm and Leinen 1987). In the subtropical South Pacific, on the Lord Howe Rise (NW of New Zealand), calcareous oozes at sites 588 to 591 DSDP contain abundant smectite in the clay fraction, associated with various amounts of kaolinite (0-35%), illite (0-25%), irregular mixed-layers (0-10%), and chlorite (0-5%) (Stein and Robert 1985). The relative abundance of smectite decreases from middle Miocene to Pleistocene, which is attributed to the decrease in chemical weathering and increase in physical alteration in Australia. The aeolian dust supplied from Australia was progressively depleted in pedogenic smectite, and enriched in illite, chlorite, and kaolinite reworked from old rocks and soils, because of an increased aridity, also documented by coarser and more abundant quartz. As the diminution of smectite abundance started earlier in the North (site 588, middle Miocene) than in the South (site 591, late Miocene), R. Stein and C. Robert infer that the increased aridity was determined by the northward migra-

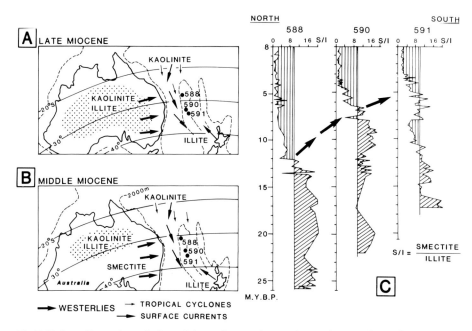

Fig. 7.13. Late Cenozoic evolution of clay sedimentation on the Lord Howe rise. (After Stein and Robert 1985). **A** Late Miocene situation. **B** Middle Miocene situation. **C** smectite to illite ratio

tion of the Indian Plate in the subtropical arid belt (Fig. 7.13). The time shift observed in the decrease of smectite abundance also argues in favor of a detrital origin for the mineral.

In the late Cretaceous and earliest Tertiary, the proximity of Africa to South America could have determined, at least partly, African sources for aeolian material deposited in the Eastern Pacific (Leinen and Heath 1981). Data from North Pacific cores 576 and GPC3 suggest deposition of relatively coarse aeolian grains and therefore noticeable wind activity (Rea et al. 1985). At DSDP site 463, West of Hawaii, Rea and Janecek (1981) study a nearly complete Upper Cretaceous section. Quartz and clay minerals are considered as essentially wind-supplied, despite low pole-to-equator temperature gradients, and therefore the probably less intense atmospheric circulation. The aeolian mass accumulation rates appear to have been higher at Aptian and Albian times than in the later Cretaceous, which is attributed to the very dry climate on land (see Frakes 1979). Smectite-dominated aeolian materials are referred to periods of volcanic activity during Aptian to early Albian and during early Maastrichtian time. The rest of the late Cretaceous series displays high amounts of illite, interpreted as issuing from continental wind-borne dust. In general the aeolian mass accumulation rates, that range over two orders of magnitude in illite-dominated sediments (500 mg/cm^2/10^3 yr in late Albian, 5 mg/cm^2/10^3 yr in Coniacian), appear to reflect global sea-level changes. Low mass accumulation rates correlate to the Cretaceous transgressions. This correspondence, coupled with probable sluggish and minimally contrasted oceanic and atmospheric circulation (e.g., Kennett

1982), indicates that the rate of aeolian sedimentation during the late Cretaceous depended on the amount of dust available rather than on the transportation process.

7.5 Conclusion

1. The importance of the aeolian contribution to deep-sea sedimentation has been *realized fairly recently* by the scientific community. Aeolian dust in pelagic sediments is now considered as one of the significant markers of paleoclimatic changes, beside the microfaunal and microfloral assemblages, the accumulation rate of biogenic particles, and the oxygen and carbon isotopic composition of planktonic and benthic foraminifera (e.g., Rea et al. 1985). Aeolian dust may provide information about the direction and intensity of atmospheric circulation, and about the aridity on source land masses. The aeolian dust production depends mainly on climate characteristics; abundant and regular dust haze may subsequently modify climate characteristics by increasing the albedo and diminishing solar radiations. Illite, associated with quartz, is the most widespread clay species supplied by the wind in modern oceans. Additional minerals may comprise all other clay and nonclay species, including abundant smectite, palygorskite, and mixed-layers.

Despite the fast progress of investigations, the knowledge of the real importance and effects of aeolian contribution to the world ocean sedimentation is still meager. Sustained efforts, quantitative approaches, and confrontations between sedimentologists, atmospherists, and modelers are necessary to better delineate the consequences of present and past wind activity. Such an aim is complicated by the difficulty of separating with certainty the aeolian particles from those issued from long-distance transport by marine currents. Much progress is expected from electron-microscope and elemental chemical investigations, but the difficulty increases when minerals have small grain sizes, as do most clay minerals and associated silicates and oxides. The distinction between wind-supplied and water-supplied smectite or illite will remain a very difficult task for a long time, even if the particle aggregation in surface sea water was demonstrated to occur extensively (see 6.3). This dramatically handicaps estimations of aeolian mass accumulation rates in the many regions potentially subjected to both aeolian and water transportation of clay- and silt-sized particles.

In addition, it is a common scientific tendency to overestimate for a while a mechanism recently identified. The temptation is often great to generalize an explanation that has been proven to be correct in a given area. For instance, the close correspondence observed in the northeastern Atlantic clay mineralogy between aerosols, suspended matter in sea water and underlying sediments invites considering aeolian supply as the exclusive transportation agent (7.7.2). Indeed, this is possible but not automatically true, since wind- and water-transported materials may have the same mineral composition, as demonstrated, for instance, off southeastern American coasts (Scheidegger and Krissek 1982). We are further tempted to interpret differences in dust and sediment composition, as in the South Atlantic, in terms of authigenic growth, and to neglect the possibility of sources and transport paths other than aeolian (see Chap. 18.2). Much progress and more balanced interpretations are expected from current research and methodologic improvements.

2. *The geographic extension and quantitative importance of aeolian input in the deep-sea clay sedimentation* is particularly demonstrated in two areas: (1) the tropical Northeastern Atlantic submitted to low river input and to frequent wind outbreaks from Sahara regions; (2) the Northeastern Pacific, predominantly exposed to Asian westerlies. Both areas are characterized by abundant illite and quartz, the latter minerals being considered as primarily indicative of wind supply; kaolinite, dolomite, and palygorskite also reflect Saharan origins. Other deep-sea regions, especially the Atlantic and Indian Oceans, often display a competition between aeolian, river, ice-raft, and marine current influences, the respective importance of which is difficult to decipher. Most Pacific Ocean regions appear suitable for favoring aeolian sedimentation, since Peri-Pacific trenches tend to trap the terrigenous matter supplied by rivers and coastal discharge, and also because of the long distances of transportation more easily covered by high atmosphere dust. The most favorable areas for accumulation and identification of aeolian dust are situated on submarine heights like oceanic ridges and seamounts, in areas located far from continental sources or devoid of significant river input, and in marine regions marked by little biogenic productivity or strong shell dissolution. These conditions again fit with large areas of Pacific deep-sea sediments, where aeolian supply may nevertheless compete with authigenic processes (see Chap. 12).

Indirect proof of the importance of aeolian contribution to deep-sea sedimentation arises from the close correspondence between dust load and sedimentation rates in regions where river supply is insignificant. For instance, dense dust loads thrown out from Northwest Africa correspond to fairly high sedimentation rates in the tropical Eastern Atlantic (ca. 1 cm/10^3 yr), while light dust loads that cross the North Pacific eastwards correlate with very slow inorganic deposition (ca. 1 mm/10^3 yr).

3. *The aeolian activity and inferred climate characteristics during past geologic times* are giving way to active research and intense discussions. Many periods of geologic history have been dustier than the present one, especially in regions located "downwind" of source land masses. It is now generally agreed that the last Pleistocene glacial stage corresponded to more arid conditions and higher atmospheric activity than the Holocene period, which correlated with a larger extension of continental dune fields (Sarnthein 1978); note that at 6000 yr B.P. more humid conditions temporarily prevailed in the Sahara. Glacial aridity nevertheless does not seem to have worldwide extension, as documented by North Pacific data. Some parts of North America and Eastern Asia appear to have experienced humid glacial and arid interglacial conditions (Janecek and Rea 1985). Even within individual regions, the pattern of continental humidity-aridity was not necessarily the same from one glacial-interglacial cycle to the next. The question of the migration of latitudinal arid belts in the course of the time has been strongly debated, especially off Northwestern Africa, where changes in orbital forcing appear to dominate the aridity cycles. Clay assemblages represent a

very useful tool to identify such climatic migration, as documented in the Eastern Atlantic and Western Pacific Oceans.

The late Cenozoic appears to be widely characterized by increased aeolian fluxes, in relation with the onset of North Hemisphere glaciation. However, the real influence of continental aridity on the late Cenozoic sedimentation is again difficult to quantify, because sea-level drop, augmentation of physical weathering, and river discharge also contributed to strongly increasing the terrigenous input. The importance of aeolian supply in the Pre-Neogene deep-sea sedimentation is very little known. First investigations suggest that important dust sedimentation may have occurred at some periods, especially during the Cretaceous and earlier Mesozoic times, where frequent sedimentation of huge evaporite bodies indicates high continental aridity. Much progress is expected from current research in this field.

Chapter 8

Terrigenous Supply in the Ocean

8.1 Climatic Control

As early as 1939, C.W. Correns reported the abundance of kaolinite in surficial sediments of the Equatorial Atlantic, of smectite in those from Cape Verde area, and of mica-illite in North Atlantic deposits. Correns (1939) came to the conclusion of independence between clay mineral formation and characteristics of the marine environment. Subsequent studies abundantly supported these first data. For instance Heezen et al. (1960) and Yeroshchev-Shak (1961) recognized the antagonistic latitudinal zonation of kaolinite and illite in the Atlantic Ocean, that led Nesteroff and Sabatier (1962) to favor the terrigenous input in deep-sea clay sedimentation. Gorbunova (1963) and Griffin and Goldberg (1963) reported the abundance of chlorite at high latitudes in the Pacific Ocean. Goldberg and Griffin (1964) showed the existence of a clay mineral zonation in the South Atlantic, while Nesteroff et al. (1964) indicated that Arctic sediments are characterized by the association of illite, chlorite, and quartz. All these first results indicated the existence of a latitudinal control of the distribution of clay assemblages in recent sediments, and therefore suggested a climatic influence.

Exhaustive data on clay distribution in the world ocean sediments were successively provided by Biscaye (1965), Griffin et al. (1968) and Rateev et al. (1968, 1969). These data, augmented by subsequent information published by different authors, were summarized and discussed by Windom (1976), who proposed average percentage values for the four major clay minerals identified in surficial deposits of world main basins (Table 8.1). H.L. Windom also provided distribution maps that, despite their frequent qualitative and quantitative imprecision, constitute a useful large-scale reference. Let us recall the main results obtained, with some additional remarks from more recent investigations.

8.1.1 Distribution of Kaolinite and Chlorite

Kaolinite forms abundantly in soils of intertropical land masses characterized by warm, humid climate (2.2.1; Fig. 2.7). The distribution of kaolinite in marine sediments reflects this dominant climatic control (Fig. 8.1), which led Griffin et al. (1968) to call kaolinite the "low-latitude mineral". *Kaolinite abundance increases toward the Equator in all oceanic basins, and therefore expresses a strong*

Table 8.1. Average content of major clay minerals in the <2 μm fraction of surficial sediments from world ocean basins. Values summed to 100% after Griffin et al. (1968), Goldberg and Griffin (1970), Venkatarathnam and Biscaye (1973) (in Windom 1976)

	Chlorite	Illite	Smectite	Kaolinite
North Atlantic	10	55	16	19
Gulf of Mexico	18	25	45	12
Carribbean Sea	11	36	28	25
South Atlantic	11	47	26	16
North Pacific	18	40	34	8
South Pacific	13	26	53	8
Indian Ocean	10	29	46	15
Bay of Bengal	14	29	45	18
Arabian Sea	18	45	28	9

climatic dependence controlled by the intensity of continental hydrolysis. In the equatorial Eastern Atlantic Ocean, kaolinite abundance reaches values as high as 50% of the clay fraction, because of the Niger and Congo river input, and of wind supply from South Saharan regions (Chap. 7). The eastern Caribbean area displays more abundant kaolinite than the western Caribbean area, that is less directly submitted to the kaolinite-rich Amazon river input. The western part of the equatorial and south-tropical Pacific Ocean receives much more kaolinite than the eastern part, because intense chemical weathering characterizes the Western Pacific coasts, in contrast to the relatively arid eastern coasts. The south equatorial zone of Indian Ocean also collects abundant kaolinite, especially in the eastern area where old lateritic soils are actively removed by Western Australian winds (Griffin et al. 1968; Fig. 7.9). Marine kaolinite derived from intertropical soils is generally associated with abundant iron oxides (mostly goethite and subamorphous components), and often with gibbsite (e.g., Equatorial Atlantic; Biscaye 1965).

Detrital chlorite mainly results from the erosion of plutonic and metamorphic rocks preserved from noticeable chemical weathering. Such rocks crop out widely on continental shields from high latitudes, where soils hardly develop and physical weathering predominates (2.2.1). The distribution of chlorite in marine sediments typically reflects these climatic conditions (Fig. 8.2), in both clay and nonclay fractions. *Chlorite abundance increases toward cold latitudinal zones parallel to the decrease of continental hydrolysis,* and tends to concentrate in polar regions of the world ocean (e.g., Berry and Johns 1966). Cold and dry or frozen regions usually correspond to the association of abundant chlorite and of noticeable amounts of quartz, mica, feldspars, amphiboles, pyroxenes, and other dense minerals, directly removed from crystalline substrates and supplied within the ocean (in Windom 1976).

Chlorite varies in opposition to kaolinite, and was called the "high-latitude clay mineral" by Griffin et al. (1968). This is especially well expressed in the Atlantic Ocean, where the chlorite percentage increases symmetrically northward and southward of the equatorial zone. The kaolinite to chlorite ratio decreases by two

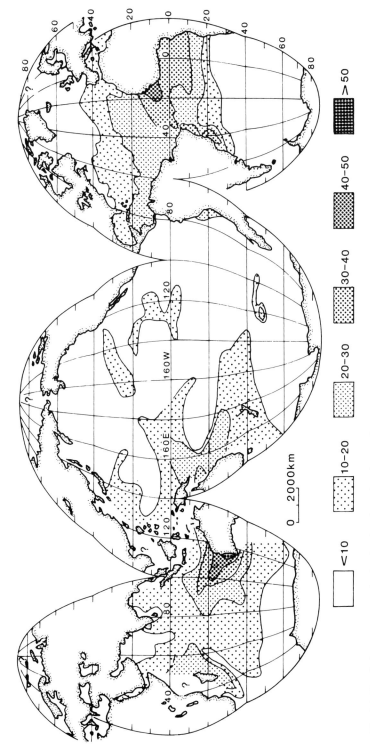

Fig. 8.1. Kaolinite percentages in the clay fraction of surface sediments in the world ocean. (After Windom 1976)

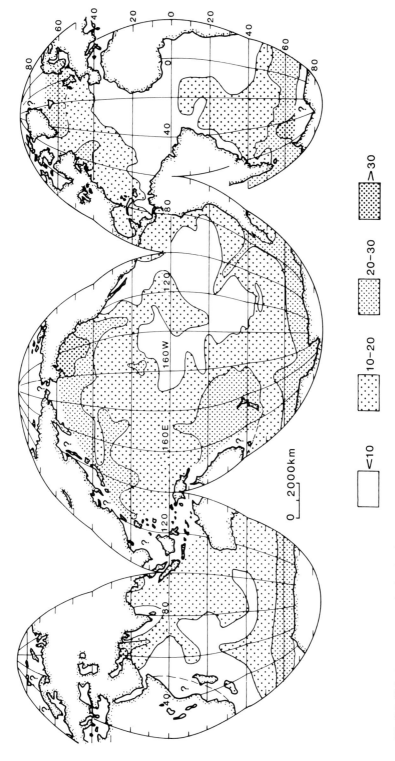

Fig. 8.2. Chlorite percentages in the clay fraction of surface sediments in the world ocean. (After Windom 1976)

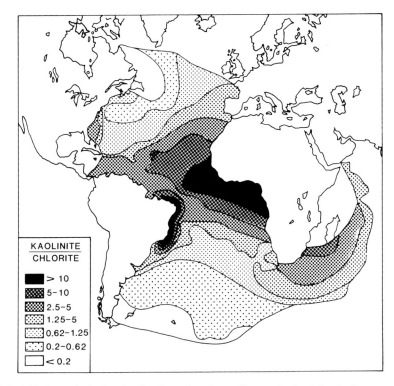

Fig. 8.3. Kaolinite/chlorite ratio in the clay fraction of surface sediments in the Atlantic Ocean. (After Biscaye 1965, in Kennett 1982). Values correspond to the peak-area ratio of 3.58 over 3.54 Å X-ray diffraction reflections

orders of magnitude from low- to high-latitude regions (Fig. 8.3). In the Pacific Ocean, abundant and well-crystallized chlorite proceeds from the physical weathering and erosion of both North American and North Asian regions. A belt of chlorite-rich sediments outlines the Antarctic continent in the South Pacific, as in other oceans. Very arid zones, that are also marked by few hydrolytic processes (2.2.1), often determine a significant supply of rock- derived chlorite toward the ocean. This is especially the case in the Southwest Pacific Ocean, East of the Australian desert, and in the Northwest Indian Ocean off the Arabian peninsula (Fig. 8.2).

8.1.2 Other Clay Minerals

The relative abundance of illite tends to increase toward high latitudes parallel to chlorite, which reflects the decrease of hydrolytic processes and the increase of direct rock erosion under cold climatic conditions. This is particularly well expressed in the North Atlantic Ocean and in the Southern Pacific and Indian Oceans (Fig. 8.4), where illite commonly exceeds 50% of the clay fraction. Data

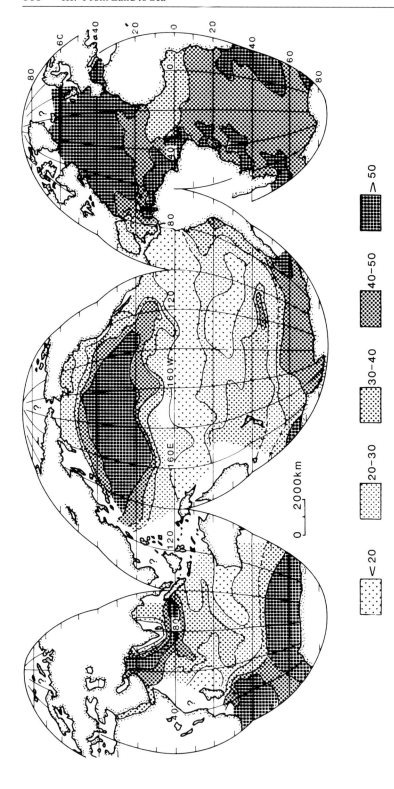

Fig. 8.4. Illite percentages in the clay fraction of surface sediments in the world ocean. (After Windom 1976)

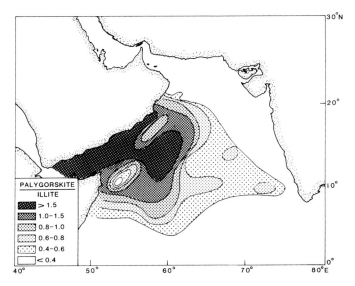

Fig. 8.5. Palygorskite/illite ratio in the clay fraction of surface sediments from the Arabian Sea. (After Kolla et al. 1976). Values correspond to the peak-height ratio of 10.5 over 10.0 Å X-ray diffraction reflections

published by Nesteroff et al. (1964), Berry and Johns (1966), and more recently by various other authors (see 8.2), show that marine illite also abundantly derives from the erosion of cold climatic regions surrounding the northernmost Pacific and the Arctic Oceans. In addition, abundant detrital illite characterizes the oceanic areas tributary either to *high-altitude cold climate regions* like the Himalaya mountains (e.g., Indus and Ganges deep-sea fans; Venkatarathnam and Biscaye 1973, Kolla et al. 1976, Bouquillon and Chamley 1986), or to *desert climate* regions like eastern continental Asia (e.g. North Pacific illite-rich belt fed by northern westerlies; Fig. 8.4.; Chap. 7.3).

Very arid conditions are sometimes characterized by abundant detrital *palygorskite*, removed from calcareous pedogenic crusts or evaporitic sediments (2.2.1, 4.2.3, Chap. 9) and transported within the ocean. This is especially the case in the northwestern Arabian Sea adjacent to the Arabian peninsula (Fig. 8.5). There is good correlation between the maximum of palygorskite percentage and the location of topographic heights, which reflects the importance of aeolian transportation in this region (Kolla et al. 1976).

The relative abundance of *Smectite* largely displays a distribution that does not parallel the zonal distribution of main weathering processes (Fig. 8.6). This indicates the accessory control of climate, and the dominance of other allochthonous and/or autochthonous processes (see 8.2.3, and Chaps. 12, 13). The increased amounts of marine smectite recorded off the *temperate to subarid regions* surrounding the South Atlantic, central North Atlantic, Central to Southeast Pacific, and Northeast Indian Oceans, as well as the Gulf of Mexico, nevertheless suggest that the mineral partly reflects conditions intermediate between those of cold-dry and warm-humid climate. These conditions cor-

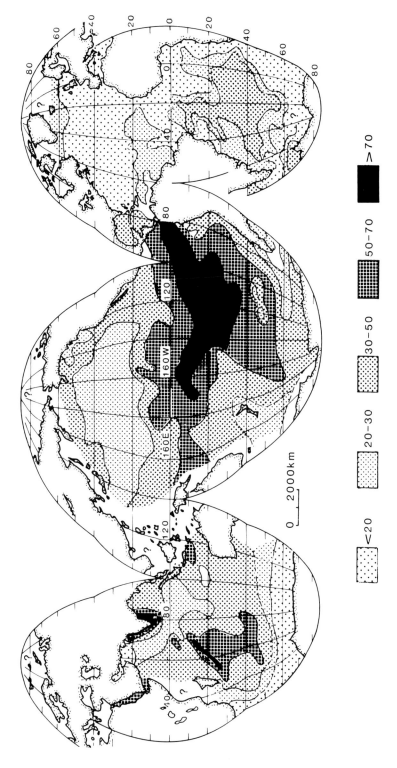

Fig. 8.6. Smectite percentages in the clay fraction of surface sediments in the world ocean. (After Windom 1976)

respond to the formation of continental brown to vertisolic soils, in which smectite develops through degradation or neoformation processes (Chap. 2). *Random mixed-layer clays* are particularly well represented in *temperate to temperate-cold latitudinal zones* (Biscaye 1965), and essentially result from the erosion of late Quaternary weathering complexes (Chap. 2).

8.2 Petrographic Control

8.2.1 Polar Regions

As cold land masses are almost exclusively submitted to physical weathering, all mineral species contained in outcropping rocks can theoretically be removed by erosion and transported within the polar oceans. The composition of detrital mineral suites depends mainly on the petrographic nature of source rocks. As Antarctica and Arctic land masses largely consist of crystalline rocks, mica-illite, and chlorite associated with quartz, feldspars and various dense minerals constitute the most frequent species of adjacent marine sediments, and therefore characterize high-latitude associations. However, various other minerals can be supplied to cold oceans, including species formerly originating under warm-humid climate and locally stored in ancient sedimentary rocks or soils.

The presence of kaolinite in Arctic sediments was first reported in the Beaufort Sea by Naidu et al. (1971), who thus cast doubt on the paleoclimatic use of the mineral as an indicator of intense weathering. Subsequent studies confirm the frequent occurrence of kaolinite in Seas of the northern hemisphere. Darby (1975) identifies average kaolinite contents of 26% throughout the Amerasian half of the Arctic Ocean, and reports mineral amounts up to 45% in the clay fraction of some core levels as old as 2.8 – 3.0 million years. Expandable mixed-layered minerals of an illite-smectite type are often associated with kaolinite. D.A. Darby suggests that Arctic kaolinite derives from old shales and paleosols of northern Alaska and Canada. Such an interpretation is supported by the presence of abundant kaolinite (up to 50% of clay) in Cretaceous and Tertiary shales and other sedimentary rocks from the Mackenzie river delta and Ellesmere islands (Bayliss and Levinson 1970, Bustin and Bayliss 1979, Datcharry 1987). Similar results arise from investigations performed in Barents Sea (Wright 1974, Bjørlykke and Elverhøi 1975), Norway Sea (Chamley 1975a), Baffin Bay (Boyd and Piper 1976, Hein and Longstaffe 1985), Gulf of Alaska (Hein et al. 1979a), etc. As kaolinite abundance does not correlate with glacial/interglacial alternations, all authors agree to consider the mineral as reworked from adjacent land masses, and devoid of direct climatic significance. Kaolinite sources consist mainly of Jurassic and Cretaceous sedimentary rocks, that crop out especially on the American part of Peri-Arctic land-masses.

Kaolinite and other clay and nonclay minerals, actively reworked through glacial and peri-glacial erosion, often display a specific distribution that allows the *identification of petrographic provinces and sources*. For instance, Wright

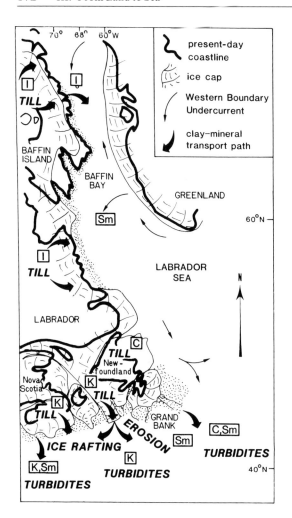

Fig. 8.7. Provenance and transportation of late Pleistocene clay minerals on the eastern margin of Canada. (After Piper and Slatt 1977). *C* chlorite; *I* illite; *K* kaolinite; *Sm* smectite and other expandible minerals

(1974) recognizes three clay mineral facies in the *South Barents Sea*. The southern facies is characterized by well-crystallized mica-illite, chlorite, amphibole, CaO, MgO, Mn, Ni, Cu, by increased amounts of smectite toward the Southeast, and is mainly attributed to rock-flour materials transported by Atlantic waters. The northern facies is marked by poorly crystalline chlorite, mixed-layered illite, kaolinite, Fe_2O_3, Zn, and results from transportation by Arctic waters. The Bear island-Spitsbergen bank facies contains abundant dioctahedral illite, Al_2O_3, K_2O, and Zr, and is of a local provenance. Piper and Slatt (1977) compare the clay mineralogy of tills and other terrestrial deposits and of marine sediments from the *eastern continental margin of Canada* (Baffin Bay to Nova Scotia), during late Pleistocene and Holocene times. They show the dominant influence of source lithologies and local input on clay associations, and the negligible influence of climate (Fig. 8.7). An illite and chlorite assemblage characterizes Pleistocene terrestrial sediments on Baffin island, Labrador, Newfoundland, and eastern Nova Scotia, as well as on the inner shelf of Labrador and Newfound-land. Illite abundance increases in areas underlain by plutonic and high-grade metamorphic rocks, while chlorite is more abundant where parent rocks have undergone low-grade metamorphism. In addition to illite and chlorite, marine sediments in Baffin Bay and on the outer continental margin of Labrador and Newfoundland contain kaolinite and smectite that derive from offshore Mesozoic-Tertiary coastal plain sediments. Kaolinite and smectite are also present on the margin of Nova Scotia and New Brunswick, where they result from the erosion of both Carboniferous-Triassic red beds and subsequent coastal plain formations.

Comparable reconstructions of petrographic sources have been made in *Peri-Antarctic seas*, and extend largely in Pre-Quaternary series thanks to Deep Sea Drilling Project sections. For instance, Piper and Pe (1977) investigate the Cenozoic clay mineralogy from Legs 28 and 29 DSDP holes on the continental rise and abyssal plain of the Australia-New Zealand sector of Antarctica. Clay associations are dominated by illite and chlorite, especially in Quaternary sediments, but kaolinite is often present (up to 13% of the clay fraction) and smectite abundance ranges between 5 and 50%. All clay minerals issue from the Antarctic continent, including smectite (see also Moriarty 1977). During the Oligocene and Miocene, Wilkes Land provided illite and some chlorite to the ocean, while Victoria Land was responsible for the supply of illite, ferriferous chlorite, and smectite. In the Plio-Pleistocene, an increase in kaolinite and smectite content off Wilkes Land is related to glacial erosion and subsequent reworking of latest Cretaceous and early Tertiary sediments cropping out on the continental shelf and accumulated under past hydrolyzing conditions.

Similar deductions on the identification and variations of southern detrital sources arise from other studies on sediments from Legs 28 (Ross Sea, Robert et al. 1988) and 113 (Weddell Sea, Barker, Kennett et al. 1987), indicating the common reworking from Antarctica of smectite, kaolinite, illite and chlorite. *A large field of investigations has been open*ed by using clay mineral successions of cold regions to recognize the terrigenous sources, their modifications in the course of the time, the location of land areas protected from active erosion by ice blankets, the effects of river influx and sea-level changes, and the variations in marine paleocirculation.

8.2.2 Arid Regions

Arid and desert warm regions, nearly devoid of noticeable chemical weathering (Chap. 2), like high-latitude regions, permit the use of mineral associations to trace the petrographic sources and their successive changes. Such a survey can be performed even with minerals reputed to be fragile, like *palygorskite* or *sepiolite*. A good example is given by the North Arabian Sea and adjacent basins. Heezen et al. (1965) posed the question of the origin of palygorskite (attapulgite) in surface sediments cored between 1000 and 2500 m water depth in the Gulf of Aden and Southern Red Sea. The abundance of palygorskite increases toward the East, off the Hadramout coast, where Müller (1961) previously identified palygorskite and sepiolite in early Tertiary sediments. G. Müller showed that fibrous clays are transported by seasonal rivers toward the Gulf of Aden, palygorskite appearing to be better preserved than sepiolite in downstream deposits. A fluvial source for fibrous clays is also demonstrated by Estéoule et al. (1970) in the submarine alluvial fan of the Rud-Hilla river, northeastern part of the Persian Gulf. Hartmann et al. (1971) identify palygorskite in most sedimentary types of the Persian Gulf and clearly demonstrate its terrigenous origin, largely dependent on river influx. Other studies indicate the importance of aeolian sources in the same region (see also 8.4). Goldberg and Griffin (1970), following Gorbunova's (1966) first investigations, show that palygorskite abundance decreases eastward in the Arabian Sea North of the Equator, which is attributed to a decreasing influence of African winds. Palygorskite occurs in both aeolian dust and marine sediments of the Gulf of Oman (Stoffers and Ross 1979, Khalaf et al. 1982), where its distribution helps to characterize petrographic provinces (e.g., Faugères and Gonthier 1981). The peculiar abundance of palygorskite on submarine hills of the Northern Arabian Sea (Kolla et al. 1976) confirms the essential role of aeolian transportation in open sea environments (8.1; Fig. 8.5).

8.2.3 Recognition of Volcanic Sources

Volcanic rocks weather preferentially into smectite, regardless of climate conditions if there is sufficient water to allow hydrolytic processes (2.2.2). The identification of volcanic terrigenous sources is therefore often easy from clay mineral investigations, especially if other sources consist of acid plutonic or metamorphic rocks that are usually devoid of smectite. This is illustrated by several studies performed in *the North Pacific Ocean and South Bering Sea.* Gardner et al. (1980) identify three dominant mineral and chemical associations in sediments deposited on the outer continental shelf of the South Bering Sea. The most significant contribution consists of coarse felsic sediment derived from the quartz-rich rocks of the Alaskan mainland, and marked by illite, K-feldspar, metamorphic rock fragments, quartz, garnet, epidote, Si, Ba, and Rb. The second group corresponds to andesitic sediments issued from the Aleutian islands, that comprise smectite, vermiculite, clinopyroxene, glass, volcanic debris, and relatively high amounts of Na, Ca, Ti, Sr, V, Mn, Cu, Fe, Al, Co, Zn, Y, Yb, and Ga. A third and local group forms basaltic sediments derived from rocks of the Pribilof islands in the Northwest, and comprises smectitic clay associated with U, Li, B, Zr, Ga, Hg, C, and S. The weakness of present-day currents, and the presence of the Bering submarine canyon between the Aleutian Islands and the outer continental shelf and slope suggest that the distribution of the mineral-chemical provinces identified mainly results from sediment dynamics occurring during late Pleistocene low sea-level stages. The favored production of smectite to marine sediments by weathering of Aleutian volcanic rocks characterizes both recent and ancient sediments. Hein et al. (1975) study six DSDP sites from Leg 19, in the Northern Pacific Ocean. They show that abundant illite and kaolinite inherited from the Kamtchatka-Koryak regions are diluted eastward and replaced by smectite and accessory chlorite derived from the Aleutian region. The volcanic activity, marked by increased amounts of smectite, appears to have been particularly active during late Miocene and Pleistocene times.

Naidu et al. (1982) and Naidu and Mowatt (1983) investigate 700 sediment samples from the major rivers and *marginal seas of Alaska*, and from clay mineral markers identify the major sources, transport trajectories, and depositional sites of land-derived fine particles (Fig. 8.8). Illite- and chlorite-rich sediments characterize most parts of the Gulf of Alaska, with slightly higher amounts of expandible clay minerals in the central area. The Alaska peninsula provides abundant smectite and associated expandible material inherited from volcanic terranes and distributed northward along the Alaska peninsula in the Bering Sea, the South Chukchi Sea, and the Southern Arctic Ocean. The Yukon river is responsible for the input of illite, chlorite, and expandible minerals, and also of significant amounts of kaolinite eroded from Mesozoic sediments and fluvial paleodeposits. Diversified clay suites from the Yukon river system are carried by major currents toward the North, where they mix with simple, illitic suites

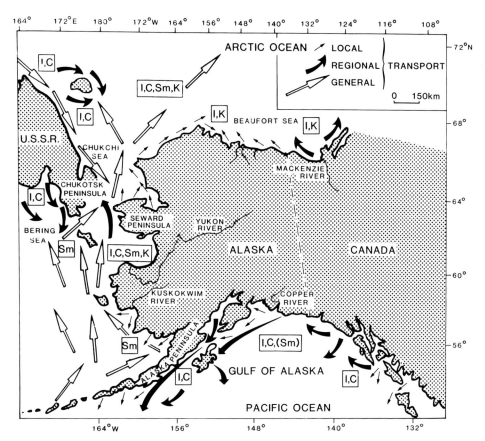

Fig. 8.8. Major transport trajectories and depositional sites of clay-sized particles in the marginal seas of Alaska. (After Naidu and Mowatt 1983). Most characteristic clay minerals are indicated. *C* chlorite; *I* illite; *K* kaolinite; *Sm* smectite and other expandible minerals

transported southward along the Chukotsk peninsula (Siberia). The Beaufort Sea displays a patchy distribution of clay minerals, with variable amounts of the four main species. This appears to result from the high seasonal sedimentary regimes as well as from ice-covering, ice-rifting, and ice-gouging effects.

Many *other examples* allow the distinction of terrestrial volcanic sources from other detrital sources. For instance, the Caribbean basin located between the Lesser Antilles Arc and South America receives weathered products issued from both volcanic islands (Fe-montmorillonite, Mg-rich beidellite, halloysite) and crystalline Venezuelan mainland (illite, kaolinite, ferriferous beidellite) (Parra et al. 1986, Gandais 1987). The Philippine and South China Seas show the antagonistic influence of volcanic archipelagoes (Philippine Islands, Indonesia) that supply abundant smectitic materials, and of the Asian continent marked by complex clay assemblages dominated by illite, chlorite, and locally kaolinite (Huang and Chen 1975, Murdmaa et al. 1977, Chen 1978, Kolla et al. 1980b). Sediments around the Indian peninsula contain abundant smectite (Fig. 8.6), derived from Deccan basaltic traps and diluted both eastward and westward by illitic clays derived from the Ganges, Indus, and Arabia (Kolla and Biscaye 1973, Kolla et al. 1976).

Note that *the high concentrations of smectite recorded in the central South Pacific Ocean* are classically attributed to volcanic influences (in Windom 1976 ; Fig. 8.6). Such a volcanic influence could be largely of subaerial nature. Considering the very low sedimentation rates in this region ($< 1.0 \, \text{mm}/10^3$ yr) and the very low continental supply, smectite could derive from the progressive erosion of volcanic archipelagoes actively submitted to equatorial-tropical chemical weathering. The difference observed in smectite abundance between North and South Pacific (Fig. 8.6) could therefore partly result from the *higher number of volcanic islands* in the latter region as opposed to the importance of dilution by continental aeolian input in the former (Chap. 7). An additional argument arises from the depleted amounts of smectite in the western South Pacific, a region exposed to westerlies carrying kaolinite, illite, and quartz from Australia (Figs. 7.9, 7.10, 8.4). The explanation of a detrital volcanic source for the smectites abundantly present in Central and South Pacific sediments is compared in Part Four (Chaps. 12, 13) with other explanations that involve authigenic processes. In addition, we should stress the relatively high contents of smectite (30-50% of clay) off Antarctica coasts in Southeast Pacific sediments, compared to Southwest Pacific deposits (Fig. 8.6). This peculiar feature could result at least partly from the erosion of late Cretaceous and younger sedimentary rocks located on the Antarctic continental margin between 60° and 170°W (UNESCO, world geological map), that are often rich in smectite (e.g., Leg 35 DSDP, Bellingshausen Sea; Zemmels and Cook 1976) and actively eroded by glaciers.

8.2.4 *Mixed Geologic and Pedologic Sources*

In most regions of the world ocean, clay detrital assemblages reflect the combined influences of land petrography and continental climate. This is systematically the case in temperate regions, where chemical weathering is weak enough to allow the recognition of geologic substrates, and strong enough to permit the mineralogic expression of pedologic processes. The petrographic influence of rocky substrates is often reflected by the so-called primary minerals, i.e., illite and chlorite, while the petrographic input from pedogenic blankets is mostly marked by irregular mixed-layers, kaolinite or halloysite, and smectite (see Chap. 2). The latter mineral group may nevertheless also derive directly from parent rocks, especially in regions where sedimentary substrates have stored products from ancient soils or chemical sediments. This is also the case of fibrous clays, of pyrophyllite (e.g., Dunoyer de Segonzac and Chamley 1968) or talc, and of other clay minerals, that are able to trace the petrographic sources in various modern and ancient sedimentary environments.

Hundreds of references have become available in the literature over the last few decades, and we quote here only very few of them. *The seas around Japan* have been extensively investigated since the first studies of Kobayashi et al. (1964). A summary is provided by Aoki et al. (1974a), Yin et al. (1987), and Chamley and Debrabant (1989). Illite and chlorite occur mainly in the Southern Japan Sea and North Philippine Sea, because of the widespread occurrence of crystalline and metamorphic rocks in South Japan (Fig. 8.9). The high amounts of smectite in northern basins result from the presence of various volcanic and sedimentary rocks. Kaolinite and mixed-layers proceed chiefly from the weathering of crystalline rocks. The supply of illite from Japan is complemented in the Japan Sea by the influence of the Tsushima warm current issuing from the East China Sea. The different rock and soil sources can be recognized in sediments as old as the Miocene, as documented in Shikoku basin from DSDP data (Chamley 1980a).

A detailed study is presented on surficial shelf sediments from *Western North Island, New Zealand*, by Hume and Nelson (1986). Marine clay assemblages are very similar to those of adjacent river and harbor sediments, which allows the identification of five north-south clay petrographic zones, characterized by specific concentrations of rock- and soil-derived minerals (Fig. 8.10): (1) Hamilton shelf sediments comprise abundant smectite and common poorly crystallized illite, issuing from Oligocene mudstones and weathered Quaternary volcanic rocks. (2) North Taranaki shelf deposits contain abundant well-crystallized illite and common chlorite and irregular mixed-layers, issuing from Miocene mudstones and overlying soils. (3) The central Taranaki shelf is characterized by detrital illite and chlorite derived from South Island via the Westland current, and by secondary smectite formed by weathering of Cape Egmont andesite. (4) South Taranaki shelf sediments mainly include well-crystallized illite and chlorite inherited, via the Westland current, from northwestern South Island rivers that drain granite and schists. (5) North Crook Strait basin deposits contain well-crystallized illite and chlorite, with associated mixed- layers, derived both from South Island via the D'Urville current and from Plio- Pleistocene mudstones of the Wanganui region.

The western continental shelf of India displays a close correlation between the sediment clay mineralogy and the petrography of adjacent land masses, which indicates the dominant influence of local source rocks rather than that of southwest

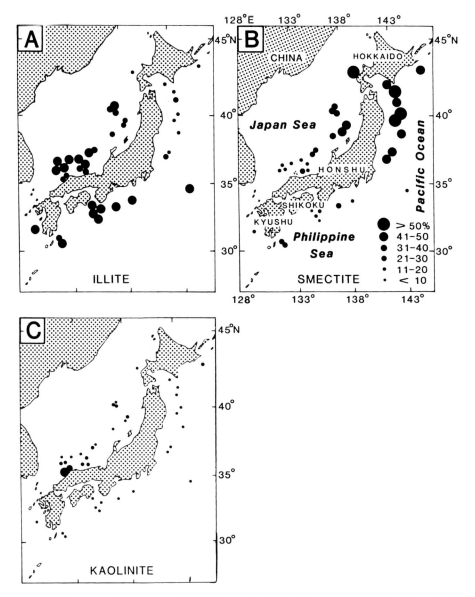

Fig. 8.9. Percentage distribution of illite (**A**), smectite (**B**) and kaolinite (**C**) in the <2 μm fraction of surface sediments around Japan. (After Yin et al. 1987)

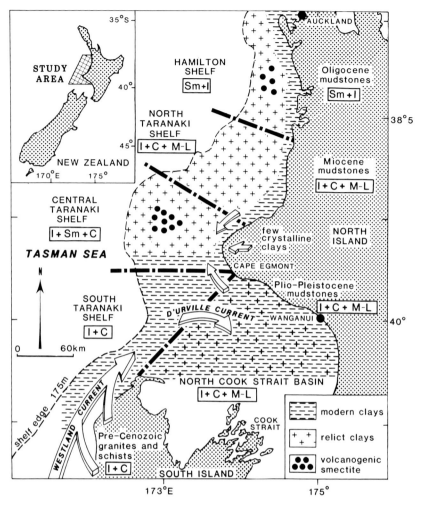

Fig. 8.10. Nature and origin of dominant clay associations in surficial shelf sediments of western North Island, New Zealand. (After Hume and Nelson 1986). *C* chlorite; *I* illite; *Sm* smectite; *M-L* mixed layers

monsoon drift along the coast (Nair et al. 1982). Four clay mineral zones are successively recorded from North to South: a smectite-kaolinite- illite-chlorite zone off the Gulf of Kutch marked by both the Indus river and local basalt influences; a smectite-rich zone off the Gulf of Cambay corresponding to widespread weathered Deccan Trap basalts; a transition zone with a southward decrease of smectite correlated with the change in coastal outcrops from principally basaltic to dominantly acid rocks; a kaolinite- and gibbsite-rich zone on the southern shelf determined by the close proximity of strongly weathered granites, gneisses, and schists.

Many studies performed in the *Mediterranean* domain demonstrate close correspondence in the clay mineralogy of river and marine sediments, and the general correlation existing between the main petrographic areas on land and mineral provinces in the sea (e.g., Chamley et al. 1962, Rateev et al. 1966, Chamley 1971, Venkatarathnam and Ryan 1971, Nir and Nathan 1972, Shaw 1978, Maldonado and Stanley 1981). The most striking fact is the strong opposition existing between western and eastern basins. The Western Mediterranean is characterized by a fairly homogeneous clay sedimentation, dominated by illite that is largely provided by the Rhone and Po rivers. Illite is associated with chlorite and various types of mixed-layers. Kaolinite and smectite often constitute accessory minerals. Increased amounts of smectite occur notably off North African coasts and in the Southwestern Adriatic basin; an obvious volcanic impact is recognized only around the island of Sicily. Unusual minerals like sepiolite are locally supplied in large amounts by North African rivers (Froget and Chamley 1977), while aeolian palygorskite is more widely scattered in small amounts (Chap. 7). The Eastern Mediterranean basins display complex and heterogeneous clay associations. Smectite often predominates, because of its abundance in Nile river materials and its reworking from various igneous and sedimentary rocks as well as from soils. Kaolinite is common and derives mostly from African soils and paleosols. African and European inputs are respectively marked by smectitic and illite-chlorite assemblages. The aeolian supply from the South is still poorly known but probably important, and characterized by both palygorskite and kaolinite. Unusual minerals, such as pyrophyllite or volcanogenic smectite, are locally reworked from specific areas, and constitute precise markers of source rocks (e.g., Chamley 1971).

From a more general point of view, *the large abundance of illite in the North Atlantic and North Pacific Oceans* (Fig. 8.4), associated with high amounts of quartz (Fig. 7.10) and fairly abundant irregular mixed-layered clays, correlates with the wide extension of continental shields in the Northern Hemisphere. These regions largely consist of granitic and acid metamorphic rocks rich in micas and quartz, the weathering of which under cool to temperate climate gives way to more or less crystalline illite, vermiculitic and smectitic mixed-layers, and quartz. These mixed rock- and soil-derived suites are easily supplied within the basins because of the relative narrowness of the Atlantic Ocean, and of the active transport by jet streams over the North Pacific Ocean. Remember also that rock-derived chlorite and soil-derived kaolinite are carried out into the Southwestern Pacific Ocean by westerlies (Figs. 8.1, 8.2), together with illite and quartz (Figs. 8.4, 7.10).

8.3 Offshore Hydrodynamic Control

Clay associations supplied from land to sea often experience further transportation through marine currents or reworking processes. Let us consider some examples about present-day marine transportation in the proximity of continental sources (see also Chap. 18.2).

8.3.1 Offshore Currents

Despite their high ability to be carried by the water masses, terrigenous clay minerals are fairly seldom employed as markers of currents in shelf and upper slope sediments. This is essentially due to the frequent difficulty in ascertaining the different water origins from the complex clay associations, to the common existence of differential settling processes in shallow and turbulent waters (see Chap. 6), and to the mixing effects between different sources. Several attempts are nevertheless presented in the literature, especially about areas characterized by strong petrographic differences on land, like the Alaskan range (8.2.3; e.g., Hein et al. 1979a, Naidu and Mowatt 1983; Fig. 8.8) or the Northern Gulf of Mexico (6.2.1; e.g., Doyle and Sparks 1980). Other examples concern continental shelves and slopes exposed to active longshore currents, whose intensity obliterates the clay sorting effects. For instance, McMaster et al. (1977) study the clay mineralogy of suspended matter off Sierra Leone and Liberia, equatorial East Atlantic Ocean. The surficial Canary current carries mainly smectite toward the Southeast, the mineral being supplied from West Africa by the Harmattan wind. By contrast, the underlying water masses, below 30 m depth, constitute a counter-current that moves to the Northwest and conveys high amounts of kaolinite issued from the African river runoff. Zimmerman (1972) shows that the green-gray Holocene sediments deposited at the foot of the New England margin, Northeast America, contain fairly abundant smectite, the distribution of which reflects the path of the Western Boundary Undercurrent issued from the Labrador Sea. The late Pleistocene sediments of the same area consist of red-brown silty clay, whose illite-rich assemblage derives from the Permo-Carboniferous red-bed area of the Canadian maritime provinces, and expresses the same southward current as Holocene sediments.

Karlin (1980) considers the different sediment sources, the annual sedimentary budget, and the clay mineral distribution *off the Oregon and North California coasts*, Northwest America. He first shows that the Columbia river, despite its very large drainage area mainly active during the summer, provides moderate amounts of terrigenous material to the ocean, relatively to the south coastal rivers that discharge 80-90% of their sediments during winter. The major silt and clay sources are the Eel river in North California (24.8 10^6t/yr), the Columbia (14.3), the Klamath (10.9), the Rogue (4.7) and the Umpqua (3.2) rivers (Fig. 8.11). The Columbia river is characterized by abundant smectite, mostly supplied from downstream arid zones of the drainage basin (Griggs and Kulm 1970), while the South Oregon and North California rivers provide abundant chlorite and associated illite. In a transect along the coast, R. Karlin determines that chlorite abundance decreases and smectite abundance increases going south to north (Fig. 8.11A, B), while illite displays only a slight northward decrease. The strong seasonal opposition observed in the river discharge between the northern Columbia river and southern coastal rivers, as well as the distinct composition of clay minerals in both river groups, allow to infer dispersal pathways of the terrigenous components of hemipelagic deposits. In winter, discharge of the North California and South Oregon coastal streams is maximum and determines the supply of abundant chlorite to the ocean, that is advected

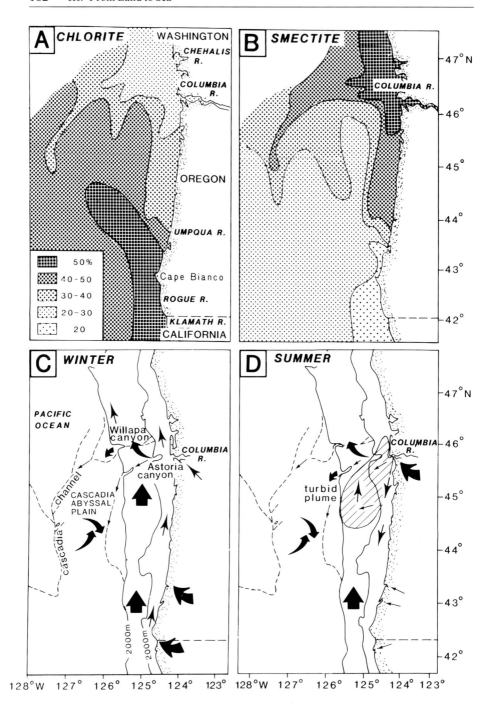

Fig. 8.11. Distribution of chlorite (**A**) and smectite (**B**) in surficial sediments off the Oregon coast. Transport pathways inferred from the clay mineral distribution patterns, in winter (**C**) and summer (**D**). (After Karlin 1980)

northwards along the Oregon slope via the poleward California Undercurrent (Fig. 8.11C). In summer, Columbia river discharge is maximum and determines a smectite-rich turbid plume off the mouth. This plume extends partly south-westwards (Fig. 8.11D), perhaps because of surface currents associated with seasonal upwelling. Another part of smectite-rich suspended matter is funneled into submarine canyons, deep-sea channels (e.g., Willapa and Cascadia channels; Fig. 8.11; see also Griggs and Kulm 1970), and abyssal zones to the North. In winter smectitic materials previously accumulated on the continental shelf off the Columbia river mouth are actively reworked and transported both toward offshore deep-sea channels and to the North on the Washington slope.

8.3.2 Resedimentation Processes

Clays and associated minerals often represent suitable indicators of reworking processes, especially if active differential settling determines the deposition of distinct assemblages in the various sectors of a given area. This is the case of numerous common shelves and slopes like the eastern margin of Canada (e.g., Piper and Slatt 1977; Fig. 8.7), but preferentially characterizes continental margins located off large river mouths, where clay supply, sorting and erosion are often important (see Chaps. 5, 6). Van der Gaast and Jansen (1984) study the shelf of Cabinda-Congo-Gabon and the *deep-sea fan of the Zaire river in Holocene and late Pleistocene times.* The Zaire river supplies mineral assemblages dominated by soil-derived kaolinite, associated with poorly crystallized illite, gibbsite, quartz, and rare poorly crystallized expandible minerals including smectite. This suite is largely distributed both westwards in the deep-sea fan and northwards on the continental shelf during the Holocene (Fig. 8.12A), apparently

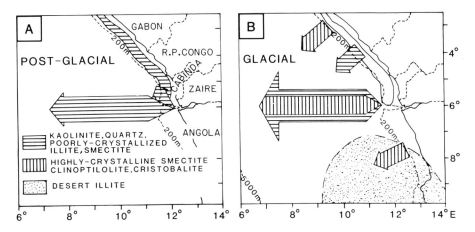

Fig. 8.12. Summary of mineral distribution on the Cabinda-Congo-Gabon shelf and Zaire deep-sea fan during Holocene (**A**) and Late Pleistocene (**B**) times. (After Van der Gaast and Jansen 1984). *Arrows* turbidites

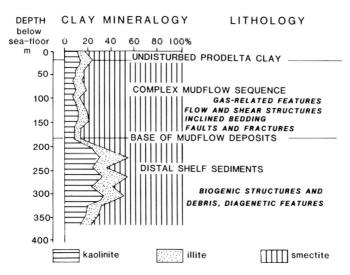

DEPTH   CLAY MINERALOGY   LITHOLOGY
below
sea-floor   0   20   40   60   80 100%
  m   0

Fig. 8.13. Clay mineralogy and lithology on the Mississippi river delta front, from borehole data on south pass area. (After Roberts 1985)

without particular sorting. By contrast, low sea-level stages of the Pleistocene reveal the incorporation in deep sediments of well-crystallized smectite from Tertiary or older material cropping out on the shelf and on the canyon walls (Fig. 8.12B). The submarine erosion and reworking are expressed by the clay mineral composition. Resedimented smectite is often associated with cristobalite and zeolites of a clinoptilolite type, also reworked from the shelf and upper slope. In addition, most stages of low sea level correlate with increased supply of aeolian illite from South African deserts.

Roberts (1985) compares the lithology (X-ray radiographs) and the clay mineralogy in a borehole drilled on the distal shelf of the *Mississippi river delta front*, in the south pass area (Fig. 8.13). The lower part of the section, between 370 and 180 m below the sea floor, contains approximately as much kaolinite plus illite as smectite (50% of the clay fraction for each group), a composition quite normal for the distal shelf sediments of Mississippi river (e.g., Griffin and Parrot 1964). The corresponding deposits display numerous biogenic structures (burrows) and debris, as well as a fine-grained, compact, and homogeneous matrix. These features indicate normal deposition processes, without any noticeable reworking. By contrast, the upper part of the section is characterized by very abundant smectite (about 80% of the clay fraction) and little kaolinite and illite. This composition is usually typical of the shallow delta front sediments, and correlates with sedimentary structures indicative of reworking processes: gas-related features, flow and shear figures, inclined bedding, micro-faults, and fractures. H.H. Roberts interprets the upper series as a complex mudflow sequence reworked in the Mississippi distal shelf area, from the shallower proximal zone. Clay minerals are therefore considered as a useful tool to identify resedimentation processes in submarine deltas, a phenomenon not necessarily reflected by physical properties like undrained shear strength values (Roberts 1986). Similar results arise from a study performed in the submarine delta of the Var river (Southeast France), where late Pleistocene clays enriched in illite and chlorite crop out on the Northwestern Mediterranean continental slope, because of the downward slumping of overlying Holocene clays enriched in smectite (Sage and Chamley 1977).

In the Ganges deep-sea fan, Northeast Indian Ocean, turbidites rich in Himalaya- derived illite and chlorite contrast with calcareous hemipelagites, the latter being characterized by differential settling processes and the progressive southward concentration of Al-Fe and Al smectites (Bouquillon and Chamley 1986). The southward progradation of the deep-sea fan in the course of the time is clearly demonstrated by the more or less regular progression of illite- and chlorite-rich turbiditic sediments above smectitic calcareous and siliceous pelagites of the Central Indian basin, during the Neogene and Quaternary (Bouquillon and Debrabant 1987). Similar deductions are presented by Holmes (1987) about the deep-sea fan developed during the early Cretaceous on the lower continental rise off *North Carolina*, Northwest Atlantic Ocean (DSDP site 603). Illite- or kaolinite-enriched sediments corresponding to sandy input are inserted between smectite-rich hemipelagites. M.A. Holmes suggests that smectite was transported in the deep sea mainly by decantation processes, while illite and kaolinite corresponded to the reworking by turbidity currents of respectively shallow marine and continental deposits.

Fruitful developments may be expected from comparable investigations on unstable areas like accretionary complexes, marked by mud volcanoes, shale diapirs and other sediment mixing and pressure effects (e.g., Barber et al. 1986). Note that *artificial reworking and consequences on the environment* may also be traced by clay mineral assemblages. Roux and Vernier (1977) show that dumping areas located in the Fos bay, Northwestern Mediterranean Sea, are characterized by smectitic sediments very different from the normal illite- and chlorite-rich mud deposited by the nearby Rhone river. The subsequent reworking of dumped sediments can be followed easily through mineralogic analysis. In the German Bight, Southeast of Helgoland, Eastern North Sea, Irion et al. (1987) identify a 3 meter-thick accumulation of muddy, polluted sediments, marked by more kaolinite and less smectite than sediments usually supplied by the Ems, Weser, and Elbe rivers. These sediments appear to have accumulated during the last 100 years because of active dredging and dumping, above a large mud body of 500 km^2 and 21 m thick that is marked by a normal clay composition.

8.4 Combination of Terrigenous Supply and Transportation Agents

Most oceanic basins reflect the existence of various controls on the distribution of terrigenous clays. The climate, land petrography, and near-shore hydrodynamic constraints intervene dominantly in turn, depending on the location of the different parts of a given basin. In addition, the long-distance transportation processes by marine currents and by the wind may strongly

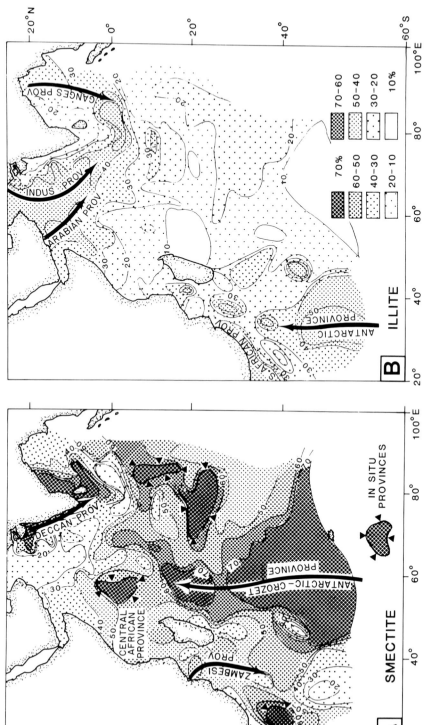

Fig. 8.14. Distribution of smectite (**A**) and illite (**B**) in the Western Indian Ocean, and inferred sources. (After Kolla et al. 1976). Percentages from X-ray diffractograms peak area

modify the distribution of detrital assemblages with respect to the terrestrial production zone. Finally, some in-situ formations of clay and associated species occur in certain basins (see Part Four), and complicate the final distribution of mineral suites on the sea floor. A lot of work has still to be done to understand the ultimate causes of the actual repartition of fine-grained silicates that constitute a large part of modern and past marine sediments. Some useful attempts have nevertheless been performed. This is especially the case of the *Western Indian Ocean*, actively studied by V. Kolla and his colleagues.

The clay mineral distribution in the Western Indian Ocean is largely influenced by the climate and the land geology, but is also controlled by physiographic patterns, submarine volcanism, and marine currents (Kolla et al. 1976). These combined constraints determine a severe alteration of the classical latitudinal distribution of clay minerals (see 8.1), which is clearly demonstrated by the repartition of the two dominant species, smectite and illite (Figs. 8.14A, B). Some areas, devoid of important sediment transport and mainly located in the central ocean, contain abundant smectite, essentially attributed to the in-situ halmyrolysis of submarine basalts. A second group of areas reflects the dominant control of continental climate and petrography, whose effects are inflected by several modes of long-distance transportation (advection): (1) Smectite-rich sediments in the Crozet and Madagascar basins. (2) Illite-rich deposits in the Indus deep-sea fan, the Agulhas basin, and the zone West of Agulhas plateau in the Southwest. (3) Illite- and palygorskite-rich areas in the westernmost and northernmost Arabian Sea (see Fig. 8.5). (4) Illite- and kaolinite-rich sediments around Madagascar and adjacent to central Africa. All these areas depend on the action of either Antarctic Bottom Water (AABW) movements, turbidity currents, or aeolian input (Fig. 8.14). The AABW appears to be responsible for the northward advection of dominantly illitic clays close to 35° E, and of predominantly smectitic clays around 60-65° E. The illitic suite is supposed to be eroded from the Antarctic continental margin, and the smectitic one from southern volcanic areas. The boundaries between the different mineral provinces are very progressive, as in all oceanic basins, because of the large dispersion properties of most clay particles.

A more detailed study in the *Arabian Sea* leads Kolla et al. (1981) to better identify the role of climate, petrographic sources, and currents on the distribution of clay minerals and quartz. Smectite-rich clays occur along the Indian margin South of the Indus river; they result from the climate weathering of Deccan basalt traps, are associated with quartz issued from local Precambrian metamorphic rocks, and appear to be primarily dispersed southerly, and to some extent northerly, by surface currents. Illite-rich clays dominate in most of the rest of the Arabian Sea, but are associated with various other minerals according to the sectors considered: little abundant chlorite in the Indus river deep-sea fan; abundant chlorite off Iran-Makran coast; palygorskite and chlorite South of Arabia; palygorskite and smectite East of Somalia (Fig. 8.15). The transportation agents appear to consist of surface currents and turbidity currents off the Indus river area, and mainly of winds in other areas. The effect of arid to desert climate in the northern and western areas favors the dominant control of land petrography on detrital associations (8.2) as well as the essential action of wind,

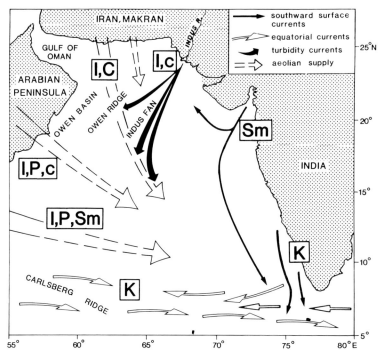

Fig. 8.15. Main environmental controls of the clay mineral distribution in surface sediments of Arabian Sea (interpretation after Kolla et al. 1981). *C* chlorite; *I* illite; *Sm* smectite; *K* kaolinite; *P* palygorskite

that probably carries palygorskite and other minerals from the Arabia-Somalia region to the Indian margin and even farther eastwards (e.g., Bouquillon and Debrabant 1987). In the southernmost area of the region, kaolinite-rich sediments probably derive from intertropical soils of Africa, Madagascar, and South India, and are mainly distributed latitudinally by north equatorial surficial currents.

Another study performed in the *Mozambique fan and adjacent areas* shows the large importance of reworking by turbidity currents, combined with the effects of petrographic sources, climate weathering, surface and bottom circulation (Kolla et al. 1980a; Table 8.2; Fig. 8.16). Four mineral provinces are identified. (1) The Zambesi province corresponds to smectite and kaolinite association in the fine sedimentary fraction, and to a hornblende and garnet association in the sand fraction. It results from the erosion of tropical soils issued from severe weathering developed at the expense of East African Precambrian metamorphic complexes, volcanics, and Paleozoic to Cenozoic sediments. This assemblage has been transported mainly by turbidity currents via the Zambesi canyon and numerous channels within the Mozambique basin since the Pleistocene. (2) The Limpopo river is responsible for an illite-rich marine province, issued from the erosion of South African substrates submitted to a drier climate than the North. The corresponding sediments are mainly distributed by surface oceanic currents.

Table 8.2. Major land and sea controls on the distribution of mineral provinces in the Mozambique fan and adjacent areas (interpretation after Kolla et al. 1980a)

Mineralogical provinces	Land petrography	Marine volcanics	Climate, soils	Surface currents	Bottom currents	Turbidity currents
Zambesi	+		+ +			+ +
Limpopo	+ +		+	+ +		
Madagascar	+		+ +	+ +		
Southern	+ +	+			+ +	

+ + Dominant control, + secondary control.

(3) Madagascar province, located on the ridge area South of the large island, is characterized by abundant kaolinite and smectite, that are produced in Madagascar soils and probably distributed by surface currents. (4) The southern province sediments comprise abundant illite and smectite, fairly abundant chlorite, that derive from high-latitude terrigenous sources, as well as perhaps from submarine volcanic sources. These clay minerals are advected over long distances by the Antarctic Bottom Water along the western margin of the Mozambique basin, as documented by the existence of current lineations and transverse bedforms on the sea floor.

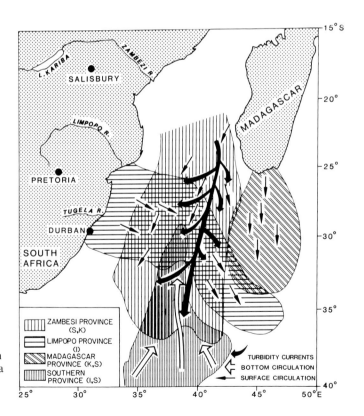

Fig. 8.16. Mineral provinces and modes of sediment dispersal in the Mozambique fan and adjacent areas, Southwestern Indian Ocean. (After Kolla et al. 1980a). *I* illite; *S* smectite; *K* kaolinite

8.5 Conclusion

1. *Terrigenous supply represents the dominant agent responsible for the constitution of clay suites in most recent sediments of the world ocean.* A long way has been covered since the first debate that opposed Correns's ideas (1939) on the major land control and Dietz's opinion (1941) on the major sea control. The debate is not concluded, but is less essential and has shifted. Most real questions arising today concern the precise nature and relative importance of authigenic processes in surficial sediments at the boundaries of the wide terrigenous clay provinces, as well as the extension of detrital influence in past sedimentary series. The main steps that led to the modern conceptions of the marine clay origin are the global maps on mineral distribution drawn by Biscaye (1965), Griffin et al. (1968) and Rateev et al. (1968), in the findings of the absence of isotopic and chemical equilibrium of marine clays with the oceanic environment (e.g. Harriss 1967), in the datings pointing to the correspondence existing between the age of most marine clays and the average age of rocks cropping out on the adjacent continent, and in the evidence of the very high particulate terrigenous flux to the ocean relatively to the dissolved flux (e.g., Lisitzin 1972). Countless papers published over the last decades testify to the correlation observed in the distribution of continental soil and marine sediment clays from the same climatic zones, to the importance and diversity of the clay influx by rivers (3.3), to the absence or scarceness of clay mineralogic changes at the land / sea transition, to the essentially physical character of sorting and settling processes during marine transportation, and to the very frequent similarity of fine-grained mineral composition in soils, river sediments and seston, aerosols, marine-suspended matter and surface sediments of a given region (Chaps. 5 and 6). The geographic limits of the marine zones unquestionably characterized by the dominant influence of terrigenous input have been extended in several basins, like the North Pacific Ocean essentially controlled by aeolian supply (e.g., Leinen and Heath 1981, Blank et al. 1985), or the Eastern Indian Ocean controlled by Ganges river supply at least as far as South of Equator (Bouquillon and Debrabant 1987). The silicate sedimentation in the South Pacific Ocean itself depends partly on aeolian supply from Australia and South America, as well as probably on bottom water supply from Antarctica regions.

2. *Illite, chlorite, associated quartz, feldspars and various dense minerals, commonly called "primary minerals", have long been considered typically terrigenous species*, and most discussions about a possible "deep-sea illitization of smectite" (Dietz 1941) appear to be definitively closed. Less common sheet silicates like talc, pyrophyllite, and serpentine also proceed essentially from the erosion of continental rocks. In addition almost all authors agree to consider kaolinite, random mixed-layers and vermiculitic minerals as characteristic products of chemical weathering and pedogenesis developed on exposed landmasses.

A much-debated question during the last decades concerns the origin of marine smectites and fibrous clays (palygorskite, sepiolite). *Smectites in marine sediments do not systematically and easily form in marine environment, and do not automatically result from the alteration of volcanic rocks.* Smectites are supplied to the sea by rivers and coastal runoff from lowest to highest latitudes, and can issue from various soils, weathering complexes, sedimentary rocks, plutonic and volcanic rocks. Smectites occur in fluvial deposits as different as those from the Congo, Mississippi, Nile and Yukon rivers, and are widely distributed in various basins such as the Equatorial Atlantic and the Arctic Ocean. Smectites display varying chemical composition according to climatic weathering conditions and parent-rock nature (Chap. 2), which considerably complicates the characterization of their precise origin. Smectites may form from alteration of volcanic rocks in both subaerial and submarine environments, the former favoring a large clay production, and the subsequent active erosion by running water or wind (e.g., smectites issued from Deccan basalts or Southern Ocean volcanic archipelagoes). Smectites do not suffer from river and marine transportation, and are often carried in the marine environment farther than other minerals because of their high buoyancy (Chap. 5). Both nonvolcanogenic and volcanogenic smectites derived from terrigenous sources add to smectitic minerals originating in the marine environment (Part Four), from which they can be distinguished only by detailed geochemical and micromorphologic investigations.

Fibrous clays, especially palygorskite, tolerate transport constraints much better than classically believed, and can constitute detrital components of various marine sediments. This is the case in coastal muds (e.g., Southwestern Mediterranean, Persian Gulf), turbidites (Gulf of Oman, Eastern Atlantic), hemipelagites (East Atlantic, Mediterranean) and pelagites (Arabian Sea). The high vulnerability of fibrous clays to erosion and transportation is largely a myth, determined by the relatively rare occurrence of these minerals at the Earth's surface, by the difficulty in identifying them in small amounts by X-ray diffraction techniques, and by the frequent necessity of checking their presence by electron-microscope observations. Palygorskite bundles survive reasonably both aeolian and subaqueous transport. Like smectites, fibrous clays set a problem of boundary between detrital and authigenic sources in marine environments (Chaps. 9 and 13).

3. *The basic zonation of terrigenous clays in the ocean is controlled by the climate.* Kaolinite forms mainly by chemical weathering in soils of humid low-latitude regions, while chlorite and illite mostly derive from the physical weathering of crystalline rocks that crop out widely in high-latitude and desert regions. Irregular mixed-layers, vermiculitic minerals and poorly crystallized smectite mainly characterize temperate regions, while well-crystallized Al-Fe smectites largely form in soils of warm-arid and poorly drained regions (Chap. 2). This soil- and climate-driven zonation is particularly well reflected by Atlantic Ocean sediments, which indicates the accessory influence of North-South currents on the clay distribution relatively to latitudinal influences (compare, for instance Figs. 2.7 and 8.1). Such a zonation has to been kept in mind when attempting to reconstruct the past climatic evolution from the study of terrigenous series (Chap. 17).

4. Recent studies show that the latitudinal zonation of marine clays allows many exceptions, which demonstrates that *climate constitutes only one of the factors responsible for the distribution of terrigenous minerals.* The clearness of the Atlantic model led to exaggerating the importance of the climatic control in the world ocean. *The average petrography of source regions becomes essential as soon*

as chemical weathering is very low, which determines the theoretical erosion and deposition in the sea of any minerals occurring in exposed rocks. This explains why kaolinite, smectite, and other so-called low-latitude minerals may be supplied in noticeable amounts within Arctic or Peri-Antarctic Seas, through the erosion of ancient sediments and soils where they have been stored during past hydrolyzing periods. The same reason explains the active supply of paleosol-derived kaolinite in marine basins located off present desert or arid regions (e.g., Saharan range; Chap. 7).

The transportation by nearshore surface or density currents also modifies the original clay zonation induced by climate (see also Chap. 18). Finally *long-distance advection by marine or aeolian currents* may severely alter the climatic zonation, as shown for instance in the Indian and North Pacific Oceans. The Western Indian Ocean represents a convincing example of the combined control of climate, petrography, hydrodynamics and aerodynamics on the large-scale distribution of deep-sea clay associations. Note that the different clay and associate mineral species may equally reflect these different influences. For instance, the Antarctic Bottom Water is traced chiefly by illite in the westernmost South Indian Ocean and by smectite toward the East, while aeolian trajectories in the North are identified by illite and either chlorite, palygorskite, or smectite. A large field of investigations is open to better understand how present-day environmental factors combine with each other to determine the distribution of clays in marine basins. Such a task may greatly help to identify the different factors responsible for the characteristics of both continental and marine past environments.

5. *Terrigenous supply is sometimes complemented or relieved by the autochthonous formation of clay minerals and other species,* especially in marine regions marked by active volcanic and hydrothermal activity, and by very low continental influx. The smectite- and oxide-rich sediments of the South Pacific and South Indian Oceans appear to represent the more suitable places for favoring mineral authigenesis (Fig. 8.6). The characterization of in- situ processes often involves a more complex methodologic approach than that of allochthonous processes, namely because newly formed minerals are not necessarily very different from those originating on and close to land masses. In addition, the regions where transitional phenomena occur are often still little investigated in detail. The autochthonous formation of sheet silicates in the marine environment is considered in Part Four and, when possible, replaced relatively to allochthonous processes. Note that the increased effort presently paid to quantify the flux variations of each different mineral of a given sediment will probably greatly help in distinguishing the terrigenous and in-situ controls on present and past marine clay sedimentation.

Part IV

Clay Genesis in the Sea

Chapter 9

Alkaline, Evaporative Environment

9.1 General Features

Alkaline and evaporative sedimentation usually corresponds to lacustrine or restricted marine environments, characterized by the active formation of carbonate or saline minerals. In Quaternary and present-day sediments, these environments display mainly detrital clay suites, as documented by most lakes, coastal marshes, lagoons, and basins (4.4.1; Chaps. 5, 8). Only few, sometimes questionable examples of in-situ clay formation are reported in modern nonvolcanic saline lakes and deltaic areas (4.4.2, 5.3.4), and no apparently convincing cases arise from studies on recent evaporative marine environments.

The present-day situation strongly differs from that reported about several geologic periods. The sedimentary genesis of clay minerals has been classically described in evaporative formations of Permo-Triassic and Paleogene from Western Europe and Northwestern Africa, and is extensively summarized and discussed by Millot (1964, 1970). Authigenic minerals quoted include mostly palygorskite, sepiolite, Al-Mg smectite, chlorite and corrensite, and sometimes illite, talc, and pyrophyllite. As clay genesis results from similar mechanisms under marine and nonmarine conditions, both evaporative environments are considered together (see also 4.2.3).

Permo-Triassic evaporative sediments were first studied by Millot (1949) in the Paris basin, by Jeannette and Lucas (1955) in Morocco, by Füchtbauer and Goldschmidt (1959) in the German Zechstein, and are investigated in detail by Lucas (1962) in France, Spain, and Morocco. J. Lucas observes a correlation between the different clay associations identified and the lithologic, geographic, and stratigraphic distribution of sediments. This is especially well expressed in the Triassic series of Jura, Northeastern France, where a transition from poorly crystallized illite, kaolinite and random mixed-layers into corrensite and well-crystallized chlorite and illite, is recorded both from the borders to the center of the basin and from basal detrital sandstones to superimposed saline sediments. This evolution is attributed to an in-situ *transformation (aggradation) of continentally-weathered illite and random mixed-layers into either regular chlorite-smectite (= corrensite) and then chlorite, or well-crystallized illite* (see 20.4). Palygorskite and sepiolite occur less extensively, and are considered as neoformation products. Additional or subsequent studies on Permo-Triassic sediments and associated magnesian authigenic clays concern especially the Western USA (e.g., Grim et al. 1960, Bodine 1978), Great Britain (e.g., Jeans 1978a) and

Northeastern France (e.g., Lucas and Ataman 1968, Geisler-Cussey and Moretto 1984). As most data and interpretations outstandingly summarized by Millot (1964, 1970) are still valid and roughly apply to more recent studies, we refer the reader to this textbook. An additional discussion about the significance of chloritic minerals developed in evaporative formations and submitted to diagenetic effects is presented in Chapter 20.

Early Tertiary alkaline and saline sediments display numerous examples of the genesis of clays that mainly consist of smectite, palygorskite, and sepiolite. Lacustrine alkaline clays have been known in Western Europe at least since 1822 (Brongniart: "magnesite" = sepiolite), and are listed by Millot (1949). By contrast, marine alkaline clays were probably first described only in 1954 (Capdecomme and Kulbicki, Senegal). G. Millot and several colleagues have abundantly investigated the basin of Paris and peri-continental basins around Western Africa from Morocco to Zaire (in Millot 1964, 1970), whose deposits include abundant limestones, dolomites, cherts, phosphates, glauconites, and clays. *A sequence characterized by the successive occurrence of kaolinite, smectite (montmorillonite), palygorskite, and sepiolite develops from coastal to basin facies and from regressive to transgressive environments*, especially during the early Eocene. G. Millot correlates this mineral sequence with a chemical sequence marked by less aluminous and more magnesian minerals when going from land toward the marine basin, which corresponds to a more and more chemically confined environment. These peculiar deposits are referred to as *"alkaline chemical sedimentation"*, permitted by strong hydrolysis processes in the hinterland, concentration of alkaline ions in the basin itself, and local tectonic quiescence preventing strong detrital input. Millot (1964, 1970) attributes the genesis of smectite and fibrous clays to a true mineral neoformation, developed from dissolved chemical elements, progressively accumulated and concentrated in marine or lacustrine water. More recent data allow to better know the conditions of genesis of fibrous clays and associated minerals in Tertiary alkaline environments (e.g., Trauth 1977, Weaver and Beck 1977, Singer and Galan 1984, Pozzuoli 1985, Galan et al. 1987). These data and inferred interpretations are summarized below (9.2).

An overview of the age, distribution, and depositional environment of palygorskite-sepiolite clays is presented by Callen (1984). The fibrous clays originating in soils,

Table 9.1. Major palygorskite-sepiolite-rich periods. (After Callen 1984). Occurrences reported on land include soil, lake and peri-marine environments

Oceans	Continents
Plio-Pleistocene	*Plio-Pleistocene* (mainly soils)
Middle Miocene – Late Oligocene	Middle Miocene (mainly) – Late Oligocene
Eocene (mainly early and middle)	Eocene (mainly)
Late Cretaceous (mainly Campanian, Albian)	Late Cretaceous
	Triassic, Late Permian, Carboniferous, Late Devonian, Cambrian

lakes, or very shallow seas appear equally to depend on semi-arid and warm climate. From available data, such conditions occurred especially during the late Devonian and Carboniferous and late Permian to Triassic in the Northern Hemisphere, and in the late Cretaceous, early and late Eocene, late Oligocene, and late Neogene in both hemispheres (Table 9.1). There is no obvious or systematic correlation between volcanic effusive phases, regression-transgression cycles, salinity and palygorskite events. R.A. Callen reports that pre-Mesozoic occurrences are essentially concentrated within 30° of the Equator in shallow landlocked seas of central and western Russia; the most extensive deposits developed from Devonian to Permian times, with associated platform dolomites and other carbonates, and could partly be related to Paleozoic volcanism. The first major widespread non marine deposits are early Eocene in age, and correlate with extensive peri-marine deposits. Evaporative sediments rich in palygorskite and sepiolite strongly decreased after the middle Miocene, the latest being lacustrine (4.2.3). Most Plio-Pleistocene fibrous clays formed in pedologic environments, especially in semi-arid regions characterized by calcareous crusts (2.2.1, 2.2.3). Note that alkaline and non evaporative sediments, such as the carbonate nodules precipitated in late Cenozoic deep-sea sediments of the Eastern Mediterranean Sea and Japan Trench (Milliman and Müller 1973, Debrabant and Chamley 1982a, Wada and Okada 1983), are lacking any clay formation.

9.2 Evaporative Clay Sedimentation, Paleogene in France

9.2.1 Basin of Paris

During the early Tertiary, the Paris Basin, Northwestern France, was submitted to an alternately marine, brackish, and continental sedimentation, under shallow-water conditions and subarid-hot climate (Pomerol 1967). The basin was protected from strong terrigenous input by flat relief and great distance from alpine orogenic areas. Sediments mainly consist of fine-grained marls, clays, limestones, dolomites and gypsum. Four main clay mineral zones follow one another from late Paleocene (Thanetian) to Oligocene (Stampian) (Trauth et al. 1969):
 – a lower aluminous zone with kaolinite, Al-smectite, and kaolinite-smectite mixed-layers;
 – a ferriferous zone marked by Fe-smectite and glauconite;
 – a magnesian zone characterized by the presence of palygorskite (attapulgite), sepiolite, illite, and smectite;
 – an upper aluminous zone with kaolinite, illite, and smectite.
 A detailed study has been performed by Trauth (1977) in the *magnesian zone developed during late Eocene* (Ludian episode of Bartonian stage), where massive sulfate deposits (Trois Masses Formation) are vertically and laterally associated

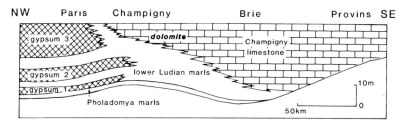

Fig. 9.1. Schematic section of Late Eocene deposits in the central basin of Paris. (After Mégnien 1974; Trauth 1977)

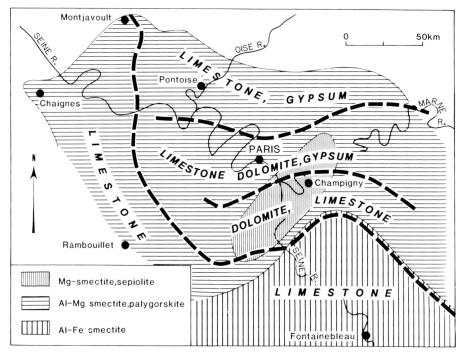

Fig. 9.2. Geographic distribution of major rock and clay mineral types in the Trois Masses Formation, Late Eocene, Paris basin. (After Trauth 1977)

with various carbonate and marl deposits (Fig. 9.1). From the periphery (SE) to the central part (NW) of the sedimentation area, the petrography progressively passes from almost exclusive limestone to dolomitic limestone, then dolomite, gypsum, and limestone, and finally very abundant gypsum forming three bodies (= "Trois Masses") with limestone (Fig. 9.2). Marls are interbedded in various amounts within carbonate and sulfate deposits, especially in the central part of the sedimentation area.

N. Trauth identifies illite, kaolinite, and different smectitic and fibrous clays in these various rocks. Illite and kaolinite usually represent minor minerals.

Table 9.2. $MgO/Al_2O_3 + Fe_2O_3$ ratio (mol) of major clay minerals of the Trois Masses Formation, late Eocene of the Paris basin. (After Trauth 1977)

Mineral	$\dfrac{MgO}{Al_2O_3 + Fe_2O_3}$
Wyoming smectite (Al, Fe)	0.13
Cheto smectite (Al, Mg, Fe)	0.75–0.87
Palygorskite (Al, Mg, Fe)	1.60–1.92
Mg-smectite (Mg, Al, Fe)	3.64–4.04
Sepiolite (Mg, Al, Fe)	4.67–5.79

Smectites comprise three dominant varieties determined by differential thermal analysis (Table 9.2): Al-Fe smectite of wyoming type (Al-Fe beidellite) characterized by the dominance of iron over magnesium in octahedra; Al-Mg smectite of cheto type (beidellite), whose octahedra contain more magnesium than iron; Mg-smectite with very abundant magnesium in octahedra. Fibrous clays include both palygorskite and sepiolite, the former species containing less magnesium than the latter and also than Mg-smectites (see Chap. 1). The geographic distribution of smectites and fibrous clays correlates partly with the petrographic zonation (Fig. 9.2). Al-Fe smectite predominates in limestones deposited on the southern border of the sedimentation area, Mg-smectite occurs mainly at the boundary of dolomitic and gypsiferous zones, and Al-Mg smectite and palygorskite characterize all other areas to the East, North, and West. The most obvious correspondence concerns magnesian minerals and rocks (Mg-clays, dolomite), the maximum amounts of which occur in the same area, South of the city of Paris.

The clay mineral distribution in late Eocene deposits of the Paris basin appears to express different modalities of the *transition from detrital to chemical processes* (Trauth 1977).

1. The Al-Fe smectite-rich and fossiliferous limestones in the southern part of the basin represent the combination of *terrigenous input* and organic activity in a mainly lacustrine environment. Al-Fe smectite, associated with little kaolinite and illite, proceeds from the erosion of soils developed in flat areas located South of the basin and submitted to warm-humid climate (Thiry and Trauth 1976b, Thiry 1981).

2. Palygorskite- and Al-Mg smectite-rich sediments that occupy the largest part of the basin correspond to intermediate environments submitted to both terrigenous and evaporative conditions. Al-Fe smectite and other detrital minerals comprise chemical elements like aluminum, silicon and titanium that combine with dissolved elements (Mg, Si), which allows their *transformation* into palygorskite and cheto-type smectite. Note that the direct terrigenous signal is

partly preserved, as shown by the frequent presence of illite and kaolinite (especially in the northern part of the basin), and that saline minerals (gypsum) may massively precipitate in these transitional environments (Fig. 9.2).

3. The sepiolite- and Mg-smectite-rich dolomitic limestones and gypsum indicate typical azoic and chemical environments, where both clay minerals and host rocks constitute true *neoformation* products (Fontes et al. 1970, Trauth 1977). Mg-clays form under strong evaporative conditions, as do gypsum, limestone, and probably dolomite, the chemistry of the brines being dominated in turn by different alkaline elements (Ca, Mg).

9.2.2 Southeastern France

During the Paleogene, several small lacustrine basins developed under alkaline conditions in western Provence and eastern Languedoc, at the periphery of the Rhone valley. These basins display relatively simple and complete examples of chemical clay sedimentation, and therefore contribute to the understanding of larger peri-marine evaporative basins, whose environmental control on clay genesis is mostly comparable and not dependent on salinity.

The basin of Mormoiron, in western Provence, experienced between Turonian and Oligocene times a continental sedimentation, characterized by two successive cycles (Triat and Trauth 1972, Trauth 1977). The first cycle (= lower unit, Table 9.3) comprises successive conglomeratic sand and clay, overlaid by sandy claystone and siliceous limestone. These formations are devoid of fossils and respectively marked by the occurrence of kaolinite-illite-wyoming smectite, wyoming smectite-palygorskite, and almost exclusive palygorskite. The corresponding lithologic and mineralogic suite is attributed to the progressive passage from detrital to moderately chemical conditions, the upper palygorskite-rich levels resulting from the partial transformation of terrigenous minerals and subsequent short-distance transportation by currents. The second cycle (= upper unit, Table 9.3) displays successive sandy marl, calcareous claystone, limestone, dolomite, and gypsum; it corresponds to the transition from typically detrital to strongly chemical conditions, which is traced by the progressive replacement of Al-Fe smectite (wyoming) by Al-Mg smectite (cheto), sepiolite and true Mg-smectite (stevensite, saponite) (Fig. 9.3). The structural formulas of smectites identified in successive lithologic formations clearly reflect the development of alkaline conditions that disfavored Al-Fe species and favored magnesian minerals and sediments (Table 9.4). As in the basin of Paris, cheto smectite and palygorskite are considered as transformation products of pre-existing detrital clays, while sepiolite, stevensite, and saponite are referred to true neoformation minerals, crystallized from ionic solutions. Mg- clays formed in restricted, strongly evaporative water ponds, filled by periodic flooding and submitted to hot and subarid climate. The lake completely desiccated sometimes, as demonstrated by the presence of sun cracks, prints of roots, and vertebrate feet. Some erosion processes occurred in the chemical basin in relation with strong rainfall episodes, and were responsible for the local resedimentation of magnesian clays and other materials.

Table 9.3. Mineral and chemical composition of Post-Cenomanian and Pre-Miocene deposits of the Mormoiron basin, Southeastern France. (After Triat and Trauth 1972; Trauth 1977)

| | Formation (thickness, m) | Lithology | Clay minerals | | Main Chemical Elements | | |
| | | | Major | Minor | Bulk material | | Clay fraction |
					Major elements	Trace elements	
Upper unit	Blauvac evaporitic complex (30–40)	Gypsum Dolomite with celestite Limestone	Stevensite Sepiolite Al-saponite Al–Mg smectite (cheto)	Palygorskite Illite Chlorite	Ca, Mg C, S (Si)	Sr	Mg, Al (Fe)
	Mormoiron green clay detrital complex (40–50)	Calcareous claystone Sandy marl	Al–Fe smectite (wyoming)	Illite	Al, Fe, Mg Si C	V, Cr, Zn, Ga, Ti, B, K	Al, Fe (Mg)
Lower unit	Jocas limestone (6–8)	Siliceous limestone	Palygorskite		Ca C (Si)		Al, Mg
	Terre à foulon (5–25)	Sandy claystone	Palygorskite Wyoming smectite	Illite Chlorite	Al, Fe, Mg Si	V, Ni, Ca, Cr, Zn, Ga, Ti	Al, Fe (Mg)
	Heterogeneous complex (reworked) (1–10)	Black claystone Clayey sand Conglomerate	Wyoming smectite Kaolinite	Illite	Al, Fe Si		Al, Fe

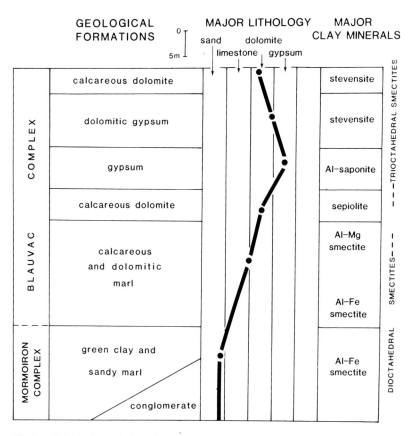

Fig. 9.3. Lithologic and mineral sequences in the upper unit of Mormoiron basin, Southeastern France. (After Trauth 1977)

The basin of Sommières, northeastern Languedoc, comprises a well-known sepiolite deposit in Oligocene lacustrine formations (Stampian s.l.), that are industrially exploited in Salinelles. Sepiolite occurs in carbonate deposits interbedded between coarse clastic sediments. Close correlations exist between lithologic and mineral successions (Trauth 1977; Fig. 9.4). The coarse detrital facies, that are probably related to tectonic instability in the hinterland, contain illite, Al-Fe smectite (wyoming) and little kaolinite. The interbedded carbonate formations include various amounts of Al-Mg smectite (cheto), Al-saponite, palygorskite, stevensite (Mg-smectite), and sepiolite, with frequent illite associated. The more magnesian the carbonates, the more abundant are sepiolite and stevensite in the clay fraction. Sepiolite is concentrated in argillaceous horizons, while stevensite accumulates rather in dolomitic beds. N. Trauth estimates that sepiolite precipitated by neoformation in shallow lacustrine ponds submitted to alternate flooding-desiccation cycles, and was periodically eroded and resedimented as thin argillaceous lenses.

Table 9.4. Smectite chemistry in the upper unit of Paleogene deposits, Mormoiron basin. (After Triat and Trauth 1972; Trauth 1977)

Genetic Process	Mineral	Structural formula	$\dfrac{MgO}{Al_2O_3+Fe_2O_3}$	Li (ppm)
Neofor-mation	Stevensite	$(Si_{3.97}Al_{0.03})(Al_{0.25}Fe_{0.07}Mg_{2.29}Li_{0.29})$ $Ca_{0.08}K_{0.06}Na_{0.08}$	4.68	3000
	Al-saponite	$(Si_{3.79}Al_{0.21})(Ti_{0.02}Al_{0.78}Fe_{0.24}Mg_{1.35})$ $Ca_{0.12}K_{0.13}$	0.77	640
Transfor-mation	Al-Mg Smectite (cheto)	$(Si_{4.03})(Ti_{0.02}Al_{1.16}Fe_{0.24}Mg_{0.60})$ $Ca_{0.12}K_{0.16}$	0.30	300
Detrital supply	Al-Fe smectite (wyoming)	$(Si_{3.92}Al_{0.08})(Ti_{0.03}Al_{1.21}Fe_{0.40}Mg_{0.30})$ $Ca_{0.16}K_{0.21}$	0.10	230

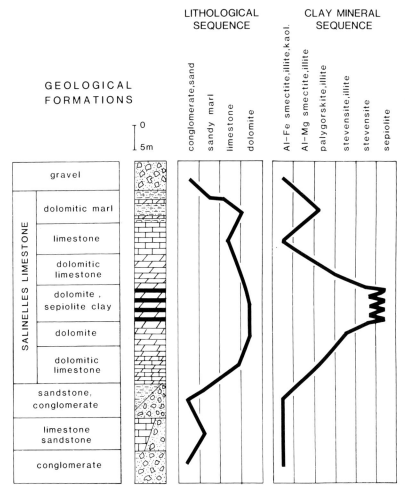

Fig. 9.4. Lithologic and mineral sequences in Stampian formations of Salinelles, Sommières basin, Southeastern France. (After Trauth 1977)

9.2.3 Geochemical Implications

The different cases investigated in the basin of Paris and in Southeastern France show that *early Tertiary peri-marine and lacustrine conditions gave way to similar lithologic, mineralogic and geochemical successions*, that occurred either laterally or vertically or both. *Nearshore facies are strongly influenced by terrigenous input*, and contain soil-derived Al-Fe smectite (Al-Fe beidellite), often associated with illite, kaolinite, and other detrital minerals. *Transitional facies correspond to interactions between detrital minerals and ionic solutions*, determining *transformation* processes and the development of Al-Mg smectite (cheto smectite, Al-saponite) and palygorskite. *Central basin facies represent true chemical environments* where minerals precipitated directly by neoformation from ionic solutions. The resulting minerals are globally called *evaporites* by Trauth (1977). They include sulfates (gypsum, celestite), carbonates (calcite, dolomite) and silicates (stevensite, sepiolite). The crystallization mechanism of Mg-clays is still little known; experiments done by Decarreau (1981) suggest it could result from the aging of silico-metalliferous coprecipitates in warm and ion-concentrated water.

Al-Mg and Mg clays evolve through chemical processes toward either sheet minerals or fibrous minerals, both suites being mutually exclusive (Fig. 9.5). The type of clay suite is conditioned by the Si/Mg ratio in ionic solutions. If this ratio is low, sheet minerals of the smectite group preferentially crystallize, while fibrous clay formation is favored by high ratios. The neoformation of evaporitic clays determines a *relative augmentation of argillaceous sediments*. It is accompanied by subtraction of silicon, magnesium, lithium and sodium from ionic solutions. Such a subtraction leads to a *chemical purification* of water and favors the precipitation of gypsiferous deposits.

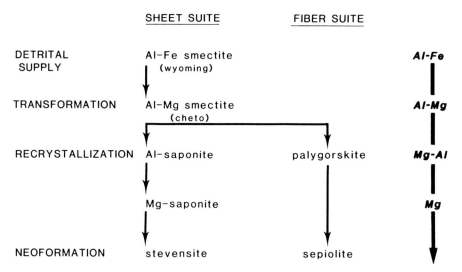

Fig. 9.5. Clay mineral sequences in alkaline-evaporative shallow-water environments, from data on Paleogene of France. (After Trauth 1977)

9.3 Other Marine Environments

9.3.1 Early Miocene, Southeastern United States

One of the most conspicuous examples of *peri-marine alkaline and evaporative clay sedimentation* is the early Miocene of Southeastern North America (Weaver and Beck 1977, Weaver 1984). Sediments deposited in a fairly stable hinge area separating the Atlantic Ocean and the Gulf of Mexico, and presently located in the North Florida-South Georgia region. During the late Oligocene, the sea transgressed over this formerly exposed area marked by an eroded karst surface, and tidal to brackish sediments formed. The general transgression continued during much of the early Miocene, and was followed by a regression culminating in the development of an extensive soil and of reworked sediments near the beginning of the Middle Miocene. While Al-Fe smectite is the dominant clay mineral in most Tertiary deposits of this region, palygorskite and sepiolite formed abundantly and specifically during the early Miocene in *brackish lagoon and tidal environments*. The distribution of palygorskite-rich shallow-water sediments extends over a distance exceeding 400 kilometers from the Northeastern Gulf of Mexico to the western margin of tropical North Atlantic Ocean. The climate was hot and responsible for drastic evaporation. The formation of fibrous clays ceased at the end of the early Miocene, when cooler climate conditions disfavored strong evaporation. Palygorskite and sepiolite were then reworked by erosion, and accumulated northeastwards in great amounts during the middle Miocene, within an adjacent marine trough where they constitute commercial clay deposits.

A nine meters core from the La Camelia palygorskite mine, North Florida, shows the presence of two minor cycles of transgression and regression, successively marked by marine, tidal, and lagoonal environments (Fig. 9.6). The lower cycle is more complete and ends with supratidal and soil materials. Palygorskite represents the predominant clay in the section, with smectite second in abundance, followed by sepiolite and illite. *The clay mineral suite is closely related to lithology.* Open-sea marine sand and continental soil contain very abundant smectite of a wyoming type (Al-Fe smectite; Table 9.5), which suggests a common origin, i.e., smectite could proceed from the erosion of early Miocene smectite-rich soils. *Palygorskite abundance increases parallel to the development of the regression*, and becomes very high in tidal, lagoonal, and supratidal deposits. Sepiolite occurs in the lower cycle only, and reaches maximum concentration at the bottom of the soil zone. In general, the Al_2O_3 and MgO contents in the clay fraction are inversely related and express the antagonistic variations of smectite

Table 9.5. Structural formulas of marine and soil smectites, early Miocene of North Florida, USA. (After Weaver 1984). Compare with Table 9.4, Al-Fe smectite of lacustrine Paleocene from SE France

Marine sediment	$(Si_{3.89}Al_{0.11})$	$(Al_{1.46}Fe_{0.28}Mg_{0.28})$	+	exchangeable cations
Soil	$(Si_{3.97}Al_{0.03})$	$(Al_{1.51}Fe_{0.25}Mg_{0.22})$	+	exchangeable cations

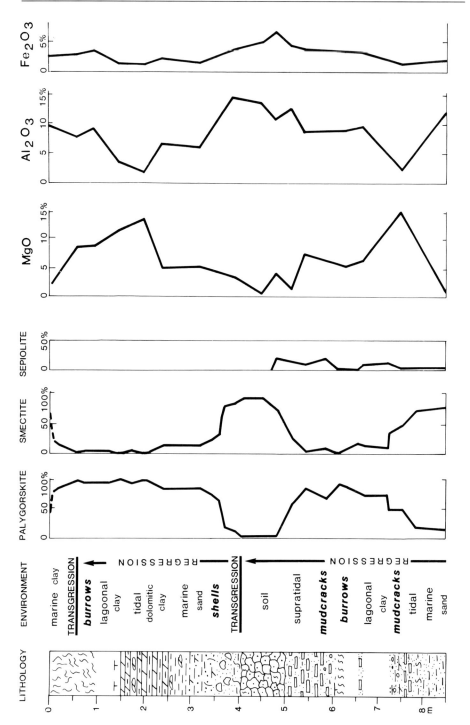

Fig. 9.6. Lithology, clay mineralogy and geochemistry of MC-1 core, La Camelia mine, Early Miocene of Florida, USA. (After Weaver and Beck 1977; Weaver 1984)

and palygorskite abundance (Fig. 9.6). Fe_2O_3 follows Al_2O_3 variations, and presents a maximum value near the base of the soil, where it forms an accumulation horizon. The K_2O values (0.5 to 1.7%) reflect illite abundance. They parallel Al_2O_3 and smectite abundances in marine sediments, suggesting a common detrital origin. The smectite identified in the palygorskite-rich clay beds appears to belong to the stevensite group (Mg-smectites), which correlates with the frequent occurrence of dolomite and resembles data collected on French Paleocene sediments (9.2).

Chemical calculations done by C.E. Weaver and K.C. Beck indicate that *most of the palygorskite formed by direct transformation of Al-Fe smectite*, with stevensite-like material being a by-product. The Al and Fe remained in constant abundance during the clay crystallization, Si, Mg, and H being provided by shallow marine and brackish water submitted to severe evaporation. Silica was largely provided by diatom frustules, strongly dissolved in warm tropical water. The conversion of smectite into palygorskite is presumed to have occurred when the pH of Mg-Si solutions was in the range of 8 to 9. Palygorskite contains an average of 24 ppm Li, suggesting it formed in waters of less than normal salinity. There is a mutual antipathy between palygorskite and clinoptilolite, with palygorskite being the fresher-water mineral and clinoptilolite the more saline equivalent. By contrast, sepiolite is concentrated shoreward of the palygorskite, and was apparently formed under less saline conditions. Sepiolite probably originated from neoformation in the near absence of smectite, from ionic solutions characterized by high pH and temperature.

Electron-microscope pictures show that most of the clay occurs as thin, parallel laminae, suggesting a periodic supply of detrital smectite to the lagoon (Weaver and Beck 1977). Two types of fibrous clay are identified. Fairly short fibers (about 1 μm) constitute the bulk of the palygorskite-sepiolite clay. They extend around smectite sheets, replace quartz and calcitic fossils, or coalesce from small opaline spheres. In addition, long fibers (> 10 μm) occur in small areas with desiccation features, and probably grew from residual fluids when dehydra8tion was almost complete. Long fibers develop in a matrix of short fibers, especially in the soil horizons where they form mats and vein filling. Almost pure monoclinic palygorskite is also reported in silicified rocks from the Miocene of Tampa Bay, Florida, where both fibrous clays and opal developed in "box-work geodes" under strong evaporitic conditions (Strom et al. 1981).

Dolomite commonly formed contemporaneously with both palygorskite and sepiolite. Dolomite and palygorskite replaced calcite of marine shells in sands underlying the lagoonal clay. They probably resulted from the downward seeping of Mg-rich waters. In addition, some apatite coprecipitated with the clay in sepiolite-rich areas where diatoms provided abundant P and Si.

The case of early Miocene environments of Florida-Georgia, supported by an extensive bibliographic survey, leads Weaver and Beck (1977) to consider that *sedimentary fibrous clays of various geologic ages chiefly formed in shallow water areas*, characterized by subtropical temperature and strong evaporation. C.E. Weaver and K.C. Beck cast doubt on the real possibility of forming palygorskite or sepiolite in non-hydrothermal deep-sea deposits. This opinion, already presented by some authors (e.g., Isphording 1973) and refuted by others (e.g.,

Couture 1978), is discussed in Chapters 12 and 13. Note that alkaline and evaporative environments, that characterize some past subtropical lakes and peri-marine basins, are hardly comparable with open deep-sea environments.

9.3.2 Uppermost Miocene, Mediterranean Sea

During Miocene times, fibrous clays frequently formed in alkaline evaporative sediments or in calcareous soils of the Mediterranean periphery (e.g., Singer and Galan 1984, Pozzuoli 1985, Galan et al. 1987). Most sedimentary environments were lacustrine, and only few corresponded to peri-marine conditions. Close to the end of the Miocene, in the Messinian period, the Mediterranean Sea became semi-closed and tended to desiccate, which determined the precipitation of giant saline evaporite bodies (see for instance Hsü, Montadert et al. 1979). Thick series of halite, gypsum, anhydrite, and other saline minerals, associated with various clays, marls, sand, some carbonates, and diatomites, accumulated in the Mediterranean basins, that were in turn submitted to the antagonistic influence of evaporation under arid climate and of water discharge from the adjacent Atlantic and land masses.

Extended mineralogic investigations on cores of the Deep Sea Drilling Project and on exposed sections from North Africa and South Europe have been performed on Messinian sediments (e.g., Chamley et al. 1978b, Chamley and Robert 1980, Rouchy 1981). Marine to brackish deposits usually contain little or no fibrous clays (maximum amounts 20% of < 2 μm fraction), that generally consist of short and broken bundles of palygorskite. If fact, the extensive formation of chain silicates appears to have vanished during the upper Miocene, except in local lacustrine basins exposed to strong evaporation and associated with pedogenic calcareous crusts (e.g., Chamley et al. 1980b, Galan et al. 1985, Leguey et al. 1985). Only very few cases could partially refer to peri-marine conditions (e.g., Chamley et al. 1978b, Galan et al. 1985). *Fibrous clays identified in Messinian marine sediments are considered as principally reworked from pre-Messinian alkaline evaporative sediments*, and secondarily from Messinian soils and lacustrine sediments.

In a few DSDP holes, namely in the Southwestern Tyrrhenian Sea (Western Mediterranean), large amounts of chlorite are identified, which resembles the clay mineral characters of some Triassic evaporitic series (e.g., Lucas 1962, Millot 1970, Jeans 1978a; see 9.1). However the sediments characterized by abundant chlorite in late Miocene sediments of the Mediterranean range have experienced strong heatflows since Messinian time, and appear to result from both evaporative and diagenetic influences (see Chap. 20).

The most striking peculiarity of Messinian sediments is the almost *systematic increase of smectite*, relatively to preceding Tortonian and subsequent Pliocene deposits (e.g., Fig. 9.7). As smectite abundance does not depend on lithology, and may be equally abundant in gypsum or halite clayey interbeds, terrigenous clays and marls, marine to brackish carbonates, coarse sand, diatom-rich sediments, etc., it is considered as fundamentally reworked from Peri- Mediterranean land

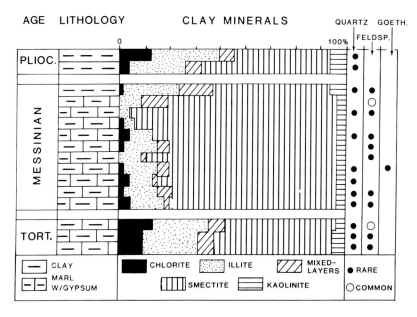

Fig. 9.7. Clay mineralogy in Late Miocene-Early Pliocene sediments of Brisighella, Northeastern Apennines, Italy. (After Chamley and Robert 1980)

masses (e.g., Chamley and Robert 1980). *Smectite probably formed abundantly in soils* developed under temperate-subarid climate around the nearly desiccated Mediterranean Sea. This question is considered in more detail in Chapter 17, from a paleoenvironmental point of view.

9.3.3 Illitization by Wetting and Drying

The presence of abundant illite is reported in some Quaternary volcanic lakes where the mineral could proceed from either in-situ chemical exchanges or sub-aerial processes (4.2.2). Almost exclusively illitic minerals also occur in some ancient lakes devoid of noticeable volcanic activity, as for instance in Oligocene lakes of the Massif Central, France (e.g., Jung 1954, Gabis 1963; see 4.1.2). Generally rich in iron, these minerals were labeled ferriferous illite or glauconitic illite, and were mostly attributed to the autochthonous evolution of pre-existing smectite. The possibility of forming illitic minerals from smectite in exposed areas appears to have been first proposed by Watts (1980) for South African calcretes submitted to alternately humid and dry conditions during Quaternary times. A similar explanation is proposed by Robinson and Wright (1987) about an early Carboniferous paleosol in South Wales, Great Britain. Laboratory experiments of Mamy and Gaultier (1975), and particularly of Eberl et al. (1986), support such a hypothesis. *K-smectites submitted to repeated wetting and drying cycles fix potassium irreversibly and progressively transform into random illite-smectite mixed-layers*, with the concomitant development of tridimensional ordering of the formerly turbostratic smectite structure. K-exchange on low-charge smectite (wyoming type) determines a collapse toward illite spacings of only a few percent of smectite layers, while high-charge smectite (cheto type) transforms into mixed-layers containing up to 50% illite layers. Potassium may issue from solutions of potassium-bearing minerals like feldspars or micas (Eberl et al. 1986, Środoń 1987). Most transformations occur during the first twenty cycles, and have been followed during about one hundred cycles by D.D. Eberl and colleagues. The first step of transformation from wyoming smectite to illitic minerals is described by Andreoli et al. (1987). It consists of the grouping and parallel alignment of elementary smectite particles, leading to the reversible formation of thick pseudocrystals similar to those of mica particles.

The possibility of illitization in peri-marine evaporative and alkaline environments is investigated by Deconinck and Strasser (1987) and Deconinck et al. (1988) *for Purbeckian sediments (early Berriasian) of the Swiss and French Jura mountains*. Purbeckian deposits formed close to the Jurassic-Cretaceous boundary in the subtidal to supratidal environments of a wide carbonate shelf located on the northern margin of the Tethys Ocean. Widespread limestones are interbedded with gray-greenish marls that preferentially occur at the top of small shallowing-upward sequences and were probably exposed for a long time. Marl interbeds contain either fresh-, brackish- or marine-water fossils, indicating both continental and marine influences. Mineral, chemical, and micromorphologic analyses show that the clay fraction of greenish marls chiefly consists of *dioctahedral illitic minerals*, characterized by high iron content in octahedral sheets, fairly abundant potassium in interlayers and high total layer charge (0.65 to 0.8). These minerals evoke K-smectites by the aspect of differential thermal curves, by the presence of some smectitic layers, and by the small size and fleecy outlines of clay particles. J.-F. Deconinck and colleagues suggest that nearly exclusive illitic clays in Purbeckian marls proceed from the transformation of Al-Fe smectites, abundantly reworked from Mesozoic continental soils. *Smectites probably accumulated in depressed areas of the periodically submerged and exposed carbonate platform, and were most likely submitted to wetting and drying cycles under hot alkaline and evaporative conditions.* Additional arguments arise from the unlikelihood of supplying almost pure detrital illite from the various rocks and soils cropping out on the hinterland, from the absence of significant diagenesis with the depth of burial, and from the increase of illite content on the outer shelf close to marine potassium sources. The Purbeckian example possibly represents an evolved stage of smectite illitization by wetting and drying alternations, the first steps of which are indicated by experiments of Eberl et al. (1986).

Further field and laboratory investigations are needed to test the possibility of extensive illitization processes by wetting and drying under warm climate in past peri-marine and lacustrine environments. Such a mechanism could explain the occurrence of illite-rich series in some regions devoid of active detrital input (tectonic quiescence, rarity of crystalline parent rocks). The illitic beds issued from evaporative peri-marine conditions could represent useful mineralogic and geochemical markers of past sea levels.

9.4 Conclusion

1. *Present-day coastal and shelf environments marked by alkaline and evaporative conditions do not display conspicuous or important clay formation*, from the bibliographic data available (see also 5.3.4). Such a lack of modern examples does not facilitate the understanding of past peri-marine environments, that present *massive clay formation at different periods placed between the late Devonian and the Tertiary*. Furthermore, clay chemical formation occurred during some geologic times under conditions as different as those typical of *peri-marine and lacustrine sediments*, which does not make it easier to understand the genesis of sedimentary processes. Finally, chemical clays like *chain silicates (palygorskite, sepiolite) also formed through pedologic processes*, some of them having taken place in Quaternary times and being perhaps still active (2.2.1, 2.2.3). Such a *convergence of clay formation in very different geographic situations*, most of them having disappeared, invites caution when interpreting environmental and geochemical conditions of clay formation.

One of the most intriguing questions to be solved is the *distinction of alkaline clays derived from either pedologic or sedimentary processes*. As documented by literature on recent and past sediments (e.g., Chap. 8), fibrous clays can easily be reworked from continental and marginal formations within the deep sea, and both soils and sediments may be responsible for such a supply. Palygorskite and sepiolite generally constitute only small amounts of calcareous soils, and therefore could hardly participate in the formation of thick Mg clay-rich deposits. But this is somewhat difficult to speculate for ancient geologic times, especially during warm and hydrolyzing periods, when pedogenic processes were probably much more extensive than in the Quaternary (e.g., late Cretaceous, Paleogene). Much progress is expected from the detailed characterization of paleosol and sediment structure, microstructure, and geochemistry. An interesting approach arises from micromorphological and grain-size investigations (e.g., Fedoroff et al. 1987, Verrechia and Freytet 1987). For instance, Leguey et al. (1985) show that the sizes of sepiolite and palygorskite aggregates in Tertiary formations from Central Spain are 1 to 5 µm in paleosols, less than 1 µm in lacustrine deposits, and 2 to 50 µm in diagenetic series.

2. *The convergence observed in the nature of pedologic, lacustrine, and peri-marine clays sheds light on some genetic conditions that must have been similar within the three types of environments. Salinity appears to have played a minor role, while strong alkalinity, water stagnation, intense evaporation, and warm climate appear to have been critical.* This implies *poor drainage and flat relief*, in order to prevent the evacuation of ionic solutions and permit chemical concentration.

The respective role of climate and chemically confined environment is illustrated by the chronological distribution of alkaline clays during Cenozoic times. *Peri-marine, lacustrine, and soil deposits developed widely during Paleogene times*, when the climate was hot. Peri-marine basins constituted particularly suitable geochemical traps, where fibrous clays could accumulate in turn with sulfates, carbonates, phosphates and organic matter (e.g., Lucas and Prévôt 1975). The formation of peri-marine alkaline clay nearly disappeared close to the middle Miocene, when *the increase of world cooling precluded active evaporation and ion trapping in shelf and coastal environments.* Freshwater chemical clays were able to form locally as late as during the uppermost Miocene, because small lacustrine basins still experienced strong evaporation and presented suitable ionic concentrations. The development of the late Neogene glaciation in turn provoked the near disappearance of lacustrine Mg-clays, and the possibility of their common formation only persisted in evaporative alkaline soils of calcrete type. Such a *climate, morphologic and geochemical control* of the formation of Mg-clays in evaporative conditions casts doubt on the possibility of crystallization of the same minerals in nonhydrothermal deep-sea environments characterized by cold, slightly alkaline, and mobile water.

3. *Minerals preferentially originating in alkaline evaporative environments include Al-Mg beidellite, Al- and Mg-saponite, stevensite, palygorskite, and sepiolite.* Some precision has been provided since the fundamental work of G. Millot and colleagues (see Millot 1964, 1970). A large part of peri-marine and lacustrine smectites consist of *Al-Fe beidellite* and are *detrital*; they chiefly accumulated in sediments deposited on the borders of the evaporative basins. These minerals were *transformed into Al-Mg smectites and palygorskite* within intermediate sedimentary environments. *Stevensite and sepiolite appear to proceed from true neoformation* under evaporative conditions that prevailed in the central part of basins or in the upper zone of chemical sequences. The genesis of magnesian sheet- and chain-silicates seems to follow two independent geochemical paths. It typically characterizes *restricted peri-marine rather than open-sea environments*, and can be referred to the neoformation of real evaporitic clays (Trauth 1977).

Other evaporative clay minerals are reported in the literature, and include mainly Permo-Triassic chlorite and corrensite, that appear to depend on hypersaline conditions and perhaps also on diagenetic control. Fruitful developments arise from experiments on the possibility of illitization through evaporation by repeated wetting and drying cycles. Such a mechanism could explain the formation of nearly true illitic clay assemblages in some fossil peri-lacustrine environments and at the surface of some ancient platforms, submitted in turn to water influx and to strong evaporation.

Chapter 10

Ferriferous Clay Granules and Facies

10.1 General Features

Iron-rich clay granules present in sediments traditionally comprise glauconite, berthierine grains, and oolitic ironstone. As the term glauconite is confusing, since it may refer either to a green granule or a specific mineral, we will follow Millot's suggestion (1964, 1970) and differentiate the granules by the term glauconie or glaucony (see also Odin and Matter 1981). As recent data show that true berthierine granules appear to occur very rarely in modern sediments, the expression "berthierine granules" will be replaced by the term verdine to designate light-green shallow-water granules (Odin 1985). In addition, the submarine alteration of volcanic rocks may give way to the formation of celadonite, a specific ferriferous clay mineral forming green vesicles or filling veins. Consequently *iron-rich clay granules and associated facies are considered as comprising glaucony, verdine, ironstone, and celadonite.*

The major mineral varieties of ferriferous clay granules are the following:

- *Glauconite or glauconitic* mica represents an iron- and potassium-rich 10 Å illite-type, characteristic of evolved glaucony ($Fe_2O_3 > 20\%$, $K_2O > 4\%$). It predominates in Paleozoic glaucony peloids but also occurs in Mesozoic and Cenozoic ones. The green ferric illites encountered in some freshwater lakes contain less than 10% Fe_2O_3 and are not glauconitic minerals (Kossovskaya and Drits 1970).

- *Glauconitic smectite* forms a mixed-layered group made of varying proportions of smectitic-, glauconitic- and sometimes illitic-type layers. It characterizes most of the Mesozoic and Cenozoic glaucony granules.

- *Green smectite* is an iron-bearing dioctahedral smectite, whose composition is intermediate between montmorillonite and nontronite. It frequently occurs in late Cenozoic glaucony granules.

- *Phyllite V* constitutes the dominant component of verdine, that forms light green granules characteristic of recent sediments from shallow marine environment (Odin et al. 1988). It essentially comprises a poorly crystallized 7.2 Å mineral of the serpentine group, marked by fairly strong magnetic properties, by the abundance of ferric iron (18-25%) and by a combined di- and tri-octahedral structure. This mineral is chemically different from both berthierine and ferriferous chlorite (Table 10.1). The 7 Å phyllite V, that appears to fill a vacant niche in the clay mineral classification, is associated in the oldest grains with small amounts of minerals marked by 14 Å and even 10 Å spacings.

Table 10.1. Geochemical data for phyllite V, berthierine, phyllite C, and chlorite. (After Odin et al. 1988)

	Phyllite V		Berthierine		Phyllite C	Chlorites Average range
	Range	Mean	Range	Mean		
SiO$_2$	33 –39	35.8	19–27	23.3	39.1	21–34
Al$_2$O$_3$	5.5–12	9.8	18–28	22.1	9.2	12–26
Fe$_2$O$_3$	18 –25	21.2	0– 5.5	3.2	16.2	0–13
FeO	5 – 9	6.4	30–37	34.8	3.8	0–40
MgO	8 –14	10.7	1– 8	3.5	10.4	1–36

- *Phyllite C* constitutes an accessory component of verdine, characterized by a dominant peak near 14.5 Å, by a ferric state of iron like phyllite V, and by fairly low magnetic properties (Odin et al. 1988). Its chemical composition differs from that of all known types of chlorites, including iron-rich types like chamosite, but is very close to that of phyllite V (Table 10.1). On X-ray diagrams the 14 Å phyllite C green clay mineral shows a behavior intermediate between those of smectite and of swelling chlorite.

- *Berthierine* is an aluminous-ferrous 7 Å trioctahedral serpentine, that has been mistaken for chamosite in some publications dealing with recent sediments. Berthierine predominates in Mesozoic peloids and ooids, but is also reported in Paleozoic ironstones (e.g., Courty 1981) and in some recent green granules (see Odin and Matter 1981).

- *Chamosite* represents an iron-rich 14 Å trioctahedral chlorite, that clearly differs from 7 Å minerals like berthierine and phyllite V by an often small but characteristic peak at 14 Å on X-ray diagrams. Chamosite is the most common clay mineral in Paleozoic oolitic ironstones, especially those having undergone late diagenesis or low-grade metamorphism (Van Houten and Purucker 1984).

Many of papers have been published over the past few decades on green granules. This results from the discovery of numerous clay granule deposits on ocean margins, which determined various applications to the understanding of ancient glaucony and ironstone formation. The renewal of publications on recent clay granules started in the early sixties (e.g., Burst 1958, Ehlmann et al. 1963, Giresse 1965, Lamboy 1968). The most striking results were probably those of Porrenga (1966, 1967), who studied the Orinoco shelf (Western Atlantic) and especially the *Niger submarine delta* in the Gulf of Guinea (Eastern Atlantic). D.H. Porrenga showed that green to blackish-green granules constitute from less than 1% to more than 60% of sediments in the Niger delta, and consist mainly of fecal pellets and fillings of foraminifera and other carbonate shells. Berthierine, called chamosite by D.H. Porrenga and corresponding to the verdine of Odin (1985), dominates between 10 and 50-60 m water depth. Glaucony is the main constituent at depths ranging from 125 to 250 m, and goethite at depths shallower than 10 m (Fig. 10.1A). "Berthierine" is poorly ordered and has a typical 7 Å basal reflection. Glaucony contains more than 70% expandable layers and resembles a Fe-rich smectite. Both minerals comprise much more iron, potassium and magnesium than the clay matrix and associated gray pellets, that are ter-

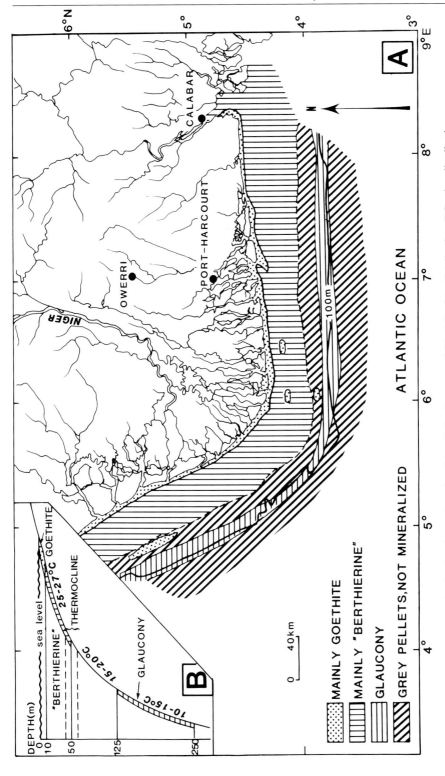

Fig. 10.1. Mineral composition of pellets in Niger submarine delta. (After Porrenga 1967). A Surface distribution. B Section distribution

rigenous and marked by differential settling processes (see 6.2.1, Fig. 6.1). Considering the water temperature at different depths, Porrenga (1967) observes that the outer boundary of "berthierine" occurrence coincides with the thermocline (Fig. 10.1B). He suggests that light green "berthierine" granules form preferentially in restricted to shallow marine environments of tropical regions, where bottom water temperature is higher than 20 °C. By contrast, temperature lower than 15 °C tends to favor the formation of dark green glauconitic granules, that consequently may develop in deeper zones and under a wider range of latitudes than "berthierine".

Several review papers have been published on the distribution and genesis of ferriferous clay granules and associated facies (e.g., Galliher 1935, Hower 1961, Borchert 1965, Triplehorn 1966, McRae 1972, Odin 1975, Odin and Matter 1981, Van Houten and Purucker 1984). In addition, the synthesis of Odin (1988) provides an outstanding and renewed reference, that is largely used in the following pages.

10.2 Nature and Distribution of Recent Green Clay Granules

10.2.1 Habits

The morphology of glaucony has been described in detail by Cayeux (1897, in Millot 1964, 1970) and Triplehorn (1966). The aspect of clay granules depends on the nature, grain size and pore size of the *substrates* submitted to greening (= "verdissement" of G.S. Odin), and on the evolution stage reached by the grains within the sediment. Odin (1975, 1988) proposes five main groups of *granular habits*, for glaucony or verdine developed on millimetric substrates: (1) *Internal molds of microfossils* show a wide geographic distribution. Foraminifera represent the most ubiquitous host substrates for glaucony and verdine, but may by replaced by shells of ostracods, bryozoans, small molluscs, and even radiolarians or sponge spicules (e.g., Kerguelen Plateau, Odin and Fröhlich, in Odin 1988). (2) *Fecal pellets* constitute the main substrate of many glaucony and verdine deposits, especially on internal continental shelves marked by the presence of abundant mud eaters or filter-feeding animals (e.g., Gulf of Guinea, Giresse and Odin 1973). (3) *Calcareous and sometimes siliceous bioclasts* may locally constitute important substrates for glauconitization (e.g., Lamboy 1968), but are not important in verdine facies. Glaucony grains fill the voids of the shells similarly to microfossil chambers, or replace partly the biogenic debris themselves. (4) *Mineral and rock debris* substrates may consist of mica, feldspar, quartz, calcite, dolomite, phosphate, chert, volcanic glass and various other materials. Usually not dominant in a given deposit, mineral and rock fragments turn green mostly along cleavage plans and cracks. (5) *Unrecognizable grains* correspond to highly evolved substrates whose original shape is obliterated by glauconitization, and mostly characterize ancient green clay deposits.

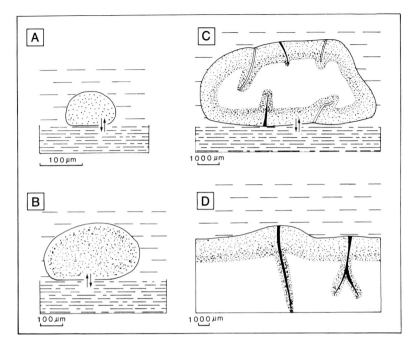

Fig. 10.2. Distribution of glauconitization process according to the size of host substrate. (After Odin and Matter 1981). **A** Fine sand. **B** Medium sand. **C** Coarse sand. **D** Boulder or hardground. The *dot density* expresses the intensity of glauconitization

In addition to the granular habit, glaucony sometimes displays a *film habit*, developed on the surface of substrates larger than a few millimeters. Substrates consist either of *boulders*, large calcareous bioclasts, flint and phosphate nodules, or of *hard grounds* extending sometimes over large distances (e.g., top of Jurassic limestones or Cretaceous chalk, Northwestern Europe). When the film habit extends downwards within the rock on a decimetric to metric thickness, it corresponds to a *diffuse habit*. Verdine facies, known in very recent sediments only (Holocene, late Pleistocene) do not present film or diffuse habits.

The extension and intensity of glauconitization usually depends on the size of host materials (Odin and Matter 1981; Fig. 10.2). Grains smaller than 100 μm are commonly little evolved and present a light green color (A). Grains of a few hundred micrometers often experience a homogeneous glauconitization (B). Millimetric grains generally display a little- or not-evolved central zone, a dark green intermediate zone and a light green external zone (C). Larger substrates (D) present film habits with the same distribution of glaucony as for coarse sand. The optimal size for obtaining pure dark-green clay from a given substrate is less than 200 μm for mica or quartz, 200 to 500 μm for fecal pellets, and about 500 μm for carbonate debris (Odin 1988).

10.2.2 Zonal and Bathymetric Distribution

Latitude. Glaucony occurs in surface sediments deposited on continental shelves and topographic highs of all oceans, from about 50°N to 50°S (Fig. 10.3A). Glaucony has been reported at latitudes as high as 63°N off Norway (Bjerkli and Östmo-Saeter 1973), 50°N off Vancouver island in Western Canada (Bornhold and Giresse 1985), and 50°S on Kerguelen Plateau (Odin and Fröhlich, in Odin 1988). The question arises about the age of outcropping glaucony, since most deposits located on Figure 10.3 concern areas marked by very low sedimentation rates. Glaucony discovered on the shelf of Vancouver island is Holocene in age (Bornhold and Giresse 1985), but many high-latitude deposits appear to be relict (e.g.,

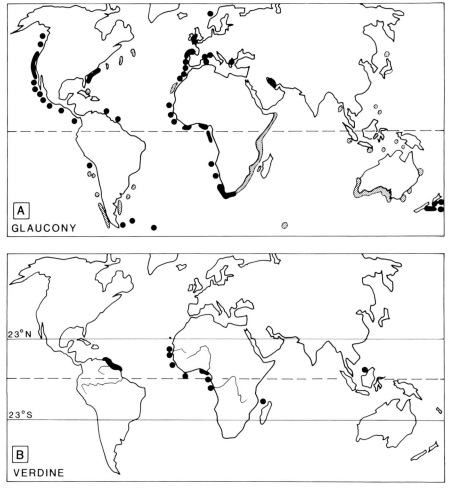

Fig. 10.3. Distribution of glaucony (**A**) and of verdine (**B**) in surface sediments of the world ocean. (After Odin 1988). *Hatched figures* areas without detailed mineralogical data. N.B. Verdine granules mainly contains 7 Å clay minerals formerly attributed to berthierine (see text)

South Africa, Kerguelen Plateau, Chatham Rise and Scotia Ridge in the Southern Ocean). It is likely that middle- to high-latitude glaucony preferentially formed during interglacial stages of Plio-Pleistocene times. In any case, glaucony formation seems to adapt easily to temperatures lower than 15°C (see Porrenga 1967) and perhaps as low as 7°C (Bjerkli and Östmo-Saeter 1973). By contrast, *verdine is restricted to intertropical sediments* (Fig. 10.3 B), which fits the former hypothesis of Porrenga (1967) on the thermal distribution of "berthierine" (Fig. 10.1B). Verdine development, that does not appear to be older than 20,000 years, necessitates average temperatures higher than 20°C and probably close to 25°C (Odin 1988). Note that true berthierine is reported in estuarine sediments of Western Scotland, at about 56°N (Rohrlich et al. 1969). Iron-rich berthierine, labeled "chamosite" by V. Rohrlich and colleagues, occurs as poorly crystallized 7 Å minerals in sand-sized fecal pellets from recent, organic-rich, sandy mud of Loch Etive.

Water Depth. Verdine may occur between less than 5 meters water depth (e.g., Casamance estuary, Senegal) and about 200 meters, but most commonly *characterizes areas located between 20 and 60 meters* (e.g., Fig. 10.4). Verdine granules encountered deeper than 100 m probably correspond to late Pleistocene deposits (before the Holocene transgression) or to reworked sediments. By contrast, magnetic grains of the *glaucony* facies mainly occur deeper than 50-80 meters, and *present the highest abundance between 150 and 300 meter depth* (Fig. 10.5). At water depths shallower than 60 meters glaucony is absent or reworked from older deposits, and is sometimes replaced by verdine. Glaucony occurs commonly

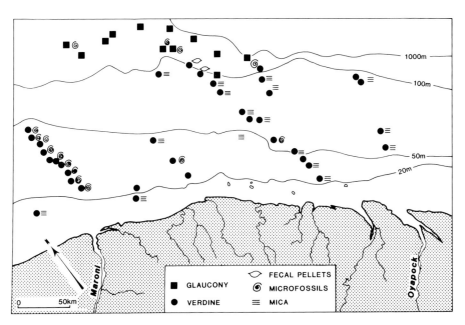

Fig. 10.4. Bathymetric distribution and nature of substrate of green clay granules, upper continental margin of French Guiana. (After Odin 1988)

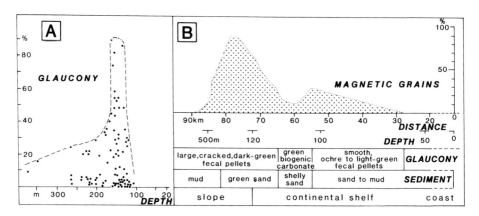

Fig. 10.5. Bathymetric distribution on glaucony of the upper continental margins off: **A** Vancouver island. (After Bornhold and Giresse 1985). **B** Congo. (After Odin 1988); grains shallower than 110 m are younger than 18,000 a B.P., deeper grains are older

down to 500 meters water depth (e.g., Fig. 10.5B). At greater depths it is often considered as reworked from shallower deposits.

The question of the possible formation of *glaucony in deep-sea sediments* is still little investigated. Various deposits are reported in the literature, as for instance on Scotia Ridge East of Patagonia (Bell and Goodell 1967), on Chatham Rise East of New Zealand (Cullen 1967), off Japan (Karig, Ingle et al. 1975), on the lower margin of Western North and Central America (Odin and Stephan 1981), and off West Africa (Giresse 1987). Deep-sea glaucony occurs in some recent sediments located as deep as 2000-3000 meters (e.g., Chatham Rise, Scotia Ridge). Many deposits appear to correspond to materials reworked from the upper margin (e.g., Eastern Atlantic Ocean). Deep-sea glaucony occurring off tectonically active regions like Japan and Mexico seems to result from successive shallower-water formation and bottom subsidence. True genesis of glaucony appears to occur down to 1000 meters water depth, mostly on topographic heights characterized by little terrigenous supply, strong bottom currents and suitable substrates (microfossils, fecal pellets). The possibility of a deeper formation is still questionable. Note that deep-sea glaucony is quoted for some old deposits. This is the case with Albian-Cenomanian sediments of the Armorican margin, Northeastern Atlantic (DSDP Leg 80), where Gillott (1984) envisages a genesis of granular and diffuse glaucony on submarine hills as deep as 1000 to 1700 meters.

Local Environment. Both verdine and glaucony facies develop in areas characterized by low sedimentation rates. Verdine, that usually forms closer to detrital sources than glaucony, occurs either beyond fast accumulating zones of river deltas (e.g., Congo river), or off swamps and mangroves where terrigenous particles are actively trapped (e.g., Ivory Coast). Glaucony is especially abundant at the shelf-slope transition, where bottom currents prevent fast sedimentation by winnowing fine terrigenous particles. Most glaucony deposits correlate with hiatuses or very slow sedimentation. *Both verdine and glaucony form essentially*

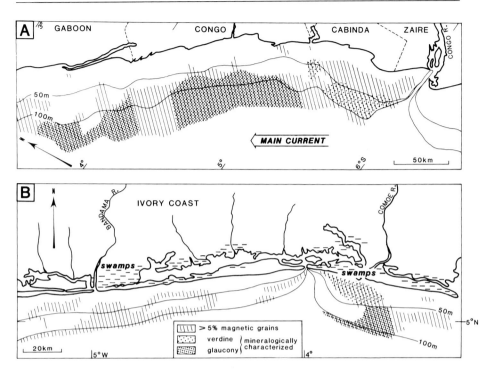

Fig. 10.6. Distribution of green clay granules off Zaïre to Gaboon (**A**) and Eastern Ivory Coast (**B**). (After Odin 1988). Note that most granules off river mouths appear to be verdine, and all granules deeper than 80 m belong probably to glaucony facies

close to the sediment-sea water interface, and hardly grow after burying. This is the reason why glaucony is considered a reliable marker of surface deposits. Accordingly potassium-rich glaucony may be used to date the time of deposition by radioactive isotopes (Odin 1982, 1988).

The formation of verdine seems to depend principally on two specific conditions (Odin 1988). First, *the oxygen content of sea water must be high* enough to determine a positive Eh in pore waters and the formation of minerals with ferric iron (phyllites V and C, see 10.1). This explains why verdine mainly develops in sandy deposits and is usually associated with goethite. Second, light green ferric granules appear to occur preferentially in the vicinity of large rivers. This is well demonstrated along the western coast of Africa, where verdine accumulates close to river mouths or on the shelf in the direction of prevailing current downstream the mouths (Fig. 10.6). *The largest verdine deposits usually correlate with large continental water supply.* The cause of such a correspondence probably lies in the high amounts of *dissolved iron* supplied by river waters, especially in intertropical regions where chemical weathering is very active. This explains the close relation existing between the latitudinal distribution of verdine and the freshwater input. Note that abundant iron may also be provided by the leaching of weathered volcanic rocks, as off New Caledonia or the Comoro islands (Odin 1988). By contrast with verdine, glaucony does not depend on the close proximity of iron-rich fluvial input, which correlates with its formation in deeper, cooler, and less restricted latitudinal conditions. In addition, glaucony commonly forms in sediments muddier than verdine deposits, which determines less oxidizing pore waters.

10.3 Genesis of Green Clay Granules

Three main mechanisms have been involved to explain the formation of glaucony granules. (1) *The layer lattice theory* (Burst 1958, Hower 1961) consists of a transformation of a degraded 2:1 layer silicate lattice (TOT clay) into an iron- and potassium-rich 2:1 layer silicate of the illite group. This is supposed to occur under reducing conditions like those provided by the interior of fecal pellets or foraminiferal tests. The glauconitization process supposes a progressive substitution of iron for aluminum in octahedral layers, the incorporation of interlayered potassium balancing the concomitant charge increase, and therefore the synchronous fixation of Fe and K within the structure. (2) *The epigenetic substitution theory* (Ehlmann et al. 1963) considers that glauconite layers form through solution of preexisting minerals, by adding ions present in sea water. This explanation was applied to the replacement by glaucony of calcite and various silicates (feldspar, quartz, pyroxene, olivine). (3) *The precipitation-dissolution-recrystallization theory* (Odin 1975, Odin and Matter 1981, Ireland et al. 1983) involves successive processes leading to a true neoformation, and therefore implies an independence between the nature of substrate and the new iron-rich clay minerals.

The layer lattice theory was widely accepted in the sixties (see Millot 1964, 1970), while recent interpretations confirm rather the reality of the precipitation-dissolution-recrystallization (pdr) theory. The epigenic substitution theory has been less discussed in the literature than the two others, and supposes neoformation processes similar to those implied by the pdr theory. Odin and Matter (1981) estimate that the layer lattice theory is incompatible with several facts observed during glauconitization: (1) The greening of detrital mica takes place beween mica sheets and not in place of them, and therefore does not require the initial architecture or ions of 10 Å layered minerals. (2) Greening often develops on calcareous substrates (shells, hard grounds) devoid of any clay lattice structure. (3) A given type of substrate, like fecal pellets, may give way to either glaucony or verdine (e.g., Gulf of Guinea), both types of green grains being characterized by very different authigenic clays. (4) Most glauconitization processes clearly show a break in the iron enrichment (i.e., strong augmentation from 10 to 15% of total Fe_2O_3), which suggests a dissolution-recrystallization instead of a progressive transformation of 2:1 clay. Despite these arguments, the transformation of preexisting layer silicates could apply to specific cases of glauconitization, as documented by experiments and modeling (e.g., Tardy and Touret 1987).

Glauconitization of fecal pellets on the upper margin of the Gulf of Guinea provides a good example of actual processes, since different stages of evolution

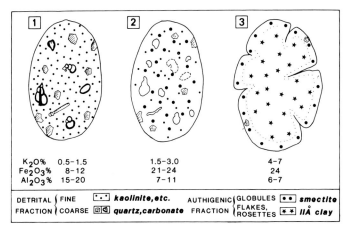

Fig. 10.7. Schématic evolution of fecal pellets in Gulf of Guinea margin through glauconitization. (After Odin 1988). Chemical data from microanalyses

are recognized on grains that have never been buried (In Odin 1988). (1) Initial fecal pellets consist of *grey mud* rich in detrital kaolinite, with minor amounts of other *terrigenous clay minerals*, biogenic carbonate shells and debris, and quartz. The mineral and chemical composition of the grains is identical to that of sediment matrix (Fig. 10.7-1). (2) Some pellets turn into *ochre to light-green grains*, and show both a strong dissolution of fine-grained calcareous particles and the appearance of *Fe-smectite* beside remains of kaolinite and quartz (Fig. 10.7-2). The organic carbon content tends to decrease relatively to the surrounding sediment, which suggests intense microbial activity (Cahet and Giresse 1983). The dissolution of carbonate debris determines an increase in porosity in pellets. Silt- and clay-sized carbonates are replaced by authigenic Fe-clays, that constitute small globules, sometimes arranged in caterpillar-like structures. Light-green granules contain much more iron and somewhat more potassium than gray pellets, and much less aluminum. (3) A third group comprises *dark-green grains*, whose initial ellipsoidal shape is difficult to recognize because of many surficial *cracks* due to an *increase of the initial volume* (Fig. 10.7-3). The internal zone of the former pellets consists of rosette- or flake-like microcrystals made of *11 Å glauconitic minerals* strongly enriched in potassium. Most terrigenous particles have disappeared within the pellets, except some kaolinite sheets in the outermost zone and a few grains of quartz. The previous porosity has nearly vanished, which determines a homogeneous structure.

The recognition of comparable evolutionary stages arises from detailed studies of clay granules sampled on various continental margins of the world ocean, whatever the nature of substrates and the initial mineral composition (e.g., Giresse et al. 1980, Odin and Matter 1981, Bornhold and Giresse 1985, Odin 1988). Nearly all cases investigated show that during glauconitization initial grains have successively experienced a dissolution of biogenous and terrigenous materials, a precipitation of Fe-smectitic clay, and a recrystallization of poorly ordered Fe- and K-glauconitic illite associated with an increase of volume. *Ver-*

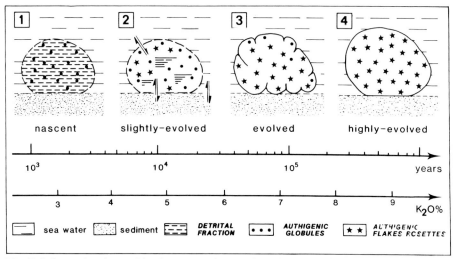

Fig. 10.8. Successive glauconitization stages of a granular substrate. (After Odin 1988)

dinization also appears to proceed from a trueneoformation, since various minerals and not only kaolinite may give way to the 7 Å phyllite V that constitutes most verdine granules. Glaucony and verdine apparently form similarly at the sediment-sea water interface, the main differences consisting of distinct neoformation products, evolution stages, chemical and chronologic constraints.

Odin and Matter (1981) and Odin (1988) consider that *the mechanism of glauconitization consists of four successive stages, that follow one another at the sediment-water interface* if suitable conditions persist (Fig. 10.8).

1. *The nascent stage* corresponds to the first development of iron-rich glauconitic minerals (mostly Fe-smectite) at the expense of detrital material. The K_2O content ranges between 2 and 4%, while detrital clays usually contain no more than 1.5% K_2O. This first stage is strongly dependent on porosity, that allows ion migration and chemical reactions.

2. *The slightly evolved stage* is characterized by the near disappearance of detrital minerals and the presence of inferred pores that are progressively filled with authigenic clays containing between 4 and 6% K_2O. Iron is still abundantly incorporated, as in the nascent stage; it probably results from Fe^{2+} mobilized at depth within the reducing sediments, and migrating upwards close to the more oxidizing sea-water interface (Ireland et al. 1983). Glauconitic clay usually consists of poorly crystallized 11-12 Å minerals, arranged in globular, caterpillar-like or blade-like particles. In this stage, Fe-smectite is supposed to be destroyed and to be replaced by the glauconitic minerals, strongly enriched in potassium and iron. Electron-microscope and microchemical observations made by Parron and Amouric (1987) suggest that glauconitic mica could result either from a siliceous gel incorporating Fe and K after destabilization of a Fe-montmorillonite, or from a ferriferous gel incorporating Si and K after destabilization of a nontronite.

3. *The evolved stage* results from a series of successive recrystallizations, tending to obliterate the initial structure. The original shape of substrates progressively disappears. Clay minerals show X-ray diffraction peaks closer to 10 Å. The clay growth occurs preferentially and more rapidly in the central zone of the grains, which determines an augmentation of the initial volume and the formation of cracks in the outer zone. K_2O content ranges between 6 and 8%.

4. *The highly evolved stage* corresponds to the filling of cracks with authigenic minerals, which determines smooth outlines. K_2O content exceeds 8% of the total granule. Clay consists of still poorly crystallized glauconitic minerals, that do not evolve significantly if the green granules remain at the sediment-sea water interface. *A crystal reorganization and ordering of glauconitic minerals may occur after burying, during early diagenesis* (Odin 1988). This probably explains why green clay granules cropping out on the sea floor and formed during Quaternary times are mineralogically different from many ancient glaucony deposits that consist of almost pure, better-crystallized glauconite. The environmental conditions of glaucony ordering after burying are still hardly known. If glaucony grains are not buried, they may either remain stable and constitute relict materials, or evolve toward goethite or phosphatized glaucony (e.g., Lamboy 1976). Goethitization of glaucony occurs in very oxidized conditions, which are favored by a sea-level drop (regression). Phosphatization of glaucony occurs rather in a more reducing environment, and often correlates with transgressive conditions.

The intensity and rapidity of glauconitization strongly depends on the nature and size of initial substrates (Giresse et al. 1980). Very small micas scattered within non evolved fecal pellets may undergo a K and Fe enrichment in less than 10 years, as reported by Bayliss and Syvitski (1982). The slightly evolved stage (2) may be reached in 10^4 years, and the evolved stage (3) in about 10^5 years, if granules are not significantly buried (Fig. 10.8). When cracks form, the granules derived from microfossils are often broken because tests are destroyed and internal volume increases, which gives way to smaller glaucony debris. Large microfossils, carbonate shells, quartz, micas, pebbles, and hard grounds are often too large to allow extensive glauconitization. Evolved glaucony may form from silt- to fine sand-sized micas, and from the surficial zone of large carbonate debris. Very small particles, like silty fecal pellets and fine calcareous debris, do not allow suitable chemical exchanges between grains and microenvironment, which precludes glauconitization process further than the nascent stage (1).

The different types of substrates suitable for glauconitization are often deposited at statistically different water depths, in a given region. On the Western African margin, for instance, planktonic foraminifera settle mainly on the outer margin and preferentially evolve toward glaucony on the upper slope-outer shelf, while verdine forms on the inner shelf principally from fecal pellets. *During a eustatic transgression, successive substrates may be glauconized in turn*, which provides a good indication of the modalities of sea-level rise (e.g., Giresse 1985, Odin 1988). A transgression may be evidenced on a given section by the landward occurrence of younger glaucony granules successively developed from planktonic tests, benthic bioclasts, fecal pellets, mineral and rock grains, and coastal shell bioclasts (Fig. 10.9). The presence of glauconized coastal shell bioclasts indicates

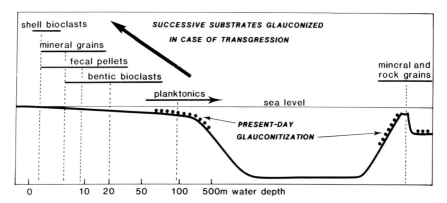

Fig. 10.9. Successive substrates potentially submitted to glauconitization during a sea-level rise. (After Odin 1988)

an important transgression in the absence of subsequent reworking, while dominant glauconized planktonic tests suggest rather a sea-level stability or even a regression. Note that glaucony deposits frequently correlate with transgressive sequences in the geologic record (e.g., Cenomanian transgression in Western Europe, Eocene (Lutetian) transgression in Paris Basin, Holocene transgression on the Congolese continental shelf; see 10.4). Transgressive conditions are not necessary to allow massive glauconitization. Nevertheless, they may favor extensive greening processes because shallower substrates are successively available, and because the sea-level rise determines a decrease in terrigenous supply, preventing rapid burying that would stop the green clay development.

The microenvironment responsible for glaucony and verdine genesis is unquestionably confined and chemically concentrated, since open-marine conditions do not modify terrigenous minerals in an appreciable manner, especially in shallow water (see Chaps. 6 to 8). A confined microenvironment is necessary to accumulate the cations used for the formation and volume increase of green granules, and to allow the slow chemical exchanges needed by the growth of green layer silicates. The existence of a confined environment is proved by the development of greening inside microtests and not outside, by the absence of greening on too small, unconfined grains, and by the darker greening and stronger mineral growth in the central zone of grains compared to their periphery (Fig. 6.7). The confined conditions necessary for chemical exchanges must nevertheless not be too strong, since greening does not develop in too closed chemical systems (e.g., buried sediments, central part of large substrates), and since a renewal of ions is necessary for granule growth. This led Odin and Matter (1981) to define a *semi-confined microenvironment* as the most suitable for glauconitization. Such an environment allows ions to slowly enter and leave the grains. *A microenvironment marked by exchanges at the boundary between reducing and oxidizing conditions, close to the sediment-sea water interface*, is also deduced from experiments done by Harder (1980), who stresses the importance of silica availability in controlling the formation of either 7 Å or 10 Å green clays. Ions are mainly provided by sea water, as demonstrated by the stopping of greening after the grains are buried. Additional ionic supply issues from the underlying sediment (especially iron; see Ireland et al. 1983) and from the substrates themselves. Note that the displacement of grains by bottom currents or bioturbation is not essential, but it may favor the incorporation of ions on all faces of a given granule.

10.4 Greensands, Ironstones

Greensands, green sandstones and ironstones identified in the geologic record consist mainly of iron-rich granules that contain a noticeable fraction of iron-rich clay. This resemblance with modern glaucony and verdine ("berthierine") deposits has led to numerous comparisons. Let us briefly consider some similarities and differences between recent and ancient Fe-rich clay granules. *Morphologically, most modern green clays* present spheroidal to ellipsoidal or ovoid, lobate or botryoidal shapes, mainly derived from fecal pellets and calcareous bioclasts. These particles mostly *refer to glaucony*, more rarely to verdine, and generally lack specific radial or concentric structures. *By contrast, ancient iron-rich granules comprise both peloids and ooids*, the latter being often more abundant than the former (Van Houten and Purucker 1984). Ooids are usually deposited in a muddy matrix. They display ellipsoidal grains made of concentric sheats of iron oxides and clays, whose origin has been much discussed and is still controversial. Ooids could result either from a layer-by-layer accretion, from a replacement of carbonate or ferric-aggregate layers by clay, from an intra-sedimentary concretion, or even from a nonoolitic growth (e.g., oncolites). The absence or scarcity of recent oolitic Fe-rich granules stimulates the discussions about the intriguing problem of past Fe-ooids, whose origin may have been diverse according to environmental conditions (see Odin 1988).

The mineralogy and geochemistry of Fe-granule clays differs notably in recent and ancient deposits. Late Cenozoic glaucony peloids contain poorly crystallized glauconitic minerals marked by broad X-ray reflections between 10 and 11 Å. By contrast, much *pre-Cenozoic glaucony*, characteristic of greensands and sandstones, *contains true glauconite* with a well-defined 10 Å peak. Often not stressed in the literature, this difference implies specific geochemical conditions of deposition for both modern and ancient glaucony, or a *crystal ordering* after burying through diagenesis and aging. The second hypothesis is more likely, since the crystalline arrangement does not usually correlate with appreciable chemical modifications. Note that J. Louail, in late Cretaceous sediments of Western France, envisages the existence of a stage intermediate between terrigenous clays and authigenic glauconite, marked by the development of lathed Fe-smectite (Louail et al. 1979, Louail 1984). This question remains open, since lathed Fe and Al-Fe smectites may occur in both glauconitic and nonglauconitic marine environments during late Phanerozoic times (e.g., Holtzapffel 1984, Hotzapffel et al. 1985; see Chap. 14). Note also that some recrystallization of ancient glaucony granules may occur if greensand deposits are exposed above sea level and submitted to meteoric water (Morton and Long 1984). Such a recrystallization may induce a modification of the isotopic composition and offers an opportunity to date sea-level changes.

Oolitic ironstone clays are mainly composed of berthierine or chamosite, two trioctahedral ferrous minerals sometimes associated with swelling chlorite. Berthierine, a 7 Å serpentine-like mineral, occurs mostly in Mesozoic ironstones, while chamosite is a 14 Å chlorite mineral represented mainly in Paleozoic ironstones. *By contrast, recent green granules apparently never contain true chamosite, and only very seldom true berthierine* (see 10.1). In particular, the so-called berthierine granules, that were re-named verdine by Odin (1985, 1988), contain phyllite V, a typically dioctahedral 7 Å ferric clay. This difference, combined with the absence of oolitic structure, diffuse habit, muddy facies and pre-late Pleistocene age for verdine, does not favor the interpretation of a verdine precursor for ironstone clays, especially for berthierine. Berthierine could result from either a primary or secondary origin, the latter consisting namely of a transformation of soil-derived kaolinite (see discussions in Courty 1982, Bhattacharyya 1983, Van Houten and Purucker 1984, Odin 1988). Perhaps diagenetic modifications have occurred and permitted the transformation of ferric minerals to ferrous ones. A diagenetic modification could also account for explaining the preferential occurrence of 14 Å ferrous minerals (chamosite) in deposits older than those rich in 7 Å ferrous minerals (berthierine). Chamosite could proceed from the transformation of berthierine, the process being favored by aging and low metamorphism. But this hypothesis lacks arguments, as for instance the existence of intermediate minerals (7–14 Å mixed-layers).

There is general agreement about the fact that *most glaucony granules and ironstones were deposited in rather shallow-water seas submitted to low terrigenous input, under fairly warm, humid climate.* Chamositic deposits formed in nearshore facies either seaward of low-energy deltas or along interdeltaic coasts of stable margins, or in inland seas, as well as in foredeeps and intracratonic basins. Shallow-water ferric oxides and chamositic ooids accumulated most commonly above upward-shoaling siliciclastic sequences, sometimes in superimposed sand-waves (e.g., Jurassic minette of Eastern France, Teyssen 1984). Many glaucony greensands deposited in open-sea areas along stable margins or on high ocean platforms, from shallow to deeper offshore environments. In general *ancient glaucony and chamositic granules accumulated in more varied environments than modern ones.* Depositional facies range from intertidal and subtidal inner shelf to deep offshore, include carbonate shelves (see Van Houten and Purucker 1984), and even mixing zones between freshwater and saline water (Sorokin et al. 1979, Whiteside and Robinson 1983). V.I. Sorokin and colleagues observe the mineral composition of ancient glaucony may vary from nearly pure smectite (high Al content) in coastal environment to glauconitic mica (low Al content) in open-sea environment.

If ancient glaucony granules appear to have formed in semi-confined micro-environments similar to recent ones, specific conditions probably characterized the genesis of oolitic ironstones. *The formation of ferrous (trioctahedral) minerals like berthierine and chamosite implies low Eh and reducing conditions,* compared to the genesis of ferric (dioctahedral) verdine or glaucony. The confined character of ironstone deposits agrees with the frequent abundance of clayey matrix, disfavoring easy oxidation. *On the other hand, the oolitic structure of most ironstone granules supposes a gently turbulent and rather oxidizing environment.* This apparent paradox has been much discussed in the literature. Some interpretations involve a two-step formation. They include for instance the post-sedimentary reduction and replacement of initial carbonate oolites, or the repeated alternation for each oolite sheet of physical accretion in oxidizing conditions and of replacement in reducing conditions (Gygi 1981). Other interpretations imply the post-sedimentary accretion of ferrous granules within reducing deposits and not in free water, or accretion processes at the sediment surface in calm and oxygen-depleted water. Some authors envisage that ooids of oolitic ironstones resulted from in-situ chemical weathering in lateritic soils and ferriferous crusts (e.g., Nahon et al. 1980), and are not shallow-marine sediments. Such a hypothesis can usually hardly apply, since most ironstones display obvious sedimentary structures, and ooids have sediment-born tangential-concentric internal fabric instead of radial-concentric (Bhattacharyya and Kakimoto 1982). A last explanation consists of a diagenetic reduction of ferric minerals into ferrous ones, as quoted above (see discussion in Odin 1988).

The geologic record of glauconitic peloids and chamositic ooids shows some parallelisms and differences (Van Houten and Purucker 1984). Chamositic and glauconitic peloids have been known since early Precambrian time, chamositic ooids appeared in late Precambrian time. Two major episodes of accumulation of glauconitic and chamositic (chamosite, berthierine) granules occurred during Phanerozoic time, the first in the early Paleozoic and the second in middle-late Mesozoic (Fig. 10.10). Both episodes were characterized by dominantly temperate to warm climate, widespread sea-level rise and transgression of cratons, dispersed cratonic blocks and open ocean gateways, and frequent decreased influx of terrigenous sediment. After the breakup of the Precambrian supercontinent about 600 my ago, the Cambrian transgression on large Laurasian cratonic blocks determined widespread glauconitic deposits. About 50 my later chamositic deposits began and developed on smaller blocks during the late Cambrian-early Ordovician high sea-level periods. After the breakup of Pangaea in late Triassic time about 200 my ago, chamositic deposits formed widely during Jurassic on Laurasia and Gondwana dispersed blocks bordering the opening Tethys. Glauconitic deposits reached a maximum 50 my later, especially along continental margins of the opening Atlantic basins. Abundant glauconitic greensands persisted until the middle Cenozoic, while oolitic iron-rich deposits strongly decreased after the late Cretaceous.

Despite the general correlation observed between Phanerozoic high stands of sea level and the development of iron-rich clay and oxide granules, there are no very close connections between both phenomena, especially as far as ironstones are concerned. For instance, some oolitic ironstones accumulated when cratonic blocks were consolidated and global sea level was relatively low, as in the middle Permian, late Triassic, earliest Jurassic and late Cenozoic time (in Van Houten and Purucker 1984). Paleozoic chamositic ooids preferentially correlated with fairly high sea level and high latitudes, while Mesozoic berthierine ooids corresponded rather to middle sea-level stands and low latitudes. Glauconitic greensands depend more closely on high sea-level periods. Note that several

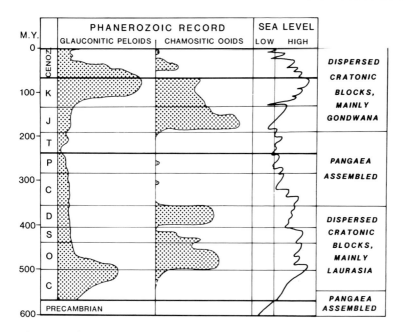

Fig. 10.10. Phanerozoic record of glauconitic peloids and chamositic ooids. Estimations collected by Van Houten and Purucker (1984)

questions remain unsolved. For instance, we do not know why during favorable episodes the development of glauconitic peloids was widespread but mostly marked by scattered granules, while chamositic ooids were less common but usually concentrated in thick ironstones. Ironstones developed mainly in Northern Europe and Northwestern Africa during early and middle Paleozoic time and in Northwestern Europe during Jurassic time. Additional questions, among others, concern the possible correlation of ironstone sequences and Milankovitch cycles. Van Houten (1986) reports in ten major sedimentary basins the repeated occurrence of 250,000- to 500,000-year asymmetric units of oolitic ironstone, developed locally between upward shoaling sequences. Such sequences could correspond to the 400,000-year Milankovitch climate cycles.

10.5 Celadonite Facies

Celadonite is a 10 Å sheet-silicate rich in octahedral ferric iron and interlayered potassium, forming green clayey accumulations in marine environment. *Celadonite* therefore *resembles glaucony in several aspects.* Even the chemical composition of individual particles of a given population may present partly overlapping domains between both minerals, as documented for instance by microprobe analyses on celadonite facies from Cyprus (Duplay et al. 1986). Let us briefly compare the characters of celadonite and glaucony. The following summary comes mainly from the review of Odin et al. (in Odin 1988) on the celadonite-bearing facies.

Celadonite is typically associated with volcanic rocks, especially lavas. It may present *five morphologic features*: filled vesicles, replaced phenocrysts, diffuse habit, films, veins. Celadonite never forms grains like glaucony. In its different habits, celadonite is always *accompanied by other minerals* that often constitute a genetic sequence marked by five steps: (1) Submarine weathering of lava (palagonitization) and precipitation of *silica*. (2) Genesis of *saponite* (Mg-smectite) in a still oxidizing microenvironment. (3) Celadonite formation, leading to reduced porosity. (4) Second genesis of clay, richer in magnesium and Al-depleted. (5) Precipitation of *calcite* and *zeolites*, determining the near closure of porosity. Other associated clay minerals may include *nontronite* (Fe-smectite) and sometimes chlorite or chlorite-smectite mixed-layers. *Celadonite is only a very accessory constituent of most altered basalts*, relatively to other clay minerals, especially saponite.

Celadonite is better defined and more constant than glauconitic minerals, from morphologic, crystallographic and geochemical points of view. Celadonite usually consists of small, fairly broad and well-shaped laths. X-ray diffraction diagrams display sharp and stable peaks, indicating a much better-ordered structure than glauconitic minerals. In particular, the 10 Å basal peak is significantly narrower for celadonite than for glauconite. The chemical composition of celadonite is marked by almost exclusively silicic tetrahedra, a low content of aluminum in octahedra, a fairly high amount of octahedral magnesium and a high abundance of potassium (Table 10.2), which differs notably from glauconite (see also 1.3.4). A characteristic formula of celadonite is:

$$K_{0.85} (Si_{3.95} Al_{0.05}) (Fe^{3+}_{0.9} Fe^{2+}_{0.25} Mg_{0.6} Al_{0.25}) O_{10} (OH)_2.$$

The formation of celadonite depends on exchanges between mineral products and sea water in a *semi-closed microenvironment*, as for glaucony granules. But the sea water involved in celadonite genesis has migrated deep within the fissured

Table 10.2. Mineral and geochemical differences between celadonite and glauconitic minerals. (after Odin et al. in Odin 1988)

	Celadonite	Glauconitic minerals
X-ray patterns	Sharp peaks	Broad peaks
Crystal shape	Well-shaped laths	Globules to irregular blades
Crystal size	Length 15 to 30 μm	Diameter 5 to 10 μm
Tetrahedral Al	<0.2/4 sites	0.25–0.5/4 sites
Octahedral Al	0.6–0.8/2 sites	0.3–0.5/2 sites
Interlayered cations	High content	Low to high content
Maximum $K_2O\%$	9.5–10.0	8.5–9.0
SiO_4 range %	52–56	47.0–50.0
Al_2O_3 range %	0.5–6.0	3.5–11.0
Fe_2O_3 range %	16–28	19–27
MgO range %	5–7	2.6–4.6
K_2O range %	7.0–9.5	3.0–8.5

and altered basalts before reaching the voids where clay growth takes place. *The chemical composition of water is* therefore modified relatively to normal sea water, and noticeably *enriched in volcanogenic elements.* This explains the frequent occurrence of a complex mineral sequence associated with celadonite formation. In addition, the temperature of sea water, measured from oxygen-isotopic composition of celadonite, ranges from a few degrees to a maximum of about 50°C. Such a temperature is often higher than that reported for glaucony, although it remains below the values usually attributed to high-temperature hydrothermal alteration processes (50 to 700°C).

Other differences between celadonite and glaucony formation concern time constraints. Celadonite probably results from a *one-step process,* contrary to glauconitic minerals that have undergone several phases of precipitation, dissolution and recrystallization (10.3). By contrast, glaucony grains display the characteristic green color and contain glauconitic smectite after a few thousand years of evolution, while *a million years or more is apparently necessary to allow celadonite formation from a submarine basalt altered by sea water.* All these characters point to the usual existence of *clear differences between two families of ferric green clays forming in both volcanic and nonvolcanic surficial marine environments.*

10.6 Conclusion

1. Great progress has been accomplished over the past few decades in the knowledge of green clay granules, especially those of recent sediments. Modern green granules represent typical marine, ferric particles, forming close to the sediment-water interface in the upper part of continental margins and submarine topographic heights, in a semi-confined microenvironment. Green granules appear to form in regions marked by reduced terrigenous supply, essentially through neoformation processes. The genesis of green granules is favored by high sea-level stands, but does not correlate necessarily with transgressive stages, sedimentation gaps, or turbulent water. Iron-rich clay granules develop from various substrates, the most suitable being calcareous microfossils and fecal pellets. The rapidity and intensity of greening noticeably depends on the nature and size of host substrates and on the microchemical environment.

2. Green clay granules in modern sediments comprise two types. Verdine (Odin 1985, 1988), that corresponds to the formerly-named berthierine nonoolitic granules, consists mainly of a 7 Å ferric clay mineral named phyllite V, and is apparently identified in late Pleistocene-Holocene sediments only. Verdine develops in shallow-water zones (maximum 50-80 m depth) of intertropical regions, close to river mouths providing abundant dissolved iron. The second type of green granules is referred to as *glaucony,* a group characterized by TOT minerals of various types: Fe-smectite, glauconitic smectite and mixed-layers, and glauconite (Fe-K illite). Glaucony statistically develops in marine areas deeper and cooler than verdine. Green granules could have formed during Holocene time at least at latitudes as high as 50° and at water depths of 1000 meters; nevertheless the maximum of glaucony production occurs between 150 and 300 meters in temperate-warm to equatorial regions. Glaucony formation proceeds in several stages, marked by a strong enrichment successively in iron and potassium, by an increase of initial volume determining external cracks, and by the disappearance of the original shape of the substrate. Glaucony also comprises film and diffuse habits, in contrast to verdine.

3. *The chemical evolution of green clay granules stops either after a long exposure at the sediment surface* (about 10^5-10^6 years for glaucony), *or after significant burying* (usually a few decimeters). Surprisingly, ancient glaucony minerals often display better crystallinity than modern ones, without any noticeable chemical difference. Depositional conditions are therefore different in ancient and recent times, or more likely a crystal ordening occurred during diagenesis (post-sedimentary processes). This example illustrates *the difficulty frequently encountered in applying the knowledge of modern green granules to the understanding of genesis of ancient glauconitic peloids and oolitic ironstones.* Most past iron-rich clay granules appear to have formed like recent ones in semiconfined environments located at the sediment-sea water interface, but many differences appear and many questions still remain (see Van Houten and Purucker 1984, Odin 1988).

Ancient clay-rich granules formed in more varying environments than recent ones, and are reported from freshwater-saline water mixing zones to deep offshore. Some oolitic ironstones are even attributed to the pedologic evolution of argillaceous sandstones in lateritic ferriferous crusts (Nahon et al. 1980). The near-absence of modern oolitic clay granules and of recent berthierine and chamosite, the apparent antagonism existing between the reducing environment necessary to generate ferrous minerals and the oxidizing-agitated environment necessary to generate oolites still complicate the explanation of ironstone genesis. Numerous hypotheses have been proposed and knowledge progresses slowly. We still hardly understand the diagenetic relations that may exist between verdine and berthierine granules, or between 7 Å berthierine, mostly Mesozoic in age, and 14 Å chamosite, mainly Paleozoic. If glauconitic peloids and chamositic ooids roughly correlate with high sea-level stands and dispersed cratonic blocks in the geologic record (Fig. 10.10), we know little about the constraints that may exist between granule genesis and climate characteristics, eustatic cycles, and Milankovitch cycles. We hardly understand the reason of the wide but fairly scattered distribution of glaucony deposits as opposed to the more local but massive deposition of ironstones. In addition, much has still to be investigated in the relations occurring between glauconitization and the formation of goethite, phosphate, dolomite, and pyrite, as well as in the microbial and organic control of clay granule genesis in both recent and ancient environments.

4. *The formation of green clay granules represents the first important step of incorporation of the terrigenous iron in mineral structures within the marine environment.* Continental shelves and upper slopes marked by glauconitization and accessorily by verdinization serve as geochemical filters during the transfer of dissolved elements from land to sea, a similar role being played in the deeper ocean by metalliferous sediments (Chap. 12). *Some convergences appear to exist between the formation of recent iron-rich clay granules and other surficial processes characterized by iron and often potassium fixation.* For instance, *gray pellets* may ex-

perience rapid iron and potassium enrichments within micaceous constituents in the absence of any greening (Bayliss and Syvitski 1982), and finally be dissociated among common muddy sediments; the resulting clays present a chemical composition slightly different from that of pure detrital clays. Nongranular enrichment of iron and potassium also seems to characterize some ancient intertidal or lagoonal *sediments submitted to wetting and drying cycles* (9.3.3). Finally the *celadonite-bearing facies*, despite its mineral, chemical, morphological, and environmental peculiarities related to volcanic dependence, presents iron and potassium enrichment, semi-confined conditions, and greening mechanism comparable to those typical of iron-rich clay.

Chapter 11

Organic Environment

11.1 Influence of Organic Activity on Clay

The activity of benthic organisms living above or within sediments may have strong effects on physical properties like sorting, particle orientation (i.e., fabric) and aggregation (Rhoads and Boyer 1982). The influence of bioturbation on the fabric of argillaceous sediments, and the dependence of ichnofacies on different types of sedimentary substrates have been especially investigated for consolidated deposits. For instance, O'Brien (1987) considers the macrofabric (X-radiography) and microfabric (scanning electronmicroscopy) of 50 samples of shales from five stratigraphic units: Jurassic of England, Permian of western Texas, Pennsylvanian of Iowa, Pennsylvanian of Illinois, Devonian of western New York state. The fabric of clay particles within the shales appears closely linked to the intensity of bioturbation:

1). Black shales formed under anaerobic conditions and devoid of bioturbation structures display fine lamination and a primary fabric characterized by parallel clay flakes.

2). Gray to dark-gray shales correspond to more oxic conditions and comprise either extensively or only partly bioturbated facies. The first facies shows a homogeneous aspect on X-radiographs with few recognizable burrows, as well as individually, randomly oriented particles similar to unbioturbated, flocculated clay; however, there are no the cardhouse structures typical of flocculated clays. The second facies includes both areas of randomly oriented particles and areas of parallel to subparallel particle orientation, which suggests a less important sediment mixing by benthic species.

N. O'Brien reports that some consolidated shales comprise both randomized fabric in burrows and preferred orientation of clay particles in interburrow areas. The coexistence of both fabrics near one another in old shales indicates that compaction and lithification do not noticeably realign clay particles, perhaps because of the existence of organic binders (e.g., mucus) in infilled burrow sediment.

Note that Reynolds (1987) finds no systematic correlation between bioturbation and random orientation, or between pelagic settling and preferred orientation, in shaly sediments from the Los Angeles basin, California. S. Reynolds observes that some bioturbated sediments possess a similar clay fabric inside and outside the burrows, and that much of the preferred orientation in recent sediments may be due to silt-sized micas rather than to clays. This points to the necessity of not oversimplifying the relationships existing between bioturbation and clay fabric.

The problem of clay mineral and chemical modifications by dwelling and feeding activity of marine organisms is still scarcely investigated. *The burrowing infauna may significantly modify the distribution of some mobile and soluble chemical elements, but does not seem to alter the original clay composition.* Piper et al. (1987) study haloed burrows in box-cored sediments from the equatorial North Pacific (151°W). Early Tertiary dusky-brown siliceous clay is blanketed by less than 5 centimeters of Quaternary sediments of the same nature. Burrows comprise a dark yellowish-brown central zone of Quaternary filling, a pale yellowish-orange surrounding zone (the halo) of Tertiary sediments, and an outermost zone of metal-oxide precipitate. Several metal elements like Mn, Ni, Cu, Co, Zn, Sb, and Ce have been leached from the light-colored halo. The major oxides, rare earth elements other than Ce, and several metals (Cr, Cs, Hf, Rb, Sc, Ta, Th, U) are not removed from the halo zone. The metal-oxide zone is enriched in Mn-oxides of todorokite type (up to 16% MnO_2). As the chemical composition of todorokite resembles that of the metal deficit in the halo and of surface ferromanganese nodules, the bioturbation is thought to be responsible for both redistribution of metals within pelagic clay and accretion of nodules on the sea floor. By contrast to these metal-element exchanges, the clay mineralogy,

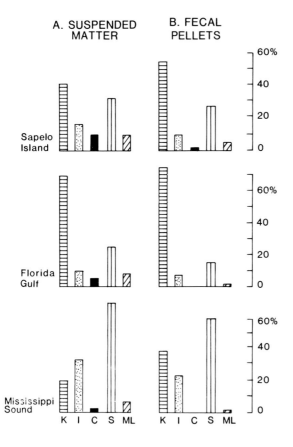

Fig. 11.1. Clay mineral assemblages (%) in suspended matter (A) and in fecal pellets of decapod *Callianassa* (B). (After Pryor 1975)

dominated by smectite in Tertiary and by illite, smectite, kaolinite and chlorite in Quaternary time, is the same in both burrows and enclosing sediments.

The alteration of clay mineral assemblages in digestive tracts of benthic organisms was probably first investigated by Anderson et al. (1958), who fed mixtures of smectite, kaolinite, and mixed-layers to oysters, clams, and fish. A.E. Anderson and others observed the partial destruction of mixed-layered clays, which was attributed to the stomach acidity of organisms.

Pryor (1975) studies the feeding activity and excretory products of the decapod *Callianassa major* and of the annelid *Onuphis microcephala*, in shallow marine sediments of the Southern Atlantic Ocean and Eastern Gulf of Mexico. These filter-feeding organisms can locally remove large amounts of argillaceous particles from suspension, and determine the deposition of fecal pellet mud-layers as thick as 4.5 mm/year. They could theoretically produce as much as 12 metric tons/km^2/yr of dry pelleted mud. W.A. Pryor shows that the sand-sized pellets consist of 80-90% clay minerals, 5-10% undigested organic debris and little sand and silt grains. The digestive process tends to alter and destroy chlorite and mixed-layers, and to disorganize illite and kaolinite structures. Smectite does not seem to be significantly modified (Fig. 11.1). *Onuphis* appears to alter chlorite and mixed-layers more intensively than *Callianassa*. These mineralogical changes apparently start in a very short time, since individual *Callianassa major* may produce an average of 2,480 pellets per day in shoreface environments of Southeast United States (Pryor 1975). The process of alteration, still little known, involves harsh and alternating conditions of acidity along the digestive tract, as well as digestive enzymes and intestinal bacteria. Clay pelleted as feces can enter the coprophagic cycles, passing through one digestive system after another, what may determine further clay-mineral alteration. Finally the organic-rich fecal pellets may serve as precursors of glaucony pellets, providing the semi-confined microenvironment necessary for glauconitization processes (see Chap. 10).

Mineral transformations by *zooplankton* are investigated experimentally by Syvitski and Lewis (1980) about *Tigriopus californicus*, a marine intertidal harpacticoid copepod species. Copepods with cleaned digestive tracts are fed by different mineral standards during a 48-hour experiment. Fecal pellets collected after 3 weeks are studied by X-ray diffraction and energy dispersive scanning

Table 11.1. Changes induced by the ingestion of mineral particles by the copepod *Tigriopus californicus*. (After Syvitski and Lewis 1980)

Initial Standard	Pellet Residue	
Minerals	Minerals	Chemical change
Microcline	Microcline	None
Illite	Illite	None
Pyrite	Pyrite	None
Clinochlore	Mg-chamosite	Slight decrease in K
Smectite, kaolinite, quartz	Vermiculite, illite, kaolinite, quartz	Increase in Mg, Ca, Fe; slight decrease in K
Tremolite	Tremolite, chamosite	Increase in Al; decrease in Ca
Vermiculite – biotite mixed-layer	Not discernable	Decrease in Mg, K, Ca, Fe

electronmicroscopy. Clinochlore appears to alter to Mg-chamosite with a reduction in Mg and an increase in Fe. Tremolite partially alters to chamosite with a reduction in Mg and Ca. Smectite apparently changes into vermiculite and mica-illite with an increase in Mg, Ca and Fe (Table 11.1). Gypsum forms in some pellets. All these modifications are attributed to mechanical and chemical mecanisms within the acidic digestive tract of the copepod. J.P.M. Syvitski and A.G. Lewis think such phenomena could in part be responsible for some chemical modifications of clay mineral assemblages in the marine environment.

11.2 Late Cenozoic Sapropels, Mediterranean Sea

11.2.1 Sapropel Characters and Environment

Organic-rich layers interbedded between hemipelagic oozes commonly occur in late Cenozoic sediments of the Eastern Mediterranean Sea, and have been studied extensively since their first description by Kullenberg (1952). Synthetic studies on Mediterranean organic-rich layers include those of Ryan and Cita (1977), Kidd et al. (1978), Sigl et al. (1978), Thunell et al. (1984). The term *sapropel* refers to discrete layers 1 cm to more than 100 cm thick and containing more than 2% by weight organic carbon. *Sapropelic marls* consist of similar sediments but with 0.5 to 2% by weight organic carbon. Sapropels comprise hemipelagic sediments mostly found on topographic heights among normal nannofossil clayey oozes, and turbiditic sediments characterized by resedimentation structures and rather associated with the sedimentary fill of trenches and depressions.

Sapropels are basically characterized by a *black to black-greenish color and a high organic content*. Organic carbon and total nitrogen contents of Pleistocene sediments from the Eastern Mediterranean (Leg 13 DSDP) range between 0.08 and 20-21% and between 0.016 and 1.275%, respectively (Calvert 1983). Average contents of carbon, nitrogen, sulfur, and carbon dioxide clearly differ in sapropels, sapropelic marls, and marls (Table 11.2). Thunell et al. (1984) report that late Pleistocene sapropels from the Ionian and Levantine basins comprise from 2.2 to 7.8% organic carbon (4 to 14% total organic content). Data from various DSDP sites show values of organic carbon content from 2.5 to 7.0% for early Pleistocene sapropels, 3.2 to 5.5% for late Pliocene sapropels and 1.7 to 3.8% for early Pliocene sapropels. Sapropels are often characterized by numerous parallel *laminae* consisting of alternating coccolith-rich and clay-rich muds, that may represent seasonal cycles (upwelling, overturn of surficial water masses). This varve-like lamination tends to be well preserved since most sapropelitic environments were devoid of significant benthic activity and bioturbation because of oxygen depletion on the sea floor. Other frequent characters of Pleistocene sapropels include the abundance of *diatom* and silicoflagellate debris, of *pyrite-marcasite* and diffuse iron sulfides, of minor *chemical elements* like barium, copper, molybdenum, nickel and zinc (Table 11.2). Sapropels and

Table 11.2. Mean chemical characteristics of Pleistocene sapropels, sapropelic marls and marls from Eastern Mediterranean Sea. (After Calvert 1983). Data from sites 125 to 130 of Leg 13 DSDP

	Sapropels	Sapropelic marls	Marls
%			
C	6.10	1.12	0.22
N	0.41	0.10	0.03
S	2.27	1.31	0.44
P	0.07	0.04	0.04
CO_2	14.61	15.08	21.27
ppm			
Ba	1269	856	639
Cu	129	57	69
Mo	73	14	3
Ni	130	77	49
Zn	117	81	73
As	7	15	3
Mn	1052	1266	954
Pb	5	6	5
Rb	47	55	52
Sr	646	677	820
Zr	92	112	190
CO_3Ca	1%	to	74%
C/Norg	16.8	to	2.4

sapropelic marls often display lower carbonate contents and higher C/N_{org} ratios than those in surrounding marl oozes. Major elements usually show only small differences between sapropels and common oozes (Calvert 1983).

The respective contribution of terrestrial and marine matter to the organic content of sapropels remains partly in question. The high C/N ratios combined with the presence of fragments of higher plants, the spectra type of *n*-alkanes, the composition of humic fractions, the nature of pyrolysis products, as well as carbon isotopic data, suggest that a large part of the organic matter is terrigenous (see Calvert 1983, Thunell et al. 1984). However, Calvert (1983) estimates that these arguments are not definite proofs, and demonstrates from calculations on sedimentation rates that the production rate of organic carbon in Quaternary sapropels should have been much higher than today to allow the formation of organic-rich layers. Such a production could have been favored by a high marine productivity, which suggests a noticeable contribution of marine organic matter to the sedimentation during sapropel deposition. Sutherland et al. (1984) study the most recent sapropel (Holocene) in eastern Ionian Sea, and conclude from carbon isotopic values ($\delta^{13}C = -2‰$) and C/N ratio (about 10) that organic material is essentially marine. ten Haven et al. (1987) investigate four late Quaternary sapropels in two piston cores from the Southeastern Mediterranean. The recognition of diagnostic organic compounds indicates that the relative terrigenous contribution does not vary significantly from one sapropel to another. By contrast, the marine contribution is characterized by relatively abundant dinoflagellates in the most recent sapropel (Holocene), while older sapropels display numerous remains of either diatoms or prymnesiophyte algae and planktonic cyanobacteria.

The deposition of sapropels occurred periodically throughout most parts of the Eastern Mediterranean Sea, including Adriatic basin, during the Plio-Pleistocene (see Thunell et al. 1984). Late Pleistocene times are peculiarly well documented. Sapropels appear to be restricted to water depths below 700-900 m, and are lacking in cores from the strait of Sicily. Although most sapropels are typical of eastern basins, a few organic-rich layers locally occur in Pleistocene sediments of the Tyrrhenian Sea (e.g., legs 13, 42A DSDP, leg 107 ODP). Biostratigraphic and oxygen isotopic data show that *sapropels and sapropelic marls can be chronologically correlated in all eastern basins* and partly even in Tyrrhenian Sea (in Thunell et al. 1984). Nine of the ten *sapropels deposited during the last 350,000 years in the Eastern Mediterranean Sea correlate with climate warmings or interglacial stages.* In addition Rossignol-Strick (1983) suggests from palynological data that late Pleistocene sapropels correspond to periods of unusually high summer insolation in the Northern Hemisphere and intensified monsoonal conditions in Africa.

Late Pleistocene-Holocene sapropels and sapropelic layers are generally attributed to *depositional conditions in an anoxic basin caused by the existence of a surficial low-salinity layer and of resulting density stratification.* The present-day equivalent could be the situation existing in the Black Sea (Fig. 11.2A), whose light surficial water issues from the Danube river, that is separated from the Mediterranean by the Bosphorus sill. The sill limiting oxygenation processes in the Quaternary Mediterranean was probably represented by the strait of Sicily. The low-salinity surface water preferentially developed during interglacial, often pluvial stages. It probably resulted from fresh-water supplied from Black Sea during high sea-level periods and/or from Nile river drainage basin during flooding episodes (references in Cita and Grignani 1982, Thunell et al. 1984). Additional arguments for the climate and high sea-level control arise from the large $\delta^{18}O$ depletions and euryhaline fauna associated with most sapropels.

Calvert (1983) tentatively proposes another explanation from data obtained on carbon accumulation rates and comparisons with the Black Sea. Rather than a direct consequence of the formation of stagnant bottom water during glacial retreat, sapropels would result from an increased flux of carbon to the sea floor caused by a higher primary production rate. S.E. Calvert envisages the possibility of increased planktonic productivity, and inferred organic carbon accumulation in sediments, as the result of a reversal in circulation in the Mediterranean Sea caused by excessive flood water runoff via the Nile river. During periods of flooding, the surface water would have flowed out from the Mediterranean, determining *upwelling* in the eastern basins and high productivity. A present-day corresponding situation would be the western margins of Africa submitted to upwelling of nutrient-rich deep water (Fig. 11.2C). Such a hypothesis, that also involves a climate dependance, appears somewhat questionable, since no definite arguments are given in the literature about the possibility of a current reversal in Gibraltar strait at least during the Pleistocene-Holocene transition (see ref. in Cossement et al. 1984). In addition, high productivity conditions do not easily explain the limitation of sapropels below about 800 meters water depth.

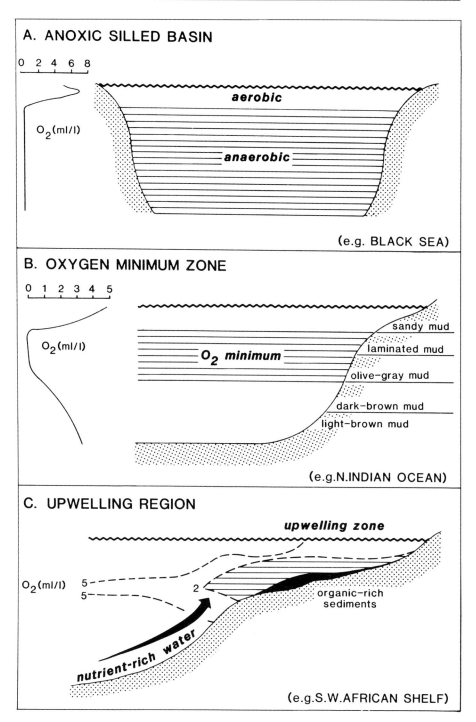

Fig. 11.2. Models for the deposition of organic-rich sediments. (After Thiede and Van Andel 1977 and Demaison and Moore 1980, in Thunell et al. 1984)

The density-stratification model appears therefore most suitable for explaining the development of sapropels in the Eastern Mediterranean Sea during late Pleistocene, early Pleistocene, and latest Pliocene time. In contrast, early and mid-late Pliocene sapropels correspond to particularly small isotopic signals, warm-type planktonic foraminiferal faunas, and unusually warm and saline bottom waters expressed by benthic foraminifera (Thunell et al. 1984). These data suggest sluggish circulation and low oxygen contents in bottom water, due to stable, warm climate conditions in the Eastern Mediterranean Region. Such stable conditions prevented the ventilation of eastern basin and water exchanges with the Western Mediterranean.

11.2.2 Clay Evolution

Sapropels often display specific clay mineral characters, relatively to under- and over-lying common sediments. Hemipelagic oozes contain diversified clay assemblages, whose constituents vary in nature and abundance according to the dependence on terrigenous sources (Rateev et al. 1966, Chamley 1971, see Chap. 8). For instance, smectite abundance increases in the Levantine basin toward the

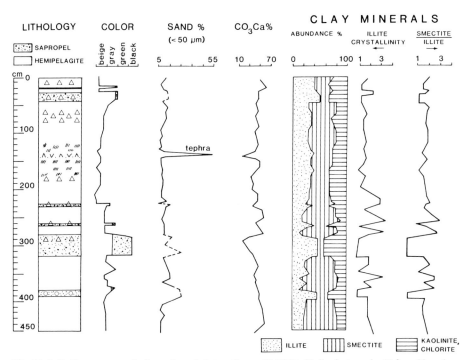

Fig. 11.3. Sedimentary and clay mineral data of core 3MO67, Hellenic trench. (After Chamley 1971). Illite crystallinity: width at half-height of 10 Å peak (1/10 θ, glycolated sample). Smectite/illite: 17/10 Å peak-height ratio (glycolated sample). *Arrows* increasing crystallinity or abundance

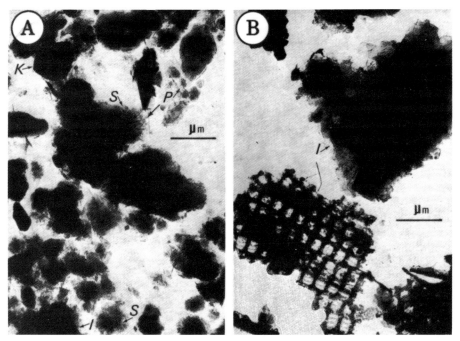

Fig. 11.4. Transmission electron micrographs of late Pleistocene sediments from core 3MO67, Hellenic trench, Eastern Mediterranean. (After Chamley 1971). **A** Hemipelagic ooze, 270 cm. Flaky smectite and mixed-layers with varying sizes (*S*), well-shaped illite and chlorite (*I*), hexagonal kaolinite (*K*), short and broken palygorskite fibers (*P*). **B** Sapropel, 282 cm. Exclusive large-sized and thick mica-illite and chlorite sheets with corroded edges (microcavities), diatom fragments

East, while illite and chlorite characterize the North Ionian basin in the West, and kaolinite and palygorskite the regions located off Libya and Tunisia. By contrast, *the sapropels tend to contain less diversified clay associations and to be especially depleted in smectite* (Fig. 11.3) and in palygorskite. Chlorite and mixed-layer abundance sometimes appears to increase (Cita et al. 1977, Sigl et al. 1978, Lunkad 1986). The crystallinity of illite-mica increases in sapropels marked by a strong diminution of smectite, and is stable or slightly decreasing when mixed-layers are more abundant. All these changes are confirmed by X-ray experiments on various grain-size fractions of sapropels, before and after removing the organic matter.

On transmission electron micrographs, common sediments display varying types, shapes, and sizes of clay particles; smectites show common flaky shapes and occur in grain sizes from 0.1 to more than 2 μm (Fig. 11.4A). By contrast numerous sapropels tend to lack small-sized particles, particularly of flaky smectites, and to mainly consist of large, usually well-outlined minerals. In some sapropels, especially those issuing from deep depressed areas like the Hellenic trench, illite and chlorite edges show microgulfs of corrosion and microcavities; diatom frustules often occur abundantly (Fig. 11.4B). This trend, particularly well documented on late Quaternary sapropels, also characterizes numerous early Pleistocene, Pliocene, and even Tortonian organic-rich sediments (Fig. 11.5). Messinian sapropel-like layers constitute an exception, since they usually show no clay mineral difference with surrounding common sediments (Sigl et al. 1978).

The cause of the strong mineralogical changes recorded in many sapropels has been discussed from sedimentological, mineralogical, micromorphological and geochemical data collected on piston and drilling core sediments from different parts of the Eastern Mediterranean (Fig. 11.6. Chamley 1971, Cita et al. 1977, Sigl et al. 1978). Two major hypotheses can be proposed.

A drastic change in terrigenous supply could be responsible for a simplification of clay assemblages and a selection of large, well-shaped micaceous particles. For instance, a tectonic rejuvenation or strong climate cooling could determine an increased supply of large, rock-derived detrital illite and chlorite. In fact sapropel deposition mostly corresponds to interglacial and pluvial conditions, characterized by high sea level and rather strong hydrolysis processes on land. Such conditions favor the production of abundant soil-derived products marked by various minerals often characterized by small-sized smectites. By contrast, the removal of large rock-derived illite and chlorite particles is rather disfavored under active chemical weathering conditions. In addition, smectite is easily transported over great distances within the sea, and high sea-level conditions favor the settling of its small particles . The hypothesis of a major modification in terrigenous input therefore appears unlikely.

The second hypothesis involves *a chemical change within the sapropel environment*. Clay minerals would be partly degraded and even destroyed under strongly reducing conditions, which would determine a mineralogic simplification. Several arguments support this hypothesis of a submarine degradation. Among minerals that preferentially disappear are smectite and palygorskite, two families characterized by small grain size, high specific surface, high exchange capacity, and low interlayer charge: these characters favor chemical attacks and reactions. Electron microscope observations show that particles preferentially preserved are rather coarse sized and display frequent microcorrosion figures. These large particles mainly consist of mica-illite and kaolinite, two clay families marked by fairly high layer charge. In addition, the clay-mineral simplification preferentially occurs within sapropels deposited in most confined marine areas, where organic deposition is high and acidic conditions well established. For instance, pH measurements made on sapropels from the deepest part of the Hellenic trench show values lower by one unit than the surrounding common sediments (e.g., pH 7 instead of 8, measured in a suspension of 10 cc sediment in bi-distilled water; Chamley 1971).

The agents responsible for the mineral destruction most likely consist of *organic acids*, well known for their ability to attack silicate minerals under experimental conditions (e.g., Huang and Keller 1971, 1972). The bacterial activity probably also participated in this submarine hydrolysis (halmyrolysis), that partly resembles mechanisms occurring on land in acidic soils of the podzol group (Chap. 2) and in underclays deposited in reducing swamps (Millot 1964, 1970). The cations subtracted from clay mineral layers were probably trapped by the diffuse organic matter acting as an absorbing complex (see Chester 1965), which

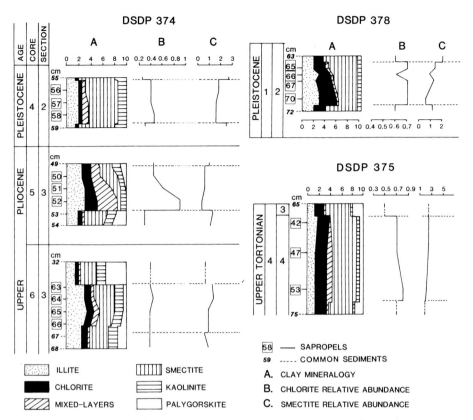

Fig. 11.5. Clay mineralogy of selected sapropels and associated common sediments from leg 42A drill sites. (After Sigl et al. 1978). Chlorite relative abundance: 4.7/5.0 Å peak-height ratios, natural sample. Smectite relative abundance: 17/10 Å peak-height ratio, glycolated sample

prevented the chemical saturation of interstitial water and permitted the continuation of subaqueous hydrolysis processes. The intensity of clay mineral alteration in the Pleistocene series depends neither on the age of sapropels nor on the depth of burial. Some sapropels showing strongly degraded clay assemblages lie above others that display very little alteration (e.g., site 374 DSDP, cores 6 and 5, Fig. 11.5; site 376, cores 6 and 1; site 378, cores 3 and 1). This indicates that *the clay degradation in sapropels probably occurred very close to the sediment-sea water interface and not after significant burying.*

Studies performed on Tortonian, Pliocene, and Pleistocene sapropels and sapropelic marls from various Eastern Mediterranean sites (Fig. 11.6) reveal that *the submarine clay alteration presents different modalities according to the initial mineral composition, the abundance of organic matter, the geographic and bathymetric position, and the sample location in a given organic horizon* (Fig. 11.7). The nature of the degradation appears to first depend on the composition of original terrigenous assemblages. For instance when palygorskite content is noticeable (> 10% of clay fraction), the other clay minerals seem to be to some

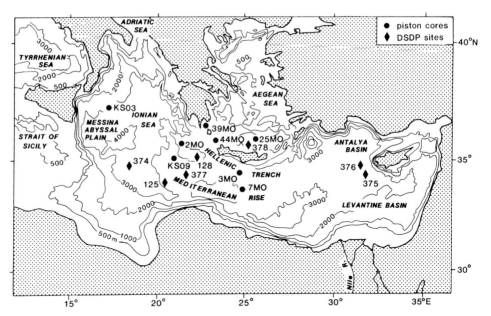

Fig. 11.6. Location of piston cores and DSDP sites studied for clay mineralogy of sapropels and sapropelic marls. (Chamley 1971, Sigl et al. 1978)

extent protected from degradation, at least until a large part of the fibrous clay is destroyed. When palygorskite is rare or absent, smectite is altered first. A good example of such *mineral barriers and steps in the clay degradation* is given in the study of Pliocene-Pleistocene sapropels of site 374, located in the central part of the Messina abyssal plain. Palygorskite, abundantly supplied during Pliocene time, is strongly destroyed in corresponding sapropels, while smectite is protected from alteration (Fig. 11.5). During Pleistocene time, palygorskite is rare or absent and smectite is preferentially degraded.

Nearly all clay minerals brought into the sapropelitic environment may be degraded, except perhaps kaolinite, whose abundance appears to remain unchanged. The resistance of kaolinite crystals probably results from their good structural organization and high layer charge, and from their relative stability in low pH conditions (Keller 1970). It is not known if kaolinite could even be formed in some sapropels, as in some continental organic-rich sediments such as tonsteins and underclays (Millot 1964, 1970). By contrast, palygorskite represents the most fragile mineral. A fairly close correlation exists between the content of organic carbon, the order of minerals successively affected by submarine degradation, and therefore the successive steps of the mineralogic simplification (Sigl et al. 1978, Lunkad 1986). The more abundant the organic carbon content, the more degraded the clay assemblages tend to be. *The simple minerals successively affected by degradation are distributed in the following order: palygorskite, smectite, chlorite, illite, kaolinite.*

Transitional minerals develop during clay halmyrolysis in sapropels. The most common transitional products identified in Eastern Mediterranean areas consist

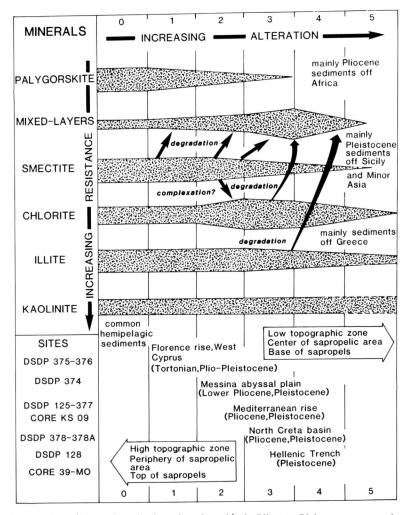

Fig. 11.7. Interpretation of clay-mineral submarine alteration in Pliocene-Pleistocene sapropels of the Eastern Mediterranean. (After Sigl et al. 1978)

of *irregular mixed-layers* (e.g. Nesteroff 1973, Sigl et al. 1978; Fig. 7.5), which resembles chemical weathering processes occurring in continental soils submitted to hydrolysis in temperate-humid regions (see Chap. 2). Smectite-vermiculite, chlorite-vermiculite, chlorite-smectite, illite-vermiculite and illite-smectite mixed-layered types may develop from the successive degradation of smectite, chlorite, and illite, and present temporarily increased proportions in clay assemblages (Fig. 11.7). If the organic influence is very strong, degradation is replaced by destruction, mixed-layered minerals do not form abundantly, and illite crystallinity displays high values because of the absence of intermediate structures.

Another type of transition consists in the development of *chlorite-like minerals*, that occurs when the organic attack is weak or moderate, and always lower than that required to produce the greatest development of mixed-layers (Fig. 11.7). Increased amounts of chlorite-like minerals in sapropels are reported on the Mediterranean rise, on the Florence rise West of Cyprus, and partly in the Messina abyssal plain (Cita et al. 1977, Sigl et al. 1978). This mineral stage apparently results from a complexation effect on smectite caused by organic matter being adsorbed in clay interlayers, bringing about a pseudo-chloritization. Such processes have been experimentally demonstrated. For instance Heller-Kallai et al. (1973) observe that the adsorption of molecular amines by montmorillonite can determine the protonation of amine groups, the retention of exchangeable cations, and the difficulty for the mineral to collapse by heating. Moinereau (1977) shows that the adsorption of organic acids by pedogenic smectites induces the resistance of clay sheets against collapse at temperatures as high as 500°C; organo-mineral complexes form as shown by differential thermal and infrared absorption curves, and disappear after peroxidation. Similar phenomena probably occurred in some Quaternary sapropels of Eastern Mediterranean close to the time of deposition. An argument is provided by the fact that sapropels with unusually abundant chlorite-like minerals in the clay fraction include in their coarse fraction terrigenous chlorite particles whose abundance is common and similar to that of surrounding organic-poor sediments.

In the sapropels themselves, *the intensity of clay degradation often decreases toward the top of organic-rich layers* (Fig. 11.5). This trend, hardly recognizable when alteration is weak or sapropels thin, becomes significant in the case of noticeable mineralogic changes (e.g., sites 374 and 378 DSDP, piston core KS09). The base of sapropels usually displays a sharp mineralogic contact with common hemipelagic marls, contrary to what occurs at the top. The upward diminution of clay mineral peculiarities usually correlates with a more or less progressive return to normal chemical composition (organic matter, trace elements) and sedimentary features (Robert 1974). This points to the rapid onset of stagnant conditions on the sea floor, and to the more progressive reversion to normal oxygenated conditions or to important downward bioturbation in upper horizons.

The location of piston and drilling cores in the ancient stagnant basins partly controls the type of alteration encountered. *Terrigenous minerals are less weathered if the sampling area is located at the periphery of the sapropelic domain,* where water exchanges were easier than in central parts. This is obvious on the edge of the Messina abyssal plain (core KS03), where slight oxidation processes may have occurred. Similar conditions characterize the Nile cone in the easternmost Mediterranean Sea (Dominik and Stoffers 1979).

Another important feature consists in the topographic situation of organic-rich layers. *In deep and depressed areas like Hellenic trench* (cores 2MO, 3MO), *water stagnation was favored, which allowed the destruction of most simple minerals* except kaolinite and little illite. *By contrast, cores from submarine elevations or slope areas experienced weak or moderate clay degradation,* probably because of less stagnant and acidic conditions due to some water exchanges. This is well documented by sites 125 and 377 DSDP on the Mediterranean rise, and by piston cores 25MO and 44MO North of Creta (Sigl et al. 1978; see also Dominik and Stoffers 1979). Sapropelitic zones located both on submarine heights and at the basin periphery show very little clay degradation, as well as the growth of both sulfides and sulfates probably from the H_2S produced by microbial activity (Robert and Chamley 1974). This is the case of cores KS03, Messina submarine cone, and 25MO, North Creta basin.

In zones marked by weak or even no in situ submarine degradation, clay associations in sapropelic layers may serve as useful markers of terrigenous sources and of peculiarities in transport conditions. Dominik and Stoffers (1979) study the clay mineralogy of sediments from three piston cores located at the southwestern part of the Mediterranean rise, on the rise itself South of Creta, and in the eastern part of the Nile river cone (Fig. 11.8). During stagnation periods, minerals derived from local, nondistant sources became relatively abundant. This is the case of illite and chlorite South of Peloponnesus in the Ionian Sea (core 22M48), of smectite and illite on the Mediterranean rise South of Creta (PC6), and of smectite and kaolinite off the Nile river (PC32). During nonsapropelitic episodes these local minerals were relatively dispersed among mineral species derived from remote or accessory sources. Distant-source minerals include smectite and kaolinite from Sicily and North Africa in the Ionian Sea, and illite and chlorite from Peloponnesus on the Mediterranean rise and in the Levantine basin. Inter-sapropelitic stages correspond to periods of relatively active water exchange and of both horizontal and vertical circulation, favoring distant supply. The main sources of terrigenous clays remained the same during stagnant and nonstagnant stages, but

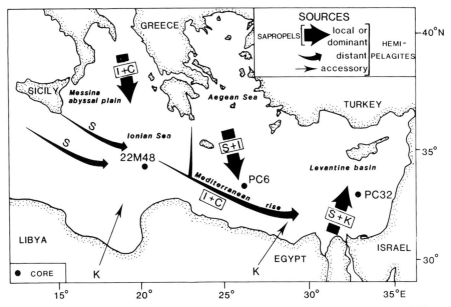

Fig. 11.8. Clay mineral sources and transportation features in Eastern Mediterranean Sea. (After Dominik and Stoffers 1979). The sapropels investigated, characterized by weak or no in situ chemical degradation, express only local or dominant sources, while common sediments depend on both local and distant or accessory sources. *C* chlorite; *I* illite; *S* smectite; *K* kaolinite

distant or secondary sources were probably of lesser importance during sapropel deposition because of density stratification and reduced circulation. Perhaps the wind activity was also weaker during interglacial and humid stages marked by sapropel development, which could have played a part in the diminution of distant supply from peri-Mediterranean land-masses into the basin.

To summarize, the following succession of clay alteration stages may be proposed in the depositional environment of Eastern Mediterranean organic-rich sediments, during the late Cenozoic (Fig. 11.7):

0. No alteration. Clay associations are exclusively terrigenous and express mineral sources, environmental conditions on land and transport conditions.

1. Very little alteration: moderate degradation of palygorskite, little development of smectitic irregular mixed-layers.

2. Weak alteration: strong degradation of palygorskite, moderate halmyrolysis of smectite leading to smectite-chlorite and smectite-illite mixed-layers, appearance of transitional chlorite-like minerals.

3. Moderate alteration: disappearance of palygorskite, increased abundance of various mixed-layers derived from active degradation of smectite and discrete degradation of chlorite and illite, formation of transitional complexation-chlorite.

4. Strong alteration: development of various mixed-layered types issued from single mineral degradation, destruction of most smectite and of small-sized and poorly crystallized chlorite and illite.

5. Very strong alteration: destruction of most clay mineral species with the exception of kaolinite and large-sized mica-illite, development of corrosion gulfs and microcavities on residual particles.

Few detailed studies appear to have been performed on organic-rich Quaternary sediments outside the Mediterranean Sea. Tompkins and Shephard (1979) investigate the central part of the Orca basin, an intraslope area marked by anoxic and hypersaline conditions in the Northwest Gulf of Mexico. Sediments usually contain abundant smectite (about 60% of the clay fraction), mainly derived from the Mississippi river (Chap. 3). A sharp decrease in smectite content occurs at 950 cm depth in a core located in the central part of the basin, which correlates with an increase in illite and kaolinite relative abundance. R.E. Tompkins and L.E. Shephard tentatively attribute this mineralogical change to a degradation of smectite in an acidic organic environment. Arguments supporting this hypothesis consist in the abundance of organic matter, the occurrence of gas expansion features immediately below 950 cm suggesting organic decomposition, and an increase in the pore water of calcium content possibly due to carbonate and smectite degradation. Other systematic studies on the clay content of recent organic-rich marine sediments do not appear to be abundant, although numerous isolated analyses exist and usually indicate the absence of significant mineral changes.

11.3 Cretaceous Black Shales, Atlantic Domain

11.3.1 Black Shale Environment

Active deposition of organic-rich black shales is known from different periods of the earth's history, especially from Paleozoic, Jurassic, and Cretaceous times. Black shales represent the source beds of many major hydrocarbon accumulations, and the origin and evolution of organic carbon in these shales largely controlled the generation of oil and gas. *Cretaceous black shales appear to be an essential source for economic hydrocarbons*, since about 60% of the world's known oil is Albian to Coniacian in age (in Schlanger and Cita 1982). The removal and burial of massive amounts of organic matter during Cretaceous time determined noticeable modifications in the global carbon budget. For instance Ryan and Cita (1977) estimate that at least 80.10^{18} grams of organic carbon were stored up in Cretaceous black shales, which is an order of magnitude greater than all identified coal and petroleum reserves. These facts, combined with the availability of numerous black shale sections from the Deep Sea Drilling Project, and with the development of new geochemical techniques, led to an active renewal of research on Cretaceous series since the mid 1970's (see review papers by Schlanger and Jenkyns 1976, Ryan and Cita 1977, Thiede and van Andel 1977, Tissot et al. 1980, de Graciansky et al. 1982, Schlanger and Cita 1982, Arthur et al. 1984a, Stein et al. 1986).

Cretaceous black shales developed in the Atlantic, Pacific, Indian and Tethyan Oceans, with a maximum extension in Atlantic basins. Deposition of organic-rich sediments in the Atlantic domain affected not only true oceanic areas but also major epicontinental seas. Black shales characterize mainly the late Jurassic-Aptian period in the southwesternmost Atlantic (Falkland area), the Hauterivian-Santonian period in North Atlantic, and the Aptian-Santonian period in South Atlantic. The shifts existing in the onset and disappearance of black shales in different areas chiefly depend on the chronology of ocean opening and widening in North and South Atlantic basins (see Kennett 1982). Three major episodes of organic-rich sedimentation, so-called "oceanic anoxic events" (Schlanger and Jenkyns 1976), occur during the Cretaceous: late Barremian to late Albian, late Cenomanian to early Turonian, and early Coniacian to late Santonian. The deposition of black shales tends to correlate with transgressive phases, especially during mid-Cretaceous time.

Diverse *mechanisms* have been proposed to explain the deposition and preservation of Cretaceous black shales. They refer to models like *stratified anoxic basins* as in the present-day Black Sea, expanded mid-depth *oxygen minimum zone* as in the Northwest Indian Ocean, and continental margin *upwelling* as on the Namibia margin, Southeast Atlantic Ocean (Fig. 11.2). Other hypotheses involve the removal of organic-rich sediments formed on continental slopes, and their transportation by *turbidity currents and slumps* within the deep sea (e.g., Robertson and Bliefnick 1983). All these mechanisms may have occurred during Cretaceous time, but some of them have been more widely distributed than others. A large part of the Cretaceous basins have probably experienced oxygen depletion in intermediate waters because of high productivity and temperature. Anoxic conditions in bottom water developed less frequently, as proved by the widespread occurrence of alternating oxygenated-bioturbated and organic-rich sediments suggesting Milankovitch-like cycles (see Berger et al. 1984).

The main factors causing the accumulation of organic carbon in the Cretaceous oceans include *the supply of terrestrial organic matter, the supply of marine organic matter, the oxygenation of deep water, and the sedimentation rate* (Stein et al. 1986). In the Atlantic Ocean, the preservation of organic matter was enhanced by the relative narrowness of the basins and equable warm climate precluding active circulation, by high eustatic sea-level favoring marine productivity and land vegetation, and by warm and saline bottom water determining reduced amounts of dissolved oxygen. In general, the depositional environment was rather oxygenated and the preservation of organic matter largely depended on sedimentation rate. Increased supply of terrigenous organic matter was especially important during Hauterivian-Albian times in Northwest and northernmost Northeast Atlantic, and during late Jurassic-Aptian in southernmost South Atlantic (Stein et al. 1986), because of high river sediment supply and humid climate on land. Anoxic deep-water conditions due to sluggish circulation and density stratification were probably restricted to relatively small and isolated basins of the South Atlantic Ocean at some periods between the late Jurassic and Albian. The only widespread episode of anoxia in the whole North Atlantic and adjacent Tethan Oceans is recorded close to the Cenomanian-Turonian boundary. Accumulation of marine organic matter due to high productivity in upwelling areas occurred locally in the south Atlantic in Aptian-Albian and in the Northeast Atlantic at Albian-Turonian times. The disappearance of organic-matter accumulation during Santonian time resulted from increased deep-sea circulation inducing ventilated deep-water conditions and erosion processes on the sea floor.

The origin of organic matter in Atlantic black shales is largely terrestrial, especially during the early Cretaceous. This fact, already shown from palynological data (Habib 1982), is strongly supported by analyses on maceral components, hydrogen index (e.g., Tissot et al. 1980, Stein et al. 1986), bitumens and kerogen pyrolysates (Simoneit 1986). *Land-supplied organic matter predominantly deposited in both North and South Atlantic Oceans from Berriasian to Albian* (Fig. 11.9A). *Marine organic matter was more frequently deposited during late Cretaceous* (Fig. 11.9B), especially in late Cenomanian-early Turonian time. The lipid matter at most DSDP sites is catagenetically *immature*, indicating a low geothermal history and a good preservation, whatever the paleoenvironmental conditions (Simoneit 1986).

Atlantic black shales comprise *various lithologic types*. Organic-rich, blackish claystones and mudstones are frequently bioturbated or laminated, and commonly alternate with or are replaced by light-colored muds, marls, marly chalks, limestones or calcareous sandstones with or without burrows and laminations. The thickness of black shales varies from a few ten to a few hundred meters. These characters differ greatly from late Cenozoic Mediterranean sapropels (11.2.1). Lithologic alternations appear to extend over very large distances and to correlate sometimes on an ocean-wide scale. The organic carbon content mostly

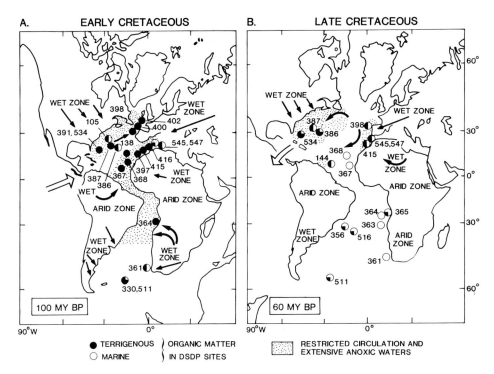

Fig. 11.9. Major origins of organic matter and restricted circulation areas in Atlantic Ocean during early (**A**) and late (**B**) Cretaceous, with main terrestrial climate areas. (After Simoneit 1986)

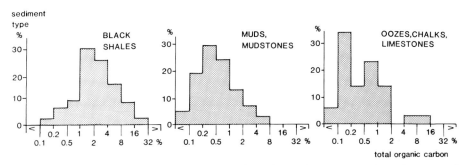

Fig. 11.10. Frequency histograms of mean organic carbon values per stage in major types of Cretaceous DSDP sediments. (After Stein et al. 1986)

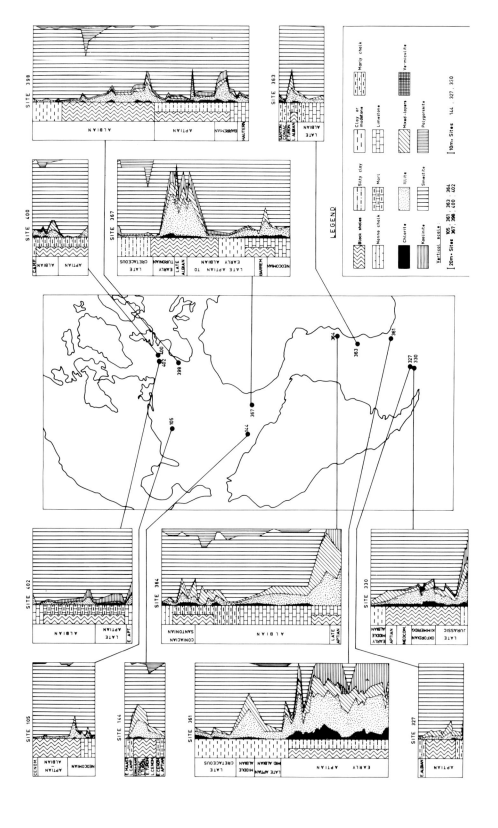

ranges between 1 and 4% (Fig. 11.10), and displays average values lower than those of many Mediterranean sapropels. The higher values recorded correspond to small, anoxic basins like the Angola basin during Albian time.

11.3.2 Clay Sedimentation

The clay mineralogy of Atlantic black shales has been investigated on numerous DSDP sites (e.g., Chamley 1979, Robert et al. 1979, Chamley and Robert 1982, de Graciansky 1982, Jacquin 1987). *Smectite* is the predominant and most ubiquitous species encountered in the clay fraction of both organic-rich and non-organic facies, commonly forming 50-80% of clay minerals (Fig. 11.11). Smectite usually presents banal flaky particles (Fig. 11.12), devoid of any degradation sign and closely resembling ordinary soil-derived minerals. *Various other clay species occur in varying amounts* beside smectite in Cretaceous black shales. They include chlorite, illite, irregular mixed-layers of illite-smectite, chlorite-smectite and rarely illite-vermiculite types, vermiculite, kaolinite, and palygorskite (Fig. 11.11). Associated non-clay minerals mainly comprise quartz, feldspars, calcite, pyrite, opal CT, clinoptilolite, and sometimes siderite, rhodochrosite or barite, diversely distributed in the sediments.

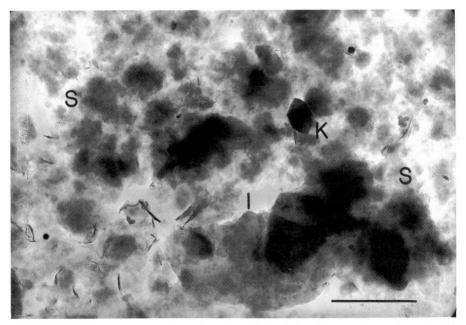

Fig. 11.12. Electron micrograph of middle Albian black shales, Hatteras Formation, site 534 DSDP, 843 m below sea floor, Blake-Bahama basin, Northwest Atlantic Ocean. Flaky smectite (*S*) with associated hexagonal kaolinite (*K*) and well-edged illite (*I*)

◄

Fig. 11.11. DSDP location in a late Cretaceous geography, and clay successions identified in black shales and associated sediments. (After Chamley and Robert 1982)

The composition of clay associations in black shales and associated sediments varies considerably according to age, but usually depends neither on the general lithology, the abundance of organic matter, or the depth of burial (e.g., Fig. 11.11). Some black shales contain very abundant smectite (e.g., Aptian at site 327, South Atlantic; Albian at site 398 off the Iberian peninsula), some others are rich in illite (Albian at site 367, Cape Verde basin), kaolinite (early Aptian at site 361, Cape basin) and even palygorskite (late Cretaceous at site 364, Angola basin). Some black shales present a clay mineralogy very similar to that of over- or under-lying organic-poor limestones or marlstones (e.g., site 105 in the Hatteras basin, site 364 in the Angola basin). The boundaries between successive clay mineral zones rarely correspond with major lithologic units (Fig. 11.11). These results are confirmed by geochemical data, that do not reveal any noticeable difference in the clay mineral composition between black shales and surrounding sediments (Chamley and Debrabant 1984a; see Chapter 19). As a consequence *clay associations in Cretaceous black shales of the Atlantic Ocean usually indicate the absence or weakness of mineral modifications linked to the organic environment.* The stability of clay minerals in black shales, that agrees with the usually moderately abundant and immature organic matter (Tissot et al. 1980, Simoneit 1986), differs greatly from what occurs in some Mediterranean sapropels during the late Cenozoic (11.2.2). The opposition is peculiarly well illustrated by the behavior of smectite and palygorskite, two minerals that tend to be destroyed in most organic sapropels and to be preserved in black shales (Figs. 11.11, 11.13). While sapropels tend to display a close correlation between lithology and clay mineralogy, black shales tend to lack such correspondence.

Local cases of clay modifications in Atlantic black shales are envisaged by Jacquin (1987) at site 364 DSDP, Angola basin. In organic-rich deposits from

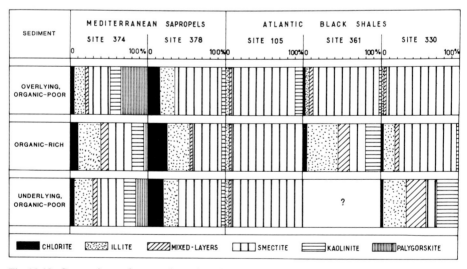

Fig. 11.13. Comparison of mean clay-mineral assemblages in organic-rich and common sediments from Eastern Mediterranean Sea (Quaternary) and Atlantic Ocean (Cretaceous). (After Chamley and Robert 1982)

early-middle Turonian time, smectite and illite appear less abundant and less crystalline than in surrounding sediments. But the existence of an actual in situ degradation remains questionable, since the fragile palygorskite fibers are not affected. The early Albian black claystones at the same site contain abundant large-sized and vermiculated kaolinite, that could result from an autochthonous formation in acidic conditions during early diagenesis (see also Chap. 16). But this process probably does not occur specifically close to the sediment-sea water interface. Autochthonous modifications clearly identified in Atlantic black shales concern minerals less stable than clays, such as pyrite, clinoptilolite, opal CT, as well as calcite, siderite, rhodochrosite, and are not specifically restricted to black shale facies (see Chamley 1979, Chamley and Debrabant 1984a). Note that small but significant differences exist in the clay mineralogy of alternating dark and light layers that characterize numerous Atlantic black shales (H.C., unpubl. res.) Calcareous horizons often contain more smectite and quartz than clayey horizons that are enriched in illite, chlorite, and quartz. These differences probably result from climate changes related to Milankovitch-like cycles (Berger et al. 1984; see Chap. 17).

11.4 Conclusion

1. The activity of benthic organisms may have strong effects on the physical properties of sediments and on the dissolution, migration and precipitation of metal oxides or sulfides. By contrast, living organisms appear to only rarely or slightly modify the mineral composition of clay associations by burrowing or feeding activity. The examples collected concern the degradation of chlorite and mixed-layered clays through sediment ingestion by some shallow-water benthic crustaceans and annelids, and the transformation of mineral standards through digestion of some copepods living in nearshore sea water. In fact observations and experiments arailable are still limited, and much has to be done on the precise nature of mineral modifications by organic ingestion, on the chemical interactions occurring in digestive tracts between minerals and organic acids, and on the actual impact of organic activity on terrigenous minerals. *The only significant alteration of Al-Si detrital minerals by living organisms appears to occur in fecal pellets, to be restricted to shallow-water sediments, and to be of little quantitative importance* since marine clay assemblages fundamentally resemble those supplied from adjacent land masses (Chaps. 3.3, 5, 8).

2. Sapropels and sapropelic marls that developed temporarily during late Cenozoic time in the Mediterranean Sea, especially in eastern basins, display a wide range of relations between organic and mineral sedimentary constituents. Some organic-rich sediments are devoid of any clay modifications, which allows the reconstruction of past detrital sources and transportation conditions. In some other sapropels clay minerals tend to be diversely degraded according to the initial mineral composition, the abundance of organic matter, and the bathymetric and topographic situation. Clay degradation tends to increase toward the central and deepest parts of sapropelitic areas, in depressed zones and at the bottom of

organic layers, and to successively affect palygorskite, smectite, chlorite, mixed-layers, and illite. Kaolinite appears to be rather stable in Mediterranean sapropels. Five alteration stages have been recognized, marked by the existence of transitional minerals like irregular mixed-layers and complexation-chlorite, and by correlative changes in inorganic geochemistry and in other sedimentary constituents such as organic matter, sulfides, biocalcareous and biosiliceous remains. *Late Cenozoic Mediterranean sapropels represent unusual marine environments where a clay degradation occurred close to the sediment-water interface mainly through the action of organic acids*, in a way similar to that characterizing continental podzolic soils (2.2.1) and some underclays.

Many questions remain unsolved and need further investigation, especially on the respective importance of terrestrial and marine organic matter, on the nature and process of organic attack, on the role of bacterial activity, on chemical fluxes that have occurred between mineral particles, organic matter, interstitial water and open-sea environment, and on the comparison between clay degradation developed in both subaerial (hydrolysis) and submarine (halmyrolysis) organic-rich environments.

3. *Cretaceous black shales in the Atlantic domain contain diversified clay associations, that mostly vary independently of the organic matter content*, and reflect environmental changes occurring on land rather than in the sea. Smectite, that represents the most common mineral, is well preserved in all Cretaceous sedimentary facies, as are other clay species including palygorskite and chlorite. The low chemical reactivity of Cretaceous black shales contrasts with the varying degradation stages recorded in late Cenozoic Mediterranean sapropels (Fig. 11.13), and seems to result mainly from differences in the composition and evolutionary state of organic matter. Atlantic black shales differ from Mediterranean sapropels by their great extension and thickness, their small dependence of submarine bathymetry and topography, their diverse lithologies, their often terrigenous, moderately abundant and immature organic matter, as well as by the near absence of in situ clay alteration. The dark color of black shales appears largely to be of a continental pedologic origin, and to have been preserved in fairly little oxygenated and rapidly buried Atlantic sediments (see Chap. 19). In situ modifications mainly concern mineral groups less stable than clay such as carbonates, sulfides, zeolites, and silica.

4. Little is known about synsedimentary relations existing between organic and clay mineral materials in non-Atlantic Cretaceous black shales and in older or younger organic-rich geological series. Most present-day organic-rich marine sediments apparently show little clay mineral differences with surrounding organic-poor deposits. Like evaporative environments, *organic environments lack of clear present-day cases of active chemical modifications of clay assemblages*, which complicates the understanding of past processes. Some significant progress is expected from comparisons between marine sediments like sapropels or black shales and continental podzolic soils or swamp underclays. Confrontations should also be envisaged between essentially synsedimentary phenomena occurring in organic-rich marine sediments and mainly postsedimentary (diagenetic) phenomena characteristic of coal deposits and tonsteins (Chapter 16).

Chapter 12

Metalliferous Clay in Deep Sea

12.1 Deep-Sea Clay Environment

12.1.1 Introduction

Metal-bearing deep-sea sediments correspond to various environments, and range from terrigenous clays deposited in oxidizing conditions to hydrothermal accumulations. *This chapter focuses on metal-bearing argillaceous deposits characterized by an autochthonous formation in the deep sea of metallic minerals and associated clays close to the sediment-sea water interface, apart from the direct influence of volcanic or hydrothermal activity.* Such a formation occurs in *hydrogenous environments.* Hydrogenous authigenic materials mostly form through inorganic reactions, although biological factors, especially those related to bacterial activity, may play a significant role in the formation of metallic minerals (e.g., Ehrlich 1981, Karlin et al. 1987). Hydrogenous deposits result either from *halmyrolysis* that implies reactions between sediment constituents and sea water, or from *precipitation* that corresponds to the direct removal of sea-water elements to form primary inorganic compounds without chemical intervention of pre-existing sediments (Elderfield 1976). Hydrogenous materials differ from other deep-sea metal-rich deposits, issued either from close hydrothermal activity (see Chap. 13) or from diagenesis after significant burying (Chaps. 14 to 16) (Fig. 12.1). Of course all transitional situations exist, natural environments being diversely submitted to the various influences responsible for the production of metalliferous clayey sediments.

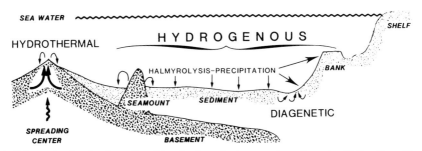

Fig. 12.1. Schematic distribution of authigenic deposits in oceanic sediments. (After Bonatti 1981)

Metalliferous clays, classically labeled as *"deep-sea red clays", mostly consist of reddish-brown to chocolate colored, very fine-grained sediments, widely distributed in the deepest parts of the oceans (> 4000 m) and deposited at a very slow rate (< 1 mm/10³ yr).* Reddish to brown clay, first discovered by the Challenger expedition in 1874 (Murray and Renard 1891), crops out widely in deep marine basins, especially in most of North Pacific Ocean, in central to eastern Equatorial and South Pacific Ocean, in Central and Southeastern Indian Ocean, and in some of the deepest areas in Atlantic Ocean. *Brown clay constituents mainly comprise Fe-Mn oxides and clay minerals, and little biogenic remains.* The calcium carbonate content is usually low, because most calcareous debris have been dissolved before reaching the deep and cold bottom waters (see Kennett 1982). Despite the fairly constant color of sediments, major and accessory components of clay minerals and oxides may vary in both nature and proportion, according to latitude, distance from terrigenous sources, planktonic productivity, wind activity, bottom circulation, or submarine volcano-hydrothermal activity. Most common associated components comprise metallic nodules and micro-nodules, zeolites (mainly phillipsite), volcanic glass, quartz, micas, feldspars and various dense minerals, fish and marine mammal teeth and bones, and cosmic spherules. Because of the slowness of deposition, the thickness of brown clay is usually small and rarely exceeds a few tens of meters; early Neogene and even Paleogene sediments can often be sampled with a few meters-long piston corers. For the same reason, and because of the paucity of organic remains and of bioturbation effects, high-resolution studies are difficult to perform on deep-sea clay.

The extremely low sedimentation rate of deep-sea brown clay favors long- lasting chemical exchanges between sediment surface and sea water. As bioturbation may determine the repeated exposure on the sea floor of slightly buried deposits, brown clay may have experienced both synsedimentary and post-sedimentary chemical changes, and it is often impossible to distinguish strictly hydrogenous processes from early diagenetic ones. We consider that processes involving subcontinuous active exchanges with the open-sea environment belong to syngenesis, whatever the duration and depth of burial.

The most obvious chemical process occurring in deep-sea brown clay consists in an *enrichment of metallic cations like Fe, Mn, Co, Ni, Cu, and Zn.* The way in which these elements accumulate is still under debate. Many authors estimate that particulate matter concentrates trace elements from sea water into hydrogenous complexes, as happens around nuclei to form manganese nodules; arguments arise from the similarity existing in the chemistry and especially the manganese / trace element ratios of the hydrogenous constituents of deep-sea clays with those of manganese nodules (see Elderfield 1976, Rao 1987). Other authors think that due to the short residence time in sea water of metallic elements like Cu, Co, Ni, the largest part of hydrogenous accumulation results from cations trapping by marine organisms and subsequent release and fixation within sediment through the dissolution of tests (e.g., Boström et al. 1978). Notice that significant differences exist in the concentration of chemical elements in hydrogenous sediments relatively to typically hydrothermal deposits. Hydrogenous materials are usually enriched in cobalt, nickel, copper, zinc, lead, and relatively depleted in iron, manganese, barium, and uranium (in Bonatti

1981). Most hydrothermal sediments display higher Si/Al ratios than hydrogenous ones. Selenium appears to characterize some hydrothermal sediments but to be hardly preserved in old deposits (Fournier-Germain 1986).

12.1.2 Metalliferous Nodules

Deep-sea red clays are associated or not with metal-rich nodules, that *consist of blackish-brown agglomerations of ferromanganese oxides concentrically arranged in millimetric layers.* The size of "manganese nodules" ranges from about 20 μm to more than 15 cm, the smaller being called micro-nodules. Metal-rich concretions develop around nuclei of various types (detrital minerals, fish teeth), and usually grow at a slow average rate (1 to 5 mm/10^6 yr)with intermittent interruptions due to burying or change in bottom oceanographic conditions. Some nodules grow much faster (up to 200 mm/10^6 yr), especially in conditions of suboxic accretion (Dymond et al. 1984). The formation of metalliferous nodules depends on both inorganic and organic processes. The organic activity intervenes through bacteria, rhizopods, and other protozoa that concentrate metal oxides (e.g., Monty 1973, Riemann 1983), and through benthic animals displacing the concretions on the sea floor (e.g., von Stackelberg 1984). Nodules occur preferentially in a zone located beneath the carbonate compensation depth, in areas marked by active bottom currents. Deep-sea nodules are extensively studied because of their scientific interest and economic usefulness (e.g., Cronan 1976, Glasby and Read 1976, Glasby et al. 1982, Académie des Sciences 1984). We present here only a few data on their mineralogic and chemical composition.

Metalliferous nodules mainly consist of subamorphous to crystalline manganese and iron oxides, associated with significant amounts of cobalt, nickel, and other metals, and including various amounts of detrital mineral grains, volcanic glass, biogenic debris, P and Ba minerals, and authigenic silicates. Manganese oxides mostly comprise todorokite and δ-MnO_2. Studies performed on Eastern Pacific sediments show that hydrogenous precipitation produces mainly δ-MnO_2, while synsedimentary evolution leads to Cu-Ni rich and stable todorokite under oxic conditions, and to unstable todorokite evolving toward a birnessite-like mineral under suboxic conditions (Dymond et al. 1984). Todorokite-rich nodules in the Pacific occur in regions of greater average depths than δ-MnO_2 rich nodules, the latter being more widespread in elevated submarine volcanic seamount areas (in Cronan 1976). Ni and Cu occur more abundantly in nodules rich in todorokite, whereas Co and Pb rather characterize those dominated by δ-MnO_2. Iron oxides are much less well known than manganese oxides, and appear to consist mainly of goethite- and hydrogoethite-like subamorphous minerals. The morphology and chemical composition of nodules often differ in their upper and lower parts, if they have remained stable on the sea floor for a sufficiently long time. The seaward (upper) sector frequently displays a smooth surface and relatively high iron and cobalt contents, while the lower sector is rather rugged and enriched in manganese and copper. This points to the existence of active chemical reactions between nodules and both overlying sea water and underlying sediment. Note also that Pacific nodules usually contain more Mn, Ni, Co and Cu, than Atlantic nodules, the latter being frequently enriched in Mn. The Indian Ocean nodules present an intermediate composition.

Silicate minerals in "manganese nodules" mainly comprise phillipsite, a Fe-Si zeolite, and smectites. These silicates are not often abundant, and sometimes are absent. *Smectite mostly comprises iron-rich types of various compositions associated or not with other clay minerals,* indicating diverse possible origins. In the central Equatorial Pacific, Calvert et al. (1978) identify Fe-rich smectite, interpreted as an autochthonous mineral whose hydrogenous and early diagenetic development possibly determined the formation of Fe-poor ferromanganese nodules. Toth (1980) shows that some Fe-Mn crusts deposited close to the ocean ridge in the Atlantic Ocean contain iron-rich, aluminum-poor nontronite, a mineral originating under strong hydrothermal influence. Mn-, Cu- and Ni-rich nodules from the northern Central Indian Ocean basin contain smectite, illite, and chlorite, an association very similar to that of associated sediments (Rao 1987) and issued from both terrigenous and autochthonous sources (Chap. 8, Fig. 8.14). In the southern part of the same basin, nodules are relatively enriched in Fe and Co, and contain only very little smectite beside illite and chlorite, which suggests a stronger terrigenous contribution.

12.1.3 Allochthonous Versus Autochthonous Origin

Before considering the clay formation in deep-sea oxidized environments, we should recall that *a large part of clay associations in metal-bearing deposits results from terrigenous supply* (see Chaps. 7 and 8, Figs. 8.1, 8.2, 8.4, 8.6, 8.14). Deep-sea brown clays from the Atlantic Ocean, from a large part of the Indian Ocean, and from the largest part of the North Pacific Ocean proceed predominantly from long-distance transportation by wind, bottom currents, and marine water masses. Most chlorite, illite, kaolinite, irregular mixed-layers in oceanic sediments have a detrital origin, like most quartz, feldspar and mica particles (Griffin et al. 1968, Kastner 1981). A large proportion of smectite derives from the erosion of continental soils, of weathering complexes on exposed volcanic rocks and of ancient sedimentary rocks. The terrigenous control on North Pacific clay sedimentation is particularly demonstrated by the similarity existing between mineral assemblages present in aeolian dust and oceanic sediments (e.g., Blank et al. 1985, Chap. 7), and between detrital series located at various distances from continents independently of lithology or bathymetry (e.g., Griffin et al. 1968, Lenôtre et al. 1985). It is significant that Pacific Ocean regions marked by the highest amounts of fine-grained sediments and therefore of clay mineral particles are located in northern basins (Fig. 12.2), which contain typically detrital clay assemblages (abundant illite and chlorite. Figs. 8.4, 8.2). Coarser particles characterize Central and South Pacific sediments, whose metalliferous facies largely consist of Fe-Mn oxide aggregates and are often poor in clay minerals (e.g., Lisitzin 1972, Cronan 1986). In addition terrigenous brown clays of the North Pacific correspond to the highest contents of fresh to altered volcanic glass in this ocean

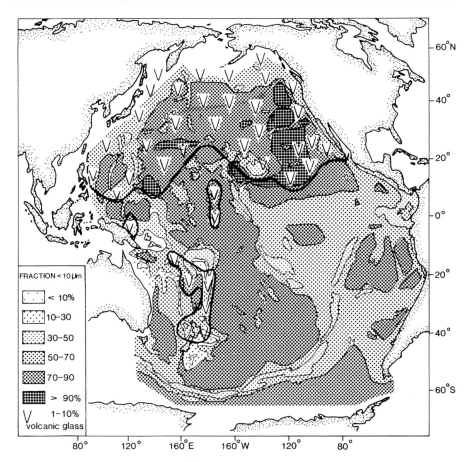

Fig. 12.2. Percentage distribution of clay-rich fraction (<10 μm) in surface sediments of the Pacific Ocean, and location of regions rich in colorless and brown volcanic glass (V) in the coarse sedimentary fraction (50–100 μm). (After Lisitzin 1972). Compare with Figs. 8.4 (illite) and 8.6 (smectite)

(Fig. 12.2), which indicates the absence of systematic correlation between clay formation and submarine alteration of volcanic material (see Chap. 13). Smectite, whose relative abundance increases in sediments of the South Pacific depleted in pyroclastic supply (Fig. 8.6), does not therefore appear to predominantly depend on the presence of volcanic material.

The importance of terrigenous influence in deep-sea sediments, including a large proportion of metal-bearing brown clays, results from the *very slow equilibration of detrital clay minerals in sea water,* a fact proved by the mineral distribution maps (Chap. 8) and by various measurements on the absolute age and isotopic composition of minerals (see Kastner 1981). Many observations show that there are few sediment-pore water interactions before and after burying in the deep-sea environment other than dissolution of most soluble phases such as silica, metal oxides, and carbonates (e.g., Bischoff and Sayles 1972, Gieskes

1983). The absence of strong chemical gradients argues against the classical geochemical balance concepts on large-scale ionic fluxes and mineral equilibrium. The most reactive components in the deep-sea environment are volcanic glass and basalt, whose abundance often appears to be little correlated with clay formation in common sediments (Chap. 13). Modern studies emphasize the fact that rather than equilibrium reactions between sedimentary particles and sea water or interstitial water, the formation of new minerals is mainly controlled by kinetic conditions (in Kastner 1981).

The main clay mineral species a priori able to form in significant amounts in metal-bearing deep-sea sediments are smectite and fibrous clays (= hormites). Some authors estimate that about half to two-thirds of smectite in surface sediments from the southern parts of the three main oceans is authigenic. Such an interpretation is supported by the non-climatic zonation of smectitic clays in southern oceans (Fig. 8.6), which should be at least partly balanced by the importance in the peri-Antarctic range of smectite-rich potential source rocks, of erosion processes, of northward bottom currents, and of wind activity (Chaps. 7, 8). The in situ formation of smectite and fibrous clays in deep-sea sediments is discussed below (12.3). In most cases, *clay minerals seem to form much less easily in deep-marine environments than authigenic silica, zeolites, and even feldspars.* The in situ growth of the latter minerals mainly occurs during diagenesis and is discussed in review papers like those of Stonecipher (1977), Sand and Mumpton (1978), and Kastner (1981). Note that most zeolites form through solution-precipitation phases in various environments. While deep-sea phillipsite mostly seems to grow from basaltic glass precursors, clinoptilolite mainly occurs in non-volcanic deposits from shallow- to deep-water environments. Clinoptilolite represents a silica-rich zeolite that may develop from the evolution of biogenic opal CT (cristobalite-tridymite) in common pelagic oozes or in alkaline evaporative sediments (Chap. 9). Analcite is a sodium-rich zeolite whose formation is favored by high Na/H ratios and by the diagenetic evolution of clinoptilolite.

One of the most constraining conditions to fulfill for the submarine genesis of aluminosilicates lies in the availability in sufficient amounts of aluminum. This very little soluble element has been envisaged to form X-ray amorphous compounds issued from subaerial weathering, and to be transformed into smectite, illite, and chlorite within the marine environment (Mackenzie and Garrels 1966a and b); but no evidence for such reactions arises from studies performed on marine sediments (in Kastner 1981). The concentration of dissolved aluminum in sea water is close to 1 µg/kg, which represents about 0.2% of its solubility at the average temperature of deep water ($2°C$). This indicates that aluminum is highly reactive in the marine enmvironment, but also very little available for mineral growth. A large part of dissolved aluminum present in sea water seems to be adsorbed on biosiliceous tests, and could subsequently serve for aluminosilicate growth in sediments (see Stoffyn-Egli 1982). But many deposits are poor in diatom frustules or radiolarian tests, and there is no tight correlation between deep-sea clay formation and biosiliceous sediments (see maps in Kennett 1982). In addition, calculations show that the total influx of dissolved aluminum by rivers is largely insufficient for the formation of marine clinoptilolite and, a fortiori, for that of clayey aluminosilicates (Kastner 1981). As a consequence, *river-derived*

particulate matter, as well as volcanic minerals and glass, probably represent the
major sources for the aluminum necessary to form deep-sea aluminosilicates.

12.2 Smectite Formation

12.2.1 Preliminary Data

Smectite is widespread in deep-sea brown clays, where it often constitutes the
dominant clay mineral (Griffin et al. 1968, Fig. 8.6). *Authigenic smectite in metal-*
liferous clays is characterized by high contents of iron and little aluminum. Most
studies performed until the mid-1970's concluded that the largest part of smectite
and of associated phillipsite formed from submarine alteration of volcanic
material, through low-temperature reactions involving sea water and often
biogenic silica (see numerous references in Kastner 1981). Some few studies
quoted by M. Kastner discussed the geochemical evolution of volcanic material,
or envisaged a distant hydrothermal influence or a hydrogenous origin. Fe-
montmorillonite was classically considered as the common alteration product of
basaltic glass, while Fe-Al-Mg smectite was rather attributed to the alteration of
nonbasaltic volcanic material. The authigenic formation of smectite appeared to
extensively occur close to the sediment-sea water interface. Further development
of the mineral was supposed to occur within the sedimentary column through
burial reactions, as suggested by the downward decrease in dissolved Mg and K,
the corresponding increase in dissolved Ca, and the ^{18}O depletion (in Kastner
1981, Gieskes 1983).

   The systematic contribution of submarine altered volcanic glass to the forma-
tion of deep-sea smectite appears highly questionable when considering a few facts
about Pacific Ocean sediments that display the highest contents of the mineral. (1)
The areas rich in smectite extend principally in Equatorial and South Pacific, a
domain marked by both comparatively few clay-sized sedimentary fractions and
little volcanic material (Lisitzin 1972, Fig. 12.2 to compare with Fig. 8.6). The
alteration stage of volcanic debris scattered within pelagic deposits does not appear to significantly affect the abundance of authigenic smectite, since smectite-
rich sediments may contain either fresh or altered volcanic material, which does
not differ from smectite-poor sediments (see Chap. 13). (2) The clay minerals
originating on the sea floor by alteration of volcanic glass (halmyrolysis) should
systematically display the negative cerium anomaly characteristic of sea water,
and a relative enrichment of heavy rare earth elements (Eu–Lu group) typical of
basalts (Piper 1974, Bonnot-Courtois 1981). In fact, these rare earth elements
characters are sometimes present, particularly in various parts of the Eastern
Equatorial and Southeastern Pacific Ocean, but are also frequently lacking; in
particular the negative Ce anomaly is not recorded in most North Pacific
Quaternary and Tertiary smectites, which recalls characteristics of detrital illites
and indicates a predominantly terrigenous influence (Piper 1974). (3) The chemical composition of smectite in Pacific surface sediments is highly variable ac-

Table 12.1. Chemistry of smectites from surface metalliferous deposits of the Pacific Ocean. Smectite (1) is considered as exclusively detrital (North Pacific) and smectite (5) as exclusively hydrothermal (metalliferous mounds field, Galapagos rift)

	North	NE equator Domes	N equator	SE equator Bauer	E equator Galapagos
	30°N–160°W (1)	14°N–126°W (2)	17°N–155°W (3)	10°S–102°W (4)	2°N–90°W (5)
%					
SiO_2	65.6	45.7	51.3	52.3	48.0
Fe_2O_3	9.9	9.2	14.6	29.7	36.7
Al_2O_3	18.3	14.1	7.4	2.4	0.02
MgO	3.5	5.2	4.2	6.1	2.4
CaO	1.1	0.9	0.5	0.5	0.7
Na_2O	1.8	2.8	1.3	0.6	1.6
K_2O	na	2.2	0.7	2.0	2.1
MnO	na	0.4	0.2	0.8	0.3
TiO_2	na	0.7	1.0	0.03	–
ppm					
Zn	na	1300	na	160	28
Ba	na	500	na	269	75
Ni	na	500	na	865	3
Cu	na	1100	na	400	2

na: not analyzed.
(1) Weliky, 1982, in Leinen 1987.
(2) Hein et al. 1979 b.
(3) Aoki et al. 1974 b.
(4) Dymond and Eklung 1978.
(5) Corliss et al. 1978.

cording to the geographic situation (Table 12.1). Diverse intermediate compositions exist between typical terrigenous smectite deposited in northern basins (see Leinen 1987) and typical hydrothermal smectite like that developed in metalliferous mounds close to Galapagos rift (Corliss et al. 1978, see also Butuzova et al. 1977). These different data strongly suggest that *diverse influences and mechanisms may intervene in the growth of smectite within deep-sea sediments, the halmyrolysis of volcanic glass being only one of them.* Let us consider a few specific cases in order to investigate this question further.

12.2.2 Domes Area, Northeast Pacific

Domes area is located between Clarion and Clipperton fracture zones in the Equatorial North Pacific, at about 2400 km West of the East Pacific ridge (14°N, 126°W). Surface and core samples consist of fine-grained brown-chocolate sediments (e.g., Bischoff et al. 1979, Hein et al. 1979c). Semi-opaque, subamorphous Fe-Mn oxyhydroxides constitute the dominant fraction of deposits (40% and more). Associated compounds include biosiliceous debris, very few carbonates (about 3% CO_3Ca), and a few clay minerals. Clay minerals comprise smectite, illite, and kaolinite. The origin of smectite has been little discussed. A hydrothermal origin related to the Clarion fracture zone to the North is envisaged by Bischoff and Rosenbauer (1977) for explaining metal-rich components, while Hein et al. (1979c) rather involve a metal enrichment through dissolution of marine organic tests. Metalliferous oxides appear to be superimposed to non-metalliferous components. Note that clay associations similar to those of the Domes area are identified in typically hydrothermal zones of the Northeast Pacific ridge, and mainly issue from detrital North American sources (e.g., Fournier-Germain 1986. Chap. 13).

12.2.3 Central South-Equatorial Pacific

The northern flank of the Marquises fracture zone, North of French Polynesia, is covered by a brown homogeneous mud a few meters thick. Hoffert (1980) and Bonnot-Courtois (1981) study several cores from this area, as well as from different other parts of the central to eastern South Equatorial Pacific (Fig. 12.3). Core TKS.16, 380 cm long, is located at 4725 m water depth (10°28′N – 144°00′W). The sediment, topped by metalliferous nodules, contains more than 60% clay-sized particles and less than 2% sand (Fig. 12.4). The CO_3Ca content is low (<8%) and decreases below the surface levels (0–3%). The >63 µm fraction

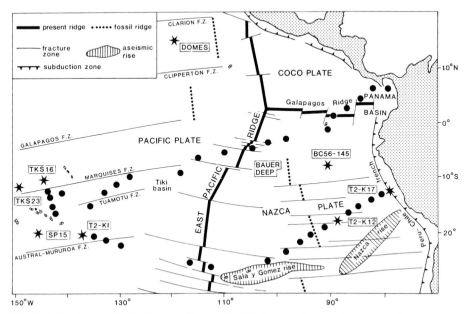

Fig. 12.3. Study area in the Eastern Central Pacific Ocean. *Stars* main cores and zones considered in Chapter 12; *dots* transects of stations studied by Hoffert (1980), Bonnot-Courtois (1981) and Tlig (1982a)

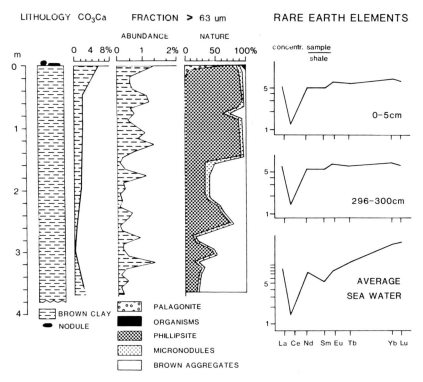

Fig. 12.4. Sedimentological and rare-earth elements data on core TKS 16, Central South-Equatorial Pacific Ocean. (After Hoffert 1980)

mainly consists of phillipsite in the upper 160 cm, and of phillipsite and brown aggregates below. Volcanic glass and palagonite, as well as some siliceous debris, are identified at the uppermost level of the sediment. Radiolarian and calcareous debris allow dating the top of the core from Plio-Quaternary time, the underlying sediments being of unknown age. The bulk mineralogy is dominated, in addition to subamorphous metal oxides, by phillipsite and clay minerals. Minor components include ubiquitous calcite and quartz, occasional feldspars, pyroxenes, barite, and apatite. Clay minerals comprise abundant smectite (about 80% of the <2 μm crystalline fraction) and accessory illite and kaolinite (about 10% each). The clay mineral nature and abundance are constant throughout the core. Thermal differential, and chemical analysis indicates that smectite is rich in iron and of a dioctahedral type (Hoffert 1980):

$$Ca_{0.46}K_{0.17}Na_{0.11}(Fe^{3+}_{1.12}Mg_{0.46}Al_{0.39})(Si_{3.37}Al_{0.63})O_{10}(OH)_2, nH_2O .$$

This structural formula resembles that of nontronite, however with more abundant octahedral magnesium and aluminum.

Rare earth elements in bulk sediment are much more abundant than those from common nonmetalliferous pelagic oozes (570 to 645 ppm instead of <100 ppm; see Bonnot-Courtois 1981). Distribution curves normalized to average continental shales display a strong negative anomaly in cerium, and do not vary from the top to the bottom of the core (Ce, Fig. 12.4). Such an anomaly is characteristic of sea-water composition (Piper 1974), and indicates that sedimentary components of core TKS.16 have undergone an authigenic formation influenced by the open marine environment. As planktonic compounds are little represented in sediments, Hoffert (1980) concludes that *metal oxides, phillipsite and smectite are newly originated by precipitation of chemical elements close to the sediment-sea water interface.* The neoformation of clay minerals in surface sediments is confirmed by the $^{87}Sr/^{86}Sr$ ratio of the < 2 μm fraction, that is close to that of sea water (0.70873 ± 0.00012 and 0.70895 ± 0.00010, respectively. Clauer 1979, Clauer et al. 1982). Smectite is therefore considered as a typically hydrogenous mineral. Volcanic debris appears to be not or only very little involved in the neoformation process. Illite and kaolinite, probably as well as quartz, represent terrigenous minerals possibly supplied by wind and scattered among authigenic sedimentary products. Relatively to hydrogenous (syngenetic) reactions, diagenetic (postsedimentary) reactions appear to develop much more slowly, and mainly consist of a downward increase in total Fe content, a progressive equilibration of strontium isotopes (increase of $^{87}Sr/^{86}Sr$ ratio in interstitial water), and a particle agglomeration in brown aggregates.

Hoffert (1980) studies the sedimentology, mineralogy, and geochemistry of a few ten surficial sediments distributed in several areas of the South Equatorial Pacific between 80° and 150° W: Panama basin, Galapagos ridge, Nazca basin North of Nazca and Sala y Gomez rises, Tiki basin West of the Pacific ridge, and French Polynesia (Fig. 12.3). Most sediments investigated consist of metalliferous brown clays, characterized by fairly abundant oxyhydroxides and clay minerals. Clay associations usually reveal the abundance of smectite, that comprises various types, with diverse chemical composition and crystallinity. Smectites are often of a ferriferous type, are associated with varying but generally small amounts of illite, kaolinite, chlorite, and rarely palygorskite. Smectites appear fairly unstable, since their flaky particles may display radial growths of peripheric laths (see 14.2.2). Phillipsite occurs in variable amounts (0-30%). Biogenous debris, often not very abundant, tend to concentrate close to the sediment-water interface, and to disappear downwards by dissolution. Volcanic glass is usually not very abundant and often comprises hydrated particles (palagonite). M. Hoffert considers that sedimentary components are almost exclusively neoformed in the sediment by hydrogenous processes, including small quartz grains and occasional minerals like palygorskite or ankerite. Hoffert (1980) distinguishes *four main types of metalliferous deep-sea clays,* marked by specific mineralogical characters, and diversely distributed according to depth, planktonic productivity, volcanic activity and sedimentary environment:

1. Brown clay deposited *close to biosiliceous oozes* preferentially forms in deep areas located beneath high productivity zones. It contains abundant 14 Å smectite, that frequently forms laths or curls and is associated with abundant brown aggregates and micronodules. Phillipsite is absent.

2. Brown clay deposited *close to biocalcareous oozes* characterizes relatively shallow submarine areas (<4000 m). It contains very abundant Fe-Mn

oxyhydroxides associated with poorly crystallized Fe-smectite and single crystals of phillipsite.

3. Brown clay deposited *close to volcanic areas* (rifts, transform faults) contains 12 Å Fe-rich smectite, monocrystalline or polycrystalline phillipsite, micronodules and palagonite.

4. Other deep-sea brown clays are devoid of specific environmental characters and comprise a mixture in varying amounts of the three former types. These sediments probably result from the reworking by bottom currents of pre-existing brown clay. They therefore represent *autochthonous deposits resedimented as allochthonous materials,* that can ulteriorly experience chemical reequilibration processes after being definitely immobilized and buried.

12.2.4 Nazca Plate

The Nazca plate extends in the Southern Pacific Ocean between South America and the East Pacific ridge, and between the Galapagos ridge close to equator and the Chile fracture zone (Fig. 12.5A). Numerous geochemical and mineralogic studies have been performed in this *region largely protected from terrigenous influences by the East Pacific subduction zone, and surrounded by active rift and transform fault zones.* Studies of Nazca plate have focused primarily on metalliferous deposits, that crop out broadly on the sea floor and have been attributed dominantly to either hydrothermal or nonhydrothermal influences (see references in Dymond 1981, Fournier-Germain 1986). For instance, the increased amounts of Fe, Mn and other transition elements in deposits near the East Pacific ridge, relatively to common pelagic sediments, suggest that sea-water hydrothermal systems constitute the major mechanisms for leaching metals from the crust and providing for their presence in brown clay. On the other hand, the chemical variability observed in surface sediments, independently of the distance to the rift, indicates that nonhydrothermal or nondirect hydrothermal sources intervene in a significant way; for instance the Bauer Deep, an isolated basin located a few hundred kilometers to the East of the Pacific ridge (Fig. 12.5A), comprises sediments with metal amounts comparable to those on the ridge crest, but with noticeably higher Ni and Fe contents.

Geochemistry of Surface Sediments
Dymond (1981) collects the information available on the variability in sedimentary sources of the Nazca plate surface sediments. At the western margin of the plate, hydrothermal precipitates related to the East Pacific ridge dominate the sediment budget. In the northern part biogenic, mainly siliceous tests predominate, reflecting the high planktonic productivity of the equatorial zone. To the East sediments mainly comprise terrigenous clays issued from adjacent South America. The basins located in the central part of the plate, like Bauer, Yupanqui and Roggeveen depressions, display strong enrichment in transition metals issued either from hydrothermal precipitates or hydrogenous deposition from normal sea water. J. Dymond develops a *normative composition model*

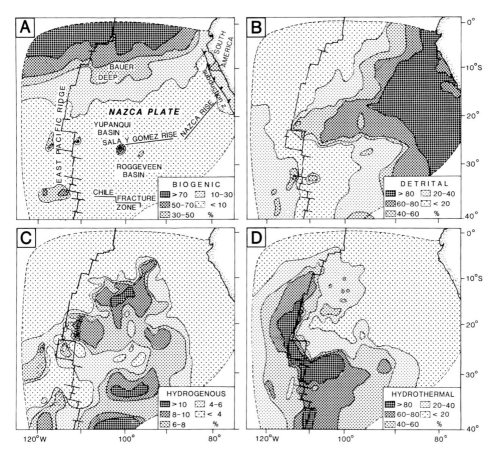

Fig. 12.5. Physiographic features of the Nazca plate and weight percents of "biogenic" (**A**), "detrital" (**B**), "hydrogenous" (**C**) and "hydrothermal" (**D**) components of surface sediments. (After Dymond 1981). Weight percents are given on a carbonate and salt-free basis

based on bulk geochemical data of 425 surface sediments from the Nazca plate. The model attempts to define the composition of pure end-member sources and to determine what mixture of sources can best account for the bulk sediment composition. Bulk chemical data are converted into quantitative estimates of the weight percent of five distinct components: (1) biogenic tests, based on the composition of biosiliceous debris and refractory organic matter, and normalized to Si; (2) detrital aluminosilicates, estimated from summary analyses of igneous and sedimentary rocks, and referred to Al; (3) authigenic hydrogenous ferromanganese precipitates, taken from metalliferous nodule and crust data, and normalized to Ni; (4) hydrothermal precipitates, evaluated from elemental ratios of samples close to the East Pacific crest and normalized to Fe; (5) dissolution residues of organisms comprising insoluble elements from the hard and soft parts of planktonic organisms, and based on Ba as an index element.

Calculations of the elemental ratio coefficients of the five components selected (Table 12.2) allow to define the *dominant sources intervening in the con-*

Table 12.2. Elemental ratio coefficients of the five components used in the normative geochemical analysis of Nazca plate surface sediments. (After Dymond 1981). Data on a carbonate-free basis

	$(E/Si)_B$	$(E/Al)_D$	$(E/Ni)_{Hg}$	$(E/Fe)_{Ht}$	$(E/Bz)_R$
Al	0.002000	*1.0000*	1.5	0.0060	0.5000
Si	*1.000000*	3.0000	4.5	0.1300	0.0000
Mn	0.000023	0.0160	30.0	0.2900	0.0070
Fe	0.001000	0.7000	15.0	*1.0000*	0.3500
Ni	0.000040	0.0015	*1.0*	0.0009	0.0065
Cu	0.000050	0.0012	0.5	0.0042	0.0160
Zn	0.000080	0.0014	0.1	0.0019	0.0040
Ba	0.002000	0.0120	0.2	0.0050	*1.0000*

E: chemical element. Si, Al, Ni, Fe, Ba: reference elements. Experimental ratios in pure biogenic (B), detrital (D), hydrogenous (Hg), hydrothermal (Ht) and dissolution residue components (R), respectively.

stitution of Nazca plate sediments (Fig. 12.5A to D), as well as their mixture in a given area (Dymond 1981). (1) *The biogenic contents* of surface deposits decrease latitudinally from about 80% *near the equator* to less than 10% South of 20°S. (2) The abundance of *detrital aluminosilicates*, mainly made of clay minerals, generally exceeds 80% in regions *within 1000 km of South America*. High concentrations of detrital components extend much more westwards between 20° and 25°S, and characterize a large portion of the Yupanqui basin. (3) *Hydrogenous authigenic components*, primarily made of metal oxyhydroxides and accessorily of aluminosilicates, mainly occur in *deep basins located East of the East Pacific ridge*, especially the Bauer Deep. Hydrogenous components never exceed a few ten percent of deposits. (4) *Hydrothermally derived components* constitute more than 80% of the sediments deposited *within 100 km of the ridge crest.* Hydrothermal precipitates appear to be carried by *bottom currents* slightly toward the West in regions North of 30°S, and strongly toward the East into Roggeveen basin South of 30°S. (5) *Dissolution residues* of biogenic matter reach a maximum abundance in regions located on the *southern edge of the equatorial high productivity zone*, a fact attributed to a decreased preservation of planktonic tests in areas of lower biogenic deposition.

Clay Mineralogy of Surface Sediments

Rosato and Kulm (1981) examine the clay mineralogy and organic carbon contents of surface and subsurface sediments at about 50 stations of the Peru continental margin and adjacent Nazca plate. Smectite, illite, chlorite, kaolinite, and mixed-layers occur in varying amounts, depending on the nature of terrigenous sources, the distance to American coast, the wind regime, the volcanic activity and the proximity of the Pacific ridge. V.J. Rosato and L.D. Kulm develop a Q-mode factor analysis, that allows the identification of three factors explaining up to 99% of the variation in clay mineral composition and organic carbon content. A first continental factor includes mixed-layered minerals, illite, chlorite, and kaolinite, and characterizes upper continental margin sediments (<2000 m water depth). A second continental factor, mainly composed of illite with subordinate kaolinite and chlorite, is accessory in most areas except on the continental shelf between 9° and 13°S. The third factor, called *oceanic factor, comprises mainly smectite derived from both authigenic and continental sources,* and accessorily wind-borne illite. The oceanic factor characterizes Nazca plate sediments. Organic carbon contents are mainly linked to equatorial and continental margin zones. The frequent increase of smectite abundance toward the deep isolated basins like the Bauer Deep strongly suggests an authigenic formation. Note that the boundary between sediments dominated by the oceanic factor and those with continental factors tended to migrate laterally seawards during low Quaternary sea-level stages, probably because of increased erosion rates and westward shift of the shoreline. Quaternary turbidites with continental clay characters occur sometimes on the seaward side of the present trench axis, which probably results from tectonic uplift after deposition (Rosato and Kulm 1981).

Mineralogy and Geochemistry, Northeast Nazca Basin
Marchig and Rösch (1983) investigate the northeastern part of the Nazca basin about 900 km South of the Galapagos ridge (5-10°S, 89-93°W. BC56–145, Fig. 12.3). Surface sediments comprise fairly abundant biogenous and terrigenous components, because of the proximity of both equatorial high productivity zone and south American continent. The area, located at about 4100 m water depth, is virtually devoid of volcanic or hydrothermal influence. Late Pleistocene to Holocene sediments, cored on an average thickness of 25 to 45 cm, are noticeably submitted to sediment-sea water exchanges. The samples marked by significant carbonate dissolution tend to present high contents of rubidium, sodium, scandium, and potassium (Table 12.3; calculations on a carbonate-free basis). The amounts of most other elements decrease parallel to the dissolution of carbonate, except Si, that remains nearly constant. The X-ray amorphous matter increases parallel to the decrease of carbonates (Fig. 12.6A). Smectite abundance also increases in carbonate-depleted samples (Fig. 12.6B), relatively to other

Table 12.3. Sedimentological and chemical data on box-cored sediments from the Northeast Nazca basin. (After Marchig and Rösch 1983). Chemical data on a carbonate-free basis in <3 µm fraction

Core	Water depth m	Depth in the core m	Sand (<63 µm) %	CO_3Ca %	Na_2O %	K_2O %	Rb ppm	Sc ppm
BC 56	3570	0– 5	62	80	0.59	0.44	39	0
		5–15	77	82	0.63	0.45	40	0
		15–25	56	74	0.99	0.43	23	0
BC 80	3930	0– 7	80	67	1.23	0.54	24	18
		7–17	76	55	1.83	0.71	27	16
		17–25	67	52	2.02	1.01	40	29
		25–35	74	50	2.05	0.99	38	40
		35–44	85	40	2.52	1.48	46	38
BC 145	4114	0–10	85	28	2.48	1.58	57	32
		10–20	83	26	2.33	1.73	63	38
		20–30	84	18	1.85	1.60	59	35
		30–38	83	9	2.59	1.58	60	34

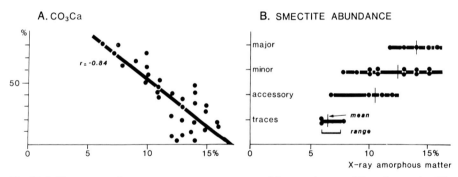

Fig. 12.6. X-ray amorphous matter contents versus calcium carbonate (**A**) and smectite (**B**) abundance. (After Marchig and Rösch 1983)

minerals (illite, kaolinite, feldspars). V. Marchig and H. Rösch attribute these mineralogic and geochemical changes to the *in situ formation of smectite*, Rb, Na, Sc, and K being possibly incorporated into the mineral interlayer positions. The origin of these elements remains unclear, as well as the absence of enrichment in Si, Al, Fe, and Mg, all elements forming the main components of smectite particles. Marchig and Rösch (1983) envisage an intermediate phase of transfer from subamorphous, metal-rich materials to new smectitic minerals. As the changes recorded do not depend systematically on the depth of burial (Table 12.3), the smectite formation may be considered as essentially hydrogenous (syngenetic), early diagenetic effects being eventually impeded or masked by bioturbation.

Mineralogy and Geochemistry, Bauer Deep
Bauer Deep sediments are known for their concentrations of trace elements roughly similar to those of the East Pacific ridge deposits, and higher than the average for Pacific pelagic clays (in Dymond 1981). Some general differences between Bauer Deep and East Pacific ridge deposits nevertheless exist, and consist mainly of lower calcium carbonate and higher smectite contents in the former relatively to the latter. Cole and Shaw (1983) and Cole (1985) discuss this question by studying two 2-meter long cores from the Bauer Deep. Sediments comprise red-brown clay to brownish marly ooze, characterized by abundant smectite (up to 70%), about 10% Fe-oxyhydroxides, 4% Mn-oxyhydroxides, and minor amounts of detrital clay minerals, quartz, feldspar, and biogenic opal. Smectite amount tends to increase in noncalcareous sediments compared to metalliferous clay, whatever the depth in the cores. Three techniques are employed to determine the type and chemical composition of smectite (Cole 1985). Infra-red and Mössbauer spectroscopy data indicate that smectite belongs to the nontronite (Fe-rich) group. The chemical composition, obtained by emission spectroscopy of five smectite separates, corresponds to the following structural formula:

$$(Ca_{0.05}Na_{0.20}K_{0.06})Fe^{3+}_{1.07}Al_{0.44}Mg_{0.54}Mn_{0.03}Ti_{0.01}Cu_{0.01})$$
$$(Si_{3.97}Al_{0.03})O_{10}(OH)_2, nH_2O .$$

This mineral contains only half the amount of octahedral Fe^{3+} required by a typical nontronite. The Fe^{3+} deficiency is compensated for by Mg^{2+} and Al^{3+}. The Bauer Deep smectite therefore corresponds to *Mg-rich, Al-rich nontronite*.

The oxygen isotopic analysis of smectite by mass spectrometry shows a uniform composition, which translates to a temperature of formation of $3° \pm 5°C$. This temperature is compatible with isotopic equilibration in normal deep-sea conditions. T.G. Cole deduces that despite the hydrothermal-type trace elements geochemistry of the Bauer Deep sediments, *smectite formation is not hydrothermal but hydrogenous*. The hydrothermal activity formerly envisaged by several authors in the Bauer Deep is not corroborated by isotopic data.

The origin of the Bauer Deep smectite is considered in terms of the source of its two dominant constituents, namely the iron characteristic of octahedral sheets and the silicon typical of tetrahedral sheets (Cole 1985). The source of aluminum, magnesium, and trace elements is not discussed, these cations being envisaged as supplied together with Fe and Si.

Since most layer silicates in metalliferous sediments consist of Fe-smectite, the origin of iron and the role of iron oxyhydroxides in the clay formation is often discussed in the literature. Many authors estimate that the dominant source of iron is hydrothermal (see Cronan 1976, Dymond 1981). Fröhlich (1980) considers that ferrous-silicic amorphous complexes, that occur abundantly in metalliferous sediments of the Indian Ocean and could have terrigenous or hydrothermal origins, represent major precursors for the growth of authigenic smectite. In the Bauer Deep, Cole (1983) deduces from oxygen isotope analysis that iron incorporation does not proceed from a hydrothermal mechanism. Iron is therefore considered to have a distant origin. This origin could be terrigenous or hydrothermal, or both. T.G. Cole estimates that owing to the geographical situation of the Bauer Deep (Fig. 12.3) and to the elemental chemical composition of its ferriferous sediments, *iron essentially derives from hydrothermal sectors of the East Pacific ridge. Fe-amorphous oxyhydroxide flocs are supposed to be transported by bottom currents from the ridge crest, and subsequent ponding would determine the accumulation of iron in Bauer Deep sediments.*

The most available source of silicon consists in biosiliceous tests. Several studies have shown that smectite development is morphologically associated with siliceous debris, especially in Pacific deep-sea brown clays (e.g., Perry et al. 1976, Hoffert 1980, Fournier-Germain 1986). In Bauer Deep sediments, Cole (1985) observes *dissolution figures on siliceous microfossils (mainly radiolarians) and a correlative growth of smectite- like particles inside and outside the tests.* The dissolution of siliceous tests and the development of smectite appear to increase in a parallel way below the uppermost core levels, which suggests a combination of syngenetic and early diagenetic processes. From geochemical and scanning electronmicroscope data, *authigenic smectite formation appears to consist in a transformation reaction of surface-active biogenic opal, in the presence of surface-active iron oxyhydroxide.* The smectite synthesis probably proceeds by coordination of Fe^{3+} from the oxyhydroxide to silanol (SiO^-) groups on the hydrolyzed surface of the dissolving opal (Cole 1985). Fe (III)-silica complexes could constitute intermediate steps for smectite development. Experiments have shown that an amorphous Fe (III)-silica precipitate could be generated by coprecipitation of silica with iron hydroxide under oxidizing conditions, and could evolve into a crystalline iron-rich clay mineral if aged under less oxidizing conditions (Harder 1978). The physico-chemical environment of the Bauer Deep pore waters is compatible with experimental conditions defined by H. Harder (Cole 1985).

Since atomic proportions of Si and Fe in the lattice are about 4 and 1, silicon is consumed at a greater rate than iron. T.G. Cole therefore envisages that *biogenic opal availability constitutes the reaction-limiting factor of smectite formation at a given location.* Nevertheless, the relative lack of smectite in the East Pacific ridge sediments, relatively to the Bauer Deep sediments, seems to be related to the excess of calcium carbonate rather to a depletion in silica. Topographic elevations like the ridge area favor the deposition of calcareous tests, which determines a higher porosity due to coarser grain-size (e.g., foraminifera tests). The dissolution of biogenic silica could be enhanced in such porous environments, and the rapid release of silicon preclude the formation of smectite (Cole 1985).

12.3 Fibrous Clay in Deep-Sea Sediments

The possibility for fibrous clay to precipitate in the deep-sea environment, especially in metal-bearing oozes protected from active terrigenous input and exposed to long-lasting chemical exchanges, has been debated since palygorskite and sepiolite were identified in Cenozoic sediments of Atlantic Ocean (e.g., Peterson et al. 1970, Rex and Murray 1970), Pacific Ocean (e.g., Zemmels and Cook 1973) and Indian Ocean (e.g., Matti et al. 1974; see other references in Couture 1977, Kastner 1981). *Experimental and theoretical investigations suggest that sepiolite can easily precipitate from sea water.* For instance, Siffert (1962b) obtains sepiolite and smectite at alkaline pH from magnesium-and silica-rich solutions. Wollast et al. (1968) precipitate poorly crystalline sepiolite by adding Na_2SiO_3 to O.1 mole $MgCl_2$ solution at 25°C and 1 atmosphere total pressure. R. Wollast and colleagues calculate that sepiolite should form spontaneously from sea water at pH 8.5 with a concentration of 6.7 ppm of dissolved silica, or at pH 7.8 with a silica concentration of 60 ppm. *By contrast, palygorskite, that contains more aluminum than sepiolite, has never been synthesized from laboratory experiments* (see Kastner 1981).

On the basis of these experimental results and calculations, sepiolite should constitute a common authigenic mineral in hydrogenous, siliceous deep-sea sediments, while palygorskite should be absent or very uncommon. Surprisingly, the opposite occurs in natural deep-sea deposits. *Palygorskite is fairly commonly quoted in recent metalliferous, biosiliceous, and other pelagic oozes. Sepiolite is, by contrast, almost never identified in recent to late Cenozoic deep oceanic deposits.* Fibrous clay minerals particularly characterize late Cretaceous to Paleogene times, in both oceanic basins and in peri-continental alkaline evaporative basins (see Chap. 9). The possibility of an in situ formation of palygorskite in the deep ocean therefore sets up a difficult and unexpected problem. Sepiolite should form easily in an oceanic environment where it is usually absent, while palygorskite can hardly form in such an environment but is often present.

Contrarily to smectite, palygorskite and sepiolite do not seem to have been described as forming in recent, nonhydrothermal, deep sediments close to the sediment-sea water interface, which makes more difficult the question of their possible hydrogenous formation during past geologic times. This problem has been investigated by Couture (1977), who observes that owing to their chemical composition, both sepiolite and palygorskite should be stable in deep-sea sediments. Palygorskite should be stable in lower dissolved Si and Mg concentrations and lower pH than sepiolite, because of its higher aluminum content. R.A. Couture calculates that palygorskite is stable with respect to smectite at pH 7.5 and at dissolved silica concentrations greater than $6.2.10^{-4}$ moles. These values are compatible with the chemical composition of interstitial water extracted from several types of deep-sea sediments. Couture (1977, 1978) deduces from these calculations that both sepiolite and palygorskite could have easily formed during Cenozoic time in deep-sea sediments, either from Mg-rich smectite and biosiliceous opal, or from biogenic opal and mainly silicic ash. Beck and Weaver (1978) think that Couture's assessments on interstitial water data are difficult to use in order to interpret the fibrous clay origin, since little reliable information is available on the processes and equilibrium constants of palygorskite formation. K.C. Beck and C.E. Weaver consider that the natural stability of both fibrous clay minerals in the deep-sea environment does not constitute an argument allowing further reasoning on a possible authigenic genesis, in the absence of convincing observations on in situ chemical equilibration and kinetics. Kastner (1981) also considers that not enough geochemical information is available to explain the formation of fibrous clays in oceanic environment, either from the evolution of sedimentary components or from direct precipitation.

The major arguments developed by Couture (1977, 1978) and by Tazaki et al. (1986) to justify the possible in situ formation of palygorskite and probably of sepiolite in nonhydrothermal, common deep-sea environments, are presented below with some comments.

1. *Palygorskite is preferentially associated with silica-rich minerals of undoubted authigenic origin, like clinoptilolite.* This is mostly true, but close investigations on data from the Deep Sea Drilling Project show numerous examples of the occurrence of one mineral group without the other or of irregular, nonparallel variations. Clinoptilolite also occurs abundantly in various pre-Pliocene sediments of the world ocean devoid of significant amounts of palygorskite (Stonecipher 1976, Nathan and Flexer 1977, Lisitzina and Butuzova 1980). *The presence of clinoptilolite in many palygorskite-containing sediments may simply reflect its widespread occurrence* (Beck and Weaver 1978). In some regions palygorskite and clinoptilolite are even mutually exclusive (Weaver and Beck 1977).

2. *Palygorskite formation appears from calculations to be favored by bottom water temperatures higher than present ones* (25°C instead of 2°C), *a fact that fits with the coexistence in late Cretaceous-early Cenozoic times of warm world climate and of abundant fibrous clays in sediments.* But isotopic investigations suggest that bottom water never reached temperatures as high as 25°C (in Kennett 1982). In

addition, the temperature of the entire deep-sea floor would have been that of the polar regions even during warm periods, which suggests that fibrous clay formation could have occurred in oceanic deposits from low to high latitude regions. Such a hypothesis is contradicted by backtracking of palygorskite- and sepiolite-bearing deep-sea sediments, which shows that they mainly occur in subtropical to tropical belts (in Kastner 1981). In fact, *major oceanic physicochemical modifications due to climate or paleocirculation changes had their greatest effects in restricted basins near land* or even on land (Weaver and Beck 1977, Kennett 1982). Beck and Weaver (1978) notice that the reworking of peri-marine fibrous clay deposits represents a simple and logical explanation for the synchronous occurrence of palygorskite and sepiolite in the deep sea, especially when considering the fact that most fibrous clay-bearing deep-sea deposits are adjacent to large shallow-water deposits (e.g., Atlantic and Indian Oceans).

3. *Palygorskite fibers are sometimes observed as forming delicate structures growing on amorphous silica surfaces, and they frequently occur in deep areas very distant from any continent* (e.g., Central and Western Pacific Ocean). This suggests an in situ formation by reaction of smectite with opaline silica and interstitial solutions (Couture 1977). But many electronmicroscope observations point to the *fairly high resistance of fibrous clays to transportation* in both water and air. Some fine sedimentary phases carried in water may include very delicate crystalline particles, whose external edges are protected against destruction by adjacent particles (e.g., distal turbidites, slumps); this, of course, does not contradict the hypothesis of an in situ growth, but indicates this is not the only possibility and that the growth could have occurred before transportation within the deep sea. In addition, recent investigations show the *importance of aeolian supply from remote sources to deep-sea environments*, especially for Pacific deep-sea brown clay characterized by very slow deposition (Chapts. 7, 8). Palygorskite particles in deep-sea sediments often display short and broken fibers indicative of important reworking processes (e.g., Lenôtre et al. 1985). Note that electronmicroscope observations of Tazaki et al. (1986) suggesting the transformation of smectite into palygorskite in one sample of North-Pacific Cenozoic deep-sea brown clay are not definitely convincing. Some palygorskite fibers may have been mistaken for lathed smectite (see Chap. 14). In addition, most palygorskite particles shown on micrographs appear morphologically independent from sheet minerals, or are superimposed to smectitic flakes rather than developing at their periphery.

To summarize, no definite arguments arise from the literature to undoubtedly support the hypothesis of a hydrogenous formation of palygorskite or sepiolite in open-sea, oxidized sediments. Fibrous clays obviously form or formed in calcareous, evaporative, terrestrial soils (Chap. 2), in peri-marine alkaline basins (Chap. 9), and in hydrothermal environments (Chap. 13), but do not seem to significantly develop in metal-bearing deep-sea surface deposits, that are usually devoid of marked alkaline character and of magnesium concentration.

12.4 Transitional Environments

A very difficult question consists in *the distinction, in a given marine basin, between clay particles of a given mineral group inherited from land and those authigenically originating in the deep ocean*. Such a question may become nearly unanswerable if the minerals considered form at the same time either in terrestrial or in submarine environments. This is particularly the case of *smectite minerals*, especially in deep-sea metalliferous clays whose very slow sedimentation rates allow the sedimentary expression of both distant aeolian input and hydrogenous neoformation. For instance, in Central and South Pacific sediments marked by fairly little crystalline clay fraction but abundant smectite percentage in this fraction, what is the proportion of smectite derived from Antarctica and enhanced by smectite-rich outcrops and strong wind activity, and the part of the same mineral group newly formed close to the sediment-sea water interface? To make progress in such a problem presumes sophisticated technical investigations, involving grain-size fractioning and the combination of different geochemical analyses. Let us take two examples.

12.4.1 Pacific Ocean

In the south equatorial part of the Eastern Pacific Ocean, Hoffert (1980) identifies abundant smectite in deep-sea brown clay, from the Panama basin and Eastern Nazca plate to French Polynesia (see stations location on Fig. 12.3). Rosato and Kulm (1981) show that the easternmost regions of this domain are characterized by significant amounts of smectite derived from both continental and marine sources. What is the respective proportion of allochthonous and autochthonous smectite in these regions, and how is terrigenous smectite diluted among hydrogenous smectite when going westwards? Is this dilution progressive or disturbed by aeolian and marine supply of smectite? Aoki et al. (1979) investigate a core from an abyssal hill Southeast of Tuamotu archipelago (SP15, Fig. 12.3), by X-ray diffraction, thermal analysis, infrared spectroscopy and chemical analysis. They identify two types of smectite (Table 12.4). One is an iron-rich smectite, not abundant, and attributed to an autochthonous formation. The other corresponds to an aluminum-rich smectite, similar to common land-derived minerals, and possibly formed by weathering of volcanic material and transported from exposed volcanic areas. These data suggest that deep-sea

Table 12.4. Chemical composition (%) of two smectite types from deep-sea sediments of Tuamotu region. (After Aoki et al. 1979)

	SiO_2	Al_2O_3	Fe_2O_3	MgO	Na_2O	TiO_2	K_2O
Fe-smectite	50.17	7.11	11.46	5.41	2.86	0.77	0.57
Al-smectite	49.39	14.39	4.73	3.61	2.12	1.18	0.34

Pacific smectites result from both autochthonous and allochthonous sources. In addition, different smectite varieties and associated sedimentary particles, originating in different parts of the ocean under distinct hydrogenous conditions, may be reworked by bottom currents from one metalliferous clay environment to another (12.2.3. Hoffert 1980).

Hoffert (1980) and Bonnot-Courtois (1981) observe from total sediment analyses that rare-earth elements display an increase in the cerium-negative anomaly from the Peru-Chile trench area to French Polynesia (Fig. 12.7A). This trend indicates the more authigenic, hydrogenous character of the deep-sea sedimentation when going westwards in the Pacific Ocean, and the correlative decrease of the continental influence. If smectite constitutes the most abundant mineral species in the brown clay, one may assume, despite the large distance between successive sampling stations, that the rare earth elements spectra obtained from bulk sediment reflect the increasing authigenic character of the clay mineral at increased distance from the South American continent. But if smectite represents only a minor sedimentary component, as is the case in many deep-sea metalliferous deposits rich in metal oxides, opal and/or zeolites, the Ce-negative anomaly could be mainly determined by nonclay authigenic minerals.

Tlig (1982a) studies the rare earth elements distribution in eleven different grain-sized fractions between > 125 μm and < 0.2 μm in surface sediments of stations regularly distributed on an East-West transect in the South Equatorial Pacific Ocean (Fig. 12.7). At the French Polynesian and Western Nazca plate stations, a negative anomaly in cerium frequently characterizes the bulk sediment as well as all grain sizes, from coarse fractions rich in iron oxides and zeolites to fine fractions enriched in smectite (e.g., TKS.23, Fig. 12.7D). By contrast, spectra obtained close to South America in the Peru-Chile trench area often display negative anomalies in the coarse fractions that contain opal or calcite, while fine clay fractions present flat spectra typical of continental influence (e.g., T2.K17, Fig. 12.7B). This result confirms the general trend established by Hoffert (1980) and Bonnot-Courtois (1981). But *some sediments of the central East Pacific basin,* as at station T2.K1 located Northeast of Mururoa island in the French Polynesian region, *display a strong Ce-negative anomaly in coarse fractions made mainly of goethite, opal, and phillipsite, and no anomaly in fine fractions dominated by smectite* (Fig. 12.7C). A negative Ce anomaly also characterizes the bulk sediment at the same station, which obliterates the signal of fine, accessory fractions. Such data pose the question of a possible terrigenous contribution of clay minerals, for instance through aeolian supply, to the deep-sea sedimentation in this area. Comparable rare earth elements spectra, marked by strong Ce-negative anomalies in both bulk sediments and coarse fractions, and by slight or no anomaly in the clay-rich fractions, occur at other stations, and apparently more frequently in the easternmost part of the transect studied by S. Tlig (e.g., station T2.K12, Central Nazca plate; location on Fig. 12.3). The question is to be sure that the cerium distribution in smectitic fractions is always reliable and significative of either a continental origin (no anomaly) or a marine origin (negative anomaly). Some marine authigenic materials, such as metalliferous nodules, display positive Ce anomalies (Piper 1974, Bonnot-Courtois 1981, Tlig 1982a), and the behavior of cerium is not perfectly known in all geochemical environments.

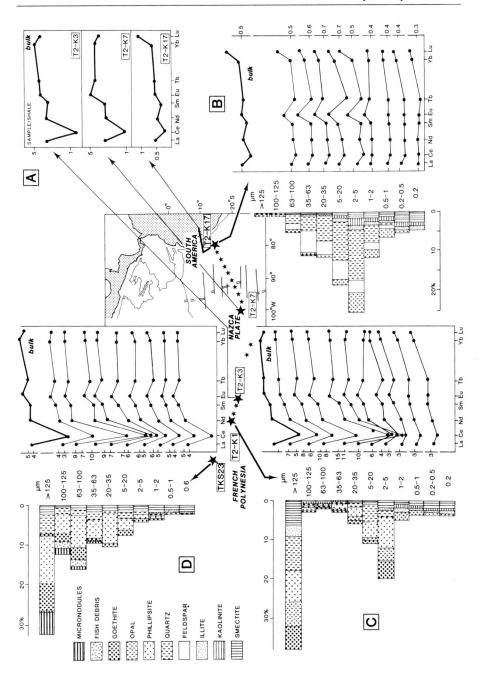

Fig. 12.7. A Rare-earth elements distribution in bulk sediments of three stations from a South America to French Polynesia transect, South Equatorial Pacific Ocean. (After Hoffert 1980). **B, C, D** Mineral and rare-earth elements distribution in bulk sediment and different grain-sized fractions from three stations in the same domain. (After Tlig 1982a). Rare-earth elements curves normalized to continental shales composition

Some answers may be expected from studies combining data on rare earth elements and on isotopes sensitive to marine environment like the rubidium and strontium system.

12.4.2 Indian Ocean

Large areas in the Indian Ocean contain abundant smectite in the clay fraction of surface sediments (Fig. 8.6). The mineral can a priori derive from continental soils and rocks, from submarine alteration of volcanic material, from hydrothermal activity and/or from hydrogenous formation. The proportion of each phenomenon is difficult to identify. Let us take one example. The smectite abundance in the Northeast Indian Ocean increases from the proximal zone of the Ganges deep-sea fan in the North to the Central Indian basin in the South. This increase, that ranges from about 20% of the < 2 μm fraction in the North to more than 50% in the South, does not correlate with a significant change in the chemical composition of smectite (Table 12.5). *Smectite belongs to the group of Fe-Al beidellite both in typical terrigenous facies related to direct input of the Ganges river and in distal pelagic facies* (Bouquillon 1987, Bouquillon et al. 1989). A comparable composition characterizes smectites from the brown siliceous and metalliferous oozes located southwards in the Central Indian basin (Tlig and Steinberg 1982). An average composition for these smectites is the following: SiO_2 55.7%, Al_2O_3 20.0%, MgO 4.9%, Fe_2O_3 13.3%, TiO_2 0.5%, K_2O 2.7%, Na_2O 1.5%, CaO 0.7% (Bouquillon 1987). Only very minor changes occur from North to South, such as, for instance, a moderate increase of octahedral and tetrahedral aluminum in some smectite particles from the distal facies relatively to more proximal facies. No increase of iron abundance, that would be indicative of smectites formation in metal-bearing clays, is recorded. Bouquillon and Chamley (1986) and Bouquillon (1987) interpret the southward increase of smectite abundance and the minor changes in its chemical composition as the result of differential settling processes, and possibly of variations in terrigenous sources

Table 12.5. Average structural formulae of smectites in the middle to distal part of the Ganges deep-sea fan (microprobe data, after Bouquillon 1987), and in the adjacent Central Indian Basin. (After Tlig and Steinberg 1982)

	Tetrahedra		Octahedra			Interlayer			
	Si	Al	Al	Fe^{3+}	Mg	Ti	K	Na	Ca
Ganges deep-sea fan									
Proximal									
Sandy mudstone	3.46	0.54	0.94	0.71	0.33	0.02	0.30	0.54	0.02
Calcareous ooze	3.51	0.49	1.06	0.58	0.49	0.02	0.26	0.19	0.03
Distal									
Brown siliceous clay	3.39	0.61	1.34	0.59	0.06	0.02	0.03	0.15	0.01
Central Indian basin									
Brown siliceous ooze	3.7	0.3	1.0	0.7	0.45	–	0.15	0.1	0.02

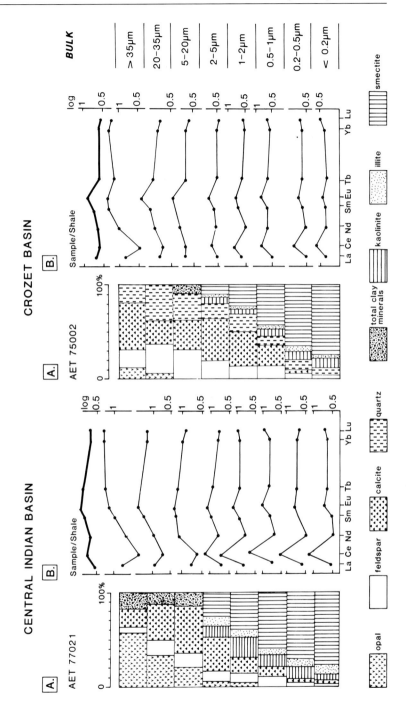

Fig. 12.8. Mineral (**A**) and rare earth elements (**B**) distribution in bulk sediment and different grain-sized fractions from Central Indian basin and Crozet basin. (After Tlig and Steinberg 1982). Rare-earth elements curves normalized to continental shales composition

and aeolian input. No in situ formation of smectite can be deduced from mineralogic and geochemical investigations of recent sediments.

It is very intriguing that brown metalliferous oozes in the Central Indian Ocean, whose abundant ferrous-silicic iron oxyhydroxides are considered as potential precursors for active smectite growth (Fröhlich 1980), contain smectite minerals with a chemical composition almost identical to that of smectites directly issued from the Ganges river. Tlig (1982b) and Tlig and Steinberg (1982) study the rare earth elements distribution in siliceous and metalliferous brown oozes in the Central Indian Ocean. Bulk materials display a fairly flat curve, only marked by a positive anomaly in europium attributed to the presence of magmatic-derived feldspar (Fig. 12.8). There is no apparent anomaly in cerium content for the total sediment. In fact, the nearly flat curve results from the *combination of coarse fractions depleted in Ce and of fine fractions enriched in Ce.* The Ce depletion is related to biogenic silica, while the Ce excess is determined by abundant smectite. The Ce enrichment in clay-sized fractions seems to be independent of both location and lithology (Tlig and Steinberg 1982).

A positive Ce-anomaly is known in iron-rich oceanic nodules, but smectites studied in the Indian Ocean lack of significant iron enrichment. Some cases of Ce-enrichment in continental environment are reported in the literature (in Tlig 1982b). In addition, the same rare earth element pattern characterizes the smectitic clays from the Central Indian Ocean that could have preferentially formed authigenically, and smectites from the Crozet basin in the South that are largely attributed to a southern detrital origin (Chap. 8, Fig. 8.14). Finally, one could envisage a volcanic influence, some volcanogenic smectites being able to concentrate cerium (Desprairies and Courtois 1980); but most sediments studied by S. Tlig do not contain appreciable amounts of altered volcanic debris. Tlig (1982b) and Tlig and Steinberg (1982) conclude that the Ce excess in fine fractions of Indian Ocean sediments is probably of continental origin, but could also derive, at least partly, from the submarine alteration of former volcanic glass or from specific oxidation processes in the marine environment. One could also envisage that the Ce excess in smectitic fractions corresponds to the subaerial weathering of volcanic materials subsequently carried by winds in the deep ocean. Whatever the cause, this type of study stresses *the difficulty in distinguishing the respective part of allochthonous and autochthonous influences in the genesis of phyllosilicates from some deep-sea metalliferous clays.*Similar problems arise from oxygen isotope data, which suggest that most quartz in the Indian Ocean is detrital, most clinoptilolite authigenic, and smectite possibly in part authigenic (Tsirambides 1986).

12.5 Ancient Deep-Sea Metalliferous Clays

Although most red, iron-bearing clayey sediments in the geologic record are of a continental origin, equivalents occur in some marine deposits of various ages. Due to the complex synsedimentary and postsedimentary history possibly undergone by old sediments, and knowing the difficulty in precisely characterizing the hydrogenous clay formation in some recent metalliferous deposits, it is a difficult task to search for the identification of similar authigenic processes that could have occurred in ancient times close to the sediment-sea water interface. The very limited data available attest to this difficulty.

In the North Atlantic Ocean, Lancelot et al. (1972) investigate multicolored banded sediments deposited during the late Cretaceous-early Tertiary in the western basin (Hatteras basin) and belonging to the Plantagenet and Bermuda Rise formations (Sheridan, Gradstein et al. 1983). The presence of goethite-rich layers leads Y. Lancelot and colleagues to compare the mineralogy and geochemistry of these dominantly reddish sediments with those of multicolored deposits linked to hydrothermal hot brines in the Red Sea (Degens and Ross 1969). The nature of iron oxides is partly similar in both cases, but the geochemistry considerably differs. For instance, Hole 105 DSDP variegated sediments display relatively to the Red Sea hot brines much more Si (x 2.8), Al (x 6.8), Mg (x 1.7), K (x 5.2), Cr (x 2.9), Ni (x 8.9) V (x 25), and much less Fe (: 5.9), Ca (: 11.1), Mn (: 5.9), Zn (: 25), Cu (: 60) and Pb (: 26). Lancelot et al. (1972) stress the abundance of Al_2O_3 and K_2O in Atlantic sediments, which they attribute to the abundance of terrigenous minerals such as illite and kaolinite. This strongly contrasts with the occurrence in Red Sea deposits of abundant metal oxides and sulfides, and of rare terrigenous clay. The authors envisage the possibility that variations in the respective supply of remote hydrothermal material and of terrigenous minerals could have determined the deposition of banded varicolored clays. Robertson (1983) studies comparable series in the same domain (Blake-Bahama basin), and attributes the relative increase of metals and the reddish color to a slow deposit in an oxidizing environment rather than to a hydrothermal influence. Whatever the cause of the coloration, clay assemblages in reddish banded sediments do not differ in a significant manner from those of over- and under-lying common non-metalliferous sediments, which precludes the hypothesis of a hydrogenous origin (Lancelot et al. 1972, Chamley et al. 1983a).

In the South Atlantic Ocean, Karpoff (1984) studies five DSDP sites located on the eastern flank of the mid-oceanic ridge, at the latitude of the Walvis rise. Red clays occur during the Miocene and represent dissolution facies of calcareous oozes. Condensed sediments contain fairly abundant iron, manganese, and trace elements, which typically resembles metalliferous clays. A.-M. Karpoff envisages from chemical and electronmicroscope data the possibility of an in situ formation of smectite in slowly deposited metal-bearing clays, especially during late Miocene time. This formation is tentatively attributed to a hydrogenous process independent of volcanic or hydrothermal activity, and similar to that reported in some South Equatorial Pacific deep-sea brown clays (12.2.3). As the clay mineral record does not show significant smectite increase in most dissolution facies compared to common sediments (Karpoff 1984), the authigenic process should necessarily be quantitatively very limited. In addition, more general studies performed on the clay mineral stratigraphy in South Atlantic basins show that smectite abundance does not significantly depend on the presence or absence of metal-bearing condensed series (Robert 1987).

In the Pacific Ocean a hydrogenous genesis is often quoted for iron and manganese oxyhydroxide in Cenozoic deposits (e.g., Kadko 1985, Barrett et al.

1986, Leinen and Heath 1987), but much more rarely for clay minerals in the same series. In the North Pacific, most clay minerals, including volcanogenic smectites, appear to derive from various sources dominated by terrigenous input, but little from hydrogenous processes (e.g., Leinen and Heath 1981, Lenôtre et al. 1985). In the Southeastern Pacific, leg 92 DSDP was devoted to the study of metalliferous sediments deposited on the western flank of the oceanic ridge, along a transect at 19°S. Surprisingly, the sediments deposited from Oligocene to Pleistocene appear to contain very few clay minerals (Barrett et al. 1986, Kastner 1986). The noncalcareous fraction of clayey nannofossil oozes to zeolitic clays consists of red-brown to yellow-brown semiopaque oxides rich in goethite. Goethite is locally associated with phillipsite, apatite, and minor quartz. Most crystalline to amorphous minerals are attributed to hydrogenous processes, but they do not include appreciable amounts of clay minerals. In particular, Fe-smectite, that occurs in noticeable amounts within recent metal-bearing clays elsewhere in the Eastern and Central Pacific (12.2), is absent from older sediments at 19°S from about 110° to 130°W. T.J. Barrett and colleagues report that this result has no methodological cause. These data are difficult to explain. Perhaps Fe-smectites did not form for some reason in this specific area, or were unstable, like Fe-smectite described in some hydrothermal environments (see Chap. 13), and were destroyed during burying.

At about 2000 km to the North, Fe-smectite constitutes an abundant component of the metalliferous phase in claystones sampled at leg 85 DSDP sites (Jarvis 1985). The mineral is attributed to a low-temperature interaction of Fe-oxyhydroxides and biogenic silica with other major cations being extracted from sea water, as described in recent sediments of the Bauer Deep (12.2.4). Note that South of the Leg 92 drilling area, a brown ooze sampled at 29°S and 131°W on the western flank of a seamount reveals the presence of Fe-smectite, goethite, phillipsite, and δMnO_2. The brown ooze, that displays a strong Ce-negative anomaly, is attributed to a hydrogenous formation (Stoffers et al. 1985). The sediment, 190 cm thick, is superimposed to green nontronitic clay of a hydrothermal origin (13.3.2).

To summarize, little undisputable information is available so far on the hydrogenous formation of clay minerals at the sediment-sea water interface in old deposits from the Pacific Ocean. As such a formation does occur recently in some basins of this domain, it probably also occurred in Cretaceous-Cenozoic times. But this was rarely demonstrated so far, or was obliterated by diagenetic changes. A special effort should be made to better separate the clay fractions from subamorphous oxides, and to search for the occurrence of discrete phyllosilicates.

A preliminary state of knowledge also characterizes the mechanisms of hydrogenous clay formation in Cenozoic brown deep-sea sediments of the *Indian Ocean*. Ferriferous smectites are recorded in several cores and probably are autochthonous (e.g., Fröhlich 1982), but the genetic process is still poorly understood. Other brown siliceous clays, like the Plio-Pleistocene deposits of Central Indian basin, contain Al-Fe smectites (e.g., site 215 DSDP, Bouquillon 1987), which resembles the composition of common nonmetalliferous sediments and suggests rather an allochthonous origin. Finally, little is known about hydrogenous processes in old metal-bearing marine clays from *land sections*. For instance some of the red clayey to radiolaritic deposits characteristic of Triassic-Jurassic times in the Tethyan domain are referred either to terrigenous supply or to diagenetic effects (e.g., Steinberg et al. 1977, Baltuck 1982, Holtzapffel and Ferrière 1982), but almost never to exchanges between surface sediments and open sea.

12.6 Conclusion

1. Deep-sea metalliferous clay represents, like organic-rich deposits, a group of sediments in which a usually minor component becomes important and determines a particular color. In both cases, the rather uniform color does not imply a single genetic process. *Red-brown clays, like black shales and sapropels, may arise from both allochthonous and autochthonous origins, especially as far as clay mineral fractions are concerned.* Unlike most organic-rich and also alkaline-evaporative sediments, deep-sea iron-rich deposits display obvious examples of clay genesis during recent times, which resembles the case of iron-rich granules in shallower-water sediments (Chap. 10). *Both deep metalliferous clays and glaucony granules appear to form essentially close to the sediment-sea water interface, and therefore result largely from hydrogenous processes.* But as the former do not usually constitute specific granules, are associated with much poorly crystallized to amorphous Fe-Mn oxyhydroxides, and correspond to extremely low sedimentation rates, the precise identification of genesis mechanisms is particularly difficult. In addition, *the source of iron* can be much more diverse in deep-sea clays than in shallow-water ferriferous granules, and *may notably include an important hydrothermal component.*

2. *Authigenic aluminosilicates in deep-sea brown clay mainly comprise Fe-smectites, that contain minor but significant amounts of aluminum and magnesium.* Zeolites of the *phillipsite* type also frequently form in this environment. By contrast, very little evidence exists for the hydrogenous formation in appreciable amounts of palygorskite or sepiolite in nonhydrothermal deep-sea sediments.

Despite their frequent occurrence, *authigenic smectites often constitute only an accessory component of metal-bearing deep-sea clays, major components consisting of iron and manganese oxyhydroxides* that form either isolated particles, aggregates, micronodules or nodules. Some "brown clays" are even almost devoid of clay minerals, particularly in the South Pacific and South Indian Oceans where metal oxides, zeolites, and biogenous debris represent the main sedimentary constituents. Some subsurface metalliferous clays in South Pacific Ocean virtually lack of clay minerals (e.g., Paleogene-Neogene deposits from leg 92 DSDP), which is different from surface deposits in the same domain and poses the question of the stability of Fe-smectites facing burial constraints. Surprisingly, the deep-sea brown oozes affected by intense authigenic processes contain less clay minerals, less fine fraction, and less volcanic materials than deep-sea brown oozes submitted to dominantly terrigenous input. This opposition clearly arises when comparing the mainly allochthonous North Pacific and the mainly autochthonous South Pacific brown clays. As a consequence, and despite the wide sea-floor surfaces covered by metalliferous deposits, *the authigenic formation of Fe-smectite in deep-sea sedimentary environments probably constitutes an accessory process, from a quantitative point of view.* This idea is supported by the very low sedimentation rate of deep-sea brown clay (often < 1 mm/10^3 yr), by its small thickness in the ocean (less than a few tens of meters), and by its few true equivalents in the geologic record relative to nonhydrogenous clayey deposits.

3. The mechanism of smectite hydrogenous formation in deep-sea brown clays is only partly understood. The most probable phenomenon reported in the Pacific Ocean consists of a *low-temperature interaction between biogenic silica providing silicon and perhaps a proportion of aluminum, Fe- oxyhydroxides providing iron, and sea water providing magnesium and accessory elements.* An alternate mechanism is envisaged in some brown, siliceous clays of the Indian Ocean, where amorphous ferrous-silicic complexes represent possible precursors for smectite genesis. *Whatever the mechanism, it appears to depend very little on direct volcanic activity or on direct hydrothermal input.* Smectite formation in deep-sea metal-bearing clays does not usually result from a predominantly volcanic influence, contrary to what is often envisaged in the literature. The hydrothermal influence appears mainly responsible for the supply of iron and trace elements that could issue from transport of amorphous flocs by bottom currents. Note that such a transport sometimes seems questionable considering the actual direction of bottom currents (e.g., compare the interpretation of Dymond 1981 and Cole 1985 of Bauer Deep sediments: 12.2.4); one should not underestimate the availability of soluble iron supplied from terrestrial and other remote sources.

The respective part of synsedimentary and early diagenetic contributions in clay formation is often difficult to assess, since the extremely low rate of deposition of pelagic brown clay favors long-lasting exchanges with open-sea environment, as well as bioturbation. The early-burying effects generally appear to be of little importance, since most old brown clays from typical authigenic environments apparently show no significant downward increase in smectite abundance (e.g., South Equatorial Pacific and South Indian Oceans). Smectite enrichments recorded in many old Cenozoic sediments do not specifically characterize metal-bearing deposits (see Part Six).

4. In addition to boundary problems existing between synsedimentary (hydrogenous) and postsedimentary (diagenetic) influences, *deep-sea metalliferous clays pose difficult questions about the transition between autochthonous and allochthonous influences.* As smectites, for instance, may issue from various continental and submarine sources, and may be reworked from one marine environment into another one, the identification of their initial origin constitutes a very uneasy task. The combination of sophisticated chemical, microchemical, micromorphologic and isotopic investigations in both sedimentary and interstitial environments may greatly help to characterize genetic conditions. Fruitful approaches are provided by rare earth element measurements on selected mineral species and grain-sized fractions, or by detailed analyses on the Rb-Sr isotope system. Significant progress is also expected from studies performed on sedimentary fractions obtained after physical or chemical removal of amorphous metal-oxyhydroxides. A wide field of research is open, based on cored or drilled

transects extending from dominantly terrigenous to dominantly autochthonous metal-bearing clays (e.g., North to South Pacific or Indian Oceans), and studied by a multidisciplinary approach in both surface and subsurface sediments.

5. Deep-sea metalliferous clays constitute a possible boundary environment with volcano-hydrothermal deposits. This transitional situation results from the presence of abundant metals in both environments, and from the frequent vicinity of basalt oceanic crust, of metalliferous crust deposits and of metal-bearing clay. The characters of these different materials are nevertheless by far not all superimposed. If all hydrothermal deposits are metalliferous, all metalliferous deposits have not undergone a hydrothermal influence. *Many deep-sea metal-bearing clays simply correspond to slow and oxidizing conditions of deposition, with no or only distant control by hydrothermal fluids.* Chapter 13 provides additional information about hydrothermal effects in deep-sea metalliferous deposition processes, relative to hydrogenous effects.

Chapter 13

Hydrothermal Environment

13.1 Submarine Alteration of Volcanic Rocks

13.1.1 Low-Temperature Alteration

Hydrothermal cooling of oceanic crust represents a widespread phenomenon at mid-oceanic ridges, and significantly contributes to the geothermal heat lost by the earth. Exchanges of chemical elements between basalt and water during hydrothermal cooling contribute to both the control of sea-water composition and to the formation of metalliferous deposits. Modelling of these processes is mostly based on the assumption that hydrothermal exchange is proportional to spreading rate (e.g., Berner et al. 1983, Sleep et al. 1983), while measurements on the East Pacific ridge suggest rather a correlation with tectonic reorganizations (Lyle et al. 1987). Ridge jumps and changes of ridge orientation in the course of the time appear to substantially increase hydrothermal activity by fracturing the basalt and allowing sea-water penetration to deep heat sources. Whatever the precise causes and chronology of hydrothermal fluxes, the study of basaltic materials attests that they have undergone some reaction with sea water. The surface of volcanic rocks presents modifications of various types and intensities, which results from a subaqueous alteration (= halmyrolysis). The products of this alteration may subsequently be reworked through tectonic activity and erosion, and feed the marine sedimentation. The characters of basalt alteration depend mainly on the temperature of hydrothermal fluids reaching the uppermost crust.

Low-temperature alteration results from interactions at temperatures close to or moderately higher than those of the bottom water, and is usually referred to basalt "weathering". The basalts altered at low temperature generally differ only slightly from the original igneous precursors. The principal changes recorded consist of a *surficial oxidation and hydration of volcanic rocks*, and of the *formation of Mg-smectite*. Associated secondary minerals may comprise celadonite (see 10.5), zeolite (mostly phillipsite), calcite, aragonite, and various Fe-Mn oxyhydroxides. The oxidation usually determines a color change at the basalt surface from blackish to reddish-brown, which led Cann (1979) to propose the term *brownstone facies* for designating low-temperature altered basalts. *The temperature* responsible for the formation of brownstone *ranges from normal bottom-water values (O-3°C) to 50-100°C* (Cann 1979, Thompson 1983). Amorphous to subamorphous basalts (= hyaloclastites) generally alter more deeply and faster than both the glassy rinds and the crystalline inner parts of basaltic pillows of flows. It is therefore necessary to distinguish the alteration processes and products of glassy and crystalline basalts.

Basaltic Glass

The subaqueous alteration of noncrystalline basalt (or sideromelane) at low temperature is called *palagonitization*. The process is described in detail by Honnorez (1981), who identifies *three stages*.

1. *The initial stage* of palagonitization is characterized by the preservation of large unaltered zones in the submarine volcaniclastic rock, and the absence of secondary minerals in the altered, oxidized residual glass (= "palagonite"). Authigenic minerals are almost exclusively restricted to the spaces located between the glass granules and their vesicles. This preliminary alteration results in the coexistence of fresh glass relicts with residual glass and with alteration minerals. Secondary minerals comprise K- and Mg-rich smectite of the *saponite* type, and Ca-poor, *Na ≥ K phillipsite*. The hydrated and oxidized glass is enriched in K, Na and Mg, and depleted in Ca. The concentration of all other elements remains unchanged.

2. In *the mature stage*, all of the fresh glass becomes altered, and both zeolite and smectite largely replace in situ the residual glass. This stage therefore corresponds to the absence of fresh glass and to the association of altered glass with both inter- and intra-granular authigenic minerals. Smectite and phillipsite largely develop. Phillipsite selectively exchanges its Na for sea-water K, which determines a *K > Na phillipsite*. Smectite is Mg- and K-rich, and comprises either *Fe-Mg saponite* or *Mg-bearing nontronite*.

3. *The final stage* of palagonitization corresponds to a complete replacement of residual minerals where initial contours between inter- and intra-granular zones are obliterated. The hyaloclastite has been completely changed into an intimate mixture of Ca-poor, *K > Na phillipsite*, various K-Mg-Fe rich *smectites* and *Fe-Mn hydrous oxides*.

The chemical budget of halmyrolysis during palagonitization consists of an uptake from sea water into altered rock of K, Mn, and Na (decreasing order), and in a release from glassy rocks into sea water of Ca, Mg, and Si (Honnorez 1981). The intensity of chemical exchanges strongly depends on the duration of sea-water circulation within the tholeiitic glass submitted to palagonitization. Noack (1981, 1983) reports the following morphological, mineralogical and chemical changes (Fig. 13.1): increase of "palagonite" thickness, evolution from trioctahedral (Mg) to dioctahedral (Mg, Fe^{2+}) smectite, increase of the oxidation rate, decrease of the Na/K ratio in phillipsite, hydration and K_2O uptake, change in the Na_2O budget, closure of cracks and porosity. As a part of Na is incorporated in phillipsite and another part is released in sea water, Y. Noack estimates that global alteration budgets in the ocean are very difficult to calculate accurately.

Experiments performed by Trichet (1970) on artificially-granulated basalts of Hawaii, submitted to freshwater alteration at 40-50°C during 3 years, confirm the chemical changes determined by extensive palagonitization. Fe and Ti ac-

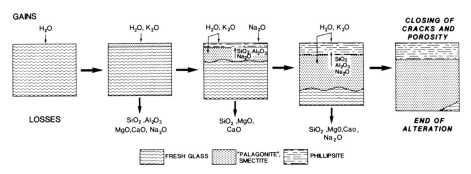

Fig. 13.1. Schematic evolution of chemical changes during palagonitization of tholeiitic glass. (After Noack 1983)

cumulate in the residual glass, whereas most K, Na, and Mg are extracted as well as 95% of the initial Si, 91% of Ca and 85% of Al. Authigenic minerals consist of Mg-rich smectite. By labeling the surface of fresh basalt by Ni (1 ppm in solution), Thomassin et al. (1983) show that initial palagonitization results both from an in situ evolution of hydrated residual glass and from a precipitation of hydrotalcite crystals (Mg_6 Al_2 CO_3 $(OH)_{16}$ $4H_2O$) followed by an Al-serpentine epigenization. These transient minerals form before Mg-smectite, and take over from one another much faster at 50°C than at bottom sea-water temperature (Crovisier et al. 1983). At 50°C saponite may replace hydrotalcite after 480-600 days; at 3°C a very thin alteration layer only forms after 600 days, and consists mainly of Fe-hydroxide (akaganeite, ßFeO OH) with associated K illite-like particles and amorphous components.

Crystalline basalts

The low-temperature alteration of crystalline basalts is usually less evolved than that of glassy basalts. Different authors have shown that *dredged basalts* cropping out on the sea floor at increasing distances from mid-ocean ridges correlate with increasing alteration stages (in Thompson 1983). As soon as submarine basalts are covered with significant amounts of little permeable sediments like clay, halmyrolysis drastically decreases. Many basalts sampled below sedimentary blankets thicker than a few meters present "weathering" characters not basically different from those reached during the time before they were buried. The more common alteration minerals identified in dredged basalts are listed in Table 13.1A. They chiefly comprise *smectite*, that is often not abundant within the rock and poorly crystallized. Associated minerals include iron oxides, typical of oxidation processes on the sea floor. Zeolites are rare or absent. Surficial hydration of basalt is ubiquitous. The alteration correlates with an uptake of K, Cs, Rb, B, Li, and ^{18}O, and with a frequent loss of Ca, Mg, and Si. Fe, Mn, Na, Cu, Ba, and Sr are either gained by or lost from the rock. Al, Ti, Y, Zr, and heavy rare earth elements usually show little or no change.

Drilled basalts present a wide range of secondary minerals, dominated by Mg- to Mg-Fe *smectites, phillipsite, celadonite, calcite* and *pyrite* (Table 13.1B). This

Table 13.1. Mineral products of submarine low-temperature alteration of dredged (A) and drilled (B) basalts. (After Honnorez 1981)

	Zeolites	Clay minerals	Carbonates	Others
A.	Dredged basalts			
	Chabazite	SMECTITE	Calcite	Hematite
	Natrolite	Chlorite		Limonite
		Sericite		Goethite
		Chlorite-smectite		Epidote
		Talc		
B.	Drilled basalts			
	PHILLIPSITE	SMECTITE	CALCITE	PYRITE
	Analcite	(saponite, Fe-saponite)	Aragonite	Marcasite
	Chabazite	CELADONITE	Mg-Calcite	Goethite
	Natrolite	Chlorite		Birnessite
		Celadonite-smectite		Fe-Mn oxides
		Talc		K-feldspar
				Amphibole

Capitals: frequent; lower case: rare.

mineralogical diversity led numerous authors to the conclusion that low-temperature alteration taking place deep in the crust may differ from that occurring close to the surface (see ref. in Honnorez 1981, Thompson 1983). The main difference could result from reactions under low water/rock ratios and nonoxidative conditions. Extensive studies on igneous materials from the Deep Sea Drilling Project provide much information, and lead to the identification of different types or stages of alteration. For instance, Bass (1976) studies basalt alteration products at site 321 (leg 34 DSDP), drilled down to a depth of 11 m in the 40 million-year old basement of the Nazca plate. The detailed mineralogic study of veins and vesicle filling allows the recognition of various secondary minerals including clays, carbonates, zeolites, iron and manganese oxides, and sulfides. M.N. Bass distinguishes four types of submarine alteration:

1. A deuteric or late magmatic alteration marked by the formation in small amounts of biotite, amphibole, chlorite, talc, smectite.

2. A preburial, variously oxidative alteration characterized by the formation of yellow or blue-green smectite at the expense of glass and olivine, and by the precipitation of amorphous Fe-Mn oxides, todorokite, and birnessite (Mn-oxides).

3. A postburial, nonoxidative alteration during which Al-bearing Fe-Mg smectite formed with minor celadonite, as well as Mg-calcite, aragonite, pyrite, and marcasite in the veins, and phillipsite in the glasses of pillow rinds. The system was probably closed and the pH alkaline (Seyfried et al. 1978).

4. A postburial, oxidative alteration in an open system, marked by the alteration of residual olivine into iron oxides, the oxidation of smectite and sulfides, a loss of Mg from calcite, the precipitation of phillipsite, metal oxides and some celadonite.

Alteration types (2) to (4) correspond to low-temperature alteration, and are presented by Bass (1976) as a sequential series of reactions and replacements.

Seyfried et al. (1978) think the different types may have occurred simultaneously at different places in the basalt environment.

After reviewing various papers, Honnorez (1981) suggests that *most buried basalts may have experienced four alteration stages*: (1) A precocious alteration, where only igneous titanomagnetite is altered into maghemite, the basalt appearing to remain fresh. (2) A slightly oxidative alteration, characterized by the precipitation of celadonite and often of K-rich nontronite and Fe-oxides in rock vesicles. (3) A strongly oxidative stage that constitutes the most important step and corresponds to the formation of the greatest part of smectite (mainly saponite), carbonates, phillipsite and metal oxides. (4) A slightly oxidative to nonoxidative stage of smaller intensity and corresponding to the formation of sulfides (mainly pyrite), Fe-rich smectite, Mg-calcite and aragonite. The timing and succession of these various stages are still open to discussion. For some authors the strongly oxidative stage occurs before lavas are buried by younger flows or sediments. For others, it starts or at least continues after burying. Whatever the timing, the final result of low-temperature alteration of oceanic basalt consists in an *increasing heterogeneity of the upper oceanic crust* and the *production of significant amounts of secondary clays, zeolites, metal oxides and sulfides, and carbonates*. According to Staudigel et al. (1981), low-temperature alteration of basalt has taken place to depth of at least 500 meters in the oceanic crust. Most reactions, especially the clay formation, have occurred within a time period shorter than 3 million years, at least for Atlantic-type crustal regimes characterized by low spreading rates.

An example of recent *low- to medium-temperature alteration under both submarine and subaerial conditions* is provided by the study of the volcanic Island Surtsey, located off the south coast of Iceland (Jakobsson and Moore 1986). Surtsey island, created in 1963-1967, consists of basaltic tephra and lava. A 181-m-deep hole drilled in 1979 permits the study of the glass after 12 years of hydrothermal alteration. The cause of heating of the tephra and of the active hydrothermal system was the intrusion of dikes below sea level, about 12.6 years before drilling. At present, the hottest zone of the hole, at a maximum temperature of 150°C, is cooling at about O.9°C per year. The section cored in basaltic tuff extends 122 m below sea level and 59 m above (Fig. 13.2). The dominant alteration process consists of the *palagonitization of sideromelane glass*. The formation of the "palagonite" rind, rich in *nontronitic smectite*, develops at a rate doubling for every 12°C increase in temperature. At 60°C, <40% of the glass is palagonitized; above 100°C, >90% is palagonitized, and above 120°C olivine crystals show external replacement by nontronite. At least ten hydrothermal minerals have crystallized in the basal tephra at $25 - 150°C$. *Nontronite, analcite, phillipsite, and tobermorite* ($Ca_5 Si_6 O_{17}, 5H_2O$) constitute the dominant species. Other minerals include halite, opal, calcite, chabazite, xonotlite ($Ca_6 Si_6 O_{17}, 5H_2O$), anhydrite, and gypsum. Only moderate mineralogic differences occur above and below sea level, probably because the subaerial tephra is saturated with sea salt. Analcite, phillipsite, and tobermorite tend to grow larger below sea level, at a given temperature. Analcite appears at lower temperature (55°C) above sea level than below (75°C). Anhydrite occurs more abundantly near the bottom of the hole, where inflowing,

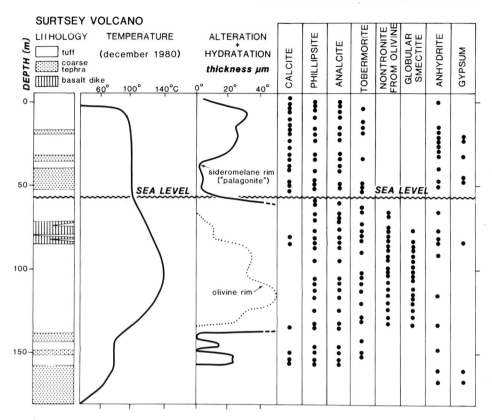

Fig. 13.2. Lithology, temperature, alteration, and secondary minerals in a drill-core from recent volcanic island Surtsey, Iceland. (After Jakobsson and Moore 1986)

cool sea water precipitates sulfate minerals more easily than in a warmer environment.

Silicic Glass

Silicic glass ($SiO_2 > 65\%$) occurs much more *rarely* in the marine environment than basaltic glass. It presents a stronger resistance against alteration because of much higher concentrations of the network-forming elements Si and Al (Fisher and Schmincke 1984). First alteration stage of silicic glass apparently involves a process of diffusion-controlled hydration and alkali ion exchange, but with minor extensive chemical changes. Secondary minerals mainly comprise *Al-smectite* rich in Mg and poor in Fe, and sometimes clinoptilolite. Note that hydrothermal reactions with silicic rocks mostly occur on land, and are responsible for the development of commercial deposits of kaolinite, smectite, serpentine, palygorskite, talc, pyrophyllite, and other species (see Millot 1964, 1970, Grim and Güven 1978, Sudo and Shimoda 1978, Keller 1982, Singer and Galan 1984, Guilbert and Park 1986).

13.1.2 High-Temperature Alteration

High-temperature alteration of volcanic rocks mainly develops in fairly deep crust, and presents characters similar to those typical of deep diagenesis and metamorphism (Chap. 15). But metamorphosed volcanics also form closer to the surface when hot hydrothermal fluids reach the uppermost part of the crust. This peculiarly occurs in slower-spreading oceanic ridges and largely faulted transform zones (Humphris and Thompson 1978), where *metamorphosed basalts* form through interaction with sea water. Metamorphosed basalts undergo major mineralogic and chemical changes relative to their precursors. While low-temperature alteration essentially involves the transformation of glass and accessorily of olivine and plagioclase, high-temperature alteration determines strong chemical reactions and modification of all initial minerals (Fig. 13.3). *With increasing temperature, the brownstone facies is progressively replaced by zeolite-, greenschist- and amphibolite-facies* (Table 13.2), before the deep magamatic evolution into gabbro takes place:

1. *The zeolite facies* corresponds to the appearance beside saponite of other Mg-rich layer silicates like *talc and sub-regular mixed-layers of smectite-vermiculite-chlorite types. Na-Ca zeolites* (analcite, heulandite) also develop.

2. *The greenschist facies* is by far the most frequent type of metamorphosed basalts recovered by dredging the sea floor, and clearly results from both

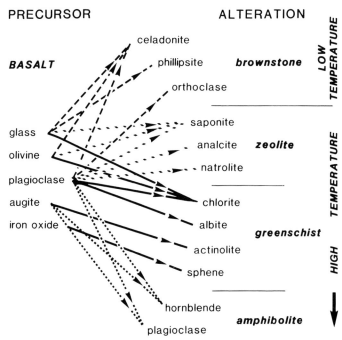

Fig. 13.3. Schematic congruent mineral reactions in different alteration facies of the oceanic crust. (After Cann 1979). The different reactions determine considerable variation in the elemental flux from facies to facies, and the formation of other minerals (see Table 13.2)

Table 13.2. Hydrothermal alteration facies of basalts and secondary minerals commonly encountered. (After Cann 1979, Thompson 1983).

Facies	Alteration minerals
Low temperature	
Brownstone	*Saponite*, "palagonite", *phillipsite, celadonite*
High temperature	
Zeolite	*Mixed-layer* chlorite-smectite-vermiculite
	talc, saponite
	analcite, heulandite
	natrolite, scolectite
Greenschist	*Chlorite, epidote*
	albite, actinolite, sphene
	talc, nontronite, quartz, *serpentine*
Amphibolite	*Hornblende, plagioclase*
	actinolite, tremolite, olivine, iron oxide,
	epidote, quartz, chlorite, sphene

hydrothermal circulation and reaction with sea water (Humphris and Thompson 1978, Thompson 1983). Main secondary sheet silicates comprise *chlorite* in basic rocks, and *talc* and possibly *serpentine* in ultrabasic rocks. Metabasalts rather include Mg-rich minerals dominated by chlorite under reducing conditions, and rather Ca-rich minerals dominated by *epidote* under oxidizing conditions.

3. *The amphibolite facies* is characterized by *hornblende and plagioclase* in metagabbros and dolerite, and by the appearance of tremolite in peridotites (Cann 1979, Cronan 1986). *Actinolite* also frequently characterizes this facies.

Comparison with active geothermal systems (Krismanndottir 1975) indicates that temperatures $\geq 230°C$ are required for the formation of epidote (mainly greenschist facies), and $> 280°C$ for actinolite (mainly amphibolite facies). Note that sulfides, dominated by pyrite, commonly contribute to alteration facies in Mg-rich metabasalts.

Specific alteration facies and minerals may develop under certain conditions. For instance, Haymon and Kastner (1986) study the alteration products of tholeiitic basaltic glass and plagioclase submitted to hydrothermal venting at the crest of the East Pacific Ridge at 21°N. Alteration crusts mainly comprise *aluminum-rich clay minerals* (19-25% Al_2O_3), which is very uncommon for phyllosilicates produced by submarine basalt evolution. Clay species include dioctohedral smectite (Al-beidellitic smectite), randomly interstratified Al-rich chlorite-smectite, minor chlorite, X-ray amorphous aluminosilicate material, and possibly minor amesite (serpentine). Oxygen-isotope analyses indicate an equilibration temperature of 290-360°C for the formation of beidellitic smectite, and of 295-360°C for that of chlorite-smectite. R.M. Haymon and M. Kastner envisage this unusual assemblage could result from high-temperature interaction between basalt glass, plagioclase and Mg-poor, acidic hydrothermal fluids, with Mg possibly provided by bottom sea water. Aluminum could have been mobilized from basalt by low-pH hydrothermal fluids and incorporated into new clay minerals. Aluminous clays could also represent residual phases following drastic alteration of glass and plagioclase by hydrothermal fluids. A high-temperature hydrothermal alteration is also envisaged by Besse et al. (1981) to explain a vertical zonation of zeolites and smectites in a 400-m-thick pyroclastic series drilled on the Ninety-East ridge, Indian Ocean (site 253 DSDP). The downward succession of phillipsite - clinoptilolite, clinoptilolite – mordenite to analcite – clinoptilolite, parallel to that of Fe-beidellite to saponite, appears to result dominantly from circulation in these porous materials of sea water heated by hydrothermal fluids (geothermal gradient of about 200°C/km).

Note that high-temperature alterations of volcanic rocks involving exchanges with sea water are also recorded in old marine, nowadays exposed series. This is, for instance, the case of the upper Triassic alkaline basalts of the Alakir Cay valley, Turkey, that have been hydrothermally metamorphosed under zeolite facies conditions (Malley et al. 1983). Zeolites of various types (natrolite, analcite, heulandite, gonnardite, thomsonite, gmelinite, stilbite, laumontite), smectite, celadonite, calcite, K-feldspar, and even chlorite have extensively developed in a more than 650-m-thick series of pillow lavas outflown in the Triassic sea. No downward evolution in the zeolite facies is recorded, indicating smooth geothermal gradient. The local development of chlorite and albite, beside smectite and zeolite, indicates an occasional transition to the greenschist facies.

13.2 Basalt Intrusion in Sediment

Hot basalt sills intruding into soft sediments, or pillow-lava flows rapidly blanketed by sedimentary deposits may determine *contact metamorphism* processes in the submarine environment. Such phenomena, often marked by the combination of thermal and hydrothermal effects, are especially reported from studies performed on Deep Sea Drilling Project materials. For instance, Einsele et al. (1980) and Fournier-Germain (1986) study the consequence of a diabase intrusion in metalliferous sediments at site 471 in Guyamas basin (Northeastern tropical Pacific, leg 63). The Miocene sediments located at the precise contact with the altered diabase are modified on a thickness of about two centimeters. They contain sulfides, abundant *chlorite and plagioclase*, and accessory quartz and illite. Chlorite and plagioclase are attributed to a slight thermal metamorphism during diabase intrusion, and to a hydrothermal alteration through rapid migration of interstitial water in heated sediments.

Desprairies and Jehanno (1983) and Rangin et al. (1983) study the mineralogy, geochemistry, micromorphology, and isotope chemistry of late Cenozoic baked sediments sampled above basalts at DSDP sites 456 (Mariana trough) and 485 (off Gulf of California). Metamorphosed sediments contain *corrensite* (regular chlorite-smectite mixed-layer) and *analcite* off the Gulf of California, and *chlorite, illite-smectite, zeolites, quartz, apatite, and epidote* in the Mariana trough (e.g., Table 13.3). Superimposed sediments are characterized either by the hydrothermal genesis of Fe-Mg smectite (site 456) or by the reworking of weathering products from exposed volcanic or common terrigenous areas (both sites). Basalts and associated sediments express an alteration at a temperature

Table 13.3. Mineralogy of the clay fraction of Pleistocene sediments above basalt at site 456 DSDP, Mariana trough. (After Desprairies and Jehanno 1983)

Samples	Chlorite %	Illite %	Random mixed-layers %	Smectites Al-Fe %	Fe-Mg %	Zeolites	Others	Estimated temperature
Common nonbaked sediments	25	30	Trace	45	–	–		*Bottom seawater*
Indurate sediments								
456 12-2-43/45	10	10	–	15	65	Clinoptilolite analcite		*90–200°C*
456A 10-3-73/75	90	–	10	–	–		Quartz	*200–230°C*
456 14	80	–	20	–	Trace	Mordenite	Quartz Calcite	*200–250°C*
456 16	80	–	20	–	–	Wairakite	Quartz Calcite Apatite Epidote	*>250°C*

between 90 and 250°C. The succession of alteration products suggests a rapid upward decrease of temperature above the basalt - sediment contact. Most Si and Al derive from basalt, while Na and Mg are mainly removed from sea water. Note that abundant corrensite also occurs in the green claystone topping Callovian basalt at site 534 DSDP in the Blake – Bahama basin, Northwestern Atlantic, and is referred to contact metamorphism (Fig. 13.4). As in the Pacific Ocean, baked sediments of the Atlantic basin extend only in a few centimeter thickness above or below intruding basalts, and display very limited thermal effects (Coulon et al. 1985).

In some other cases, sediments close to basalt intrusions do not reflect appreciable metamorphic change. Chamley and Bonnot-Courtois (1981) investigate the transition between basalt and intercalated or overlying sediments in Northern Phillipine Sea (DSDP leg 58) and in Northwestern Atlantic (leg 11). Altered basalts contain well-crystallized Mg-smectite characterized by a relative enrichment in heavy rare earth elements (europium, terbium, yterbium, lutetium), and by the absence of negative anomaly in cerium. These characters suggest an alteration in a tholeiitic environment protected from noticeable sea-water penetration. By contrast the associated sediments display various clay assemblages and rare earth-elements spectra, reflecting either an influence of basalt alteration, some exchanges with sea water, or a dominantly-terrigenous input. For instance at site 105 (4966 m water depth) in Hatteras basin (leg 11) the limestones intercalated between basalts display a negative Ce anomaly and an enrichment in heavy rare earth elements, and contain abundant sepiolite and lath-shaped smectite in the clay fraction (Fig. 13.5). These characters suggest a submarine alteration under the influence of both basalt and sea water (see 13.4), but no evidence of baking

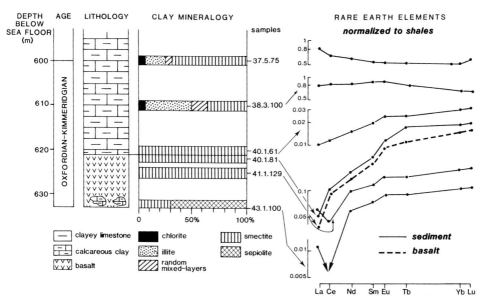

Fig. 13.4. Clay mineralogy and rare-earth element distribution at the basalt-sediment contact of hole 195 DSDP, Hatteras basin, Northwestern Atlantic. (After Chamley and Bonnot-Courtois 1981)

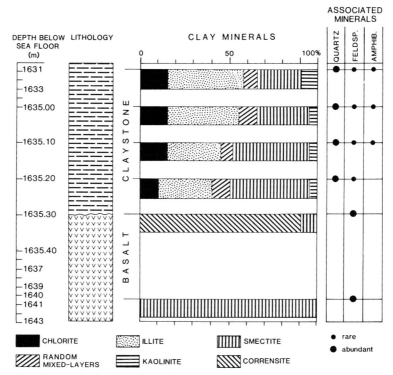

Fig. 13.5. Clay mineralogy at the basalt-sediment boundary, DSDP hole 534A, Blake-Bahama basin. (After Chamley et al. 1983)

exists. The reddish calcareous clay located immediately above the basalt comprises the same type of smectite as basalts, and probably results from the reworking of alteration products formed in the absence of sea-water influence and of contact metamorphism. The red clayey limestones deposited above volcanic rocks during the late Jurassic contain diversified clay assemblages, enriched in light rare earth elements and devoid of volcanic or marine influence, which indicates an essentially terrigenous influence (see Chaps. 18, 19).

The influence of basalt intrusion in adjacent sediments is sometimes recognized in old *exposed series*. Brauckmann and Füchtbauer (1983) investigate the Turonian – Coniacian sandstones and siltstones of the Atane formation in West Greenland. The unaltered rocks, that have been buried less than 1 kilometer, mainly contain kaolinite, quartz and feldspar in siltstones, and kaolinite, illite-smectite, illite and smectite in the sandstone matrix. Adjacent to thin basalt dykes and sills albite, quartzine, smectite, and goethite formed at the expense of kaolinite matrix, at temperature probably of less than 200°C. Adjacent to thicker basalt sills, albite of different types developed in sandstones, with associated ondulose quartz, muscovite, chlorite, epidote, nontronite and anatase; siltstones comprise secondary muscovite, chlorite, pyrophyllite and albite. These various minerals are attributed to a strong thermal effect during basalt intrusion, at temperatures ranging from 250 to 500°C. The intensity of metamorphic evolution was probably enhanced through convective heat transfer by pore water. Discussing the origin of chemical elements needed for mineral growth, F. Brauckmann and H. Füchtbauer attribute Mg, Ti and Fe to intruding basalts, Si and K to potassium feldspar, and Na to either episodical sea-water or groundwater influx.

13.3 Hydrothermal Deposits

13.3.1 Main Precipitates

The importance of hydrothermal circulation in the oceanic crust was first suggested by the discovery in the 1960's of iron-rich deposits at the surface of basalt rocks. The frequent occurrence of metalliferous crusts and sediments recognized as boundary layers between oceanic basalts and pelagic calcareous, siliceous or argillaceous oozes constitutes the basic evidence that *common deep-sea sedimentation is often preceded by hydrothermal exhalation and precipitation.* Reviews on the subject are provided by Rona (1978), Bonatti (1981) and Thompson (1983). *Metal deposits resulting from hydrothermal activity may occur as layers or discontinuous lenses at the bottom of the sedimentary column in oceanic basins, as crusts or mounds in ridge axes, transform zones, or intraplate volcanic areas, and as various sediments in marginal, back-arc, and inland volcanic basins.*

Typically hydrothermal deposits differ from other metalliferous sediments by high Fe, Mn, Ba and U contents and high Si/Al ratios (Bonatti 1981; see 12.1.1). The main hydrothermal phases are controlled by the *abundance of iron or manganese*, that constitute either metal oxyhydroxides, metal silicates, or metal sulfides. *Metal oxyhydroxides* range in composition from Fe-rich, Mn-free end members to Mn-rich, Fe-free end members, with various intermediate compositions. They represent the most abundant and ubiquitous components of hydrothermal sediments. Fe predominantly occurs as poorly crystalline $Fe(OH)_3$ or *goethite*, while Mn is mainly included in *todorokite* ($(Na, Ca, K, Ba, Mn^{2+}) Mn_5 O_{12}, 3H_2O$), *birnessite* ($4MnO_2, Mn(OH)_2, 2H_2O$) and MnO_2. *Metal silicates* represent common accessory minerals. They chiefly comprise iron-rich smectites of the *nontronite* type. This differs from the direct hydrothermal alteration products of basalts that usually consist of Mg-smectite (saponite; see 13.1.1). *Metal sulfides* occur more rarely, but locally constitute spectacular concentrations developed as metal mounds and chimneys on the sea floor. Sulfides mostly include *pyrite, chalcopyrite* ($CuFeS_2$), *wurtzite* [$(SnFe)S$] and *sphalerite* (Zn sulfide), and are usually associated with iron hydroxides. Various elements other than Fe and Mn can precipitate from hydrothermal solutions on the sea floor and form new minerals. It is the case of Si (opal), Ba (barite), Mg (talc, chrysolite),Pb (galena), Ca (anhydrite, gypsum), etc.

Massive sulfide deposits represent one of the most diversified precipitation facies deposited on the ridge axes. Such accumulations are especially known on fast-spreading ridges. Detailed descriptions are reported from the East Pacific, Galapagos and Juan de Fuca ridges (e.g., Styrt et al. 1981, Bischoff et al. 1983, Davis et al. 1987). Occurrences also arise from the East African rift (Vaslet et al. 1987). The sulfide mineralogy seems to be mainly related to the fluid temperature and vent history. At 21°N on the East Pacific ridge, high-temperature fluids (about 350°C) deposit large amounts of chalcopyrite, while lower temperature

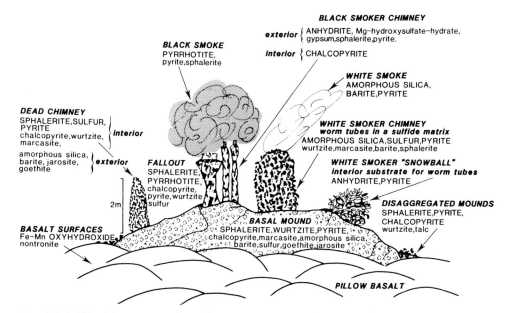

Fig. 13.6. Sulfide-rich structures observed at 21°N vent area on the East Pacific ridge; major and minor minerals associated. (After Haymon and Kastner 1981)

fluids (260–270°C) precipitate abundant wurtzite. Active vents contain wurtzite and copper-rich sulfides, while extinct vents are enriched in pyrite and sphalerite (Styrt et al. 1981). Haymon and Kastner (1981) present a schematic distribution and history of the various minerals precipitated in different sulfide-rich deposits of the 21°N vent area (Fig. 13.6). Metal-rich sea water leached from the underlying basalt discharges on the sea floor at temperatures up to 350°C. The mixing of this hot, sulfide-rich, acidic water with cold, oxidized, bottom sea water determines the precipitation of *wurtzite – sphalerite – pyrite – pyrrhotite minerals* in association with Ca-Mg sulfates (mainly anhydrite, gypsum). These precipitates constitute the outer walls of the chimneys. Within the walls, pyrrhotite is converted to pyrite and marcasite. Aging of chimneys determines the thickening of the walls, whose outer cooler parts become encrusted with worms. Adjacent to vents, deep-sea autotrophic clam and worm communities may develop. Hot vents tend to produce black smoke and cooler vents white smoke, the latter containing less sulfides and more barite and opal. When the temperature of the outer walls drops below approximately 130°C, anhydrite dissolves. As the vents stop working, pyrrhotite and wurtzite are rapidly oxidized and replaced by *Fe and Zn oxyhydroxides*. Subsequently, the chimneys often collapse and constitute basal mounds. Various other minerals precipitate, transform, or disaggregate around the chimneys and on the adjacent basalt surfaces (Fig. 13.6). *Fe-Mn oxyhydroxides* precipitate from the venting solution over a broad area, while *nontronite* develops more locally (Fig. 13.15).

13.3.2 Pacific Ocean

East Pacific Ridge

The clay sedimentation associated with hydrothermal activity differs considerably in nature and origin according to the location. This is illustrated by various studies performed in different rift and transform zone areas. McMurtry and Yeh (1981) study the clay minerals of surface metalliferous sediments sampled on the crest of the East Pacific ridge at *6 and 10°S*. Foraminifera oozes contain a clay fraction essentially made up of poorly crystallized smectite. Chemical analysis indicates a *dioctahedral smectite of an iron-rich type* (Table 13.4). Oxygen-isotope formation temperatures range between 30 and 50°C in the 10°S region and between 2 and 35°C in the 6°S region. Smectite is considered as formed in situ by low-temperature hydrothermal processes. The mineral could result either from cooling and oxidation of instable, high-temperature sulfide assemblages (380°C), or from the percolation of hydrothermally altered sea-water solutions through underlying basalt and sediment. Authigenic Fe-smectite appears to be widespread in surface sediments, possibly as a result of colloidal transport of hydrothermal materials eroded from metal mounds by bottom currents. McMurtry and Yeh (1981) think that some smectite of the adjacent Bauer Deep could have a similar origin, which is supported by the similarity of chemical composition in both areas. Note that Cole (1983) deduces from oxygen-isotope investigations that the Bauer Deep smectite, only slightly depleted in Fe relatively

Table 13.4. Examples of smectite composition in surface calcareous ooze, East Pacific ridge, 10 and 6° S. (After McMurty and Yeh 1981)

Location (water depth)	Structural formula									Temperature formation °C	
	CO_3Ca	Tetrahedra		Octahedra			Interlayers				
	%	Si	Al	Fe^{3+}	Al	Mg	Ca	Na	K	Min.	Max.
10°32'S 110°04'W (3111 m)	85.0	3.83	0.17	1.21	0.47	0.34	0.03	0.29	0.07	32±1	51± 5
5°48'S 107°00'W (3157 m)	74.6	3.96	0.04	1.24	0.31	0.55	0.05	0.45	0.04	2±2	35±18

to Pacific Ridge smectite, results from hydrogenous processes with indirect and distant hydrothermal influence (see 12.2.4). This exemplifies the difficulty arising in distinguishing hydrothermal and hydrogenous effects in transitional deep-sea environments, especially when the crystalline clay fraction is poorly represented. Walter and Stoffers (1985) show that hydrothermal deposits on the East Pacific ridge between 2°N and 42°S are dominated by Fe-Mn hydroxides, with generally little clay mineral fraction. The increasing spreading rate from North to South is reflected by a higher hydrothermal input of subamorphous metal oxyhydroxides, with almost no crystalline species.

Fournier-Germain (1986) investigates the deposits of the East-Pacific ridge axis, sampled at *12°50'N* by diving, dredging and coring. Sediments mainly consist of brown-reddish oozes rich in metal oxides, with associated foraminifera, radiolarians, and clay minerals, and very little volcanic glass. X-ray diffraction analysis of the bulk sediments reveals abundant metal hydroxides, some quartz, calcite and feldspar, and few clay minerals. The less than 2-μm fraction, studied by X-ray diffraction after removing the amorphous components, is composed of illite, kaolinite and smectite, with accessory chlorite, quartz, feldspar, and goethite. Despite the close hydrothermal environment, this *clay assemblage* is *mainly detrital*, and corresponds to the terrigenous influx from Central America. Similar data are reported by Blaise et al. (1985) on the clay fraction of the northern segment of the Juan de Fuca ridge, at 51°N in the Northeastern Pacific. B. Fournier-Germain reports a decrease of amorphous oxyhydroxides West of the East-Pacific ridge axis, and the weak dependence existing between volcanic glass content and clay associations. The only arguments for an in situ evolution of clay exists in the presence of some lathed and crumpled smectite particles, and of very small kaolinite hexagonal sheets, possibly of hydrothermal origin. The hydrothermal impact concerns predominantly the formation of Fe-Mn oxyhydroxides, and very little that of clay minerals.

Galapagos Rift Mounds

The Galapagos *hydrothermal mounds field* is located 18–32 km south of the Galapagos rift axis at 86°W and 0°25'-0°50'N (Fig. 13.7). Hydrothermal deposits

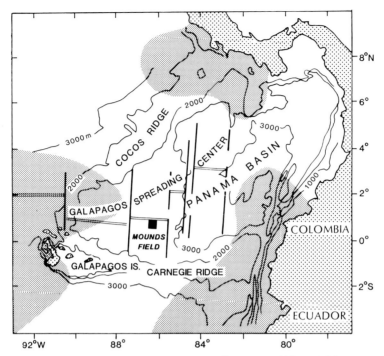

Fig. 13.7. Location map of Galapagos spreading center. (After McMurtry et al. 1983). *Shaded areas* correspond to surface sediments whose smectite is of probable detrital origin. (Heath et al. 1974, in McMurtry et al. 1983)

occur within a broad band of high conductive heat flow, as a linear system of faults and fissures extending subparallel to the spreading axis. Corliss et al. (1978) identify a nontronitic mineral as a major component of the mounds, that largely consist of green argillaceous sediments. J.B. Corliss and colleagues suggest by analogy to Harder's experiments (1976) that nontronite precipitates from interstitial fluids containing 20 ppm monomeric SiO_2 and 4-7.5 ppm Fe, at temperatures of 3 to 20°C. The environment should be less reducing than inferred from H. Harder's experiments (Eh = − 0.1 to − 0.8 V), since abundant Mn oxides precipitate in mound deposits together with clay. Corliss et al. (1978) suggest that the hydrothermal formation results largely from percolation of sea water within underlying basaltic crust and sediments, and from subsequent upward migration.

The Galapagos hydrothermal mounds are known in some detail following leg 54 of the Deep Sea Drilling Project (Hekinian et al. 1978). A hydrothermal mound displays the following sequence from top to bottom, on a thickness of about 30 meters: Mn-oxide cap, Fe-rich green clay layers intercalated with pelagic sediments, normal pelagic sediment and basement basalt. Oxides mainly comprise todorokite, birnessite, and amorphous compounds, with a relative depletion in Fe and accessory transition elements (Ni, Co, Cu). *The green clays consist of Fe- rich minerals* with geochemical and morphological characters of *celadonite* and *Fe-beidellite to nontronite* (Hoffert et al. 1980, Rateev et al. 1980, Schrader et al. 1980).

Table 13.5. Smectite composition of green clay and siliceous ooze, Galapagos hydrothermal mounds field. (After McMurtry et al. 1983)

Samples	Tetrahedra			Octahedra			Interlayers			
	Si	Al	Fe^{3+}	Al	Ti	Mg	Ca	Mn	Na	K
A. DSDP 424A-2-1-118/ 124 (00°35.33′N– 86°07.81′W) green hydrothermal clay	3.94	0.06	*1.59*	0.03	–	0.38	0.03	0.01	0.02	0.36
B. Southtow 18PG-2/7 (00°35.1′N–86°11.1′W) siliceous carbonate-bearing ooze	3.97	0.03	0.48	*1.12*	0.03	0.37	0.09	0.02	0.06	0.11

McMurtry et al. (1983) compare the geochemistry and oxygen-isotope composition of purified clay minerals from the green layers sampled in the mounds, and from pelagic oozes cored in the vicinity of the hydrothermal field. The major component of the hydrothermal mounds at DSDP site 424 is a *dark green nontronite*, whose formation temperature ranges *from 25 to 47°C*. The nontronite is *very poor in aluminum* both in octahedral and tetrahedral sheets (Table 13.5A), a character attributed to a typical hydrothermal deposition under low-temperature conditions. In addition to nontronite, the <2-μm fraction of the Galapagos green clay contains some amorphous iron-oxide and silica compounds. These amorphous phases are interpreted by G.M. McMurtry and colleagues either as "surplus" hydrothermal reactants or as pre-existing sedimentary components in excess to those used for nontronite formation.

In contrast to the specific nontronitic character of hydrothermal mounds, the adjacent pelagic sediments contain a much more common smectite of a Fe-montmorillonite to Fe-beidellite type (Table 13.5B). Assuming an exclusively authigenic origin, McMurtry et al. (1983) calculate from oxygen isotopic data that this mineral would have formed at a temperature of 27 to 39°C. Such a temperature is incompatible with the sedimentary environment marked by pelagic siliceous, carbonate-bearing ooze. In addition, the chemical composition of the mineral resembles that of smectites from some metal-bearing deep-sea clays submitted to terrigenous to hydrogenous influence (Chap. 12). According to McMurtry et al. (1983), the apparent high isotopic temperature of the Fe-smectite could reflect (1) a detrital origin of the mineral, (2) a hydrothermal formation at the spreading axis followed by a dispersal by bottom currents, (3) a mixture of authigenic and detrital smectites. This result points to the difficulty arising in identifying the precise origin of minerals that do not clearly belong to end-members.

McMurtry et al. (1983) plot the chemical data of various smectites from the Eastern Pacific Ocean (Fig. 13.8). Nontronite from the Galapagos mound field represents a Fe end-member characterized by very abundant iron and little aluminum and magnesium. Al-beidellite of detrital, mainly pedogenic origin con-

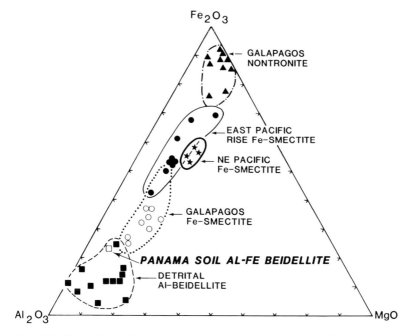

Fig. 13.8. Al_2O_3-Fe_2O_3-MgO variation diagram for deep-sea smectites. (Bibliographic compiling after McMurtry et al. 1983)

stitutes an Al end-member with abundant aluminum and fairly little iron and magnesium. *Intermediate Fe-smectites with noticeable amounts of Al and Mg occur largely in metalliferous sediments of the East-Pacific ridge area and adjacent regions, where they could characterize predominantly hydrogenous processes or a mixture of different origins.* This is the case with Al-Fe smectites from pelagic oozes of the Galapagos spreading centre. Walter and Stoffers (1985) show that the noncarbonate fraction of the Galapagos rift sediments are only locally of a strictly hydrothermal origin, and that the terrigenous influence on the clay composition decreases westwards. One may imagine that the hydrogenous formation of smectite increases westwards parallel to the diminution of terrigenous input, with an accessory contribution of iron from distant hydrothermal vents. In the Galapagos hydrothermal mounds field itself, the terrigenous contribution is probably still noticeable, as suggested by the similar chemical composition of smectites from pelagic sediments in this area and from exposed soils from Panama (Fig. 13.8).

Southwest Pacific
Singer et al. (1984) and Stoffers et al. (1985) describe a 354-cm-long core recovered on a transect Tahiti – East Pacific ridge at 29°28.08′S, 131°36.81′W in 4250 m water depth. The hole is located on the western flank of a small seamount. Dark brown, fine homogeneous, metalliferous sediments constitute the upper part of the core. They are replaced below 190 cm by dark-gray coarse-

sand aggregates rich in Mn-oxides. The lower part of the core, below 215 cm, displays an olive-green granular clay with few oxide layers. Some clay granules are coated by birnessite.

The upper sediments are essentially composed of Fe-smectite, goethite, phillipsite, and poorly crystallized δMnO_2. The rare earth-elements distribution in the $<63 \mu m$ fraction presents a strong negative Ce anomaly indicative of a hydrogenous formation in contact with sea water. The coarse layers mainly consist of well-crystallized todorokite. The *green granular clay* in the lower part of the core comprises *almost pure, well-crystallized nontronite, characterized by very little or no aluminum*, high SiO_2 and MgO contents, extremely low rare earth-elements concentrations and a high Fe_2O_3/FeO ratio. The chemical composition of nontronite is close to the Fe end-member of the nontronite – beidellite serie:

$$(NH_4)_{0.34}K_{0.15})Fe^{3+}_{1.66}Mg_{0.40}) (Si_{3.67}Fe^{3+}_{0.33})O_{10}(OH)_2$$
(core catcher $<0.6 \mu m$)

Oxygen-isotope measurements indicate a formation temperature of 21.3-22.8°C for nontronite. The rare earth elements in green clay comprise more heavy components than the brown clay of the upper unit and a strong negative Ce anomaly, both characters typical of hydrothermal deposits (Corliss et al. 1978, Bonnot-Courtois 1981). Singer et al. (1984) and Stoffers et al. (1985) conclude that *nontronite formed hydrothermally at low temperature*.

Euhedral authigenic quartz occurs in noticeable amounts in the lower part of the core. Oxygen-isotope measurements indicate that quartz formed at a temperature of about 25°C. The coexistence of hydrothermal nontronite and quartz sets the problem of the age of the formation of each component. Experiments done by Harder (1976) show that one of the critical conditions for the formation of nontronite consists in relatively low concentrations of silica, which does not fit with the presence of quartz. In fact, the quartz crystallites present at 29°S are found embedded in the clay matrix, and even coating clay granules. This suggests that nontronite precipitation has preceded the quartz crystallization. Strontium isotope investigations indicate a minimum age of formation for the nontronite of 12 million years, the real age being possibly of 20 to 30 million years. Considering the northwestward migration of the Pacific plate since the Miocene, and the approximately 25-million-year-old formation of the crust in the sampling area, Singer et al. (1984) envisage that nontronite could have precipitated in situ in the vicinity of the active East Pacific ridge. Such a timing is compatible with the inferred sedimentation rate in this region, close to 1 mm/10^3 yr. The quartz grains probably formed in a subsequent phase of hydrothermal activity, under a similar thermal regime. Note that hydrothermal deposits close to active vents are reported East of Tahiti in the Souwestern Pacific, where reddish encrustments contain authigenic vivianite (Fe-phosphate), aragonite, antigorite, and perhaps some little kaolinite and smectite (Hoffert et al. 1987).

13.3.3 Mid-Atlantic Ridge

As early as 1971, Copeland et al. envisaged a local hydrothermal or volcanic origin for a coarse iron-rich smectite identified in a sample dredged at 22°38′N in the ridge axis. Fe-smectite occurs beside fine-grained terrigenous smectite and greenstone-derived chlorite. Hoffert et al. (1978) study two hydrothermal fields discovered and sampled with the submersible Cyana in the *FAMOUS area* of the Mid-Atlantic ridge, at 37°N in transform fault "A". Both fields lie 200 m apart at a water depth of 2670 – 2690 m, each of them covering an area of about 600 m² in a zone of active left-lateral strike-slip faulting. Hydrothermal deposits develop above pelagic sediment and form small, asymmetrical, East-West elongated ridges of about 40 m long and 15 m width, with an average thickness of 10 to 100 cm. The mound thickness increases close to the small, fissure-like vents from which they apparently issue (Fig. 13.9). The major deposits encountered are *black Fe-Mn concretions and clay-rich material*, both being often intermixed. The concretions comprise mainly todorokite, birnessite, rancieite, and MnO_2, with fairly high contents of Mn, Ba, V, Ni, Co and Cu. Samples near to the vents are enriched in MnO_2, those precipitated at a greater distance contain relatively more rancieite. The clay-rich material, especially abundant close to the vents, is mainly composed of *nontronite, celadonite, and Fe-Mn concretions*, with relatively abundant Fe, Si, and K. An example of structural formula of the FAMOUS nontronite is the following (after Hoffert et al. 1978):

$$(Ca_{0.01}Na_{0.25}K_{0.32})\,(Fe^{3+}_{1.75}Mg_{0.33})\,(Si_{3.51}Al_{0.02}Fe^{3+}_{0.47})O_{10}(OH)_2 .$$

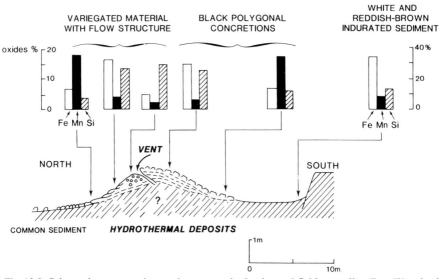

Fig. 13.9. Schematic cross-section at the western hydrothermal field extending East-West in the Famous area, 37°N, Mid-Atlantic ridge. Abundance of Fe, Mn and Si oxides in different samples. (After Hoffert et al. 1978). Si oxides mainly represent clay minerals, Fe and Mn oxides mainly metal concretions

Table 13.6. Mineralogy and chemical composition (%) of major types of hydrothermal deposits in TAG area, 26°N, Mid-Atlantic ridge. (After Thompson et al. 1985)

	Crystalline laminated birnessite	Green earthy nontronite	Black earthy birnessite	Red earthy amorphous iron oxide
Fe	< 0.1	*32.35*	9.57	*41.52*
Mn	*52.30*	0.67	*41.54*	2.43
Si	0.06	*12.15*	1.24	5.88
Ca	1.56	0.17	1.39	0.92
Al	0.03	0.05	0.09	0.05
Mg	0.92	1.86	1.05	0.62
Ti	0.31	0.94	0.37	0.65
K	0.43	1.34	0.51	0.90
Co	< 0.02	< 0.01	< 0.01	< 0.01
Ni	< 0.01	< 0.01	0.03	< 0.01
Cu	0.01	< 0.01	0.01	< 0.01
Zn	< 0.01	< 0.01	0.01	< 0.01
Mo	0.09	< 0.01	0.10	< 0.01
Ignition loss	17.52	6.85	19.25	13.55

In addition to concretions and clay material, Fe-Mn coatings on basalt pillow lavas occur around the hydrothermal fields. These coatings display more abundant transition elements and lower Fe/Mn and Si/Al ratios than the clay-rich deposits, a character suggesting a hydrogenous influence through sea water in addition to hydrothermal precipitation.

At 26°N in the Mid-Atlantic ridge exists another hydrothermal area extensively studied since the late 1970's, the *TAG area* (Rona 1980). The deposits include surficial metal-rich staining on the surface of carbonate oozes, as well as massive layered deposits. The stained sediment patches are mainly made up of black Mn-oxides, locally forming crystalline birnessite. The massive deposits range in size from less than 1 m² to about 15 x 20 m. Their nature and composition vary considerably and display four major types (Thompson et al. 1985; Table 13.6): (1) *Black crystalline laminated birnessite* forming discrete plate-like layers several centimeters thick, with a rough upper surface and a smooth lustrous under surface conforming in shape to the underlying deposit. (2) *Green earthy nontronite* very rich in iron and strongly depleted in aluminum, precipitated as loose friable earthy powder intermixed with other deposits. (3) *Black earthy, impure birnessite* consisting of massive loose friable powder, interspersed with the black crystalline laminated layers. (4) *Red earthy*, disorganized or layered deposits of iron oxide, interspersed with black earthy birnessite and green earthy nontronite, and forming *Fe-amorphous oxyhydroxides* without crystalline components.

G. Thompson and colleagues report that the variations in morphology, fractionation and crystallinity of the different deposits in the TAG area are mainly controlled by local Eh conditions and by the hydrodynamics of circulation in the basalt upper crust. Hydrothermal deposits probably result from the *precipitation of low-temperature solutions.* Most of the metal-rich waters do not reach the surface; they may react with oceanic crust at depth at higher temperature and precipitate metal-rich sulfides within the crust. The local presence of Cu-Fe-Zn-rich layers and of scattered clam shells nevertheless suggests the occasional existence of high-temperature black smoke-type venting. Thompson et al. (1985) estimate that the best analogy to the TAG-type hydrothermal process could be the Troodos massif in Cyprus, marked by Mn- and Fe-rich low-temperature "umbers" in the upper surface, and by metal-rich sulphides at depth (see 13.3.5).

13.3.4 Red Sea

The Red Sea represents a young, narrow (<300 km in width) oceanic basin whose formation started during the late Eocene. After a long quiescence period of about 30 million years marked by the massive deposition of evaporites and associated detrital sediments, accretion processes started again in the Pliocene, and are still continuing. The oceanization of the Red Sea is more evolved in the southern part than in the northern part. The axial valley is characterized by the presence of numerous depressions, where hot brines issuing from evaporite dissolution and hydrothermal activity accumulate. The most important depressions occur in the central part of the Red Sea, and have been called the Kebrit, Gypsum, Vema, Nereus, Thetis, Atlantis II, Shagara, Erba, Sudan, and Suakin Deeps, from North to South. Owing to its geologic peculiarities, to its tropical situation provoking strong evaporation and to its narrowness determining noticeable terrigenous supply, the Red Sea displays a large array of sedimentary types. For instance, Bischoff (1969) identifies seven facies in the Atlantis II Deep: detrital, iron-smectite, goethite-amorphous, sulfide, manganosiderite, anhydrite and manganite. The most spectacular deposits consist of iron-rich accumulations, extensively studied since the mid-1960's (e.g., Miller et al. 1966, Degens and Ross 1969, Bignell 1978, Thisse 1982).

Bignell (1978) synthesizes the data obtained for sediments from the different deeps, and recognizes five major groups of authigenic facies: *oxides, sulfides, sulfates, carbonates, and silicates* (Table 13.7). Most precipitates form out of the brine column that overlies the deposits. Bischoff (1969) discusses the main processes of mineral precipitation within the Atlantis II Deep, at about 21°N. Red Sea water probably circulates through the mainly evaporitic sediments and the basalts that constitute the substrates of hydrothermal deposits. The resulting brines are heated while moving toward the central part of the Red-Sea rift zone, and finally percolate upwards, through fractures, to the sea floor. Dense, metal-rich brines issue in the Red-Sea water mass, where they progressively cool and may precipitate solids at a rate of about 40 cm/10^3 yr. Goethite and other iron-oxide phases mainly precipitate in a 44°C brine that separates the denser and warmer lower brines and the upper normal Red-Sea water, as indicated by the marked depletion of dissolved iron in this intermediate layer. Owing to its limited distribution and to its high Fe content, iron-smectite is presumed to precipitate in a zone of decreased oxygen availability, probably lower in the brine column than

Table 13.7. Major facies groups of metalliferous sediments from Red-Sea Deeps. (After Bignell 1978)

Mineral group	Minerals	Accumulating metals
Oxides		
Fe-oxides	Limonite, goethite, lepidocrocite, hematite	Fe, Zn, Cu,Co, Pb, Hg, V, Ba
	magnetite	Fe, Cu, V, Ba, Ti, Zn, Co, Pb, Hg
Mn-oxides	Manganite, groutite, woodruffite, todorokite	Mn, Fe, Zn, Cu, Pb, Hg, Sr, Ba
Sulfides		
Mixed sulfides	Sphalerite, chalcopyrite, marcasite	Fe, Cu, Zn, Ag, Ba, Cd, Hg, Ga
Pyrite	Pyrite	Fe, Cu
Sulfates	Anhydrite, gypsum	Ca
Carbonates	Siderite	Fe, Ca, Zn, Cu, Pb
	manganosiderite	Fe, Mn, Ca, Zn
	rhodochrosite	Mn, Ca, Zn
Silicates	Smectite	Fe, Zn, Cu, Pb, Hg, Ag, Ba
	chamosite	Fe, Mg, Al, Zn, Cu, Hg

for the ferric hydroxide. Anhydrite apparently results from mixing of SO_4 from the overlying sea water with Ca provided by the brine. Most sulfides may have precipitated at a coalescing front between a metal-rich, sulfide-poor brine and a sulfide-rich, metal-poor brine, both coming along the same subterranean channel (see discussion in Bischoff 1969, Bignell 1978, Zierenberg and Shanks 1983).

The origin of clay associations in Red-Sea sediments varies considerably, according to the respective nature and importance of detrital input and of hydrothermal activity. *Many sedimentary areas are characterized by an essentially terrigenous supply* providing illite, chlorite, kaolinite, smectite, palygorskite, and even sepiolite from exposed sedimentary rocks and weathering formations (e.g., Schneider and Schumann 1979). In particular, palygorskite, previously envisaged as forming through hydrothermal process in the Southern Red Sea (Heezen et al. 1965, Bonatti and Joensuu 1969), has never been identified so far as really originating in metalliferous deposits. On the other hand, palygorskite is known to outcrop largely on adjacent land-masses (Chaps. 7, 8) and to be air-transported over the basin (Tomadin et al. 1987). *The hydrothermal formation of clay minerals occurs principally in some depressions within the rift valley*, and varies in both nature and importance. In northern depressions like the Shaban Deep at 26°N, the sedimentation is dominated by common, non metalliferous sediments rich in detrital kaolinite, smectite, illite, feldspar, quartz, mica, calcite, and dolomite (Fournier-Germain 1986). Only local metalliferous accumulations occur, marked by abundant amorphous Fe-oxides and goethite, and by accessory calcite, dolomite, smectite, and illite. The origin of these local, metal-rich interlayers appears essentially hydrothermal.

As at the Atlantic and Pacific ridge axes, hydrothermal precipitation in Red-Sea deeps is characterized by *Fe-rich smectites of the nontronite type* (e.g., Miller et al. 1966, Bischoff 1969, Goulart 1976, Butuzova et al. 1979, Cole 1983).

Mineralogic, geochemical, and oxygen isotopic data point to a precipitation of Fe-smectite from hydrothermal fluids. The mineral mostly corresponds to a *dioctahedral ferric smectite*, whose composition does not vary significantly throughout the sedimentary column.

The existence in Red Sea hydrothermal deposits of di- to tri-octahedral smectites was envisaged by Bischoff in 1972. This question is investigated in detail by Badaut et al. (1985) for ferriferous sediments sampled at the sea-floor surface in the southwestern part of the Atlantic II Deep. The sediment consists of a water-saturated black mud, whose color rapidly changes to dark red under laboratory conditions. The material is therefore unstable and oxidizes very quickly. The mineralogy and geochemistry are determined with preservative methods by transmission electronmicroscopy coupled with energy-dispersive spectrometry, by transmission X-ray diffraction and by Mössbauer spectrometry. All analyses indicate the presence of a *trioctahedral iron clay* of the following approximate structural formula:

$$(\text{Interlayer}_x)(\text{Si}_{4-x}\text{Al}_x)\text{Fe}_3^{2+}\text{O}_{10}(\text{OH})_2 .$$

In the Mg – Fe series of trioctahedral smectites, this mineral represents the iron end-member (= *ferrous stevensite*), not clearly identified in nature so far. Note that Decarreau et al. (1987) synthesize from silico-ferric coprecipitates a pure dioctahedral smectite ($\text{Si}_4\ \text{Fe}_{1.83}^{3+}\ \text{O}_{10}\ (\text{OH})_2\ \text{Ca}_{0.26}$), growing slowly at 100-150°C. The mineral identified by D. Badaut and colleagues is quite different from the synthetic mineral, but is *very unstable*. If removed from the sea floor and not rapidly frozen, it oxidizes spontaneously into silico-ferric, more dioctahedral materials, and tends to draw nearer to the mineral synthesized by Decarreau et al. (1987). Badaut et al. (1985) suggest from electronmicroscope observations that oxidation of octahedral Fe^{2+} into Fe^{3+} starts at the periphery of clay particles and progresses centripetally. This leads to the formation of a dioctahedral area at the outer part of the trioctadedral mineral. Such a chemical evolution could explain the di- to tri-octahedral character of Fe-smectites formerly identified by Bischoff (1972) in the Red Sea. With continuing oxidation, the excess iron released from the lattice migrates and accumulates around the clay particles as iron oxyhydroxides.

The ferrous smectite of Atlantis II Deep precipitates very close to hydrothermal springs (Badaut et al. 1985). This fact agrees with thermodynamic calculations of Zierenberg and Shanks (1983) on the possible formation of ferrous saponite at *high temperature* (150°C) in the lower brines of Red Sea Deeps. The ferriferous clay precipitating at a greater distance from the vents is rather dioctahedral (nontronite), which also fits with the average temperature of 60°C calculated for the hydrothermal formation of ferric smectite (Zierenberg and Shanks 1983).

The evolution of hydrothermal activity in the course of time and the post-sedimentary evolution of hydrothermally originating clay are studied by Singer and Stoffers (1987) for a 11.91-m-long core sampled in the southwestern basin of the Atlantis II Deep. The sediment is dominated in turn by iron-oxide, smectite, sulfide and anhydrite facies, the two former being the more frequent (Fig. 13.10).

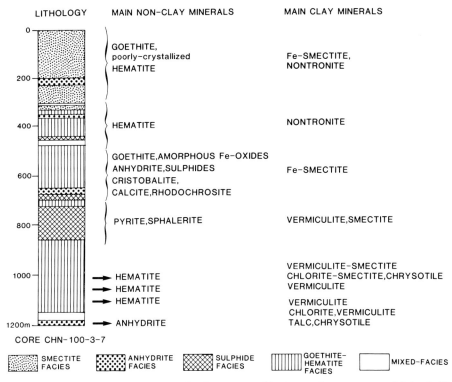

LITHOLOGY   MAIN NON-CLAY MINERALS   MAIN CLAY MINERALS

GOETHITE,
poorly-crystallized   Fe-SMECTITE,
HEMATITE   NONTRONITE

HEMATITE   NONTRONITE

GOETHITE,AMORPHOUS Fe-OXIDES
ANHYDRITE,SULPHIDES   Fe-SMECTITE
CRISTOBALITE,
CALCITE,RHODOCHROSITE

PYRITE,SPHALERITE   VERMICULITE,SMECTITE

                             VERMICULITE-SMECTITE
HEMATITE   CHLORITE-SMECTITE,CHRYSOTILE
HEMATITE   VERMICULITE
HEMATITE   VERMICULITE
                             CHLORITE,VERMICULITE
ANHYDRITE   TALC,CHRYSOTILE

CORE CHN-100-3-7

SMECTITE FACIES   ANHYDRITE FACIES   SULPHIDE FACIES   GOETHITE-HEMATITE FACIES   MIXED-FACIES

Fig. 13.10. Lithology and mineralogy of sediments cored in the southwestern part of Atlantic II Deep, Red Sea. (After Singer and Stoffers 1987)

The iron-oxide facies predominates in the lower three quarters of the core, and the smectite facies in the upper 3 meters.

In the crystalline clay fraction, *talc* dominates at the bottom of the core, near the brine discharge vent. At 11.83 meter depth, clay minerals consist of *vermiculite*, well-shaped *chlorite*, and *chrysotile*, which is attributed to either the submarine alteration of hydrothermal talc or to a direct hydrothermal precipitation due to changes in the brine composition. At 11.70 meters smectite minerals occur, then progressively alter to vermiculite, while at 10.25 meters chlorite, talc, and chrysotile again become prominent. The rest of the core displays an upward increase of Fe-rich smectite of the *nontronite* type. Smectite is especially abundant in the three upper meters. This succession shows that the hydrothermal brines of the Atlantic II Deep have given rise to a large variety of authigenic minerals, *iron oxides and Mg- to Fe-clay minerals* being the most abundant.

The magnesian clays identified in the lower part of the sedimentary column, close to the basaltic basement, recall the largely magnesian character of hydrothermally altered materials formed within the oceanic crust (13.1), and also the chloritic nature of many sediments in old evaporitic series (9.1. Chap. 20). The combination of the proximity of the basalt, hydrothermal exhalations, and hot saline brines appears to be essential for the genesis of talc, chlorite, and chrysotile. This mineral association evokes a high-temperature alteration similar to that responsible for alteration products of metamorphosed basalts. By contrast the preferential development of Fe-smectite in the upper sediments resembles the low-temperature hydrothermal processes occurring in more open environments, like those typical of most vent areas from mid-oceanic ridges (13.3.2, 13.3.3).

13.3.5 Old Hydrothermal Deposits

The brown to reddish, sometimes encrusted deposits identified at the *base of the sedimentary columns at Deep Sea Drilling Project sites* are often attributed to hydrothermal activity. Mostly recovered in the Northeastern and Central to Eastern Equatorial Pacific Ocean and in the Philippine Sea, these deposits contain abundant *subamorphous Fe and Mn oxyhydroxides (= red-brown semi-opaque oxides)*. The crystalline phases are usually not abundant and composed of *goethite, Fe-smectite, phillipsite, and feldspar*. The pelagic carbonate fraction is often not abundant, as well as micronodules, palagonite, and volcanic glass. Considering the location of these deposits at the boundary between oceanic basalt and common sediments, and owing to their mineralogic and chemical composition very similar to that of metalliferous sediments of the East Pacific ridge, most authors consider that they mainly result from *low- temperature hydrothermal precipitation* (e.g., von der Borch and Rex 1970, von der Borch et al 1971, Drever 1971b, Cronan 1973, Anderson et al. 1976, Boström et al. 1976, Bonatti et al. 1979, Leinen 1981, Leggett 1982, Leinen, Rea et al. 1986b, etc.). A post-sedimentary evolution is sometimes suggested by the development of lathed-smectite and crystalline iron oxides. Contact metamorphic effects are locally deduced from the presence of hematite, subregular illite-smectite mixed-layers, serpentine, pyrophyllite, or chlorite (Chamley 1980b, Fournier-Germain 1986). The hydrothermal impact is usually limited to sediments immediately overlying the basalt, and does not significantly affect the superimposed deposits.

Hydrothermal deposits are also reported in several *exposed formations*. For instance, the *Troodos ophiolite complex* in Cyprus, that formed at an active spreading center in late Cretaceous, includes about 90 massive sulfide deposits overlain by heavy-metal-enriched sediments. Sea water originally reacted with the basalts of the Troodos complex at high water/rock ratios (see Thompson 1983). *Sulfide-rich deposits* are overlain by ochreous sediments considered as secondary alteration products (Robertson and Boyle 1983). "Umber earth" deposits locally cover the ophiolite formations. They contain abundant *Fe-Mn oxyhydroxides and aluminous celadonite* (greenish earth) or *magnesian to ferriferous smectites* (typical "umbers"). "Umbers" closely resemble modern metalliferous deposits related to oceanic ridges. Associated pyroclastites display regular chorite-smectite and illite-smectite mixed-layers (Desprairies and Lapierre 1973, Guillemot and Nesteroff 1979). All these metal-rich deposits are referred to high- to low-temperature hydrothermal activity. Other hydrothermal deposits dominated by metal oxides, sulfides, or more rarely by iron-rich clays are reported in the literature. This is the case with deposits associated with ophiolites from the late Cretaceous to Tertiary in Oman (Juteau 1984), late Jurassic in the Appenines (Bonatti et al. 1976) and Western Alps (in Lemoine et al. 1982), or early Paleozoic in Newfoundland (Kay 1975).

13.4 Palygorskite and Sepiolite Deposits

A volcanic to hydrothermal influence is sometimes envisaged as explaining the fibrous clay occurrence in recent or ancient marine deposits (e.g., Callen 1984). Most examples arise from the Atlantic Ocean, which a priori complicates the interpretations since fibrous clay deposits crop out largely on adjacent continents and may be widely reworked. Let us briefly list the present-day marine deep-sea deposits, the main facts and interpretations. Hathaway and Sachs (1965) first identify sepiolite in whitish material recovered by dredging at about 4000 m water depth in the Central Equatorial Atlantic Ocean, South of St.Paul Island, on the western flank of the ridge. The sediments contain clinoptilolite, smectite, coccoliths, and various other minerals in the same dredge. Referring to the experimental synthesis of sepiolite by Siffert (1962b), J.C. Hathaway and P.L. Sachs consider that the Atlantic sepiolite formed by *reaction of silica from submarine, perhaps hydrothermal alteration of acidic volcanic rocks, with dissolved magnesium from sea water*. Clinoptilolite is attributed to the same alteration of silicic glass. Serpentine, also present in the dredged material, is thought to have a distinct origin, owing to its morphological independence from sepiolite and clinoptilolite.

Siever and Kastner (1967) describe an association of sepiolite, phillipsite, talc, and smectite in a sediment cored close to the Mid-Atlantic ridge at 22°N, and envisage a submarine alteration. Bonatti and Joensuu (1968) study a 410-cm-long core sampled on the Barracuda escarpment, Eastern Puerto-Rico trench (17°N, 58°W). The sediment mainly comprises common detrital minerals (illite, smectite, chlorite, kaolinite, quartz, feldspars). At 90 cm below the sea floor occurs a 10 cm-thick brown layer rich in semi-indurated nodular, subspherical aggregates up to 3 cm in diameter. The nodules comprise either light brown-yellowish grains rich in palygorskite with little sepiolite, or light yellow grains composed of clinoptilolite, palygorskite, smectite, serpentine, and quartz. The authors envisage an *alteration on the sea floor of smectite by Mg-rich solutions provided by hydrothermal sources having migrated through basaltic material.* E. Bonatti and O. Joensuu report the presence of hydrothermally altered basalt at the same station, and the existence of hydrothermal activity in the Barracuda fracture zone a few miles from the location of the palygorskite-containing core. *The role of smectite as a fibrous- clay precursor* is suggested by an intergrowth of palygorskite and smectite forming microaggregates. Bonatti and Joensuu (1968) think that the reaction of smectite and silica submitted to Ca- and Mg-hydrothermal solutions leached from basalt could be responsible for the formation of clinoptilolite in addition to palygorskite, and also of sepiolite and serpentine. Note that Hoffert et al. (1980) envisage a similar origin for nodular palygorskite and phillipsite encrusted with Mn-oxides and sampled in the Cape Verde basin, tropical Northeast Atlantic.

Bowles et al. (1971) investigate 34 dredge hauls from several regions of the Eastern and Western Atlantic. Five of the sampled materials, systematically originating from fracture zones and tectonic escarpments, contain a salmon-colored clay partly coated by a ferro-manganese crust. Clay minerals consist of almost exclusive palygorskite forming parallel-oriented crystals, and accompanied by little quartz and locally calcite or dolomite. F.A. Bowles and colleagues estimate that the relative purity of mineral composition excludes a terrigenous origin. They also stress the fact that the noticeable alteration of submarine glass into palygorskite and even into smectite is rarely unquestionably demonstrated in the literature. Bowles et al. (1971) therefore propose a *primary precipitation of palygorskite on the sea floor, without smectite or volcanic glass precursor, by direct reaction between the sea water and magnesium-rich hydrothermal solutions* emanating from underlying fissures.

Kossowskaya et al. (1975) and Lomova (1975) study the smectite- and palygorskite-rich sediments deposited from late Cretaceous to Oligocene off Western Africa in the Eastern Atlantic Ocean, and sampled during leg 2 of the Deep Sea Drilling Project. The abundance of both minerals and the frequent presence of glass and ash in sediments lead these authors to envisage a massive clay formation in the deep-sea environment by *direct alteration of volcanic material reworked from Cape Verde, Canary and Madeira islands*. The hydrothermal activity could intervene as an accessory phenomenon. Such a hypothesis is also claimed by Gorbunova (1979) to explain the presence of palygorskite in DSDP leg 6 materials (Northwestern Pacific). Lomova's hypothesis replaced the former interpretation of Peterson et al. (1970), who envisaged that sepiolite and palygorskite in the Cape Verde sediments resulted from heavy magnesium-rich brines originating in nearshore areas and percolating downslope into the ash-bearing basin sediments. The explanation proposed by O.S. Lomova and colleagues supposes a widespread transformation of volcanic materials in East Atlantic basins, which implies a huge and subcontinuous erosional activity on the rather small volcanic archipelagoes. Such an interpretation appears all the more questionable, since palygorskite deposits widely formed in Cretaceous-Paleogene times under alkaline evaporative conditions in West African peri-marine basins (Millot 1964, 1970), from where they were easily reworked during marine instability periods (Mélières 1978, Timofeev et al. 1978, Chamley 1979; see Chap. 19).

In the Hatteras basin, Northwestern Atlantic, independent pieces of clayey limestones intercalated within late Jurassic basalt flows contain abundant sepiolite, made up of long and flexuous fibers, and associated with coarse-lathed smectite (Chamley and Bonnot-Courtois 1981). The rare-earth element distribution displays an enrichment of heavy components suggesting the *influence of basalt*, and a depletion in cerium suggesting the *influence of sea water*. The intervention of hydrothermal activity is not quoted by the authors. The fact that sepiolitic sediments depend on basalt from their stratigraphic position and chemical composition, but are independent from their location and lithology, is nevertheless compatible with a hydrothermal influence.

Table 13.8. Structural formula of palygorskite and iron-beidellite close to the basalt-sediment contact, Mariana trench, Western Pacific Ocean. (After Desprairies 1981)

Palygorskite (460-7-cc-5/6)	$(Si_{7.87}Al_{0.13})(Al_{1.38}Fe^{3+}_{0.49}Mg_{1.87})K_{0.14}Ca_{0.02}O_{20}(OH_2(OH_2)_4$
Iron-beidellite (459B-60-1-10/11)	$(Si_{3.73}Al_{0.27})(Al_{1.13}Fe^{3+}_{0.62}Mg_{0.19})K_{0.43}Na_{0.15}Ca_{0.02}O_{10}(OH)_2$

The Pacific Ocean provides a few evidences on the hydrothermal formation of fibrous clays. Church and Velde (1979) study a 240-cm-thick palygorskite-rich horizon limited by two Mn-crust layers, in a core from the eastern equatorial basin, West of the oceanic ridge (7°N, 117°W). Palygorskite occurs together with clinoptilolite and smectite. After being concentrated by mechanical treatment, palygorskite reveals an oxygen-isotope composition much lower than that expected from minerals formed in equilibrium with either sea water or earth's surface conditions, which suggests a *hydrothermal origin*. $^{87}Sr/^{86}Sr$ ratios are higher than those of sea water, indicating a *nondirect precipitation* on the sea floor and the existence of some mineral precursors. These data provide additional arguments against the hydrogenous formation of fibrous clays in deep-sea, metal-bearing, open-marine sediments (see 12.3).

Desprairies (1981) identifies abundant palygorskite in late Eocene claystone to mudstone of DSDP hole 460, drilled on the inner wall of the Mariana trench in Western Pacific (18°N, 148°E). Palygorskite also occurs in lesser amounts at nearby site 459, in sediments blanketing the basalts and even within basalts themselves. Energy-dispersive analyses of palygorskite by X-ray techniques indicate a structural formula fairly similar to that of Fe-beidellite (Table 13.8), with of course more magnesium and trace elements. As Fe-beidellite also occurs in the same environment and displays, like palygorskite, a Ce depletion, A. Desprairies thinks that smectite could be a precursor for the fibrous clay formation. Palygorskite is attributed to the *hydrothermal alteration of iron-smectite and possibly volcanic glass by high-temperature sea- water fluids marked by high Mg/Ca ratio*. Saponite, that typically derives from the low-temperature alteration of basalt within the volcanic rock (13.1.1), does not display any cerium depletion and therefore clearly proceeds from a distinct genetic process.

To summarize, various hypotheses have been proposed in the literature to explain the presence of palygorskite or sepiolite in marine sediments associated with volcanic and hydrothermal activity: exclusive alteration of volcanic glass, exclusive precipitation on the sea floor from hydrothermal fluids, alteration of volcanic glass or smectite by hydrothermal fluids. We suggest that *deep-sea fibrous clays preferentially form from mineral precursors in sedimentary environments located close to volcanic rocks, under the combined influence of sea water and hydrothermal activity*. Such a genesis, that should be tested by further investigations, is inferred from the following observations:

1. Fibrous clays often develop in sediments located close to volcanic rocks, from chemical elements of partial volcanic origin as indicated by the rare-earth element distribution. In addition, the magnesian character of palygorskite and

sepiolite resembles that of Mg-smectite (saponite) formed at proximity by low-temperature alteration of basalt. But fibrous clays apparently do not commonly form in the igneous rocks themselves. Volcanic glass or ash preferentially alters to Mg-smectite, high-temperature sheet-silicates or phillipsite.

2. A temperature significantly higher than that of bottom sea water is suggested by isotopic measurements, and by the frequent vicinity of fibrous clays to basalt intrusions and tectonically unstable crustal zones.

3. The influence of sea water is indicated by the negative anomaly in cerium, a character often not encountered for strictly basalt-altered Mg-clays like saponite.

4. The existence of a mineral precursor is necessary since fibrous clays are not in equilibrium with sea-water composition. Iron-smectites appear to represent particularly suitable precursors. The Fe-smectites themselves could result from former low-temperature hydrothermal deposition.

5. The hydrothermal formation of fibrous clays probably corresponds with semi-confined chemical conditions, as suggested by their high magnesian character different from open-sea authigenic minerals, and by their frequent nodular, vein, systematically local habit.

13.5 Sedimentary Impact of Volcano-Hydrothermal Activity

13.5.1 Lateral Extension

Knowing the areal extension of the sea floor submitted to hydrothermal influence at a given period is very difficult, since a continuum exists between highly concentrated deposits and common metal-bearing deep-sea clays (Chap. 12). As Fe-Mn oxyhydroxides represent the most abundant and widespread products of hydrothermal activity on the sea floor, it is convenient to distinguish the "concentrated" deposits with high $Mn + Fe$ contents and the "diluted" deposits where hydrothermally derived minerals are largely replaced by terrigenous, biogenous, or hydrogenous material (Bonatti 1981). Boström (1975) proposed the use of the $Al/Al + Fe + Mn$ ratio ($=$ index D) to distinguish metalliferous from non-metalliferous sediments. Metal-bearing sediments occur widely in the deep ocean (Fig. 13.11). Values below 0.4 generally indicate some anomalous metal enrichment, that could be related to hydrothermal activity. If such values are correct, *hydrothermally affected deep-sea sediments appear mainly located in the vicinity of oceanic ridges* (Fig. 13.11), and are much less widely distributed than other metal-bearing sediments. The use of Boström's index for various time slices in Atlantic Ocean deposits suggests that metallization due to hydrothermal activity was fairly constant since the late Jurassic and never exceeded two to three hundred kilometers on either side of the ridge crest.

In situ measurements on the present-day *dispersal of hydrothermal plumes* in marine water provide useful information on fluid fluxes (see Thompson 1983), on the actual impact of fluid exhalations, and on the possible transportation of metal flocs from oceanic ridges to adjacent basins (see 12.2.4). Data on the sea-water

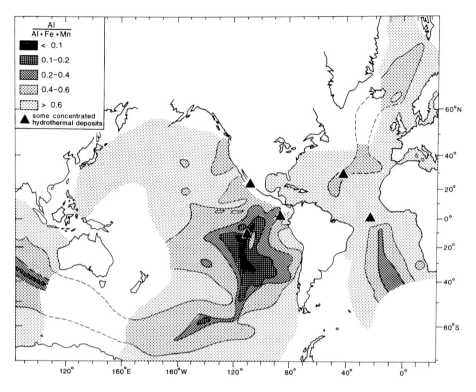

Fig. 13.11. Ratio Al/Al + Fe + Mn in oceanic sediments. (After Boström 1975, Bonatti 1981)

chemistry in the water column above the ridge axis and at proximity exist for several regions (e.g., Trefry et al. 1985, Klinkhammer and Hudson 1986). For instance, manganese profiles over the fast-spreading East Pacific ridge between 22°15′S and 19°25′S reveal the existence of a series of hydrothermal plumes formed by buoyant emanation of hot vent fluids (Klinkhammer and Hudson 1986). The Mn-anomaly maximum can occur several hundred meters above the sea floor. The hydrothermal plumes are dispersed laterally as far as 2000 km from the spreading axis. At about 20°S, the plumes migrate mainly westwards, which reflects the main direction of mid-depth marine currents. At about 30°S, the plumes are advected eastwards because of change in the direction of currents (Fig. 13.12). The asymmetrical dispersal pattern inferred from the measurements coincides with the asymmetrical distribution of metalliferous sediments on both sides of the East Pacific ridge. This demonstrates that *metal-rich sediments located close to the ridge essentially derive from hydrothermal activity*, and that the plumes generated by high-temperature venting dominantly influence the formation of these sediments.

The manganese tends to precipitate less rapidly than iron in many metalliferous deposits. The increase of Mn relative to Fe therefore allows the recognition of decreasing volcano-hydrothermal influence and of sea-floor spreading effects in recent and ancient environments (Boström et al. 1972, Debrabant and

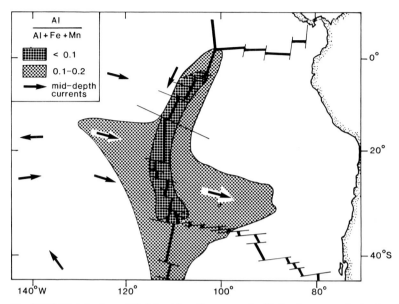

Fig. 13.12. Distribution of Fe-Mn rich sediments and mid-depth currents in the East Pacific ridge region between the equator and 40°S. (After Boström et al. 1969, and Reid 1981; in Klinkhammer and Hudson 1986)

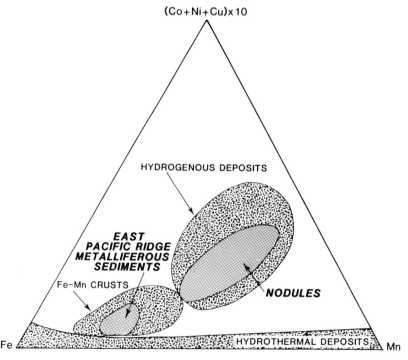

Fig. 13.13. Average distribution of Fe, Mn, and minor transition elements (Co + Ni + Cu) in hydrothermal to hydrogenous deposits. (Compiled by Toth 1980)

Foulon 1979, Debrabant and Chamley 1982b). Other chemical markers of the hydrothermal impact in ridge axes and of dilution gradients in deep-sea sediments may be used, as for instance mercury, cadmium, arsenic, antimony, and selenium (e.g., Grousset and Donard 1984, Fournier-Germain 1986).

By contrast to Fe-Mn compounds, the *strictly hydrothermally derived clay minerals appear to be restricted at the close proximity of submarine vents* (13.3.2). True nontronite or Fe-saponite, that characterize hydrothermal deposits above the oceanic crust, are mostly replaced by Fe-Mg smectites or detrital clay associations at a very short distance from the vents. Clay minerals therefore mostly constitute very local witnesses of hydrothermal activity in sediments, while oxides characterize both close and distant influences. In addition to very abundant iron and manganese oxyhydroxides and to nontronitic clay minerals, typical hydrothermal deposits are also identified by low concentrations of minor chemical elements (Co, Cu, Ni, Pb; Fig. 13.13) and rare earth elements, by low Co/Zn ratios, and by Fe and Si contents that covary (Toth 1980). The "dilution" of these characteristics at increasing distance from the ridge axes corresponds to the transition from typical hydrothermal to typical hydrogenous sedimentary environment (see also Chap. 12).

13.5.2 Vertical Extension

The duration of the hydrothermal and volcanic influence on the marine sediments successively deposited above the oceanic crust is expressed by metal oxide and elemental distribution rather than by silicate minerals. Boström et al. (1972) consider about ten DSDP sites distributed on an East-West transect in the Southwestern Atlantic Ocean, West of the mid-oceanic ridge at a latitude of 28-30°S. The chemistry of major elements is determined for eight time slices, from the uppermost Cretaceous – lower Tertiary to the Plio-Pleistocene. The basal deposits above oceanic basalt are relatively rich in Fe, Mn, and P, even those presently located far from the crest of the ridge. This is particularly well expressed by the values of the Al/Al + Fe + Mn ratio (Fig. 13.14). *The closer to the ridge the sediment within a given time unit,the lower the Al/Al + Fe + Mn ratio.* Such a result suggests that submarine ridge volcanism, and associated hydrothermal activity, have been a significant source of Fe and Mn throughout time. The volcano-hydrothermal processes appear to have been approximately constant in intensity since the uppermost Cretaceous, a result somewhat qualified by Maillot (1982).

The metal supply from the ridge decreases rather rapidly in the course of the time, but nevertheless appears to persist in South Atlantic common sediments for a few million years (Boström et al. 1972, Maillot 1982). Similar data arise from studies performed on Fe and Mn contents in North Atlantic basins (Debrabant and Foulon 1979, Debrabant and Chamley 1982b). Studies on Ca and Mg distribution in interstitial water lead to similar conclusions (Gieskes 1983; see 14.1). *In contrast to these chemical data, the clay mineral assemblages reflect a volcanic or hydrothermal influence on a very small thickness only* (usually less than a few tens of centimeters. See Fig. 13.4), *and rapidly become of a predominantly con-*

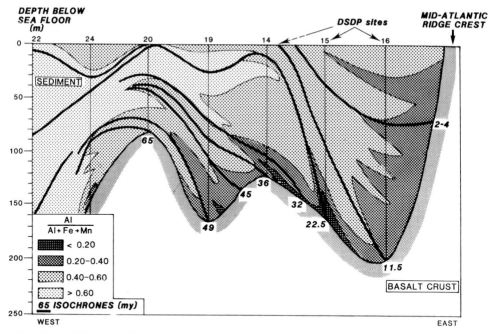

Fig. 13.14. Al/Al + Fe + Mn distribution in an 29°S transect westward of the ridge, South Western Atlantic Ocean. (After Boström et al. 1972). Isochrones in million years

tinental origin (Boström et al. 1972, Chamley and Debrabant 1984, Robert 1987 ; see Chap. 19). This indicates that the direct impact of hydrothermal activity on the deep-sea clay sedimentation is about two orders of magnitude lower than that affecting the elemental iron and manganese distribution. Comparable results arise from investigations on Pacific DSDP sediments (e.g., Desprairies 1981, and references in 13.3.5).

The hydrothermal impact from ancient oceanic crust and fracture zones on *presently exposed marine sediments* is still little investigated, especially from a clay mineralogic point of view. Lemoine et al. (1982, 1983) list the possible indicators of hydrothermal activity linked to the expanding Jurassic – early Cretaceous Tethys Ocean in the Western Alps, France and Italy. Alpine ophiolites locally display either a hydrothermal metasomatism in gabbros, a spilitization of pillow lavas, a ^{18}O depletion in ophicalcite breccias suggesting high-temperature effects, or local Cu-Fe ore deposits. Faunal anomalies like Hauterivian *Peregrinella* beds (brachiopods) and Callovian sponge and pelecypod bioconstructions occur in post-ophiolite deep-sea sediments from the North Tethyan continental margin. These faunas suggest similarities with modern abyssal-benthic communities developing close to hydrothermal fields (Lemoine et al. 1982). Such an interpretation is nevertheless not exclusive, since deep-sea benthic communities may develop in various environments characterized by fluid expulsion, many of them being independent of crust-derived hydrothermal exhalations (accretionary prisms or muddy slopes submitted to strong fluid pressure. e.g., Faugères et al. 1987).

13.5.3 Authigenic Formation and Reworking

Volcanogenic and hydrothermally altered clays can be easily reworked, like all soft sediments, especially on the submarine slopes of volcanic seamounts and islands. It may sometimes be *difficult to distinguish the clayey materials formed in situ from those reworked, particularly if local subaerial weathering is active and adds to the sedimentary contribution.* Various studies deal with this question. As early as 1963 Moberly, followed by Moberly et al. (1968), reported around the Hawaiin islands the presence of abundant fine oozes, chiefly made up of halloysite, allophane, and amorphous materials, with accessory talc. This material mainly derives from the alteration of local volcanic rocks, and appears to evolve secondarily into illitic minerals (celadonite ?) and smectite, and perhaps into chlorite and kaolinite. The respective influence of subaerial and submarine alteration is difficult to assess, as well as that of hydrothermal activity.

Monaco et al. (1979) recognize several origins in the mineral phases of the volcaniclastic sediments deposited near Vulcano island, in Central Mediterranean. Hydrothermally precipitated minerals mainly comprise iron oxides and sulfides, manganese oxides, sulfates, and perhaps some smectite and kaolinite. As terrigenous minerals also include smectite and kaolinite from both local and distant origins, it is very difficult to obviously identify the authigenic clays. Parra et al. (1985) notice the large abundance of smectite in the clay fraction of marine sediments deposited around the Faeroe Islands, North Atlantic Ocean. Chemical analyses on the less than 0.3-μm fraction lead to the identification of three types of smectite: saponite and associated celadonite issuing from hydrothermal alteration of basalts, Fe-Mg smectite derived from exposed Faeroe soils, and secondary Al-Fe smectite possibly originating in the marine environment under hydrogenous conditions. Despite the uncertainty arising from analytical procedures (i.e. measurements on polished clay surface instead of on isolated particles), such studies allow considerable progress in the knowledge of submarine clays and should be developed. Notice that most smectite-rich sedimentary sediments accumulating around exposed volcanic areas issue largely from subaerial weathering and erosion (e.g., Chamley 1980b, Desprairies 1981), and therefore constitute detrital volcanogenic clays.

13.6 Conclusion

1. *The hydrothermal environment develops at the transition zone between internal and surficial geodynamical domains.* Internal processes include the alteration of the oceanic crust by high-temperature fluids and the metamorphism of sediments intruded by basalt sills. Most hydrothermal alteration phenomena occur before the basalt is buried beneath marine sediments. External processes comprise low-temperature alteration and the dispersal and precipitation of hydrothermal material in the deep sea.

The hydrothermal alteration of basaltic glass develops faster than that of crystalline basalt. *Low-temperature alteration of glass* is called *palagonitization*, a process characterized by the successive formation of *hydrated glass, Mg-smectite (saponite), metal oxides and phillipsite.* The final stage of palagonitization corresponds to the development of exclusive authigenic minerals intermixed in complex aggregates. Low-temperature alteration of *crystalline basalt* determines the formation of various minerals, whose precise succession and timing is still imperfectly known. *Oxidizing conditions favor the formation of Mg-Fe smectites relatively depleted in magnesium (<6% MgO), of K- rich celadonite, Mg-nontronite, carbonates and Fe-Mn oxides. Reducing conditions rather lead to Mg-rich smectites (>20% MgO) and Fe-sulfides.*

High-temperature alteration may occur close to the basalt – sea water interface if hot fluids rise up in cracked and fractured rocks at oceanic ridge axes, in transform-fault zones or in other tectonically unstable crustal areas. High-temperature alteration processes nevertheless occur mainly in the deep crust. Whatever the location, they determine the formation of several petrographic facies characterized by a large number of authigenic minerals. Increasing thermal effects correspond in turn to *zeolite facies (e.g., analcite, subregular mixed-layers, talc), greenschist facies (e.g., chlorite, serpentine, epidote) and amphibolite facies (e.g., hornblende, plagioclase).* Al-smectite may exceptionally form close to the sea floor, probably under the influence of acidic hot hydrothermal fluids.

A large part of hydrothermal alteration phenomena depends on the action of *sea water migrating within the upper oceanic crust,* which corresponds to the first phases of the *submarine hydrothermal cycle* (Fig. 13.15). Cold sea water enters cracks and fissures in recharge zones of basalt and circulates downwards in more and more warm rock. The water cools the basalt, leaches chemical elements from

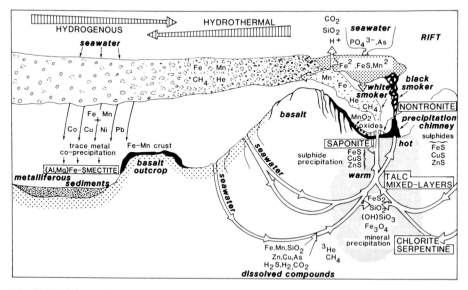

Fig. 13.15. Schematic cross-section through a ridge-crest hydrothermal system and adjacent basalt crust and hydrogenous sediments, showing the chemical exchanges and the main mineral phases formed by alteration and precipitation. (After Toth 1980, Leinen, Rea et al. 1986)

it, what determines exchange reactions. Some precipitations take place under more or less reducing conditions (mainly sulfides, but also high-temperature sheet-silicates). Sea water may circulate through the basalt as deep as 1 to 2 kilometers, depending on the extension of the fracture system and on fluid pressure constraints. Finally, the hot water enriched in metal elements migrates upwards, notably near the spreading axes, and determines the alteration of cooler basalt. Precipitation of reduced minerals, especially sulfides, mainly takes place within the rocks before the sea water emerges. The water itself progressively cools, giving rise to low-temperature alteration (palagonitization, formation of saponite and Mg-Fe smectite).

2. *Hydrothermal deposits are fundamentally characterized by the precipitation of abundant Fe-Mn oxyhydroxides relatively depleted in minor transition elements (Co, Cu, Ni).* Deposition of *massive sulfides* and other minerals may also take place near hot and warm water vents as the fluids emerge (Fig. 13.15). Clay minerals in hydrothermal deposits are often not abundant, perhaps because of an excess in iron or depletion in silica or aluminum. When noticeably present, clay minerals basically consist of *nontronite*, a typical dioctahedral iron-smectite that usually precipitates fairly close to the vent areas. Unstable trioctahedral iron-smectite (= *ferrous saponite*) locally occurs adjacent to vents in Red-Sea depressions filled with warm brines. The discovery of such unstable iron-rich smectite could perhaps help in understanding the genetic relations existing between some deep-sea amorphous ferro-silicic materials and authigenic clay minerals (see Chap. 12).

Fibrous clays of both palygorskite and sepiolite types may form in hydrothermal environment. They appear to result preferentially from the transformation of smectite-rich sediments located in the immediate vicinity of basalts, and submitted to the combined influence of hydrothermal fluids, of basalt and of sea water reacting under semi-confined conditions. Peculiar clay formation characterizes the pelagic sediments intruded by or rapidly blanketing hot basalt sills. The deposits are baked on a few centimeters thickness and experience a very local *contact metamorphism*. The resulting products consist of regular mixed-layers clays like *corrensite* (chlorite-smectite), or of chlorite and high-temperature zeolites.

3. Hydrothermal fluids and associated particulate mineral material issued from ridge axes and other fractured crustal zones rise as buoyant plumes or diffuse in the open sea, and are advected away by deep- to mid-water circulation (Fig. 13.15). The particulate matter is oxidized and adsorbs trace metals from sea water (especially Co, Cu, Ni, Pb), and finally settles and precipitates to form metalliferous sediments. *The progressive dilution of hydrothermal plumes in the sea water therefore determines a transition from hydrothermal to hydrogenous environments.*

It is very difficult to recognize the dominant influence, and therefore to precisely identify the genetic processes occurring in *sediments transitional between true hydrothermal and true hydrogenous or terrigenous environments*. The great abundance of iron and manganese relative to aluminum, and the depletion of Co, Cu, and Ni relative to Zn, constitute fairly reliable indicators of dominantly hydrothermal impact (Figs. 13.11, 13.14), relative to hydrogenous influence

Table 13.9. Tentative correlation between genetic environment and chemical composition of smectites from Equatorial to South Pacific metalliferous sediments. [1] Singer et al. 1984; [2] McMurty et al. 1983; [3] Cole 1985; [4] Hoffert 1980

Type of smectite	Tetrahedra			Octahedra			Interlayers			
	Si	Al	Fe^{3+}	Al	Fe^{3+}	Mg	Ca	NH_4	K	Na
Pure old hydrothermal (Southwest Pacific[1])	3.67	–	*0.33*	–	*1.66*	0.40	–	0.34	0.15	–
Pure recent hydrothermal (Galapagos mounds field[2])	3.94	0.06	–	0.03	*1.59*	0.38	0.03	–	0.36	0.02
Hydrothermal and hydrogenous (Bauer Deep[3])	3.97	0.03	–	*0.44*	1.07	0.54	0.05	–	0.06	0.20
Hydrogenous > hydrothermal (Galapagos spreading centre[2])	3.97	0.03	–	1.12	0.48	0.37	0.09	–	0.11	0.06
Pure hydrogenous (?) (North Marquises fracture zone[4])	3.37	*0.63*	–	*0.39*	1.12	*0.46*	0.46	–	0.17	0.11

characterized by more abundant minor metals, and to terrigenous influence marked by less amorphous metal oxides. But the crystalline minerals possibly forming in these different environments may vary greatly in nature and abundance. One should therefore be especially prudent in qualifying the origin of sedimentary components in metal-rich deposits, particularly the clay minerals that are dominated by smectites of very different possible composition and genesis.

Schematically, *pure hydrothermal smectite is very rich in iron (nontronite), while pure hydrogenous smectite contains noticeable amounts of aluminum and magnesium (Al-Mg bearing Fe-smectite*; Table 13.9). *By contrast terrigenous smectite is frequently of an Al-Fe beidellite type, especially in warm climatic regions, and is often accompagnied by various other clay minerals.* But such a distribution characterizes *end-member clays. Various other situations may occur.* For instance, some active hydrothermal zones display abundant metal oxides associated with nearly pure terrigenous clay minerals (e.g., northern part of the East Pacific ridge, Juan de Fuca ridge). Other areas like the Galapagos rift center contain in close vicinity deposits as different as pure hydrothermal nontronitic mounds, hydrogenous clayey muds and terrigenous - biogenous oozes. Other regions are marked by abundant smectitic deposits reworked from exposed volcanic archipelagoes (e.g., Philippine Sea). Such a diversity of situations stresses the prudence needed in interpretation.

4. *The geographic and chronologic impact of hydrothermal influence on the common deep-sea sedimentation may be identified by measuring the Fe + Mn abundance.* This influence extends fairly broadly on either side of oceanic ridges (at least a few hundred kilometers), and largely depends on the major direction of

bottom or mid-depth currents. As manganese statistically tends to precipitate farther than iron, it is possible to follow the timing of hydrothermal influence from a given spreading axis. The volcano-hydrothermal signal expressed by metal oxides above basalt may be recognized in sediments of at least a few tens of meters thick. *By contrast, the hydrothermally derived clay minerals are restricted to the lowermost part of sedimentary columns, and in the proximity of hydrothermal vents.* Hydrothermal clay minerals are replaced by hydrogenous and terrigenous clays at short lateral and very short vertical distances from the venting areas, except in some restricted basins characterized by hot brines, active exhalation, or rapid precipitation (e.g., Red Sea). On the other hand, the reworking of volcanogenic and clays originating hydrothermally often affects wide submarine surfaces and large time intervals, especially if minerals originating subaerially are added to material formed in the marine environment.

Part V

Clay Diagenesis

Chapter 14

Early Processes

14.1 Introduction

Several books are largely devoted to clay diagenesis processes (e.g., Kisch 1983, Singer and Müller 1983, Parker and Sellwood 1984, Velde 1985). Our purpose is to summarize the major factors controlling the post-sedimentary evolution of clay associations and to present some recent research trends, in order to interpret in a reliable manner the clay mineral successions displayed by ancient sediments. *The diagenetic history of a sediment is conventionally considered as starting after the deposit is buried and definitively removed from the influence of the open aqueous environment.* Diagenetic (= post-sedimentary) effects therefore clearly differ from syngenetic (= syn-sedimentary, hydrogenous) effects, the latter being characterized by free exchanges between solid and liquid environments (i.e., sediment-water interface, sediment reworked by bioturbation, highly-porous sediment; see Part 4). Of course, *the boundary between syngenetic and diagenetic processes is transitional*, since sediments successively buried are only progressively subtracted from the influence of the open aqueous environment.

Early clay diagenesis conventionally concerns buried sediments whose clay associations are protected from noticeable thermodynamic effects. The clay mineralogy of these sediments does not significantly depend on temperature and pressure inluence linked to increasing depth of burial, which usually corresponds to sedimentary columns of less than two kilometers' thickness (in Kisch 1983, Singer and Müller 1983). Sediments buried more deeply than two to three kilometers may experience clay mineralogic changes determined by high temperature and pressure, what characterizes late diagenetic and metamorphic domains (Chap. 15). Early- to late-diagenetic clay-mineral changes controlled by specific lithologic variations, or by tectonic or hydrothermal activity, are considered separately (Chap. 16).

The possibility of forming or transforming clay minerals in slightly buried marine sediments was first envisaged by Arrhenius (1954, in Arrhenius 1963) about some Pacific cores. Arguments supporting the existence of such reactions arose from the downward increase of silica dissolved in interstitial water, the high content of boron associated with 1Md micas, the higher crystallinity of ancient smectites compared to recent ones, and the presence of aluminosilicate-like overgrowths in partly dissolved siliceous skeletons. Some additional studies scattered in the literature concluded with the same possibility, most interpretations involving the syn-sedimentary alteration of volcanic glass into clay (in Kastner 1981;

see also 14.2, 14.3). First experimentations performed on the stability of clay minerals in the marine environment also suggested some in situ changes. For instance, Whitehouse and McCarter (1958) report a noticeable alteration of smectite to chlorite and illite in sea water under laboratory conditions. Mackenzie and Garrels (1966a) observe in similar conditions the release of silicon from smectite and other clay minerals, which indicates a crystal instability. Russell (1970) deduces from the uptake of magnesium by smectite that small amounts of the mineral could transform into chlorite-like lattice. In addition, theoretical considerations led Mackenzie and Garrels (1966b) to estimate that in order to maintain steady state conditions in the sea about 7% of the clay minerals in marine sediments need to be formed authigenically. Recalculations by Berner (1971) led to about 4%.

By contrast to these first studies and theoretical considerations, *most investigations performed since the mid-1960's come to the conclusion of the general stability of clay associations in shallow-buried common sediments*. Arrhenius (1963) already noticed that synthesis experiments suggest the formation of zeolites rather than of phyllosilicates from the precipitation of aluminum species in marine environment. Price (1976) and Kastner (1981) summarize the data available and induce lack of evidence about possible changes in the structure of clay minerals in ordinary marine sediments such as hemipelagic clay or mud and biogenic ooze. The only significant clay mineral modifications appear to occur close to the sediment-sea water interface (Part 4), and in volcanogenic or hydrothermally affected sediments. Most chemical exchanges involving sedimentary clays concern interlayer cations (K, Na, Ca, Mg) rather than octahedral or tetrahedral elements (Carroll and Starkey 1958, Roberson 1974, Dilli and Rao 1982). For instance, Lerman et al. (1975) calculate that silica dissolution rates at distances of 100–300 m below the sediment-water interface range from $0.5.10^{-3}$ to $1.5.10^{-3}$ mg SiO_2/liter/yr. These very low values apply to sediments as old as 1.10^7 years.

If dissolution or formation of clay minerals occurs in slightly buried common sediments, it appears to account for a very small part of the total clay only, which is generally not recognizable by most chemical and mineralogic techniques. Mackin and Aller (1984) and Mackin (1986) envisage from measurements of dissolved Al and Si in China Sea and New-England shelf sediments that very small amounts of aluminosilicates like chlorite could form through an alteration / precipitation process. Similarly, De Lange and Rispens (1986) suggest from dissolved Fe and Si data on recent sediments from the Nares abyssal plain, Western North Atlantic, that nontronite could locally precipitate and constitute about 0.1% of the clay fraction.

Early diagenetic processes occurring within the first hundred meters below the seafloor generally concern carbonates, phosphates, silica, zeolites, sulfides and sulfates, metal oxides, and organic matter, much more than clay minerals (e.g., Larsen and Chilingar 1979, 1983; Berner 1980). This is particularly well expressed by chemical measurements on interstitial waters of deep-sea sediments (e.g., Gieskes 1983). Most depth profiles display a downward increase of calcium and a correlative decrease of magnesium, which is attributed to a release of Ca and a removal of Mg by underlying basalts submitted to alteration. Sites with conservative Ca

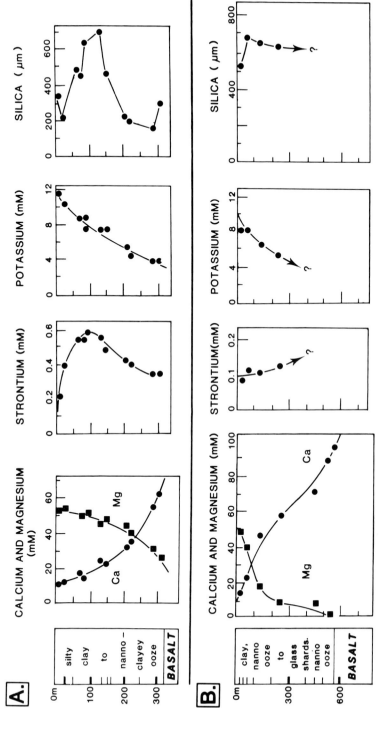

Fig. 14.1. Concentration-depth profiles of some chemical elements dissolved in interstitial water. (After Gieskes 1983). **A** DSDP Site 239 (Mascarene basin, Indian Ocean), conservative Ca and Mg concentration profile. **B** DSDP Site 285 (South Fiji basin, Southwestern Pacific Ocean), non-conservative Ca and Mg concentration profile. *mM* millimoles; *µm* micromoles

and Mg concentration-depth profiles (Fig. 14.1A) imply that the gradients exclusively result from the mass transport of these elements through the sediments between basalt and the ocean. Sites with nonconservative profiles are less frequent (Fig. 14.1B); in these cases the differences occurring between Ca and Mg concentrations imply that there have been reactions in the sediment column, magnesium being more rapidly removed than calcium is supplied. This could account for a slight formation of smectite within the sediment (Gieskes 1983). Notice that the alteration of underlying volcanic material is also suggested by the downward depletion in the $\delta^{18}O$ and in the $^{87}Sr/^{86}Sr$ ratio of the interstitial waters. Whatever the type of concentration-depth profile and its significance, *the convective flow through the sediments stops when sediment thicknesses are greater than 100–150 m, and remains quantitatively very low.* Gieskes (1983) estimates that the calcium flux associated with the average gradient is less than 1% of the river flux, and that the magnesium flux is about 2%. In addition to Ca and Mg gradients, measurements on interstitial water composition generally show a downward decrease of potassium, suggesting an uptake into the sediment, a local maximum of strontium presumably caused by carbonate recrystallization, and variable amounts of dissolved silica mainly controlled by the local abundance of siliceous tests (Fig. 14.1).

14.2 Smectite and Early Diagenesis

14.2.1 General Features

Among the clay minerals buried in common marine sediments, *smectites constitute the group the most frequently quoted as possibly subject to post-sedimentary modifications.* While illite, chlorite, kaolinite, vermiculite, irregular mixed-layers, and even palygorskite or sepiolite are mostly considered as stable in sediments buried at distances exceeding one kilometer below the sea floor, smectites are often attributed to early diagenetic effects. Such interpretations mainly result (1) from the unexpected abundance of smectite in many Mesozoic and Cenozoic marine deposits, (2) from both the diversity and continuum of chemical composition of smectites implying possible in situ modifications, (3) from the sensitivity of smectitic minerals to cation-rich environments, and (4) from the well-known dependence of smectite on volcanic or hydrothermal activity.

Various discussions arise from the literature about the possible diagenetic formation of smectite in common sediments (see Kastner 1981). For instance Brown et al. (1969) report the presence of abundant smectite associated with clinoptilolite in Jurassic, Cretaceous and Paleocene sediments from Southeast England. This mineral association, often combined with glauconite and amorphous silica, suggests an authigenic origin apparently without the participation of volcanic glass as source material. G. Brown and colleagues envisage that zeolite- and smectite-rich sediments could result from chemical elements diagenetically assembled after the dissolution of silica from flint of biogenic opal, the removal of alkalis from glauconite, micas or feldspars, and the dissolution of calcium carbonate. Kossowskaya et al. (1975) and Lomova (1975) identify large amounts of smectite and palygorskite in late Cretaceous to Oligocene Atlantic sediments deposited off Northwestern Africa. This mineral association is thought to result from the massive submarine alteration of volcanic material, which is locally scattered in the sediments (see 13.4). Jeans (1978b) studies South England deposits comparable to those investigated by Brown et al. (1969), but focuses on Cretaceous series. Abundant smectite and silica (opal CT) are interpreted as precipitated from the pore waters of their host sediment at some depth (10 m or more) below the seafloor. The mineralogy of silica and smectite appears to have been controlled by the SiO_2 concentration of the pore water and more locally by pH gradients. In a later study, Jeans et al. (1982) consider that Cretaceous smectite of Southern England and Northern Ireland mainly results from the diagenetic evolution of volcanic material, the sediments being probably of volcanic origin. Similar interpretations, involving the post-sedimentary alteration of volcanic glass dispersed within the deposits, are presented for some early Tertiary series, such as the deep-sea brown clays of the Northwestern Pacific Ocean (Schoonmaker et al. 1985), and the clays and limestones of Blanche Point formation in South Australia (Jones and Fitzgerald 1984, 1987). *To summarize, late Mesozoic and early Cenozoic sedimentary smectites are often attributed to post-sedimentary processes, the clay minerals being supposed to develop in common deposits with or without the participation of volcanic materials.*

Cretaceous to Paleogene sediments of the Atlantic domain often contain high amounts of smectite in the clay fraction. Site 534 of the Deep Sea Drilling Project (Blake-Bahama basin) represents a good example of the clay stratigraphic record commonly encountered in this domain (Fig. 14.2). *Atlantic ordinary sediments constitute favorable materials to test the hypothesis of an early- diagenetic origin of smectite,* for the following reasons: (1) they usually endure a moderate depth of burial (less than 2 km); (2) they are submitted to moderate geothermal gradients (about 30°C/km); (3) they mostly consist of biogenic to clayey materials devoid of abundant volcaniclastic or hydrothermally derived constituents; (4) they were deposited at noticeable sedimentation rates (usually ≥ 1cm.10^3yr), which precluded appreciable hydrogenous genesis at the sediment – sea water interface; (5) their richness in smectite allows a precise identification of its mineralogy, geochemistry, and morphology. Let us consider the main characteristics of the nature and distribution of smectite in these Atlantic sediments.

Mineralogy. Thousands X-ray diffraction analyses performed on Cretaceous-Paleogene sediments from the Atlantic domain indicate that most smectites contain few illite-like layers, and present a rather narrow basal reflection at *17 to 17.5* Å on glycolated samples. The 8.9 Å peak is often present after glycolation, and the (060) reflection is close to 1.502 Å, indicating a dioctahedral mineral. X-ray diffraction characters vary little from one area or geologic period to another. On transmission electronmicrographs, smectites mostly display *flaky outlines* very similar to those encountered in continental soils (e.g., Chamley 1979, 1981, Chamley et al. 1979a, 1983a, Robert 1987; see Fig. 19.1). Some laths occur sometimes at the periphery of smectite flakes, as is discussed below (14.2.2).

Chemical Composition. Microprobe investigations on isolated particles show that smectites from common Cretaceous-Paleogene clays to limestones consist of

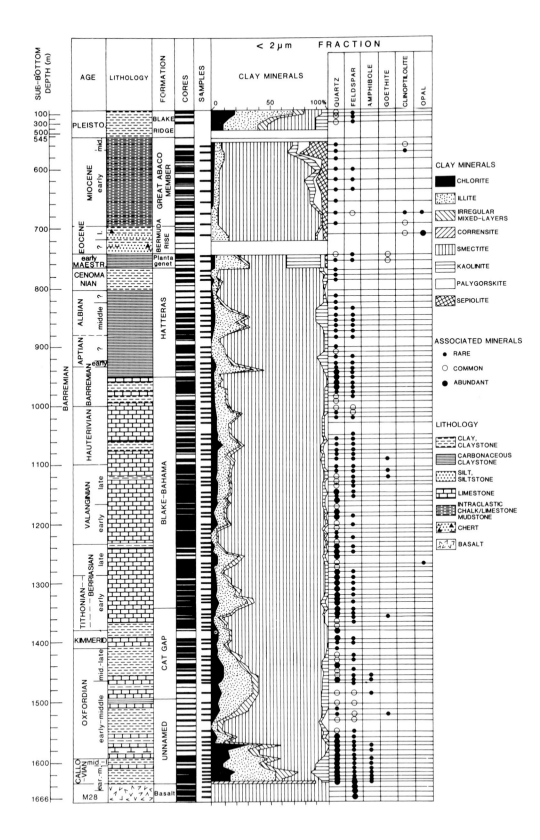

Table 14.1. Structural formulae of smectites from clayey to calcareous sediments of the late Cretaceous of tropical Northeastern Atlantic. (After Chamley et al. 1988). Comparison with Peleogene detrital Al–Fe smectite (Wyoming) of the Mormoiron Basin, SE France (after Trauth 1977), and with smectites from recent West African soils. (After Paquet 1970, Duplay 1982)

Samples	Tetrahedra		Octahedra				Interlayers			
	Si	Al	Al	Fe^{3+}	Mg	Ti	Na	K	Mg	Ca
Senegal Basin, Kafoutine										
N° 620, Maastrichtian	3.12	0.88	1	0.50	0.30	0.02	0.62	0.42	–	0.33
N° 1190, Santonian	3.22	0.78	1.31	0.44	0.23	0.02	0.32	0.30	–	0.20
Cape Verde Basin, DSDP 367										
19-1-100, late Cretaceous	3.74	0.26	1.23	0.43	0.44	0.02	0.03	0.17	–	0.06
22-6-91, late Albian	3.31	0.69	1.07	0.69	0.32	0.02	0.05	0.40	0.05	0.07
23-2-46, Albian	3.57	0.43	1.47	0.38	0.18	0.02	0.04	0.16	0.08	0.04
Mormoiron Basin, SE France										
Al–Fe smectite (wyoming)	3.92	0.08	1.21	0.40	0.30	0.03	?	0.21	–	0.16
West African Soils (parent rock)										
M-5D (gneiss w/amphibole)	3.67	0.33	1.07	0.58	0.47	0.03		?		
GB 92 (granite)	3.38	0.62	1.43	0.49	0.21	0.06		?		
LIV-2-3 (Eocene clay)	3.61	0.39	1.37	0.46	0.24	0.05		?		
Godola (granite)	3.30	0.70	1.29	0.68	0.19	0.05	?	0.09	?	0.01

beidellites, marked by the presence of aluminum in both tetrahedra and octahedra, and by a partial substitution of ferric iron and magnesium for octahedral aluminum (Table 14.1). The relative amounts of Al, Fe, and Mg may vary in a moderate manner, which characterizes a rather well-defined group of *Al-Fe beidellites*. This composition is comparable to that of smectites originating in continental soils from warm regions or reworked from such soils in pericontinental deposits (e.g., Paquet 1970, Trauth 1977). By contrast, the chemical composition of most Atlantic smectites differs greatly from that of smectites derived from the alteration of volcanic glass or basalt, of from hydrothermal or hydrogenous precipitation, that are much richer in magnesium or iron (Chap. 12, 13).

Rare Earth Elements Composition. Sediments from the Eastern Atlantic Ocean rich in clay fraction (60-90%) and containing abundant smectite within this fraction have been studied by neutron activation from both stratigraphic and geographic points of view (Courtois and Chamley 1978, Bonnot-Courtois 1981). The vertical distribution of rare earth elements displays a very homogeneous pattern in most sediments, from early Cretaceous to Quaternary times (Fig. 14.3).

Fig. 14.2. Blake-Bahama basin, Northwestern Atlantic ocean. Lithology and mineralogy of the <2 µm sedimentary fraction of DSDP Hole 534 A. (After Chamley et al. 1983a), with reference to site 391 for the Pleistocene. (After Pastouret et al. 1978a)

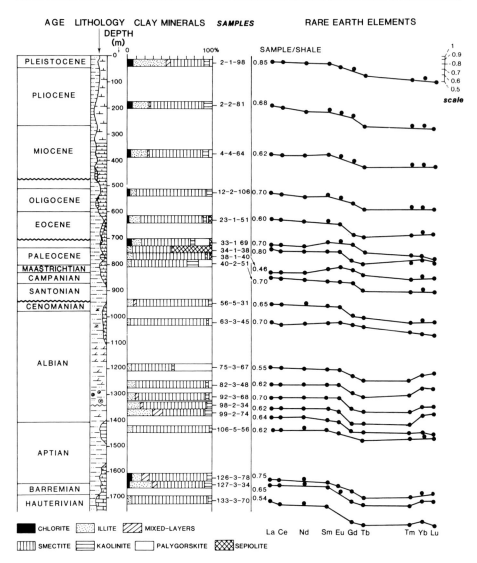

Fig. 14.3. Lithology, clay mineralogy and rare earth elements distribution at DSDP site 398, off Iberian Peninsula, Northeastern Atlantic. (After Courtois and Chamley 1978, Bonnot-Courtois 1981)

When normalized to continental shales, the curves appear rather flat. *The rare earths are devoid of both enrichment in heavy elements* (europium, gadolinium, terbium, tamarium, yterbium, lutetium) *and depletion in cerium.* A slight enrichment in light elements (lantanium, cerium, neodymium, samarium) is usually recorded. All these characters suggest a strong continental influence and the absence of noticeable impact of volcanic activity or sea-water influence on smectite formation. Rare earth elements in Cretaceous-Cenozoic deposits of the Eastern Atlantic are not equilibrated with sea-water composition, suggesting that smectitic

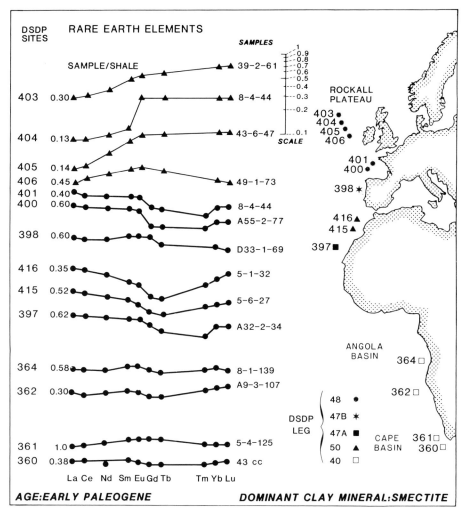

Fig. 14.4. Distribution of rare earth elements in smectite-rich sediments of the Eastern Atlantic margin during Paleocene time (mainly Eocene). (After Courtois and Chamley 1978)

minerals represent reliable markers of past continental environments. Similar results are provided by Mesozoic sediments drilled in the Western Atlantic Ocean, where rare earth elements are studied on the fine, carbonate-free sedimentary fraction (Chamley and Bonnot-Courtois 1981; see Fig. 13.4).

The horizontal distribution of rare earth elements spectra in a given time slice closely resembles the vertical distribution (Fig. 14.4). An exception occurs in the Eocene in the northernmost part of the Eastern Atlantic Ocean, where sediments display a relative enrichment in heavy elements (Eu, Gd, Tb, Tm, Yb, Lu). The corresponding DSDP sites are located on the Rockall Plateau, an area submitted to intense tholeiitic volcanic activity during this period, in relation with the opening of the Norway Sea. Sediments contain abundant volcanic materials, the alteration of which led to the formation of smectite rich in heavy rare earth elements. Notice that the absence of a Ce negative anomaly in sediments from the Rockall Plateau suggests that most alteration processes occurred subaerially, before the volcanic materials were eroded and transported into the marine environment (Courtois and Chamley 1978, Desprairies et al. 1981).

Strontium Isotope Composition. Pure smectites of very small grain size (<0.2 μm and 0.2–0.4 μm) are studied by Clauer et al. (1984) from Cenomanian and Paleocene sediments of the North Atlantic Ocean. For instance, Paleocene oozes at DSDP site 386, Southeast of Bermuda, display $^{87}Sr/^{86}Sr$ ratios of 0.70860 ± 0.00044 for the 0.2–0.4 μm fraction, and of 0.70899 ± 0.00043 for the <0.2 μm fraction. These values are much higher than that characteristic of Paleocene marine depositional environments (0.70783 ± 0.00028). In addition, the apparent age indicated by strontium isochrones and by K–Ar dating ranges between 167 ± 5 and 200 ± 6.10^6 years, instead of 56 ± 3.10^6 years, which corresponds to the Paleocene stage. Both groups of results indicate that *Atlantic smectites are not in equilibrium with the marine environment and are dominantly inherited from old, presumably terrigenous parent-rocks.*

Associated Minerals. Smectite in Atlantic sediments is often accompanied by clinoptilolite and opal CT (cristobalite – tridymite series), which could suggest a mineral paragenesis developed through diagenetic processes. In fact, the distribution of the three mineral groups is largely independent. If clinoptilolite- and/or opal-rich sediments frequently contain abundant smectite, *smectite-rich sediments are often devoid of clinoptilolite and opal CT* (Fig. 14.2. Chamley 1979; see also the Initial Reports of the Deep Sea Drilling Project, namely volumes 39, 40, 41, 44, 47, 48, 50, 76, 77, 79, 80, 81). Clinoptilolite and opal develop mainly in late Cretaceous to Paleogene sediments (Stonecipher 1976), while smectite-rich deposits occur from late Jurassic to late Paleogene. In addition, the smectite composition and relative abundance do not depend on the presence or absence of zeolite or opal. . There is therefore no systematic correlation between silica-rich minerals and smectites, which suggests the absence of a real paragenesis.

Volcanic Constituents. Volcanic glass is rarely identified in significant amounts within late Jurassic to Cenozoic sediments of the Atlantic domain where smectite abundantly occurs. It is somewhat intriguing that marine sediments whose abundant smectite is attributed to the diagenetic transformation of volcanic glass are virtually devoid of undisputed volcanic remains (e.g., Jeans et al. 1982, Jones and Fitzgerald 1984, 1987). The only possible volcano-derived materials sometimes recognized consist of quartz (e.g., Jones and Fitzgerald 1984), that could also have a detrital aeolian or subaqueous origin, and could in any case hardly reflect oceanic tholeiitic volcanism. Most authors supporting the hypothesis of a volcanic origin for smectite in common sediments imply that initial volcanic glass has been altered and destroyed through smectite formation. The often total absence of glass is nevertheless difficult to understand, knowing that the alteration potential of volcanic material greatly differs with size, shape, porosity and composition, and that non-altered volcanic debris usually still exist in deposits rich in unquestionably volcanogenic smectite. In addition, the lower part of Atlantic sedimentary columns above oceanic crust is characterized by Fe–Mn and Ca–Mn migrations directly related to the basalt proximity (Boström et al. 1972,

Debrabant and Chamley 1982b, Gieskes 1983), but usually lacks any increase in smectite, clinoptilolite, or opal abundance. On the contrary, the sediments blanketing the basalt crust generally reveal a decrease in smectite abundance relatively to illite, chlorite, and kaolinite (e.g., Fig. 14.2; see also Chamley 1979, Robert 1987). In a similar way, Albian sediments deposited close to the North American continent contain abundant smectite, while contemporary deposits located close to the Atlantic ridge and in the volcanically active Cape Verde basin are depleted in this mineral (Chamley and Debrabant 1984a).

Lithologic Distribution. In most Atlantic series, the relative abundance of smectite does not depend on the main lithologic successions (e.g., Fig. 14.2). The mineral content varies independently of the dominance of either reddish clay, organic-rich clay, limestone, chalk, chert, silt, pelagic, or hemipelagic ooze. This general independence, combined with the absence of control of smectite abundance by the depth of burial, suggests that the mineral formation and evolution are not noticeably controlled by early diagenesis. Of course this general behavior does not exclude the possibility of minor lithologic control, such as local degradation in organic environment, or dependence on short-scale changes like limestone-marl alternations (Chap. 17).

To summarize, the characters and distribution of most late Mesozoic to Cenozoic Atlantic smectites appear to depend little or not at all on the depth of burial, lithology, associated minerals and volcanic activity. On the contrary, these characters and distribution are largely compatible with land-derived minerals. This suggests that smectite is little controlled by post-sedimentary processes and notably issues from land-masses adjacent to the Atlantic Ocean (see Part 6). One cannot exclude that authigenic processes could have been responsible for the formation of smectite in past common sediments, but no reliable arguments for such a possibility arise from data provided by the present-day sedimentation or by the study of moderately buried series. Similar conclusions could be presented about various other Mesozoic and Cenozoic successions in clayey to calcareous sediments, such as those of the Western Tethyan domain (e.g., Giroud d'Argoud et al. 1976, Deconinck et al. 1985).

14.2.2 Lathed Smectites, Clay Mineral Overgrowths

The existence of smectites marked by a fine lathed morphology is locally reported in brown metal-bearing deep-sea clays (e.g., Karpoff et al. 1981, Lenôtre et al. 1985, Tazaki et al. 1986), *in common pelagic or hemipelagic oozes* (e.g., Clauer et al. 1984, Robert et al. 1985, Chamley and Debrabant 1988), *and even in pericontinental sediments* (e.g., Trauth 1977). The origin of these singular structures is sometimes attributed to hydrogenous authigenesis in metalliferous environment (e.g. Hoffert 1980), sometimes to pre-glauconitization growths (e.g., Louail 1984), or to the early diagenetic transformation of volcanic materials (e.g., Jeans et al. 1982). Lathed smectites and associated overgrowths are especially investigated in the Atlantic and Indian Oceans. Let us summarize the main facts and interpretations proposed in both domains.

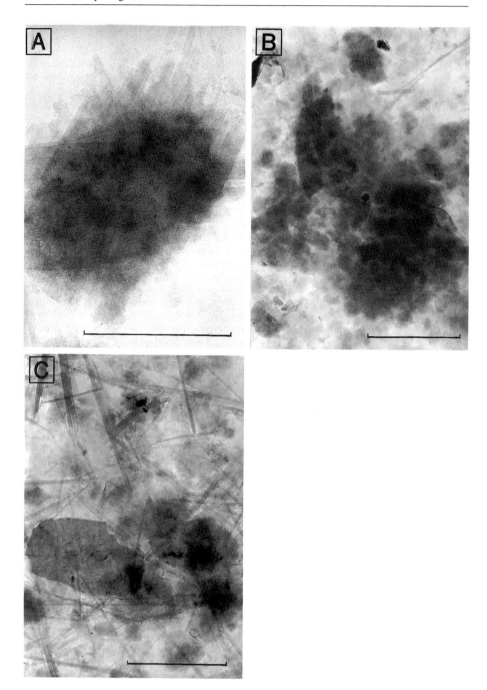

Fig. 14.5. Smectite laths (**A**), smectite flakes (**B**) and detrital palygorskite (**C**). Micrographs T. Holtzapffel, H. Chamley (bar: 1 μm). **A** Lathed smectite with flaky center (Albian, North Atlantic Ocean). **B** Flaky smectite, with rare palygorskite fibers (*P*) (early Cenozoic, NW Pacific Ocean, Site 578 DSDP). **C** Flaky smectite (*S*) with abundant reworked palygorsky fibers (*P*) and illite sheets (*I*) (late Cretaceous, NW Pacific Ocean, Site 576 DSDP)

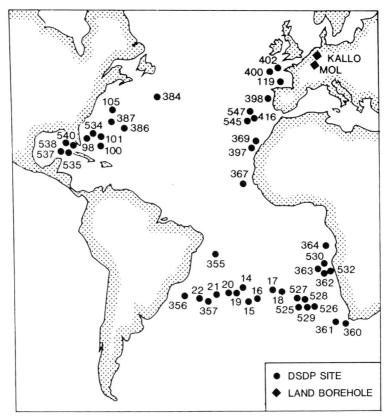

Fig. 14.6. Location of drilling sites devoted to investigations on lathed smectites. (After Holtzapffel and Chamley 1986)

Atlantic Ocean

Lathed smectites frequently occur in clayey to calcareous sediments deposited since the late Jurassic in the North and South Atlantic Ocean (Holtzapffel et al. 1985, Holtzapffel and Chamley 1986). Lathed smectites usually consist of very fine (about 0.02 to 0.1 μm wide), elongated (about 0.05 to >1.0 μm long) particles assembled in *bundles oriented at 60° from each other* (Fig. 14.5). The laths constitute either the totality of the smectite particles (= *lathed smectite ss*) or, more frequently, only the outer part (= *transitional smectite*). Common *flakes of smectite* often dominate the clay fraction where laths occur, or are even exclusive.

The main characters of Atlantic lathed smectites, investigated by T. Holtzapffel and colleagues on 225 samples issued from 39 DSDP and land-drilling sites (Fig. 14.6), are the following:

1. *All transitional morphologies can occur, in a given sample, from exclusive flaky to exclusive lathed particles*, what suggests a genetic relation between the different smectite types. The lath abundance systematically increases in smallest grain-size fractions (<0.5 μm) compared to larger sizes (0.5–2 μm), reflecting the sensitivity of high-surface particles to environmental influence.

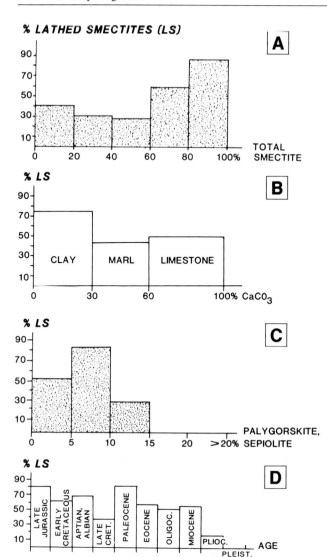

Fig. 14.7. Quantitative distribution in Atlantic drilling sites of lathed smectites relatively to total smectite percentage (**A**), major lithology (**B**), fibrous clay abundance (**C**) and sediment age (**D**). (After Holtzapffel and Chamley 1986)

2. *The percentage of total smectite in a given sample does not depend on the abundance of laths relatively to flakes.* Some samples made up of more than 90% smectite in the clay fraction are devoid of laths, and may be next to samples containing as much smectite with 10 to 40% laths. This implies that the formation of lathed smectite does not appreciably modify the clay mineral proportion. Nevertheless, *laths often tend to occur more abundantly in sediments rich in smectite* (Fig. 14.7A), which probably results from the ability of smectite to favor lath formation. Notice that illite or kaolinite particles display lath overgrowths much more rarely than smectite.

3. *The chemical composition of flake-rich and lath-rich smectites,* determined by atomic absorption spectrometry or microprobe, *is identical or very similar.*

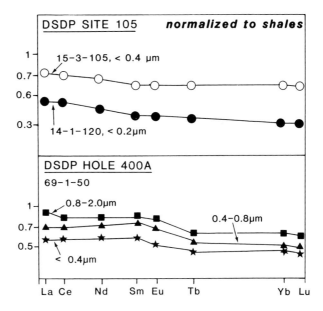

Fig. 14.8. Rare earth elements distribution in smectite-rich (>80% of clay), lath-rich grained size fractions of Albian black shales, North Atlantic Ocean. (After Holtzapffel et al. 1985)

Both types belong to the group of Al-Fe beidellites, without significant differences (Holtzapffel et al. 1985). In addition, *the rare earth element distribution strongly resembles that of terrigenous clay*, whatever the abundance of lathed smectite or the grain-size fraction considered. There is neither a heavy-element enrichment nor a Ce depletion (Fig. 14.8). These data indicate the general stability of clay chemistry whatever the importance of lathed structures. Such a result implies that the terrigenous signal of smectites, suggested by previous studies (see 14.2.1), is mineralogically and chemically not noticeably modified by the development of laths. *Lathed smectites basically appear to result from a morphological reorganization of detrital smectite flakes, with little chemical modifications*. Notice that no chemical argument arises about a specific volcanic control on lath formation.

4. *The development of lathed-smectites is favored in argillaceous sediments relatively to calcareous ones* (Fig. 14.7B), *and in fine nannofossil- rich deposits* relatively to coarser foraminifera-rich ones. This suggests that lath formation develops rather in little permeable sediments, which agrees with the lack of noticeable chemical modifications and therefore the likelihood of restricted micro-environmental exchanges. Such conditions are supported by the fact that sediments deposited very slowly (<6 m/10^6 yr), i.e., that have experienced long-lasting exchanges with the sea-water environment, are very poor in lathed smectites. On the other hand, *lath formation appears to be unfavorable in too impervious micro-environments*, when pore spaces are filled with authigenic opal CT or clinoptilolite. Similarly, the presence of more than 15% of fibrous clays (palygorskite, sepiolite) strongly inhibits the presence of smectite laths (Fig. 14.7C), perhaps also because of too low-permeability conditions. Notice that the presence of laths in deep-sea Atlantic sediments never appears to corre-

Table 14.2. Apparent age of Paleocene smectites relatively to the actual age, from Rb-Sr data on DSDP site 119,
South Biscay Bay, Northeastern Atlantic. (After Holtzapffel et al. 1985)

DSDP 119-30-5-60	
Grain-size fraction	Apparent age (10^6 yr)
0.8–2.0 µm	205 ± 11
0.4–0.8 µm	201 ± 8
<0.4 µm	141 ± 4
Real age	*56 ± 3*

late with the presence of undoubtedly volcanogenic materials or with the
proximity of volcanic activity.

5. Lathed smectites mainly occur in sediments deposited at *moderate
sedimentation rates* (about 10 to 20 m/10^6 yr), and are disfavored when the
deposition exceeds 30 m/10^6 years. This result indicates that a rapid burying impedes the development of laths, which therefore should preferentially occur at a
very early diagenetic stage. This hypothesis is supported by strontium isotope
data, that show a slight reequilibration toward sea-water composition and therefore the *proximity of the open marine environment* (Clauer et al. 1984, Holtzapffel
et al. 1985). Rb-Sr data obtained on the smallest smectitic fractions enriched in
lathed particles express a younger apparent age than for larger grain sizes
(Table 14.2), pointing to the more-evolved character of lath-rich materials.

6. The abundance of lathed smectites does not depend on the depth of burial,
and the real age of lath formation estimated from strontium isotope data is very
close to the actual deposition time (Holtzapffel et al. 1985). Both these results
support the hypothesis of a *very early diagenetic change* from flakes to laths. *The
lath abundance varies strongly with the age of deposition*, the most favorable
periods occurring in the late Jurassic, Aptian-Albian and Paleocene, and the
most disfavorable at late Cretaceous and Plio-Pliocene (Fig. 14.7D). These variations appear to be controlled mainly by lithologic and kinetic factors.

Scanning transmission electronmicroscope (STEM) investigations allow better understanding of the ion migrations and chemical modifications associated
with the formation of laths (Rautureau and Steinberg 1985). The STEM study of
the 0.4–2.0 µm fraction of an Albian gray mudstone from DSDP site 105
(Northwestern Atlantic, core 15.3.100) is performed by Steinberg et al. (1984,
1987) on about ten smectite particles displaying a transition from central flakes
to outer laths. The flakes present a composition of Al-Fe beidellite, with variable
contents of Fe and Al suggesting a diversity of terrigenous sources. The lathed
part of the particles tends to be slightly enriched in silicon and aluminum, and
depleted in iron, which determines a composition closer to that of montmorillonite. M. Steinberg and colleagues conclude that a *discrete chemical reorganization and homogeneization*, as well as a slight silication probably take place in the
closed interstitial microenvironment. The quantitative importance of the ion
migration and rearrangement seems too small to be identified by X-ray diffrac-

tion and common chemical investigations. A few illite particles also display short overgrowths, that appear to be enriched in Si and Fe and depleted in Al relatively to the central part (Steinberg et al. 1987). Illite overgrowths therefore tend to be closer to montmorillonite composition.

Indian Ocean

Lathed smectites frequently occur in Cenozoic sediments of the Northeastern Indian Ocean, especially in the Ganges deep-sea fan and adjacent areas (Bouquillon and Chamley 1986, Bouquillon 1987, Bouquillon et al. 1989). Overgrowths also develop on the edges of illite and kaolinite particles, apparently more frequently than in the Atlantic Ocean. Detailed investigations performed by A. Bouquillon show *a few characters similar to those identified in Atlantic basins*: the laths (1) present all transitional morphologies between pure flaky to pure lathed particles, (2) display an arrangement of bundles oriented at angles of 60° to each other, (3) preferentially develop and are larger around smectite flakes than around illite or kaolinite sheets, (4) decrease in abundance when palygorskite exceeds 10% of clay minerals, (5) do not depend on the depth of burial or on the presence of volcanic materials, and (6) present no or very slight chemical differences when rare or abundant in equally smectite-rich sediments.

Some other characters differ radically from those of Atlantic lathed smectites:

1. *Laths occur both in ancient and recent sediments* from Eocene to Pleistocene age, in highly variable amounts (from 0 to about 70% of total smectite content). Lathed smectite is even identified in noticeable amounts in late Pleistocene sediments (< 18,000 yr BP) from the proximal zone of the Ganges deep-sea fan.

2. *The presence of laths depends little on lithology.* Lathed smectites equally develop or not in terrigenous clayey sand and silt, in biocalcareous oozes, and are independent of carbonate-rich / clay-rich alternations. Laths nevertheless are little represented or absent in the organic-rich mud of the Sri Lanka basin and in the metalliferous deep-sea clays of the Central Indian basin.

3. *The lath formation is not controlled by the sedimentation rate.* For instance, lathed smectites occur in similar amounts both in rapidly deposited siliciclastic sediments off the mouth of the Ganges (≥ 100 m/10^6 yr) and in slowly deposited foraminifera oozes of the ninety-east ridge (10.2 m/10^6 yr).

4. Preliminary STEM investigations done by M. Steinberg suggest an *increase of silicon and iron* instead of silicon and aluminum, from central to outer parts of transitional smectite particles (Bouquillon 1987). Conversely, aluminum and potassium contents tend to decrease from central flakes to peripheral laths.

5. The laths appear to include within the mineral particles some *diffuse organic components*, as indicated by the more or less intense degradation of lathed structures submitted to organic reactants (methanol, K_2O, toluene). These organic compounds could partly control the lath growth.

6. Calculation of structural formulae and electron microdiffraction analysis suggest a *poor crystal organization* of lathed smectites, and no systematic structural improvement relatively to pure smectite flakes.

To summarize, the formation of lathed smectites and other clay overgrowths on clay particles in common deep-sea sediments depends on varying and complex mechanisms, still poorly understood in many cases. Notice that lathed smectites develop in Pacific metalliferous clays characterized by very slow sedimentation rates and abundant iron (Hoffert 1980), whereas they appear not to be favored in comparable metalliferous or siliceous reddish sediments from the Northern Indian Ocean (Bouquillon 1987). In addition, some spectacular lath-smectitic facies occur in shallow-water environments next to iron-rich sediments where glaucony granules developed (e.g., late Cretaceous on the Brittany margin, Northeastern Atlantic. Louail 1984). The environmental constraints on lath formation defined in the deep-sea Atlantic environments appear to apply little to the other environments studied so far. Much remains to be investigated to better understand the mineral and organic control on lathed-clay formation, which undisputably belongs to the early diagenetic domain.

14.3 Specific Environments

Early Effects of Burial

The possibility of weak but irreversible modifications of the clay mineral associations contained in soft sediments submitted to moderate depth of burial (< 1–2 km) is periodically discussed in the literature. For instance, Chamley (1968) described a downward increase of illite crystallinity in 30 piston-cores from the Northwestern Mediterranean Sea, at a sediment depth of about one to two meters. This increase was attributed to a fixation of potassium in the interlayers of altered illite, and therefore interpreted as the very first expression of diagenetic clay reorganization. Subsequent studies performed on larger cores and on drilling materials showed that illite crystallinity does not increase further in more deeply-buried sediments, and even tends to decrease or to present apparently random variations (Chamley 1971, 1975b). In fact, the variations recorded appear to reflect mainly changes occurring in weathering conditions on land, before illite was eroded from alterations and soils, and reworked within the sea. Well-crystallized illite of a few meter-deep sediments probably corresponds to the late Pleistocene cold climate on land, leading to less-weathered minerals than those supplied later in post-glacial sediments (Chap. 17). Early diagenetic effects on illite crystallinity are therefore lacking or are insignificant in shallow-buried deposits of the Mediterranean Sea.

In Paleocene to Recent sediments of the Indus deep-sea cone, Northwestern Indian Ocean, Matter (1974) identifies a diversified clay assemblage mainly inherited from Indus river (illite, chlorite, kaolinite), from India margin by turbidity currents (smectite), and from North or West by wind (palygorskite). Notice that A. Matter envisages that palygorskite could also have an autochthonous origin. Parallel to the progressive compaction of sediment under increasing overburden pressure, Matter (1974) observes from 300 to 1300 meter depth a slight decrease of illite abundance, a concomitant augmentation of 17 Å minerals (smectite and expandable mixed-layers) and a narrowing of the 10 Å illite peak. A. Matter interprets these changes as reflecting an early diagenetic breakdown of illite, a diminution of expandable layers in altered illite, and a neoformation of smectitic minerals. As a consequence, the clay mineral diagenesis is envisaged to start at much shallower depths than usually believed. In fact, the real influence of the depth of burial appears questionable in Indus cone sediments, since the increase in illite crystallinity occurs on air-dried samples, but not on the usually more reliable glycolated samples. In addition, the amelioration of illite crystallinity does not correlate with an increase in the abundance of illite but of smectite, a trend opposite to that usually recorded in clay assemblages modified by depth influence. As smectite abundance generally decreases with increased depth of burial (see Chap. 15), A. Matter's interpretation sets up a difficult question.

Similar questions arise from the study of some DSDP sites located in the Western Pacific Ocean off the Japan margin (Mann and Müller 1980). The possible fixation of potassium within smectite layers, leading to smectite-illite mixed-layers, is suggested from the downward-increasing sharpness of the (001) smectite peak after sample glycolation. As illite does not show a comparable trend, as the change recorded in smectite crystallinity is discrete and correlates with an augmentation in smectite abundance, and as various paleoenvironmental factors obviously control the clay distribution in the studied area (Mann and Müller 1980), it seems difficult to come unquestionably to the conclusion of diagenetic effects. The variations observed in smectite crystallinity could result from changes in sources due to the site migration on the Pacific plate, or from changes in the climate or tectonic context.

As a consequence, *early effects of depth of burial on clay associations appear questionable or very weak in most common marine sediments buried at less than one to two kilometers.* Notice that stronger early diagenetic modifications may occur in continental sediments, especially those characterized by high initial permeability and subject to active groundwater circulation. This is the case of Cenozoic first-cycle desert alluvium from Southwestern United States and Northwestern Mexico (Walker et al. 1978). In addition to mechanical infiltration of detrital clay in heterometric porous sediments, some mineralogic changes result from partial removal of framework grains of feldspars and ferromagnesian silicates, and from precipitation of authigenic potassium feldspar, zeolite, smectite, quartz, hematite and calcite. But these changes depend only partly on the depth of burial, and are largely controlled by lateral migration of groundwater at variable depth levels (see Chap. 16).

Chlorite, Palygorskite Formation

The genesis of chlorite in slightly buried sediments was discussed by Swindale and Fan (1967) for the Hawaii region, in the central North Pacific Ocean. Soil-derived gibbsite, issuing from strong hydrolysis on adjacent islands, is thought to be unstable in the marine environment and to react with dissolved silica, magnesium, and potassium. Chlorite was described to progressively replace gibbsite crystals by successive peripheral haloes. Carroll (1969, 1970) envisaged the diagenetic formation of chlorite in Quaternary deposits of the central North Pacific and Arctic Oceans. The increase of the mineral abundance and crystallinity below surface sediments, locally associated with an augmentation of vermiculitic minerals, was attributed to authigenic processes. Very few similar interpretations arise from the recent literature. It seems that *the chlorite increase in pre-Holocene sediments could largely proceed from increased land supply* under the colder and less-hydrolyzing late-Pleistocene conditions, compared to Holocene ones (Chap. 17).

The early diagenetic formation of palygorskite and sepiolite is sometimes envisaged (e.g., Kossowskaya et al. 1975, Lomova 1975, Gorbunova 1979, Tazaki et al. 1986), but *does not seem to be indisputably demonstrated* so far. The unlikelihood of deep-sea hydrogenous genesis of palygorskite at the expense of smectite or by direct precipitation at the sediment-sea water interface is commented in Chapter 12.3. The reality of fibrous clay formation in some restricted hydrothermal sediments located close to basalt rocks is reported in Chapter 13.4. The fibrous clay-rich deposits identified in Atlantic margin and abyssal plain sediments of late Cretaceous and early Cenozoic times do not display any systematic correlation with the depth of burial, the lithology, the abundance of smectite or the presence of volcanic glass (e.g., Weaver and Beck 1977, Mélières 1978, Timofeev et al. 1978, Chamley 1979, Debrabant et al. 1984, Jacquin 1987, Robert 1987; see also the Initial Reports of the Deep Sea Drilling Project, namely volumes 39, 40, 41, 44, 47, 48, 50, 76, 77, 79). Palygorskite occurs in sediments as different as black shales, limestones and chalks, zeolite-bearing varicolored clays, chert-bearing sediments, hemipelagic oozes and clays, turbidites and other resedimented deposits. Palygorskite often displays on electronmicrographs short and broken isolated fibers, or disorganized and altered bundles. We therefore tend to attribute most fibrous clay occurrences in common deep-sea sediments to the reworking of peri-marine, shallow-water sediments where the minerals formed authigenically (Chap. 9). The search for possible diagenetic overgrowths of palygorskite on smectite particles or for other kinds of fibrous clay genesis in common deep-sea sediments should be pursued through further micromorphological and microchemical investigations, namely by focusing the analyses on deposits containing particles with easily available magnesium and silica (e.g., Mg-calcite, basaltic glass debris, biosiliceous tests).

Diagenesis in Slightly Buried Volcaniclastic Sediments

The ability of volcanic glass dispersed in deep-sea sediments to alter in clay minerals is much debated in the literature, and still imperfectly understood (see Fisher and Schmincke 1984). Arrhenius noticed that although glass shards in Mesozoic sediments may be unaltered, similar volcanic debris in Quaternary deposits are sometimes transformed into smectite or phillipsite. Kennett and Thunell (1975) interpret the increased abundance of volcanic glass in late Cenozoic sediments as the result of a global augmentation in explosive volcanism with possible global-climate implications (Kennett 1982); one could also envisage that this increase simply corresponds to a lesser alteration due to the recent age of glass particles (e.g., Hein and Scholl 1978). Some ash and glass accumulations are reported as being either abundantly transformed in smectite (e.g., Hein and Scholl 1978, Jeans et al. 1982), or as hardly evolving into secondary minerals (e.g., Keller et al. 1978, Palais 1985, Imbert and Desprairies 1987). The downward decrease of dissolved Mg and increase of Ca in interstitial water within sediments located above basalt basement (Gieskes 1983) suggest the diagenetic formation of smectite, a possibility argued by Perry et al. (1976) for Miocene sediments of the

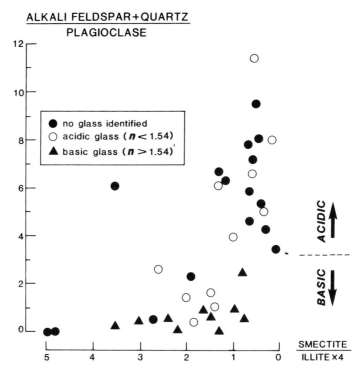

Fig. 14.9. Relation of clay minerals to volcanic debris in Southeastern Pacific sediments. (After Peterson and Griffin 1964)

Southeastern Pacific. But such a clay formation could proceed from the diffusion of chemical elements issued from the alteration of either the sedimentary volcanogenic debris or the basalt itself. In addition, the Ca and Mg opposite gradients are generally too small to allow a significant clay formation (Gieskes 1983). Many sediments blanketing altered basalts contain abundant terrigenous clay minerals instead of volcanogenic ones (e.g., Chamley 1979, Debrabant and Chamley 1982b). Is seems therefore difficult to exclude the possibility for pre-Pliocene smectite to be reworked from either subaerially weathered volcanic rocks or common smectite-rich soils.

The indisputable dependence of smectite on volcanic glass in deep-sea sediments was first demonstrated by Peterson and Griffin (1964) for surficial deposits of the Southeastern Pacific Ocean (Fig. 14.9). Dominantly basic volcanic debris identified by refractive indices greater than 1.54 and by abundant associated plagioclase grains clearly correlate with large proportions of smectite. Smectite appears to form soon after glass deposition in slightly buried sediments. On the other hand, rather acidic volcanic debris is associated with small proportions of smectite, and with fairly abundant alkali feldspar, quartz, illite and chlorite. The latter minerals are considered as inherited from exposed land-masses through long-distance transportation. Griffin et al. (1967) also demonstrate a close correspondence between volcaniclastic debris and smectite abundance in the Lau

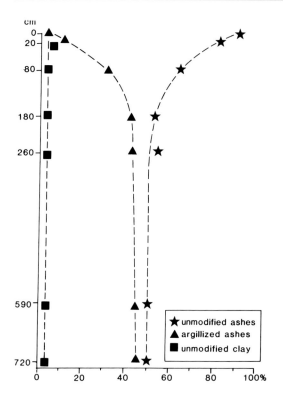

Fig. 14.10. Diagenetic evolution of pumice particles into smectite in volcaniclastic, diatom-bearing sediments from Santorini collapsed caldeira, Aegean Sea, Eastern Mediterranean Sea. (After Chamley 1971)

basin, Southwestern Pacific. The sediments virtually lack illite and quartz, but contain some chlorite that is independent of igneous particles and could have either a terrestrial or volcanic origin.

An early diagenetic formation of smectite occurs in *volcanic-rich oozes deposited near Santorini archipelago*, that collapsed into the Eastern Mediterranean Sea about 1400 years B.C. (in Chamley 1971, Keller et al. 1978). A 720-cm-long core sampled at 280 m water depth within the submerged caldeira of Santorini shows a rapid downward change in the nature and abundance of clay minerals, diatom frustules and volcanic debris (Figs. 14.10, 14.11). Surface sediments contain abundant ash particles made up of either sharp shards or porous silt- to clay-sized pumices. Clay minerals are rare and consist of comparable amounts of terrigenous illite, chlorite, and smectite. Diatom tests occur abundantly and are rather well preserved. With increasing depth in the sediment, the smectite abundance and crystallinity increase, while illite and chlorite remain constant. On electronmicrographs, the glass shards are unmodified, while small pumice particles progressively display filmy clay overgrowths. The proportion of argillized pumices reaches 5% in surface sediments, 35% at 80 cm depth, 40% at 260 cm and 50% at the base of the core. Simultaneously, the diatom frustules appear progressively dissolved; chemical anlyses show that the total amount of silica remains constant throughout the core, relatively to aluminum. At about 180 cm below the sediment surface all biogenic silica is dissolved, at a level where the development rate of smectite around volcanic ashes decreases (Fig. 14.10).

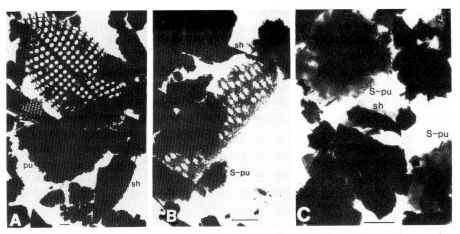

Fig. 14.11. Santorini caldeira, Eastern Mediterranean. Core 29MO67, electronmicrographs. (After Chamley 1971). Bar = 1 µm. **A** 0–5 cm. Well-preserved diatom frustules, glass shards (*sh*) and pumices (*pu*). **B** 180 cm. Strongly dissolved frustule, partly argillized pumices (*S-pu*), unaltered glass shards (*sh*). **C** 180 cm. Strongly argillized pumices (*S-pu*), unaltered glass shards (*sh*)

This evolution is attributed to a *diagenetic alteration of unstable pumice particles into smectite, the clay formation preferentially developing by using the silica provided by easily soluble diatom frustules, and accessorily that provided by the glass itself.* Chamley (1971) concludes from comparisons with data on Eastern Mediterranean volcaniclastites that the transformation of volcanic glass to smectite is favored by the presence of very small glass particles, of porous ash and of associated biosiliceous debris.

Various other cases of smectite formation in volcanic-rich sediments are reported in the literature. For instance, Karpoff et al. (1980) investigate the late Cenozoic volcaniclastic sediments deposited on the Emperor seamounts, Northwestern Pacific. The halmyrolysis of volcanic material induced the successive formation of smectites, phosphates, and iron-manganese oxides. Smectites formed from altered glass during early diagenetic stages marked by low sedimentation rates. Phosphates developed in shallow water and metal oxides in deeper, oxidized environments, both groups being favored during non deposition periods. Desprairies (1981) describes the diagenetic formation of smectite and zeolites in volcaniclastic tuffs and mudstones accumulated above basaltic basement around the Mariana arc and trench. Chamley (1980a), Chamley et al. (1985a) and Chamley and Debrabant (1989) report the formation of different types of smectites in various glass-bearing sediments, South and East of Japan (Fig. 14.12): large and broad laths of saponite close to basalt basement (A); tufts at the periphery of fine pumice fragments (B); small, transparent, well-edged but nonpolygonal sheets of saponite (C); ruffled hair-like particles rich in silica, magnesium, and iron (D); intergrowths in mica particles of aluminum-rich Mg-beidellites (E); filmy, transparent veils of Si-Al-rich smectites devoid of tetrahedral substitutions and depleted in magnesium, growing at the periphery of little-altered, whitish siliceous clay to siltstone (F). *The diversity in composition and morphology of smectite minerals formed subaqueously at the expense of vol-*

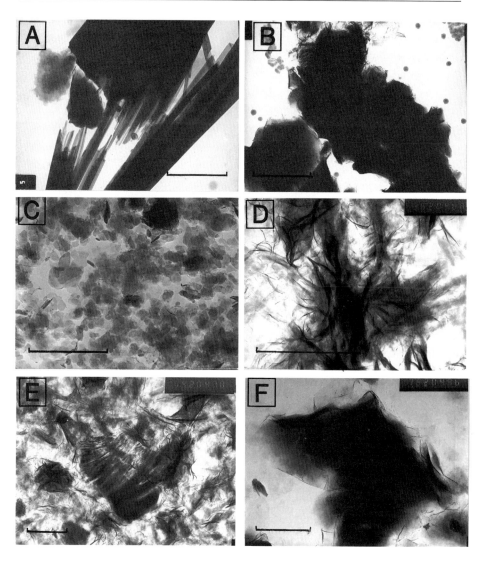

Fig. 14.12. Electron microfacies of volcanogenic West Pacific smectites. Bar: 1 μm. **A** Shikoku basin, DSDP 444A-23-1-50, middle Miocene reddish-brown nannofossil clay. Broad spaonite sheets, with lathed endings. **B** Daïto basin, DSDP 446-13-4-39, Oligocene white ash. Smectite tufts at the periphery of porous glass. **C** Mariana basin, DSDP 585A-17-1-15, Aptian, volcanogenic mudstone. Very small, transparent, subpolygonal Mg-smectite. **D** Nankai trough, South Zenisu ridge, Kaiko project (1-4-3W), whitish to blackish volcaniclastic mudstone. Ruppled, hair-like smectite having entirely replaced volcanic glass. **E** Nankai trough, South Zenisu ridge, Kaiko projet (1-4-3W), whitish to blackish volcaniclastic and micaceous sandy mudstone. Biotite altered into smectite and hair-like smectite. **F** Kuril trench inner wall, Kaiko project (3.6.6), white-greenish tuffaceous volcaniclastic mudstone to siltstone. Filmy veils of Al-smectite developed at the expense of glass

canic debris appears very large. The diverse smectite types encountered mostly differ morphologically and chemically from the dominantly flaky smectites commonly supplied from exposed land masses.

14.4 Conclusion

1. *Diagenetic processes are conventionally considered to start after the sediment is slightly buried and protected from direct exchanges with the sea water. Early diagenetic processes on clay minerals concern sediments that are not buried deeply enough to experience noticeable pressure and temperature effects.* For common geothermal gradients (about 30°C/km) this corresponds to depths of burial of less than about two to three kilometers.

Most clay minerals contained in common deep-sea sediments are not very sensitive to slight or moderate depths of burial. This is particularly the case of illite, chlorite, irregular mixed-layers, and kaolinite. A few studies suggest that smectite, palygorskite, chlorite, and even illite could experience some early-diagenetic degradations or aggradations, but this is difficult to establish obviously. Most sedimentary columns drilled in the deep sea or sampled on exposed land-masses display clay mineral changes that mainly depend on paleoenvironmental conditions (see Part 6) rather than on diagenetic constraints. The clay diagenetic modifications affecting sediments buried less than 2 kilometers are usually much less important than the modifications induced by changes in tectonic activity, climate, paleocirculation or detrital sources.

2. *Some smectite-bearing sediments of various ages show the presence of lath systems oriented at 60° from each other.* Laths preferentially develop at the expense of smectite flakes, especially the smallest ones, that are sometimes entirely replaced. Laths may also grow at the periphery of other minerals, namely illite and kaolinite, but this is much less frequent and leads to short overgrowths only. The proportion of lathed minerals never correlates with increased depth of burial. Micromorphologic and microprobe investigations indicate that laths mainly consist of smectites, whose chemical composition differs only little from that of initial flakes. In adddition, the percentage of total smectites in a given sediment does not depend on the lathed smectite / flaky smectite ratio. *The development of smectite laths therefore does not appreciably modify the mineral and chemical composition of the pre-existing sediment.*

Lathed clay minerals are especially studied in Meso-Cenozoic deposits of the Atlantic Ocean and in Cenozoic deposits of the Northeastern Indian Ocean. Atlantic lathed smectites appear to be favored by little porosity and moderate sedimentation rate, and to result from almost iso-chemical recrystallization of detrital smectite flakes in slightly buried sediments; lath formation is hindered by the presence of abundant foraminifera, opal, clinoptilolite or palygorskite, and does not easily occur in post-Pliocene deposits. In the Ganges deep-sea fan and central Indian basin, smectite laths are much less dependent on lithology, age, or sedimentation rate, and appear to be partly controlled by organic compounds. Further studies are needed to better understand the intimate nature and genesis of secondary lathed clays, that unquestionably belong to the early diagenetic domain.

3. *The large abundance of smectite in many ancient, common marine sediments, peculiarly those of late Mesozoic and early Cenozoic age, poses an intriguing problem.* Where does smectite come from, how did it form ? It is enticing to solve this problem by attributing smectite to the transformation of volcanic material, since smectite is known to develop easily through glass and ash alteration. Some authors readily attribute the abundance of smectite in past clays, muds, clayey carbonates or calcareous oozes to volcanic activity. Detailed investigations performed on late Jurassic to Quaternary sediments from the Atlantic and Tethyan domains show that such a correlation is rarely demonstrated. *In most common deposits, smectite consists of Al-Fe beidellite, whose morphology and major, minor, and rare earth element distribution does not resemble that of Mg- or Fe-rich smectites typically related to alkaline volcanism or to hydrothermal or hydrogenous activity.* Smectite-rich common sediments often contain very few volcanic remains or do not correlate with the abundance of such remains, which differs from indisputably volcano-derived smectitic sediments. In addition, the huge abundance of smectite in Meso-Cenozoic sediments of Atlantic and Tethyan regions would suppose a massive and subcontinuous explosive volcanism in the ocean, a possibility hardly compatible with the mainly effusive character of volcanism in oceanic domains; the submarine alteration of basalt lavas usually determines only a very local increase of smectite. At the present stage of knowledge, a systematic or even common volcanogenic origin of past marine smectites does not appear easily convincing.

In opposition to these apparent inconsistencies, a good agreement exists between the characters of many Meso-Cenozoic sedimentary smectites and a terrigenous origin. *The mineralogy, chemical composition, and shape of most marine smectites closely resemble those of smectites presently formed in surficial soils of warm, poorly drained regions. The ubiquity of smectite in ancient sedimentary facies, its independence of lithology and depth of burial, its presence in both contemporaneous continental and marine sediments* (e.g., late Cretaceous-Paleogene: Sittler 1965a, Chamley et al. 1971, Redondo 1986), *and its correspondence with the warm climate of late Mesozoic-early Cenozoic times, suggest that the mineral is mainly reworked from exposed land masses, as are most other clay minerals.* In the absence of further arguments supporting an autochthonous genesis, we are inclined to favor an allochthonous origin for the abundant smectite contained in most ancient, common deep-sea deposits. The frequent augmentation of smectite abundance in offshore sediments and during transgression periods could largely result from differential settling (6.2) and from the development of poorly drained continental soils.

4. *An early diagenetic formation of smectite occurs in some marine sediments characterized by effectively abundant volcanic glass of explosive origin.* Such sediments, particularly investigated in South and West Pacific Ocean and in Eastern Mediterranean Sea, display a clear correlation between the development of smectite and the occurrence of volcanic glass. Synsedimentary smectites appear to hardly form from massive glassy debris, and to be favored by the combination of small-sized, pumiceous ash, and of easily available biogenic silica. Volcano-derived submarine smectite may present various chemical compositions and shapes depending on the nature of initial volcanic materials, but appears to differ from most common soil-derived terrigenous clay minerals.

Chapter 15

Depth of Burial

15.1 Basic Evolution

15.1.1 Introduction

The clay mineral composition of slightly to moderately buried sediments commonly displays a wide diversity and variability, which reflects the various paleoenvironmental controls typically acting under conditions of the Earth's surface. *At depths of burial usually exceeding two kilometers, clay assemblages tend to become simpler, and to be more or less progressively enriched in illitic and chloritic minerals. Smectite tends to disappear downwards, irregular mixed-layers tend to be less expandable and to include subregular to regular types.* Kaolinite is preserved deeper than smectite, but also tends to diminish in abundance. This general evolution expresses the increase of temperature and pressure in deeply buried series, and corresponds to the *transition from early to late diagenetic conditions.* The precise depth of the downward passage to thermodynamically affected sediments primarily depends on the geothermal gradient, secondarily on pressure, lithology, permeability, tectonics, or hydrothermal circulation (see Chap. 16). Common geothermal gradients average 30°C/km, and significantly affect the clay mineralogy beyond 2.5 to 3.0 kilometers of burial, at temperatures exceeding 80°C. At greater temperature and pressure, anchimetamorphic conditions replace the diagenetic ones. Anchimetamorphism is characterized by an increase in illite crystallinity, the appearance of transitional minerals like pyrophyllite or paragonite, the development of a slaty cleavage and the passage from clay-sized to silt- and sand-sized mica and chlorite. Further evolution leads to the upper zone (= epizone) of the metamorphic domain, with successive laumontite-, pumpellyite/prehnite-, and greenschist-facies.

A lot of research and review papers have been published since the late 1960's about the late diagenesis of clay and associated minerals. They all point to a downward clay mineral simplification (e.g., Fig. 15.1), with various modalities and local peculiarities. Most thick series investigated contain abundant argillaceous deposits devoid of important vertical or lateral secondary ion migrations. The reader is especially referred to Dunoyer de Segonzac (1969), Scholle and Schluger (1979), Kisch (1983), Singer and Müller (1983), Durand (1984), Parker and Sellwood (1984), Robinson (1985), Velde (1985), Frey (1987), as well as to the special volumes on diagenesis published in the journals Clay Minerals (e.g., no. 19, 3, 1984; 21, 4, 1986) and Clays and Clay Minerals (e.g., no. 34, 2, 1986).

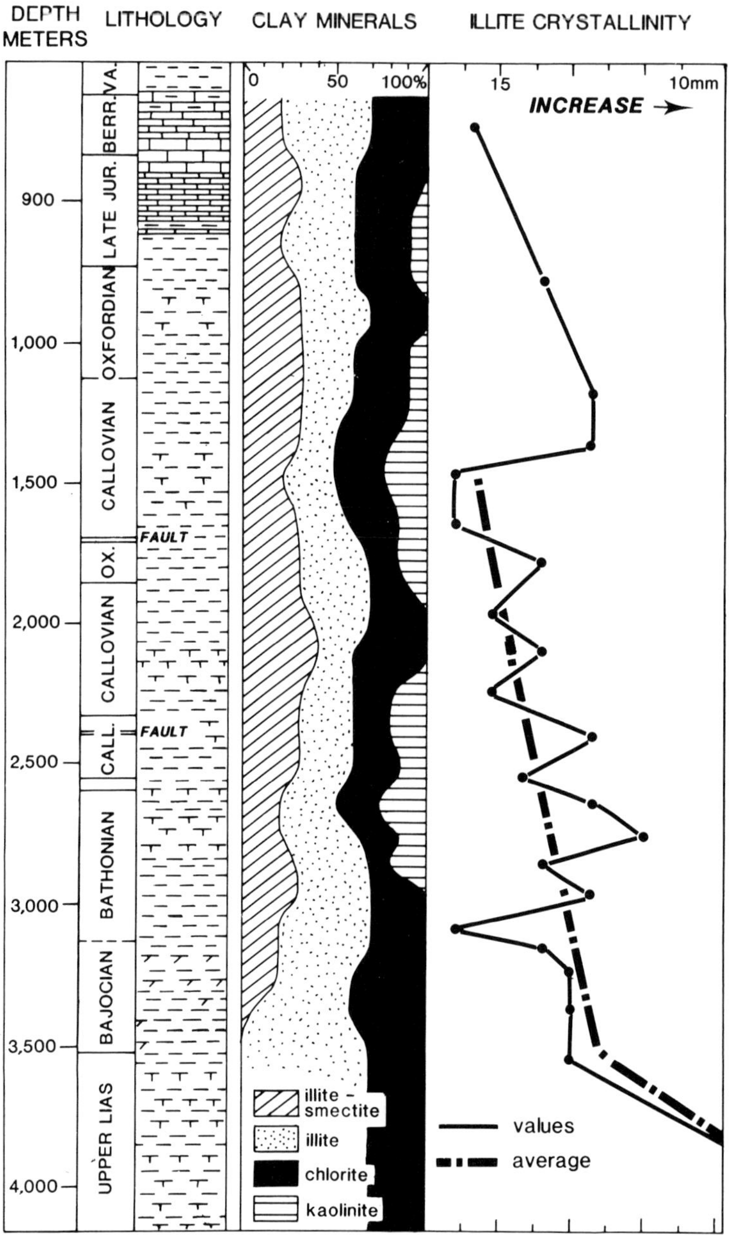

Fig. 15.1. Clay mineralogy and illite crystallinity of Montagne-de-Lure borehole sediments, Southeast France (after Dunoyer de Segonzac 1969)

15.1.2 An Example: Gulf of Mexico Coast Sediments

The sediments deposited from *Eocene to Pleistocene* off Mexico to Florida coasts comprise *shale-rich series* extending over several kilometers thickness and extensively drilled by oil companies. The sediments chiefly belong to Anahuac and especially Frio formations. As early as 1959, Burst noted a progressive modification in X-ray diffraction pattern of smectite and the subsequent disappearance of the mineral with increasing depth of burial. Burst (1959, 1969) successively explained the smectite change as due to a gradual fixation of potassium and magnesium to form illite and chlorite, and to a simple dehydration to one water layer without noticeable chemical conversion. Powers (1959) considered the decrease of smectite with depth in Gulf Coast sediments off Texas as a transformation of Mg for Al in the octahedral sheet, with the consequent fixation of interlayer potassium.

First quantitative data on the mineral and chemical evolution in Gulf Coast sediments are reported by Perry and Hower (1970, 1972). The smectite-rich layers dominate the clay mineralogy in shallow-buried deposits and are progressively replaced by illite-rich layers, down to 5500 meter depth. *The decrease of expandability of illite-smectite minerals* (sensu Reynolds 1980, 1983) *ranges from about 80% to 20% smectite layers with increasing depth.* Discrete illite, kaolinite, and chlorite phases are variously present and considered as detrital. The bulk rock chemical composition lacks systematic variation with depth, except for a decrease in calcium and magnesium due to carbonate solution. The clay fraction shows a downward increase of potassium attributed to an interlayer fixation of the element in illite-smectite as the proportion of illite layers increases. The diagenetic reaction depends on the depth of burial mainly through temperature effects, not on the geologic age or stratigraphic boundaries. Reynolds and Hower (1970) report from various examples the existence of three types of mixed-layering: random, ordered (allevardite-like organization), and superlattice units (three illite and one smectite layers, labeled as type IISI; see Fig. 15.8). Illite-smectite minerals with 40 to 100% smectite layers belong to random mixed-layers, and mostly characterize sediments buried at less than 3.5 kilometers. Allevardite-like and superlattice varieties correspond to highly evolved sediments poor in expandible layers (about 10%) and often buried deeper than about 6 kilometers.

A detailed examination of burial diagenetic reactions in Oligocene – Miocene Gulf Coast shales is performed by Hower et al. (1976), Aronson and Hower (1976), and Yeh and Savin (1977). These authors consider the mineralogy, the inorganic chemistry and the radiogenic and stable isotope chemistry of bulk materials and different grain-size fractions from the 1250 to 5500 m depth-interval of the well CWRU-GC6. *The major mineralogic changes, mainly marked by an increase in illite layers within smectite-illite, occur over the depth interval 2000 to 3700 meters, after which no significant modifications are detectable.* The reduction of illite-smectite expandability is more important for very small grain-sized fractions than for larger fractions (Fig. 15.2A). Between 2000 and 3700 meters, calcite abundance decreases from about 20% of the rock to almost zero, disappearing from progressively coarser fractions with increasing depth. Potas-

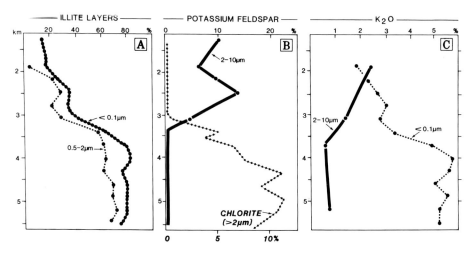

Fig. 15.2. CWRU Gulf Coast well 6, late Oligocene-Miocene. Vertical distribution of **A** illite layers in illite-smectite; **B** chlorite and K-feldspar percent; **C** K₂O in different ignited, CaO free fractions (% weight). (After Hower et al. 1976)

sium feldspar decreases to zero, while chlorite appears to slightly increase in amount (Fig. 15,2B). The distribution of chemical elements shows, in addition to previous data (Perry and Hower 1970), an opposite behavior of potassium in coarse and fine fractions (Fig. 15.2C), suggesting a transfer from sandy minerals to newly formed illitic clay-layers. The strong increase of potassium in the smallest size-fraction (<0.1 μm) is associated with an increase in aluminum and a decrease in silicon, while quartz abundance increases in coarser fractions. The potassium and aluminum appear to be derived from the diagenetic decomposition of potassium-feldspar (and of mica?), and the excess of silicon probably allows the formation of quartz. The atomic proportions closely approximate the following reaction:

$$smectite + Al + K = illite + Si,$$

which can be interpreted as follows:

$$smectite + K\text{-}feldspar\ (+\ mica?) = illite + quartz\ (+\ \text{chlorite?}).$$

Hower et al. (1976) consider *that the shale acts as a closed system for all components except* H_2O, *CaO*, Na_2O *and* CO_2. The reality of the diagenetic process is confirmed by isotopic dating (Perry 1974, Aronson and Hower 1976). *The whole-rock apparent K-Ar ages decrease with increasing depth* (Fig. 15.3), pointing to the existence of dominantly detrital minerals derived from old formations in the upper part of the well, and of diagenetically rejuvenated minerals in the lower part. Argon is lost from K-feldspars and micas during their diagenetic destruction to yield the potassium used in converting smectite to illite. This conforms the nearly isochemical nature of the diagenetic evolution.

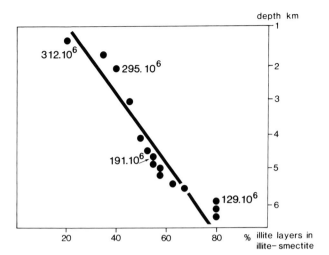

depth km

312.10⁶

● 295. 10⁶

191.10⁶

129.10⁶

Fig. 15.3. Apparent ages of illite-smectite in Gulf Coast sediments, from K-Ar dating. (After Perry 1974)

% illite layers in illite– smectite

Yeh and Savin (1977) investigate the oxygen-isotope composition in three Gulf Coast wells, including the one studied by J. Hower and colleagues. Their results indicate that the Cenozoic sediments do not constitute isotopically equilibrated systems, even those buried at 5600 meters where temperatures are as high as 170°C. The oxygen-isotope disequilibrium among the clay fractions decreases as burial temperature increases. The isotope exchange between clay and pore water becomes more important in the temperature range 55 to 90°C, in which the most extensive conversion of smectite layers to illite layers occurs within mixed-layer phases. Yeh and Savin's data (1977) suggest that Gulf Coast sediments are still subject to diagenetic changes, which tends to determine a better equilibrium between mineral isotope composition and interstitial environment.

Such a conclusion is contested by Morton (1985), who studies the Rb-Sr composition and $^{87}Sr/^{86}Sr$ ratio in <0.1 and <0.5 μm fractions of Gulf Coast sediments (no. 2 Pleasant Bayou well, Texas). If Rb-Sr dating confirms the age decrease with depth, the isochron calculations suggest that illite-smectite mixed-layering equilibrated about 23 million years ago at a depth of burial close to 2.5 kilometers. About 3 kilometers of additional post-Oligocene burial apparently has not resulted in the formation of younger illite-smectite. J.P. Morton proposes a model of episodic or "punctuated" diagenesis, whereby rapid diagenesis is initiated by a change in pore-water chemistry. According to this model, clay reaction occurs temporarily only, over a range of temperature such that the deeper clays are transformed to 80% illite layers, and the shallower clays less completely. Such a model conflicts with the prevailing concept of a gradual illitization of smectite with progressive burial. Morton (1985) concludes that there is no evidence that clay diagenesis still acts in Oligocene sediments of the Frio formation. J.P. Morton's assessments set up *the question of the duration of diagenetically efficient temperatures in buried series,* and of the temperature interval necessary for allowing a significant increase of illitization process. Yau et al. (1987a) do not observe an appreciable change in the characteristics of a smectite hydrothermally treated at 200 to 460°C for 71 to 584 days. Ramseyer and Boles (1986) show that

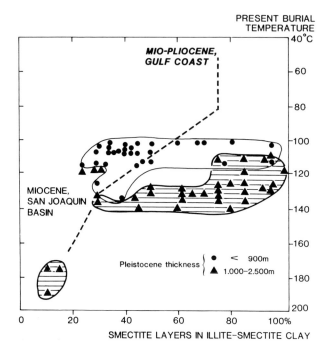

Fig. 15.4. Present-day burial temperature vs percentage of smectite layers in <2 μm illite-smectite from Miocene rocks of the San Joaquin basin, California. (After Ramseyer and Boles 1986). Gulf Coast reference curve from Perry and Hower (1972)

series buried less than 900 meters may experience clay diagenetic changes much more important than contemporary series buried more than 2500 meters (Fig. 15.4), if the residence time at critical temperatures is significantly larger. The same authors estimate that highly expandable smectite-illite changed to slightly expandable illite-smectite over a narrow temperature interval (10–20°C) acting during a sufficient time.

15.1.3 Other Cases

An extensive study of the diagenetic evolution of clay minerals during burial is performed by Dunoyer de Segonzac (1969) on two- to five-kilometers thick boreholes and sections from Cameroon, West Africa, France and Italy. One of the most relevant series studied consists of the Logbaba well in *Douala basin, Cameroon*, where G. Dunoyer identifies an increase of illitic layers in illite-smectite mixed-layers down to 4000 meter depth. This trend correlates with an increase of temperature (from about 40 to 150°C), of potassium content in the <2 μm fraction (from 1 to 6% K_2O), and of nonvolatile organic carbon compared to total carbon (carbon ratio from 0.1 to 0.6), as well as with a decrease in the cation exchange capacity (from 20 to 10 milliequivalent/100g). The smectite abundance suddenly decreases at 1200 meter depth and the mineral disappears near 1500 meters, which is hardly attributable to the diagenesis only (see Chap. 21). Chlorite, kaolinite, and irregular mixed-layers display noticeable quantitative variations along the borehole, which probably reflects transitional situations between strictly paleoenvironmental and stricly diagenetic constraints. From a general point of view, Dunoyer de Segonzac (1969) characterizes the *late diagenetic domain* by the *development of illite and chlorite, the absence of true smectite, the instability of kaolinite, and the frequent presence of allevardite* (regular illite-smectite mixed-layer). Smectite is considered as having been transformed either to chlorite in a magnesian environment with a corrensite-like transitional stage, or to illite in a potassic environment with an allevardite-like transitional stage. Kaolinite is supposed to recrystallize into dickite under acidic conditions, and to be destroyed and serve for chlorite and illite aggradation under basic conditions. The clay mineral transformations are irreversible and lead to the illite/chlorite facies.

Numerous other examples from deeply buried series point to a downward change from smectite-rich to more illitic clay assemblages, similar to what is described in Gulf Coast wells. This is the case with Mesozoic basins from Papua, New Guinea (van Moort 1971), of the Rhine graben in Germany (Heling 1978), the Northwestern Canada basins (Foscolos et al. 1976, Powell et al. 1978), and of Central Poland (Środoń 1984a). In Tertiary clayey silts of the Vienna basin, the less than 0.2 µm sedimentary fraction essentially contains mica, kaolinite, and illite-smectite (Johns and Kurzweil 1979). With increasing depth illite-smectite changes from a random mixed-layer phase with about 75% smectite layers to an ordered mixture with about 20% smectite layers. The illitization process does not involve mica and kaolinite, whose relative abundance remains constant throughout the series. *The potassium and aluminum needed for illitization* therefore *derive from coarser, nonsheet silicate sedimentary constituents, such as K-feldspars*. A similar conclusion arises from the study of the Precambrian belt supergroup, Clark Ford, Idaho, for the diagenetic domain (Eslinger and Sellars 1981). In the underlying metamorphic domain, the contribution of K-feldspars to further illitization appears to be much less important or even absent.

Middle Jurassic to Oligocene sediments of the northern North Sea undergo progressive *illitization* starting around 2000–2500 meter depth and being *stabilized around 3200–3800 meters at a smectite 20% – illite 80% stage* (Pearson et al. 1982, Pearson and Small 1988). Based on an average present-day geothermal gradient of 3°C/100 m, the temperatures at the onset of smectite evolution average 65-80°C and reach 100-200°C in the illite-smectite ordering zone. Clay ordering and dehydration occur in Paleocene – Eocene deposits in the northern part of the basin (Viking graben), and in older and more buried deposits when moving southwards. The level of organic maturity, located below the clay dehydration level, also deepens to the South. This trend probably results from late tectonic activity rather than from differences in geothermal gradient.

Cretaceous and Tertiary argillaceous sediments from 13 deeply drilled wells in Japan display a *parallel, nonsynchronous evolution of clay minerals, zeolites, and silica during burial diagenesis* (Aoyagi and Kazama 1980). Smectite is

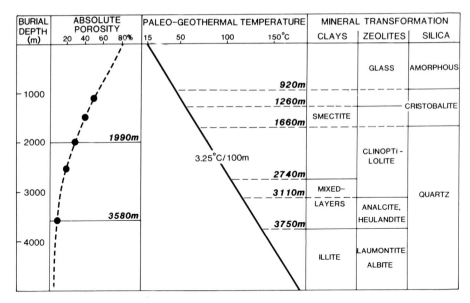

Fig. 15.5. Average relationships between depth of burial, porosity, temperature and mineral transformations in Neogene clayey sediments of Japan. (After Aoyagi and Kazama 1980)

replaced by smectite-illite mixed-layers beyond about 2700 meter depth, while illite-rich layers become abundant below 3750 meters (Fig. 15.5). The transformation from smectite to mixed-layers implies a pressure of approximately 900 kg cm^{-2} and a paleotemperature of about 100°C, and the transformation from smectite to illite 920 kg cm^{-2} and 140°C. The degradation of kaolinite requires much higher pressures and temperatures than for smectite. Zeolites proceed from the diagenetic transformation of volcanic glass, and show the following succession: clinoptilolite at 330 kg cm^{-2} and 60°C, heulandite and/or analcite at 860 kg cm^{-2} and 120°C, and laumontite with sometimes albitic plagioclase at 930 kg cm^{-2} and 140°C. The transformation from amorphous opal to low-cristobalite (250 kg cm^{-2}, 50°C) and then to quartz (660 kg cm^{-2}, 70°C) takes place at lower pressures and temperatures than for clays and zeolites.

Specific lithologies or geothermal gradients may modify or accelerate the basic clay evolution with increasing depth of burial. Cambro-Ordovician deposits drilled on the continental margin of western Newfoundland show clay suites successively dominated by smectite, corrensite and illite (early to middle Ordovician), by expandable chlorite and illite (early Ordovician), and by typical chlorite and illite (middle Cambrian to lowermost Ordovician. Suchecki et al. 1977). The two upper zones are attributed to the burial evolution of Mg-rich volcanic glass, while the lower zone is devoid of glass and is considered to contain mainly detrital minerals.

Plio-Quaternary shales of the Salton Sea region, northern end of the Gulf of California in the delta of the Colorado river, are submitted to a very high geothermal gradient (about 30°C/100 m) because of active crustal spreading. Detrital minerals undergo very rapid reactions from the earliest stages near the

Table 15.1. Weight % of some major minerals at increasing depths in shales from the Salton Sea geothermal field. (After Yau et al. 1987 b)

Depth of burial (m)	256	348	439	540	732	983
Temperature (°C)	115	150	200	220	260	300
% Smectite in illite-smectite	40	20	10	0	0	0
Illite, illite-smectite	38	36	46	53	47	17
Chlorite, kaolinite	6	5	6	7	9	10
Albite	2	3	7	3	5	15
K-Feldspar	0.2	1	1	1	2	4
Dolomite, ankerite	11	13	0	0	0	0
Quartz	32	34	27	25	27	37
Calcite	12	8	13	12	10	16

surface, through low-grade greenschist-facies metamorphism at less than 1 kilometer depth (Muffler and White 1969, McDowell and Elders 1980). Due to geothermal activity and probable hydrothermal exhalations, phyllosilicates progress through zones of illite-muscovite (115-220°C, < 540 m depth), chlorite (220-310°C, 1259 m) and biotite (>310°C). At increasing depths the original detrital phases (e.g.,smectite) appear to have reacted directly to precipitate secondary minerals without intermediate phases (Yau et al. 1988). The increase of illite layers in illite-smectite is very fast, the conversion being completed at about 220 meter depth (Table 15.1). Minerals newly formed beyond 200 meters include particularly albite, K-feldspar, and chlorite. Dolomite and ankerite seem not to be favored in Salton Sea shales (Yau et al. 1987b), whereas in other regions like the Central USA, some argillaceous carbonates appear to use the Mg released by smectite-to-illite conversion for producing postcompactional dolomite cements (McHargue and Price 1982).

The geothermal wells drilled in the Imperial Valley, also located in the Plio-Pleistocene delta of the Colorado River, allow a comparison of temperature and kinetic controls on the late diagenetic-early metamorphic evolution of clay minerals (Jennings and Thompson 1986). *Below approximately 175°C, the duration of reactions significantly influences the mineral evolution*, which is essentially characterized by the progressive transformation of illite-smectite to pure illite (70 to 210°C) and by the correlative decrease in the amount of detrital potassium-feldspar. *Temperature above 175°C*, rather than time, *appears to mainly govern the reactions*, that consist of the probable appearance of chlorite (180 to 194°C) and of the loss of kaolinite (210°C). Similar conclusions arise from Ramseyer and Boles's study (1986) on Tertiary sandstones and shales of the San Joaquin basin, California (Fig. 15.4). Sedimentary series of similar lithology, age and temperature (e.g., Miocene shales at 120-140°C) can present in distinct areas either high (>80%) or low (<20%) amounts of expandable layers, the less buried sediments often displaying the lower expandability. Such a result can only be explained by a *different resident time at high temperature*. Areas containing illite-smectite with high expandabilities present a time-temperature index much lower than areas marked by low expandabilites.

The comparison of clay mineralogy, organic matter characteristics and borehole measurements sometimes allows the reconstruction of past tempera-

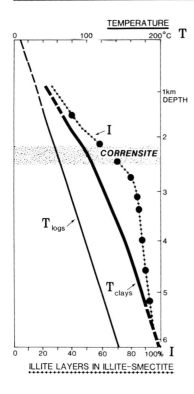

Fig. 15.6. Paleotemperature gradient as determined by clay mineral geothermometers (T clays) and present-day thermal gradient from incorrected borehole temperature (T logs). (After Pollastro and Barker 1986). Clay mineral temperatures from changes in illite-smectite ordering and composition, and from the appearance of corrensite. (After Hoffman and Hower, in Scholle and Schluger 1979)

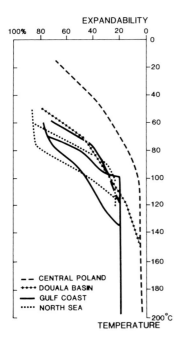

Fig. 15.7. Smectite to illite evolution in shales from different sedimentary basins, as a function of temperature. (After Šrodoń and Eberl 1984)

tures and post-diagenetic history of a given region. Pollastro and Barker (1986) investigate the 1500-5500 meters sedimentary interval of the Green River basin, Wyoming. Late Cretaceous to early Tertiary rocks were drilled in the Pinedale anticline. Paleotemperatures calculated from changes in the composition and ordering of illite-smectite and in the mean vitrinite reflectance reached 200°C at 5500 meters. The present-day temperature measured at this depth is about 135-150°C (Fig. 15.6). These data suggest that the maximum temperature responsible for the inorganic and organic evolution was about 50°C higher than present-day borehole temperature. The apparent temperature decrease can be explained by the erosion of about 1700 meters of the section, a fact documented by vitrinite reflectance data and implying that the geothermal gradient has remained constant since maximum burial. The erosion itself resulted from uplift of the Pinedale anticline during the Neogene. Notice that the fluid homogenization temperature from aqueous inclusions in quartz and calcite veins at 5200 meters conforms to the present temperature regime, indicating that mineralization processes occurred after the uplift and erosion took place.

To summarize, the most systematic evolution recorded in series buried more than 2 kilometers and submitted to temperatures higher than 60 to 80°C *consists of the progressive illitization of smectite minerals* (Fig. 15.7). Subregular illite-smectite mixed-layers and allevardite-like minerals usually develop above 100°C in series buried more than 3 kilometers, but higher temperature may allow a clay ordering in much less deeply buried sediments (Fig. 15.8). The last stages of clay

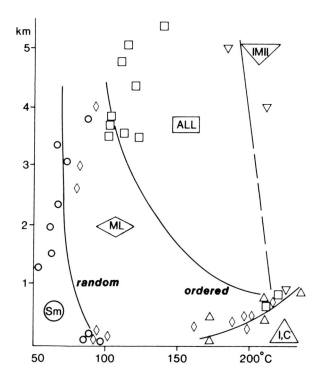

Fig. 15.8. Depth-temperature plot of clay mineral characteristics from deep boreholes and wells in hydrothermal areas. (After Velde 1985). *Sm* 70–100% smectite in illite-smectite; *ML* 40–60% expandable layers in ordered illite-smectite; *All* allewardite, superstructure phase, regular illite-smectite; *IMII* <10% expandable layers, superlattice ordered group; *I,C* illite, chlorite facies

mineral evolution under increasing thermodynamical constraints are characterized by the superlattice ordering defined by Reynolds and Hower(1970; Type IISI, three illite and one smectite layers), and by the illite/chlorite facies. The latter facies, that appears to depend on high temperatures more frequently than on high pressures, correlates with an increase of mineral grain size and crystallinity.

15.2 Mechanisms of Evolution

15.2.1 Open Versus Closed Systems

First studies on deeply buried shales suggested that some cations like potassium and magnesium, necessary for allowing the evolution of smectite to illite or chlorite, were supplied through long-distance transportation processes. For instance, Weaver and Beck (1971) proposed from calculations on mineralogical and chemical budgets that K and Mg were added to diagenetic minerals by fluids migrating upwards along the sedimentary column. By contrast, most subsequent investigations concluded to *short-distance migrations* only, the elements needed for illite formation being provided by smectite itself and by adjacent K-feldspars and micas. For instance, Hiltabrand et al. (1973) performed experiments on recent marine muds of the Louisiana coast, whose clay composition is similar to that of Gulf Coast recent series. Sediments immersed in sea water at 100 and 200°C display the formation of illitic and chamositic minerals, and the diminution of feldspars and expandable mixed-layers. This shows that a clay-mineral evolution comparable to that observed in Gulf Coast Tertiary sediments can occur in a closed system without addition of new ions. Aronson and Hower (1976) observed a decrease with depth of whole-rock apparent K-Ar ages, which confirms the *isochemical nature of the diagenetic transformation* and also suggests a closed chemical system.

The dominance of very short-distance exchanges in shaly sediments is confirmed by high-resolution transmission electronmicroscopy, that shows the growth of small packets of illite layers within subparallel layers of smectite matrix (Ahn and Peacor 1986. 15.2.2). Even at 5500 meters non-equilibrated assemblages of relatively reactive clay minerals exist. The coexistence, at a distance of a few tens of angströms, of smectite and illite layers and of structural imperfections implies that ionic homogenization is very limited. As water is the principal medium for allowing ion transport, this means that diagenetic sediments have not been affected by important water migration, at least not since their initial compaction. Ahn and Peacor (1986) conclude that *the argillaceous Gulf Coast deposits behaved essentially as a nearly closed system, water being present in very small amounts compared to solid materials, and acting as a catalyst for short-distance ion transport and for local clay reconstruction at the reaction interfaces.*

By contrast to Gulf Coast shales, *open systems may occur in more permeable or more active geothermal environments.* This is the case of the shales from the *Salton Sea geothermal field* (Yau et al. 1987b). Individual, euhedral to subhedral illite crystals develop in open pore space during burial diagenesis, and probably result from a direct crystallization from solutions. The components derive partly by dissolution of detrital smectite, whose abundance decreases with increasing depth. As smectite is not observed by transmission electronmicroscopy in the immediate proximity of neoformed illite, the ions necessary for illite growth issue at least partly from relative distant sites. This supposes a high relative water/rock ratio, high proportions of pore space, and active flow of solutions. In the shallower sediments of Salton Sea geothermal field detrital smectite is also observed to be in part directly replaced by illitic layers, similarly to what happens in nearly closed microenviromnents of Gulf Coast sediments (Yau et al. 1987b). This points to the existence of *intermediate situations* between closed and open systems. Pollastro (1985) describes another transitional case in two shale-to-sandstone series of Wyoming and Colorado, Rocky Mountains. Late Cretaceous and early Tertiary deposits display both individual illite laths and illite-smectite mixed-layer clays, whose abundance simultaneously increases downwards. This implies the probable coexistence of short-distance and larger-distance ion migrations.

Other evidence for the mainly local character of chemical exchanges in deeply buried sediments arises from the existence of *local microenvironments where specific mineral formations take place*. Pye et al. (1986) perform backscattered elctronmicroscopic observations and energy-dispersive X-ray analyses on Gulf Coast CWRU no.6-well sediments already studied by Hower et al. (1976). Above 2750 meter depth, calcite foraminifera tests are partly filled with authigenic pyrite, kaolinite and chlorite, with some unoccupied pore space. Such new minerals do not occur in other shale constituents adjacent to the tests. Between 2750 and 3400 meters the remaining voids are filled with authigenic chlorite or illite, while the calcareous tests themselves are dissolved. In the same depth interval iron-rich chlorite replaces zoned carbonate rhombs formerly formed in sulfate reduction and fermentation zones. K. Pye and colleagues conclude the usefulness of foraminifera tests and their authigenic mineral infillings as indicators of diagenesis in soft, fine-grained shales. In a further study of Texas Gulf Coast sediments, Burton et al. (1987) confirm the *local authigenesis of kaolinite and chlorite in different pore environments marked by specific textures and chemical characters*. In all occurrences kaolinite appears to have formed before chlorite, both minerals probably being able to have precipitated directly from solution. Primmer and Shaw (1987) also report the existence of *microenvironmental diagenetic changes superimposed to the general illitization of mixed-layer clays with depth*, in both Gulf Coast Tertiary and North Sea Mesozoic shales. In Gulf Coast sediments the neoformations concern mainly kaolinite and chlorite. In North Sea deposits a large variety of diagenetic reactions occurs in different microenvironments, involving phyllosilicate, feldspar, carbonate and sulfide phases. The types of secondary minerals identified depend primarily on the nature of original sediments in relation with differences in detrital sources, and accessorily on the depth of burial.

15.2.2 Transformation Versus Solution-Precipitation

Illitization

Transformation through mixed-layering. Several mechanisms have been put forward to explain the progressive passage from smectite-rich to illite-rich minerals at increased depths of burial. J. Hower and colleagues (in Hower et al. 1976) first proposed a precise mechanism of *mineral transformation* for explaining the apparent increase in the proportion and ordering of illite layers at depth. The destruction of detrital potassium-feldspars and perhaps of mica was suggested to provide the potassium and aluminum required for the progressive transformation of smectite to illite. The substitution of octahedra-derived Al^{3+} for Si^{4+} in the tetrahedral layers of smectite was envisaged to be responsible for the fixation of K^+ in the interlayer positions by increasing the layer negative charge. The correlative release of Si^{4+}, Fe^{2+} and Mg^{2+} was supposed to participate in forming other new minerals like quartz and illite.

*Transformation with mixed-layering and cristallization.*Pye et al. (1986) observe that main depth-controlled mineralogical changes occur in the zone of organic matter decarboxylation. The morphological, mineralogical and chemical variations recorded by punctual microscopic and geochemical analyses lead these authors to propose a five-step illitization process, starting in the decarboxylation zone: (1) reduction of pH due to decarboxylation, and simultaneous iron reduction; (2) carbonate dissolution; (3) dissolution of detrital K-feldspars and micas; (4) appearance of negative octahedral layer charge in mixed-layer clays, due to reduction of Fe^{3+} to Fe^{2+}; (5) fixation of interlayered K^+ and substitution of Al^{3+} for tetrahedral Si^{4+}, the supplementary K and Al being derived from non-clay detrital silicates. In addition, a direct precipitation of illite and chlorite is reported in pore space, as well as a replacement of kaolinite by chlorite. K. Pye and colleagues therefore envisage the *coexistence of transformation and dissolution-precipitation processes*, the former being dominant and responsible for the downward decrease of expandability in illite-smectite layers. A similar combination of mechanisms occurs in Plio-Quaternary sediments of the Salton Sea geothermal field, where solution-precipitation reactions dominate transformation reactions, particularly in the more deeply buried sediments subject to high water/rock ratios (Yau et al. 1987b). *The respective importance of each type of process appears to largely depend on the sediment permeability and water flows,* which also closely influences the importance of short-distance versus larger-distance exchanges (15.2.2).

An alternative mechanism of *illitization, intermediate between transformation and dissolution-precipitation*, is referred to as "*smectite cannibalization*" and documented from field studies and laboratory experiments by Boles and Franks (1979), Inoue (1983) and Pollastro (1985). For instance R.M. Pollastro investigates the late Cretaceous-early Tertiary interbedded sandstones and shales in the Green River basin, Wyoming, as well as the late Cretaceous chalks to chalky shales of the Niobrara Formation, Colorado. Scanning electronmicroscopic observations on deeply buried rocks reveal the presence of illite-rich, ordered illite-smectite minerals growing on a substrate with a typically smectitic morphology. These textural characteristics suggest the *combined inheritance and solution-recrystallization of smectitic minerals*, with only slight chemical modification. This process partly resembles the growth of smectite laths at the periphery of smectites flakes sometimes recorded during early diagenetic history (14.2.2), but here involves a real mineral change from smectite to illite. Pollastro (1985) envisages that a part of illite-smectite clay is destroyed under high temperature conditions, through a selective cannibalization of smectite layers. This partial, structural destruction would provide some or all of the chemical elements needed to build illite-smectite with a higher illite content. As a result, the number of clay layers in a given particle would tend to decrease in the course of time and with increasing depth, by progressive destruction of smectitic layers and correlative feeding of illite layers. The initial Al and K would serve primarily for illitization, while Si, Ca, Fe and Mg could participate in the formation of discrete illite, chlorite, quartz and dolomite.

Dissolution-precipitation without mixed-layering. An exclusive solution-recrystallization process for burial illitization is proposed by Nadeau et al. (1984a and b, 1985). The question was first posed by McHardy et al. (1982), who observed that a mineral displaying on X-ray diffraction diagrams an illite-smectite pattern looked on electronmicrographs like dioctahedral illite made up of very thin particles (about 3 nanometers thick). The explanation proposed was that smectite interlayers of the "illite-smectite" minerals were in fact the spaces existing between individual platy or lathed crystals of very thin illite. *Within the areas where such small illite crystals overlap, ethylene-glycol or water can be adsorbed and swelling can take place as within true smectite interlayers.* W.J. McHardy and colleagues therefore suggested that *interparticle planar contacts may act as smectite interlayers*, even if the samples studied do not contain true smectite particles. These ideas are tested on <0.1 μm-sized fractions by P.H. Nadeau and colleagues, who experimentally produce a wide range of materials that yield X-ray diffraction patterns typical of mixed-layer minerals simply by making mixtures of elementary particles of different minerals in various combinations and proportions. Nadeau et al. (1984a, b) think that many common mixed-layered clay minerals really consist of the superimposition of a few very thin particles of a single mineral species, that behave on X-ray diffraction diagrams like mixtures of this mineral and of smectite. A sedimented aggregate of 2 nm elementary illite particles would appear as regular illite-smectite, and of 2-4 nm elementary chlorite as corrensite.

As a direct consequence, *the illite-smectite mixed-layers identified during burial evolution of smectitic shales are considered to be in fact extremely small illite particles that display interparticle diffraction effects* (P.H. Nadeau). These small and thin illite particles are believed to result from the dissolution of detrital smectite followed by a recrystallization. According to the *interparticle* diffraction concept of intertratified clays (Nadeau et al. 1985), random interstratified illite-smectite is composed of one to three elementary illite and smectite particles (Fig. 15.9). When smectite particles are entirely dissolved, the remaining population would essentially consist of two to five elementary illite particles, that appear through X-ray diffraction to be regularly interstratified illite-smectite with at least 50% illite particles. With increasing depth of burial and diagenesis, the thickness of fundamental illite particles increases within the population. When

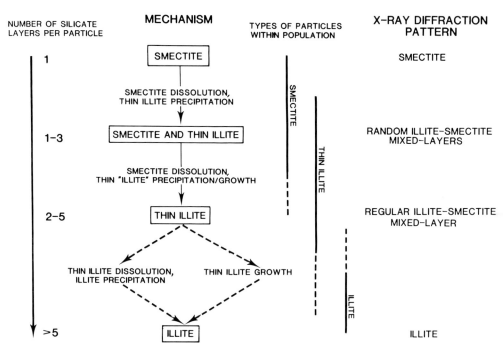

Fig. 15.9. Neoformation mechanism for the conversion of smectite to illite, after the interparticle diffraction concept. (After Nadeau et al. 1985)

the particles become thick enough to allow dominant *intraparticle* diffraction effects (more than five stacked elementary layers), the clay material is identified by X-ray diffraction as illite. P.H. Nadeau and colleagues therefore think that *the diagenetic conversion of smectite to illite constitutes a recrystallization mechanism, where smectite dissolution and illite precipitation occur within a population of extremely small crystals without intermediate stage.* Nadeau's concept is particularly supported by Corbato and Tettenhorst's calculations (1987), that suggest the rarity of illite-smectite clays among all crystalline materials, since their X-ray diffraction size is equal to or larger than the size of their constituent physically separable particles. Wilson and Nadeau (1985) suggest that the interparticle concept should be tested in environments other than diagenetic ones, especially in weathering and soil continental profiles.

Transformation without mixed-layering. Ahn and Peacor (1986) study by high-resolution transmission and analytical electronmicroscopy three argillaceous sediments located at 1750, 2450 and 5500 meter depth in the Gulf Coast CWRU no.6 well already investigated by Hower et al. (1976). Punctual chemical analyses show distinct compositions for smectite and illite (Table 15.2), and support J. Hower's interpretation on the release of Si from smectite and the uptake of external Al and K. *Smectite particles present a more variable chemical composition than illite ones, which probably results from the diversity of detrital smectites supplied within the basin during sedimentation, and from the homogenization of illitic minerals through burial diagenetic history.*

Table 15.2. Average analytical electronmicroscopic analyses of illite, smectite and chlorid minerals form Gulf Coast buried sediments. (After Ahn and Peacor 1986, 1987b). Range values in brackets

	Smectite 1750 m	Illite 5500 m	Chlorite 5500 m	7–14 Å mixed-layer chlorite 5500 m
Tetrahedra				
Si	3.84 (3.66–3.88)	3.40 (3.31–3.48)	2.80 (2.70–2.90)	2.70 (2.70)
Al	0.16 (0.12–0.34)	0.60 (0.52–0.69)	1.20 (1.10–1.30)	1.30 (1.30)
Octahedra				
Al	1.41 (1.16–1.74)	1.69 (1.46–1.85)	1.50 (1.40–1.70)	1.40 (1.30–1.60)
Fe^{2+}	0.34 (0.21–0.63)	0.19 (0.10–0.28)	2.80 (2.60–3.00)	3.20 (3.00–3.30)
Fe^{3+}	0.34 (0.21–0.63)	0.19 (0.10–0.28)	0.30 (0.30)	0.40 (0.30–0.40
Interlayers				
Mg	0.25 (0.04–0.31)	0.12 (0.00–0.29)	1.10 (1.00–1.20)	0.80 (0.80–0.90)
K	0.14 (0.00–0.32)	0.63 (0.52–0.72	–	–
Na	0.17 (0.00–0.25)	0.01 (0.00–0.06)	–	–

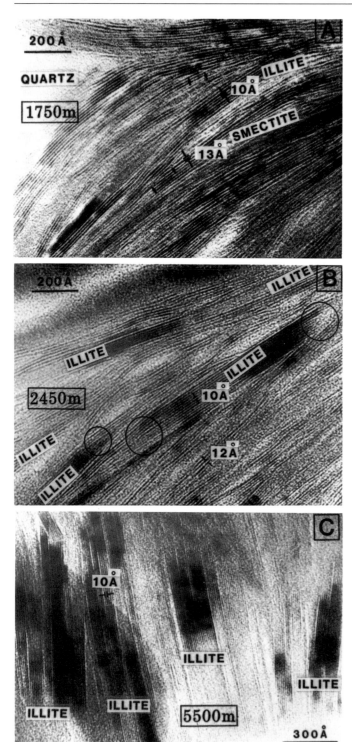

The micromorphological study of thin sections whose plane is normal to the clay layering provides spectacular new informations (Fig. 15.10). Smectite presents a texture of wavy 10- to 13-Å layers with a high density of edge dislocations, while illite is made up of relatively defect-free straight 10-Å layers. The microstructures of smectite and illite are not continuously parallel to (001) planes at smectite-illite interfaces. The relations between both mineral species are therefore characterized by the *widespread occurrence of chemical, textural and structural discontinuities, and by the presence of separate discrete domains of either smectite or illite. These data suggest the absence of true mixed- layered clay minerals,* the random or regular alternation of illite and smectite layers being replaced by small packets of either illite or smectite texturally discontinous. Similar results arise about the depth-zoning of phyllosilicates in geothermally altered shales of Salton Sea, California (Yau et al. 1988). Notice that Ahn and Peacor's observations (1986) contradict Nadeau et al.'s assessments (1984a and b, 1985) on the interparticle diffraction concept, since high resolution microscope observations show that illite packets are usually thicker than five layers (instead of $\leq$ 3; see Fig. 15.9) in the 2450 meter depth-sample which gives a X-ray diffraction pattern of randomly interstratified illite-smectite mixed-layer. Similarly, illite packets are thicker than ten layers in the 5500 meter-depth sample whose diffraction pattern indicates an ordered illite-smectite with 80% illite layers; at this level, illite packets should contain $\leq$ 5 layers according to Nadeau's concept (1985. Fig. 15.9). Ahn and Peacor (1986) suggest that the discrepancy between both types of data could result from experimental causes, P.H. Neadeau's observations being based on size-fractionated samples (<0.1 μm) instead of unsegregated original shale samples. The indirect observations could have provoked the disarticulation of original packets of illite and smectite to thinner units even down to a thickness of one or two layers, presumably rearranged during specimen preparation and possibly giving artificial interlayer stacking relations of illite and smectite.

Whatever the real cause of discrepancy between direct and indirect electron-microscopic observations, both Nadeau (1985) and Ahn and Peacor (1986) refute the existence of true illite-smectite mixed-layers in buried samples whose X-ray diffraction patterns suggest their presence. The mechanism involved for illitization nevertheless strongly differs in both cases. Instead of a dissolution-precipitation process, J.H. Ahn and D.R. Peacor propose a *transformation process* acting in the following way (Fig. 15.11):

1. *Initial smectite contains anastomosing layers with variable thickness and abundant dislocations,* forming continuous structural units referred to as megacrystals. *By contrast, illite consists of well-defined, straight and relatively*

Fig. 15.10. High-resolution transmission electron micrographs of Gulf Coast buried sediments. (After Ahn and Peacor 1986). Lattice fringe images. **A** 1750 m-depth. Discontinuous and wavy smectite layers having variable orientation. Edge dislocations indicated by arrows. **B** 2450 m-depth. 50–100 Å packets of illite occurring between subparallel smectite layers. **C** 5500 m-depth. Abundant, thick and subparallel packets of illite, locally coalescing

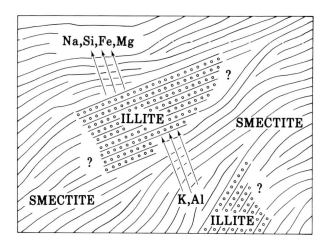

Fig. 15.11. Interpretation of the direct transformation of smectite to illite in Gulf Coast argillaceous sediments, from high resolution transmission electron microscopic observations. (After Ahn and Peacor 1986)

defect-free layers arranged in packets within the subparallel imperfect smectite matrix.

2. *Layers of illite grow discontinuously with layers of smectite in the along-layer direction.* This growth is determined by the *diffusion of K and Al through the smectite matrix to the transition boundary*, and by the diffusion of Na, Si, Fe and Mg away this boundary. *The abundant defects in smectite are thought to serve as pathways for ion transport.* The transformation probably involves the partial breakdown of both octahedral and tetrahedral sheets, facilitated by the presence of interstitial and interlayered water. As a consequence, chemical arrangements occur in both octahedral and tetrahedral sheets and lead to relatively defect-free illite layers.

3. As the reaction progresses, *the boundaries of illite packets advance until they coalesce*, as indicated on high-resolution electronmicrographs. Because illite inherits the layer orientation of the smectite where it growths, and because initial smectite layers are divergent, layers of illite packets become subparallel where they intersect, resulting in small-angle grain boundary-like features (Ahn and Peacor 1986). The final texture at 5500 meter depth is dominated by thick illite packets with subparallel orientations.

J.H. Ahn and D.R. Peacor's observations provide a convincing example of an *illitization mechanism devoid of both true mixed-layer stages and complete smectite dissolution*. This model, based on the observation of a few samples in a shaly, almost closed environment, represents a useful tool to be tested in sediments with different lithologic and thermodynamic characteristics. Ahn and Peacor's model is supported by tests on layer-by-layer mechanisms of smectite illitization, assuming a solid-state transformation (Bethke and Altaner 1986). Notice that Bell (1986) also describes coherent smectite and illitic crystallites in South Texas Gulf Coast and New Mexico sediments, which conflicts with P.H. Nadeau and colleagues's concept. In addition, T.E. Bell observes on high-resolution electronmicrographs true mixed-layered spacings at 17 to 28 Å, suggesting the common occurrence of individual crystallites of true illite-smectite mixed-layer minerals.

Chloritization, Kaolinization

Chlorite minerals. In the Douala basin, Cameroon, chlorite is supposed to result from the downward transformation of smectite in a magnesian environment (Dunoyer de Segonzac 1969). In *Gulf Coast sediments*, chlorite is not abundant and its origin has been little discussed through first investigations. Beyond burial depths of 2500 meters chlorite is lacking in clay assemblages from shales or is very rare (Hower et al. 1976), except in some restricted microenvironments like foraminiferal test infillings (Pye et al. 1986). Chlorite occurs in shaly facies below 2500 meters and increases in abundance down to 3700 meters, below which it remains almost constant. Hower et al. (1976) envisaged that a part of magnesium and iron released during the transformation of smectite to illite could have favored the formation of diagenetic chlorite. Other authors suggested that kaolinite is a major source of Al and Si for chlorite formation during burial diagenesis (e.g., Muffler and White 1969, Perry and Hower 1970, Boles and Franks 1979).

Ahn and Peacor (1985) observe by high-resolution electronmicroscopy the texture, structure and chemical composition of chloritic minerals at 2450 and 5500 meter depth in Gulf Coast CWRU no.6-well sediments. *Chlorite occurs in small amounts as 100-150 Å thick packets intergrown with and semi-coherent with respect to the smectitic layers* in the 2450 meter samples. At 5500 meters, chlorite is more abundant and presents thicker packets. The chlorite packets are sub-parallel to each other and vary in thickness. The composition corresponds to a *Fe-rich, trioctahedral chlorite*, with moderate amounts of Mg only (Table 15.2). The average chemical formula, that resembles brunsvigite, is the following:

$$Fe_{3.1} Mg_{1.1} Al_{2.7} Si_{2.8} O_{10} (OH)_8.$$

The richness of iron relatively to magnesium in Gulf Coast diagenetic chlorite correlates with chemical data of illite-smectite clays (Hower et al. 1976) that show a greater decrease in Fe than in MgO with increasing depth and illite proportion. This chemical change suggests that smectite layers lose more Fe than Mg during smectite-to-illite conversion. Ahn and Peacor (1985) therefore envisage that Fe and Mg from smectite represent the major source for chlorite formation. This interpretation agrees with the fact that chlorite layers are intergrown parallel to layers of surrounding illite-smectite. *Chlorite therefore appears as a direct byproduct of the smectite-to-illite transformation, utilizing iron and magnesium released from smectite, diffused at proximity and reprecipitated* together with silicon supplied from smectite or other detrital silicates. The formation of a trioctahedral chlorite from mainly dioctahedral smectite particles implies a strong *structural reorganization*, involving more drastic disruptions in octahedral and tetrahedral layers than for the smectite- to-illite conversion. Notice that J.H. Ahn and D.R. Peacor identify beside new chlorite particles few amounts of 7 Å berthierine and 7-14 Å mixed-layered chlorite, whose chemical composition closely resembles that of coexisting chlorite with a slightly higher iron content (Table 15.2). *Berthierine minerals are considered as metastable precursors of chlorite*, and are suggested to represent diagnostic markers of the diagenetic environment.

Kaolinite. Beside detrital particles (Hower et al. 1976) and secondary test infillings (Pye et al. 1986), kaolinite is reported in Gulf Coast sediments as thin packets of layers contained exclusively within a matrix of smectite or illite (Ahn and Peacor 1987). High-resolution transmission electronmicrographs show that kaolinite packets are commonly interstratified with illite packets, and generally extend subparallel to the surrounding illite or smectite layers. With increasing depth of burial, the interfaces between kaolinite and interstratified illite tend to be more coherent. Where kaolinite layers terminate, they are discontinuous and do not show any transition with illite layers. All these textural relations suggest that *kaolinite and illite crystallized simultaneously at the expense of smectite.* Such a diagenetic origin for kaolinite is further supported by the increase in the thickness of the mineral packets with depth, although the relative abundance of kaolinite remains low.

In contrast to the smectite-to-illite conversion, micromorphological observations imply that *kaolinite formation probably results from a dissolution- crystallization mechanism* rather than from a transformation. The absence of structural and textural continuity between kaolinite and smectite supposes the destruction of major structural units of the preexisting particles (i.e., the combined octahedral and tetrahedral smectite sheets). Eberl and Hower (1977) already envisaged that some smectite layers could dissolve to produce the elements necessary for the development of both illite and kaolinite. The cause of the formation of kaolinite together with illite, instead of illite alone, is still little documented. Ahn and Peacor (1987b) suggest, on the principle of a limiting factor (Hoffman and Hower 1979), that kaolinite formation could be favored by a depletion in potassium availability within the diagenetic microenvironment.

15.3 Modification of Physical Properties

15.3.1 Compaction

The marine clays accumulated in *surficial sediments* usually display flocculated domains of particles arranged in a edge-to-face, *house of cards, random pattern.* This is the case with most recent coastal, marginal and deep-sea deposits (e.g., Rieke and Chilingarian 1974, Ferrell 1987). *With increasing depth of burial, most clayey sediments progressively change from a random-oriented to a more or less oriented fabric.* For instance, the Mississippi delta sediments display between the mudline and 85 meter depth three successive stages (Bennett et al. 1981): (1) edge-to-face and face-to-face contacts of particles with relatively high-void ratio (> 2.5); (2) particle-to-particle packing, with medium-void ratio (1.5–2.5) and coexistence of dominantly random-oriented domains and of chains; (3) oriented domains of particles associated with thin, long voids and very low-void ratio (< 1.5). Similar results arise from the study of DSDP site 515 in the southern Brazil basin, where a preferred particle orientation starts at about 50 meter depth, when porosity is reduced to about 61% and pressure reaches 6 kg/cm^2

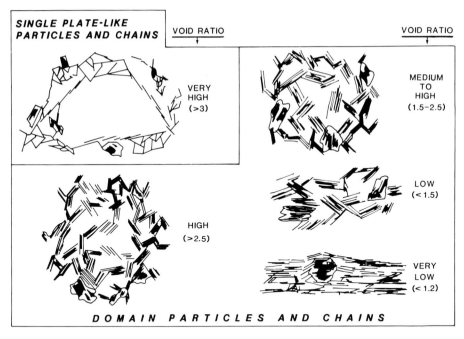

Fig. 15.12. Model of clay fabric evolution with increasing depth of burial. (After Bennett et al. 1981)

(Faas and Crocket 1983). Bennett et al. (1981) propose a model of clay fabric that accounts for the main physical changes experienced by clay particles submitted to an increasing sedimentary load (Fig. 15.12). Of course at such burial depths clay minerals are not affected by chemical diagenesis, and display reorganizations of particles only, not of layers (see Chap. 14).

With increasing depth of burial and appearance of chemical modifications, the clay layers tend in turn to present physical reorganizations (see Rieke and Chilingarian 1974). Powers (1967) already suggested that during burial diagenesis and compaction of mudrocks the "open" structure of smectite transformed first to give free pore water and denser illite, and second to provoke a greatly reduced volume when water is expelled and illite itself compacted. Ahn and Peacor (1985, 1986, 1987b) show from high-resolution electronmicroscopic observations that the conversion of smectite to illite, and the related chlorite and kaolinite formation, tend to determine a *parallel-to- subparallel orientation of newly formed clay layers* (15.2.2). Keller et al. (1986) document the parallel changes of illite-smectite morphology and of the proportion of expandable layers in deeply buried smectites and hydrothermally altered volcanic rocks from Japan. The most clearly marked morphological change in the smectite-to-illite conversion occurs in the range 45-30% expandable layers, which correlates with the transition from random to oriented mixed-layering on X-ray diffraction diagrams. In this range the clay fabric changes from a sponge-like or cellular to a platy or ribbon-like arrangement. At the same time *a modification in layer stacking occurs, from turbostratic to rotational ordering.* The rotationally ordered structure, that leads to an almost precise juxtaposition of quasihexagonal oxygen surfaces from adjacent layers, favors a more crystalline regularity and a more plate-like or sheet-like habit.

15.3.2 Application to Oil Exploration

The physical, mineralogical and chemical modifications experienced by clay minerals during burial diagenesis determine fundamental changes in the water and organic matter content, and in the potential of fluid migration. The nature and abundance of clay minerals in a given rock may significantly affect its reservoir or cap-rock properties. Clay minerals are therefore considered since a long time as useful tools in search for oil (e.g., Weaver 1960, Burst 1969, Sarkisyan 1972). A huge amount of investigation is being continuously performed on the subject with emphasis applied to petroleum generation, exploration and production (e.g., approximately 20,000 references for the period 1978–1986, Roaldset 1987). New techniques develop through both borehole and laboratory investigations in order to better understand the control of burial constraints on organic-inorganic relations (e.g., Clay Minerals 1986, 21, 4; Bell 1986, Herron 1986). Clay research concerns either argillaceous sediments possibly acting as parent rocks and transitory reservoirs, or sandy rocks serving as final reservoirs where clay coatings or infillings can take place (Chap. 16). In addition to information on the petroleum history, clay mineral studies may supply data on the paleogeographic evolution of oil-related basins, which may possibly facilitate further exploration.

Argillaceous materials submitted to increasing depth of burial fundamentally experience a change in fabric due to compaction (15.3.1). This progressively tends to destroy the porosity and permeability down to a certain depth (generally about 3 km). Beyond this depth the chemical environment often becomes corrosive; some minerals dissolve with consequently the development of a secondary porosity (Schmidt and McDonald 1979). Such an opposite evolution with increasing depth mainly characterizes sandy rocks subject to acidic groundwater migration (Curtis 1987a), but may also affect more argillaceous series marked by relatively high pore pressures and water flows. Surdam and Crossey (1987) indicate that main diagenetic reactions involving inorganic constituents in sediments correlate with the maximum concentrations of organic solvents (Fig. 15.13). Such a correlation emphasizes the *interdependence existing between mineral diagenesis and petroleum history.*

The late diagenetic evolution of clay minerals consists essentially of the conversion of smectite to illite. This evolution is accompanied by a water explosion from smectite interlayers. As early as 1967, Powers suggested that *release of water during illitization is important in flushing out hydrocarbons from source rocks and transporting dissolved ions for the dissolution- precipitation of cements in adjacent reservoir sandstones.* In addition, the water expulsion could determine *overpressuring and hydro-fracturing* in the rocks submitted to clay diagenesis; however, the increase in fluid pressure could inhibit further dehydration until the water released through illitization migrates out of the rock (Colten-Bradley

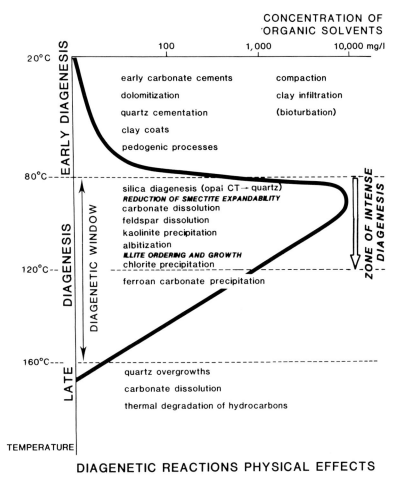

CONCENTRATION OF
ORGANIC SOLVENTS

Fig. 15.13. Correlations between diagenetic reactions and organic acids versus temperature, in oil field waters. (After Surdam and Crossey 1987)

1987). In fact, the role of water as a carrier of organic and inorganic materials within the rocks depends on rock permeability but also on geothermal gradient, pore pressure, interlayered water density, geological age, etc. Burst (1969) and Perry and Hower (1972) proposed a *shale-dewatering* history marked by two or three stages succeeding one another during burial (Fig. 15.14). After initial connate-water loss due to compaction, E.A. Perry and J. Hower suggested the existence of two stages of dehydration related to the clay-mineral evolution: one stage would correspond to a random collapse of smectite layers; the other, deeper stage would correlate with the transition from random to ordered interlayering. The clay dehydration stages apparently develop at temperatures higher than 80-90°C and characterize burial thicknesses of more than 2 to 3 kilometers. The importance of water release from clay for oil formation is proved by the *frequent coincidence observed in the decrease of smectitic expandable layers and the*

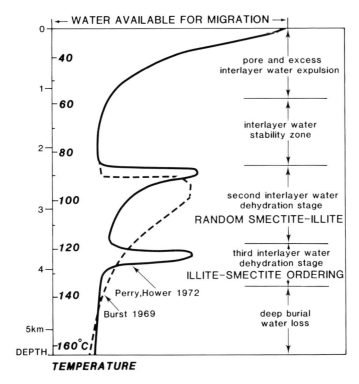

Fig. 15.14. Water escape curves from smectitic sediments during illitization upon burial

hydrocarbon appearance. Edman and Surdam (1986) observe that the diminution of expandable clay minerals correlates with a pulse of organic acids and the subsequent formation and migration of most liquid hydrocarbons, in late-Cretaceous Ericson sandstones from Green River basin, Wyoming. Odigi (1986) reports in Tertiary sediments from the eastern Niger delta the synchronous loss of smectite, increase of iron and hydrocarbon occurrence, over the 3250 to 3720 meter-depth interval where illite-smectite varies from 32 to 82% illite layers. Colten-Bradley's calculations (1987) indicate the coincidence occurring between the temperature range for interlayer water loss by smectite under differential pressure conditions, and the beginning of the smectite-to-illite conversion and hydrocarbon generation (see also Surdam and Crossey 1987, Pearson and Small 1988).

   The presence of abundant smectite in sedimentary rocks submitted to burial effects appears to noticeably favor the production and migration of hydrocarbons. First smectite interlayers may *incorporate huge amounts of organic products* that constitute potential precursors for hydrocarbons. E. Roaldset (pers.commun.) calculates that one metric ton smectite could theoretically adsorb as much as 550 kg proteins, 250 kg polypeptids, 200 kg carbohydrates or 150 kg amino acids. Second smectitic clays act as important *water reservoirs*, that can provide through diagenesis the carrier necessary for hydrocarbon migration (Powers 1967). In addition, the water adsorbed in smectite interlayers may help in *resist-*

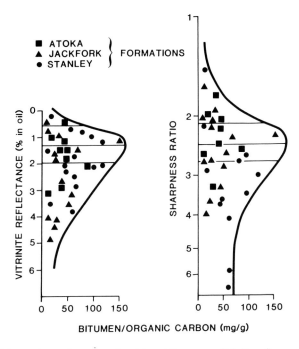

Fig. 15.15. Relationships between bitumen/organic carbon (mg/g) and the mean vitrinite reflectance, the illite crystallinity. (After Guthrie et al. 1986). Illite crystallinity calculated from Weaver's (1960) sharpness ratio = diffractogram peak height at 10 Å divided by peak height at 10.5 Å. Horizontal lines designated as onset (*1*) and end (*2*) of oil generation, and end of oil preservation (*3*). (Tissot and Welte 1978)

ing burial compaction and permeability decrease of clays and other sediments; this could explain why smectite-bearing calcareous rocks commonly exhibit higher porosity than kaolinite-containing rocks (Aoyagi and Chilingarian 1972). Finally, smectite constitutes a possible *catalyst for decarboxylation*, a very important step in petroleum generation. Johns and Shimoyama (1972) experimentally demonstrate that the presence of expandable minerals facilitates the liquid-hydrocarbon cracking reactions; smectite catalyzes the decarboxylation of fatty acids and the production of long-chain alkanes. Taguchi et al. (1986) estimate that the thermocatalytic effect of the smectite-to-illite conversion is responsible for changes occurring in the lipid materials, the fixation of nitrogen and the carbon isotope fractionation in Neogene rocks from Northeastern Japan. Crossey et al.(1986) observe a coincidence in time, temperature and space of smectite-to-illite conversion with peak concentrations of organic acids in Texas Gulf Coast oil field brines. This suggests that the mineral evolution favors the generation of disfunctional carboxylic acids. Such acids could pass through sandstones adjacent to shales just prior the hydrocarbon generation and allow an increasing permeability of reservoirs by dissolving carbonates and some aluminosilicates.

Notice that conversely to the organic-maturation control by clay minerals, *the organic matter possibly acts as a catalyst in clay dissolution processes*, which emphasizes the importance of organic-inorganic interrelations during burial diagenesis, and explains why oil generation and migration are often connected with clay evolution. Kharaka et al. (1986) list the chemical aspects of dissolved organic species that favor their intervention in mineral diagenesis: source or sink for hydrogen and correlative control of pH; reduction agents and control agents for sediment-oxidation processes; productors of decarboxylated phases; source for organo-mineral complexes. Edman and Surdam (1986) investigate the late Cretaceous sandstones of Green River basin, Wyoming. Illite-smectite minerals were degraded by organic solvents that mainly consist of carboxylic acids and phenols. During maturation stages the complexation of aluminum determined the dissolution of clay, resulting in a porosity enhancement. J.D. Edman and R.C. Surdam convert the interpreted organic-inorganic interactions and associated porosity increases into diagrams using a clastic reaction-pathway flow chart. The construction of such flow charts, that involve the four component system of CO_2, organic acids, carbonates and aluminosilicates, represents a useful step in predicting regions of maximum enhanced and preserved porosity in the subsurface.

The crystallinity of illite generally increases in the lower part of the late diagenetic zone only (Kubler 1968, Dunoyer de Segonzac 1969), but sometimes displays an earlier amelioration (e.g., Fig. 15.1). *Significant and irreversible increase of illite crystallinity statistically correlates with an increase of the hydrocarbon potential at depth* (Weaver 1960). Guthrie et al. (1986) study organic and inorganic parameters in diverse Carboniferous formations of the Ouachita mountains, Oklahoma and Arkansas. They demonstrate that illite crystallinity is significantly related to vitrinite reflectance, a very useful index in the search for oil within deeply buried series. Plots of bitumen/total orbanic carbon ratio versus vitrinite reflectance and 10 Å-illite sharpness ratio reveal hydrocarbon generation-preservation curves that allow to define submature, mature and supermature zones with regard to a liquid hydrocarbon window (Fig. 15.15). J.M. Guthrie and colleagues suggest that in the absence of vitrinite in a given buried series, illite crystallinity can be used quantitatively in order to estimate the levels of thermal maturity, and even to approximate hydrocarbon generation-preservation stages.

15.4 Conclusion

1. *A great deal of progress* has been made the last few years in the knowledge of clay diagenesis with increasing depth of burial. This progress proceeds mainly from the *application of new high-resolution electronmicroscopical and microchemical techniques and from the development of in situ borehole measurements.* Some of the more significant results are listed below:

– *The successive steps in the decrease of expandability of smectite minerals represent the more reliable and easy marker of depth effects*, than can be observed on X-ray diffraction diagrams. Illitization appears to mainly result from micro-environmental, chemical exchanges involving the potassium and aluminum released by the degradation of smectite and accessorily of coarser-sized feldspars. This supposes that smectite minerals are effectively present in the initial detrital sediments. If non- or slightly buried deposits mainly comprise illite and chlorite minerals, like in some peri-tectonic areas, the identification of burial effects may be more difficult and needs micromorphological investigations.

– *The modifications that affect the clay minerals usually occur at burial depths exceeding 2 kilometers, and appear to be less progressive than formerly believed.* In series marked by normal geothermal gradients (about 30°C/km) the major ordering processes develop between 2.5 and 3.5 kilometers, and do not progress beyond as deep as 5 to 6 kilometers. This suggests *the existence of fairly narrow temperature domains at which major changes take place, or the occurrence at certain depths of specific organic-inorganic interactions permitted* by the downward chemical evolution.

– *The residence time of a given temperature in a buried series appears to be more important than the absolute, instantaneous temperature values.* Borehole experiments show that some deeply-buried sediments are less affected by clay diagenetic changes than less-deeper sediments of the same lithology and age, submitted to comparable temperature. The duration of diagenetically efficient temperatures may have been longer in the former series compared to the latter, which can be documented by the study of secondary minerals formed through successive diagenetic stages. This points to the *dominant control of kinetics relatively to chemical conditions.* The *question arises of the real age of diagenetic changes in a given buried series,* and of the continuous or discontinuous history of burial diagenesis with increased age and depth.

– *The physical reorganization of clay particles and aggregates submitted to progressive compaction during early diagenesis is followed by a physico-chemical reorganization during late diagenesis.* Both types of modifications converge to determine a subparallel to parallel orientation of clay particles and clay layers. Such changes, that are sometimes associated with punctual clay solution-precipitation phenomena in pore voids and test chambers, lead to a *complete modification of the original mineral arrangement in deeply buried series,* which considerably modifies the physical properties and paleomagnetic characteristics.

2. *Some classical concepts on burial diagenesis of clay should be modified or abandoned* owing to recent results:

– *Chlorite does not usually result from the transformation of smectite in a magnesian environment.* Diagenetic chlorite is usually not abundant (less than 10% in Gulf Coast sediments) and of a ferriferous type; magnesian chlorite develops commonly during metamorphism only. Smectite does not appear to easily alter directly into chlorite, the latter mineral being more probably a byproduct of illitization derived from important dissolution and recrystallization.

– *Kaolinite often does not noticeably decrease in the late diagenetic zone, and may even form* anew under either common or restricted conditions. *Kaolinite does not progressively alter to chlorite or illite with increased depth;* it is either stable, growing or simply destroyed, according to the chemical environment.

– *True regular mixed-layers like allevardite or corrensite do not frequently form as transitional minerals during the burial history of smectite.* Regular mixed-layers are only rarely quoted in sediments buried at less than 4 to 6 kilometers, and often appear to be related to specific lithologies (Chaps. 16, 21). Regular mixed-layers occur mainly in the low-temperature metamorphic zone (anchizone) and mostly consist of illite-smectite (= allevardite. In Velde 1985).

   – *The concept of mixed-layering during the illitization process appears questionable* according to high-resolution electronmicroscopic data. Mixed-layered spacings occur only locally on ultra-thin clay sections cut normal to the layer planes (e.g., Bell 1986, Klimentidis and Mackinnon 1986). Many observations indicate the existence of packets of single clay minerals (smectite, illite, chlorite, kaolinite) rather than irregular or regular layer alternations. *Large discrepancies therefore arise between X-ray diffraction data* suggesting the development at depth of transitional, more and more ordered mixed-layered minerals, *and microscopic observations* pointing to the absence or rarity of such minerals.

   – *Diagenetic processes do not depend on the absolute age of buried series*, contrarily to what is often believed (in Weaver 1967b, Sudo and Shimoda 1978). Some very old sediments contain clay assemblages very diversified and similar to those of recent detrital deposits (e.g. late Precambrian early Paleozoic of Western Africa; Chamley et al. 1978a, 1980a), while late Tertiary series may have experienced drastic diagenetic changes (e.g., Ramseyer and Boles 1986). The statistical increase of illite and chlorite contents with increasing geological age results simply from the higher chance for old series to have been submitted to burial, tectonics and metamorphism, relatively to young series. More important than geological age are certainly the geothermal gradient and residence time of diagenetically active temperature.

   3. *The mechanisms of clay evolution during burial diagenesis largely differ according to lithology, fluid pressure and geothermal gradient*, and are still imperfectly known. The major control appears to consist in the *sediment permeability*, that determines the water/rock ratio and the importance of ion exchanges. In most sediments *chemical migrations occur on short distances* only (i.e., less than a few meters). In almost-closed systems such as argillaceous series, illitization of smectite chiefly proceeds from the *transformation* of adjacent detrital smectite, while chlorite and kaolinite may locally form in small amounts through phenomena evoking *dissolution-precipitation*. In more open systems allowed by coarser lithology or high interstitial water flows, diagenetic illite rather newly forms after dissolution of smectite and K-feldspars. *Intermediate situations* may exist, with, for instance, the synchronous growth of both lathed illite on detrital smectite (transformation) and of independent euhedral illite crystals (neoformation). All transitional stages probably occur between what happens in impervious clay and in porous sandstones (see Chap. 16).

   4. *The diagenetic history of buried clayey series closely parallels that of organic matter*, which determines extensive investigations to increase knowledge of the mineral characteristics and evolution in relation with oil generation, migration, accumulation, preservation and exploitation. *The abundance of smectite in initial sediments favors the production of hydrocarbons* because this mineral is able to adsorb abundant organic compounds, to release high amounts of water acting as a possible carrier, to increase the permeability through its degradation, and to act as a catalyst. The major stage of smectite-illite ordering during clay diagenesis often occurs fairly shortly before oil generation and migration, indicating close

organic-inorganic interactions. Organic acids produced during decarboxylation may in turn intervene in the diagenetic degradation of smectite. In addition, the values of illite crystallinity parallel the vitrinite reflectance evolution with depth, which may help in the identification of thermal maturity and hydrocarbon formation.

5. *Some new questions arise from recent investigations.* They concern, for instance, the inorganic-organic interrelations during burial history, the influence of temperature variations along the time on lithologically distinct buried series, or the precise interactions occurring between detrital and secondary minerals at increasing depths. But one of the most intriguing questions consists of the *real significance of X-ray diffraction patterns attributed to random or ordered mixed-layered minerals.* Some authors estimate that they result from *interparticle* instead of intraparticle *diffraction effects* (Nadeau et al. 1984b, 1985), very small and thin illite particles reacting with ethylene-glycol like normal smectite particles. The interparticle diffraction concept leads to a *dissolution-recrystallization* process for explaining illitization at depth. Other authors show on high-resolution micrographs the existence of coexisting *smectite and illite packets,* the latter increasing at depth and *intergrowing* at the expense of the former (e.g., Ahn and Peacor 1986). Such a smectite-to-illite conversion process, that in convincing since based on non destructive observations, characterizes a *transformation* rather than a neoformation. Whatever the mechanism, these data set up the question of the true development or not of mixed-layered clay minerals during burial diagenesis, and of the concept of interlayering itself. High-resolution electronmicroscopic and chemical analyses should be performed on types of buried sediments other than those commonly investigated (namely the Gulf Coast sediments). They should also be applied to other clayey materials characterized by abundant mixed-layer minerals, such as weathering profiles, soils, common slightly buried sediments, evaporitic and glauconitic deposits, etc. Such investigations would greatly help in a better understanding of both environmental and diagenetic histories of past deposits.

Chapter 16

Tectonic, Lithologic and Hydrothermal Constraints

16.1 Diagenesis Through Tectonics

16.1.1 Lateral Effects

Tectonics and metamorphism are often associated. Intense deformations determine an increase in pressure and temperature that induces various chemical and mineralogical changes. Metamorphic zones develop in close connection with most active orogenic areas. Laterally *the sedimentary formations involved in the tectonic structuration undergo attenuate thermodynamic efforts, and diagenetic changes can take place. Clay minerals tend to become less diversified and better crystallized, isotopic reequilibrations occur.* Many examples are provided by the literature. For instance, the eastern Provence, Southeast France, experienced during Albo-Cenomanian time pre-alpine deformations, which determined an increase in illite crystallinity and a rubidium/strontium homogenization (Bonhomme et al. 1969). In the Atlas basins, Morocco, illite and chlorite developed during Hercynian orogeny at the expense of kaolinite, smectite, mixed- layers, at the same time as chemical, isotopic and grain-size homogenizations took place (e.g., Robillard and Piqué 1981, Wybrecht et al. 1985). Diagenetic changes in Atlas mountains largely resulted from local temperature and pressure rather than from burial effects, since some less-buried series are much more affected than deeply buried ones (A. Piqué, pers. comm.).

At the periphery of mountain chains, the clay associations in sediments may or may not reflect the tectonic structuration. *Transitional situations may laterally occur between non diagenetic and diagenetic sediments.* This is particularly well documented in the subalpine and outer alpine range, *Western Alps,* where sediments present slight to moderate tectonic deformations. Dunoyer de Segonzac et al. (1966) and Dunoyer de Segonzac (1969) study a West-East transect in Callovian-Oxfordian marls (= Terres Noires), from the city of Die in the subalpine zone to Barcelonnette in the inner Alps. When moving eastwards, the sediments become progressively darker, harder and marked by a slaty cleavage. An obvious schistosity appears East of Gap city, at the same time as kaolinite disappears from chlorite- and illite-rich clay assemblages. The illite crystallinity progressively increases eastwards, and its values display a narrower range (Fig. 16.1). *The lateral evolution resembles that recorded vertically downwards in deeply buried series* (Chapter 15). G. Dunoyer and colleagues attribute the mineralogical evolu-

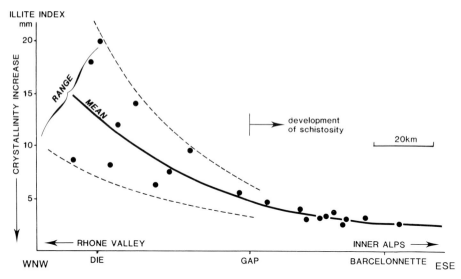

Fig. 16.1. Lateral variation of illite crystallinity in the Callovian-Oxfordian marls of Western Alps, Southeast France. (After Dunoyer de Segonzac et al. 1966). The values correspond to the width at mid-height of the 10 Å illite peak on X-ray diagrams (mm). City location: see Fig. 16.3

tion recorded to late diagenetic- early metamorphic processes induced by the alpine tectonics. Frey (1970) decribes similar changes in late Triassic-early Jurassic pelitic rocks of the alpine border region in Switzerland. When passing from unaffected clays and marls to anchimetamorphic phyllites, the illite crystallinity increases, while kaolinite and irregular illite-smectite mixed-layers are replaced by pyrophyllite, phengite and Al-rich chlorite. In the southwestern French Alps, Rhetian marl-carbonate alternations also show an eastward reorganization of illite structure, associated with the development of chlorite, paragonite and albite (Dunoyer de Segonzac and Abbas 1976).

Chlorite represents a peculiarly useful marker of post-sedimentary mineral changes in late Mesozoic calcareous sediments of the French subalpine range. As this mineral virtually lacks in non diagenetic sediments, its appearance obviously characterizes post-sedimentary modifications. Illite, smectite and kaolinite occur in variable amounts in both non diagenetic and diagenetic formations, and their percentage variations may reflect either changes in detrital sources and climate (see Part 6), or thermodynamic constraints. Ferry et al. (1983) report that chlorite abundance increases progressively in the Vocontian basin from West to East, within isochronous Valanginian limestone-marl alternations. This change is attributed to a smectite-to-illite transformation, determined by a thermal diagenesis whose intensity increases toward the major alpine thrust zones. A similar variation is recorded for different early Cretaceous periods by Deconinck and Chamley (1983) and Deconinck (1987), from the outermost to innermost parts of the subalpine range. The eastward increase of chlorite abundance correlates both with increasing temperature and sedimentary overburden. *The chlorite amount increases eastwards much more rapidly in limestone beds than in marl inter-*

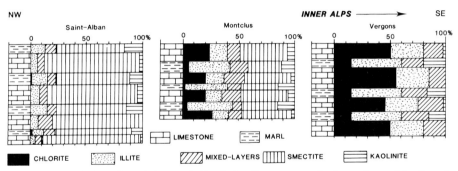

Fig. 16.2. Lateral clay mineral changes in Berriasian limestone-marls alternations of the French subalpine range, from western, nondiagenetic to eastern, diagenetic zones. (After Deconinck and Chamley 1983). Mineral changes in alternations of the left column result from climate variations (see Chap. 17.4)

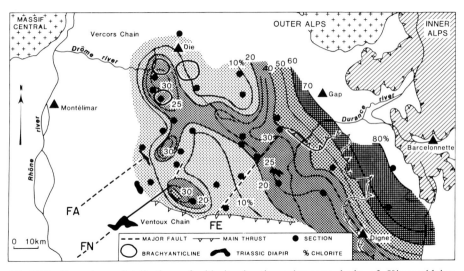

Fig. 16.3. Percentage distribution of chlorite in the calcareous beds of Kimmeridgian isochronous limestone-marl alternations from the Vocontian basin, Southeastern France. (After Levert and Ferry 1988). *FA* Aygues fault. *FE* Eygalayes fault. *FM* Ménée fault. *BA* Brachyanticlines

beds (Fig. 16.2), which is attributed to higher permeability and stronger heat transfer in calcareous levels. Marly beds appear to keep the paleoenvironmental signal of detrital clays much longer than calcareous beds. Diagenetic chlorites display an iron- rich type, which resembles the composition of chlorite deeply buried in Gulf Coast sediments (Hower et al. 1976) and could result from an iron release by pre-existing smectites (see 15.2.2). The illite abundance increases parallel to chlorite, but much more in marl than in limestone layers, suggesting a strong lithologic control (see 16.2.2).

Levert and Ferry (1988) confirm the eastward increase of Fe-chlorite abundance in four isochronous horizons of the Vocontian basin. Oxfordian, Kimmeridgian, Valanginian and Aptian bundles of limestone-marl alternations show a progressive enrichment of chlorite toward the inner Alps, where calcareous beds may contain up to 80% of the mineral in the $< 2\,\mu m$ fraction (Fig. 16.3). The mineralogical change correlates with an increase in the maturation state of organic matter. In addition, local increases in more magnesian chlorite occur in the western part of the basin. The geographic situation of these anomalous chlorite enrichments varies according to the period, but preferentially characterizes some areas marked by fault systems, Triassic diapirs or brachyanticlines. J. Levert and S. Ferry suggest that the local Mg-chlorite augmentations, which sometimes correlate with a kaolinite enrichment, could result from hydrothermal mineralizations linked to fluid migrations along fault systems or above Triassic salt diapirs.

Other examples arise from the literature on lateral clay diagenetic modifications induced by the tectonic activity. For instance, Chennaux et al. (1970) correlate the westward appearance of pyrophyllite in Silurian-Devonian claystones from the Western Sahara with the increased influence of the Hercynian orogeny. Leikine and Velde (1974) interpret the northward increase of illitic layers in illite-smectite mixed-layers from Senonian deposits of Northeast Algeria as metamorphic effects linked to the pre-alpine tectonics.

16.1.2 Tectonic Overburden

If sedimentary series are tectonically covered by thick geological formations, diagenetic effects may develop similarly to what occurs during progressive burial. *The pressure and temperature due to the tectonic overburden may determine a clay mineral simplification and reorganization.* Such a possibility was envisaged by Dunoyer de Segonzac and Bernoulli (1976) for Rhetian deposits of the Austrian-alpine domain. If the superimposed tectonic units are ulteriorely removed by erosion, the autochthonous or para-autochthonous underlying formations may crop out and keep in their clay mineralogy the evidence of the former covering. Such a possibility is documented by Deconinck and Debrabant (1985) in the northwestern part of the French subalpine domain (Fig. 16.4). Typical detrital clay assemblages occur in late Cretaceous limestones of the western sector, marked by dominant beidellitic-smectite associated with various amounts of illite and little random mixed-layers, chlorite or kaolinite. When crossing the Arcalod fault to the East, these assemblages in contemporaneous sediments are suddenly replaced by clay associations rich in well-crystallized illite and chlorite, with associated sub- regular mixed-layers. This change is thought to result from the overburden effect of former tectonic nappes, that issued from the inner Alps and invaded the eastern part of the subalpine range during Paleogene time. Such a local overburden effect would explain the lack of lateral gradient in clay-mineral characteristics. J.-F. Deconinck and P. Debrabant suggest that the Arcalod fault results from a collapse of the eastern sector due to an excess of sedimentary load.

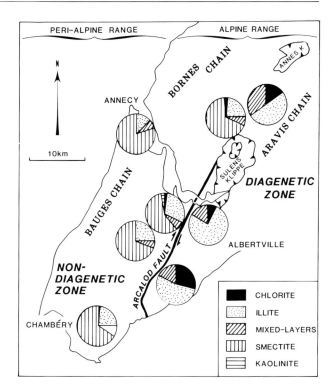

Fig. 16.4. Clay mineral distribution in the late Cretaceous sublithographic limestones of the subalpine range, Savoie, Southeast France. (After Deconinck and Debrabant 1985)

Further erosion determined the disappearance of the allochthonous formations, except in some local areas where isolated klippes are preserved. Clay mineral assemblages therefore appear as useful markers of past tectonic overburden in regions marked by active erosion.

The precise identification of clay-mineral characters in buried series from tectonically active regions may help to *date the time of diagenesis relatively to the time of tectonic structuration.* Bouquillon et al. (1985) study two deep boreholes drilled down to 4000 and 5000 meters in the Devono-Carboniferous series of Northern France. Two structural units are superimposed and separated by a major thrust contact of Hercynian age, the Midi fault. The illite crystallinity tends to increase downwards in both structural units, suggesting a burial effect; but the crystallinity displays higher values in the upper unit than in the lower one. This indicates that the clay diagenesis took place before the final tectonic structuration, probably in the Paleozoic basin where sediments progressively accumulated.

16.2 Chemically Restricted Environment

Deeply Buried Smectite
The possibility for smectite to exist in late diagenetic to epimetamorphic environments was probably first reported by Yashkin (1967) for late Proterozoic rocks of

the Dniestr region, USSR. Wilson et al. (1968) document the epimetamorphic formation of saponite in late Precambrian to Cambrian (= Dalradian) metalimestones of the Banffshire coast, Northeast Scotland. Saponite occurs as irregular, often streaked out or folded aggregates associated with either chlorite, talc, phlogopite or chlorite-smectite mixed-layers. Other minerals comprise very abundant calcite, frequent dolomite and quartz, and various accessory minerals including kaolinite and tremolite. Referring to experimental data on hydrothermal syntheses, M.J. Wilson and colleagues suggest that saponite formed under high temperature and pressure conditions by some reaction such as:

$$dolomite + kaolinite + quartz + H_2O = calcite + saponite + CO_2.$$

Dunoyer de Segonzac and Abbas (1976) report in the metamorphic Briançonnais zone, Western Alps, the presence of an aluminous montmorillonite that preferentially occurs in dolomitic rocks of Rhetian age. Smectite constitutes up to 80% of the clay fraction, which permits proposing approximate structural formulae such as the following (Ascension section):

$$K_{0.35}Na_{0.05}(Al_{1.40}Mg_{0.43}Fe^{3+}_{0.16}Ti_{0.05})(Si_{3.85}Al_{0.15})O_{10}(OH)_2, nH_2O.$$

Due to its high crystallinity, its specific occurrence in a metamorphic zone and its preferential association with carbonate facies, the Briançonnais montmorillonite is attributed to a low-metamorphic authigenesis in a possibly hydrothermal environment.

Studer and Bertrand (1981) describe the formation of trioctahedral smectites and mixed-layers in the Moroccan Atlas, within marly sediments intruded by basic igneous dykes. The smectite-rich sediments are attributed to cooling processes following metamorphism, when gas-rich aqueous fluids interacted with metamorphic minerals. Bouquillon et al. (1985) identify smectitic minerals in Carboniferous series buried at 3000 to 4500 meter depth in two boreholes drilled in the North of France. Visean limestones and dolomites locally contain very abundant (up to 90% of the clay fraction), highly crystalline saponite. This mineral is accompanied by other magnesian sheet-silicates like talc and corrensite (Table 16.1), as well as by celestite, some swelling chlorite and abundant strontium (3950-5350 ppm in the bulk rock, compared to about 100-400 ppm in surrounding rocks). The close connection of smectite and corrensite with

Table 16.1. Structural formula of expandable clay minerals from deeply buried Carboniferous dolomitic limestone (Epinoy-1 borehole, 3 200 meter depth. After Bouquillon et al. 1985)

Saponite
$$(Si_{3.73}Al_{0.27})(Mg_{2.59}Al_{0.15}Fe^{3+}_{0.04})K_{0.29}Ca_{0.09}Na_{0.05}O_{10}(OH)_2$$

Corrensite
$$(Si_{3.18}Al_{0.82})(Mg_{1.06}Al_{0.81}Fe^{3+}_{0.47}Ti_{0.01})K_{0.39}Mg_{0.13}Ca_{0.06}Na_{0.05}O_{10}(OH)_2$$

Talc
$$(Si_{3.85}Al_{0.15})(Mg_{2.89}Fe^{2+}_{0.03})Na_{0.10}Ca_{0.08}K_{0.05}O_{10}(OH)_2$$

evaporative sediments suggests that Mg-clays result from the *combination of alkaline sedimentary and chemically restricted diagenetic conditions*. Blaise and Mercier (1988) report a similar occurrence of Mg-smectite, talc and corrensite in late Proterozoic, intensely diagenized rocks of the Yukon Territory, Western Canada. The Mg-clays are restricted to black, organic-rich dolomites and appear to represent preserved facies of past evaporative conditions.

The occasional formation or preservation of smectite in deeply buried sediments is compatible with some hydrothermal syntheses and other experiments at high temperature and pressure (e.g., refer. in Wilson et al. 1968). Sass et al. (1987) construct isothermal, isobaric activity diagrams for the system K_2O - Al_2O_3 – SiO_2 – H_2O. The illite-smectite assemblage appears metastable at temperatures between 110 and 200°C, but may coexist stably with either kaolinite or microcline above 200°C. The smectite stability is namely controlled by the quartz solubility. Yau et al. (1987a) show that Wyoming bentonite is not affected by hydrothermal treatment at 200 to 460°C for 71 to 574 days, probably because calcite has stabilized the initial expandable clay. This could explain why deeply buried smectite preferentially remains and eventually becomes better crystallized in carbonated rocks than in others.

Limestone – Marl Alternations

When passing from non diagenetic to diagenetic series in early Cretaceous limestone-marl alternations of the French subalpine range, both lithologic types show an eastward increase of illite and chlorite content in the clay fraction (16.1.1. Fig. 16.2). *Calcareous beds display a preferential increase of chlorite abundance, while argillaceous interbeds show a preferential increase of illite abundance* (Deconinck and Debrabant 1985). The absence of any chemical gradient from limestones to marls in a given set of lithologic alternations points to the lack of appreciable vertical ion migration. The potassium necessary for the illite development probably results from the diagenetic degradation of K- feldspar and mica, two mineral species that occur in marly interbeds more commonly than in calcareous beds. The mechanism of illitization would be similar to what is described at increasing depths of burial in Gulf Coast sediments (15.1.2). The magnesian character of the diagenetic chlorite preferentially developed in the calcareous beds could result from the evolution of low- magnesian to nonmagnesian calcite. Microprobe analyses reveal that chlorite is also enriched in iron. Both iron and magnesium could be partly provided by the initial smectite, the illitization of which would have determined an incorporation of potassium and aluminum and a release of iron and magnesium.

A similar control of the diagenetic evolution by lithology is reported by Bouquillon et al. (1985) about the Devonian shale-sandstone alternations from Jeumont-1 borehole in the North of France. Iron-rich diagenetic chlorites formed preferentially in sandy beds compared to argillaceous beds. All these mechanisms involve *different permeabilities and fluid migrations in alternating lithologic types*. Similar phenomena characterize some barite- or phosphate-nodules scattered among Cretaceous marls of Northern France (Holtzapffel 1987), where original smectite may diagenetically evolve into either illite, chlorite or vermiculite.

16.3 Highly-Porous Rocks

16.3.1 Introduction

The ability of clay minerals to form diagenetically in the voids of permeable rocks is suspected, since accordeon-like, vermicular or blocky kaolinite has been described from thin-section studies in various sandstones (e.g., Termier 1890, Ross and Kerr 1931; see Millot 1964, 1970) These delicate particles can hardly result from a genetic process similar to that responsible for the supply of the typically-detrital minerals among which they occur (quartz, mica, feldspar,etc). *Because their high porosity and permeability favor the post- sedimentary migration of fluids, the sandstones represent particularly suitable environments for the diagenetic formation of secondary minerals* like carbonates, silica and clays. Dapples (1979) identifies three major stages that characterize the diagenetic evolution of sandstones: (1) The redoxomorphic stage occurs during early burial and is dominated by oxidation and reduction reactions developing at the time of initial compaction and fluid ejection. (2) The locomorphic stage corresponds to mineral replacements and precipitations in the pore spaces, which leads to primary cementation and to the development of chalcedony, siderite, calcite, etc. (3) The phyllomorphic stage concerns lithified sandstones and results chiefly in the formation of clay minerals and, in the most advanced stages, of mica.

The three stages of sandstone diagenesis tend to follow each other at increasing depth of burial. For instance, Galloway (1974) reports in Tertiary basins of Northwestern America the successive formation at increasing depth of calcite pore-filling cement, of authigenic clay rims and coats around detrital grains, and of authigenic phyllosilicate and laumontite pore-filling cement. But as lateral migration of chemically active groundwaters may occur at various stratigraphic levels and geological periods, *the depth of burial constitutes only one of the factors controlling the formation of secondary minerals in sandstones.*

Most sandstones contain, beside the typically detrital mineral species like quartz, some sheet-silicates that also derive from terrigenous sources. This is obvious for the sand- or silt-sized micas, but also concerns clay-sized minerals such as kaolinite reworked from lateritic blankets, or illite and chlorite eroded from crystalline rocks. It is therefore difficult to distinguish a priori the primary and secondary clay minerals in sandstones. For a long time it was believed that secondary clays in sandstones occurred in very small amounts only, and that most sheet-silicates were of a detrital origin. The search for new oil- reservoir rocks, that has developed dramatically since the early 1970's, favored extensive laboratory investigations. A lot of new results arose, which showed that secondary clay minerals may represent an importat proportion of the total clay fraction. The development of clay diagenetic growths in sandstones is in some cases even considered as one of the major ways to form the graywacke matrix (e.g., Galloway 1974).

Detailed optical studies help greatly to distinguish primary clay minerals from secondary species. One of the first relevant studies was presented by Wilson and Pittman (1977), from data on thin sections and scanning electronmicroscopic

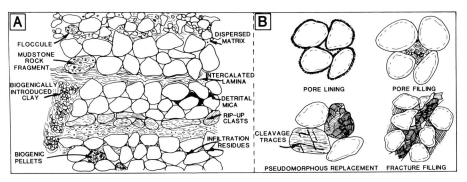

Fig. 16.5. Modes of occurrence of allochthonous (**A**) and autochthonous (**B**) clay in sandstones. (After Wilson and Pittman 1977). The size of individual flakes and aggregates is exaggerated

preparations. *Autochthonous clays* are reported as pore linings, pore fillings, pseudomorphic replacements and fracture fillings (Fig. 16.5B). Arguments for an authigenic origin consist mainly in the delicacy of clay morphology, the absence of pore-lining clay only at grain contacts, and a mineral composition different from that of typically allochthonous minerals. By contrast, *allochthonous clays* mainly occur as dispersed matrix, sand-sized floccules, sand- to cobble-sized mud, shale clasts, specific laminae, or as a result of post-depositional infiltration or bioturbation (Fig. 16.5A). Many papers are available on this subject, as well as on the genetic processes involved in the secondary formation of clay in sandstones. A summary only is presented here. For more details, the reader is referred to specialized publications [e.g., Larsen and Chilingar 1979, 1983; Clay Minerals 1982 (17, 2), 1984 (19, 3), 1986 (21, 4)].

16.3.2 Diagenetic Clay Minerals in Sandstones

Kaolinite

The most frequent authigenic clay minerals in sandstones comprise kaolinite, illite and chlorite. Some mixed-layered clays are also attributed to a post-sedimentary formation. Kaolinite is widely distributed in sandstones, and may derive from either allochthonous sources or autochthonous genesis. For instance, the Carboniferous series of Scotland contain variable amounts and types of kaolinite according to the age and lithology (Wilson et al. 1972). Late Carboniferous sediments largely consist of shales and limestones; they contain abundant detrital kaolinite, that is small-sized, moderately ordered and mainly derives from the erosion of a tropically-weathered lateritic cover. Underclay horizons contain a very poorly-ordered form of kaolinite, that is thought to have been degraded in an acid swamp environment before burying. Early Carboniferous sediments are rich in dolomites and sandstones; the latter lithologic type is characterized by a highly crystalline, morphologically- perfect kaolinite that formed during diagenesis.

In most cases, the chemical elements needed for kaolinite formation, namely silicon and aluminum, are thought to derive from the leaching of some minerals pre-existing in the sandstones. *Potassium feldspars and micas constitute the more probable Si and Al sources.* Arditto (1983) discusses the formation of well- ordered kaolinite in the Jurassic Pilliga sandstone, Southeastern Australia, in terms of chemical equilibrium and detrital mineral stability. Kaolinite is the major stable mineral phase in contact with bore and surface waters in the local aqueous system, while K-feldspars and micas tend to alter under the same conditions. The alteration of detrital feldspars and micas is documented by thin-section observations. As K-feldspar constitutes up to 20% of the total detrital material and biotite up to 3%, both mineral species represent easy sources for the chemical elements necessary for kaolinite growth. P.A. Arditto demonstrates that chemical exchanges within the hydraulic system essentially occur in a horizontal way and not vertically. Kaolinization, that develops at the expense of extensively deformed and crushed feldspars, clearly postdates significant compaction and belongs to the late diagenetic history.

According to Curtis (1987b), the dissolution of source minerals, that is favored by acidic conditions and the presence of relict organic matter, is followed by a pH rise due to the acid consumption. This would determine kaolinite precipitation, since the solubility of the mineral decreases sharply as neutral pH values are reached. Another way to raise the pH and precipitate kaolinite from organic-rich solutions consists in the decarboxylation stage that characterizes the hydrocarbon maturation.

Dickite sometimes occurs in sandstones, with kaolinite or alone. Kaolinite is often considered as mainly detrital and deposited during sedimentation, while dickite represents a diagenetic mineral formed at depth. Ferrero and Kubler (1964) report the dominance of dickite over kaolinite in Cambrian sandstones from Hassi-Messaoud, Sahara. Dickite constitutes large hexagonal sheets (20 – 60 μm long) and is widely distributed in most coarse facies. True kaolinite particles, that hardly exceed 2 μm in length, are restricted to fine sandstone and siltstone beds. Cassan and Lucas (1966) confirm that dickitization occurs during late diagenesis in Hassi-Messaoud sandstones, after two silication phases. Dickite seems to develop by transformation of pre-existing kaolinite, and to correlate with a corrosion of quartz grains. Loughnan and Roberts (1986) discuss the intimate association of dickite with ordered and disordered kaolinite in Triassic sandstones from the Sydney basin, Southeast Australia. Dickite appears as the most stable phase, and as the only unequivocally authigenic kaolin polytype. Dickite probably precipitated very slowly from migrating groundwaters because of unusually low concentrations of silica.

Illite

Illite is recognized as a possible diagenetic mineral in sandstones since it has been identified as growing at the expense of kaolinite particles (e.g., Kulbicki and Millot 1960). *Hairy illite* represents the most convincing diagenetic facies for illite, since the fragile and long mineral needles developing at the periphery of quartz or other substrates may hardly have experienced significant transportation (e.g., Bailey 1980a). Güven et al. (1980) report the frequent occurrence of authigenic lathed illite in the pores of many sandstone reservoirs from the United States, such as those existing in the Eocene of Texas (Wilcox formation), the Paleocene of Wyoming (Fort Union formation) and the Jurassic of Mississippi (Norphlet formation). The laths develop perfect morphologies, and are up to 30 µm long, 0.1 to 0.3 µm wide and up to 200 Å thick (0.02 µm). Electronmicroscopic analyses show that hair-like illites are associated with irregular blob-like cores, the chemistry of which is very similar to that of laths. This resembles the chemical similarity of lathed and flaky smectites in some Meso-Cenozoic argillaceous diagenetic environments (Chap. 14). N. Güven and colleagues stress the fact that the presence of hairy illite in sandstone pores increases the microporosity and pore tortuosity, and therefore decreases the permeability, which may significantly modify the reservoir properties.

Morad and AlDahan (1987a) identify hair-like, filamentous illites in sandstones as old as those of the late Proterozoic Visingsö group, Southern Sweden, which indicates the stable nature of this facies. Visingsö sandstones also contain flaky diagenetic illites. Both facies appear to derive from a dissolution- crystallization process, the source minerals probably consisting of potassium feldspars. Some K-feldspars show noticeable dissolution and albitization features, and the newly formed illite or albite grains may display parallel orientation in two directions, possibly representing the cleavage planes of the replaced mineral. The reason for the preferential development of laths or sheets appears difficult to assess. The cause of the preferential formation of illite instead of kaolinite seems to be related to the chemical composition and pH of migrating groundwaters, illite apparently developing under less acidic conditions.

A specific case of sandstone illitization under the action of *warm migrating waters* is documented by Whitney and Northrop (1987) for the late Jurassic Morrison formation in the southern part of the San Juan basin, New Mexico (USA). Mineral and isotopic investigations allow to infer the mineral-fluid reactions over a distance of at least 60 kilometers. A 100 meter-thick serie of massive, homogeneous, hydrologically continuous fluvial sandstones has served as a primary conduit for fluids in the course of the time. The initial smectitic minerals express a diagenetic alteration that varies independently of the lithology. The proportion of illite layers ranges from 0 to 90%. The sandstone is less illitic at the upper and lower contacts of the sand body and becomes more illitic toward the center. The same trend occurs from the margin to the center of the basin. Oxygen and hydrogen isotope compositions of illite suggest a fluid temperature above 100°C. Illitization through warm migrating fluid was accompanied by some chloritization processes. G. Whitney and H.R. Northrop infer from the tridimensional mineral zonation that warm, evolved fluids migrated up dip from the center of the basin,under the influence of a regional hydraulic head. Mineral-fluid reactions probably resulted from a brief, tectonically- induced pulse of fluid, which is consistent with the Tertiary movements of petroleum-related fluids in overlying rocks from the San Juan basin.

Chlorite

The possibility for chlorite to form diagenetically in sandstones is envisaged by Smoot and Narain (1960) for Cambro-Ordovician and Mississippian formations from Illinois, USA. Kulke (1969) reports the presence of dioctahedral chlorite (= sudoite) and of both random and ordered chlorite-smectite mixed-layers in late Triassic (Keuper) sandstones of South Germany; chloritic minerals are considered to form under late diagenetic conditions, at depths where kaolinite is unstable and dissolves.

Diagenetic chlorite appears to consist chiefly of *coatings* around detrital grains (e.g., Hayes 1970). Curtis et al. (1984) study the composition of chloritic coatings in Cretaceous sandstones from the Tuscaloosa formation in Louisiana, USA (burial depths 2390 and 5500 m), and from the Fahler formation in Alberta, Canada (1876 m). Scanning transmission electronmicroscopic analyses indicate the dominant presence of true chlorites, the chemical composition of which varies relatively slightly and corresponds to a magnesian chamosite (= Mg-bearing Fe chlorite). In addition, some coatings include swelling chlorite-like minerals that contain a vermiculitic component. Chloritic coatings can be so continuous within the sandstone that they may prevent further overgrowth of framework grains from saturated pore waters, thereby preserving the initial pososity.

Replacement chlorite also exists in some sandstones. For instance, Morad and AlDahan (1987b) report the common substitution of detrital microcline and albite by chlorite in the Visingsö group of the late Proterozoic from Southern Sweden. Authigenic quartz and sometimes illite or feldspar are associated with the diagenetic chlorite. The feldspar is usually dissolved before to be replaced, the iron and magnesium needed for the chlorite growth being derived from biotite and from metal oxide grains or pigments. A tentative reaction is proposed by S. Morad and A.A. AlDahan:

$$\text{microline (108.7 cc)} + \text{reactants} \qquad \qquad =$$
$$\text{KAlSi}_3\text{O}_8 \qquad \quad +0.4\text{Fe}^{2+}+0.3\text{Mg} \quad +1.4\text{H}_2\text{O} =$$
$$\text{chlorite (62 cc)} \qquad \qquad \qquad \qquad +\text{quartz} +\text{products}$$
$$0.3(\text{Fe}_{1.4}\text{Mg}_{1.2}\text{Al}_{2.5})(\text{Al}_{0.7}\text{Si}_{3.3})\text{O}_{10}(\text{OH})_8 + 2\text{SiO}_2 \quad +\text{K}+0.4\text{H}$$

The fairly homogeneous chemical composition of chloritic rims in sandstones suggests noticeable migrations of the fluids responsible for diagenetic exchanges and reactions (Curtis 1987b, Morad and AlDahan 1987b, Whitney and Northrop 1987). The formation of chlorite appears to be favored by reducing and late diagenetic (= noticeable burial depth) conditions.

16.3.3 Diagenetic Processes and History

The detailed genetic relations existing between the different secondary and source minerals in sandstones are only partly understood. Kulbicki and Millot (1960) observe in *Cambro-Ordovician formations from the Central Sahara* that diagenetic processes preferentially occur in sandstone facies and are inhibited in impervious shales where detrital assemblages prevail. Illite always forms after kaolinite, since the former mineral develops at the expense or at the periphery of the latter. G. Kulbicki and G. Millot propose a two-step history: (1) *degradation of detrital micas, leading to coarse-sized vermicular and accordeon-like, diagenetic kaolinite* that coexists with small-sized, detrital kaolinite; (2) *transformation of diagenetic kaolinite, leading to secondary illite* that may ulteriorly evolve to diagenetic mica. The kaolinization is associated with some silication of detrital quartz, and is considered to occur during early diagenesis in relation with acidic fluid migration controlled by surface climate conditions. By contrast, the illitization of kaolinite grains is attributed to the late diagenetic influence at depth of saline groundwaters. Similar conclusions arise from Sommer's investigations (1975) of a Jurassic reservoir from the North Sea. The kaolinization phase corresponds mainly to the formation of diagenetic dickite under the action of acidic freshwater; dickite forms at the expense of both mica and plagioclase and is associated with Ca- and Fe-cements. The illitization of dickite occurs during late diagenesis and is associated with the formation of not abundant chlorite, while dickite dissolves under the action of saline groundwaters.

Detailed mineralogical and electronoptical studies show that clay diagenesis in sandstones may be much more complex than previously described. For instance, Huggett (1984) investigates the fluvial sandstones and associated siltstones, shales, coals, and occasional marine sandstones, from the *Westphalian Coal Measures* in East Midlands, UK. The sandstone bodies contain 10 to 15% clay minerals, that consist of kaolinite, chlorite and illite (Fig. 16.6).

1. *Kaolinite* may be blocky or platy, some of the grains showing cleaving figures that apparently result from compaction and pressure due to other crystals growth. Kaolinite, considered as entirely authigenic, is overgrown by quartz, feldspar and illite, which suggests an early diagenetic formation. Locally, however, almost all the kaolinite is blocky and occurs in open pores on top of chlorite rims, indicating a late formation. The preferential formation of kaolinite in Carboniferous sandstones is clearly related to leaching by freshwater and to the lack of brackish- or marine-water influence (Huggett 1986). Some kinetic factors probably also intervene, as suggested by Huang et al.'s experiments (1986) on the preferential formation of kaolinite with high fluid / rock ratios.

2. *Diagenetic illite* displays plate-, curved lath- and straight blade habits, that clearly differ from tangentially arranged, grain-lining, ragged plates of detrital illite. Illite occurs as pore-filling and pore-rimming clay, and as inclusions in mineral overgrowths. The distribution of detrital illite is controlled by the deposition facies, most clay particles accumulating in the channel abandonment deposits. The diagenetic replacement of feldspar by illite is ubiquitous, that of muscovite is favored in clean, washed sands, and that of kaolinite is neither abundant nor facies-related. The alteration of feldspar to illite instead of kaolinite appears to be favored by ion-enriched, less acidic pore waters. Such conditions conform Huang et al.'s experiments (1986) which show that mass nucleation and growth of illite platelets on albite surfaces are favored by low fluid/rock ratios. The radial illite grain-rim cements have no immediately adjacent precursor phase and represent true precipitation minerals. Illite and quartz authigenesis occurred concurrently, the development of the former species probably being favored by the presence of K-feldspar. Unlike illite blades, illite laths are observed overgrown by quartz, which suggests they have formed before the blades. Finally, J.M. Huggett's observations indicate that physical and

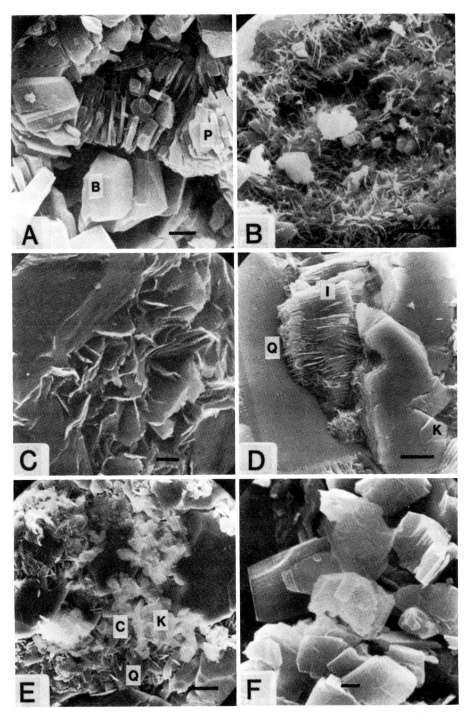

Fig. 16.6. Scanning electronmicrographs of diagenetic clay minerals in sandstones. (Courtesy J. M. Huggett. See Huggett 1984). **A** Mixture of blocky (*B*) and platy (*P*) kaolinite. Bar = 5 μm. **B** Radial illite laths. Bar = 5 μm. **C** Illite plates. Bar = 1 μm. **D** Illite (*I*) replacing kaolinite (*K*) and overgrown by quartz (*Q*). Bar = 5 μm. **E** Chlorite coatings (*C*) overgrown by quartz (*Q*) and blocky kaolinite (*K*). Bar = 0.5 μm. **F** Blocky pseudohexagonal chlorite. Bar = 1 μm

chemical constraints are much more important than depth constraints. Nevertheless illite tends to form predominantly under later diagenetic conditions than kaolinite. The precise reasons for the preferential development of a given illite habit are still poorly understood.

3. *Chlorite* is both detrital and authigenic, the latter type being only locally significantly abundant. Diagenetic chlorite consists of either small grain rims or blocky-pseudohexagonal crystals. The first type is the most frequently encountered. The presence of kaolinite-overgrowing chlorite suggests a mostly early-diagenetic formation. The local abundance of chlorite could be related to the local presence of ferromagnesian minerals (e.g., biotite) or of marine to brackish environment. Backscattered electron imaging supports the hypothesis of the replacement of biotite by chlorite (Huggett 1986).

Huggett (1984) proposes from his investigations a tentative timing for the mineral cement formation, the authigenic clays being situated among carbonate and other silicate precipitates (Fig. 16.7).

Other studies sometimes conclude fairly different genesis and timing conditions for the clay diagenesis in sandstones. Kantorowicz (1984) compares the authigenic clay minerals in *middle Jurassic sandstones* from the Ravenscar group in Yorkshire (UK) and from the Ninian Field Brent group in the North Sea. In the Ravenscar group *early diagenetic conditions* induce the formation of pore-lining *illite*, pore-lining *chlorite* and pore-filling vermiform *kaolinite*, all minerals excluding mutually each other because of different depositional pore-water chemistry: sea water determines the formation of illite, anoxic freshwater that of chlorite, and oxygenated freshwater that of kaolinite. Similar conditions characterize the Brent group, but vermiform kaolinite also occurs in the marine sand-

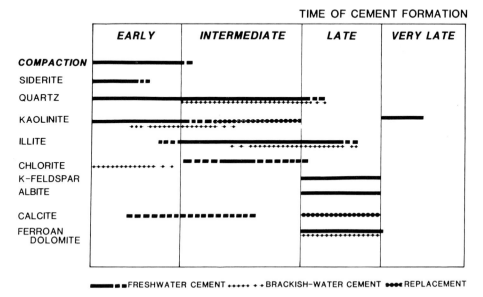

Fig. 16.7. Tentative timing of authigenic cement formation in Coal Measures sandstones, Westphalian of East Midlands, UK. (After Huggett 1984)

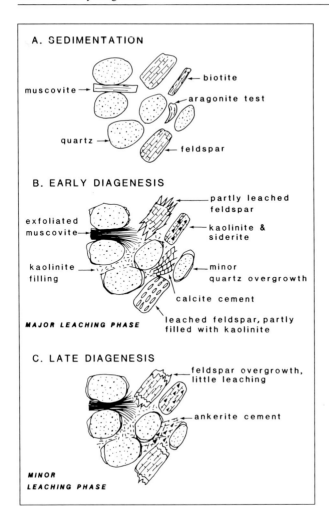

A. SEDIMENTATION

muscovite →

biotite

aragonite test

quartz →

feldspar

B. EARLY DIAGENESIS

partly leached
feldspar

exfoliated
muscovite →

kaolinite &
siderite

kaolinite
filling

minor
quartz overgrowth

calcite cement

MAJOR LEACHING PHASE

leached feldspar, partly
filled with kaolinite

C. LATE DIAGENESIS

feldspar overgrowth,
little leaching

ankerite cement

**MINOR
LEACHING PHASE**

Fig. 16.8. Main diagenetic changes in the middle Jurassic sandstones of the Brent group, Statfjord field, North Sea. (After Bjørlykke and Brendsdal 1986)

stones, perhaps due to temporary freshwater-table development following the sedimentary progradation. Whatever the precise causes, the local geographic, climatic and hydrologic context appears to strongly control the types of early diagenetic clay minerals. *Late diagenetic conditions* correspond to the precipitation of *dickite* and other kaolinite-like minerals in both groups, but subsequent illite formed in the Brent group only, perhaps from alkaline formation waters and at burial depths not reached by the Ravenscar group rocks. Finally, secondary vermiculite is identified in some sandstones of the Ravenscar group, which probably results from recent subaerial alteration of chlorite.

Bjørlykke and Brendsdal (1986) study the middle Jurassic Brent group in the Statfjord field, North Sea, and show that early diagenesis of sandstone determines mainly the kaolinite, siderite and calcite formation. Late diagenetic processes chiefly correspond to feldspar overgrowths and to ankerite cementation (Fig. 16.8).

In the *early Permian Rotliegendes sandstones* of the Rough Gas field, North Sea, *early* environmentally related *diagenesis* is characterized by few mineral formations, grain displacement, and corrosion (Goodchild and Whitaker 1986): (1) growth of anhydrite and fine rhombic dolomite within aeolian sandstones; (2) infiltration of detrital clay and possible formation of mixed-layered clays and coarse dolomite within fluvial sandstones; (3) moderate alteration of feldspars within all facies. *Late diagenetic features* are superimposed on early cements: illitization of pre-existing clays, development of chlorite as quartz rims, formation of ferroan dolomite cements. During *later diagenesis*, early anhydrite and both former dolomite phases were partially dissolved, probably under the action of acidic pore waters generated by the maturation of organic matter within the underlying Carboniferous formation. This late dissolution episode, that created a secondary porosity, was followed by kaolinite, gypsum and minor pyrite precipitation. Gas emplacement from the late Cretaceous onwards halted further diagenetic reactions (Fig. 16.9).

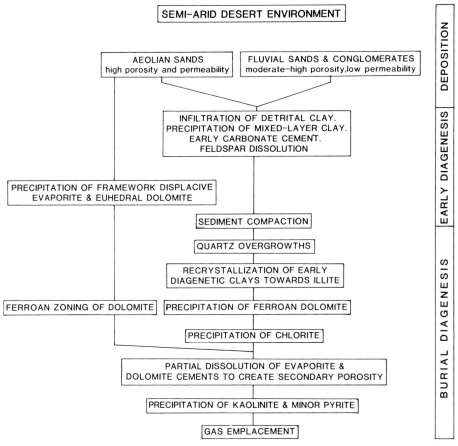

Fig. 16.9. Interpretation of the diagenetic history of the Rotliegendes sandstone in the Rough Gas field, North Sea. (After Goodchild and Whitaker 1986)

16.3.4 Clay Diagenesis and Reservoir Properties

Many sandstones from productive oil wells contain diagenetic kaolinite only, while unproductive wells filled by saline water show the presence of both kaolinite and illite (e.g., Kulbicki and Millot 1960, Smoot and Narain 1969, Sommer 1975). As kaolinite is mostly considered as an early-diagenetic species that tends to be destroyed through illitization, its preservation in a given sandstone reservoir is attributed to oil irruption. *The filling up of a rock reservoir by oil probably prevents late-diagenetic illitization.* Similar results arise from gas emplacement (e.g., Goodchild and Whitaker 1986). *By contrast, the filling up of the reservoir by saline water favors illitization of detrital or early-diagenetic minerals.* The study of the distribution of different diagenetic clays in a given sandstone body may therefore help to understand the chronological evolution of oil and water migration. Thomas (1986) calculates from potassium / argon dating in Jurassic sandstones of the Central Viking graben, North Sea, that the stopping of illitization by the hydrocarbon charge occurred between late Eocene (40 my-44 my BP) and Oligocene (28 my).

The development of secondary clays during sandstone diagenesis tends to decrease the permeability through pore space occupancy, and therefore to impede the hydrocarbon migration and diminish the reservoir volume. *The more evolved the diagenetic clay cementations and precipitations, the less favorable the reservoir properties.* The preservation of part of the primary permeability is favored either by chlorite coatings that often prevent further clay growth (e.g., Curtis et al. 1984, Dogan and Brenner 1987), or by a hydrocarbon migration taking place before noticeable illitization has started (i.e., during early-diagenetic kaolinization; e.g., Thomas 1986).

Bjørlykke and Brendsdal (1986) show that the leaching of feldspar and the formation of kaolinite in Jurassic Brent sandstone, North Sea, have been caused by the flow of meteoric water at moderate burial depth during early diagenesis (Fig. 16.8B), while more isochemical changes occurred during late diagenesis (Fig. 16.8C). Most of the diagenetic kaolinite results from the alteration of micas rather than of feldspar. Muscovite that altered to kaolinite expanded and flared into the pores, causing dramatic reduction in the sandstone permeability. As detrital mica preferentially settled in low-energy depositional environments, the corresponding lithofacies display the maximum reduction of permeability after the diagenetic evolution. *The recognition of the different types of depositional facies therefore greatly helps to predict the reservoir quality of sandy bodies that experienced clay diagenetic modifications.* Further three-dimensional modeling of reservoir performance by using geological, lithological and mineralogical data, combined with wireline logs (e.g., Hurst and Archer 1986), may greatly help the commercial exploitation of a given hydrocarbon-bearing field.

The production operations in oil fields result either from nature-driven energy (e.g., the expansion of the naturally compressed hydrocarbons, or the water influx from a highly pressured aquifer) or from artificial injection of water or gas. After such processes, the remaining oil may be recovered by heating the oil to lower its viscosity, or by changing the properties of the injection water. Such *production treatments may modify the reservoir properties and prevent further field development programs.* Many of the effects do not alter the mineralogy, but do change the permeability. Kantorowicz et al. (1986) investigate experimentally some of the reservoir-property modifications induced by production operations. *Acidization* to remove drilling mud dissolves siderite and chlorite without significant increase in rock permeability, because the dissolved siderite releases some previously cemented clay particles into the pore space. *Water injection* may cause movement of siderite rhombs and of clay particles as well as smectite to swell, which tends to determine a decrease of permeability and to impede the hydrocarbon extraction. *Steam injection* induces mineralogical changes, namely the dolomite and kaolinite destruction, as well as textural modifications. Such experiments allow prediction of the secondary effects of rock-fluid interaction in wells, and development of production programs without unintentional damage being caused to the reservoir.

16.4 Volcanic Environment

16.4.1 Igneous Rock Accumulations

Most volcanic rocks, especially volcaniclastites, present a *high pore and fracture permeability*, as do the sandstones. Their aptitude to favor secondary fluid migration and consequent diagenetic modifications is enhanced by the *dominantly sub-amorphous, disorganized and often porous* character of their constituents. In addition to the horizontal or oblique hydraulic conductivity, the depth of burial and the presence of *hydrothermal solutions* may commonly participate in the diagenetic evolution of volcanic rocks. The possible coexistence of all these factors explains the *high reactivity* of many volcaniclastites to postsedimentary processes. Detailed information on this subject is provided by Sand and Mumpton (1978), and by Fisher and Schmincke (1984). Let us give a few examples of clay diagenesis in different types of volcanic accumulations.

Thick accumulations of *silicic acidic glass* are particularly studied in Japan. Iijima (1978) and his colleagues investigate pluri-kilometric series of marine silicic volcaniclastites that have altered at depth in *four successive zones* (Fig. 16.10): (1) alteration of silicic glass to montmorillonite and opal A or opal CT (cristobalite-tridymite); (2) reaction of glass with groundwater to form alkali zeolites such as clinoptilolite and mordenite, additional montmorillonite and opal CT; (3) transformation of alkali zeolites into analcite, with additional change of clinoptilolite into heulandite followed by a change of heulandite into laumontite; (4) evolution of analcime into albite. Chloritic minerals and illite develop at depth within zones 3 and 4, that progressively pass to the prehnite-pumpellyite facies of the epimetamorphic zone. The general evolution corresponds to a *progressive dehydration* with depth and characterizes the *reaction series silicic glass-alkali zeolites-albite.* The study of Japanese deep oil-field wells shows that the transition between the successive zones ranges over restricted temperature intervals: 84-91 °C at 1700-3500 meter depth for the boundary zone

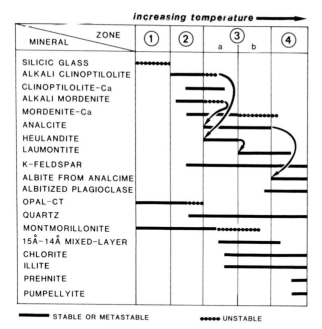

Fig. 16.10. Successive zones of authigenic silicates formed during burial diagenesis of marine silicic volcaniclastics. (After Iijima 1978). *Zone 4* grades into metamorphic domain

1-2, and 120-124°C at 2500-4500 meters between zones 3 and 4. The *temperature control* therefore appears to be much more important than the pressure control. As for other buried sediments, the *duration* of efficient temperatures probably also strongly intervenes in determining the intensity of diagenetic changes (see Fisher and Schmincke 1984).

In the Hokuroku Kuroko mineralization area, Northeast Japan, Inoue (1987) studies *silicic pyroclastic rocks* that have undergone both *burial and hydrothermal alterations favored by pore fluid migrations* (see also 16.5). A 900-meter-thick glass accumulation shows the parallel downward evolution of clinoptilolite into mordenite, analcite and laumontite, and of trioctahedral smectite into illite-smectite and especially chlorite-smectite mixed-layers. By contrast with the swelling rate of the smectite-illite serie that changes continuously from 100 to 0% expandability at increased depth, the smectite-to-chlorite conversion displays a discontinuous decrease in the percentage of expandable layers. The chlorite-smectite exhibits a tri-modal frequency in expandability throughout the drill holes, with steps at 100-80% (almost pure smectite), 50-40% (true corrensite) and 15-0% (swelling chlorite to true chlorite). The three types of chloritic minerals coexist in a depth range of about 200 meters (i.e., between 600 and 800 m). A. Inoue shows that the discontinuous diagenetic change of smectite into chlorite correlates with a continuous increase in tetrahedral aluminum, octahedral iron and exchangeable magnesium and iron, and with a continuous decrease in tetrahedral silicon and exchangeable sodium and potassium. *Discontinuous structural changes therefore correspond to nearly-continuous chemical changes.*

Alkaline rocks submitted to fluid migration and to depth of burial effects are investigated in the Pasco basin, Washington (USA), where up to 1500-meter-thick basaltic flows accumulated over 200,000 km² during Miocene times (Benson and Teague 1982). *The generalized sequence of secondary minerals with time and depth is iron-smectite – clinoptilolite – silica.* The nontronite forms at first, but is present at nearly all depths; it is probably associated with amorphous iron- bearing oxyhydroxides. Clinoptilolite appears below 350 meters. Silica occurs as quartz, cristobalite, tridymite and opal CT; quartz is ubiquitous whereas other silica types occur abundantly below 600 meters only. Minor components include various zeolites (mordenite, phillipsite, erionite, chabazite), celadonite, apatite, pyrite and gypsum. Mordenite develops below about 900 meter depth, which correlates with the beginning of clinoptilolite dissolution. According to Benson and Teague (1982), the diagenetic alteration of Pasco basin basalts does not involve significant hydrothermal processes, since clinoptilolite and quartz, that are metastable at temperatures exceeding 100°C (transformation into analcite and high-temperature silica, respectively), occur throughout the series. The chemical composition and location of groundwater masses associated with the rock texture, chemistry and permeability probably controlled most diagenetic processes, under thermodynamical conditions similar to those existing today.

The sedimentary *accumulation of fine ashes in lacustrine depressions* is sometimes responsible for the massive diagenetic formation of analcite-rich deposits called *analcimolites*. Analcimolites are mostly attributed to the post- sedimentary evolution of volcanic ash, although some analcite-rich sediments could be independent of igneous activity (see discussion in Millot 1964, 1970). In the Agadès basin, Niger, thick lacustrine sediments deposited from Permian to early Cretaceous times contain numerous analcite-rich beds, that constitute commercial analcimolites (Pacquet 1968). *Analcite developed during early diagenetic stages* from the alteration of fine, permeable and reactive volcanic ashes reworked from Devono-Carboniferous rocks of the Aïr massif. The formation of extensive analcimolites was favored by the alkaline lacustine environment, high pH values, and the availability of abundant volcanogenic silicon, aluminum and sodium. *Further diagenetic evolution of permeable volcaniclastic lacustrine rocks led to the alteration of analcite into various clay minerals*, depending on the leaching conditions permitted by the migrating groundwaters: kaolinite under active leaching, ferriferous chlorite under poor leaching and restricted chemical conditions, and aluminous montmorillonite under intermediate conditions.

16.4.2 Bentonites, Tonsteins

Thin beds of sediments extending over large areas and considered to have a major volcanic origin are classically called bentonites or tonsteins. The term bentonite originally refers to smectite-rich beds mostly derived from tephra deposits, while the term tonstein strictly designates kaolinite-rich claystone beds interbedded with coal-bearing strata (see Grim and Güven 1978, Bouroz et al. 1983, Fisher and Schmincke 1984). Bentonites usually correspond to subaqueous, mainly marine, smectitic horizons, while tonsteins rather represent con-

tinental subaqueous, kaolinitic and organic-rich horizons. In fact these defini-
tions do not easily apply to all situations encountered in nature or described in
the literature. Some so-called bentonites are not dominated by smectite in the
clay fractions, but by kaolinite or illite (e.g., Weaver 1963, Pollastro and
Martinez 1985, Teale and Spears 1986). More significant, the clay fraction of
some tonstein layers obviously passes laterally from a kaolinitic to an illitic or
smectitic composition, when the organic content of the surrounding sediments
decreases. We are therefore inclined to follow Fisher and Schmincke's suggestion
(1984), to designate by the unique term *bentonite* all *laterally widespread clay-
rich, thin beds that are of probable volcanic origin*. According to the dominant clay
species a specification should be given in the following way: *smectite-bentonite,
kaolinite-bentonite, illite-bentonite*. Notice that some bentonites are characterized
by the presence of halloysite in continental and even marine environments (e.g.,
Nagasawa, in Sudo and Shimoda 1978; Imbert and Desprairies 1987), and are
referred to halloysite-bentonites if this specific clay mineral is sufficiently
abundant in the sediment.

*Most clay minerals in bentonites undoubtedly result from the early diagenetic
alteration of vitric fallout ash*, as summarized by Millot (1964, 1970), Grim and
Güven (1978), and Fisher and Schmincke (1984). The most relevant arguments
are the following: (1) thinness usually less than 10 centimeters associated with
lateral extents exceeding tens to hundreds kilometers; (2) sharp lower and upper
contacts with adjacent sedimentary rocks; (3) presence of frequent vitroclastic
textures and of local unaltered vitric tuffs; (4) local existence of lateral gradation
from unaltered ash-rich to altered clay-rich sediment; (5) occurrence of some
relict minerals originating under high-temperature conditions ; these minerals,
that usually do not exist in adjacent common sediments, may include biotite,
rutile, sanidine, sphene, zircon and other species. In addition, high contents of
some conservative chemical elements such as Zr, Nb and Th, may participate in
the characterization of ancient volcanic layers modified by diagenetic alteration
(e.g., Pacey 1984).

Some bentonites, especially kaolinite-bentonites, are attributed to the
reworking over large distance of laterite-derived materials, for instance during
flood periods (see in Millot 1964, 1970). In fact the systematic attribution of
kaolinite-rich horizons to the in situ evolution of volcanic layers appears un-
reasonable, especially if the geographic extension of the layers is moderate (i.e.,
less than a few kilometers) and if the sediments lack any volcanic remains. The
situation is then comparable to that of some deep-sea sediments, for which the
high abundance of smectite in the clay fraction is systematically attributed to the
submarine evolution of volcanic glass (see Chaps. 12, 13). Soil-derived kaolinitic
tonsteins s.l. do exist in some coal fields. In addition, a lateritic alteration may
have affected ash beds after diagenesis if the series have been exposed in the
course of the time (e.g., Bouroz et al. 1983). Diagenetic kaolinite in tonsteins is
often distinguished from pedogenic kaolinite by its vermiform habit. Detailed
optical and chemical investigations may greatly help in distinguishing allochthonous and autochthonous causes in the mineral composition.

The dominant clay minerals in bentonites are smectites, that comprise either
the trioctahedral hectorite-saponite serie, or the dioctahedral montmorillonite-

beidellite-nontronite group. The nature of the smectite type depends on both the initial glass composition and the groundwater characteristics. For instance in the South Bering Sea, saponite-nontronite varieties of smectite and high Fe/Al ratios characterize bentonite beds derived from basaltic glass, while montmorillonite varieties with low Fe/Al and Ti/Al ratios point to silicic volcanic ash as source material (Hein and Scholl 1978). *Montmorillonite and Al- beidellite are the most frequently reported species*, which results from the dominant rhyolitic to dacitic nature of volcaniclastites. The rapid formation of smectite-bentonite is favored by a moderate magnesium content (5-10%) and discouraged by too high silicon contents ($>70\%$ SiO_2) (Grim and Güven 1978). The alteration of silicic ash to bentonite requires external MgO, which usually implies the contribution of sea-water magnesium and noticeable migration of pore water. Note that Hein and Scholl (1978) propose to distinguish volcano-diagenetic smectites from common detrital smectites by the very low percentage of illitic layers ($<15\%$) in the former minerals.

The second abundant mineral in bentonites is *kaolinite*, that typically characterizes the thin beds (0.5-2 cm thick) developed over wide distances in Carboniferous coal series. Kaolinite is mainly attributed to the *alteration of tephra layers under acidic conditions* determined by the organic-rich swamp environment. Low salinity appears to favor the alteration of vitric ash to kaolinite. Some hypotheses imply a transitional stage marked by the solution of the glass, the possible formation of a gel, and the crystallization of kaolinite from the gel in reducing microenvironments (e.g., Pollastro 1981). The alteration processes probably started soon after sedimentation at low burial depth, and typically belong to the early-diagenetic zone.

With increasing burial depth, both smectite-bentonites and kaolinite-bentonites tend to be replaced by typical late-diagenetic clays such as illite-smectite, illite and even chlorite (Fisher and Schmincke 1984, Inoue 1987). Note that other diagenetic minerals may be commonly associated to clays in bentonites, especially silica and zeolites of the clinoptilolite group (e.g., Hein and Scholl 1978, Senkayi et al. 1987).

Bentonites s.l. represent useful *stratigraphic markers* and dating tools, especially the tonstein layers occurring within continental or lagoonal sequences that present rapid lateral facies changes. The correlation of bentonite beds in the different sectors of a given basin is facilitated by distinct mineral associations and trace element compositions, the use of which is particularly relevant if diagenetic alterations are moderate or of a comparable intensity (e.g., Spears and Kanaris-Sotiriou 1979, Bouroz et al. 1983, Fisher and Schmincke 1984). In addition, mineralogical and geochemical studies greatly help to *recognize the origin of volcanic fallouts*. For instance, Spears and Kanaris-Sotiriou (1979) compare the composition of some British and other European Carboniferous tonsteins by using clay mineral suites and various elemental ratios (Ti, Cr, Zr and Ni over Al). The British tonsteins of acidic composition resemble the French and German tonsteins, suggesting a common distant origin and allowing some stratigraphic correlations. In contrast, the British tonsteins derived from basic ash contain variable amounts of detrital constituents, do not resemble other European bentonitic beds, and probably originate from local eruptions. Pacey (1984) studies the composition of about 20 thin persistent marl horizons in the post-Cenomanian chalk sequence of central Eastern England. The abundance of Mg-smectite and the anomalously high content of Zr, Nb and Th relatively to Ba, Rb and K, suggest a unique volcanic origin. The analyses and mapping of the bentonitic beds in the British Isles area indicate that the ashfalls are of the tropospheric type, and were drawn by westerly winds from a center located to the West or Northwest of mainland Britain. N.R. Pacey envisages that the Anton Dohrn seamount in the Rockall Trough, that destabilized immediately before the separation of Greenland from Northwestern Europe, is a likely center of origin for the late Cretaceous ashfalls in this region.

16.5 Hydrothermal Environment

Introduction

Warm and ion-enriched fluids migrating through rocky formations tend to influence, and often to accelerate the diagenetic reactions. For instance, the conversion of smectite to illite, that usually occurs at depths of burial exceeding 2.5 kilometers, is nearly completed at much shallower depth in the Jurassic sediments of Scotland, due to local convective hot-water circulation cells associated with Tertiary igneous intrusions (Andrews 1987. See also Yau et al. 1987b, 1988. 15.1.3). The mechanisms developing within the rocks usually preceed in a more or less continuous sequence those occurring when hydrothermal fluids reach the surface of the earth or the sediment-sea water interface (see Chap. 13). Different approaches allow estimation of the relative depth and temperature of the different hydrothermal manifestations occurring in a given rock. *Clay minerals may constitute useful markers* in such *thermal estimations.* For instance, Noack (1985) distinguishes the main characters of three common sheet silicates forming at either low or high temperature. Talc, chlorite and serpentine display distinct crystallographic, geochemical and isotopic characteristics in both environments (Table 16.2). Cathelineau and Nieva (1985) show that the geochemical characteristics of hydrothermal chlorites parallel the microthermometric data of fluid

Table 16.2. Some characters of high- and low-temperature talc, chlorite and serpentine (after Noack 1985)

	High temperature	Low temperature
Talc	$\delta^{18}O = +10‰$	$\delta^{18}O = +20‰$
	ferrous iron	ferrous and ferric iron
Chlorite	IIb polytype	Ib or Ia polytype, or
		chlorite-vermiculite mixed-layer
	Al > 40%	Al = 40%
	Fe > 20%	Fe = 40–50%
	positive Al_{oct}–Fe correlation	negative Al_{oct}–Fe correlation
Serpentine	alteration and precipitation	precipitation
	Si, Al, Fe	Si, Al, Fe

inclusions in gangue minerals from altered andesites in the active geothermal system of Los Azufres, Mexico. The abundance of tetrahedral aluminum correlates especially well with absolute temperature in the 150 to 300°C range. *Chlorite therefore represents a reliable solid solution geothermometer.* Similar results are provided by Boiron and Cathelineau (1987) for hydrothermal chlorite, illite and phengite from Carboniferous volcaniclastites and rhyolites of the Massif Central, France. Meunier et al. (1987) use the hydrothermally formed smectite-illite-chlorite phases as geothermometers in various fractured plutonic and metamorphic rocks from France, that were submitted to fluid temperatures lower than 200°C. The mineralogical suite developed at increasing temperatures is the following: montmorillonite and vermiculite, random smectite-vermiculite mixed-layers (40 to 100% smectitic layers), ordered smectite-chlorite mixed-layers (< 40% smectitic layers), illite – chlorite – corrensite or illite – chlorite. A. Meunier and colleagues deduce a succession of phase diagrams characteristic of four temperature domains situated around 30, 100, 160 and 200°C.

Hydrothermal fluxes participate in an active way in the transformation of hard silicate rocks into soft argillaceous rocks (see Millot 1964, 1970, Sudo and Shimoda 1978, Thompson 1983, Berger and Bethke 1985, Velde 1985). Deep Earth Drilling Programs developed in various countries favor rapid progress in the knowledge of hydrothermal processes within both igneous and sedimentary rocks. Most hydrothermal processes occur in rock fractures and fissures, and determine the growth of secondary minerals such as vein fillings or wall-rocks. Some massive hydrothermal alteration also exists, and develops mainly at the expense of acidic plutonic or volcanic rocks.

Vein Filling and Wall-Rock Minerals

A large variety of clay minerals may form by interaction of migrating hydrothermal fluids with host rocks. Magnesian minerals are the more widespread species, and are dominated by talc, chlorite and serpentine (see Millot 1964, 1970, Noack 1985). For instance, Girard (1985) identifies in late Precambrian shales and sandstones of Taoudeni basin, Mauritania, three hydrothermal phases occurring at decreasing temperatures with time: (1) high-temperature wall-rock metamorphism (350-600°C) leading to talc, calcite, diopsite, forsterite, phlogopite, tremolite and quartz in calcareous environment, and to biotite, phengite, quartz and titanium oxides in siliceous environment; (2) medium temperature wall-rock alteration (200-400°C) resulting in the formation of serpentine in carbonates and of chlorite, albite, kyanite in sandstones; (3) fairly low-temperature cementation (120-150°C) by various minerals such as illite, chlorite, siderite and calcite. Parneix et al. (1985) study the hydrothermal chloritization of biotite in two granitic rocks of the Massif Central, France, and demonstrate that migrating fluids supply significant amounts of the magnesium and iron necessary for the mineral change. Hydrothermal processes therefore appear characterized by the existence of *open chemical conditions* and by *non isochemical reactions*, what differs from diagenetic reactions basically controlled by the depth of burial in shaly sediments (Chap. 15).

Regular chlorite-smectite mixed-layers (corrensite group) are frequently reported in vein fillings or pore space of various rocks. Morrison and Parry

(1986) describe the formation of a dioctahedral chlorite-smectite in Permian sandstones of the Lisbon valley, Utah, through the rising of hydrothermal fluids along a major fault system. Corrensite and associated illite-smectite are attributed to the interaction of moderately warm fluids (about 100°C) with smectite and an Al-bearing phase such as K-feldspar or kaolinite. Pouteau et al. (1985) report complex pervasive and wall-rock alterations in volcaniclastic andesitic accumulations of Bossa, Haiti. Pervasive alteration at high temperature (200-280°C) mainly led to chlorite, while wall-rock alteration favored the formation at medium temperature (< 200°C) of corrensite, saponite and illite-smectite mixed-layers. Corrensite occurs in other hydrothermally affected fractured rocks such as monzonitic syenites of the Massif Central, France (Meunier 1982), or marble, quartzite, paragneiss and metavolcanics of the Precambrian from Ontario, Canada (de Kimpe and Miles 1987).

Fibrous clays constitute a third group of magnesian clay minerals frequently associated with hydrothermal processes in continental fault and fracture systems, which resembles the probable hydrothermal dependence of deep-sea authigenic palygorskite or sepiolite (13.4). Most hydrothermal fibrous clays develop in basalts, serpentinites or other related basic rocks, and are often accompanied by secondary carbonates. Numerous examples are provided by Sudo (in Sudo and Shimoda 1978), Callen, and Imai and Otsuka (in Singer and Galan 1984), and Velde (1985). Some fibrous clays accumulations of hydrothermal origin are of a commercial interest, as are some of those derived from alkaline evaporative sedimentation (Chap. 9). Note that all palygorskite-rich deposits associated with basic volcanic rocks are not necessarily of a hydrothermal origin. Siddiqui (in Singer and Galan 1984) reports that the large palygorskite fields (> 14 million tons) occurring in pockets along the margin of the Deccan Trap formation, Western India, result strictly from post-Eocene weathering of basic igneous rocks under surface conditions.

Other wall-rock clay minerals issued from hydrothermal alteration include *saponite* (e.g., Post 1984, Pouteau et al. 1985), *halloysite* (e.g. Mârza et al. 1966), *kaolinite* (e.g., Keller 1988), *vermiculite*, *illite* and various mixed- layered varieties (e.g., Meunier et al. 1987).

Vein-Filling Clay Sequences

Mineralogical and geochemical sequences often develop in rock veins and walls, the successively formed hydrothermal minerals being possibly transformed in turn, depending on changes of physico-chemical conditions in the course of the time. Such sequences are mainly documented for igneous rocks, much less for sedimentary rocks (e.g., Girard 1985, Yau et al. 1988). Many data concern *plutonic rocks*. For instance, Cathelineau (1983) identifies from metallographic, mineralogical and microthermometric data seven successive stages of clay authigenesis in vein fillings and alteration zones of five uranium hydrothermal deposits located in leucogranites from Western and Central France; (1) phengites and chlorites related to the earliest stages of granite alteration, before the hydrothermal activity really took place; (2) phengites crystallization within the granite, close to the veins; (3) illite and celadonite precipitation on the wall-rocks or as inclusions in quartz; (4) illite-smectite mixed-layer crystallization during pitchblende deposition; (5) removal by fluids of the primary uranium ores, resulting in new parageneses from the inner to the outer part of the wall-rock: illite, illite-smectite, K-smectite and adularia, chlorite, kaolinite; (6) smectite and coffinite precipitation; (7) late formation of kaolinite, preceding the meteoric alteration. M. Cathelineau demonstrates that the mineral sequence is controlled by the progressive temperature decrease of circulation fluids with time. A similar correlation of mineral sequences with temperature decrease arises from the study of Massif Central granites, in the Haut-Cézallier area (Griffault et al. 1984): (1) high-temperature, propylitic alteration (200-230°C) leading to chlorite, adularia, prehnite and calcite; (2) medium-temperature formation (175-200°C) of regular mixed-layers, kaolinite, ankerite and quartz in whitish veins; (3) low-temperature precipitation (<100°C) of kaolinite, siderite, K- smectite in brown veins.

Volcanic rocks display extensive vein hydrothermal alteration sequences in the continental environment. The *andesite* and rhyolite flows piled up in the region of Los Azufres, Mexico, are located in a commercial geothermal field where active fluids permanently migrate in a strongly fractured system. X-ray diffraction, microprobe analyses and microscospe investigations lead to the identification of three mineral zones, following each other at increasing temperature, from the surface to more than 2300 meter depth (Cathelineau et al. 1985): (1) clays (mainly smectite, kaolinite and illite) and zeolite; (2) calcite, chlorite and illite; (3) chlorite, quartz and epidote (Fig. 16.11). The mapping of the distribution of main primary and secondary minerals along about 40 active wells reveal the existence of a large geothermal fluid body, that ascends and discharges through two main

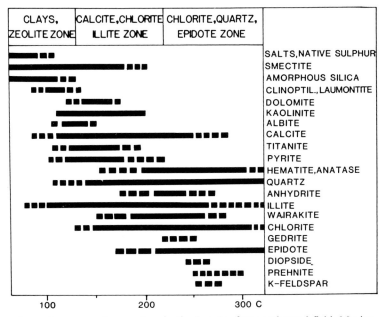

Fig. 16.11. Mineral zonation versus temperature in the Los Azufres goethermal field, Mexico. (After Cathelineau et al. 1985)

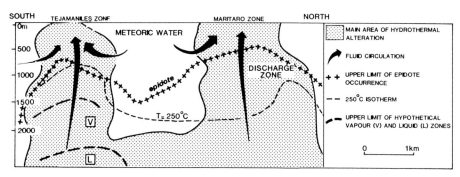

Fig. 16.12. Schematic cross-section of hydrothermal-alteration distribution in a South-North transect, Los Azufres goethermal field, Mexico. (After Cathelineau et al. 1985)

fracture systems. These two privileged circulation areas display concentric aureoles of upward-decreasing hydrothermal alteration (Fig. 16.12). The general pattern of the geothermal field corresponds to a dome structure induced by the anomalously high thermal gradient, and distorted by the local effects of the fracture systems that favor intensified fluid circulation.

Massive Alteration

Most extensive hydrothermal alterations occur in igneous rocks associated with mountain chain systems, where the continental crust is deeply fractured and submitted to internal constraints. The large ores of *kaolinite deposits* that cover *granitic areas in ancient orogenic belts* like Britain or Cornwall are primarily attributed to hydrothermal activity (e.g. Brown 1953, Nicolas 1956). The kaolinization phase often corresponds to a late hydrothermal alteration phase, that followed or accompanied other phases like tourmalinization and sericitization. Meunier and Velde (1982) show that a two-mica granite from Southern Armorican chain, Western France, experienced a low-temperature hydrothermal metamorphism that produced three mineral sequences: (1) muscovite alteration to Fe-Mg phengite, and biotite alteration to kaolinite and titanium oxide; (2) recrystallization of phengite, muscovite and feldspar into Al-Mg sericite; (3) massive alteration of feldspar to produce kaolinite, K-beidellite and Al-illite. Kaolinite is the predominant mineral species in most altered facies.

Wall-rock alteration may display massive features if host rocks are highly reactive and sensitive to hydrothermal fluids. This is particularly the case with *pyroclastic, vitreous rocks.* The Kuroko deposits in Japan represent a good example of such processes (Shirozu, in Sudo and Shimoda 1978). Kuroko deposits belong to the Green Tuff formations, that largely crop out in the northern part of Honshu island and in Hokkaido, and that consist of strata-bound polymetallic sulfide-sulfate materials related to Miocene felsic volcanism. Kuroko deposits show complex alteration facies extending on a kilometric scale. Four alteration zones extend from the lower or inner to the upper or outer: silicified zone accompanying stockwork ores, clay zone enclosing stratiform ore bodies, sericite-chlorite zone, and montmorillonite zone. Clay minerals in the clay zone consist

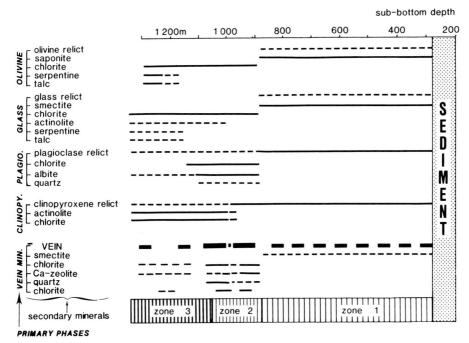

Fig. 16.13. Hole 504B DSDP, Galapagos spreading center. Major secondary minerals in relation to their primary solid phases in the basalt, and to the depth distribution of veins. (After Kawahata and Furuta 1985)

mainly of dioctahedral mica-montmorillonite mixed-layers (sericite habit), Mg-chlorite and kaolinite. The montmorillonite zone is characterized by Al-smectite, opal and zeolites, and progressively grades into the adjacent regional zeolitic alteration zone (analcite-rich deposits).

In the deep sea, basalt alterations recovered from drilled series display mineral sequences that are often dominated by *iron- to magnesium-rich minerals*. The mineral suites identified at increasing depths more or less recall the different associations forming close to the sea bottom through the action of hydrothermal fluids expelled at various temperatures (Stakes and O'Neil 1982. See also 13.1, 13.6). Kawahata and Furuta (1985) investigate Hole 504B of the Deep Sea Drilling Project, that was drilled down to 1350 meters below the sea floor in the southern crest of Costa Rica rift, Galapagos spreading center (Eastern Equatorial Pacific). Rocky formations altered through high-temperature water-rock interaction are recovered below 890 meter depth, and display secondary associations of the greenschist facies. Three mineral zones are identified (Fig. 16.13): (1) 234-890 m: fresh basalt, minor alteration to saponite, serpentine, talc. (2) 890-1050 m: strongly altered basalt marked by abundant veins filled with secondary chlorite, epidote, laumontite, quartz and actinolite. Some samples are enriched in Mn and Zn. (3) 1050-1350 m: moderately altered basalt, less abundant veins filled with the same minerals as in zone 2. The estimated alteration temperature changes sharply at boundary zone 1-2. It is less than 100°C in zone 1, and more than 200°C in zones 2 and 3. This probably results from a time gap. Intensive hydrothermal activity most likely occurred during the formation of oceanic crust in the rift valley, corresponding to zone 3 and 2 basaltic flows. Much later, another lava flow corresponding to zone 1 covered the rocks at some distance from the ridge axis, and subsequently experienced much less active hydrothermal circulation.

16.6 Conclusion

1. *The influence of anomalous temperature, pressure or interstitial fluids in buried sediments may interfere with the basic effects of either ion exchanges during early diagenesis or depth of burial during late diagenesis.* Such interferences tend to be enhanced in the case of *high permeabilities* or *rapid lithologic contrasts.* The effects of strong geothermal gradients or active fluid circulation may be cumulative with those of normal burial, and determine important diagenetic changes over a sediment thickness much lower than usually expected (e.g., illitization of smectite over a few hundred meters instead of a few kilometers; see also Chap. 15). Most frequently, however, the effects of actively migrating fluids determine important local changes, which tend to obliterate the burial effects. This is especially the case for permeable sandstones, reactive volcaniclastites and hydrothermally affected formations. The recognition of the boundary zone between sedimentary and diagenetic constraints on mineral associations is often complicated by such situations, that prevent reliable paleogeographic applications (see Part 6).

2. Lateral clay diagenetic changes in mountain-belt areas reflect the attenuate effects of metamorphism linked to orogeny. Peculiarly well documented in the Western Alpine range, *the clay-mineral modifications recorded in surficial deposits from outer to inner parts of mountain chains roughly parallel the modifications observed vertically at increasing depths of burial.* Clay mineral suites tend to become simpler, better crystallized, and to be dominated by illite and chlorite. The respective influence of temperature and pressure on the clay mineral changes is still little documented, but thermal effects appear to play an especially significant role. Local paleogeographic or hydrothermal peculiarities may complicate the progressive lateral obliteration of the paleoenvironmental messages by the diagenetic constraints. *The overburden of sedimentary formations by tectonic nappes may also induce mineral modifications similar to those resulting from progressive burial.* The diagenetic imprint in such overloaded sediments is preserved permanently, which may help to recognize the geographic extension of past tectonic covering even if the nappes are subsequently removed by erosion.

3. *Less permeable and chemically confined sediments may relatively resist the diagenetic constraints induced by depth of burial or tectonics.* This is the case with the *limestone-marl alternations* submitted to vertical or lateral diagenesis, in which the *marly interbeds* are preserved much better and longer than calcareous beds. In addition the marly interbeds preferentially evolve toward illite-rich clay suites by using the potassium released by detrital feldspars and micas, while more permeable and Mg-bearing limestones are enriched in chlorite. Another case of

resistance to thermodynamical changes concerns *smectites*, that may locally be preserved and display enhanced crystallinity in alkaline, often magnesian environments of the late diagenetic-early metamorphic zones. Smectites may even have formed in such deep and warm environments, as confirmed by laboratory experiments. *Metamorphic smectites* appear to be of a highly crystalline, magnesian type (saponite), to be restricted to calcareous- dolomitic sediments, and to be frequently associated with corrensite, talc and perhaps swelling chlorite.

4. *Highly permeable sediments favor the horizontal circulation of groundwaters and induce specific diagenetic changes that interfere strongly with burial effects.* This is especially the case with coarse to medium-sized sandstones, the major constituents of which are detrital, but may have been modified or complemented by minor amounts of various secondary minerals. Clay minerals constitute the most frequent species diagenetically formed in permeable sandstones. They mainly comprise kaolinite, illite and chlorite, and accessorily mixed-layers and other species. The habits, mineral suites and chemical compositions of diagenetic clays in sandstones display large diversity, and only the major phenomena are clearly understood so far. One of the prime difficulties arises from the similar nature and composition of both detrital and diagenetic clay minerals, which necessitates the use of sophisticated optical and chemical investigation methods.

Kaolinite predominantly consists of large hexagonal sheets stacked as vermiform aggregates, but may also occur as blocky particles or overgrowths. Kaolinite derives mostly from the early diagenetic alteration of detrital micas or feldspars, under the action of meteoric freshwater migrating in more or less oxygenated sediments. Mica-derived kaolinite is often favored in interchannel sediments where detrital sheet-silicates abundantly deposited under low-energy conditions. Late diagenetic kaolinite occurrences are also reported in some deeply buried sandy environments (e.g., North Sea reservoirs), where the mineral may be replaced by dickite. *Illite* mainly presents a hairy habit, and results from late diagenetic replacements or overgrowths on kaolinite, quartz and other silicate grains. The diagenetic formation of illite is favored by saline, ion- enriched, organic-poor migrating groundwaters. Other habits include pore- fillings of blocky illites. A similar diversity characterizes *chlorite*, that occurs less frequently and constitutes mostly coatings of fine particles around coarse detrital grains. Diagenetic chlorite appears to be favored by the presence of anoxic and non saline groundwater.

Diagenetic clay minerals in permeable sedimentary rocks are actively studied from environmental, textural and kinetic points of view, since their formation or inhibition dramatically affects the *hydrocarbon reservoir properties*. The occupancy of pore space by clay diminishes the rock permeability, which is of a great importance if diagenetic processes occur in an early phase. The infilling of sandstones by oil or gas prevents further argillization, whereas the migration of saline waters favors late diagenetic illitization.

5. *Volcaniclastic sediments are particularly sensitive to diagenetic modifications* because of their usually high permeability, the high reactivity of their amorphous constituents to chemical exchanges, and their frequent association with hydrothermal exhalations. Accelerated diagenetic changes often occur in such environments, which favors the formation of *zeolites* and little- hydrated silicates

rather than of clay minerals. A mineral sequence commonly encountered at increasing depth comprises smectite, *clinoptilolite, heulandite and analcite* successively. Analcite also constitutes local accumulations (= *analcimolites*) derived from the early diagenetic alteration of fine vitric ash in alkaline lacustrine deposits, and ulteriorly possibly altered into various clay minerals. In a general way, the types of clay minerals originating from the alteration of volcanic materials may be very diverse, and depend greatly on the chemical composition of parent components.

Bentonites represent thin, widespread beds of clay-rich sediments that chiefly derive from the early diagenetic subaqueous alteration of fine volcanic ash. Bentonites are characterized by either *smectite* (= bentonites s.s., or S-bentonites), *kaolinite* (K-bentonites) or *illite* (I-bentonites). Smectite-rich bentonites are particularly frequent, and mostly result from a submarine alteration. The types of secondary smectite vary from Al to Mg and Fe end-members, depending on the initial composition of the glass. Kaolinite-bentonites mainly characterize continental, organic-rich environments, where they constitute brown, persistent horizons (= *tonsteins*). Kaolinite mostly derives from the early-diagenetic evolution of fallout glass, but may locally have a pedogenic origin (e.g. reworking of lateritic soils, or post-sedimentary weathering). All types of bentonites may serve as useful stratigraphic and source markers.

6. *Hydrothermal processes affect buried series in both continental and marine environments.* Warm, chemically concentrated fluids migrate upwards within the rocks along the fracture and pore systems, and determine various types of alteration. Most alteration phenomena concern *igneous rocks* of both plutonic and volcanic origin. Sedimentary rocks are more rarely affected, or less well studied from this point of view. The minerals originating from hydrothermal fluid-rock interactions present a very large diversity, depending on the composition of both migrating fluids and primary solid constituents. The minerals most frequently encountered in hydrothermal veins and wall-rocks are *magnesian species such as chlorite, talc, serpentine, fibrous clays, corrensite, Mg-smectite* and numerous non-clay minerals. Other types of smectite, as well as halloysite, vermiculite, various mixed-layers and kaolinite, are also reported from vein and fracture fillings, where mineral sequences often develop according to the alteration depth and temperature. The massive alteration of igneous rocks induced the preferential formation of kaolinite in granitic rocks, whereas volcanic rocks altered to diverse minerals, depending on their chemical composition and on the continental or submarine context.

Part VI

Clay Stratigraphy and
Paleoenvironment

Chapter 17

Paleoclimate Expression

17.1 Bases and Conditions

17.1.1 Fundamentals

The clay mineral composition of weathering profiles and soils largely depends on the climatic conditions existing on land at successive periods of the geological history (Chap. 2). This main climate control allows the use of paleosol clay associations in order to reconstruct past climates (e.g., Singer 1980). In many cases, however, the paleosols have disappeared through erosion, are truncated, or have undergone cumulative weathering processes, and therefore reflect poorly the ancient climate conditions.

The products of soil erosion consist mainly of clay minerals, that are less cohesive and more easily reworked than rock-derived materials. The uppermost part of pedologic profiles, that particularly tends to be equilibrated with ongoing climate, is preferentially subject to removal by erosion agents. *The soil-derived, terrigenous clay minerals statistically accumulate in a stratigraphic order within sedimentary basins, in the course of time.*

The distribution of clay minerals in recent sediments of the world ocean largely reflects the continental weathering processes and the latitudinal distribution of soils and weathering complexes on exposed land masses (Chap. 8). As clay minerals are largely dispersed in the marine environment and often experience no or only little diagenetic modification (Chaps. 14 to 16), it is tempting to consider their successive assemblages in sediments as possible indicators of successive paleoclimates. Some reviews have been presented on this subject (e.g., Chamley 1971, 1974, 1979, Singer 1984). Let us summarize the main principles and limitations of the *paleoclimatic use of sedimentary clays*, through a few examples and applications.

First data on the possible use of clay minerals as paleoclimatic markers in Quaternary sediments appeared in the late 1960's. Hallan (1966) reported from recent sediments deposited on the Sigsbee Deep, Gulf of Mexico, that smectite and kaolinite abundance increased during interglacial stages, while illite and chlorite amounts increased during glacial stages. Monaco (1967, 1971) observed a relative augmentation of chlorite and illite abundance in late Pleistocene relative to Holocene sediments on the Roussillon margin, South France. Chamley (1967) reported in two piston cores from the Northwestern Mediterranean a close, *direct correlation between the width at half-height of the 10 Å X-ray diffraction peak of illite, and the abundance of warm-water planktonic foraminifera.* As the width of the 10 Å peak is inversely proportional to the crystallinity and structural order of illite particles, the sedimentary levels deposited during warmer periods appeared to correspond to stronger hydrolysis processes (i.e., more humid and warmer climate) on peri-Mediterranean land- masses. By contrast, the higher crystallinity of illite observed in cold-water sedimentary levels indicated a better preservation of rock-derived minerals, and therefore less hydrolyzing conditions on land (i.e., colder or dryer climate). The correspondence existing between illite crystallinity and planktonic faunas suggested using the degree of "opening" of illite crystals as an index of the hydrolyzing power capacity of the environment in the continental source area. Parry and Reeves (1968a) described an augmentation of kaolinite and smectite abundance in lacustrine horizons deposited during Quaternary pluvial periods in the southern plains of Texas, USA, which was related to climate variations.

A systematic and critical approach of the paleoclimatic use of clay and associated minerals is developed for the middle to late Quaternary hemipelagic muds deposited in the Northwestern Mediterranean Sea (Blanc-Vernet et al. 1969, Chamley 1971). Clay assemblages derive mainly from the Rhone river and comprise abundant illite, fairly abundant chlorite and smectite, and accessory amounts of kaolinite and irregular mixed-layers (mainly illite-smectite, chlorite-smectite). In a general way, the sedimentary levels whose planktonic foraminifera

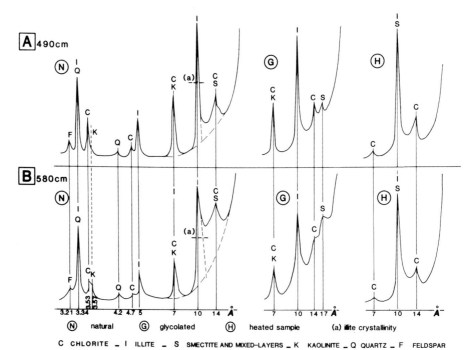

Fig. 17.1. Examples of X-ray diagrams in beige (**A**) and gray (**B**) hemipelagites from core 1MO67, Northwestern Mediterranean Sea (41°42′N–5°52′E, 2460 m-depth. After Chamley1971)

or pteropods lived in temperate to warm sea water display a relative decrease in illite crystallinity, and an increase in smectite and kaolinite abundance (Fig. 17.1A). In contrast, the levels enriched in planktonic fauna typical of cool to cold sea-water show an increase in illite crystallinity, chlorite abundance and crystallinity, and in smectite crystallinity (Fig. 17.1B). This suggests that *periods of relatively warm water in the sea correlated on land to relatively strong hydrolysis (i.e., high rainfall and temperature), responsible for the degradation of illite, the production of pedogenic kaolinite and smectite. Opposite conditions marked by low rainfall and temperature appear to have prevailed during periods of cold sea water, leading to the preservation on land of pre-existing, rather well- crystallized illite, chlorite and smectite.* As a consequence, *marine clay- mineral characteristics may be envisaged as reliable indicators of successive climates that occurred in a given continental domain.* The crystallinity of illite and other clay-mineral parameters can help to distinguish cold-dry conditions, that are frequently associated with glacial stages, from warm-humid, more hydrolyzing interglacial conditions.

In addition to clay mineral data, the levels suggesting enhanced hydrolysis on land tend to be enriched in coarse fraction, in calcium carbonate and in feldspars compared to quartz, and display yellowish to beige colors (Fig. 17.2). The coarse fraction is impoverished in fragile mineral species like feldspars, biotite and amphiboles. Geochemical analyses indicate a relative depletion of the more easily leached silicon relatively to aluminum. Transmission electronmicrographs reveal the abundance of small-sized, altered and flaky clay particles. All these characters converge to suggest that beige, degraded illite- bearing levels correspond to rather warm and humid continental climate, favoring the development of chemical alteration and of surficial soils in the hinterland. By contrast the grey, crystalline illite-bearing levels correlate with less calcareous, more siliceous sediments that show large, well-edged and well preserved clay particles. These levels are indicative of stronger physical and lesser chemical weathering conditions on land, and point to the existence of cooler, dryer climate.

X-ray diagrams obtained on carefully prepared oriented pastes of less than 2 μm, non calcareous particles allow the measurement of reliable and *reproducible parameters, that may represent quantitative indicators of hydrolysis processes on land* (Fig. 17.2). The following indices appear especially useful to characterize past climate in the Mediterranean range (the sign+or − in parentheses indicates the evolution trend of the values under increasing hydrolysis conditions):

− Chlorite relative abundance: 4.7/5 Å peak-height ratio, natural sample (-).
− Chlorite crystallinity: sharpness of the 4.7 Å peak, natural sample (-). 1 = well-defined, 3 = poorly-defined peak.
− Smectite relative abundance: 17/10 Å peak-height ratio, glycolated sample (+). The smectite augmentation mainly results from the degradation of chlorite and illite under moderately hydrolyzing conditions.
− Smectite crystallinity: angle (°) of the 17 Å peak, glycolated sample (+). This index partly includes the irregular mixed-layers, the abundance of which tends to increase with increasing hydrolysis under temperate climate.
− Kaolinite relative abundance: aspect of the 3.5 Å doublet, natural sample (+). 1=dominant chlorite particles (3.53 Å), 3=dominant kaolinite particles (3.57 Å).

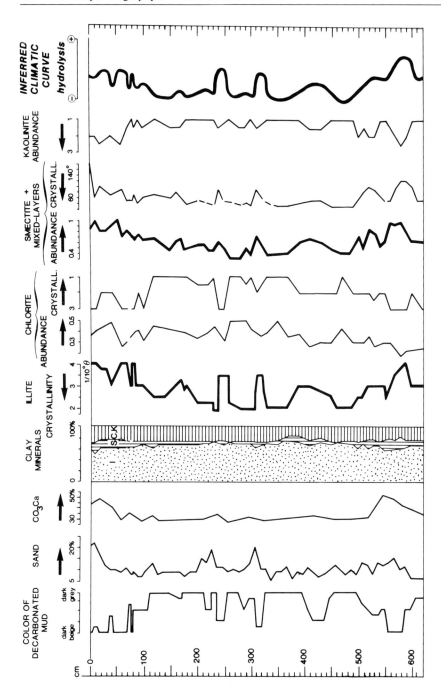

Fig. 17.2. Sedimentological and mineralogical data on late Quaternary hemipelagites of Northwestern Mediterranean Sea (core 1MO67, Balearic plain. After Chamley 1971). *I* illite; *S* smectite and expandable mixed-layers; *C* chlorite; *K* kaolinite; *arrows* indicate an augmentation or an amelioration. Other explanations: see text

The paleoclimate information deduced from clay mineral assemblages in marine sediments basically concerns the degree of hydrolysis at the surface of exposed land-masses. The respective impact of temperature and rainfall on the resulting mineral associations is difficult to assess a priori, and may be complicated by other factors like seasonality or drainage conditions. The use of clay minerals to reconstruct past climates appears particularly interesting to supplement the information provided by other indicators (marine faunas, pollen and spores, oxygen isotopes), or to replace other indicators when they are lacking or modified by sorting, dissolution, local chemical conditions, etc.

17.1.2 Conditions of Application

Clay and associated minerals formed through weathering processes during most periods of geological history, at least since the late Precambrian, when oxygenated conditions replaced reducing conditions on the Earth's surface. Clay minerals therefore a priori represent useful markers of the successive climates responsible for the soil formation in the course of time. As the terrigenous clays deposited in sedimentary basins result mainly from the surficial erosion of continental soils, they may constitute reliable indicators of successive climate conditions on land. Some limitations to such a use nevertheless occur. *The basic limitation consists in the possible existence of diagenetic changes.* The sedimentary series investigated from a paleoclimatic point of view must have experienced *moderate depths of burial* (usually less than 2.5 km), and have been *preserved from aggressive chemical conditions* (see Chaps. 14 to 16). Suitable situations exist in many natural environments, especially in *clay-rich sedimentary series deposited since the Mesozoic at a fairly high sedimentation rate in Tethys- or Atlantic-like basins.*

One of the prerequisites to using clay minerals as paleoclimatic markers is the availability of *precise, discriminant and highly reproducible techniques in sample treatment and X-ray diffraction processing.* The mineralogical variations induced by climate changes during a period of 10^3 to 10^5 years often do not exceed 10 to 20 percent of the relative abundance of a given clay group. The quantitative estimations of clay minerals on X-ray diagrams obtained from oriented preparations are sometimes not precise, and the range of values may be of the same order of magnitude as those determined by climate changes. Such limitations may be responsible for the fact that in a given region some authors identified climatic variations in late Quaternary sedimentary clays, while others did not (e.g., Gulf of Mexico: Harlan 1966, Scafe and Kunze 1971). The availability of modern equipment and the use of reliable indices on the relative abundance and crystallinity of clay minerals, associated with a statistical approach, usually make it possible to overcome these technical difficulties.

Many clay suites in marine and lacustrine sediments result from the erosion of both soils and rocky substrates. For instance the abundance of illite and chlorite in the Northwestern Mediterranean Sea reflects the widespread occurrence of crystalline rocks cropping out in the mountainous regions of the Rhone river basin, while degraded smectite, irregular mixed-layers and kaolinite predominantly represent soil-derived minerals. In the Eastern Mediterranean range, smectite proceeds from the erosion of both Meso-Cenozoic sedimentary rocks and Quaternary soils (Chamley et al. 1962, Chamley 1971). Some typically pedogenic minerals sometimes contribute to the formation of marine sediments under climatic conditions that are incompatible with their formation in surrounding soils; this is the case with kaolinite in late Quaternary sediments from the Arctic Ocean, and results from the reworking of peri-Arctic Mesozoic sediments (e.g. Naidu et al. 1971. See Chap. 8). *Paleoclimatic reconstructions should therefore be based systematically on relative variations in the abundance of the different clay minerals,* rather than on absolute values.

In some regions, weathering processes are too weak or too strong to allow appreciable differentiation in the clay composition during successive climatic periods. This is the case when very dry or very cold conditions prevail in the course of time, preventing any significant hydrolysis of rocky substrates (e.g., deserts, or very high-latitude or -altitude regions). Similarly, soils developed under strongly aggressive and hydrolyzing conditions may not register minor climatic variations. Such situations occur in intertropical regions where thick lateritic soils have developed since the late Mesozoic under predominantly warm-humid climate (e.g., African or South American shield beneath equatorial forest). The detrital clays removed by erosion from these evolved soils may be hardly sensitive to climate variations, especially if the terrigenous material includes minerals derived from the erosion of old pedologic blankets (e.g., Bellaiche et al. 1972, Lange 1982).

The use of clay minerals for climate reconstruction supposes that continental soils tend to reach an approximate equilibrium state with meteoric conditions at successive investigation periods, so that the materials eroded and deposited within the basins really express the current weathering patterns. Such equilibrium conditions imply *little tectonic activity and fairly stable continental morphology.* Strong rejuvenation processes preclude the development of soils and favor the direct erosion of rocky substrates, which leads to the supply in the basin of less evolved clay minerals suggesting less hydrolyzing climate (Chap. 18.3). Such characters are reported for early Pleistocene sediments from Zakinthos island, Western Greece, where active tectonic activity prevented the pedogenic formation of kaolinite and smectite, and determined the erosion of less weathered illite and chlorite and of vermiculitic mixed-layers (Blanc-Vernet et al. 1979). A tectonic rejuvenation may also bring about the reworking of old pedologic blankets, and give the appearance of a more hydrolyzing climate. This is the case with the Belgium basin during the Neogene. The tectonic uplift of the Ardenne massif determined the abundant supply of old pedogenic kaolinite in the sedimentation area, at a period where the world cooling actually discouraged the formation of kaolinite in soils of Northwestern Europe (Mercier-Castiaux et al. 1988).

If the sources of clay minerals correspond to distinct petrographic and pedologic features and have changed in the course of time, the climatic signal may be deformed or even obliterated. Such varying sources are reported by Hein et al. (1975) for the far Northwestern Pacific and Southern Bering Sea. Plio-Pleistocene clay-mineral suites derive either from western Kamchatka-Koryak and the outer Bering margin responsible for illite and kaolinite supply, or from the eastern Aleutian islands and Alaskan shelf rich in volcanogenic smectite and chlorite. Modifications in the importance of each source over time may have determined stronger clay mineral changes than those possibly induced by climatic variations under the fairly high local latitudes (55°N), which prevents reliable information on past weathering processes.

Another limitation factor concerns the sorting and differential settling processes. *At short distances from land masses, especially in shallow-water environments, hydrodynamic constraints may induce clay mineral segregations that are of the same order of magnitude as those controlled by climate.* If hydrodynamic conditions have changed in a given sector with time, their effects may be misinterpreted in terms of climatic variations, or may mask or exaggerate the climatic signal. For instance, strong resedimentation processes in the Northwestern Mediterranean Sea locally determine the preferential settling of coarse particles dominated by well-crystallized illite and chlorite, leading to the impression of cold, less hydrolyzing climate (Bellaiche et al. 1972, Monaco et al. 1982). In Makkowik bay, Labrador, Barrie (1983) suggests that the systematic decrease away from source in the abundance of mica during the last glacial stage results from physical sorting by size in the fresh-water layer of a highly stratified estuary. This phenomenon does not exist for post-glacial periods during which clay minerals are mixed and uniformly distributed. Under such estuarine conditions, the climatic signal tends to be preserved during interglacial stages and to be hidden during glacial stages. In a general way, *the quality and reliability of the climatic expression of clay suites appear to improve with increased distances from shore areas.*

The existence of active meridian currents in a given oceanic basin may determine the mixing of suspended material originating from different latitudes, and therefore the confusion of climatic signals borne by clay minerals. This is the case in the Vema Channel, Southwestern Atlantic Ocean, where recent sediments contain in the clay fraction fairly abundant chlorite supplied from high- southern latitudes by the Antarctic Bottom Water (Jones 1984). The percentage variations of chlorite correlate with the abundance of displaced Antarctic diatoms, which confirms the origin of the clay mineral and also indicates the allochthonous origin of biosiliceous debris. In the North Atlantic at 40°N, clay minerals appear to derive from various sources located in the temperate-cool climatic zone, while planktonic forminifera proceed from the mixing of cool, more local water masses and of the more distant Gulf Stream or Labrador currents (e.g., Rotschy and Chamley 1971, Pastouret et al. 1975). *The clay minerals constitute all the more reliable paleoclimatic markers since they are transported over limited distances by meridian currents.* Advective transport over North-South distances exceeding a few hundred kilometers reduces their interest for direct paleoclimatic reconstructions greatly. The same restriction applies to most other biogenic and non biogenic constituents of marine sediments. Note that the clay transportation by bottom or surface water masses may be impeded at certain climatic periods, either by sluggish circulation due to attenuate high-latitude exchanges, or by the develoment of large ice packs (e.g., Kennett 1982, Zimmerman 1982).

17.2 Late Quaternary

17.2.1 Direct Climatic Expression

Many reconstructions are available for the *Mediterranean Sea*, whose latitudinal arrangement, relative proximity of land, and sluggish current regime favor the use of clay minerals as climate indicators. For instance, the core Alinat-C3, sampled on a diapir structure in the Ligurian basin, *Northwestern Mediterranean* (42°47'N, 07°41'E, about 140 km West of northernmost Corsica), is considered from a triple point of view (Rotschy et al. 1972. Fig. 17.3):

1. *The clay minerals* comprise abundant illite (35 to 65% of the <2µm fraction) associated with variable amounts of smectite, chlorite, kaolinite and irregular mixed-layers (Table 17.1). Quartz and feldspars commonly occur in the clay fraction. A climatic curve that is mainly based on the *variations of illite crystallinity and* on the *relative abundance of smectite and other expandable minerals* provides information on the degree of hydrolysis in Southeastern France and Corsica.

2. *The planktonic foraminifera* are divided in two groups. One group mainly characterizes cold sea water (*Globigerina pachyderma, Globigerina quinqueloba, Globigerinoides bulloides*), the second develops rather in temperate to warm water (*Globorotalia inflata, Globorotalia scitula, Globorotalia truncatulinoides, Globigerinita glutinata, Orbulina universa, Hastigerina siphonifera, Globigerinoides ruber*). The relative abundance of each group shows a climatic curve indicative of the relative variations of *sea-water temperature*.

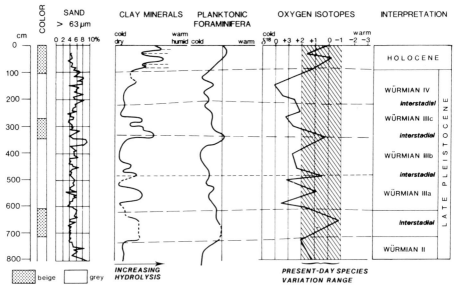

Fig. 17.3. Climatic curves deduced from clay minerals, planktonic foraminifera and oxygen isotope data in core Alinat-C3, Northwestern Mediterranean Sea. (After Rotschy et al. 1972)

Table 17.1. Clay mineral characteristics of cold-dry and warm-humid end-member sediments from core Alinat-C3, Ligurian basin, Northwestern Mediterranean Sea (afterRotschy et al. 1972)

	Illite crystallinity (1/10 °O)	Illite abundance (%)	Smectite, mixed-layers (%)	Chlorite, kaolinite (%)
Cold-dry (grey)	1.5	65	5 (M-L rare)	30 (K rare)
Warm-humid (beige)	4.0	35	45 (M-L common)	20 (K common)

3. *The oxygen isotope composition* ($^{18}O/^{16}O$) of planktonic foraminifera is measured on the calcareous tests of six species: *Globigerinoides ruber, Globigerina bulloides, Globigerina pachyderma, Globorotalia inflata, Globorotalia truncatulinoides and Orbulina universa*. Additional data are provided by benthic foraminifera. The sedimentary levels marked by excess values in the heavy isotope (^{18}O) indicate a preferential accumulation of evaporated water on continental ice-caps, and therefore tend to correlate with cold periods (see Kennett 1982). By contrast, the levels enriched in the light isotope (^{16}O) correspond rather to ice melting, suggesting climate warming. The resulting distribution curve provides information on *general temperature* at successive periods.

There is very good agreement between the climatic information provided by the different indicators, each of them bearing a slightly different signal (continental hydrolysis, sea-water temperature, general temperature, respectively). Considering an average sedimentation rate of 10 cm/10^3 years for hemipelagic oozes from the Western Mediterranean Sea, the Pleistocene-Holocene boundary (about 10,000 yr BP) approximately occurs at 1 meter depth (Fig. 17.3). The lower part of the core belongs to the late Pleistocene (different glacial and interglacial stages of the Würmian period). Note the fairly high *sensitivity of terrigenous clay minerals to climatic changes*, that appear to be expressed at a scale of about 1000 years.

Similar results arise from the study of numerous other cores scattered in both western and eastern basins of the Mediterranean Sea, where clay mineral data are considered alone or combined with data on foraminifera, pteropods, nannofossils or oxygen isotopes (e.g., Blanc-Vernet et al. 1969, Chamley 1971, Müller et al. 1974, Cita et al. 1977, Robert 1980a, Lunkad 1986, Alonso 1987). Comparisons with climatic information from continental cave fillings show close correlations in climate successions recorded on land and under the sea, especially in norhtwestern basins. *Strongly hydrolyzing periods indicated by clay suites often correlate with increasing sedimentation rates of hemipelagic oozes*, which sometimes allows to interpret the clay mineral record in terms of both temperature and humidity (Fig. 17.4). *Some shifts may occur between the data of inorganic and organic markers*, which results both from the different climatic indications provided by each marker and from the specific delay needed by a given marker to register the climatic changes. Clay minerals often express climate modifications slightly before the planktonic foraminifera that depend on progressive changes in

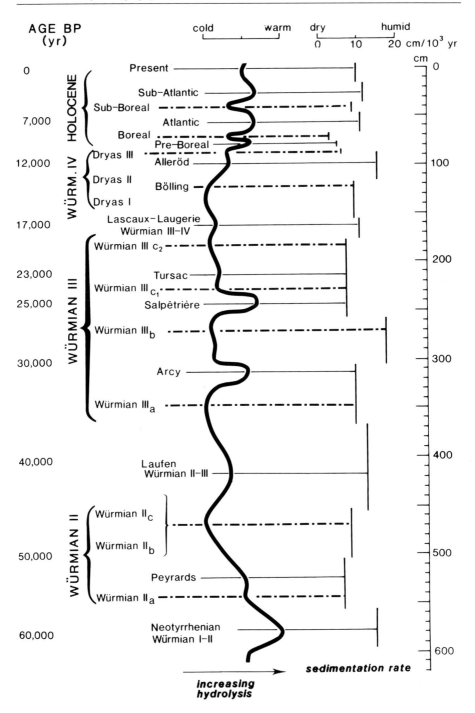

Fig. 17.4. Stratigraphic and climatic interpretation of clay mineral data, core 1MO67, Northwestern Mediterranean Sea. (After Chamley 1971; denomination of late Quaternary stages after Escalon de Fonton 1966)

the water-mass temperature (e.g., Blanc-Vernet et al. 1969, 1975), and before pollen and spores that depend on the development of new clay-bearing soils on land (e.g., Francavilla and Tomadin 1973).

Climatic reconstructions from clay mineral data exist for Quaternary sections in various continental environments such as *cave fillings* (e.g., Onoratini et al. 1973), *fluvio-lacustrine deposits* (Bornand and Chamley 1974, 1975) and *fluvio-glacial accumulations* (Tiercelin and Chamley 1975). The climatic message provided by such sections often represents a local significance only, since clay variations express changes that have occurred in a restricted area, under specific pedogenic conditions. Climate reconstructions on land sections are sometimes

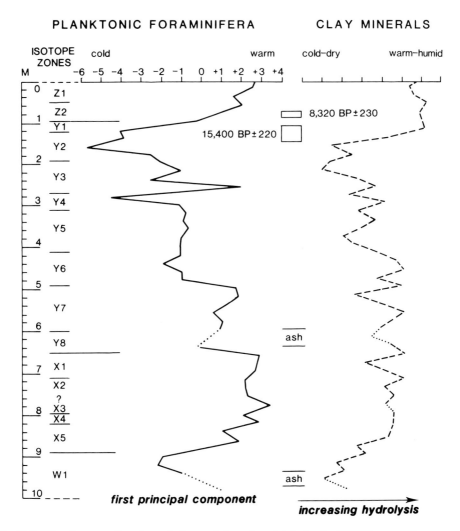

Fig. 17.5. Climatic curves from foraminifera and clay mineral data for the 135,000 years BP, core TR 126-25, Campeche Bay, Southwestern Gulf of Mexico. (After Chamley and Kennett 1976)

hardly correlatable with marine records, the latter usually bearing a more general message.

Comparable results to those obtained in the Mediterranean Sea arise from studies performed in the *Gulf of Mexico*, whose latitudinal location favors moderate weathering processes, and whose semi-closure discourages long-distance water- mass exchanges (e.g., Chamley and Kennett 1976. Fig. 17.5). In the *Atlantic Ocean, different clay parameters take turns from high to low latitudes to express the climate variations*, in the marine sectors located away from too- active meridian circulation. Degraded smectite and associated mixed-layers appear to constitute reliable markers at fairly high latitudes, like in the Norway Sea and Northeastern Ocean (e.g., Rotschy and Chamley 1971, Chamley 1975a). Illite crystallinity combines with various other parameters in temperate to temperate-warm regions (e.g., Chamley 1971, Chamley et al. 1977a), while kaolinite and soil-authigenic smectite apply better under tropical and equatorial latitudes (e.g., Pastouret et al. 1978). Off arid regions like the Sahara, the quartz abundance seems to represent a particularly reliable indicator of cold periods marked by increasing wind activity (e.g., Bowles 1975, Sarnthein et al. 1982b, Dauphin 1983. See Chap. 7).

In the *Pacific Ocean*, fairly few detailed investigations have been performed so far on the paleoclimatic expression of land-derived minerals during late Quaternary times. Quartz appears to be an especially suitable indicator (e.g., Eslinger and Yeh 1981, Morley et al. 1987). In various cores from the Eastern *Indian Ocean*, Aoki and Sudo (1973) report an increase in relative abundance of smectite and kaolinite during interglacial stages, while glacial-stage sediments are relatively enriched in illite and chlorite; these clay mineral variations are attributed to changes in the degree of continental hydrolysis, interglacial stages favoring weathering processes on Asian land masses. This interpretation closely resembles that proposed in the Mediterranean range. Similar results are reported by Bouquillon and Labeyrie (1988), who compare clay mineral and oxygen isotope data in the proximal zone of the Ganges deep-sea fan.

17.2.2 Indirect Effects

Intensity of rainfall. Under tropical humid latitudes where kaolinite forms abundantly in well-drained soils during both glacial and interglacial periods, the variations recorded in the composition of terrigenous clay input may reflect temporary changes in the intensity or geographic location of rainfall. Pastouret et al. (1978b) study from an interdisciplinary point of view a 15.15-meter-long core from the outer Niger delta, *eastern Equatorial Atlantic* (03°31'1N, 05°34'1E, 1181 m water depth). The comparison of the clay mineral suites with oxygen isotope, planktonic foraminifera and quartz data points to the importance of rainfall variations in the Niger river basin during the last 30,000 years. Abundant supply of kaolinite from about 13,000 to 4500 years BP correlates with low amounts of planktonic foraminifera, low $^{18}O/^{16}O$ ratios, and very high sedimentation rates (up to 600 cm/10^3 yr between 11,500 and 10,900 yr BP). This suggests very active precipitation over the continental river basin, especially in middle and upstream areas where kaolinite preferentially formed in Cenozoic soils. On the other hand, low sedimentation rates, foraminifera-enriched sediments and heavy-isotope-enriched periods correspond to enhanced supply of detrital smectite, chlorite, palygorskite and sometimes irregular mixed-layers, all mineral species preferentially inherited from the downstream parts of the river drainage basin or from moderate weathering processes. This second assemblage suggests either a decrease in the precipitation regime (i.e., dryer climate), or preferential rainfall on the coastal zones of North-Equatorial Western Africa. High quartz concentrations are typical for the transition between oxygen isotope stages 1 and 2 (about 15,000 to 11,000 yr BP), at the inception of heavy precipitations in the southern Sahel zone.

Ice cover. The covering of land or sea by ice sheets during glacial stages may significantly affect the nature and composition of the materials eroded and transported within sedimentary basins. Ice blankets prevent the reworking of underlying rocks and soils, whose composition may be quite different from that of outcrops in uncovered areas. *The growing, melting or displacement of ice blankets in the course of the time can determine further changes in the detrital input,* which induces indirect climate effects and not a direct control of weathering processes. Duncan et al. (1970) attribute the strong increase of the relative abundance of smectite in Holocene sediments of the Cascadia basin, *Northeastern Pacific Ocean,* to the melting of the ice cover in the upper part of West-American river basins, namely the Columbia river, where abundant smectite- bearing formations crop out. In the *Black Sea,* Müller and Stoffers (1974) show that the clay-mineral distribution in recent sediments reflects the petrography of adjacent land masses. Illite prevails in northern sediments that mainly derive from the mica-rich rocks of the Russian platform. Smectite abundance increases toward the Anatolian coast in the South, where the drainage area comprises abundant weathered volcanic rocks. The same pattern characterizes most Holocene sediments of the Black Sea. In late Pleistocene deposits, the abundance of smectite increases relative to illite in both southern and northern areas. The highest smectite amounts occur at the maximum cold event of the last glaciation (Würmian III, about 18,000 years ago) and just before the massive deglaciation (about 11,000 years). G. Müller and P. Stoffers interpret these changes as the result of the extension of ice cover and permafrost in the regions located North of the Black Sea. The increase in smectite proportion expresses the dominant supply from southern continental areas, at periods where very cold climate prevented northern detrital input. The calcium carbonate abundance roughly parallels the variations of smectite abundance, suggesting that calcite in marine sediments partly results from the erosion of southern soils and rocks. An increase in illite signifies the growing influence of northern hinterland, when ice blankets and permafrost soils melted on the Russian platform. In the *Weddell Sea* C.Robert (pers. commun.) attributes the increase of smectite abundance during late-Cenozoic glacial periods to the combination of (1) increasing submarine erosion by active currents of the smectite-bearing rocks cropping out on the Antartic continental margin, and (2) increasing protection against subaerial erosion of illite-bearing crystalline rocks by extending ice sheets.

Grousset and Duplessy (1983) study a 4-meter-long piston core from the *Greenland basin*, Northwest of the present polar front (66°36'N, 10°30'5W; 1487 m-depth), from lithologic, mineralogic and isotopic points of view. Three time units are recognized from the base to the top of the core.

   1. Before 18,000 years BP: little carbonate and coarse fraction, abundant volcanic glass, plagioclase and pyroxene. The clay fraction consists exclusively of smectite associated with abundant titanium and iron. This basal unit is attributed to the erosion of Iceland basaltic rocks and to short-distance transportation by turbidity currents, while the sea was ice-covered.

   2. From 18,000 to 9000 years BP: increase of carbonate content in marine heterogeneous mud, appearance and increase of illite, chlorite and kaolinite in the clay fraction. The iron and titanium amount decreases while rubidium, lead and thorium contents increase. F. Grousset and J.-C. Duplessy suggest that illite, chlorite and kaolinite result from the southward drift of icebergs carrying continental material. The icebergs are thought to have been transported into the basin by surface currents from Greenland, Scandinavia or the Arctic. Transport had to be surficial since submarine ridges prevented active bottom flows. This implies an early melting on Greenland in response to the increase in insolation at that time. Kellogg (1985) disagrees with such a possibility and argues that the mineralogic change between 18,000 and 9,000 years BP results from advection of continental material from the South and South-West, and not from the North. Over North America the change from ice sheet to retreat between 16,000 and 13,000 years BP caused an increased supply of clay-rich rock floor to the ocean. The long-distance currents progressively developing could have carried North American illite, chlorite and kaolinite into Norway Sea and Greenland basin. Grousset and Duplessy (1985) maintain their interpretation of an early deglaciation in the Norway Sea area by citing $^{87}Sr/^{86}Sr$ data indicating that the non car-

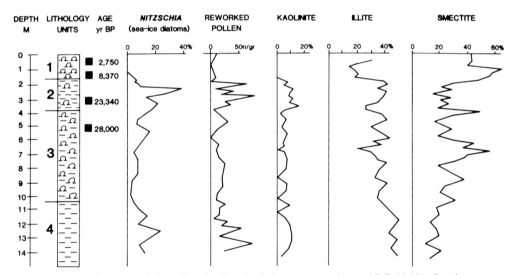

Fig. 17.6. Variation of biological and mineralogical parameters in core RC 14-121, Southern Behring Sea (54°51'N, 170°41'W, 2530 m water depth. After Sancetta et al. 1985)

bonate bulk sediments of the core are more closely related to Scandinavian than to North American continental values. Whatever the precise origin, this second unit displays clay mineral changes that result from ice melting effects and not from direct weathering processes.

3. From 9000 years to Present: increase of smectite, iron and titanium in the clay fraction of muddy sediments, that become homogeneous. This last sedimentation interval is considered to reflect the deglaciation of the smectite-rich Iceland domain.

Sea-level changes. The variations of the sea level during alternating glacial- interglacial periods may have induced noticeable changes in the composition of mineral suites derived from river drainage basins or continental shelves. In *the Behring Sea*, strong clay mineralogical changes are recorded, that are much too intense to result from variations in weathering processes under the very-cold climate conditions occurring in this region (Naidu et al. 1982). Sedimentary levels enriched in smectitic minerals correlate with high sea-level stands, which results from enhanced detrital supply by the Yukon river that drained widespread, partly ice-free continental areas. By contrast, the deposits enriched in illite appear to correspond to low sea-level stages, when local rivers and coastal erosion mainly fed the terrigenous sedimentation (see 8.2.1).

Sancetta et al. (1985) conduct an interdisciplinary approach for a 14.5-meter-long core from the Southern Behring Sea. The distribution of diatoms, pollen, oxygen isotopes and clay minerals indicates the existence of four successive zones, from the early Würmian (= early Wisconsin, isotope stage 4) to the Holocene (Fig. 17.6):

1. 14.5-10.5 meters, clayey silt. Fairly abundant kaolinite correlates with abundant diatom species characteristic of the sea-ice flora (*Nitzschia grunowii, Nitzschia cylindra*), which indirectly demonstrates that the clay mineral derives from old rocks or soils and not from contemporary weathering processes. Kaolinite is associated with abundant illite, as well as with abundant reworked pollen grains. This zone is attributed to a low stand of sea level during isotopic stage 4, when terrigenous materials were actively displaced and transported down Behring canyon from drainage of the Southeast Alaskan mainland.

2. 10.5-4.0 meters, silty diatom ooze. The low abundance of sea-ice diatoms suggests a warming, which correlates with decreased amounts of kaolinite, illite, reworked pollen and sand-sized clastics. A corresponding increase of smectite agrees with a partial melting of land ice, allowing the supply of expandable minerals from the Yukon river basin. In addition, smectite could derive from the Aleutian arc that constitutes a more local source to the South. All these characters suggest a middle Wisconsinian age.

3. 4.0-1.8 meters, diatom-bearing silt. [14]C dates indicate a late Wisconsin age, corresponding to the last glacial maximum stage and very low sea-level. The concentration of kaolinite increases markedly due to Alaskan borders erosion, while smectite slightly decreases and illite remains high. This zone corresponds again to an increase of sea-ice diatoms and reworked pollen grains. A possible increase of aeolian transport is suggested by abundant conifer pollen, and by noticeable amounts of well-rounded and frosted quartz grains in the coarse fraction.

4. 1.8-0.0 meters, diatom ooze. This unit represents the Holocene interval, during which the sea-level rise and warming determined the decrease in the *Nitzschia* flora, the decrease of coastal illite, the disappearance of coastal kaolinite, the strong diminution of displaced shelf diatoms and reworked pollen, and the abundant supply of smectite from the Yukon drainage system and from the Aleutian volcanic arc.

Other examples of the indirect control of the late Quaternary climate on clay assemblages through sea-level variations are reported in the literature (e.g., Japan Sea, Oinuma and Aoki 1977; Mauritania slope, Lange 1985). Bowles and Fleischer (1985) show that the greatest quantities of Amazon sediments, marked by kaolinite, smectite and illite, reach the *Eastern Caribbean basin* during inter-glacial periods when sea level is high; at these times Orinoco-derived materials appear diluted among Amazon terrigenous clays. By contrast during glacial periods the Amazon contribution to the Caribbean decreases, because deposition of river sediments mainly occurs near the shelf edge and causes greater amounts of sediment to be transported into the deep Atlantic via the Amazon submarine canyon. The decrease of the Amazon contribution in the Caribbean appears partially compensated by an increase in the input from the Orinoco river, marked by the presence of pyrophyllite. Terrigenous materials derived from the Orinoco river, whose mouth is closer to the Caribbean basin than the Amazon mouth, are less trapped on the shelf and in the Gulf of Paria during low sea-level stands, and enter the deep sea directly.

Marine currents. Climate changes may induce noticeable modifications in the celerity, volume and origin of water masses that carry suspended minerals and other material in the ocean. For instance, in late Quaternary sediments between the Gibbs fracture and the Greenland basin, *Northeastern Atlantic*, the distribution of clay minerals is mainly related to the relative impact of different means of transport and deposition (Grousset et al. 1982, Zimmerman 1982). Bottom currents predominantly operate during the warm interglacial stages, while ice-rafting becomes more important in cold, glacial periods. Both sources of transportation tend to determine an increase of illite relative to smectite through the dilution of locally derived volcanogenic materials. But the relative importance of each source in the course of the time strongly interferes with climate effects.

In the Vema Channel, *Southwestern Atlantic*, Chamley (1975d) observes an augmentation of chlorite and smectite abundance and a correlative decrease of kaolinite content, in the 2800-3500-meter depth interval at the transition from Holocene to late Pleistocene. He suggests this change results from an expansion of the chlorite-bearing, northward-flowing Antarctic Bottom Water relative to the southward North Atlantic Deep Water. Jones (1984) agrees with the mainly southern origin of chlorite, but thinks that kaolinite represents a laterally derived advective mineral rather than a species issued from long-distance meridian transport by the North Atlantic Deep Water (see 18.2.3. Fig. 18.4).

In the northern Cascadia basin-Juan de Fuca abyssal plain, *Northeastern Pacific*, Carson and Arcaro (1983) demonstrate that the relative enrichment of smectite and depletion of illite and chlorite in Holocene compared to Pleistocene sediments results from a size dependency. Clay-mineral changes in both abundance and crystallinity parallel textural variations in such a manner that mineralogy can be predicted on the basis of clay-mineral size dependency and grain-size distribution. B. Carson and N.P. Arcaro deduce from their experiments that local stratigraphic variations in clay mineralogy between late Pleistocene and Holocene sediments result from selective transport and deposition rather than from constrasting source areas or weathering conditions. Even in this case, a climatic dependence affects the clay distribution, but it is of an indirect nature. The control of the clay mineral distribution by hydrodynamics in past sediments is considered in more detail in Chapter 18.

17.3 Cenozoic

Cenozoic times are climatically characterized by the passage from a nonglacial to a glacial world. After preliminary cooling events in middle Paleocene and middle Eocene times, and a major warming during the early Eocene, the definite climate deterioration started near the Eocene-Oligocene boundary. The world cooling developed step by step during late Paleogene, Neogene and Quaternary times. Most important cooling phases occurred during middle and late Oligocene (oldest marine glacial packs off Antarctica), middle and late Miocene (formation of East- and then West-Antarctic ice-sheets), and middle-late Pliocene (development of north-hemisphere ice-caps). The cooling periods were separated by temporary warmings of varying intensity, especially in the early Miocene. This general Cenozoic evolution is documented by oxygen isotope and microfaunal records (e.g., in Haq 1981, Kennett 1982, Miller et al. 1987). How do clay mineral suites express this evolution?

17.3.1 Atlantic Ocean

Most Atlantic sites investigated during the Deep Sea Drilling and Ocean Drilling Projects show a *step-by-step clay-mineral evolution during Cenozoic times. In early Paleogene times, well-crystallized smectite usually occurs in large abundance within the sedimentary clay fraction.* The amount of expandable minerals decreases noticeably only in some sedimentary horizons marked by abundant palygorskite and/or sepiolite (e.g., Fig.17.7). By contrast, illite, chlorite, irregular mixed-layers, kaolinite, as well as quartz, feldspars and amphiboles are often lacking or occur in very small quantities.

    Close to the Eocene-Oligocene boundary, illite becomes ubiquitous, chlorite and irregular mixed-layers appear, quartz and feldspars occur commonly in the clay fraction, whereas smectite abundance tends to decrease upwards. This trend goes on step by step until the late Cenozoic, with enhanced increases of the illite group (illite, chlorite, mixed-layers) especially during middle-late Oligocene, middle Miocene and late Miocene (e.g., Karlsson et al. 1978, Chamley 1979, 1981, Robert 1982). Further increase of the illite group, associated with fairly abundant quartz,

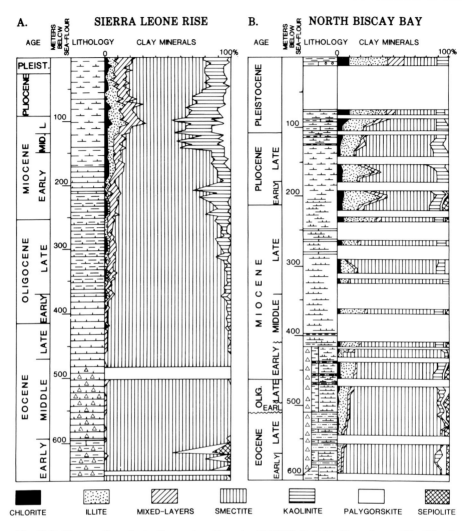

Fig. 17.7. Clay stratigraphy in Cenozoic sediments of DSDP site 366, Sierra Leone Rise (North-Equatorial, Eastern Atlantic; after Robert 1982) and site 400, North Biscay Bay (Northeastern Atlantic; after Chamley 1979 and Debrabant et al. 1979)

feldspar and dense minerals, takes place in Pliocene and Pleistocene times in the northern hemisphere (e.g., Fig. 17.7B), what correlates with the development of glaciers on Greenland, North America, Europe and Asia, as well as in the Alpine range. The illite group commonly constitutes up to 70-80% of the clay fraction in late Cenozoic sediments from middle to high latitude in northern and southern hemispheres, both domains where smectite abundance and crystallinity tend to drastically decrease. The behavior of kaolinite also depends on the latitude. In intertropical areas the kaolinite abundance increases after the Eocene and may even parallel the illite evolution (e.g., Fig. 17.7A). Under higher latitudes the kaolinite abundance tends to decrease or to present random variations. Fibrous clays usually decrease in abundance toward the late Cenozoic, except in some tropical to subtropical regions, where they may display temporary augmentation.

The abundance of smectite in early Cenozoic sediments is generally disconnected from any lithologic, diagenetic or volcanogenic control (see Chaps. 14, 15). The chemical resemblance of most Atlantic smectites with soil-forming Al-Fe beidellites, their common occurrence in Cenozoic soils and continental sediments (e.g., Sittler 1965a, Chamley et al. 1971, Rasplus et al. 1976), their frequent association with terrigenous organic matter, suggest a mainly pedogenic origin (see Chaps. 2, 8). *Early Cenozoic and Cretaceous smectites therefore appear to derive mainly from the erosion of continental soils developed on peri-Atlantic land masses.* Such an interpretation agrees with all climate reconstructions, that stress the widespread occurrence of *warm climate* from low to fairly high latitude zones in both continental and marine environments (e.g., Frakes 1979, Kennett 1982). The huge amounts of smectite supplied to marine basins probably result from the large development of soils on Atlantic borderlands, especially in coastal plains where depleted drainage favored hydromorphous conditions, ion trapping and vertisol-like soils (Fig. 17.8A). The large abundance of Al-Fe smectite in late Mesozoic-early Cenozoic deposits suggests the occurrence of a climatic regime dominated by *seasonal contrast in precipitation*, favoring both hydrolysis and hydromorphous conditions in poorly drained soils. Kaolinite, that is described in various continental sediments of similar ages, preferentially formed in upstream zones of continental drainage basins, and hardly reached the deep ocean because of too long-distance transportation.

From late Eocene time onwards, the increase of illite abundance, the correlative appearance and increase of chlorite, random mixed-layers and associated nonclay minerals correspond to the instalment and development of ice sheets on high- latitude land-masses. The world cooling discouraged chemical weathering, that was progressively reduced and replaced by physical weathering. Clay and associated minerals contained in the rocks and paleosols tended to be increasingly eroded relative to soil-forming minerals. Physically produced terrigenous minerals statistically comprise abundant mica-like and chloritic species, while transitional conditions between physical and chemical weathering favor the production of intermediate minerals of the random mixed-layered family (Chap. 2).

The widespread increase of the illite group in Atlantic sediments during Cenozoic time is therefore attributed to the decrease of hydrolyzing and hydromorphous conditions, in relation with the world cooling (Fig. 17.8B, C). Such an interpretation is supported by the correspondence existing between major cooling phases and the illite-group augmentations (Eocene-Oligocene boundary, middle-late Oligocene, middle and late Miocene, middle-late Pliocene), and also by differentiation in latitude that resulted from an enhanced increase in physical weathering processes in northern and southern regions (e.g., Chamley 1979, Latouche and Maillet 1982). The increased reworking of rocky substrates compared to that of soil blankets was favored on a global scale by the sea-level lowering, especially during late Cenozoic times, and more regionally by Alpine tectonics (see Chap. 18). The frequent low-latitude increase of kaolinite abundance

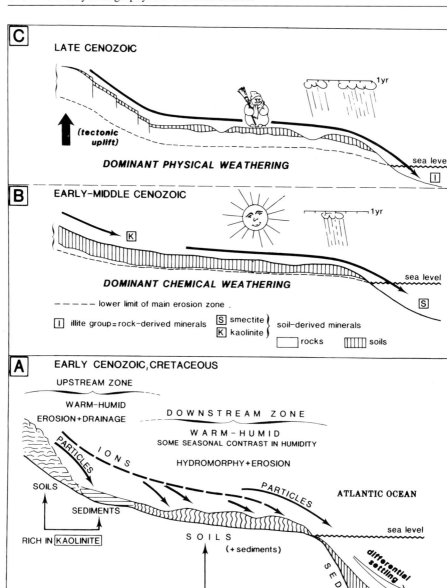

Fig. 17.8. Tentative interpretation of clay sources in late Mesozoic-early Paleogene sediments (**A**) and late Paleogene-late Cenozoic sediments (**B, C**) of the Atlantic range. (After Chamley 1979)

during Cenozoic times resulted from the development of long wet seasons with still high temperature, as well as from enhanced reworking of old lateritic soils and from the migration of the African plate across the equatorial zone. The temporary increase of fibrous clay amounts, especially of palygorskite, proceeds either from the development of peri-marine basins under transgressive and warm climate conditions, or from the reworking of old evaporative-alkaline coastal sediments (see Chaps. 9, 18, and discussions in Chamley 1979, Chamley and Debrabant 1984a, Stein 1985, Robert and Chamley 1987).

The climate record by clay suites in Cenozoic deposits of the Atlantic range is often exaggerated by sedimentary hiatuses or by additional effects due to modifications in the current regime (e.g., Robert 1980b, 1987, Latouche and Maillet 1985, Chamley 1986. Chap. 18). Of course, local control by rock petrography, soil characteristics, land morphology, water mass and wind regime, may have determined various modalities in the climatic record registered by the clay-stratigraphy patterns (Diester-Haass and Chamley 1980, Latouche and Maillet 1982, Stein and Sarnthein 1984). One of the more relevant minerals in the climatic expression consists of *kaolinite*, that is almost exclusively terrigenous and preferentially forms in soils submitted to active hydrolysis and drainage. Robert and Chamley (1986, 1987) tentatively use the relative abundance of kaolinite in 45 DSDP sites of the Atlantic and Western Pacific Oceans to estimate the major changes in continental humidity during Cenozoic times. Periods of increasing humidity expressed by kaolinite augmentations occur during late Oligocene (about 30 my ago), middle Miocene (14 my), late Miocene (10 my) and late Pliocene (3 my), which correlates with major developments of polar ice sheets. It is suggested that *high precipitation regimes, that privilege the formation of lateritic soils on low-latitude continents, favor the growth of snow fields and ice-caps in high-latitude regions.* Note that attenuate increases of kaolinite abundance are restricted to the high-latitude DSDP sites during late Miocene and late Pliocene times, both periods during which low-latitude sites display a depletion in the kaolinite supply. This opposite trend appears to result from the *late Cenozoic latitudinal differentiation in the climate,* that favored enhanced humidity toward high latitudes and more arid conditions off tropical Africa and Australia.

17.3.2 Other Oceans

In addition to the latitudinal differentiation suggested by the behavior of kaolinite, the *clay mineral suites of the Pacific Ocean often express the general world cooling that characterizes Cenozoic times,* especially in regions subject to noticeable terrigenous influx. As early as 1972, Jacobs and Hays from the study of 12 cores of the *North-Equatorial Pacific Ocean* report a decrease in smectite abundance and an increase in illite, chlorite and kaolinite abundance, from middle Miocene to Pleistocene times. This change is interpreted to indicate an increasing influx of detrital minerals issued from continental weathering, in relation with the world cooling, the shift of global wind belts toward the equator and the enhanced production of ice-derived detritus. The climatic interpretation of clay mineral changes is supported by the increase of illite crystallinity and of average sedimentation rates from lowermost Miocene to Pleistocene, which fits the type of results obtained in the Mediterranean Sea at more recent time intervals (17.2.1). Note that the *climatic signal* expressed by sediments like those studied by M.B. Jacobs and J.D. Hays may be noticeably *enhanced by the northwestward migration of the Pacific plate*, provoking the crossing of successively cooler latitude zones in the course of the time (e.g., Leinen and Heath 1981). In the *Philippine Sea*, northwestern Pacific Ocean, the late Cenozoic increase of typically rock-derived minerals is clearly recorded at the DSDP sites drilled in the Shikoku basin and on the Daito ridge (Chamley 1980a, b). The augmentation of terrigenous illite, chlorite, random mixed-layers and quartz, from late Miocene to Pleistocene times, occurs at the expense of smectite, that mainly derived in earlier Cenozoic times from the extensive weathering of surrounding volcanic rocks.

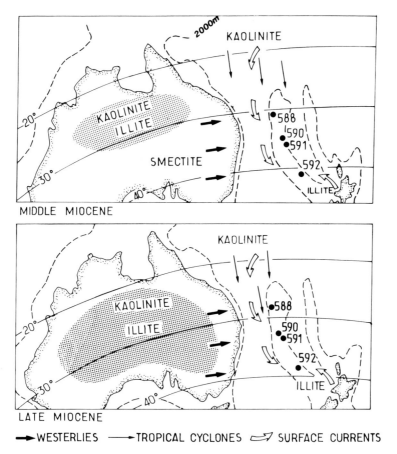

Fig. 17.9. DSDP Leg 90 site location, and interpretation of clay mineral changes recorded on the Lord Howe Rise during Miocene times. (After Stein and Robert 1985)

In the *Lord Howe rise* area, *Southwestern Pacific*, a combined study of clay minerals, grain size and terrigenous flux rates suggests that the Neogene history of desertification in Australia was controlled by both the northward drift of the Indo-Australian plate and the buildup of Antarctic ice sheets (Stein and Robert 1985). Sites 588 to 592 of DSDP leg 90 were drilled in Miocene to Pleistocene calcareous oozes and chalks deposited at 1100 to 2100 meters water depth (Fig. 17.9). Scarce but ubiquitous, the clay fraction is considered to be chiefly transported from Australia by the westerlies, and from the North by the East-Australian current. Clay minerals in the less than 2 μm fraction consist of smectite (40-100%), kaolinite (0-35%), illite (0-25%), random mixed-layers (0-10%) and chlorite (0-5%). Two main results are discussed by R. Stein and C. Robert:

1. *The relative abundance of smectite*, very high in the Miocene, *decreases during late Cenozoic* (Fig. 17.10). This change appears to result from the *progres-*

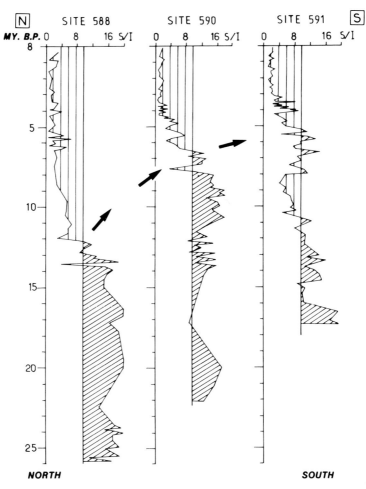

Fig. 17.10. Variations of the smectite/illite ratio on the Lord Howe Rise since the middle Miocene. (After Stein and Robert 1985)

sive extension of deserts in Northern and Northwestern Australia (Kemp 1978, in Stein and Robert 1985). The decrease of chemical weathering probably induced a decrease in soil development and a correlative increase in physical alteration and erosion. Pedogenic smectites appear to have been progressively replaced by rock-derived minerals (illite, chlorite) and by kaolinite reworked from older soils, as a result of erosion enhanced by the disappearance of vegetation with increasing dryness. The Miocene change to an arid climate is also documented by an increase in wind-supplied particles and in the amount of terrigenous input on the Lord Howe rise. It is probably related to the northward drift of the Indian Plate (i.e., migration of Australia to a dryer climate belt), and to the development of the Antarctic ice cap (see Kennett 1982). Maximum phases of Australian aridity expressed by terrigenous material roughly correlate with major cooling events around 14, 9, 5 and 3 million years ago.

2. *The decrease in smectite abundance during Miocene time in not synchronous at the different sites* drilled on the Lord Howe rise (Fig. 17.10). It started earlier to the North (site 588, middle Miocene) than to the South (site 591, late Miocene). This shift indirectly confirms the generally-allochthonous origin of clay minerals, including smectite, in the homogeneous calcareous sediments of this region. Moreover, it suggests that *the change to an arid climate started earlier in northern than in southern Australia, in relation to the northward migration of the Indian Plate* within the arid climate belt (Fig. 17.9).

In the southern Ocean, Cenozoic cooling phases were first identified from clay mineral record by Jacobs (1974). Deep-sea cores from the peri-Antarctic range indicate the transition from smectite-rich Eocene and Oligocene sediments to illite- and chlorite-rich Miocene and Plio-Pleistocene deposits. Smectite was considered as authigenic and progressively replaced by land-derived minerals, resulting from extensive cooling and glaciation on the Antarctic continent. M.B. Jacobs suggested that poorly crystalline illite in pre-Miocene sediments indicated highly hydrolyzing, aggressive climate conditions, while well- crystallized illite from late Miocene time upwards reflected low chemical weathering and dominant physical alteration due to very cold climate. Piper and Pe (1977) and Robert et al. (1988) show the importance of terrigenous supply, including smectitic material, in Cenozoic sediments of the Ross Sea. Main phases of cooling and further glaciation determined progressive changes in the clay composition, and correlate with major development stages in the marine circulation (see Chap. 18). In the Weddell Sea, ODP leg 113 sites also document the step by step replacement of smectite by illite and chlorite during Cenozoic times, in relation with the glaciation and circulation history in the Western Antarctic range (Barker, Kennett et al. 1987).

17.3.3 Mediterranean Sea

Pliocene, Pleistocene

Preliminary data on the general evolution of *Plio-Pleistocene climate* arise from clay mineral studies on DSDP sites in Western and Eastern Mediterranean

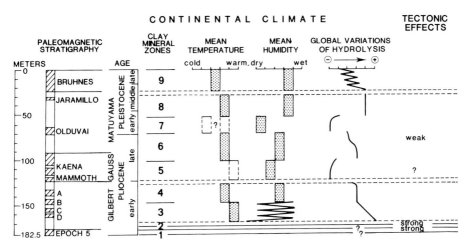

Fig. 17.11. Tentative reconstruction of the Plio-Pleistocene climate evolution in the Western Mediterranean range, based on the clay mineralogy of DSDP site 132, Tyrrhenian Sea. (After Chamley 1975 b)

(Chamley 1975b, c). One of the most complete sections consists of DSDP site 132 in the *Tyrrhenian Sea* (Fig. 17.11).

1. The *Pleistocene* successions are studied in a similar way as described for late Quaternary series (17.2.1), since the climate factors influencing the clay composition generally varied in a similar manner. Interglacial stages were of a predominantly humid and temperate type, leading to "open" illite, abundant expandable minerals and increasing amounts of kaolinite. By contrast, glacial stages usually correlated with cold and dryer conditions (more abundant chlorite and crystalline illite), especially on the European borderlands (e.g., Frakes 1979). The climate curve inferred from clay mineral data correlates quite well to that based on planktonic foraminifera (Cita 1973, in Chamley 1975b). Three successive climate intervals are proposed as follows:

  – 2 to 1.5 my BP: rather dry and temperate.
  – 1.5 to 0.7 my: temperate to temperate-warm and humid.
  – 0.7 my to Present: average temperate-humid conditions with great variations in the intensity of continental weathering due to glacial – interglacial alternation stages.

2. The *Pliocene* successions are more difficult to interpret, since some warm-temperate periods probably corresponded to low rainfall, some cold periods to high precipitation, and since foraminifera assemblages taken for comparison sometimes display slight dissolution figures. The opposite relations temporarily occurring between temperature and humidity are documented by the fact that some levels marked by very low hydrolysis (i.e., well-crystallized illite) correspond to warm-water foraminifera, and vice-versa. The comparison of clay mineral and foraminifera records therefore allows to envisage the distinction between warm-humid, warm-dry, cold-dry and cold-humid conditions during continental weathering. Apart from two short episodes marked by tectonically

controlled reworking of palygorskite, smectite, kaolinite and iron oxides, the
Pliocene can tentatively be subdivided into four climate intervals as follows:

– 5 to 4.1 my BP: subarid, warm with periodic fluctuations in continental
humidity.

– 4.1 to 3.4 my: transitional, temperate-warm and fairly humid.

– 3.4 to 2.7 my: fairly dry, probably temperate-warm with increasing
hydrolysis upwards.

– 2.7 to 2.0 my: transitional, temperate-warm, fairly humid with an upward
decrease of hydrolysis before Pleistocene time.

Site 653 of ODP leg 107 was drilled very close to site 132 in the Tyrrhenian
Sea. Its statistical clay-mineral study suggests the existence of more non climatic
paleoenvironmental changes than those previously recorded (de Visser and
Chamley 1989). In the *Pliocene Trubi facies of Southern Sicily*, de Visser et al.
(1988) investigate in details the rhythmic *marl-limestone alternations* that crop
out in the Punta di Maiata section. The alternations consist of gray and beige
layers that appear to mainly result from the alternating predominance of terrigenous supply and of marine organic-carbonate production. The gray, organic-
rich layers display relatively high amounts of smectite in the clay fraction,
abundant land-derived palynomorphs, a rather fine grain-size, a depletion of
aeolian silt and a decrease in ^{18}O compared to ^{16}O, indicative of an augmenta
tion in runoff. All these characters suggest that the gray layers probably formed
under relatively humid and warm conditions with increased discharge of
nutrient-rich, sediment-loaded Sicilian river water. By contrast, the beige, most
carbonate-poor layers are enriched in palygorsksite and kaolinite as well as in
aeolian silt, and correspond to higher $^{18}O/^{16}O$ ratios, indicative of less continental water supply. This suggests that beige layers correlate with relatively dry and
cool climate conditions, low sea-water fertility and enhanced aeolian dust
transport from North African regions rich in fibrous clays and kaolinite (see
Chap. 7). Intermediate climate conditions probably characterized transitional,
white sedimentary layers. J.P. de Visser and colleagues estimate that the average
duration of the rhythmites corresponds to the periodicity of Earth's precession
(about 20,000 years). *Marl-limestone alternations and inferred variations in biotic
and abiotic parameters therefore appear to result from cyclic variations in insolation through astronomical forcing.*

Messinian

*The uppermost Miocene in the Mediterranean range is characterized by the
widespread desiccation of most parts of the marine basin,* during what is called the
Messinian salinity crisis (see Kennett 1982). The Mediterranean Sea became almost isolated from the world ocean after the Tortonian, and experienced complex sedimentation patterns especially marked by the periodic deposition of huge
amounts of saline evaporites. The isolation of the Mediterranean basin resulted
both from plate movements bringing Africa into direct contact with Southern
Europe in Betic-Rif straits, and from the glacial-eustatic lowering of the ocean
sea-level (e.g., Adams et al. 1977). The clay mineralogy of the late Tortonian to
early Pliocene stratigraphic interval, that comprises the whole Messinian, has
been studied in various continental and marine sections from Western and

Eastern Mediterranean (DSDP legs 13 and 42A), Spain, Algeria, Sicily, Southern France, Northern Italy, Cyprus, etc (Chamley et al. 1977b, 1978b, Chamley and Robert 1980, Rouchy 1981). In most sections investigated, *the abundance of smectite increases strongly in Messinian deposits relatively to those of Tortonian and Pliocene age* (e.g., Fig. 9.7). The smectite augmentation, that is associated with a lower diversity in the clay mineral suites, *does not appreciably depend on the lithological variations* in most Messinian series (9.3.2). Smectite is dominantly of a Fe-Al beidellite type. Its composition and morphology resemble those of smectitic minerals born in continental soils under subarid climate conditions. Smectite-rich clayey sediments are often associated with oxygen isotope compositions characteristic of continental-water discharge. Since smectites are generally less abundant in Pre-Messinian, namely Tortonian deposits, than in Messinian sediments, it seems mostly probable that they represent the erosion products of soils that developed during the Messinian itself.

The development of smectites in peri-Mediterranean soils during Messinian times could have been favored by the *alternation of dry and wet episodes*, that were responsible for the frequent cyclicity of evaporite-marl deposition. The arid episodes would have corresponded to the preferential ion trapping in downstream areas of the desiccating basin, to the formation of major evaporite bodies in the basin proper, and to the last Cenozoic growth of palygorskite and sepiolite in some restricted, lacustrine basins of the upstream parts of drainage basins (e.g., Chamley et al. 1977b, 1980b). The humid episodes of a given cycle would have determined active hydrolysis and reworking of soils on land, and the subsequent deposition of smectite-rich terrigenous sediments on top of evaporite bodies.

The conspicuous *mineralogical changes typical for Messinian stage in the Mediterranean range are completely lacking* outside the Mediterranean at the same period. For instance, sediments from the adjacent Northeastern Atlantic show no any noticeable break in the clay mineralogy, and no strong smectite increase, from the Tortonian upwards to the Pleistocene. This is documented by DSDP site 397 off Cape Bojador (Chamley and Diester-Haass 1979), or by DSDP site 544 Southwest of Gibraltar strait (Stein and Sarnthein 1984). Only moderate variations in clay mineralogy are observed in most parts of the open ocean, indicating a fairly stable climatic regime on adjacent continents since the late Miocene, with some alternating cooling and warming phases (17.3.1). In addition, the great clay mineral differences recorded between Mediterranean and outer- Mediterranean areas appear to be strictly restricted to the Messinian period (Fig. 17.12).

The Messinian climate in the Mediterranean range was therefore probably influenced by specific conditions occurring in the Mediterranean domain itself. It is suggested that *the Messinian salinity crisis was associated to a peri- Mediterranean climate crisis*, that was controlled by a number of specific environmental conditions. These conditions may correspond to the following chain of events (Chamley and Robert 1980. Fig. 17.12): semi-closing of the Mediterranean Sea from the Atlantic Ocean; increasing desequilibrium of the water budget (excess of evaporation over precipitation and ocean-water supply); formation of saline evaporites; desiccation of the Mediterranean and peri- Mediterranean atmo-

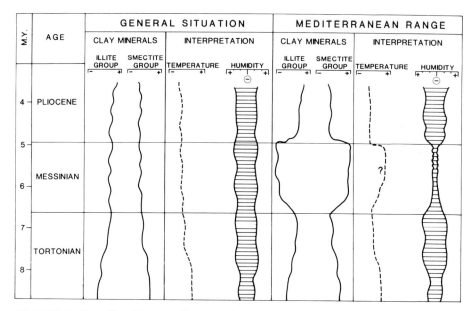

Fig. 17.12. Outline of late Neogene climate evolution inside and outside the Mediterranean area, inferred from clay mineral data. (After Chamley and Robert 1980)

sphere; relative aridification of Mediterranean borderlands; appearance of dry/humid alternating cycles of variable duration, development of smectite-rich soils and, perhaps, slight increase of the averaging temperate- warm temperature. *Basically, the Messinian climate seems to be characterized by an increase in continental aridity balanced by short and strong humid periods, which favored the developement of smectitic soils and their periodic reworking toward the Mediterranean basins.* Such an explanation agrees with most lithological and paleontological observations (see discussion in Chamley and Robert 1980). In early Pliocene times, the ingression of Atlantic water in the Mediterranean Sea determined the interruption of these specific regional conditions, leading to the reappearance of climate patterns similar to those typical of open-sea regions.

17.4 Pre-Cenozoic

Middle-Late Mesozoic

Cretaceous sediments of the Atlantic and Tethyan domains are often characterized by the abundance of Al-Fe smectites in the clay fraction of both exposed and submarine series (e.g., Chamley 1979, 1981, Robert 1981, Chamley and Debrabant 1984a, Deconinck et al. 1985). Smectites are well crystallized, and usually comprise only few illitic layers according to R.C. Reynolds's classification (see 1.3.10). The mineralogical, chemical and morphological characters of Cretaceous smectites closely resemble those of early Paleogene smectites in the same domains, and correlate with the widespread occurrence of warm, hydrolyzing climate on the Earth (see Frakes 1979, Arthur et al. 1985). In addition, the abundance of Al-Fe smectites in Cretaceous sediments is generally independent of any apparent influence of volcanic or hydrothermal activity, the result of which usually consists of the formation of magnesian or ferriferous smectites. Diagenetic modifications appear to have been very moderate and to have not appreciably modified the proportions of the different species within clay mineral associations. Finally, the warm climate conditions and tectonic stability that occurred widespread during Cretaceous times favored a priori the development of thick continental soils. All these considerations converge to suggest that *Cretaceous smectites basically derived from the erosion of Cretaceous soils massively developed under hydrolyzing climate* (Fig. 17.8). The specific abundance of smectite relative to kaolinite indicates that the Cretaceous climate was probably dominated by noticeable *contrasts in seasonal humidity*, with well-defined and rather long dry seasons. Such subarid conditions favor ion trapping and concentration, as well as a depletion in the continental drainage (see Chap. 2).

By contrast to Cretaceous series, *Jurassic sediments often present fairly high contents of kaolinite in the clay fraction.* Smectite frequently occurs in small amounts or is even lacking. Its abundance increases in late Jurassic formations only (e.g., Millot 1964, 1970, Decommer and Chamley 1981, Chamley and Deconinck 1985, Tribovillard 1988). The relative abundance of kaolinite during Jurassic times suggests the existence of *almost year-round humid conditions on land masses, combined with warm temperature.* Such an interpretation agrees with palynological interpretations by Reyre (1980), who reports the passage from warm-humid, homogeneous Jurassic conditions to more contrasted Cretaceous climate. Y. Reyre concludes from his review that seasonal contrasts and variability affected the continental humidity much more than the temperature which remained rather constant. Appreciable climatic changes are expressed by pollen and spore assemblages from the Barremian upwards only. Similar conclusions arise from Batten's investigations (1984) about the Cretaceous floral provinces in the northern hemisphere, indicating first major changes close to the Jurassic-Cretaceous boundary, and an increased differentiation in latitude climatic zonation during Cretaceous time. Notice that Hallam (1984) disagrees with such an interpretation and considers from the general distribution of major lithologic types that late Jurassic times experienced more humid climate conditions than Cretaceous times. A similar conclusion can be drawn from Sladen and Batten's investigations (1984) in Southeast England and Northwest Europe, but the climatic change is considered to be essentially local due to tectonic uplift, and not linked to global changes in climatic zones.

Limestone-Marl Alternations
The occurrence of repeated cycles of alternating carbonate-enriched and clay- enriched sediments is widespread recognized in Mesozoic sediments, and also frequently occurs in Cenozoic and even Paleozoic series (see Berger et al. 1984). Rhythmicity characterizes many pelagic carbonate formations whether deposited in shallow or deep-water environments. Most cycles appear to represent either about 20,000 or 40,000 years, and are sometimes grouped as bundles with periods

of 100,000 years. These different periodicities all suggest a *strong dependence on climate variations driven by astronomical forcing* (precession, short-period and long-period eccentricity, respectively), and correspond to Milankovitch-like cycles. The limestone-marl alternations correlate with more or less periodic changes in insolation, evaporation, wind stress and/or rainfall, causing changes in terrigenous input, water-mass stratification, planktonic productivity, deep-water oxygen content, sea-water salinity, and carbonate- dissolution rates (e.g., Cotillon et al. 1980, Arthur et al. 1984b, Mount and Ward 1986).

The clay mineralogy of Cretaceous cycles generally displays a systematic increase of smectite abundance in carbonate-rich beds, and an increase of kaolinite and often of illite in clay-enriched interbeds. Such a variation is especially documented in diagenesis-free series of Southeastern France and other Western Tethyan regions (e.g., Chamley et al. 1973, Cotillon et al. 1980, Ferry et al. 1983, Deconinck and Chamley 1983. Fig. 16.2). Similar variations occur in Cretaceous alternations of other regions like Mid-Pacific mountains (Ferry and Schaaf 1981) and the Atlantic domain (pers. unpubl. data). In the Central Apennines, Italy, Johnsson and Reynolds (1986) report increased amounts of illite-smectite mixed-layers in shale partings of late Cretaceous cycles, as well as a better ordering of mixed-layered minerals. Note that a kaolinite increase is also reported in marly interbeds of Callovian-Oxfordian alternations, Southeast France (Tribovillard 1988). Similar trends are reported in sedimentary cycles of various other ages (e.g., Pliocene of Sicily, 17.3.3).

The mineral cyclicity is generally attributed to more humid conditions during the deposition of marly interbeds, the increase of continental runoff determining stronger hydrolysis processes, the preferential development of pedogenic kaolinite and the increased erosion of both soil- and rock-derived minerals (namely upstream-supplied kaolinite, illite). Such an interpretation agrees with the higher proportion of detrital clay fraction diluting marine carbonates in marly interbeds, and with often higher sedimentation rates (e.g., Mount and Ward 1986). By contrast, *smectite-enriched calcareous beds are usually considered as reflecting dryer climate on land*, causing a lesser terrigenous influx, the preferential removal of downstream soil products, and a higher planktonic productivity.

Many questions remain about the precise paleoenvironmental signal registered by the clay fraction of limestone-marl alternations: respective importance of local and global climate on the clay signal, possible changes in sources, respective importance of soil-derived and paleosol-derived kaolinite, aeolian or diagenetic significance of the quartz often associated with smectite-enriched beds, possible partial obliteration by diagenetic imprint, etc. A large field of research has to be investigated, where clay mineral data and interpretations have to be compared to the modeling of rhythmic bedding sequences (e.g., Barron et al. 1985), and inserted in a comparison of different-scale cyclicities (e.g., Ferry and Rubino 1987).

Pre-Jurassic
Very few investigations have been performed so far on the detailed climatic significance of clay successions in old sedimentary formations. We know that a climatic signal is borne by clay assemblages in various series of early Mesozoic,

Paleozoic (Millot 1964, 1970) and even late Proterozoic ages (e.g., Chamley et al. 1978a, 1980a). As the possibility of diagenetic obliteration of paleoenvironmental messages statistically increases with increasing time, one must be especially careful in studying such old series. But the *frequent occurrence of obviously non diagenetic series in numerous regions devoid of important depth of burial, tectonic structuration and fluid migrations* (Chaps. 14 to 16), as well as the difficulty in investigating most paleosols reliably (Chap. 2), strongly suggest developing paleoclimatic approaches from clay- mineral investigations in ancient sedimentary formations.

17.5 Conclusion

1. Clay minerals preferentially form through weathering and pedogenesis at the surface of the Earth. The widespread, easy and nearly continuous erosion of soft pedogenic blankets determines a dominant participation of soil-derived clay minerals to the formation of recent and past sediments. As clay minerals experience no or only slight diagenetic modifications in the numerous sedimentary series buried less than 2 to 3 kilometers and devoid of significant volcano-hydrothermal impact, they may represent useful markers of past continental climate. The correlations observed in the variations of mineral, biogenic and isotopic markers demonstrate the frequent validity of detrital clays to reconstruct past climate successions in Meso-Cenozoic sediments. The range of variations recorded in the nature and proportions of clay assemblages from vertical sedimentary columns roughly resembles the range of variations displayed horizontally by the present-day latitudinal zonation linked to weathering on land masses.

Clay minerals basically express the intensity of weathering, and especially of hydrolysis, in the land masses adjacent to sedimentary basins. The information provided by pedogenic minerals fundamentally integrates the combined effects of temperature and precipitation, with sometimes additional data on the rainfall seasonality or drainage conditions. The climatic interpretation of clay mineral results is mostly based on the variations of illite crystallinity and of the relative abundance of smectite and other expandable species. These parameters apply especially to the reconstruction of climate variations occurring at middle latitudes or altitudes. They are supplemented or replaced in other climatic zones by distinct indices derived from measurements on X-ray diffraction diagrams. In addition to or instead of the *direct influence of climate* on the formation of past weathering profiles and soils, the clay mineral record sometimes provides useful information on *indirect climatic changes* such as those affecting marine currents, sea-level stands, and ice blankets on land or sea.

The paleoclimate expression of clay minerals is all the more reliable since the basins investigated were protected during sedimentation from important reworking of ancient soils, changes in detrital sources, differential settling processes, meridian marine circulation and tectonic activity. The statistical value of clay information about climate generally increases with increasing distance from pedogenic sources. As the climate control on clay assemblages usually determines moderate and progressive changes, it is necessary to dispose of high-quality and reproducible equipments, and to base the interpretations on relative abundance or crystallinity ratios rather than on absolute values.

2. *The middle to late Quaternary alternation of glacial and interglacial periods generally correlates with an alternation of rather physically altered and rather chemically altered clay assemblages.* Cold periods indicated by microfauna or oxygen isotope data usually correspond under temperate latitudes to sedimentary levels enriched in well crystallized illite, chlorite and smectite, and in feldspars, which indicates an average cold and dry climate on land. By contrast, warm-water periods correspond rather to enhanced amounts of kaolinite and poorly crystallized expandable minerals (smectite, random mixed-layers), and illite. These variations express the dominant supply to the sea of materials alternately eroded from rocky substrates and from currently forming soils. *Detailed climate reconstructions using the clay stratigraphic record are especially documented in mid-latitude, relatively restricted basins* such as the Mediterranean Sea, where long-distance advection processes hardly developed and where temperate climate on adjacent land masses allowed numerous transitional situations. Pedoclimatic conditions appear much more difficult to reconstruct in high-latitude basins where chemical alteration is virtually lacking during Quaternary times, as well as in low-latitude areas where permanently strong hydrolysis precludes appreciable changes in soil-derived clays.

3. *Cenozoic sedimentary series frequently display a step-by-step increase of rock-derived minerals, namely illite, chlorite, feldspars, at the expense of Al- Fe smectite and often of kaolinite.* This general trend, whose acceleration and slackening stages roughly parallel the cooling and warming periods indicated by climate markers such as oxygen isotopes, is particularly documented in the Atlantic range characterized by stable margins and little tectonic instability. The same trend occurs in various other sedimentary basins. The clay mineral change is attributed to the *transition from non glacial to glacial conditions at the Earth's surface.* The development of ice caps since the late Eocene, and the correlative latitudinal climate differentiation probably result in a decrease in chemical weathering in the course of the time, and an increase in physical alteration. Extensive and thick smectitic soil blankets tended to be eroded as aggressive hydrolysis decreased and physical weathering processes increased, especially toward high latitudes where cooling extended dramatically. Note that the variations in the relative abundance of kaolinite, a mineral that mainly depends on hydrolysis intensity, make it possible to propose general variations in the continental humidity. The major phases of ice-cap development at high latitudes during Oligocene, Miocene and Pliocene times correlate with increase in kaolinite abundance, suggesting higher precipitation periods on southern and northern oceanic domains.

The general Cenozoic trend was probably temporarily interrupted during *uppermost Miocene* times in the *Mediterranean range.* The near-closure of the sea apparently determined both a salinity crisis in the marine basin and a *climate crisis marked by subarid smectitic soils* on its borders. The occurrence of shorter, less extensive dry conditions is also suggested for other Cenozoic periods from the comparison of clay mineral with microfauna data. For instance, the combination of less weathered clay minerals (crystalline illite, fairly abundant chlorite, few random mixed-layers) and of warm-water foraminifera at some Pliocene episodes in the Western Mediterranean Sea implies a depletion of precipitation on adjacent land masses, and high temperature.

4. Pre-Cenozoic series offer a wide field of paleoclimate research, that is still little investigated. *Cretaceous times* often correlate with *smectite-rich clays*, which appears to result from the erosion of thick pedogenic blankets developed under *hot temperature and seasonally contrasted humidity*. On the other hand, Jurassic sediments usually display large proportions of kaolinite in clay fractions, suggesting *warm conditions combined with more constant annual humidity*. Both *Jurassic* and Cretaceous stages appear characterized by rather little variation in climatic conditions both in the course of time and at various latitudes. The differentiation in soil development appears noticeable only in late Cretaceous and especially Paleogene times. Small-scale climate variations are expressed by *limestone-marl alternations*, that particularly characterize late Mesozoic periods, but also occur in both pre-Jurassic and post-Cretaceous times. The *calcareous beds* of such cycles usually contain enriched amounts of *smectite*, while marly interbeds display increasing amounts of kaolinite and/or illite. This suggests an alternation of *respectively dryer and more humid conditions on land*. Detailed climate reconstructions are still rare for sedimentary series older than Jurassic time. They should be developed by studying some of the numerous sections that are devoid of noticeable post- sedimentary clay-mineral changes.

Paleoclimate reconstructions from clay stratigraphic data may be envisaged in series as old as the late Precambrian, especially in regions marked by the absence of important post-sedimentary overburden and tectonic structuration, and where clay-rich sediments actively and regularly deposited at the periphery of variously weathered land masses. Clay minerals, the climatic interest of which strongly depends on the precise knowledge of the petrographic context and geological history, generally display a *regional signal* only. They present transitional situations from climate control to source or tectonics control, which often complicates the interpretations (Chap. 18). On the other hand, clay mineral sequences present several *specific advantages for climate reconstructions*. They do not depend on biotic or evolutional factors, and often not on physico-chemical post-depositional modifications. They occur virtually in all types of sedimentary rocks from both continental and marine environments, and provide an exclusive, integrated signal for land-surface climatic conditions, with a fairly high sensitivity. Their use as a climate indicator is all the more useful, since they are considered together with other sedimentary markers, providing a synthetic expression of past successions.

Chapter 18

Paleocirculation and Tectonics

18.1 Identification of Sources

The recognition of present sources from clay mineral assemblages of surficial sediments has been broadly investigated since the 1960's. In the deep Gulf of Mexico for instance, Pinsak and Murray showed as early as 1960 that the distribution of smectite, illite, chlorite, kaolinite and mixed-layers is primarily controlled by terrigenous sources, some in situ adjustments being thought to intervene after sedimentation. Further studies indicated the almost exclusive influence of detrital sources and differential settling, and to virtual lack of autochthonous clay modifications (e.g., Devine et al. 1973). In the Mediterranean Sea the identification of clay mineral provinces related to terrigenous or volcanic sources was also made at the same period (Chamley et al. 1962, Rateev et al. 1966). Subsequent studies allowed clarification of some origins. For instance, in the Central Mediterranean, Blanc-Vernet et al. (1975) showed that illite and chlorite issue predominantly from Sardinia and from the Ionian Sea, smectite from Sicily and palygorskite from North Africa. The precise origin of fibrous clays (palygorskite, sepiolite) varies according to the regions considered from West to East. They derive from both northern and southern sources in the Alboran Sea (Cossement et al. 1984), essentially from the South in the central basin (Blanc-Vernet et al. 1975), and almost exclusively from Africa in eastern basins (Chamley 1975c, Mélières et al. 1978) ; note that X-ray reflections attributed to palygorskite by Stanley et al. (1981) locally suggest a Greek source for the mineral. Similar investigations are being performed in many regions of the world ocean, and numerous publications are issued each year that participate in clarifying the mineral sources in recent sediments. Some examples are given in Chapter 8.2, especially for very cold and very arid regions where the materials feeding the marine sedimentation preferentially derive from rocky substrates.

The application of different identification techniques on recent marine sediments sometimes demonstrates that the origin of clay minerals in a given region may be much more complicated than indicated by simple X-ray diffraction data. This is the case in the *Northeastern Indian Ocean* for smectite minerals. As early as 1973, Kolla and Biscaye explained from geographic-distribution patterns that recent smectites could derive from either the Indonesian Archipelago, the Mid-Indian Ocean ridge or the Deccan traps in India. Similar conclusions arose from the study performed in the same region on DSDP leg 22 cores of middle Miocene to Pleistocene age, smectite being systematically attributed to volcanogenic

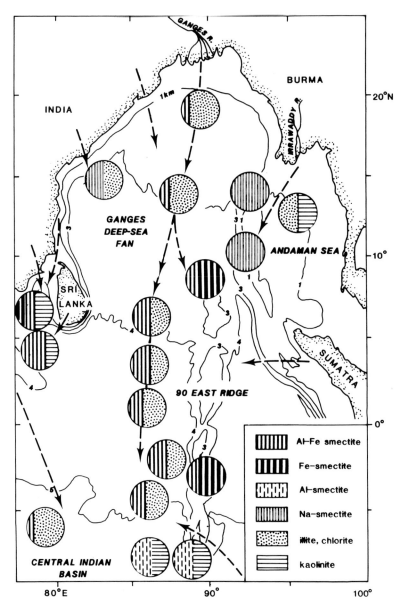

Fig. 18.1. Distribution of characteristic chemical types of smectite, and other dominant clay minerals, in recent sediments from the Northeastern Indian Ocean. (After Bouquillon et al. 1989). Minor or non-representative clay species are not represented. *Arrows* indicate the possible clay mineral sources

sources (Kolla 1974). Bouquillon et al. (1989) perform X-ray diffraction investigations, microprobe and transmission-electron microscopy analyses on numerous piston-core sediments of the same region. *Smectites comprise dominant Al-Fe types, and also Al-, Fe- and Na-types. The distribution of the different smectite types varies in recent sediments according to the location* (Fig. 18.1):

– Al-Fe smectites occur widely and seem to result primarily from the reworking of Indian soils, developed on various types of parent-rocks under strongly hydrolyzing conditions. The relative abundance of smectite tends to increase with increasing distance from river mouths, which is mainly attributed to differential settling processes (Chap. 6).

– Fairly homogeneous iron-rich smectites are identified in the Sri Lanka basin, where they derive from adjacent ferriferous soils or from the displacement by contour currents of India-derived sediments. Iron- to silicon-rich smectites also exist on the Ninety-East ridge, associated with Al-Fe smectites; the ferriferous varieties could partly result from a local authigenesis, as suggested by their association with amorphous silico-ferrous complexes and by low sedimentation rates (12.4).

– Aluminous smectites occur mainly in the more distal zone of the Ganges deep- sea fan and in the northern part of the Central Indian basin. Al-smectites are associated with kaolinite and palygorskite of a probable Australian origin through wind and current action (7.3), and probably have the same origin.

– Sodic-smectites are identified West of Sri Lanka island, as well as in the Western Andaman Sea. Bibliographic data suggest a detrital supply from southeastern India and Burma, respectively.

– In addition, well-crystallized smectites associated with high amounts of illite occur in the equatorial zone. A. Bouquillon and colleagues think this association could have a northwestern origin, and be transported by surface or wind currents from the western coast on India (materials issued form Indus river and Deccan traps).

Clay mineral associations are often used in the geological record to identify past sources. Many examples are provided in the literature, and many new references arise regularly. Let us quote a few cases of different periods. In the Black Sea, Müller and Stoffers (1974) have shown that *late Quaternary* sediments are dominated in the clay fraction either by illite, that mainly derives from the northern Russian platform, or by smectite, essentially originating from southern Turkish volcanics. Owing to these data, Stoffers and Müller (1978) interpret in the same area the increasing percentage of illite in DSDP Leg 42B *Pleistocene* sediments, relatively to *late Miocene-lowermost Pleistocene* deposits, as the result of a change from predominantly southern to predominantly northern sources. One could wonder if this upward clay change does not also partly result from the late Cenozoic climate cooling, that progressively favored the erosion of rock-derived instead of soil-derived minerals (see 17.3). In *early Miocene* sediments of North Aquitaine, Southwestern France, Alvinerie and Latouche (1967) report the northern origin of smectite and the southern origin of illite, each mineral being trapped on the corresponding side of a submarine rise and being prevented from progressing southwards and northwards, respectively. Similarly, the study of the *Paleogene* Kesan sandstone formation of Southwestern Turkish Thrace allows Ataman and Gökçen (1975) to distinguish two successive sources of detrital materials. A southern origin first prevailed, characterized by soil-derived kaolinite, mixed-layers and illite developed on a peneplain topography under tropical humid climate; a northern supply subsequently took place during late Paleogene times, and determined the input within the basin of corrensite issued from the rapid alteration of metamorphics in the Rhodope mountains.

Pre-Cenozoic series offer various source reconstructions. For instance, in the *Cretaceous* sediments of Provence, Southeastern France, the Barremian clay assemblages indicate a permanent southern source marked by degraded illite, random mixed-layers and some kaolinite, suggesting the erosion of weathered crystalline substrates (= Pyrenees – Corsica – Sardinia land mass). From late Barremian to Aptian a large supply of kaolinite and smectite indicates an increasing northwestern source correlative to the extensive phase of carbonate platform development. The changes recorded in clay mineral sources seem to be related to distant tectonic activity (Chamley and Masse 1975, Giroud d'Argoud et al. 1976). Diverse refinements of the source distribution and associated mechanisms in this region arise from subsequent studies (Deconinck et al. 1985, Levert and Ferry 1988). In the *Paleozoic* of the United States of America, numerous attempts contribute to the recognition of sources. For instance, the Upper Ordovician (Cincinnatian) formations of Ohio display a close correspondence between clay mineral suites and paleocurrent figures; the successive increase and decrease of chlorite and vermiculite abundance relative to illite and illite-smectite reflect a change from more easterly to more northerly sources (Bassarab and Huff 1969). The Pennsylvanian (Desmoinesian) black shales of the mid-continent region show a northward increase in the abundance of detrital illite, kaolinite, chlorite and illite-smectite, suggesting a source area in the Ouachita and Iowa regions (Ece 1987). In *early Paleozoic and late Proterozoic* deposits of Mauritania, clay assemblage data combined with sedimentary structures allow identification of a northern to northeastern origin for glacier bodies, and distinguishing local from distant sources according to the varying extension of African ice-sheets (Chamley et al. 1977a, 1980a).

18.2 Paleocurrents and Resedimentation

18.2.1 Introduction

The use of clay assemblages to characterize present coastal and slope water masses is well known (see Chaps. 5, 6, 8.3), *and invites using these assemblages in reconstituting past water masses.* For instance, Pierce et al. (1972), with 8.0 μm pore filters, trap the material suspended in the waters on the *continental shelf from Chesapeake Bay to Savannah, Georgia,* Northwestern Atlantic. The amount of material trapped ranges from 0.05 to 8.44 mg/liter. The clay and associated mineral suites allow identifying *four distinct water masses,* that locally intermix: (1) The Virginian coastal water contains illite, chlorite, talc, quartz, amphibole, dolomite, feldspar and rare kaolinite. This composition suggests northern affinities and a probable origin located North of Chesapeake Bay. (2) The Carolinian coastal water comprises terrigenous minerals of poorer crystallinity, marked by kaolinite and some smectite in addition to ubiquitous quartz and illite. This suite probably results from river runoff in the Carolinas and Georgia. (3) The Gulf Stream water includes well-crystallized kaolinite, quartz, dolomite and talc, with very minor amounts of chlorite. (4) The Carolinian slope water carries talc, kaolinite, illite, dolomite, calcite, feldspar and quartz. J.W. Pierce and colleagues conclude that precise mineralogy of suspended matter can be used to trace water masses in a way similar to that in which temperature and salinity are usually employed.

The relevance of clay minerals as *tracers of oceanic water masses* is a priori favored by the frequent occurrence of abundant suspended material of both organic and inorganic nature at various depths in the water column (e.g., Kennett 1982. See also Chap. 6). The most abundant particles generally occur in bottom nepheloid layers, that are frequently present in the oceans. Most of the material in the nepheloid layer occurs in a strip along the western boundary currents, as for instance in the Atlantic (Fig. 18.2) and Indian Oceans. The nepheloid water consists

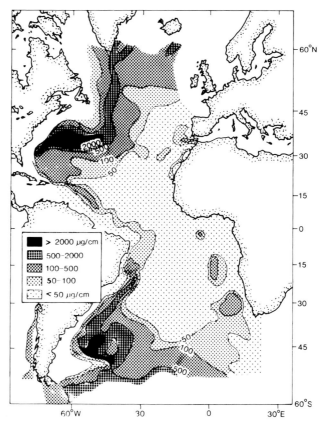

Fig. 18.2. Distribution of suspended matter in the bottom nepheloid layer in the Atlantic Ocean. (After Biscaye and Eittreim 1977). Material settling from above into the layer is subtracted

of particles maintained in suspension by constant erosion and redeposition, and carried over more or less long distances above the sea floor. Such resuspensions have been traced by clay minerals in the Western Mediterranean Sea, despite the moderate intensity of the bottom circulation (Pierce and Stanley 1975). Remember that reworking and redeposition processes linked to turbidity currents may also be identified by clay mineral changes (e.g., Rupke 1975, Monaco et al. 1982. See also 8.3.2), as well as by the clay fabric (e.g., O'Brien et al. 1980). All these considerations invite envisaging some applications to past geological times. The glacial periods, often marked by enhanced bottom circulation and resedimentation processes (e.g., Gardner et al. 1980), may be distinguished a priori by mineral suites from interglacial, high sea-level periods. Let us discuss a few examples of tentative reconstructions of past circulation from mineralogical and sedimentological investigations.

18.2.2 Mediterranean Sea

The use of clay minerals in the Mediterranean Sea as water-mass tracers is favored by distinct petrographic sources (see Chap. 8) and by fairly well-known circulation patterns. As early as 1971, Venkarathnam and Ryan identify in the *eastern basin* six mineral assemblages, from surface sediment analysis. These assemblages are tentatively attributed to six distinct dispersion agents: (1) A Nile assemblage, rich in well-crystallized smectite ($> 50\%$ of the clay fraction) with associated kaolinite (15-25%), is dispersed on the eastern Nile cone and within the Levantine basin by easterly-directed surface currents. (2) A smectite (40-60%), chlorite and illite assemblage issues from the Southeast Aegean Sea and is thought to be transported to the Mediterranean ridge southeast of Crete by Levantine intermediate water. (3) A kaolinite-rich assemblage (> 30-20%) associated with high carbonate values is recorded on the western section of the Mediterranean ridge and in the western Nile cone, as a result of transport by winds from North Africa. Further studies show that palygorskite also characterizes African-borne aeolian supply (Chap. 7). (4) and (5) Kithira and Messina assemblages are both marked by abundant illite and chlorite, that are carried respectively from south Italy and South Greece into the deep parts of the Ionian basin by bottom circulation and turbidity currents. (6) A Sicilian assemblage with kaolinite (> 30-20%) and smectite (30-50%) characterizes the westernmost part of the Ionian basin South of Sicily, and is attributed to the dispersal by easterly-moving surface waters from the Western Mediterranean.

In the *Tyrrhenian Sea*, a similar attempt to correlate clay suites from surface sediments and transportation agents is presented by Tomadin (1974). Illite and chlorite are related to bottom currents flowing from Italian and Sardinian shelves into the bathyal plain. Smectite is largely attributed to volcanic sources located in the Napoli and North Sicily areas, and is especially carried by surface and intermediate water masses to the central Tyrrhenian Sea. Kaolinite and poorly crystallized illite are thought to issue largely from North Africa by winds.

The *Alboran Sea*, in the westernmost part of the Mediterranean Sea, is particularly investigated, since strong water-mass exchanges occur with the adjacent Atlantic Ocean and may have been modified during glacial-interglacial alternations. Huang et al. (1972) first reported a slight eastward increase in smectite abundance in surface sediments, which was tentatively attributed to an eastern volcanic source located in the Alboran Sea itself. As the distribution of smectite around the Alboran volcanic islands did not justify such a hypothesis (Valette 1972), a more easterly terrigenous origin for smectite was later suggested by Pierce and Stanley (1975). In fact, extensive mineralogical studies performed on piston- and box-cores from the Western and Eastern Alboran Sea show that smectite presents rather an average increase in abundance to the West, in the direction of the Gibraltar strait and Atlantic Ocean (Auffret et al. 1974, Cossement et al. 1984. Table 18.1). This suggests an Atlantic origin through surficial water mass for Alboran smectite. Such an interpretation agrees with the fairly high abundance of smectite West of the Strait of Gibraltar in the Cadiz Bay (Mélières 1974). Moreover the only Alboran cores relatively depleted in smectite are located outside the main path followed by the Atlantic surficial water mass (Auffret et al. 1974, Cossement et al. 1984). In addition, the distribution in the same cores of planktonic foraminifera shows an eastward transportation into the Mediterranean by east-flowing Atlantic Surface Water. As the suspended particles settle through the water column, they are partly redistributed at depth by the westward-flowing Mediterranean Deep Water Mass (Auffret et al. 1974). The distribution of illite in surface sediments is opposite to that of smectite (Table 18.1), suggesting a predominantly eastern source for the first mineral. This agrees with the existence of an illite-rich clay-mineral province in Western Mediterranean basins (Chamley 1971). Note that the lithological, mineralogical or microfaunal changes occurring at the late Pleistocene-Holocene transition correspond to common climate variations (see 17.2.1), and do not indicate any current reversal in the strait of Gibraltar (see discussion in Cossement et al. 1984).

The late Quaternary sedimentation in the *Southeastern Levantine basin*, Eastern Mediterranean Sea, depends mainly on the terrigenous supply from the Nile river (Venkatarathnam and Ryan 1971). Maldonado and Stanley (1981) investigate this area and the adjacent Nile cone from 25 piston cores sampled from the Present back to 23,000 years BP. Five stratigraphic layers are recognized, that present distinct lithologic and mineralogic characters. Essentially two clay patterns are identified:

1. The recent surface sediments (Fig. 18.3A) and the 23,000 to 18,000 years BP layer display a regular northwestward decrease in smectite abundance. This results from an injection of smectite-rich Nile-river sediment by surface currents

Table 18.1. West-East variations of smectite and illite relative abundance in the clay fraction of surface sediments from the Alboran Sea, westernmost Mediterranean Sea (after Auffret et al. 1974)

	West KR 38 04°26′2W	KR 36 04°12′8W	KR 35 03°58′0W	KR 33 03°12′7W	KR 41 02°52′5W	KR 31 024°30′8W	East KR 42 00°17′9W
Smectite %	30	25	25	20	20	15	10
Illite %	40	45	45	45	45	50	55

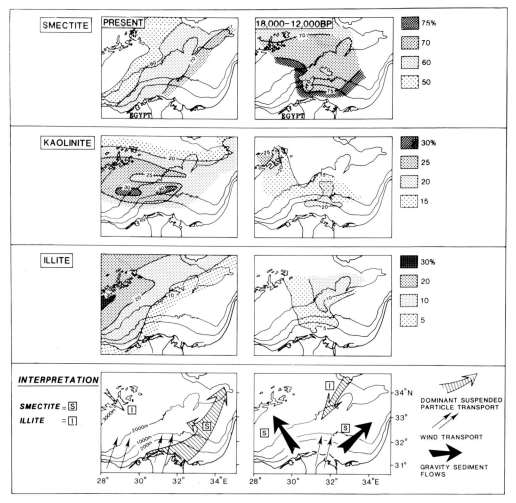

Fig. 18.3. Clay-mineral distribution and depositional interpretation in the Southeastern Levantine Sea, Eastern Mediterranean. (After Maldonado and Stanley 1981). *Left* surficial layer, recent deposits. *Right* 18,000 to 12,000 years BP stratigraphic layer

into the counter-clockwise Eastern-Mediterranean circulation. Kaolinite represents the next most abundant species, with the largest concentration on the Nile cone and on its western side. Such a distribution suggests that kaolinite derives in part from the Nile river and is also blown in northwards from the western desert. Illite and chlorite are both marked by decreasing percentages going southeastwards. These latter minerals are mainly advected from the North by intermediate to deep water masses.

2. Three intermediate layers, extending respectively from Recent to 5700 years BP, from 5700 to 12,000 yr and from 12,000 to 18,000 yr (Fig. 18.3B), show a more irregular distribution of clay minerals. In this hemipelagic to turbiditic

mud, smectite is concentrated to the westernmost part of the area, decreasing in abundance from the Southwest to the Northeast. Smectite mainly results from downslope gravity processes, that determined the direct supply of Nile-derived minerals from the distal continental margin to the upper continental slope. Mass-gravity flows were discouraged at times of higher stands of sea level, as shown by a progressive decrease in the importance of gravity flows and of smectite resedimentation from 18,000 years to Recent, a time interval marked by a sea-level rise. Kaolinite of African aeolian origin is relatively less abundant than in surface and bottom layers. It appears to be diluted among smectite-rich gravity flows. Illite and chlorite are regularly advected from northern regions during the 18,000 years-Recent time interval, with accessory variations in the intensity and direction of supply.

A. Maldonado and D.J. Stanley infer from their study that *depositional processes are more important than climate variations in determining the clay mineral distribution in the Southeastern Levantine Sea during the past 23,000 years*. No correspondence exists between the clay record and the climate evolution, which is marked in this region by a change from arid to humid conditions. This conclusion confirms the frequently poor climatic expression of clay minerals and other terrigenous material in sediments located close to coastal areas and submitted to important redeposition processes (17.1.2). The sea-level changes appear to have significantly controlled the clay distribution in the Levantine basin. Eustatic changes, however, cannot entirely explain the distribution of clay minerals, since the lower layer cored (25,000 to 18,000 yr BP), corresponding to a sea-level lowering, is similar in composition to the surficial layer corresponding to a high sea-level. The relative proportion of gravitite-sediment types available at a given period is probably also of high importance.

18.2.3 Southern Ocean

The possibility to use clay assemblages in reconstructing paleocirculation patterns is illustrated in several parts of the Southern Ocean, where important meridian movements characterize the different water masses. On the *Walvis Ridge, Southeastern Atlantic Ocean*, Robert (1982) investigates the DSDP sites drilled during leg 74. An increase in chlorite and illite abundance occurs from Eocene to Present time only at sites 525 and 526 that are located at relatively shallow water depths (2477 and 1064 meters, respectively). This change is atrributed to an increasing influence of the Benguela Current, that carries illite and chlorite eroded in higher amounts from South African deserts. By contrast, the deeper sites 527 and 528 (4438 and 3035 meter depth) do not display any important modification in the illite and chlorite content, probably because they are located aside from the direct influence of the Benguela Current.

The *Vema Channel, Southwestern Atlantic Ocean*, constitutes a narrow passage at 4565 meter depth between the Argentina and Brazil basins. The southward-flowing North Atlantic Deep Water and the underlying northward-flowing Antarctic Bottom Water go through this channel, that cuts a submarine

West-East barrier, the Sao Paulo Plateau-Rio Grande Rise. Chamley (1975d) investigated ten box-cores spanning the depth interval 1903 to 4563 meters on the western side of the Rio Grande Rise at about 31°S. Clay minerals in the less than 2 μm fraction consist of chlorite (3-20%), illite (30-50%), smectite (15-35%), kaolinite (5- 20%) and random mixed-layers (5-15%). In surface sediments, 0 to 2 centimeters below the sea water-sediment interface, kaolinite and poorly crystallized illite characterize the clay fraction in the 1903-3274 meter water-depth interval. This was attributed to a *preferential advection from the North by the North Atlantic Deep Water*, whose depth range approximately covers the 2000- to 3500-meter interval. Kaolinite and poorly ordered illite were presumed to derive from low-latitude soils subject to strong hydrolysis. Below 3500 meter water depth, the chlorite abundance noticeably increases, together with that of medium-crystallized smectite, illite-smectite mixed-layers, quartz and feldspars. Illite is less abundant but better crystallized than at shallower water depths. This change was related to the *influence of the Antarctic Bottom Water, carrying Antarctic- and South American-derived minerals to the North*. The relatively high abundance of rock-derived minerals (chlorite, quartz, feldspars), combined with the good crystallinity of illite, agreed with the occurrence of very low hydrolysis processes under high latitudes (Chap. 2). In addition, smectite appeared to have a predominantly southern detrital origin, a fact later confirmed by Robert (1982. See also 7.2.2, 8.2.1). Additional investigations in subsurface sediments led Chamley (1975d) to identify an upward shift, from 3274 to 2878 meter water depth, in the chlorite-rich/kaolinite-rich transition. This shift occurred within the box cores at a depth of about 10 centimeters below the sediment-water interface, in sedimentary levels dated from the uppermost Pleistocene (Melguen and Thiede 1975). The shift was tentatively attributed to a *volume extension and celerity increase of the chlorite-carrying Antarctic Bottom Water at the end of the last glacial period*. It was therefore proposed to use clay assemblages to follow the variations of different water masses in the course of the time (e.g., Melguen et al. 1978). Clay mineral interpretations were supported by observations on nannofossil and foraminifera assemblages (Diester-Haass 1975, Melguen and Thiede 1975).

Jones (1984) presents a detailed study on the chlorite and kaolinite distribution in 40 core-top samples from the area previously investigated by H. Chamley, on the western Rio Grande Rise in the 1430-4565 meter depth interval. The comparison between a hydrologic profile across the Vema Channel, the abundance of Antarctic-derived diatoms (*Nitzschia kerguelensis*, Jones and Johnson 1984), and the abundance of chlorite confirm the southern origin of the clay mineral (Fig. 18.4). *The chlorite percentage variations obviously correlate to the boundary zone between Antarctic Bottom Water and North Atlantic Deep Water*. Kaolinite reaches maximum concentration at mid-depth, approximately at 3500 meters. As the core of the North Atlantic Deep Water is located at about 3000 meter depth, G.A. Jones thinks that kaolinite can hardly be advected from the North by this water mass. He suggests a lateral, western source from the Brazil coast, with several transportation steps. Kaolinite could be introduced at 20°S on the Brazil shelf by the Rio Doce river, be carried southwards by the Brazil current to Cabo Frio, then be more or less incorporated into fecal pellets, be transported eastwards to the Sao Paulo Plateau, and finally be resuspended by bottom cur-

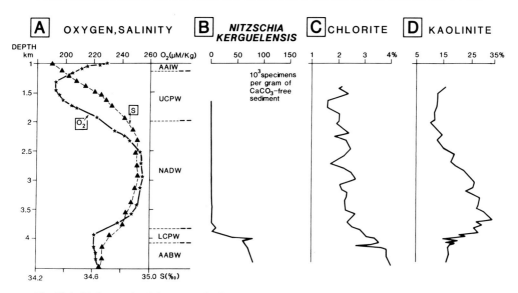

Fig. 18.4. Hydrography (**A**), Antarctic diatoms (**B**), chlorite (**C**) and kaolinite (**D**) in the Vema Channel, Southwestern Atlantic Ocean. (After Jones 1984). Oxygen content and salinity of the following water masses: Antarctic Intermediate Water (*AAIW*), Upper Circumpolar Water (*UCPW*), North Atlantic Deep Water (*NADW*), Lower Circumpolar Water (*LCPW*), Antarctic Bottom Water (*AABW*). Diatoms and clay minerals from 40 core-top samples from the Rio Grande Rise in the 1430 to 4565 m water-depth interval. Chlorite and kaolinite normalized to talc

rents and advected toward the Vema Channel and Rio Grande Rise. In addition, Jones (1984) and Jones and Johnson (1984) do not identify from Antarctic diatoms or clay minerals significant shallowing or deepening of the Antarctic Bottom Water during the glacial or interglacial intervals examined (Present, 18,000 yr, 120,000 yr, 140,000 yr BP). This constitutes another disagreement with Chamley's assumptions (1975d), that were based on different time slices (Present and about 10,000 yr BP). All these considerations point to the complexity and still preliminary character of paleocurrent reconstructions proposed from mineralogical data on deep-sea sediments.

Following preliminary investigations by Piper and Pe (1977), Robert et al. (1988) provide paleocurrent data combined with information on climate evolution about Cenozoic sediments from DSDP site 274, drilled North of the *Ross Sea* in the Australia – New Zealand sector of the Antarctic. Smectite often constitutes the dominant clay mineral in the <2 μm fraction (30-70%), and is associated with fairly abundant illite, accessory chlorite and little kaolinite and mixed-layers. Smectite is considered as essentially terrigenous, as suggested by geochemical data, the absence of relationship with dated volcanic events, the existence of smectite-rich Cretaceous rocks cropping out on the Antarctic margin, and the local presence of reworked Cretaceous pollen. A combined study of clay minerals, inorganic geochemistry and radiolaria suggests the following steps in the Cenozoic evolution of this part of the Southern Ocean : (1) southward circulation during Oligocene providing soil-forming smectite from Australia; (2) early-Miocene cooling on Antarctica responsible for a current inversion and the production of increasing amounts of rock-derived illite; (3) middle-late Miocene erosion of smectite-rich rocks cropping out on the Antarctic continental margin, correlative of enhanced circulation of cold water; (4) Plio-Pleistocene development of Antarctic Bottom Water carrying smectite, chlorite and well-crystallized illite to the North.

Notice that few studies on the hydrodynamic expression of detrital clay suites exist for continental series (e.g., Liebling and Scherp 1976), except those based on the distribution of bentonite or tonstein layers (e.g., Grim and Güven 1978, Fisher and Schmincke 1984. See also 16.4.2).

18.3 Tectonic control

18.3.1 An Example: The Pliocene of South Sicily

The renewal of relief due to tectonic uplift often results in great changes in the processes of erosion and transport. The composition of detrital clay suites may register such changes because of noticeable modifications in the erosion conditions of both exposed rocks and soils. On the southern coast of Sicily at Capo Rossello, Pliocene marine sediments of the Caltanissetta Basin comprise two successive intervals that are lithologically and mineralogically distinct (Fig. 18.5). The early Pliocene deposits consist of hemipelagic, marly limestones (= *Trubi facies*), that contain significant amounts of *palygorskite* (up to 20% of the clay fraction) and are relatively enriched in *illite* compared to smectite and kaolinite. Palygorskite occurs as short and broken fibers in Trubi hemipelagites, indicating a detrital origin. Palygorskite exists in large amounts within the Paleogene sediments blanketing the North-African shield, namely in Tunisia (in Chamley 1971, Sassi 1974), and is much less frequent in Sicily (e.g., Mascle 1973). One can infer that sediments deposited during early Pliocene in the marine basin between Sicily and Tunisia received palygorskite particles that were removed essentially from North Africa.

The Trubi limestones are overlaid with terrigenous, clayey marls (= *Monte Narbone facies*) deposited from early late Pliocene to uppermost Pliocene times. The lithologic change correlates with a rapid decrease in the palygorskite content and with a sudden increase in *smectite and kaolinite* amounts relatively to illite. Kaolinite, and especially smectite, are known to occur in large amounts in pre-Miocene Sicilian rocks (Mascle 1973. 17.3.3). The more terrigenous character of the sedimentation (i.e., increase of clay fraction, decrease of biocalcareous fraction) suggests an increasing influence of Sicilian output, because the Capo Rossello section is located on the Sicilian side of the basin. This indirectly confirms the African origin of palygorskite.

The sudden lithological and mineralogical change recorded in Pliocene sediments cropping out on the southern coast of Sicily roughly corresponds to a tec-

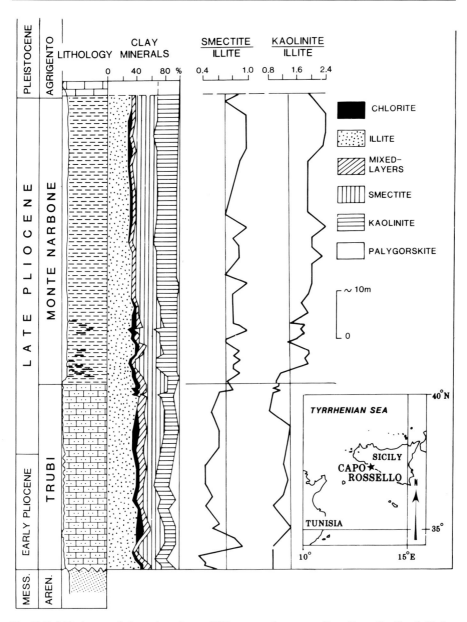

Fig. 18.5. Lithology and clay mineralogy of Pliocene sediments at Capo Rossello, South Sicily. (After Chamley 1976). Smectite/illite = 17/10 Å peak-height ratio, glycolated sample. Kaolinite/illite = 7/10 Å peak-height ratio, glycolated sample (chlorite, that constitutes a very minor part of the 7 Å peak, is neglected)

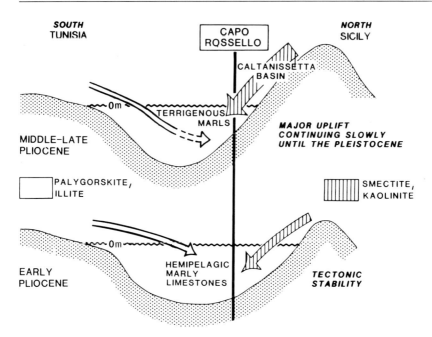

Fig. 18.6. Tectonic interpretation of lithological and mineralogical changes in Pliocene marine sediments of Capo Rossello, Caltanissetta Basin, southern coast of Sicily

tonic uplift of Northern Sicily (Mascle 1973). *The correlative change from hemipelagic marly limestones to marly clays, and from palygorskite-enriched to smectite- and kaolinite-enriched clay fraction, therefore appears to express a specific phase of tectonic activity.* During early Pliocene time the Capo Rossello area was located in an open marine basin where calcareous oozes deposited. The admixture of terrigenous components to planktonic shells was controlled by the influence of both southern and northern sources (Fig. 18.6). Various minerals were supplied, including palygorskite and illite predominantly derived from Africa, as well as smectite and kaolinite predominantly supplied from Sicily. During the early late Pliocene, a tectonic event caused the uplift of Northern Sicily, leading to a rapidly increasing supply of smectite- and kaolinite-rich clastics to the Capo Rossello area. The marine biogenic and the North-African terrigenous contributions to sedimentation were then diluted by large input of Sicilian material. The slight increase of smectite and kaolinite abundance toward the uppermost Pliocene, correlative to the almost disappearance of palygorskite, suggests a continuation of the Sicily uplift. This is proved by the final emersion of the marine basin at Capo Rossello close to the Pliocene-Pleistocene boundary. *Clay minerals in marine sediments therefore are able to reflect both major and attenuate tectonic movements on adjacent Sicily during Pliocene time.*

18.3.2 Applications

Mediterranean Range, Late Cenozoic

The Plio-Pleistocene section drilled at DSDP site 125, located on *Eastern Mediterranean Rise* in the Ionian Sea, displays clay mineral changes resembling those recorded in South Sicily. Close to the Plio-Pleistocene boundary, the amount of palygorskite decreases from about 30% to 10% of the clay fraction (Fig. 18.7). Palygorskite fibers are short and broken, indicating a detrital origin. Palygorskite is accompanied by noticeable amounts of kaolinite, a typically Africa-derived mineral in this region (Venkatarathnam et al. 1971, Maldonado and Stanley 1981). As palygorskite occurs abundantly within the early Tertiary rocks cropping out in North Africa (see Chamley 1971), its decrease in marine sediments suggests a relative diminution in the terrigenous influx from the South. The palygorskite decrease is balanced by a strong increase in the abundance of illite, chlorite, quartz and feldspars, all minerals characteristic of northern sources. By contrast to the stable African shield, the European borders of the Eastern Mediterranean were submitted during late Tertiary times to an active tectonic structuration related to Alpine orogeny. The mineralogical change observed on the Mediterranean Rise close to the Pliocene-Pleistocene boundary therefore appears to result from an *acceleration of the Peloponnesus uplift*, responsible for an increasing erosion in Southern Greece (Chamley 1975c). Similar expressions of tectonic activity during late Cenozoic times are recorded in other parts of the

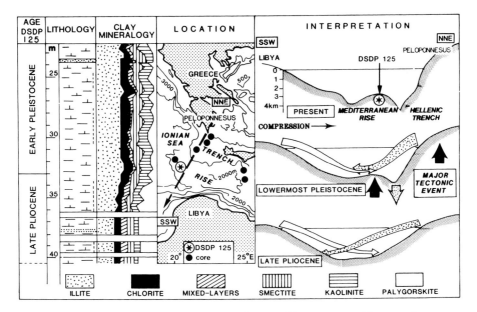

Fig. 18.7. Clay mineral stratigraphy at the Pliocene-Pleistocene transition on the eastern Mediterranean rise, Ionian Sea, and interpretation. (After Chamley 1975c)

Mediterranean range such as the Tyrrhenian Sea, South Adriatic Sea, Messina abyssal plain and West of Cyprus (e.g., Chamley et al. 1977c, Blanc-Vernet et al. 1979). In the Alboran Sea, the increasing character of the epcirogenic activity toward the strait of Gibraltar is expressed by a westward augmentation of rock-derived minerals relative to soil- derived minerals (Cossement et al. 1984).

The case of the Eastern Mediterranean rise sets up two additional questions.

1. The mineralogical modification recorded at the Pliocene-Pleistocene boundary does not correlate with any lithological change. The whole Plio-Pleistocene section is made up of homogeneous clayey calcareous oozes. *Clay minerals suites in hemipelagites are therefore able to express tectonic changes when other sedimentary components remain unchanged.* Such a property may be particularly useful to identify tectonic events in sedimentary series deposited remote from orogenic areas.

2. Complementary investigations performed in the Hellenic Trench, North of the Mediterranean Rise, reveal the presence of abundant palygorskite in Pliocene sediments, indicating the probable output of African material as far as the foot of the Peloponnesus. Such an important supply of African-derived minerals is hardly compatible with the presence of a noticeable submarine relief between Africa and the Hellenic Trench, i.e., with the existence of the East Mediterranean Rise as a significant submarine barrier. In addition, palygorskite is virtually lacking in Pleistocene sediments of the Hellenic Trench, whereas the mineral is still present in contemporary deposits from the Mediterranean Rise itself and from the basin located between the rise and Libya. This suggests that *the Mediterranean rise did not significantly exist until the late Pliocene, and experienced a major uplift phase close to the Pliocene-Pleistocene boundary.* Palygorskite was abundantly supplied North of the rise when the submarine relief was low, and was retained South of this relief after it rose. As a consequence, a major step of the Mediterranean Rise formation, due to compression between Africa and Europe, probably occurred no later than lowermost Pleistocene time and correlated to the uplift of Peloponnesus (Fig. 18.7). This interpretation is compatible with geophysical and palynological data (in Chamley 1975c). It suggests that *the clay stratigraphic record is able to express tectonic changes occurring on both exposed and submarine areas.*

New Zealand Region, Cenozoic

The leg 90 of the Deep Sea Drilling Project provides a subcontinous Cenozoic record on the Lord Howe Rise (site 592), the Challenger Plateau (site 593) and the Chatham Rise (site 594), three zones located around New Zealand (Fig. 18.8). Late Eocene to Pleistocene nannofossil oozes and chalks are investigated from clay mineral assemblages, total grain-size distribution and flux rates (Robert et al. 1985). Smectite and illite represent the major clay minerals, and are accompanied by chlorite, irregular mixed-layers, kaolinite and various non clay minerals. Strong variations occur in the clay stratigraphy, which contrasts with the generally homogeneous lithology. These mineralogical variations cannot be correlated with an in situ evolution of volcanic glass, because of the absence or only local co-occurrence of volcaniclastics with clay minerals. Most mineralogi-

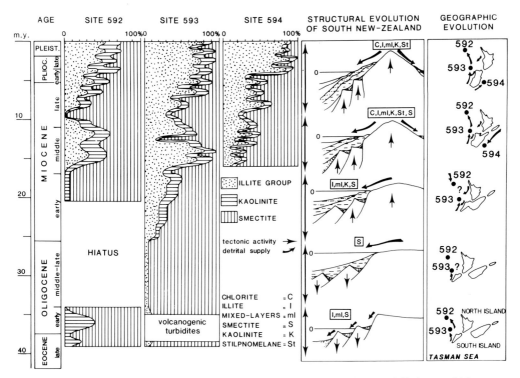

Fig. 18.8. Clay mineral data and interpretation about Cenozoic sediments drilled around New Zealand. (Leg 90 DSDP, after Robert et al. 1985)

cal changes are too intense to correspond to climate variations and are not synchronous to known climatic events. *The Cenozoic clay stratigraphy around New Zealand mainly reflects the tectonic history of this area, and accessorily the paleocirculation pattern.* Five periods arise from the clay mineral zonation (Fig. 18.8).

1. *Eocene-Oligocene boundary.* A strong increase in illite group minerals (i.e., illite, chlorite, mixed-layers) occurs at site 592, associated with abundant volcaniclastics and reworked deposits at site 593 (Kennett, von der Borch et al. 1985). This reflects major extensional tectonics and graben formation on the southern island of New Zealand (Norris et al. 1978). The block-faulting was probably accompanied by rejuvenation of the morphology in the hinterland, causing an increasing erosion of rocky substrates.

2. *Oligocene.* The Oligocene sedimentation is only recorded on the Challenger Plateau (site 593). Abundant fleecy smectite flakes characterize the clay fraction of the biocalcareous oozes, that are devoid of volcanic debris. Smectite appears to derive mainly from the erosion of soils developed under hydrolyzing conditions in the peneplaned borderlands of New Zealand, where the same clay suites occur (Hume 1978). This suggests the existence of a fairly stable period. The sedimentation gap on the Lord Howe Rise (site 592) could be related to erosion by strong currents flowing southwards (in Robert et al. 1985).

3. *Early Miocene.* An increasing supply of illite group minerals and kaolinite at site 593, as well as onshore (Hume 1978), correlates to compressive tectonic activity in the North Island of New-Zealand (Nelson and Hume 1977). The tectonic rejuvenation of the island relief probably induced the reworking of both rock products (illite, etc) and upstream soil products (kaolinite). New Zealand derived materials did not reach the Lord Howe Rise area (site 592), that was still submitted to southward-flowing currents (Stein and Robert 1985).

4. *Middle to early late Miocene.* A strong, long-lasting supply of the illite group occurs West of New Zealand. It is attributed to a new compressive phase, especially documented in the South Island by reverse faulting and deposition of coarse detrital sediments (Norris et al. 1978). The specific contribution of South New Zealand to the clay sedimentation is demonstrated by the appearance of stilpnomelane, a sheet-mineral reworked from old metamorphic schists exposed in this island. The mineralogical event is recorded on both Challenger Plateau (site 593) and Lord Howe Rise (site 592), probably because of the progressive development of northward currents in the Tasman Sea, in relation to the development of the Antarctic ice cap. By contrast, only slight changes characterize the clay sedimentation on the southeastern side of New Zealand (Chatham Rise). These slight changes are probably related to tectonic pulses as at other sites, But the effects of the tectonic activity were probably diluted by distant supply of smectite through southward currents flowing East of New Zealand.

5. *Late Miocene to Pleistocene.* A last increase in the abundance of illite and associated minerals is recorded, reflecting the Kaikoura orogeny in New Zealand (Crook and Feary 1982), again responsible for a morphological rejuvenation and increasing erosion. This tectonic activity is expressed at the three sites, pointing to the development of northward-flowing currents on both western and eastern sides of New-Zealand. The tectonic effects are probably superimposed on the effects of the late Cenozoic cooling, also responsible for an increasing supply to the ocean of rock-derived minerals relatively to soil-derived minerals (Chap. 17.3).

North Atlantic Domain, Late Mesozoic

The first sediments deposited in North Atlantic basins during late Jurassic times consist of Callovian to Kimmeridgian brown-gray claystones to reddish limestones. In the underlying oceanic crust, the clay fraction of basalts and interbedded carbonates is characterized by abundant crystalline Mg-smectite or sepiolite related to submarine alteration or hydrothermal activity (13.2. Fig. 13.4). By contrast, the sediments blanketing the basalts show a typically detrital clay-mineral assemblage (Figs. 13.4, 14.2. Chamley 1979, 1981, Chamley and Debrabant 1984a). More or less weathered illite and chlorite occur abundantly in the less than 2 µm fraction, together with random mixed- layers, poorly crystallized smectite, quartz and feldspars. Such a clay association suggests a strong erosion of both rocks and not evolved soils on adjacent land masses, and therefore a tectonic rejuvenation. The detrital character of clay suites is supported by rare-earth element data (Chamley and Bonnot-Courtois 1981). The cause of the tectonic instability on American and Euro-African borders is attributed to the early stages of Atlantic Ocean spreading, after the rifting has determined the collapse of the oceanic crust, the initiation of the young marine basin, the relative uplift of adjacent, margins and the active erosion of the continental rocks. *The strong increase recorded in the abundance of the illite group during first oceanic sedimentation stages appears to reflect the tectonic erosion of adjacent continental margins after the basin initiation.* The subsequent augmentation of Al-Fe smectite is attributed to further tectonic relaxation stages and to the development of continental soils (17.3, 17.4).

By extension, *the sudden and strong increases recorded in the illite-group abundance at different periods of the Cretaceous history in North and South Atlantic basins are thought to partly result from tectonic rejuvenation stages, sometimes associated with sea-level changes* (e.g., Chamley 1979; Chamley et al. 1979a and b, 1983; Robert 1982, Chamley and Debrabant 1984a). These mineralogical events, especially identified at Barremian, Albian and Santonian- to-Maastrichtian times, are presumed to correlate with increasing rates of oceanic spreading and subsequent structural equilibration of continental margins (see Chap. 19). The respective influence of the actual tectonic rejuvenation of ocean margins and of sea-level changes on the clay-mineral record has still to be investigated, by comparing continent-to-ocean transects in both stable and non stable regions.

Other Regions

Various other examples of the tectonic signature of clay mineral sequences are reported in the literature. For instance, in the *North Indian Ocean*, the major surrection stages of the Himalaya chain since the early Miocene are expressed in both Ganges and Indus deep-sea fans by the temporary abundance of illite- and chlorite-rich turbidites, and by the almost complete absence of smectite-bearing hemipelagic deposits (Bouquillon et al. 1989, Debrabant et al. 1989). Note that *illite and chlorite are not the only clay minerals possibly representative of tectonic rejuvenation. The nature of the minerals indicative of a tectonic instability depends on the petrographic nature of rejuvenated substrates*, and sometimes on other factors. On the Owen Ridge, Northwestern Indian Ocean, the expression of the Himalaya uplift by the illite group is followed at middle Miocene time by a strong and sudden increase of wind-supplied palygorskite. This mineralogical change results from the *submarine surrection of the Owen Ridge* itself, which precluded further direct supply of Indus-derived illitic turbidites, and favored the settling of windborne dust issuing from Arabian and Northeastern Africa (Debrabant et al. 1989).

Tectonic events may be recorded in various environments. Sebastian and Soria (1987) attribute the periodic input of paragonite in Neogene lakes of the Betic range, Southeastern Spain, to repeated rejuvenation stages of the adjacent cordillera. Chamley et al. (1985b) correlate the sudden decrease of illite, chlorite and kaolinite abundance in late Oligocene sediments from the eastern margin of Japan to the collapse of a continental land mass migrating from the North-Northwest. Deconinck et al. (1985) observe a close correspondence between the lithofacies and the clay mineralogy, in Cretaceous megasequences of the subalpine range, Southeast France. Periods of large supply of illite and chlorite, corresponding to marly deposits and fairly deep environments, are progressively replaced by smectite-rich platform carbonate facies. The first episodes are at-

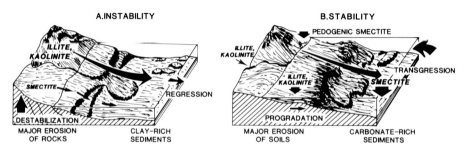

Fig. 18.9. Tectonic control on clay mineral successions in the subalpine range, Southeast France, during late Jurassic and Cretaceous times. (After Deconinck et al. 1985)

tributed to tectonic rejuvenation and instability, associated (Valanginian) or not (Aptian – Albian) with sea-level changes (Fig. 18.9A). The subsequent increases in smectite are thought to reflect a tectonic relaxation, favoring the development of continental soils and of biocalcareous platforms, as well as progradation and differential settling processes (Fig. 18.9B). Bjørlykke and Dypvik (1975) correlate major input of chlorite, plagioclase and chromite in early Paleozoic sediments of Norway to the Caledonian orogeny developed on northwestern land masses.

18.3.3 General Expression of Tectonic Activity by Clay

The clay mineralogical changes induced by tectonic activity and recorded in sedimentary series are generally of large amplitude, except in areas marked by slight, epeirogenic instability (e.g., Alboran Sea or Zakinthos basin, Mediterranean Sea. 18.3.2). *The climatic or hydrodynamic messages originating in clay minerals are usually obliterated when the tectonic control becomes significant.* The clay mineral variations caused by tectonics often *cover large time intervals.* Due to the dispersal properties of clays by currents, the tectonic phases may be *recognized in sediments located far away from unstable land masses.*

The major effect of tectonic rejuvenation on exposed land masses results in the erosion of soils and weathering complexes covering the rocky substrates. If soils are less developed, as in temperate-humid regions, the erosion products supplied after tectonic renewal within sedimentary basins do not fundamentally differ from those supplied during stable periods; clay minerals nevertheless tend to be better crystallized, enriched in fairly fragile species like chlorite or fibrous clays, and associated with more non clay minerals (quartz, feldspars, amphiboles and other dense species). As a consequence, the resulting clay assemblage displays a *mineralogical diversification.* This diversification is much more intense if the soil blanket has developed under strongly hydrolyzing conditions, such as those prevailing in present intertropical areas or in most regions at Cretaceous and Paleogene times (Chap. 17). Whilst extensive pedogenesis tends to determine a clay-mineral homogeneization toward smectite or kaolinite, the strong erosion of soils induced by tectonic rejuvenation allows virtually the exposure of various

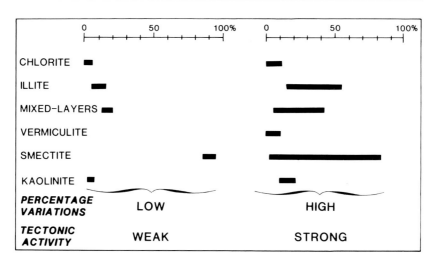

Fig. 18.10. Site 367 DSDP, Cape Verde basin. Variation range in the relative abundance of clay minerals during stable and unstable tectonic periods, late Mesozoic times. (After Chamley et al. 1980c. Late Jurassic: 19 samples. Early Cretaceous: 24 samples. Late Cretaceous: 11 samples)

parent rocks, and leads to a strong diversification of eroded materials. Generally speaking, stable tectonic conditions correlate rather to homogeneous clay suites that are able to reflect climate, transportation and deposition characteristics. In contrast, *the more active the tectonics and more hydrolyzing the climate, the more intense is the resulting clay diversification and the wider is the range of quantitative variations for each mineral group* (Fig. 18.10).

Sedimentary clay assemblages are able to express the tectonic stability or instability independently of the lithology. A less diversified and less changing clay suite, which is indicative of rather stable conditions in source regions, may correlate to a homogeneous lithology (Fig. 18.11A), for example because of regular supply of terrigenous material or of active planktonic productivity; it may also be associated to heterogeneous and changing lithofacies (Fig. 18.11B), due to local reworking or variations in marine productivity, circulation, etc. Similarly, a highly diversified and variable clay suite, indicative of unstable tectonic conditions in source areas, may correspond either to homogeneous or to heterogeneous lithofacies; the first case often characterizes series deposited far away from tectonized regions (Fig. 18.11C); the second case occurs rather in sections located close to unstable land masses and receiving abundant coarse- sized materials (Fig. 18.11D). In all situations, *the stratigraphic successions of clay mineral associations appear to constitute reliable markers of tectonic stability or instability, whatever the type and intensity of the lithologic obliteration.*

The mineralogical expression by marine sediments of the tectonic events having occurred on continental margins and exposed land masses is conditioned by the submarine topography. If the sections investigated are located beyond a submarine barrier or a submarine trough, the tectonic message can hardly be registered by deep-sea sediments. Similarly, the flanks of oceanic ridges generally do not

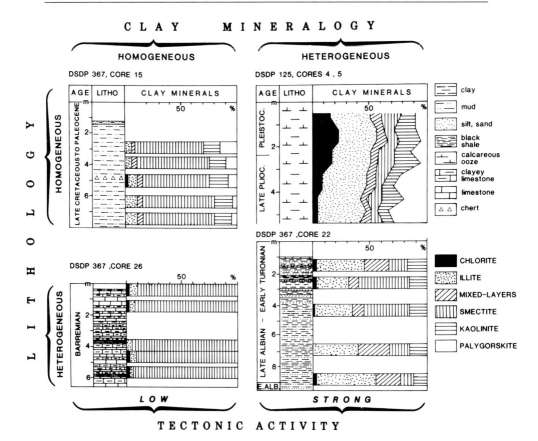

Fig. 18.11. Examples of the expression of tectonic activity or stability by clay mineralogy, compared to lithology, from DSDP data. Site 367, Cape Verde basin, East Atlantic Ocean. Site 125, Eastern Mediterranean Rise

receive terrigenous material in sufficient abundance or significance to express the tectonic changes on continents. In addition, *the submarine or subaerial position of the young oceanic ridge is critical for the clay stratigraphic successions recorded in sedimentary basins* (Fig. 18.12).

1. *If the rift is immersed,* as in most parts of the Atlantic Ocean during late Jurassic and early Cretaceous times, the margin destabilization following the ocean initiation induces a supply of various, typically rock-derived minerals dominated by the *illite group* (illite, chlorite, random mixed-layers, quartz, feldspar. See 18.3.2. Figs. 13.4, 18.2A). The Mg-smectite-bearing tholeiitic basalts are therefore blanketed by sediments characterized by a *large mineralogical diversity. After the tectonic relaxation* of continental margins, *Al-Fe smectite* forms in continental soils and is increasingly supplied to the marine basin, which tends to determine a *mineralogical simplification.*

2. *If the rift is exposed,* the subaerial weathering of its vulnerable volcanic rocks induces the massive formation of *Mg- to Fe-smectite,* that is easily

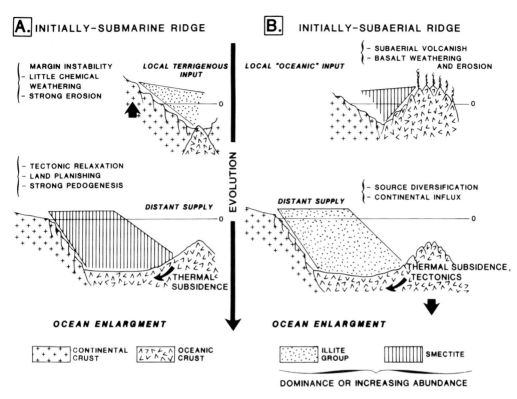

Fig. 18.12. Control of topography and ocean enlargement on clay-mineral successions deposited in a marine basin characterized by an either submarine (**A**) or subaerial (**B**) ridge. Interpretation. (After Debrabant and Chamley 1982 b)

reworked and supplied to the adjacent oceanic basin. The clay mineralogy of first marine sediments tends to be characterized by *almost exclusive smectite* (Fig. 18.12B), that is highly crystalline and often enriched in Na compared to K. Such a situation occurs during Barremian at DSDP site 384, South of Newfoundland on the J-anomaly ridge (Debrabant and Chamley 1982b). This is also the case with the Rockall Plateau, Northeastern Atlantic Ocean, after the opening of the Norway Sea at Paleocene time allowed the formation of volcanic archipelagoes between Iceland and the British Isles (Debrabant et al. 1979, Desprairies et al. 1981). A similar situation characterizes today the Eastern African rift, where abundant fluvio-lacustrine sediments rich in smectite deposit due to the massive erosion of volcanic mountains submitted to chemical weathering. *After the first subsidence phases and the rift submersion* have taken place, clay minerals tend to originate from a more distant source. This determines a *mineralogical diversification,* and especially the appearance of minerals from the *illite group.*

An opposite clay mineralogical evolution therefore characterizes the young sedimentary sequences deposited in basins, according to the submarine or the subaerial position of the rift.

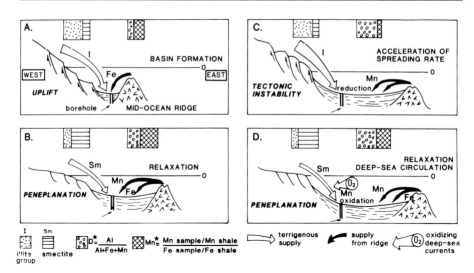

Fig. 18.13. Continental and marine influence on the inorganic sedimentation during the early history of the Northwestern Atlantic Ocean. (After Debrabant and Chamley 1982 b). Increasing values of D★ generally reflect an increasing continental influence. Increasing values of Mn★ reflect either the close proximity of the spreading axis (volcano-hydrothermal manganese; step A), or oxidized conditions on the sea floor (precipitated manganese; step D). Stages A to D explained in the text

In the North Atlantic basins marked by a permanently submarine position of the ridge, the clay mineral and inorganic geochemical successions often reflect *four successive steps in the early history of the ocean* (Fig. 18.13A to D). (A) The *tectonic destabilization of continental margins* associated to the basin initiation is marked by the abundance of the illite group indicative of active erosion of continental crust and sediments, and by abundant iron issuing from hydrothermal and volcanic activity on the near spreading axis (e.g., reddish clayey limestones of late Jurassic age). (B) The *tectonic relaxation following the ocean formation* allows the development of smectitic soils on land masses; pedogenic Al-Fe smectite therefore characterizes the sedimentary clay fraction. The continuing spreading is responsible for an increasing supply of ridge- derived iron relative to manganese, because manganese precipitates closer to the sources (e.g., early Cretaceous grayish limestones and marls). (C) A *new destabilization* of continental margins related to a spreading acceleration induces a rejuvenation of land relief and a new flux of rock-derived minerals (e.g., illite groups). At the same time the migrating site tends to be protected from the direct chemical influence of the ridge: metal precipitations are low, or even prevented by a reducing environment (e.g., early to middle Cretaceous deposition of black shales. (D) A *further relaxation* can induce a new supply of pedogenic smectite, while the ocean enlargment allows enhanced deep-sea circulation, bottom oxidation and manganese precipitation at the sediment-sea water interface (e.g., late Cretaceous varicolored clay). *Clay minerals therefore appear to constitute privileged markers of the paleoenvironmental evolution on the land masses bordering the basin, while transition metals characterize rather changes occurring in the ocean itself.*

Thrusting or complex tectonic structuration are sometimes described as being expressed by clay mineralogical changes. Hoffman and Hower (1979) relate the alteration of clays in the Montana disturbed belt, USA, to thermal insulation resulting from emplacement of thrust sheets. In the Barbados accretionary prism, in the Caribbean, Schoonmaker et al. (1986) attribute the local increase of illite layers in illite-smectite mixed-layers to circulation of fluids in a reverse faulting zone. The zig-zag trend with depth of the mixed-layer ordering could therefore express late tectonic modifications superimposed on burial diagenetic reactions. At DSDP site 541, drilled during leg 78 in late Cenozoic series of the Barbados accretionary wedge, Schoonmaker (1986) reports the occurrence of a trioctahedral smectite in disturbed sediments of the décollement zone. Trioctahedral smectites apparently do not exist in contemporary sediments drilled at the close site 542, which represents reference series that are not tectonized and are located out of the thrusting zone in the Atlantic basin. The trioctahedral smectite is referred to saponite, and is thought to result from chemical changes associated with reverse faulting. J. Schoonmaker envisages that the basal décollement zone acted as conduits for upward migration of fluids released at depth in the wedge, providing a specific chemical environment in which formation of saponite developed. The reality of such phenomena must be further documented. In the case of the Barbados accretionary prism, the identification of saponite is based on crystallinity criteria only, and is not supported so far by chemical evidence. It is still unclcar why saponite could form in certain parts of the décollement zone (Schoonmaker 1986), and ordered illite-smectite in other parts of the same accretionary prism (Schoonmaker et al. 1986). In addition, Mg-smectite originating from basalt alteration at depth could perhaps have migrated upwards together with fluids, and have been reworked instead of having formed in situ. These restrictions do not diminish the interest of the search for using clay mineral changes as markers of complex tectonic processes, especially in subduction areas.

18.4 Conclusion

1. *Changes in sources or in current activity often determine modifications of the clay assemblages that are slight* and of the same order of magnitude as mineralogical changes induced by climate variations. Such *convergences in climate and circulation effects* sometimes complicate the recognition of the factors actually responsible for the changes recorded in the clay stratigraphy. An accurate knowledge of the geological context greatly facilitates the interpretations.

A considerable literature exists about the use of clay assemblages as *indicators of detrital sources in recent and ancient sediments from various depositional environments.* Much progress has still to be made on the precise identification of terrigenous origins, since a given clay mineral group may result from different sources in a single region. For instance, smectites in recent sediments of the North Indian Ocean may derive either from common soils developed under hydrolyzing climate on the Asian continent, from soils and rocks cropping out in Australia or East Africa, from the subaerial alteration of volcanic rocks (e.g. Indonesia), from altered submarine volcniclastites or basalts, or from in situ growth. The combination of mineralogical, micromorphological and micro-chemical data helps greatly in reconstructing recent and past sources. Extensive investigations are also to be developed to use clay suites as *paleocurrent markers*. Promising studies exist on late Quaternary periods, for which advected minerals may serve as indicators of the different superimposed water masses, of their variations in volume and celerity in relation with climate changes, or of the respective influence of wind and water circulation. A large field of investigations is open for application for older periods, which definitely implies an inter-disciplinary approach.

2. *Tectonic phases generally determine important and long-lasting modifica-tions in detrital clay assemblages, which differs from most effects induced by changes in climate or currents.* As tectonic rejuvenation usually impedes the development of continental soils and favors the direct removal of material eroded from various rocky substrates, the resulting deposited clays often display a *mineral diversification*. When tectonized land masses comprise mainly crystalline rocks, the detrital clay assemblages tend to be enriched in mica-illite, chlorite and associated non clay minerals, and resemble the clay suites formed at depth through burial diagenesis or zonal metamorphism (Chaps. 15, 16). For instance, the clay assemblages derived either from burying at depths exceeding 3 to 5 kilometers, or from the direct erosion of uplifted, strongly sloped crystalline rocks, can display similar patterns. Such a *convergence between thermodynamical and mechanical effects* may complicate the intepretation of clay stratigraphic successions.

3. *Clay mineral suites in sediments constitute useful markers of tectonic activity or relaxation.* Because of their large dispersal properties, *clays may reflect rejuvenation phases far away from tectonized areas*, in sedimentary basins never reached by coarse terrigenous fractions. Because of their dependence on meteoric conditions, pedogenic clays are highly sensitive to slight tectonic changes; *they are able to reflect some epeirogenic stages that determine small morphologic modifications but noticeable changes in the development of surficial soils.* A decrease in the epeirogenic activity tends to be followed by the development of new soils, that may form in a few hundred years, and whose erosion products ac-cumulating in sedimentary basins bear modified paleoenvironmental messages.

Sedimentary clay suites reflect in a privileged manner the *tectonic events oc-curring on exposed land masses*, especially the *uplift* movements determined by compression phases or by the positive reaction of continental crust to subsidence in adjacent marine basins. Clay successions may also express *tectonic events oc-curring in the basins themselves*, such as those resulting in the formation of a sub-marine barrier (= *uplift*) or of a trough or trench (= *downlift*): in both cases, the dispersal of clays by marine currents tends to be impeded, which often modifies the composition of detrital assemblages.

Opposite mineralogical trends characterize the clay successions recorded during the early formation stages of marine basins of Atlantic type, according to the mor-phological position of the oceanic ridge. A subaerial position followed by thermal subsidence determines the successive supply of volcano-derived smectite and of diversified, illite-bearing terrigenous clays. A submarine position of the ridge followed by thermal subsidence and tectonic relaxation favors the successive supply of land-derived illitic clay and of soil-derived smectite (Fig. 18.12). Other potential ways of research concern complex structuration areas like subduction zones and accretionary prisms, where overburden and thrusting may induce *fluid migrations* and possible clay diagenetic modifications.

4. *Changes in detrital sources, in marine or aeolian currents and in tectonic activity induce modifications of clay successions that are either complementary or in opposition to those resulting from climate changes. The mineral modifications due to tectonics often obliterate the signals provided by other environmental factors.* In many cases two or more factors may intervene simultaneously on the characteristics of clay as well as of other sedimentary constituents. Deciphering the paleoenvironmental messages originating in clay stratigraphic successions therefore constitutes a difficult and fascinating task. The possibility of using sedimentary clays as a tool for integrated paleogeographic reconstructions is considered in Chapter 19. Such a purpose is based on the concept that *clay stratigraphic changes reflect the sum of different environmental factors, whose respective importance varies with time, and the most important of which are expressed only at one given period of geological history.*

Chapter 19

Paleoenvironmental Reconstruction

19.1 Chronological Evolution of Hatteras and Cape Verde Basins

19.1.1 Introduction

Clay mineral successions in terrigenous series have long been envisaged since as expressing the paleoenvironmental evolution of the land masses adjacent to sedimentary basins (see Millot 1964, 1970). Studies taking into account different aspects of the paleoenvironment and spanning large time intervals concentrate on the nearly continuous sedimentary columns recovered during the Deep Sea Drilling Project (DSDP) and the Ocean Drilling Program (ODP). The Atlantic Ocean, that progressively developed after the separation of American and Europe-African land masses, was almost permanently submitted to strong continental influences. It therefore lends itself particularly well to paleoenvironmental reconstructions involving both continental and oceanic controls.

The predominantly terrigenous character of the Atlantic clay sedimentation since late Jurassic times is documented by the following facts (see Chapts. 11.3, 13.3.3, 13.4, 14.2.1, 14.2.2, 15, 16): (1) absence of mineralogical trend with increasing depth of burial; (2) absence of systematic relationship between clay composition and lithofacies distribution; (3) rareness of chemically active hydrogenous environments; (4) only local influence of volcanic and hydrothermal influence; (5) clay overgrowths developing locally only and inducing no appreciable changes in the proportion of the different clay minerals. The high detrital input of clay material in the Atlantic domain in the course of the time is favored by the relative narrowness of the ocean especially during early formation stages, the general absence of oceanic troughs trapping terrigenous matter and thus preventing large dispersal, the presence of large river systems feeding the marine sedimentation and inducing high sedimentation rates, and the scarcity of intraplate volcanism.

Many DSDP sites have been investigated from clay stratigraphy. They provide detailed mineralogical, geochemical and micromorphological logs (e.g., Initial Reports of the DSDP, volumes 39, 40, 41, 44, 47A and B, 48, 50, 71, 72, 73, 74, 75, 76, 77, 79. See Fig. 19.1). *Sites 105 and 367* are respectively located in the *western and eastern basins of the North Atlantic Ocean* (Fig. 19.2). Site 105 was drilled in the Hatteras basin at a depth of 5231 meters and penetrated 633 meters into the sea floor (34°53.72'N, 60°19.40'W. Hollister, Ewing et al. 1972). Site 367 is located in the Cape Verde basin at 4748 meters water depth (12°29.2'N,

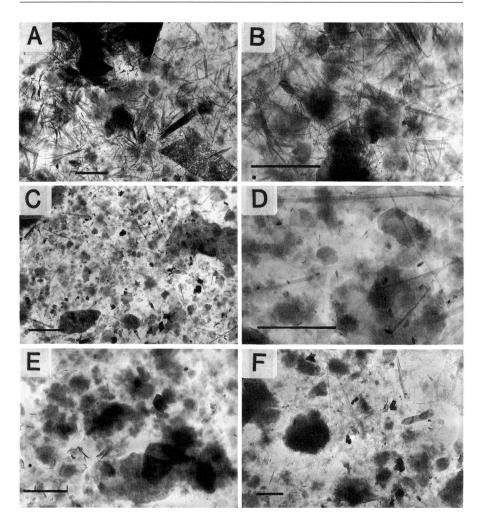

Fig. 19.1. Hole 534 A DSDP, Blake Bahama basin, Northwestern Atlantic Ocean. Transmission electron micrographs. (After Chamley et al. 1983. Bar = 1 μm. n.b. Compare with Fig. 14.2. **A, B** Early Miocene, Great Abaco Member, Blake Ridge formation (sample 4-2-100; approx. 575 m depth). Short and often bundled fibers of sepiolite, flaky smectite, few well-edged illite and kaolinite sheets, radiolarian debris. **C, D** Early Maastrichtian, Plantagenet formation (24-1-32; approx. 740 m). Abundant kaolinite hexagons and well-edged illite sheets; smectite with blurred contours; scarce short to long, often broken fibers of palygorskite and sepiolite. **E** Middle Albian, Hatteras formation (35-2-100; approx. 850 m). Abundant illite and smectite, some hexagonal sheets of smectite. **F** Early Valanginian, Blake-Bahama formation (76-3-48; approx. 1210 m). Abundant fleecy smectite, noticeable amounts of short and broken fibers of palygorskite. **G** Early Berriasian, Blake-Bahama formation (91-1-101; approx. 1340 m). Abundant flaky smectite, fairly abundant well-edged illite. **H** Callovian-Oxfordian, "Unnamed" formation (119-1-113: approx. 1565 m). Very abundant smectite with blurred contours, few illite particles. **I** Callovian, "Unnamed" formation (127-1-51; approx. 1630 m). Large, abundant, well-edged illite and chlorite; small, flaky particles of random mixed-layers and smectite. **J** Callovian, basalt-sediment contact (127-cc-10; 1635.30 m; see Fig. 13.5). Refringent, well-shaped, diversely-sized sheets of metamorphic corrensite. **K, L** Slightly-weathered basalt crust (128-1-88, 1642 m). Authigenic smectite as tufts or broad boards

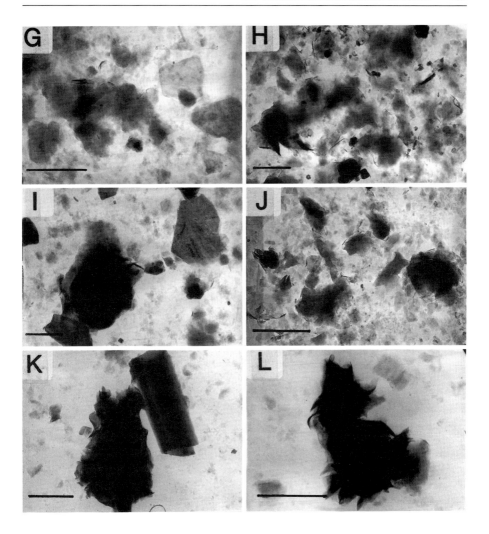

20°02.8'W; penetration 1153 meters. Lancelot, Seibold et al. 1978). One hundred samples of each site have been studied (Chamley et al. 1980c, Chamley and Debrabant 1984a). The most significant results are summarized in Fig. 19.3. They include the general lithology, the clay mineralogy (< 2 µm fraction), and selected geochemical percentages and ratios obtained on the bulk sediment:

    – CaO (%) indicative of the percentage of calcium carbonate.

    – Organic carbon (C%).

    – Abundance of manganese (Mn^*), measured as the ratio of Mn within the sediment over Mn in continental shales, relatively to iron content in the same materials (Steinberg and Mpodozis-Marin 1978)

    – Abundance of aluminum (D), measured as the ratio of Al over $Al + Fe + Mn$. D is roughly proportional to the terrigenous input (Boström et al. 1969).

    – MgO/Al_2O_3 and MgO/K_2O.

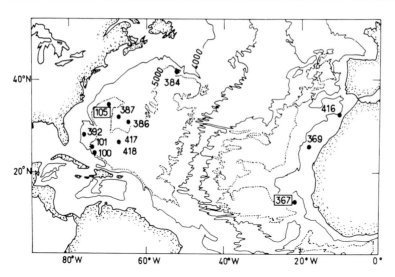

Fig. 19.2. Location of some DSDP sites in the North Atlantic Ocean

19.1.2 Major Similarities

Continental and Oceanic Influences

At sites 105 and 367, the clay mineral assemblages show no relationship to the general lithology, the presence of volcanic debris or the degree of oxidation-reduction. Except at the basalt-sediment contact and within the basalts (see 13.2. Fig. 13.5), they are not significantly influenced by volcanism, sea water or the depth of burial. *Above the oceanic basalts* (late Jurassic), the *strong supply of primary minerals* (illite, chlorite, quartz, fedspars), associated with random mixed-layers and poorly crystallized smectite, *reflects erosion and tectonic destabilization of the ocean margins*, after the rifting and initial spreading stages have taken place (see 18.3.2. Fig. 14.2). *During the same period the manganiferous and magnesian character of the sediments expresses the distal influence of oceanic volcanism* (Debrabant and Foulon 1979). When considering Fe and Mn abundance in contemporaneous marine sediments between the mid-Atlantic ridge and Site 105, it is possible to observe the passage from the proximal to distal influence of oceanic volcanism, and to deduce an approximate spreading rate (Debrabant and Chamley 1982b. See 18.3.3. Fig. 18.13). *The clay mineralogy and the inorganic geochemistry therefore represent two complementary tools in deciphering terrigenous and marine influences during the early evolution of oceanic basins.*

Cretaceous Peneplanation and Rejuvenation of Continental Relief

The increase of Al-Fe smectites in the latest Jurassic sediments, and their abundance during a large part of Cretaceous and even Paleogene times, do not depend on depositional conditions or the chemical environment. Smectites are attributed chiefly to the reworking of continental soils, developed after a sig-

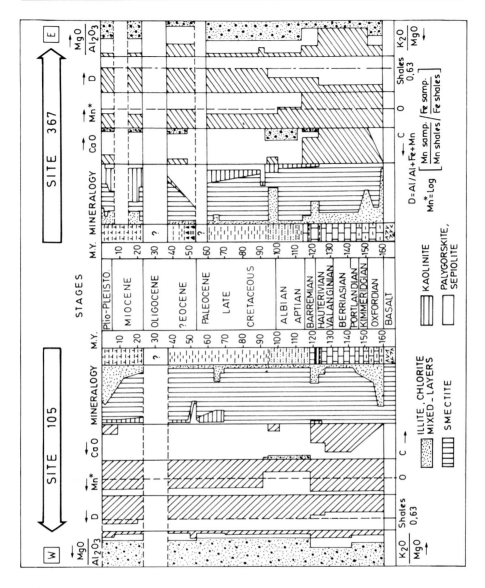

Fig. 19.3. Schematic stratigraphic distribution of clay mineral suites and geochemical parameters in Hatteras (site 105) and Cape Verde (site 367) basins, since the late Jurassic. (After Chamley and Debrabant 1984a)

nificant peneplanation of gently sloping and poorly drained areas (see 17.4. Fig. 17.8). The *smectite-rich facies correlate to tectonically stable periods,* reflecting the development and the periodic erosion of soils on the borderlands (Debrabant and Chamley 1982b). Only a limited portion of the sedimentary smectites appears to derive locally from in-situ growth and replacement, which does not appreciably modify the relative abundance of the different clay species (Holtzapffel and Chamley 1986. See 14.4.2).

The Cretaceous time is episodically marked on both North Atlantic margins by a *strong supply of illite and associated minerals,* including occasional kaolinite or palygorskite, and by an increase of the index D pointing to a more terrigenous influence. These mineralogical changes occur especially around the Hauterivian-Barremian boundary and during the Santonian-Campanian-Maastrichtian, and are accompanied by a great variability in the clay associations (Fig. 18.10). The main causes are probably the *tectonic rejuvenation of the Atlantic margins, the sea-level variations,* and the increase of water-mass exchange in relation with acceleration stages of the spreading rate. Tectonic instability of the structural blocks of the margins induced by ocean widening (e.g., Montadert et al. 1979) seems to be responsible for the strong erosion of the hinterland and the supply of detrital material derived from various soils and rocks to the ocean.

Morphological Evolution

The absence of kaolinite in *late Jurassic-early Cretaceous* sediments at sites 105 and 367 is somewhat unusual, because this mineral is well known in many continental and shallow-water sediments of the same age (e.g. Millot 1964, 1970, Robert 1987). Its absence could be due to the existence of *carbonate platforms* acting as morphological barriers to the transit of the fast-settling kaolinite toward the oceanic basin. It could also result from the *still immature morphology of the river-drainage basins* supplying terrigenous material to the young Atlantic Ocean. Kaolinite, that preferentially forms in well-drained, upstream soils in continental basins, could hardly have been removed and carried downwards onto the ocean. If such a hypothesis is correct, kaolinite would serve as an indicator of the morphological maturation of continental river-fed basins developed after the initiation of an oceanic basin.

Oxidation-Reduction Processes

The development of black-shale facies *during and after Barremian time* is accompagnied by an increase in the organic matter, a decrease in the Mn^* index and the disappearance of the calcium carbonate (CaO). Such a drastic change in sediment composition, which occurred in progressively deepening, still-narrow North Atlantic basins with restricted deep-water circulation (Berggren and Hollister 1977, Lancelot, Seibold et al. 1978), implies a strongly confined basin. Local diagenetic processes occurred, resulting in migration of Mn, Fe and Ca in the interstitial waters, and local precipitation of rhodochrosite, siderite, etc. (Debrabant et al. 1979). During the *late Cretaceous* the deep-water exchange in the rather broad Atlantic Ocean is marked by the precipitation of Mn in varicolored clays (see index Mn^*), the strong decrease of organic carbon and the appearance of goethite. Thus *geochemical parameters constitute good indicators of* *the evolution of ocean bottom conditions* through this time. In contrast, the clay mineralogy varies independently of oxidation-reduction conditions (see 11.2.2) and reflects modifications occurring in the terrigenous input. The organic matter is also largely of a terrigenous origin (e.g., Kagami et al. 1983, de Graciansky et al. 1987).

Development of Peri-Marine Confined Basins
The appearance of palygorskite in *late Cretaceous* sediments from the two North Atlantic basins occurs in lithologically variable sediments (Chamley 1979), and often coincides with strong resedimentation processes and changes in detrital characteristics (index D, Al_2O_3/K_2O). The *early Paleogene* occurrence of palygorskite and sepiolite is recorded in most parts of North and South Atlantic margins (Chamley 1979, Robert 1982), and also coincides with extensive resedimentation facies. These fibrous clays are mostly considered as minerals inherited from peri-marine basins, the depositional environment and the age of which they help to characterize. *Palygorskite and sepiolite formed under alkaline conditions in restricted basins* subject to marine transgressions, limited water exchange, warm and humidity-contrasted climate, strong evaporation, periodic subsidence and tectonic instability (see Chap. 9).

Expression of Continental Climates
The extensive occurrence of aluminum-iron smectites suggests that the *late Mesozoic to early Paleogene climate was hot with contrasting seasonal humidity*, which favored the development of poorly drained soils on gently sloping coastal plains (see 17.4). The present-day blackish vertisols may be relics of such soils which, until the late Eocene, could have characterized a large part of the non glacial world, marked by high temperatures and low climatic contrast with latitude (e.g., Frakes 1979, Kennett 1982). In contrast, *the irregular increase of the illite group*, of the D index and of the Al_2O_3/K_2O ratio, *from the early Oligocene until the Quaternary, reflects passage toward colder conditions*, during which the established pedogenic processes were partly replaced by incomplete weathering and by the direct erosion of rocks instead of soils (Chamley 1979a, Robert 1982. See 17.3.1).

19.1.3 Major Differences

Continental and Marine Influences on Sedimentation
The Hatteras basin (site 105) *experienced strong continental influence* (rather high values of the D index), even though the sedimentation rate was fairly low because the site was located on a topographic high (Hollister, Ewing et al. 1972) and favored early diagenetic processes (growth of opal CT, clinoptilolite). *In the Cape Verde basin, the marine volcanic influence seems to be fairly high*, as shown by the high values of the Mn^* index and the relatively low values of D (i.e., abundance of Fe and Mn). These conditions extend from the late Jurassic to the early late Cretaceous, and are attributed to the vicinity of different and successive volcanic

Table 19.1. Geochemical and clay mineral characteristics in Hatteras (site 105) and Cape Verde (site 367) basins during late Cenozoic times (after Chamley and Debrabant 1984a)

Site	Al_2O_3/K_2O	MgO/K_2O	D	Mn*	Illite, chlorite, mixed-layers (%)	Smectite (%)	Kaolinite (%)	Palygorskite (%)
105 (west)	6.5	0.8	0.67	+0.54	25	67	8	Traces
367 (east)	9.4	1.3	0.63	−0.32	15	52	30	3

provinces: the mid-oceanic ridge, Cape Verde and Sierra Leone rises, and Romanche fracture zone.

Expression of Major Atlantic Spreading Stages
A strong "illite event", restricted to the Cape Verde basin (site 367), occurred during the *Albian-Turonian* period. Associated with an increase in the sedimentation rate and higher amounts of organic matter in the sediments, this event also correlates to increasing contents of Fe, Zn, Ni, Cu and V. Rather than being associated with a tectonic phase on the adjacent continent which is not documented by the literature (Dillon and Sougy 1974), the mineralogical and geochemical changes probably reflect the ultimate *separation of the African and South-American continents* (Berggren and Hollister 1977). Such separation resulted in margin instability and in increase of volcanic activity in the Central Atlantic area. In the same period, the Hatteras basin was located much more northwards in Western Atlantic, and could not have been affected by the tectonic instability developing in central regions.

Bottom Conditions
Deep-water circulation temporarily increased in the Cape Verde basin at the *Barremian-Aptian transition*, which is expressed by the absence of organic carbon, an increase of transition elements, and local lithological changes (in Lancelot, Seibold et al. 1978). These changes are probably due to a *bottom oxygenation* linked to the increasing deep-water circulation, an event not recorded in the Hatteras basin. Calcareous turbidites and other resedimentation processes are frequent at site 367 during the Cretaceous, especially in the Albian-Turonian period, and less common at site 105. During the Cenozoic, the Cape Verde basin is still characterized by a negative anomaly in Mn* index (Fig. 19.3, Table 19.1), and by noticeable amounts of organic carbon and vanadium. Such patterns do not exist at site 105, and indicate the *persistence of a more confined environment in the Cape Verde basin than in the Hatteras basin*, as a result of higher sedimentation rates and more depressed topography.

Development of Peri-Marine Basins
Fibrous clays and their geochemical associations (Mg, Si) *occur much more abundantly and include more sepiolite in the Cape Verde* basin than in the Hatteras basin during late Cretaceous and Paleogene times. This difference probably

Table 19.2. Main paleoenvironmental events recorded in Hatteras and Cape Verde Basins since the late Mesozoic, from mineralogical and geochemical data. (After Chamley and Debrabant 1984a). I illite; C chlorite; Ig illite group; S smectite; K kaolinite; P palygorskite; F fibrous clays; Sr strontium; Mn manganese index (Mn*); D detrital index

Stage	General events	Regional events	
		West, Hatteras basin (Site 105)	East, Cape Verde basin (Site 367)
Plio-Pleistocene Miocene	Step-by-step world cooling (Ig, D)	Northward migration of the American plate inducing stronger cooling (Ig)	Sub-tropical conditions (K, P) Crossing of the equatorial zone (K)
Oligocene	Intensification of north-south circulation (Mn)		
Eocene	Existence of perimarine basins, transgression (F)	Cooler climate (Ig)	Wide development of confined perimarine basins
Late Cretaceous	Tectonic rejuvenation of Atlantic margins, major spreading stage (I) Increase of deep-water mass exchange (Mn)		
Albian	Existence of perimarine basins (F)	Margin stability (Sm)	Major spreading phase, definitive opening of the South-North Atlantic gateway (I)
Aptian	Continental pedogenesis and peneplanation (S)		
Barremian	Ocean deepening, reducing environment, carbonate dissolution (Mn, C)		Temporary intensification of bottom circulation (Mn)
Hauterivian	Tectonic reactivation of margins, major spreading stage (I)		
Valanginian Berriasian	Continental peneplanation (S)		
Portlandian Kimmeridgian Oxfordian	Major phase of rifting-spreading, tectonic activity on margins (I)		

Vertical annotations (upper section): Development of bottom circulation (Mn); Presence of relatively restricted conditions of deep circulation (Mn)

Vertical annotations (lower section): Warm and humidity-contrasted climate (S); Volcanic influences (Mn); Stronger dissolution processes (Sr); Persistence of moderate volcanic activity (Mn, D)

results chiefly from a *warmer and more humidity-contrasted climate* on the African side of the North Atlantic, which favored the genesis of fibrous clays in peri-marine basins adjacent to the ocean. The North American plate occupied a more northerly position after the late Cretaceous, and its eastern margins were exposed to stronger North-South circulation (Berggren and Hollister 1977, Tucholke and Mountain 1979). This favored cool climate and prevented the genesis of fibrous clays in peri-marine basins. Fibrous clays developed only in the southern part of North America, where they were occasionally reworked toward the open sea (Chamley and Debrabant 1984b). An additional reason for the differences observed between both domains could be the mountainous topography along the American side of the North Atlantic (Appalachian chains s.l.), discouraging the development of evaporative coastal basins.

Cenozoic Climate Changes
Both western and eastern basins display marked and increasing differences through time, *since the early Oligocene*. The sediments in the *Hatteras basin* show a *higher content of illite* and associated minerals, of aluminum and potassium, and of the D index, which indicates an increasing supply of rock detritus and moderately weathered soil products. This probably reflects the *northward migration of the American plate toward regions increasingly submitted to Cenozoic glaciations*. During the same time interval, *the African plate also moved toward the North and crossed the tropical and equatorial zones, which is marked by an increase of kaolinite, goethite, gibbsite and fibrous clays*. All these minerals are products of severe hydrolysis, developing in distinct morphological contexts. The strong increase in kaolinite abundance during the early Miocene at site 367, which is presently located at 12°29'N, could reflect the passage of the site area across the equatorial zone, or the temporary occurrence of a more humid climate belt (Robert and Chamley 1987). *Terrigenous clay minerals may therefore express the latitudinal movements of the lithospheric plates in the Atlantic Ocean.*

A summary of major paleoenvironmental similarities and differences recorded in Hatteras and Cape Verde basins since the late Mezozoic is given in Table 19.2.

19.2 Paleogeographic Evolution of Atlantic Regions

19.2.1 North Atlantic Basins at Albian Time

The Albian period, characterized by the deposition of black-shale facies, is considered from various DSDP sites located in both the Western and Eastern North Atlantic (legs 11, 41, 43, 44, 50 and 51-53), with special reference to sites 105, 387, 386, 367 and 369 (Chamley and Debrabant 1984a. Figs. 19.2, 19.4). The clay mineralogy of sediments from the western basin shows a *strong decrease in smectite content with increasing distance from North America* (from site 105 to site 387); this decrease is balanced by an increase in the abundance of illite, chlorite and random mixed-layers, and traces of kaolinite are recognizable eastwards.

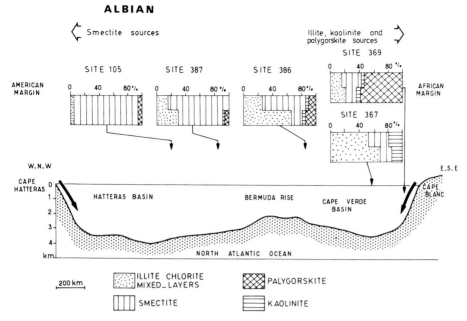

Fig. 19.4. Clay mineralogy in Albian sediments along a North-Atlantic transect. (After Chamley and Debrabant 1984a)

Such a gradient confirms the interpretation of the mainly terrigenous and non volcanogenic origin for the smectite (see 14.2.1), because the mineral abundance decreases with distance from the continent and with approach to the basaltic area of Bermuda. However, the question arises about the origin of other clay minerals, because only volcanic rocks and smectite-rich alteration have been identified on the Aptian-Albian substratum of the Bermuda Rise (Humphris et al. 1980, Pertsev and Rusinov 1980, Rusinov et al. 1980).

A mineralogical break occurs in the *Bermuda area* (Fig. 19.5). Site 386, located southeast of site 387, shows a *drastic increase of the gradient* observed in the Hatteras basin. The smectite abundance decreases strongly within a short distance, while the abundance of all other clay minerals increases. Kaolinite and palygorskite vary in parallel with the typically terrigenous illite group, which also points to a probably detrital origin. When moving more eastwards *to the African margin*, the mineralogical gradient becomes still more pronounced; *high amounts of illite, kaolinite and/or palygorskite* occur at sites 367, 369 and 416, and only little smectite. These different results suggest that *during Albian time smectites were chiefly produced on American borderlands, other clay minerals chiefly on the African side*, and that *the Bermuda area acted as a major barrier for mineral dispersal*. This implies that transportation by water masses was much more important than aeolian transportation (see 7.4.2).

Complementary data arise from studies of other sedimentary characteristics, whose major trends are schematically represented in Fig. 19.5. On the American side of the Bermuda area, compared to the African side: (1) the D index

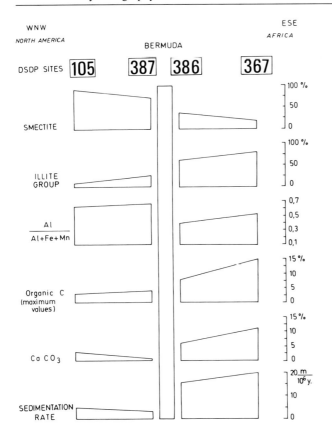

Fig. 19.5. Schematic distribution of some characteristics of Albian sediments along a West-East transect across the North Atlantic Ocean. (After Chamley and Debrabant 1984a)

(Al/Al + Fe + Mn) is higher, which indicates stronger continental influences with regard to the volcanic activity; and (2) the organic carbon, the calcium carbonate contents and the sedimentation rates are lower. The values of the last three parameters generally decrease in a direction away from both American and African margins. All these data indicate the existence of *opposing influences on either side of the Bermuda rise*. This peculiar distribution of sedimentary components is in agreement with the paleogeographical reconstruction of the North Atlantic about 100 million years ago, as proposed by Chenet and Francheteau (1980) from geophysical data (Fig. 19.6). The Bermuda area, located close to sites 386 and 387 (Fig. 19.1), appears as a topographic elevation forming a morphological barrier, extending northeast toward the J-anomaly ridge and the Grand Banks area, where the black-shale facies pass to shallow-water carbonates.

  The clay mineralogical and sedimentological record provides control for paleoenvironmental reconstruction in both marine and continental environments at Albian time (Fig. 19.7). *The Bermuda ridge appears to have been the symmetry axis of the North Atlantic Ocean*, acting as a major submarine barrier. As a consequence, a large part of the present Western Atlantic Ocean, East of the Bermuda area, was submitted to African terrigenous input. Clay minerals

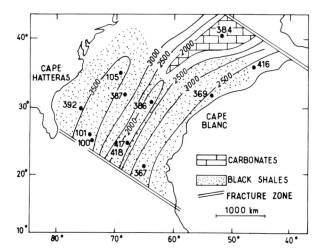

Fig. 19.6. The North Atlantic Ocean at Albian time. (After Chenet and Francheteau 1980)

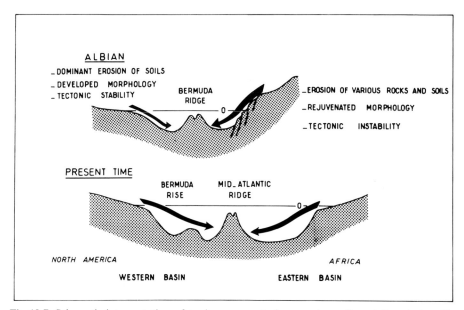

Fig. 19.7. Schematic interpretation of terrigenous control on marine sedimentation during Albian time, compared to the present-day situation. (After Chamley and Debrabant 1984a)

represent useful markers of marine dispersion of suspended matter, because of their small size and good buoyancy. The presence of small amounts of Africa-derived minerals in the Bermuda area suggests that the top of the ridge was located significantly below the sea level, permitting some exchanges of suspended matter between both basins, or that wind supply was added to water-borne supply (see 7.4.2). Note that the northernmost part of the J-anomaly ridge has probably emerged since the Barremian (Tucholke and Vogt 1979), which is expressed in the clay mineralogy by high amounts of volcanic-derived smectites formed by the subaerial weathering of tholeiitic basalts (Debrabant and Chamley 1982b. See 18.3.3, Fig. 18.12).

Strong differences in continental morphology characterize both sides of the North Atlantic Ocean, according to sedimentological data. *On the American side* the abundance of Al-Fe smectites, associated with strong detrital content (D index), rather low sedimentation rates, oxidized organic matter and few carbonates, suggests the existence of low relief. *The tectonic setting was therefore probably stable*, which is supported by Jansa's observations (1981). This precluded strong erosion, fast sedimentation and noticeable reworking of platform and slope carbonates, and favored the development of poorly drained soils in which smectites developed. *On the African side* of the Atlantic Ocean, the mixture of minerals inherited from crystalline rocks (illite group), surficial sedimentary rocks (smectite, palygorskite), and soils developed into downstream (smectite) and upstream (kaolinite) zones of the river basins, suggests strong erosion of continental areas. *Tectonic rejuvenation of morphology* is consistent with high sedimentation rates due to active erosion, with relatively abundant organic matter rapidly buried and therefore preserved from oxidation, and with occurrence of reworked carbonates in turbidites transported from the margin adjacent to the Cape Verde basin. All the above data indicate tectonic instability on the African margin, accompanied by significant volcanic activity on the submarine rises (low values of D index), and by an increasing spreading rate in the Eastern North Atlantic Ocean documented by the geochemistry. These phenomena were probably associated with the spreading stages when definitive separation of the African and South-American lithospheric plates took place.

19.2.2 North Atlantic Ocean, Cretaceous Times

De Graciansky et al. (1982, 1987) compare the paleoenvironmental evolution of the North Atlantic basins during Cretaceous times by considering the distribution of organic matter and of some other sedimentary constituents. Three main phases of deposition are identified, which are separated by two unconformities (E1 and E2): (1) The Blake-Bahama phase, extending from Valanginian to early Aptian (Bedoulian), corresponds to relatively uniform depositional conditions with alternating oxic and anoxic bottom environment. (2) The Hatteras phase, from late Aptian to late Cenomanian, reveals various bottom environments according to the regions: persistent anoxic conditions in southeastern basins, almost no anoxia in basins along the European margin, alternating oxia and anoxia in western basins, and oxic conditions over the mid-ocean ridge. (3) The Plantagenet phase, covering the Turonian-Senonian period, corresponded to homogeneous oxic conditions in the whole North Atlantic. The geographic distribution of clay mineral assemblages reflects mainly terrigenous sources, except perhaps for a fraction of smectite. As deduced from investigations on about 15 DSDP sites, *clay suites are more diverse during Cretaceous times in eastern basins than in western ones*, which extends the data obtained for the single Albian period

(19.2.1). P.C. de Graciansky and colleagues divide the *Cretaceous clay stratigraphy into four main phases. The first three reflect the asymmetry of the North Atlantic with respect to terrigenous supply. The fourth phase expresses a more homogeneous detrital input.*

1. *Prior to the earliest Aptian*, smectite dominates the clay suites in the western basin, associated with minor amounts of illite and mixed-layers. Temporary increases of the illite group minerals, especially at Barremian time, probably reflect a general destabilization of Atlantic margins (Chamley and Debrabant 1984a and b. See 19.1.1). Eastern basin sediments contain diversified minerals, including chlorite, illite, random mixed-layers, smectite, kaolinite and palygorskite.

2. *From Aptian to middle Albian*, the difference between both western and eastern basins is particularly well marked. Smectite is very abundant on the American side, and illite and kaolinite on the African side. A tectonic cause is envisaged, involving the instability of the African margin compared to the American one (see 19.2.1). A control by the continental climate could additionally intervene.

3. *From late Albian to early Cenomanian* the difference between the two sub-basins is still apparent but evolves in course of the time. In the western basin smectite-bearing sediments are progressively enriched in kaolinite and then in illite; this change is recorded in southern sites prior to northern ones. In the eastern basin an opposite trend is recorded, diversified clay assemblages being progressively replaced by smectite-rich suites. De Graciansky et al. (1987) envisage that smectite and associated palygorskite are largely reworked from continental and peri-marine sources, and could also partly derive from diagenetic changes occurring in the organic-rich environments. The latter possibility appears to be little compatible with general clay-mineral data on black-shale facies (see 11.3).

4. *From late Turonian to Senonian*, smectites are associated with abundant illite, kaolinite, and locally with chlorite and mixed-layers. This association first appeared in the Western North-Atlantic basins, and then progressively developed in the eastern basins from North to South. This resulted in both a mineralogical diversification within the clay fraction and a general homogenization of the clay composition within the different North Atlantic basins. The homogenization correlates to oxidized bottom conditions, and probably results from the development of active North-South circulation and of advection processes.

Note that the *Eastern Gulf of Mexico*, West of Florida, which formed at lowermost Cretaceous times, appears to have been marked by a continuous tectonic instability until the late Albian. This differs from the stable environments recorded on the eastern side of Florida in the Western Atlantic Ocean (Chamley and Debrabant 1984b, Debrabant et al. 1984). The transition from continental to marine environments around Berriasian time is expressed North of the Campeche Escarpment (DSDP site 537) by an evaporitic-clay stage, rich in authigenic palygorskite (Fig. 19.8C and D). The destabilization of the margin throughout the early Cretaceous is documented by the occurrence of strongly changing and mineralogically diversified clay assemblages, that are dominated by

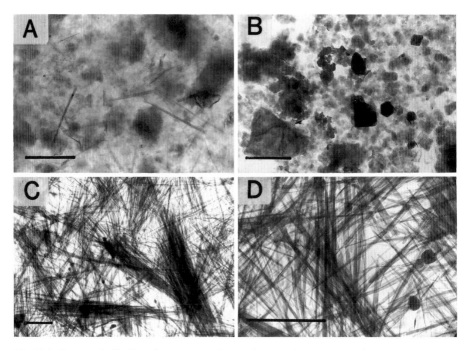

Fig. 19.8. Eastern Gulf of Mexico, Leg 77 DSDP. Electronmicrographs. (After Debrabant et al. 1984). Bar = 1 μm. **A** Site 538 A-21-3-130, North of Campeche escarpment, late Cretaceous (Coniacian-Santonian?). Abundant fleecy smectite suggesting the supply of soil-derived minerals in a period of tectonic stability. Sparse, short and broken fibers of palygorskite reworked from older evaporative facies. **B** Site 540-64-1-66, Southwestern Florida escarpment, late Albian. Illite and kaolinite facies, suggesting tectonic instability in source regions. Background of small, blurred smectite and mixed-layer particles. **C, D** Site 537-13-1-8, North of Campeche escarpment, lowermost Cretaceous (Berriasian?). Almost exclusive palygorskite as long, dense, imbricated bundles suggesting an autochthonous formation in alkaline evaporative, shallow-water environment

the illite group (illite, chlorite, random mixed-layers. Fig. 19.8B). The tectonic relaxation and the resulting development of continental soils were completed in late Cretaceous time only, as suggested by the input in great amounts of Al-Fe smectite from Cenomanian upwards (Fig. 19.8A).

19.2.3 South Atlantic Ocean

Paleoenvironmental reconstructions of the South-Atlantic Ocean history are deduced from clay stratigraphic successions by several studies, especially those conducted by C. Robert and colleagues (e.g., Robert et al. 1979, Robert 1981, 1982, Robert and Maillot 1983, Robert 1987). *Late Mesozoic clay suites express mainly variations in the tectonic activity, while Cenozoic assemblages reflect rather changes occurring in the climate and current regime.* As in the North Atlantic domain, strong and rapid increases in the abundance of illite proportions, as-

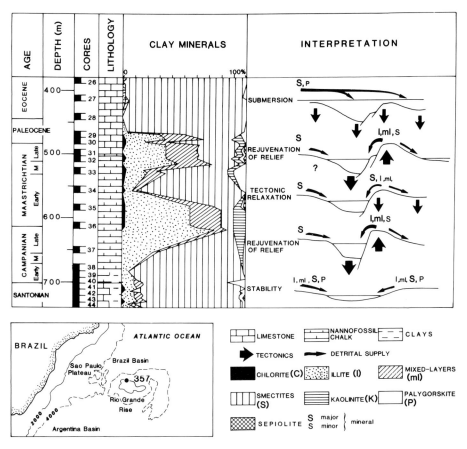

Fig. 19.9. Santonian to Eocene clay-mineral succession on the Rio Grande Rise, Southwestern Atlantic Ocean, and interpretation. (After Robert 1987)

sociated with a large mineralogical diversification, are usually referred to a tectonic destabilization of the ocean margins and/or to sea-level changes. The renewal of the land masses bordering the ocean determined in Jurassic and Cretaceous times the frequent erosion of various paleosols and rocky substrates. In contrast, tectonic relaxation stages favored the development of continental soils, the clay fraction of which appeared to predominantly comprise Al-Fe smectites during Cretaceous and Palaeogene times. Stable tectonic periods priviledged the climatic expression of terrigenous clay associations (Chap. 17). In late Mesozoic sediments abundant smectite or kaolinite, and sometimes fibrous clays (palygorskite, sepiolite) reflect the warm and hydrolyzing conditions dominating on continents. In Cenozoic sediments increasing amounts of illite, chlorite and random mixed-layers progressively occupy the clay sedimentary fraction at middle to high latitudes, reflecting the world cooling and the increasing development of physical weathering compared to chemical weathering.

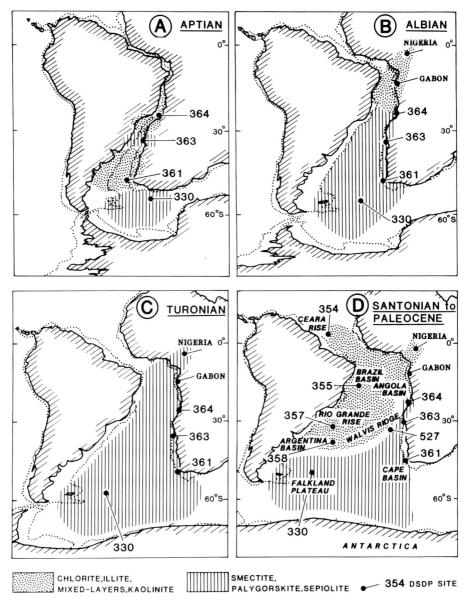

Fig. 19.10. Distribution of the dominant clay associations in South Atlantic sediments during Aptian (**A**), Albian (**B**), Turonian (**C**) and Santonian-Paleocene (**D**). (After Robert 1987)

On the Rio Grande Rise for instance, the DSDP site 357 and other sites from legs 3 and 39 reveal an *alternation of stability and instability periods, from Santonian to Eocene times* (Fig. 19.9). During Santonian and early Campanian, abundant smectites appear to derive from the erosion of widespread soils developed under stable conditions on the exposed Rio Grande Rise and South-

American land mass. Two major tectonic instability phases during late Campanian-early Maastrichtian and late Maastrichtian-early Paleocene times are deduced from strong increases in illite, mixed-layer and kaolinite abundance. These unstable periods are separated by a Maastrichtian phase of quiescence with increasing supply of kaolinite. Finally, a peneplanation completed in approximately 0.75 million years extended during the Paleocene and was followed by the definite submersion of the Rio Grande Rise.

The compiling of data from ten DSDP sites in the *whole South Atlantic Ocean* and from land sections in Western Africa leads Robert (1987) to propose *six mineralogical stages from middle Jurassic to Eocene, interpreted in terms of tectonics, peneplanation and climate.* Such interpretations are strongly supported by bibliographic data, indicating that clay successions, including smectitic ones, constitute reliable markers of the paleoenvironmental changes occurred on peri-Atlantic land masses.

1. Prior to the Oxfordian, abundant illite, random mixed-layers and kaolinite reworked from rocky substrates and ancient soils reflect the tectonic extension occurring between South Africa and Antarctica.

2. From Oxfordian to early Albian time, smectites are thought to have developed in soils of Southernmost Africa, after the land peneplanation had been favored by subsidence. At the same time chlorite, illite, random mixed-layers and kaolinite were abundantly supplied in the Cape and Southern Angola basins submitted to a distensive tectonic regime (Fig. 19.10A).

3. From the Albian period to early Turonian, abundant smectite accompanied by more or less palygorskite and sepiolite characterize the clay sedimentation from the Southern Ocean to the Southern Angola basin, while the illite group is restricted to Gabon and Nigeria areas (Fig. 19.10B), as well as at the junction zone between South and North Atlantic (see 19.2.1).

4. From early Turonian to early Campanian times smectite, palygorskite and sepiolite are widespread in South Atlantic basins, indicating general transgressive conditions, tectonic relaxation and large development of continental soils (Fig. 19.10C).

5. From early Campanian to early Paleocene, the illite group and kaolinite increase in abundance within the northern basins of the South Atlantic, in relation to compressive tectonics in Nigeria and final opening of the Central Atlantic Ocean. In contrast, the southern basins continue to receive large amounts of smectite, palygorskite and even sepiolite, indicative of more stable conditions on adjacent land masses (Fig. 19.10D).

6. After the early Paleocene smectite, palygorskite and sepiolite are widespread in the whole South Atlantic, reflecting warm climate and tectonic relaxation conditions that dominated around the ocean.

19.2.4 Adjacent Regions

Mesozoic and Cenozoic periods are investigated from clay paleoenvironmental messages in different regions adjacent to the Atlantic Ocean. In the *Norway Sea*, the Tertiary record at DSDP sites from leg 38 reflects the following successive events

(Froget 1981): early Eocene instability of continental margins in relation with the basin opening, and testified by a strong supply of illite and kaolinite; margin stabilization in middle-late Eocene time, marked by increasing amounts of soil-derived and volcano-derived smectite; progressive cooling in middle Oligocene, early and middle Miocene, and Plio-Pleistocene times, expressed by successive increases in the abundance of minerals derived from crystalline substrates and moderately weathered rocks (chlorite, illite, random mixed-layers). This tectonic-to-climatic evolution is opposite to that recorded during the same time interval in sediments from *Brittany, Western France* (Estéoule-Choux 1967b). In the latter region the removal during Paleogene time of kaolinite derived from thick, stable soils was followed during the Miocene by the formation of smectite under warm-subarid climate, and during the Pliocene by a rock- and soil-derived mineral influx due to tectonic rejuvenation.

The Mesozoic series in the Iberian Peninsula show clay successions that partly parallel and partly differ from those recovered in the adjacent Atlantic Ocean. Berthou et al. (1982) compare the Cretaceous clay sedimentation in the *Western Portuguese basin* and on the adjacent *Galicia Bank* at DSDP site 398. From Hauterivian to early Cenomanian times, illite and kaolinite prevail in the Portuguese basin, and smectite at site 398. These differences appear to result from distinct continental sources. Site 398 probably received terrigenous materials derived from poorly drained soils to the North, whereas the Portuguese basin was fed at the same time with products inherited from well-sloped and actively eroded relief to the South. A major mineralogical break occurs at the base of the late Cenomanian in the Portuguese basin. Smectite amounts rapidly increase, while no major change exists in the adjacent Atlantic basin. The two regions therefore progressively show similar clay assemblages in Cenomanian deposits. This major change is attributed to both the Cenomanian transgression and the tectonic collapse to the North of the Portuguese basin along the Nazaré fault-system. Such tectono-eustatic variations favored the development of coastal soils where smectites could have formed. Large water-mass exchanges subsequently developed and allowed lateral advection processes of suspended minerals.

In the *Betic Cordillera, South-East Spain*, the Liassic transgression is characterized by a transition from kaolinite-rich to smectite-rich clay assemblages, which is related to a change in the depositional depth (Palomo Delgado et al. 1985). Similar trends are reported in the same region for late Cretaceous black-greenish mudstones, where a smectite increase in transgressive facies is attributed to either detrital supply, volcanic activity or in situ growth (Lopez Galindo et al. 1985, Lopez Galindo 1986). Note that the chemical composition of Cretaceous smectites from the Subbetic zone, Spain (Lopez Galindo 1986), strongly resembles that of late Mesozoic Al-Fe smectites from the open Atlantic Ocean, that are typically related to a terrigenous origin by Debrabant et al. (1985).

In the *Senegal basin*, Western Africa, Michaud and Flicoteaux (1987) investigate the clay mineralogy of Cretaceous shales from 11 commercial boreholes. In pre-Turonian sediments, kaolinite and illite dominate the clay assemblages, which is attributed to the distinct effects of pore diagenesis in coarse-grained facies, of lithostatic pressure in deeply buried sediments, and of tectonic instability in younger deposits (see also Chap. 20). The abundance of smectite increases in Turonian sediments and keeps high until the uppermost Cretaceous time, as a result of the combined action of hydrolyzing climate and of tectonic stability on the continental margin and the hinterland. Note that attempts of paleoenvironmental reconstructions from clay successions also exist about series deposited *East of the African continent*. For instance the late Cretaceous to Cenozoic deposits recovered during DSDP leg 25 in the South Somali basin, Western Indian Ocean, reflect the following evolutionary phases, which resemble those identified on the western tropical margin of Africa (Vernier and Froget 1984):

– Late Cretaceous: abundant smectite reworked from hydromorphic soils in low-relief continental basins submitted to hot climate with a subarid seasonal regime.

– Campanian to Paleocene: tectonic rejuvenation related to the early stages of the Northwestern Indian Ocean opening, and responsible for a noticeable supply of chlorite, illite and random mixed-layers.

– Early Eocene: reworking of palygorskite from peri-marine, semi-enclosed basins developing on the subsiding marginal regions.

– Late Oligocene to Quaternary: slight increase of chlorite, illite and kaolinite amounts, in relation with the worldwide development of the Cenozoic glaciation. The expression of this cooling is slight in the study area, which is located in the equatorial climatic belt and was continuously submitted to hydrolyzing conditions with strong contrasts in seasonal humidity.

19.3 Specific Paleoenvironmental Successions

19.3.1 Intraplate Volcanic Environment, Late Cretaceous of the Mariana Basin

The *site 585 of DSDP leg 89* is located in the northern part of the Mariana basin, Western Tropical Pacific Ocean, at 6109 meters water depth (13°29.00'N, 156°48.91'E. Fig. 19.11). The drilling allowed to recover more than 500 meters of volcaniclastic to argillaceous sediments deposited during the *late Aptian-Maastrichtian stratigraphic interval* (Moberly, Schlanger et al. 1985). Detailed mineralogical, micromorphological (TEM), geochemical and microchemical investigations have been carried out on this sub-continuous series (Chamley et al. 1985a). The different results put together show the existence of *four successive stages* reflecting very different environmental conditions (Fig. 19.12, 19.13).

1. *Late Aptian to early Albian*. The heterogeneous sequence of volcanic sandstones, mudstones and breccias contains a highly crystallized, trioctahedral, magnesian-smectite (Na-smectite. Table 19.3), associated with celadonite, analcite, titaniferous augite, olivine, Na- to K-feldspars. Characteristic major elements of bulk materials include Fe, Mg, Na, K with numerous associated trace elements (Mn, Ni, Cr, Co, Pb, V, Zn, Ti). All lithological, mineralogical and geochemical data emphasize the importance of volcanic influences. Mélières et al.

Fig. 19.11. Location of DSDP site 585 in the Mariana basin, Western Tropical Pacific

(1981) already suggested a volcanic origin for the Aptian smectites in the Mid-Pacific mountains. The volcanic environment appears to be responsible for the peculiar clay-mineral morphology at site 585 which includes small, well-shaped and transparent smectites (Fig. 14.2C), celadonite chips and hexagonal chlorites. The large size and heterogeneity of the volcaniclastic debris demonstrate the proximity of volcanoes at the site 585 area. The volcanoes were probably mostly subaerial, as suggested by the presence of displaced shallow-water bioclastic debris (Sliter 1985), the abundance of sodium and the presence of aluminum hydroxides of a possible pedogenic origin. Furthermore, Thiede et al. (1982) stress the importance of terrestrial organic matter in Mid-Cretaceous sediments of the tropical and subtropical Pacific Ocean. In summary the first interval recorded at site 585 shows the *strong influence of local and subaerial volcanic activity on the sedimentation.*

2. *Early to middle Albian.* Volcaniclastic supply remains important, and is confirmed by the persistent occurrence of analcite, augite, and Na- to K-feldspars, and by the relative abundance of Mg, Ti, Fe, and Mn. Nevertheless, the volcanic influence strongly decreases. The fine siltstones to claystones constituting the overall lithofacies are devoid of olivine and celadonite, which are partly replaced by true detrital illite. Smectites comprise dioctahedral types with aluminum in the octahedra (Mg-beidellite. Table 19.3; Fig. 19.14, core 46), an increasing SiO_2 content and decreasing amounts of MgO. Volcanogenic clay particles, consisting of hairy and only few transparent smectites, are largely

Table 19.3. Structural formulas of smectite at site 585, from microprobe data (after Chamley et al. 1985a)

Samples	Structural formulae	Types of smectite
585-27-1-13 (Campanian)	$(Si_{3.96}Al_{0.04})\,(Al_{1.05}Fe^{3+}_{0.52}Mg_{0.47}Ti_{0.02})$ $K_{0.10}Ca_{0.09}Na_{0.03}O_{10}(OH)_2$ $(Si_{3.79}Al_{0.21})\,(Al_{0.92}Fe^{3+}_{0.73}Mg_{0.46}Ti_{0.02})$ $K_{0.11}Ca_{0.06}Na_{0.03}O_{10}(OH)_2$ $(Si_{3.66}Al_{0.34})\,(Fe^{3+}_{0.97}Al_{0.71}Mg_{0.44}Ti_{0.02})$ $K_{0.15}Ca_{0.06}Na_{0.03}Mg_{0.02}O_{10}(OH)_2$	Fe-beidellites
585-31-3-29 (Turonian)	$(Si_{3.88}Al_{0.12})\,(Al_{0.72}Fe^{3+}_{0.61}Mg_{0.53}Ti_{0.04})$ $K_{0.59}Na_{0.24}Ca_{0.04}O_{10}(OH)_2$ $(Si_{3.68}Al_{0.32})\,(Al_{0.62}Fe^{3+}_{0.68}MgO_{0.58}Ti_{0.03})$ $K_{0.69}Na_{0.31}Ca_{0.07}O_{10}(OH)_2$	Alkalines smectites
585-44-3-17 (m. Albian)	$(Si_{3.42}Al_{0.58})\,(Mg_{0.87}Al_{0.78}Fe^{3+}_{0.56}Ti_{0.06})$ $K_{0.23}Na_{0.15}Ca_{0.09}Mg_{0.01}O_{10}(OH)_2$	Mg-beidellites
585-46-2-40 (e.-m. Albian)	$(Si_{2.98}Al_{1.02})\,(Mg_{1.35}Al_{0.60}Fe^{3+}_{0.46}Ti_{0.03})$ $Mg_{0.32}Na_{0.32}Na_{0.17}K_{0.11}Ca_{0.05}O_{10}(OH)_2$ $(Si_{2.80}Al_{1.18}Fe^{3+}_{0.02})\,(Mg_{2.19}Fe^{3+}_{0.50}Ti_{0.03})$ $Mg_{0.44}Na_{0.13}K_{0.07}Ca_{0.06}O_{10}(OH)_2$	Mg-smectites with transition terms to Mg-chlorite
585A-17-1-15 (1. Aptian)	$(Si_{2.66}Al_{0.87}Fe^{3+}_{0.47})\,(Mg_{2.52}Fe^{3+}_{0.17}Ti_{0.04})$ $(Na_{1.21}K_{0.27}Ca_{0.12}O_{10}(OH)_2$	Saponite

associated with fleecy and lath-shaped smectites not typical of volcanic environments. As a consequence, interval 2 appears as a *transition period, characterized by a still persisting, but less proximal and less subaerial volcanic influence.* Note the temporary occurrence of a clay mineral suite ranging from Mg-beidellite to Mg-chlorite, including chlorite-smectite irregular mixed-layer minerals and expandable chlorite, all suggesting a local, confined and magnesium-rich environment.

3. *Middle Albian to middle Campanian.* Coinciding with the passage from volcanogenic sediments to common claystones, the zone 2 / zone 3 boundary is defined by a marked change in the bulk mineralogy and geochemistry. All the volcanic markers disappear, except the persistence of a good correlation between Na_2O and TiO_2 and the presence of Na- to K-feldspars. Augite and analcite are replaced by quartz, opal and clinoptilolite. Smectites present a silica-rich, alkaline character (alkaline smectites. Table 19.3); they are medium-crystallized and display all transitional morphologies between common fleecy types to completely lath-shaped types. Such change in smectite morphology has been described and discussed in Cretaceous and Cenozoic sediments from the Atlantic and Indian Oceans, and does not correlate to noticeable chemical changes related to diagenesis (see 14.2.2). It seems to correspond to an in situ recrystallization of fleecy to lath-shaped smectites without noticeable chemical changes. Opal CT and zeolites also proceed from in situ crystallization (see Riech and von Rad 1979). As a result, the third interval recorded in Cretaceous sediments at site 585

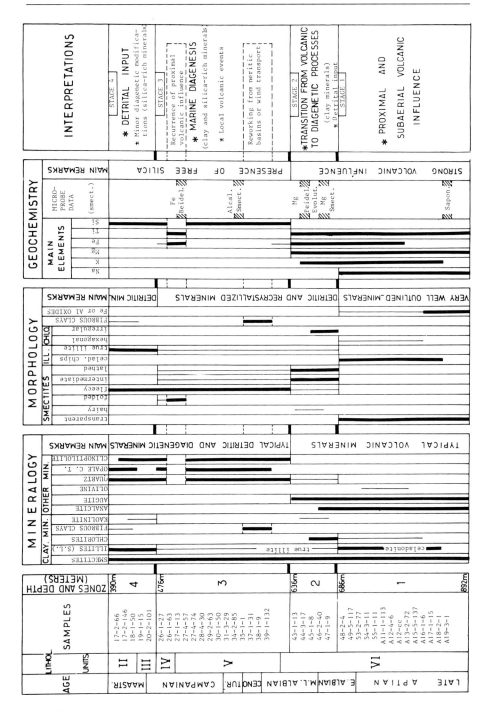

Fig. 19.13. Main data and interpretation of the late Cretaceous inorganic sedimentation in the Mariana basin (DSDP site 585. After Chamley et al. 1985a)

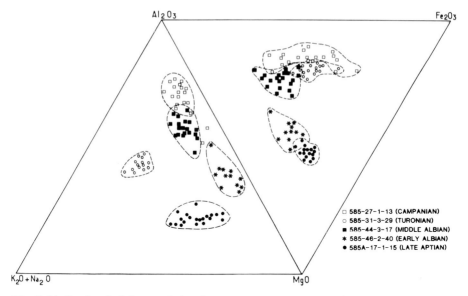

Fig. 19.14. Geochemical characteristics of smectites at site 585, from microprobe analyses. (After Chamley et al. 1985 a)

indicates a *strong decrease of the volcanic influence and the importance of marine early-diagenetic processes.*

Note the presence of a recurrence during Campanian time of volcanic input in the uppermost part of zone 3, marked by folded, Fe-rich smectites (Table 19.3, core 27), and by increasing amounts of Fe, Mn, Na, K and Ti. This event correlates with the supply at site 585 of middle-neritic to upper-bathyal foraminifera (Sliter 1985), suggesting remobilization and downslope transport linked to volcanic activity.

The late Albian to Cenomanian sediments locally contain short, more or less broken and isolated palygorskite fibers, that probably result from the reworking of authigenic sediments formed in alkaline coastal basins during the middle-late Cretaceous. The presence of this fibrous clay input could correspond to the passage of the site 585 area across a warm, subarid climatic belt where fibrous clays formed and were supplied to the ocean by marine or aeolian currents.

4. *Late Campanian to Maastrichtian.* Although no major lithological changes occurred during the latest Cretaceous, the mineralogy and geochemistry show a new important change. The volcanic influence nearly disappears, as shown by the absence of a correlation between Na_2O and TiO_2, and the decrease of Ti, Fe, K, Na. An increasing terrigenous influence is suggested by the aluminous character of the sediments (Fig. 19.12), as well as by the abundant presence of Al-Fe beidellites in the clay fraction. There is a strong mineralogical diversification in the clay assemblage, marked by the combined presence of smectite, illite, palygorskite, kaolinite and random mixed-layer minerals. The diagenetic modifications chiefly consist in the formation of opal and clinoptilolite, and apparently concern the

clay minerals very little. Smectites mainly display common fleecy particles, with only few and partly shaped ones. This major change in sedimentary conditions, which indicates a diversified detrital supply from distant sources, more dependent on input than local environment, probably reflects increasing marine circulation, as is suggested by more oxidizing conditions (index Mn*. Figs. 19.2, 19.13). This increasing exchange of water masses was essentially connected to the subsidence of the migrating Pacific plate. As a result, the last Cretaceous interval recorded at site 585 reflects the *increasing detrital input from distant sources distributed by an evolving current system, and only minor diagenetic modifications.*

To summarize, *the late Cretaceous sedimentary sequence at DSDP site 585, Mariana basin, is characterized by the combination of geodynamic and paleoenvironmental changes that are reflected by the clay mineralogy, geochemistry and morphology. As the subsidence related to the East-Pacific plate migration progressed, the inorganic sedimentation successively displayed a local, subaerial volcanic control, then more distant, submarine and diagenetic effects, and finally the influence of diversified, very distant and mainly terrigenous influences.*

19.3.2 Tectono-Eustatic Environment, Late Miocene of Sicily

The combined study of different, equally conservative components of sediments allows to propose integrated paleoenvironmental interpretations in series spanning sufficiently large and continuous time intervals. This is the case of the late Miocene deposits from South Sicily, where a permanent marine sedimentation developed *from about 14.6 to 5.3 million years BP* in the Caltanissetta basin. Giammoia and Falconara sections, located in this basin East of the Capo Rossello section (Fig. 18.5), provide a 180-meter-thick series of hemipelagic clays embracing the Serravallian to lowermost Messinian interval. Organic-rich interbeds are frequently intercalated in the hemipelagites, that pass upwards to Messinian diatomites and limestones. The investigations performed on about 90 samples comprise *X-ray diffraction on the clay fraction, determination of planktonic foraminiferal assemblages and characterization of oxygen- and carbon-isotope patterns.*

The clay minerals identified include smectite (30-70% of the < 2 μm fraction), illite (15-30%), kaolinite (10-40%), chlorite (0-10%), random mixed-layers (traces – 10%), palygorskite (0-10%) and sepiolite (0% – traces). The calculation of mineralogical ratios on X-ray diagrams obtained from glycolated samples, as well as the percentage estimates of clay and nonclay minerals, permit the distinction of *three major units*, which can further be subdivided into seven minor intervals (Fig. 19.15).

As the clay mineral associations lack any indication of diagenetic alteration in the marine environment, they can be used as indicators of the nature and intensity of pedogenesis and erosion in the hinterland. Smectite is considered to form mainly in downstream parts of drainage systems under warm climate, whereas kaolinite formation is favored in well-drained soils characterized by strong hydrolysis. The relative amount of illite roughly represents the contribu-

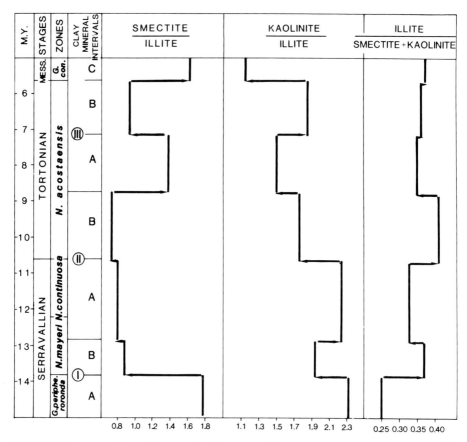

Fig. 19.15. Average values of the smectite/illite (18/10 Å), kaolinite (+ chlorite)/illite (7/10 Å), and illite/smectite + kaolinite (10/18 + 7 Å) ratios in the successive clay mineral intervals of the Giammoia-Falconara series. (After Chamley et al. 1986). Peak-height ratios measured on X-ray diagrams, glycolated samples. Foraminifera zones from Zachariasse and Spaak (1983). *G* Globorotalia. *N* Neogloboquadrina

tion of unaltered source-rocks from which smectite and kaolinite form pedogeni-cally. Thus *the successive clay mineral intervals are tentatively interpreted in terms of climate, sea-level changes and tectonic rejuvenation*, whose relative importance varies in course of the time and determines variable supply of rock- and soil-derived minerals within the Caltanissetta basin (Fig. 19.16). From early Serraval-lian to late Tortonian, the general increase of the relative abundance of illite, and the correlative decrease of kaolinite, reflect the *late Cenozoic world cooling*,

Fig. 19.16. Main interpretations inferred from the integrated study of clay minerals, plantonic foraminifera and stable isotopes (O, C) of the Giammoia-Falconara series. Tentative correlations with the ranges of sporomorph associations and some continental and Paratethyan stages. (After Chamley et al. 1986)

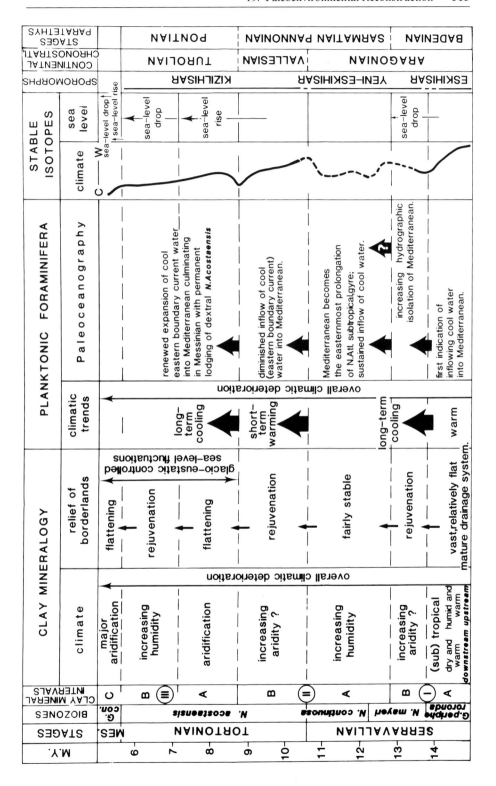

marked by the progressive replacement of soil-derived by rock-derived detrital material (see 17.3).

The high relative amounts of smectite and kaolinite and the low amounts of illite and nonclay minerals in the lowermost part of the sequence (interval IA, *early Serravallian*), suggest the existence of a *mature drainage system* providing a large supply of kaolinite from upstream, humid, and of smectite from downstream, more arid areas. Probably the Caltanissetta basin was situated in the subtropical belt and bordered by a vast, relatively flat landmass. The parallel decrease of smectite/illite and kaolinite/illite ratios at interval IA – IB boundary, caused by the increase of illite abundance, suggests a general decrease of soil formation, both in downstream and upstream areas, in the course of the Serravallian. This assumption is supported by the increase of rock-derived, non clay minerals, such as feldspars. The *increase of illite-micas* and non clay minerals may have been related to a sustained *rejuvenation of the subaerial relief* during the early Serravallian, caused either by a eustatic sea-level lowering or by tectonic uplift. The effects of the lowering of the base level of erosion obliterate any clear climatic signal that might be inferred from the changes in the composition of the clay-mineral associations in unit I.

In the *late Serravallian* interval IIA, environmental conditions were completely different from those occurring before. The pronounced drop of the illite/smectite + kaolinite ratio across the boundary between units I and II, and the high relative amount of kaolinite in interval IIA, suggest *increasing pedogenesis* and the *predominance of kaolinite-rich soil formation*. Probably the paleogeographic setting was characterized by the existence of fairly pronounced reliefs in a relatively stable hinterland where such kaolinite-rich soils could develop under *humid conditions*. More humid climatic conditions during the late Serravallian can also be deduced from palynological data (Bessedik 1984).

At about the *transition from the Serravallian to the Tortonian* (boundary IIA - IIB), the supply of kaolinite decreases, whereas the amount of smectite in the upper Serravallian interval IIA and the Tortonian interval IIB is roughly the same. This may reflect a decrease in annual rainfall. Increasing aridity under unchanged paleogeographic conditions, however, would rather favor an increase of smectite formation in downstream areas. Therefore the changes across the Serravallian-Tortonian boundary may better be explained as having been caused by a *rejuvenation of the subaerial relief*, favoring the supply of rock-derived illite relative to surficial kaolinite.

During the *late Tortonian and early Messinian* unit III the environmental conditions in the borderlands of the Caltanissetta basin had changed again. The inverse relationship between the smectite/illite and kaolinite/illite ratios (Fig. 19.15) suggests *alternating periods with development of smectite-rich soils in poorly drained and of kaolinite-rich soils in well-drained areas*. This alternation was most probably controlled by climatic fluctuations.

The conspicuous increase of smectite and the associated decrease of kaolinite and of illite-mica in the *late Tortonian* interval IIIA suggest a flattening of the relief and the development of smectite-rich soils in borderlands under semi-arid conditions. This *flattening of the relief* could have been caused either by *peneplanation or by sea-level rise*.

In the *latest Tortonian* interval IIIB there may have been a return towards *more humid climatic conditions*, as indicated by the sudden and strong increase of kaolinite and concomitant decrease of smectite at the transition IIIA – IIIB. The ensuing decrease of kaolinite in interval IIIB seems to have been roughly balanced by an increase in the supply of rock-derived illite. This could reflect a *lowering of the base level of erosion*, due to a glacio-eustatic drop of the sea level or to tectonic uplift.

At about the *transition from Tortonian to Messinian* (transition IIIB – IIIC), conditions became favorable again for the widespread development of smectite-rich soils. This could have resulted from an *aridifaction* and flattening of the relief of the borderlands, which flattening must be attributed to a relative sea-level rise. Apparently fairly similar environmental conditions can be postulated for the late Tortonian interval IIIA and the early Messinian interval IIIC. These conditions were defined by seasonal contrasts in humidity, which favored the formation of smectite in surficial, poorly drained soils, and prevented the formation of kaolinite in relatively better-drained areas. During the Messinian the development of smectite-rich soils was accentuated by the effects of the Messinian salinity crisis (see 17.3).

The boundaries between the clay-mineral intervals coincide with major, presumedly Atlantic-born changes in the abundance of planktonic foraminifera. First-order changes in the clay mineral record are also expressed in the stable isotope patterns. *The synchronism of the modifications recorded in the abiotic and biotic parameters is attributed to a complex interplay between climatic fluctuations, sea-level variations and tectonically controlled paleogeographic reorganizations in the Central Mediterranean during the middle and late Miocene.* The major changes in the physico-chemical and biotic records occurred during the Serravallian (about 13.8 and 12.8 my BP), at the transition from the Serravallian to the Tortonian (around 10.6 my), in the Tortonian (about 8.7 and 7.2 my) and at the Tortonian-Messinian boundary (5.6 my). The main events can be tentatively discussed as follows (Fig. 19.6).

1. The planktonic foraminiferal patterns primarily reflect climatic fluctuations, resulting in an *overall climatic deterioration* and a *stronger inflow of cool Atlantic water* into the Mediterranean. The clay mineral record leads to a similar conclusion, suggesting that the climatic regime of the Sicilian hinterland shifted from subtropical to temperate-warm in the course of late Miocene time.

2. The stable isotope and clay-mineral records indicate *eustatic fluctuations controlled either by climate or by tectonics*. A *first sea-level* drop is suggested during *Serravallian times* by an increase of illite abundance and higher values of the carbon isotope ratio ($\delta^{13}C$). Simultaneously, the planktonic foraminiferal record shows the *replacement of a tropical association by a subtropical one*, suggesting the onset of climate cooling. This cooling is partly obliterated in the oxygen isotope record by salinity changes. However, the stable isotope data indicate that during Serravallian the Mediterranean *water masses became increasingly stratified*. Notice that the onset of cooler conditions and the associated sea-level drop in the Serravallian correspond to a global fall in sea level (Vail et al. 1977). This change is also roughly synchronous with several extra-Mediterranean regional events, such as the isolation of the Red Sea and of the Paratethys (El-

Heiny 1982, Rögl and Steininger 1983), that are clearly correlated to the middle Miocene global cooling. In addition, the late Serravallian climatic and paleogeographic changes are reflected in the peri-Mediterranean flora, especially the sporomorph associations (Fig. 19.16).

3. At the *Serravallian-Tortonian boundary* the clay mineral changes suggest a *rejuvenation of Sicilian relief*, while the planktonics and stable isotopes indicate a *temporary climate warming*. Similar paleoclimatic and paleogeographic variations are reported from other parts of the Mediterranean domain, especially from flora and mammal data (see Benda et al. 1982).

4. Important changes in clay mineralogy, composition of planktonic foraminifera and stable isotope patterns reflect a *major intra-Tortonian event* at about 8.7 million years ago (interval III). The clay mineralogy and the stable isotope record suggest the beginning of a period marked by important *glacio-eustatic fluctuations*. Planktonics indicate a renewed *cooling* after the early Tortonian. These major changes occurred significantly before the beginning of the Messinian, a conclusion also supported by shifts in the $\delta^{13}C$ already starting close to the IIB – IIIA interval boundary. The shifts indicate an intra-Tortonian sea-level rise followed by two phases of important sea-level drop, one of which is early Messinian in age. The changes inferred from the Sicilian record correspond to a major step in the continental stratigraphy (e.g., mammal stages, Fig. 19.16), and to large-scale variations in the Paratethys paleogeography. This suggests that *the Messinian salinity crisis forms only an overprint on general paleoenvironmental conditions* originating already in late Tortonian time both in the Mediterranean and Paratethys basins.

19.3.3 Cretaceous – Tertiary Transition

The discovery in the early 1980's of a conspicuous iridium anomaly at the Cretaceous-Tertiary boundary renewed the discussions about the cause of the exceptional phenomena which marked that period. In the marine sedimentary record, the Cretaceous-Tertiary boundary levels generally correlate to high-rate extinction of coccoliths and planktonic foraminifera, the replacement in benthic faunas of suspension feeders by deposit feeders and carnivores, a negative $\delta^{13}C$ excursion, microtektite-like spherules, shocked quartz, a drop in carbonate content, in addition to enrichments in siderophile chemical elements. These *catastrophic changes*, contemporary with the disappearance of ammonite and dinosaur faunas, are the subject of *intense debate about the possibilities of extra-terrestrial and terrestrial causes* (e.g., Courtillot and Cisowski 1987, Hallam 1987). Crucial arguments in extra-terrestrial scenarios consist in the enrichment in iridium and other siderophile elements, and the presence of impact-born spherules. Current terrestrial scenarios involve an uppermost Cretaceous sea-level drop and/or intense global volcanism with major activity in the Northern Indian domain (Deccan area. See McLean 1985, Courtillot et al. 1986). Recent investigations tend to better consider the possibility of terrestrial causes, especially by the way of major disturbance in the Earth's mantle determining intense volcanism. What is the contribution of the clay mineral record to such a debate?

Large-Scale Studies

First investigations often concluded an independence between the clay mineral record and massive extinction of Cretaceous faunas. For instance, Sittler (1965b) reported the permanent expression by clay associations of warm, hydrolyzing climate across the Cretaceous-Tertiary boundary in Provence and Languedoc regions, Southeastern France, within several continental sections where dinosaurs bones and eggs suddenly disappeared.

Subsequent large-scale investigations often point to the existence of *certain changes in the clay mineral stratigraphy at the Mesozoic-Cenozoic transition.* Chamley and Robert (1979) study the Santonian to Eocene interval in *Atlantic sediments* from 12 DSDP sites. They observe from the Campanian upwards, and especially in the late Maastrichtian-early Paleocene period, an increase in the relative abundance of illite and chlorite, generally associated with more abundant random mixed-layers, quartz and feldspars. Discussing the possible causes of these mineralogical changes, the authors favor the hypothesis of a *general tectonic destabilization* associated with increasing water-mass exchanges in the ocean, and reject the possibility of a general climate cooling.

Chamley et al. (1984) conduct mineralogical, electronmicroscope and geochemical analyses on late Campanian to late Paleocene sediments from DSDP sites 525, 527, 528 and 529, that were drilled during leg 74 on the western flank of the *Walvis Ridge,* Southeastern Atlantic Ocean. The Cretaceous-Tertiary boundary is characterized by a *strong increase in the relative abundance of smectite* (Fig.19.17), that occurs among a complex paleoenvironmental history reflected by noticeable changes in the inorganic record. The main facts and interpretations are the following (Fig. 19.18).

1. The sediments intercalated between late Cretaceous salts or located immediately above them, as well as most Campanian-Maastrichtian-Paleocene deposits, contain various clay assemblages, whereas the basalt alterations are

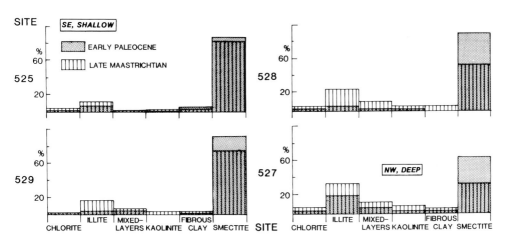

Fig. 19.17. Average clay mineral percentages at late Maastrichtian and early Paleocene times on the Western Walvis Ridge, Southeastern Atlantic Ocean. (After Chamley et al. 1984)

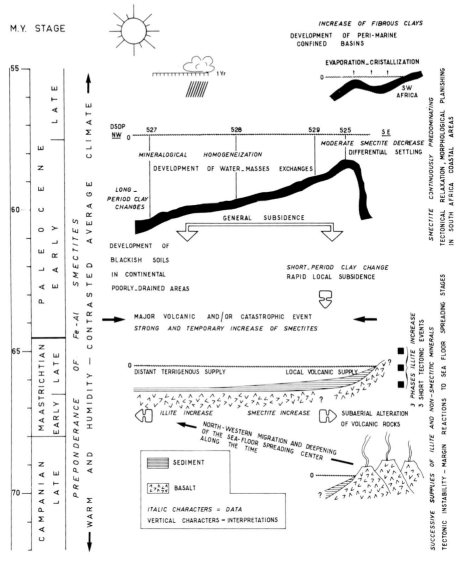

Fig. 19.18. Main data and clay sedimentation (*roman letters*) and inferred interpretations (*italics*), of the Cretaceous-Tertiary transition on the Walvis Ridge, Southeastern Atlantic Ocean. (After Chamley et al. 1984)

dominated by crystalline smectite. The lithology, mineralogy, morphology and geochemistry indicate the *predominantly detrital character of sedimentary clays*, and the only local volcanogenic origin of smectites.

2. The preponderance of Fe-Al smectite in marine sediments suggests the persistence of an average hot climate, with sharp contrasts in humidity, favoring the development of poorly drained soils in coastal continental areas (see Chap. 17).

3. During the late Cretaceous, volcanogenic smectite predominated at the shallowest site 525 (2467 m water depth), indicating the proximity of strongly-weathered, eroded subaerial basalts. Toward the deepest site 527 (4428 m) there occurs a progressive diversification of the clay suites. Combined mineralogical and geochemical data indicate a northwestward increase in the terrigenous supply from distant sources, as well as a deepening of, and an increase in distance from, the seafloor spreading center.

4. Three successive inputs of illite, chlorite, mixed-layer clays, kaolinite, and some palygorskite characterize the late Maastrichtian period. They are probably expressions of *tectonic destabilization* on the Walvis Ridge and/or the African margin, associated with the development of deep circulation. They may also correspond to *fluctuations in the sea level*.

5. The *Maastrichtian-Paleocene boundary* is marked by a strong and sudden increase in the amount of smectite, and by associated geochemical changes indicative of *volcanic influence and local subsidence*. This short-period clay change is especially recorded at the shallowest site 525, located closer to the ridge crest.

6. During the *Paleocene* a clay homogenization develops at all the sites, reflecting the *subsidence* of the Walvis Ridge and the *increase of water-mass exchanges*. Subsidence, leading to the disappearance of subaerial volcanic rocks, becomes important during the late Paleocene, as expressed by the geochemistry. At the same time a morphological peneplanation occurs on the adjacent continent, as suggested by the abundance of pedogenic-like smectite. The extensive formation of soils is associated with peri-marine basins propitious to the neoformation of fibrous clays. All these characters indicate the existence of a time of tectonic quiescence and transgressive conditions.

Cretaceous-Tertiary Boundary Layer
The precise contact between the uppermost Maastrichtian deposits and the first Paleocene sediments is often characterized by a clayey layer, known as the Cretaceous-Tertiary boundary clay. Investigations on the clay lithology, mineralogy, and geochemistry of this layer and surrounding horizons point to the existence of *significant changes* and lead to interpretations involving either extraterrestrial or terrestrial causes.

Extraterrestrial Causes
Kastner et al. (1984) investigate the boundary layer from Stevns Klint, *Western Denmark*, and from DSDP hole 465A in the *North-Central Pacific Ocean*. At Stevns Klint, the only clay mineral detected is a pure smectite. In contrast, the clay fractions above and below the boundary layer comprise illite, mixed-layer illite-smectite rich in smectite and quartz, all minerals of detrital origin. The

purity of the boundary smectite, its very local presence and the high value of its $\delta^{18}O$ content suggest an in situ formation by alteration of glass. The formation of smectite from extraterrestrial impact rather than from volcanic glass in the marine environment is supported by the high content of iridium and gold, and by the absence of a negative anomaly in cerium. M. Kastner and colleagues suggest that the pure smectite layer at the Cretaceous-Tertiary boundary at Stevns Klint formed authigenically in a low water/rock submarine diagenetic environment, most probably isochemically from the aluminum- and iron-rich aluminosilicate impact spherules. At DSDP hole 465A, the abundance of smectitic minerals in the zone of high iridium anomaly is tentatively attributed to the alteration of the same glass as at Stevns Klint.

Pollastro and Pillmore (1987) study the Cretaceous-Tertiary boundary at the top of a kaolinitic claystone layer in a sequence of coal-bearing, fluvial rocks in the east-central part of the Raton basin, *New Mexico and Colorado* (USA). The boundary layer, which is defined by the disappearance of certain fossil-pollen taxa, contains well-crystallized kaolinite and random illite-smectite mixed-layers, associated with shocked quartz grains and abundance anomalies of iridium, chromium and other elements. The authors estimate that the clay minerals have been formed by the in situ alteration of vitreous material in a coal-swamp environment. The well-crystallized kaolinite is considered as the diagenetic stable phase for original fallout material that first altered to cabbage-like microspherules of a halloysite or allophane nature. The microspherulitic fabric and subsequent crystalline kaolinite are thought to represent the distinctive character indicative of an asteroid impact at the end of the Cretaceous. Similar conclusions arise from studies by Bohor et al. (1984, 1987) and Bohor and Triplehorn (1987) about the Cretaceous-Tertiary layer in continental series from *Montana and Wyoming*, USA. In the Lance Formation at Dogie Creek, Wyoming, the boundary clay consists of a basal kaolinitic claystone as much as 3-centimeter-thick containing hollow goyazite spherules. This layer is overlain by a 2- to 3-millimeter-thick smectitic horizon containing both shock-metamorphosed minerals and an iridium anomaly of 21 ppb. The association of spherules, unusual clay mineralogy and geochemistry with a palynological break at the base of the claystone layer is tentatively attributed to an extraterrestrial impact, probably responsible for the widespread emanation of a cloud of glassy material.

Terrestrial Causes

Various authors stress the fact that the geochemical anomalies, microspherules, shocked quartz and other peculiarities recorded at the Cretaceous-Tertiary boundary could equally result from extraterrestrial and terrestrial causes, the latter being possibly related to intense and widespread volcanic activity (see Courtillot and Cisowski 1987, Hallam 1987). Several studies on the clay mineral successions across the boundary support the hypothesis of a major terrestrial event. For instance Rampino and Reynolds (1983) compare the clay mineralogy at four localities from *Western Europe and North Africa*, and conclude *large differences from place to place in the clay mineralogy of the boundary layer*.

1. The boundary sequence at *Nye Kløv, Denmark*, is very similar to the Stevns Klint sequence located at 300 kilometers in the same country and studied by

Kastner et al. (1984). Pure smectite characterizes the boundary clay, whereas underlying limestones contain common detrital illite and smectitic minerals. Smectite in the boundary clay is attributed to the common alteration of volcanogenic material, withoug the need for an asteroid impact.

2. The *Gubbio section, Central Italy*, displays similar illite- and kaolinite-rich detrital assemblages in the boundary clays and 1.73 meters below the boundary. Other investigations on the same section provide some precision. Wezel et al. (1981), Johnsson and Reynolds (1986) and Jehanno et al. (1987) report an augmentation of the relative abundance of kaolinite close to the boundary. H. Chamley (unpubl.) observes an augmentation of kaolinite from 10 to 30% of the <2 μm fraction in the thin red clay forming the boundary layer, with a rapid subsequent decrease from 30 to 15-10%; this very local kaolinite increase is accompanied by an increase in illite and random mixed-layers abundance. It occurs within an uppermost Cretaceous-lowermost Paleogene period marked by clay assemblages rather rich in kaolinite relatively to older and younger periods (from 15 meters below to 25 meters above the boundary). The cause of the kaolinite increase is attributed to either continental margin progradation and retreat, shifting depositional pattern, climatic change, sea-level lowering or new source terrains. A significant *increase in erosion conditions* is inferred from most of these possible explanations.

3,4. The boundary clay and the clays above and below it from *Caravaca,Southeastern Spain, and El Kef, Northern Tunisia*, contain comparable mineral associations. Abundant kaolinite and smectitic minerals occur, accompanied by little illite and sometimes palygorskite. M.R. Rampino and R.C. Reynolds envisage that smectite could derive from the alteration of volcanic material, and that kaolinite could represent either terrigenous detritus or bentonitic alteration. Note that Jehanno et al. (1987) report an obvious decrease of kaolinite abundance in the boundary clay at Caravaca section, only minor variations at Finestrat section located in the same area, and a slight decrease of kaolinite abundance at Bidart section located in Southwestern France close to the frontier with Spain. At El Kef section, Tunisia, C. Jehanno and colleagues report a slight increase in the kaolinite abundance at the Cretaceous – Tertiary boundary. A comparable trend is seen by H. Chamley (unpubl.), who observes an onward increase of both kaolinite and illite abundance in the boundary clay at El Kef section (from 20 to 30% and from traces to 5%, respectively). This mineralogical variation appears to result from a sea-level fall or tectonic rejuvenation rather than from an increase in continental humidity. A conspicuous increase in the kaolinite abundance also occurs at the Maastrichtian-Paleocene transition in the external part of the Senegal basin, Western Africa (Chamley et al. 1988. See Chap. 20).

To summarize, the clay mineralogy of the Cretaceous-Tertiary boundary layer is characterized by widespread modifications, the nature of which considerably differs from place to place. Some sections reveal a large increase in the relative abundance of smectite (Walvis Ridge, Denmark), some other an increase of both illite and smectitic minerals, or of kaolinite and illite, or even more complex changes. All the minerals identified are known in normal depositional or pedogenic environments and exist in various amounts within the common

Cretaceous and Paleogene sediments. This points to the absence of mineralogical variations of a unique type, that could be related to an exotic cause. An extraterrestrial control of the high-resolution clay stratigraphy at the Cretaceous-Tertiary boundary can therefore hardly be envisaged. *A terrestrial control* appears to be much more likely. A *major disturbance in the Earth's mantle*, responsible for intense volcanic activity and glass emissions, could represent a unique cause for the changes observed in the inorganic record. Such a disturbance *could account for both the secondary alteration of volcanic glass into smectitic clay in some areas, and for a widespread tectonic destabilization in some other areas* . Such a destabilization could have induced a relative drop of the sea level, the drastic erosion of exposed areas, and the supply to sedimentary basins of various detrital assemblages. This hypothesis, together with that of an extraterrestrial impact, should be further tested by performing detailed investigations on sections located close to the areas submitted to major volcanic activity during uppermost Cretaceous times.

19.4 Conclusion

1. *Stratigraphic successions of clay mineral associations represent a useful tool to reconstruct the evolution of past environments*, especially when they are combined with lithologic, geochemical and micromorphological data. The large dispersal properties of clay minerals favor their use for paleogeographic reconstructions in sedimentary areas located far away from source regions. *Clay sedimentary successions basically reflect the paleogeographic history of the land masses adjacent to the depositional basins, and accessorily the evolution of the basins themselves.* Most large-scale studies performed so far concern the Mesozoic and Cenozoic series from the Atlantic domain, an ocean strongly dependent throughout time on the history of American, European, African and Antarctic borderlands. *Clay assemblages provide information mainly on the tectonic structuration and morphologic evolution for Jurassic and Cretaceous times, due to the early formation and growth of Atlantic basins. In contrast, clay successions in Cenozoic series mainly reflect changes occurring in climate, detrital sources and transportation agents, because of fundamental climatic modifications correlating to the existence of mature oceanic basins.*

An increasing number of investigations applies to other basins such as the Tethyan, Pacific and Indian Oceans, where paleoenvironmental reconstructions may be partly impeded by diagenetic modifications. The Mediterranean region was submitted during late Cenozoic times to a complex tectonic, morphologic and climatic history, that may be reconstructed by combining mineralogical, paleontological and isotope information. Some applications exist for geological periods as old as the late Precambrian, in series preserved from noticeable burial effects, tectonic structuration and fluid migrations (see also Chaps. 17, 18, 20).

2. *Vertical movements of the Earth's crust are clearly reflected by successive clay associations.* The combination of mineralogical and inorganic-organic geochemical analyses allows better understanding of the stages of ocean forma-

tion, widening and deepening. The *post-rift stages* generally determine the tectonic destabilization of continental margins, leading to the priviledged input to the basin of substrate-derived minerals dominated by illite, chlorite and other typically continental species. *Spreading-acceleration stages*, responsible for the rapid subsidence of continental margins, apparently also induce enhanced erosion of crystalline substrates on land. The effects of such *tectonic destabilization may interfere with the effects of sea-level changes*, whose actual influence on the clay record is still poorly understood. Detailed investigations of land-to-sea transects in different geodynamical and eustatic contexts are necessary to better decipher the respective influence of tectonic rejuvenation or peneplanation and sea-level changes, for a given time slice. *Vertical movements occurring in the basins themselves*, as, for instance, the formation or disappearance of submarine barriers or troughs, ridge jumps and other intraplate deformations, can also be expressed by clay mineral changes. Note that the detrital mineral species indicative of a given tectonic or source signal may differ considerably according to the petrographic nature of continental formations. Illite and associated minerals (chlorite, random mixed-layers, quartz, feldspars, dense minerals) often constitute favored markers, but these minerals may be accompanied or replaced by minerals as diverse as smectite, kaolinite, palygorskite, sepiolite, talc, serpentine and others.

3. *Horizontal movements of lithospheric plates are often registered by detrital clay successions.* The displacement of a given sedimentary area at *increasing distance* from a spreading axis or from a land mass tends to determine differential precipitation of metal oxides or differential settling processes (see Chaps. 6, 18). The *crossing by migrating plates of successive climate belts* induces changes in the nature of the soil-derived minerals supplied to the basins and of transportation agents. The *convergence and collision of lithospheric plates*, as well as subduction mechanisms, often lead to great changes in the detrital supply. Convincing examples arise from the comparison of clay suites in Eastern and Western Atlantic basins, and promising studies exist for other oceans. For instance the northward migration of the North American plate induced an increasing input of rock-derived minerals (illite group) during late Cenozoic times, a period during which the crossing of the equatorial zone by the African plate determined massive influx of soil-derived minerals (kaolinite group). The northwestward displacement of the Pacific plate since the late Cretaceous allowed the successive influence in a given sedimentary area of changing climate, volcanic activity, aeolian influx and marine currents. As for vertical movements, the nature of minerals indicative of horizontal movements may vary greatly according to petrographic, morphologic and climatic conditions in source regions. A large field of investigations is open, especially about the climatic variations expressed by terrigenous clay suites in relation with the situation of past continental domains relatively to oceanic domains.

4. *The morphologic evolution of the continental source regions responsible for the terrigenous input to adjacent marine basins fundamentally controls the pedologic or non pedologic origin of mineral suites.* An unstable continental relief precludes pedogenic processes and favors the direct removal of materials from rocky substrates. Resulting deposits frequently contain abundant illite and diver-

sified clay assemblages. In contrast, stable landscapes favor peneplanation, the development of soils, and the precipitation of chemical sediments in continental or peri-marine basins. Stable conditions may therefore result in the reworking of little-diversified clay assemblages, dominated by either pedogenic smectite, pedogenic kaolinite or chemically deposited palygorskite and sepiolite, depending on the topography and climatic conditions. Organic and inorganic geochemical investigations greatly help to understand the significance of successive mineral associations.

5. *Specific paleoenvironmental conditions* appear to have prevailed at certain geological periods, as documented by clay sedimentary successions. For instance, *the North Atlantic Ocean*, despite its rather simple structural evolution, *was morphologically segregated during the Albian and some other Cretaceous periods*. The major submarine relief was located in the Bermuda area, in the present Western Atlantic, separating the mainly soil-derived input from America and the dominantly rock-derived input from Africa. The mineralogical and geochemical record therefore permits proposing the existence of opposite paleogeographic conditions on both Atlantic borderlands. Similar reconstructions arise from studies in the south Atlantic Ocean or the Gulf of Mexico, and many others may be envisaged from submarine or exposed series from various basins and paleobasins. *In the Central Pacific Ocean*, the volcanic activity and the subsidence phases during Cretaceous times correlate to the *succession of local volcanogenic, diagenetic, and finally distant-detrital inorganic contributions to the sedimentation*. Particular messages are expressed by the clay sedimentation in relation with geological crises. For instance, *the Cretaceous-Tertiary boundary appears from clay successions to have been globally subject to a global terrestrial disturbance associated with intense volcanism and tectonics*, rather than to an extraterrestrial impact. This is documented by the occurrence of widespread mineralogical changes, the nature of which varies according to the geological and sedimentological context. Countless applications are expected from similar investigations in the numerous sedimentary series whose siliciclastic components are devoid of appreciable post-depositional modifications.

Chapter 20

Clay and Geodynamics

20.1 Diverse Significance of Sedimentary Clay Assemblages

The previous chapters have emphasized the possibility for a given sedimentary clay mineral or clay mineral association to derive from very different origins. The awareness of such *diverse possible origins of clay minerals* has increased during the few last years, although some species are still often attributed to a single provenance. For instance, smectites, fibrous clays and associated zeolites of the clinoptilolite group are frequently referred to an authigenic volcanic or hydrogenous origin, although they may also be reworked from various environments (e.g., discussions by Kossowskaya et al. 1975, Macaire et al. 1977, Timofeev et al. 1978, Chamley 1979, Logvinenko et al. 1979, Dymond 1981, Nadeau and Reynolds 1981, Rosato and Kulm 1981, Chamley and Debrabant 1984a). This is the reason why, in order to clarify the genetic mode of a given mineral group in a sediment, it is often very useful to obtain data other than only X-ray diffraction results. Geochemical, isotope, microprobe and micromorphologic information is usually particularly helpful. The methodological approach, which is at present progressing, closely resembles those that were used a long time ago for minerals whose origins and mixtures are less complex than those of clay minerals (e.g., quartz, Eslinger et al. 1973).

Figures 20.1 to 20.3 schematically illustrate some of the more usual genetic paths existing for illites, smectites and palygorskites that occur in common marine sediments from recent and ancient ages. Various other cases have been mentioned in the previous chapters, for these minerals and other species. They all converge to *similar final products*, although the paleoenvironmental, synsedimentary or diagenetic conditions may have been fundamentally different. For instance, *illites* in a sedimentary basin may commonly derive, as well as chlorites, from the erosion and detrital supply of sloped and little-weathered outcrops of metamorphic or plutonic source-rocks, or from a diagenetic evolution related to a severe depth of burial (more than 3 km, the geothermal gradient being of a normal value of about 30°C/km). It is very difficult to differentiate the origin of both types of illite, except perhaps through isotope dating.

Smectites constitute a broad mineralogical group, whose different components may display very diverse origins. The chemistry and shape of individual particles, and some other geochemical characters (e.g., isotope composition) may help to recognize the precise origin of smectites, but such an aim is often complicated by the coexistence of mineral varieties that proceed from different sources.

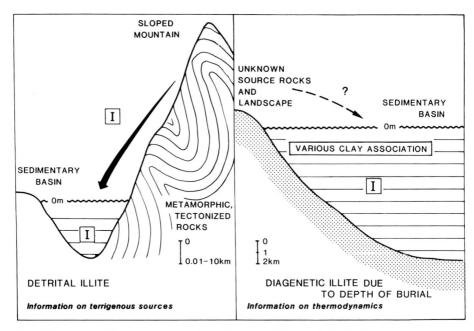

Fig. 20.1. Two possible opposite origins for sedimentary illites (and chlorites)

In addition, some smectites present a certain variability of chemical composition in a given genetic environment. It is therefore sometimes difficult to distinguish easily and accurately between the following possible origins (Fig. 20.2): reworking from soils developed under hydrolyzing conditions in poorly drained areas, or from past sediments that resulted from the reworking of such soils (mainly Al-Fe beidellites); reworking of subaerial alteration products from exposed volcanoes (Al to Mg or Fe smectites); submarine evolution and dispersal of volcanic products, through hydrothermal influence (often Mg-smectites) or not (often Fe-smectites); in situ precipitation, or reworking, of smectites from evaporative, more or less alkaline basins (diverse Mg-smectites originating in ancient lakes or peri-marine basins); preservation of smectites in some epimetamorphic rocks especially carbonate-rich types. One should add the possibility of reworking degradation smectites of various chemical compositions that have widely formed since the Miocene in soils and weathering profiles from temperate regions, as well as the occurrence in Pacific-like regions of hydrogenous Fe-smectites.

Palygorskite and sepiolite commonly derive from chemical precipitation in evaporative basins, as do some Mg-smectites (stevensite, saponite, etc.). Many studies, however, have shown that these minerals may also form under subarid conditions in calcareous pedogenic crusts, and that the fibrous clays may be easily reworked by wind or water from those subaerial, lacustrine or peri-marine environments. Fibrous clays may therefore be either authigenic or detrital in a given sedimentary environment (Fig. 20.3). In addition, some studies point to the likelihood of generating fibrous clays under hydrothermal conditions (Chap. 13).

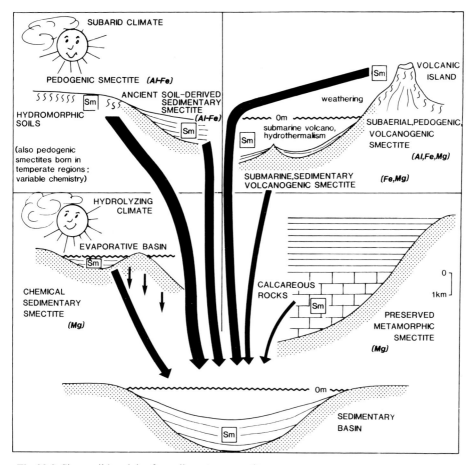

Fig. 20.2. Six possible origins for sedimentary smectites

The presence of short and broken fibers, and of dissociated bundles, often helps in the identification of detrital palygorskite and sepiolite, but the massive reworking of these minerals (e.g., by mass flows or turbidites) is often accompanied by the preservation of the initial shape, which complicates the recognition of the precise origin.

Let us consider in a few *examples* how the detailed knowledge of the clay mineral succession in a given area may serve to identify the respective part of the allochthonous and autochthonous controls, and therefore to determine what aspects of the geodynamic history are mainly responsible for the geologic evolution of this area.

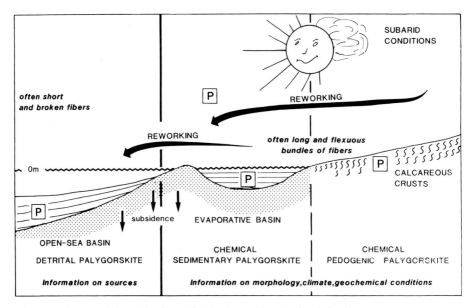

Fig. 20.3. Three possible origins for surficial palygorskites (and sepiolites)

20.2 Cenozoic Clay Sedimentation in the North Pacific Ocean

The problem of the origin of clay minerals in Pacific recent sediments is often difficult to solve, since *scattered terrigenous materials may be mixed with slowly formed authigenic products* (Chaps. 12, 13). The question becomes much more conflicting when considering the clay sedimentation during past geological times, potentially subject to global tectonic, climatic, diagenetic and other changes. Let us take some examples. As early as 1977, Steinberg et al. discussed the origin of clay minerals and feldspars in Cenozoic silica-rich sediments from the Northwestern Pacific Ocean, the Japan Sea, the California coast range and extra-Pacific regions. The rare earth element pattern led these authors to consider that most Japan Sea minerals were of a detrital origin, while those from the open Pacific Ocean and from California were authigenic. This assertion was mainly based on the existence of a negative anomaly in the cerium relative content, indicating a mineral genesis in the presence of sea water. But the rare earth measurements were mainly performed on the bulk sediments, whose major components could have been of a biogenic or hydrogenous origin (e.g., radiolaria, Fe-Mn oxides), and could not be representative of the origin of the less abundant clay mineral constituents (see Tlig 1982a).

Several U.S. investigations are devoted to the study of a few *giant cores* that have been sampled in order to plan the possible disposal of high-level nuclear waste, and whose sediments encompass the whole Cenozoic period. Corliss, Hollister et al. (1982) first summarize the data of core LL44-GPC3, sampled in the Central North Pacific Ocean at 30°N and 175°W (water depth 5705 m, reliable penetration 25 m. Fig. 20.4). Considering the migration path of the coring area in

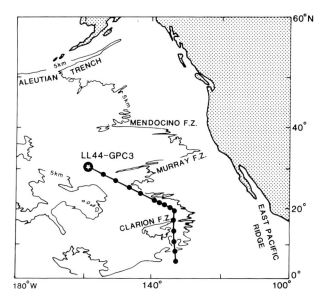

Fig. 20.4. Present location of giant core LL44-GPC3, with its paleo-backtrack path since the uppermost Cretaceous. (After Leinen 1987)

the Eastern Pacific Plate during the last 65 million years, the authors recognize two mineralogical periods. From 65 to 22 million years the sedimentation area was located South of about 20°N latitude, and was characterized by the deposition of smectite, phillipsite, feldspars, clinoptilolite and fish debris: this composition is assimilated to the authigenic assemblage found presently South of 20°N. From about 22 million years to the Present the sediments consist of quartz, illite, chlorite, kaolinite, mica, feldspars, smectite, cristobalite, fish debris and abundant manganese micronodules: such an assemblage is considered to be broadly influenced by detrital input, similarly to what is found presently North of 20°N. Using the paleolatitude positions, Corliss, Hollister et al. (1982) estimate that the transition from authigenic to detrital clay assemblages corresponds to a position of about 22 – 24°N. They therefore propose the existence of a *sedimentation model* during the Cenozoic which is simply *based on the north-northwestward migration of the Pacific Plate through different sedimentation provinces*, without temporary variations in the volcanic activity or terrigenous supply. The climatic change due to the northern hemisphere glaciation is considered to have little effect on the sedimentation area, other than to increase the sedimentation rate by a factor of two.

Kadko (1985) draws similar conclusions about the mineralogy and geochemistry of cores from three sites of the North Pacific, whose sediment age ranges over the last 20-30 million years. Similar major chemical variations are modeled by coupling the vertical and horizontal motions of the Pacific Plate with a *constant authigenic flux of elements into the sediments*. The elements with a substantial authigenic component display a marked inverse relation to the sedimentation rate, which depends on the Cenozoic motion of the Pacific Plate and climatic change. Superimposed over this flux is an *input of detrital material of relatively constant chemical composition*. Many assumptions of D. Kadko, like those of Corliss, Hollister et al. (1982) and of other authors (e.g., Aoki 1984, Schoonmaker et al. 1985), are based on the idea that almost all the smectite is authigenic and derives from volcanic, hydrogenous or diagenetic activity. The conclusions are therefore easy to draw, authigenic conditions favorable to the formation of smectite being replaced in the course of time by allogenic conditions marked by the illite group, quartz and kaolinite. Such an idea is nevertheless a priori somewhat questionable, since the relative abundance of smectite in some open-ocean areas studied by D. Kadko is lower than in regions located closer to Asia where the detrital supply is almost exclusive (e.g., Chamley et al. 1985b, Lenôtre et al. 1985).

Leinen (1987) carries out partitioning experiments on the origin of paleochemical signatures in North Pacific clays. Admitting that smectites may be significantly of detrital origin, which is confirmed by comparisons between the aerosol and sediment composition (Blank et al. 1985. 7.2.2), M. Leinen compares three techniques for partitioning the sediments chemical composition into source end-members. The core LL44-GPC3, previously studied by Corliss, Hollister et al. (1982), is especially investigated. All partitioning models indicate that *the downcore elemental distributions reflect the changing relative influence of detrital, hydrothermal and hydrogenous sediment components.* Two models suggest that

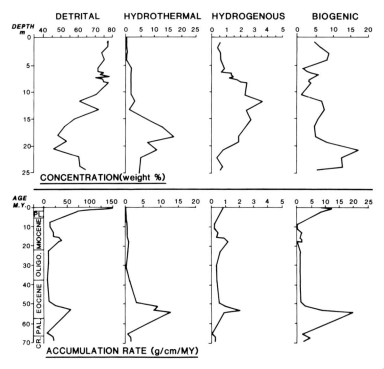

Fig. 20.5. Concentration (weight % versus depth) and accumulation rate (g/cm²/my versus age) of normative partitioning end-members in core LL44-GPC3. (After Leinen 1987, and M. Leinen, pers. commun. Note that accumulation rates are somewhat exaggerated)

biogenic sediment components also influence the sedimentation. The chemical composition of the end-members apparently did not change significantly over the last 70 million years of clay sedimentation. By contrast to previous authors, Leinen (1987) observes strong changes in the relative importance of the end-members, the detrital factor being the more important throughout the Cenozoic (45 to 80% in the normative partitioning model, Fig. 20.5).

Detrital components, which are largely aeolian, appear to have been very important between 70 and 50 million years ago, and also since the beginning of the Neogene. The influence of hydrothermal activity apparently increased between 60 and 50 million years ago, although the site was not near an actively spreading ridge at that time. This suggests that the changes in the relative importance of the chemical end-members downcore cannot be explained solely by changes in the position of the core site along its backtrack path, which contradicts previous conclusions. If the backtrack positions of the core were essential, the hydrothermal component would increase in relative importance toward the base of the core, when the sampling site was nearest the ridge crest, and there would be no increase in detritus at the bottom of the core. Hydrogenous components make up less than 5% of the sediment at site LL44-GPC3, and accumulated at a nearly constant rate through time. Biogenic components were relatively abundant when the site was located near the Equator, and accumulated at the highest rate 60-50 million years ago when the equatorial productivity zone was very wide. Leinen (1987) concludes that *important changes in sedimentation patterns took place in the North Pacific Ocean during the Cenozoic*, which fundamentally renews the previous concepts. The geochemical investigations performed through such studies would probably greatly benefit by the addition of mineralogic, micromorphologic and microprobe information. In the present state of research, they clearly show that the geodynamical evolution of the North Pacific Ocean filling depends on much more diverse causes than the lonely geographic migration of lithospheric plates.

20.3 Late Mesozoic Clay Sedimentation in the Eastern Atlantic and Western Tethyan Domains

20.3.1 Comparison Between Cape Verde and Senegal Basins

Introduction. Studies performed by Michaud and Flicoteaux (1987) on 11 commercial boreholes from the Senegal basin show the existence in Cretaceous sediments of diagenetic processes, that mainly result from fluid migration in permeable sandstones and from depth of burial. Such processes are not recognized in DSDP sites from the adjacent Atlantic Ocean (e.g., Chamley 1979, Chamley and Debrabant 1984a, Holtzapffel et al. 1985). It is therefore interesting to compare the two domains by identical techniques, in order to clarify the resemblances and differences, to appreciate the regional control and to define the geodynamical evolution of each area.

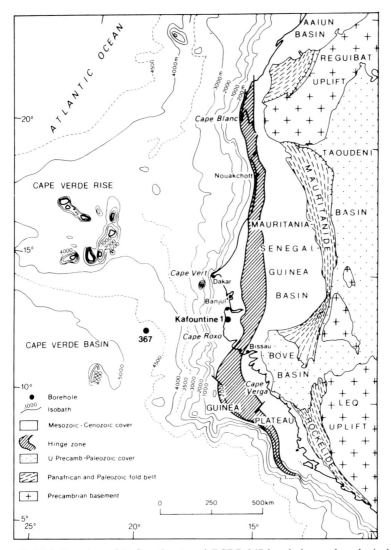

Fig. 20.6. Location of Kafountine 1 and DSDP 367 boreholes, and geological sketch. (After Chamley et al. 1988)

Let us compare the Senegal and the Cape Verde basins by an interdisciplinary approach of two drill sites, each of them being representative of one area. *Site Kafountine 1* was drilled by Elf-Aquitaine Oil Company in the Senegal basin, close to the present-day coastal line (12°55.34'N, 16°43.47'W; Fig. 20.6), to a depth of 5395 meters. *DSDP site 367* is located about 400 kilometers West of Kafountine in the Cape Verde basin, Eastern Tropical North Atlantic Ocean (12°20.2'N, 20°02.8'W; 4748 water depth, 1153 m penetration). About 100 samples of each site are studied by X-ray diffraction, with additional investigations by transmission electronmicroscopy and microprobe. Most data obtained concern the Cretaceous record, that is almost complete and comparable at both sites.

Data on site Kafountine 1. Kafountine 1 sediments are characterized by the predominance of quartz, clay minerals and calcite, the respective abundance of which changes through time. The Valanginian to early Cenomanian deposits, as well as early Cenozoic sediments, contain mainly calcite, while early-middle Cenomanian to Maastrichtian deposits are marked by a progressive increase of quartz and clay minerals. The latter mineralogical zone corresponds to a major progradation stage of the coastal basin of Senegal.

Four clay mineral zones are identified at Kafountine site from Valanginian to Oligocene (Fig. 20.7). The oldest deposits (*zone III*), from 5400 to 3900-3500 meters, contain abundant illite and chlorite, mainly consisting of large, transparent, well outlined but nonhexagonal particles. The illite and chlorite crystallinity is fairly good but much less than in typical anchizonal rocks (Kisch 1983). The heat resistance of the chlorite 14 Å peak suggests a dominantly magnesian composition or a good crystalline organization. Zone III sediments comprise some kaolinite ($<$ 5%) and noticeable amounts of irregular to subregular mixed-layers mainly composed of illite- and vermiculite-type layers. *Zone II*, that extends approximately from 3900-3500 to 2000 meters below the surface, ends in lowermost middle-Cenomanian marly limestones. It is marked by the appearance of smectite, the increase of kaolinite abundance, the decrease of magnesian, possibly ordered chlorite, and the progressive replacement of vermiculitic by smectitic mixed-layers. Electronmicroscope observations show the presence of both large and small well-shaped particles. The overlying *zone I* develops in late Cretaceous claystones and sandstones (2000-590 m), when prograding sediments deposited in the Senegal basin. A rapid increase of smectite and kaolinite abundance occurs between 2025 and 1975 meters, whereas illite and chlorite percentages decrease, as well as those of mixed-layers which mainly comprise illite-smectite and chlorite-smectite types. Well-crystallized smectite consists of flaky particles associated with fairly small illites, kaolinites, and sometimes rare palygorskite fibers. The abundance of smectite increases upwards. The Paleogene mineralogical *zone 0* extends from 590 to 138 meters depth in platform carbonates and associated facies. It is highly variable, and characterized by the temporary abundance of either flaky smectite, hexagonal kaolinite or palygorskite fibers.

Microprobe analyses of individual clay particles show moderate chemical changes with depth (Table 20.1). *Illite-like* particles have an alumino-magnesian composition and are relatively poor in potassium; the chemical nature of the more aluminous particles resembles that of regular illite-vermiculite mixed-layers (= rectorite), and occurs in zones III and II where vermiculitic transition terms are identified on X-ray diagrams. *Chlorites* present a large chemical diversity with the frequent co-occurrence of dioctahedral (Al^{3+}, Fe^{3+} in octahedra) and trioctaedral (Mg^{2+}, Fe^{2+}) types. The lack of systematic downhole change suggests the absence of progressive transformation with increasing depth of burial. Dioctahedral chlorites nevertheless occur more frequently in the lowest zone. *Smectites*, abundant in zone I, consist of aluminous beidellites. Their total deficit of charges decreases downwards (from 1.7 to 0.5 from 620 to 1640 m depth), at

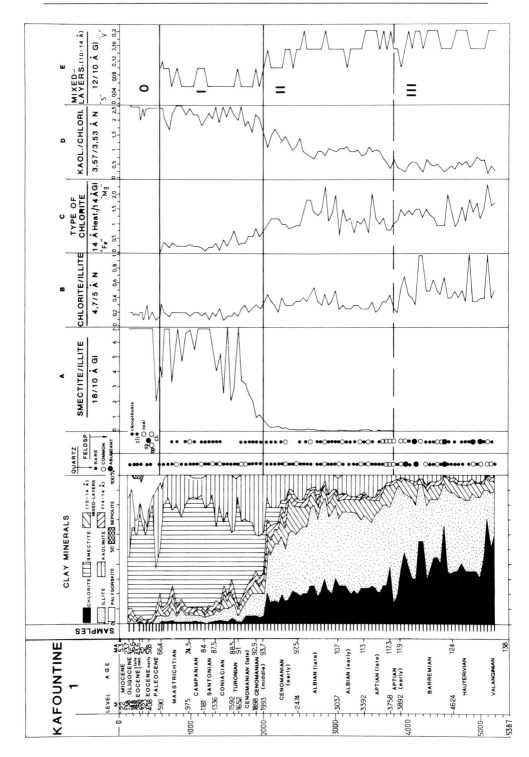

the same time as tetrahedral silicon and octahedral aluminum increase, and interlayer cations (Na, K, Ca) decrease. This evolution contradicts the possibility of a diagenetic transformation of smectite to illite at increasing depth of burial.

Data on DSDP Site 367. The Valanginian to Paleocene sedimentary column is about 750 meters thick, which corresponds to one seventh of the thickness at site Kafountine. The bulk mineralogy is characterized by the abundance of calcite in Neocomian to Aptian times, of clay minerals with variable amounts of quartz during the late Cretacous, and of clay, opal and clinoptilolite during the Paleogene.

The clay mineralogy at site 367 is marked by the abundance of well-crystallized, flaky, *Al-Fe smectite throughout the Cretaceous and Paleogene*, including the lowermost Neocomian deposits. The permanence of smectites at DSDP hole is obvious when comparing sites Kafountine and 367 by dividing the clay mineral record in successive zones marked by average values (Fig. 20.8). There is no continuous change related to the depth of burial, but a *sudden increase of certain minerals at specific periods*: illite, chlorite, irregular mixed-layers in the late Hauterivian-Barremian and during the late Cretaceous, the same species plus kaolinite and feldspars from late Albian to early Turonian, palygorskite at latest Cretaceous to Paleocene times, sepiolite and/or palygorskite, clinoptilolite and opal during the Eocene. Kaolinite is identified fairly late in the geologic record, in Hauterivian-Barremian time.

Microprobe data show the *absence of vertical composition change* at site 367, in both illite-rich and smectite-rich levels analyzed at different depths (Table 20.1). The composition of illite is fairly stable, with abundant Al, K and rather abundant Fe. Some Albian illite particles display a structural formula very similar to that of illites at site Kafountine. Smectites vary noticeably in composition at a given sedimentary level, from Si-rich to Fe-rich types; but all smectites belong to the Al-beidellite group, with noticeable amounts of Mg.

Diagenesis related to depth of burial. At site 367 no diagenetic effect related to the age or to the depth below the sea-floor arises from the clay mineral record. Similar conclusions arise from the study of most Atlantic DSDP sites (Chamley and Debrabant 1984a. 19.1, 19.2). The only evidence for a clay diagenesis consists in the local formation of lathed smectites (14.2), and does not depend on the depth of burial. The absence of burial effects is supported by the presence at any levels of clay minerals usually originating in surficial conditions (abundant smec-

---

Fig. 20.7. Clay mineralogy at site Kafountine 1. (After Chamley et al. 1988). Mixed-layers (10–14 Å): illite-smectite and illite-vermiculite; (14–14 Å): chlorite-smectite and chlorite-vermiculite. **A** Smectite relative abundance compared to illite: ratio of 18/10 Å peak heights on X-ray diffraction diagrams, glycolated samples. **B** Chlorite relative abundance compared to illite: 4.7/5.0 Å peak heights ratio, natural sample. **C** Indication of the average chemical types of chlorite: 14 Å peak heights of heated sample (490 °C, 2 h) over glycolated sample. The reinforcement of the 14 Å peak by heating is interpreted as an increase of the magnesian character of chlorite. **D** Kaolinite relative abundance compared to chlorite: 3.57/3.53 Å peak heights ratio, natural sample. **E** 10–14 Å mixed-layers average type: 12/10 Å peak heights ratio, glycolated sample. Higher values indicate more vermiculitic-like mixed-layers, lower values more smectitic ones

Table 20.1. Examples of structural formulas of clay minerals from Kafountine 1 and DSDP site 367 boreholes (microprobe data, after Chamley et al. 1988)

Borehole	Kafountine 1								DSDP 367			
Minerals	Chlorites				Illites		Smectites		Smectites			
Samples	2650 m		5202 m		2650 m	5202 m	620 m	1640 m	19-2-100 (late Tur.)	20-4-100 (early Tur.)	22-6-91 (late Alb.)	23-2-46 (Apt.-Alb.)
No. of particles	2	4	5	3	21	8	20	24	4	9	6	18
Type	Dioct.	Trioct.	Dioct.	Trioct.								
Octahedra												
Al	0.87	–	0.88	–	1.49	1.62	1.00	1.43	1.43	1.31	1.07	1.47
Fe^{3+}	0.93	–	0.85	–	0.22	0.22	0.50	0.42	0.21	0.41	0.69	0.38
Mg	–	1.89	–	1.28	0.43	0.22	0.30	0.16	0.41	0.30	0.32	0.18
Fe^{2+}	–	0.01	–	1,53								
Ti	0.01	0.01	0.01	0.01	–	0.01	0.02	0.03	0.01	0.02	0.02	0.02
Tetrahedra												
Al	0.71	0.85	0.51	1.15	0.88	1.07	0.88	0.48	0.10	0.50	0.69	0.43
Si	3.29	3.15	3.49	2.85	3.12	2.93	3.12	3.52	3.90	3.50	3.31	3.57
Brucitic layer												
Al	1.07	1.85	1.65	2.13								
Mg	1.75	0.65	0.78	0.38								
Interlayer												
Na	0.29	0.12	0.46	0.19	0.14	0.23	0.62	0.04	0.03	0.07	0.05	0.04
K	0.40	0.18	0.27	0.11	0.33	0.27	0.42	0.19	0.17	0.33	0.40	0.16
Mg	–	–	–	–	0.18	0.20	–	0.07	–	–	0.05	0.08
Ca	0.02	0.03	0.02	0.02	0.03	0.01	0.33	0.06	0.06	0.13	0.07	0.04

tite, fibrous clays, irregular mixed-layers), the independence existing between the rare earth element composition of clay and the volcanism or sea-water influence, and the absence of correspondence between the clay composition and the lithology or the occurrence of zeolites, opal, organic matter or volcanic debris.

In contrast, site Kafountine 1 appears to depend on clay diagenesis controlled by the depth of burial, in addition to local diagenetic changes due to fluid migration in permeable rocks (e.g., kaolinite increase in some sandstones. See Michaud and Flicoteaux 1987, and 16.3). The burial effects seem particularly active *below 2000 meters depth*, in mineralogical zones II and III (early Cenomanian to Valanginian; Fig. 20.7) whose clay assemblages comprise abundant illite and chlorite, and differ strongly from site 367 contemporeneous smectite-rich assemblages (fig. 20.8). The downhole relative increase of well-organized or magnesian chlorite (Fig. 20.7B), the more vermiculitic character of 10-14 Å mixed-layers that tend to shift nearer to illite (E), the decrease of kaolinite abundance (D) and the disappearance of smectite (A), simultaneously suggest significant diagenetic changes (see Kisch 1983, and Chap. 15). The downward increase of clay particle grain-size also agrees with the intervention of burial influence. The main cause of the diagenetic change lies in the thick sedimentary overburden caused by the rapid subsidence of the Senegal basin during early Cretaceous times.

Despite its probable existence, *the argillaceous diagenesis controlled by burial effects seems to be of moderate intensity at site Kafountine*. Such an assertion is supported by several observations: the crystallinity of illite displays moderate values and does not increase downwards, as shown by the width at half-height of the 10 Å peak on glycolated samples, that varies randomly between 4.5 and 6.0 mm, and even tends to average larger values in zone III than in zone II; apparent or true regular mixed-layers lack throughout the borehole; there are no particles whose chemical composition is obviously intermediate between those of smectite or kaolinite and those of illite or chlorite. Probably the lowest mineralogical zone III corresponds to a *late diagenetic environment located significantly above the anchizone* (see Kisch 1983, and 15.1). Such a deduction is consistent with a 5300 meter-thick lithostatic pressure under a moderate geothermal heat flow (about 3°C/100 m), such as that existing on the Western African margin. As a consequence, *the large amounts of illite and chlorite in the lower part of Kafountine 1 borehole cannot derive from burial effects alone*. Another cause probably intervenes.

Influence of tectonics. The mineralogical changes occurring in middle Cenomanian sediments near the bottom of mineralogical zone II do not easily fit the usual phenomena controlled by depth of burial. The downward disappearance of smectite (Fig. 20.7), and the correlative increase of the illite group, occur at a relative *shallow depth of burial* (about 2000 m) considering the local absence of high geothermal gradient (in Latil-Brun and Flicoteaux 1986). Generally a significant increase of the illite abundance does not occur below a sedimentary thickness of about 3000 meters. In addition, *the transition* from mineralogical zones I to II *is very sharp* (about 50 m thickness), in view of the lack of sedimentary hiatus and the high sedimentation rate characteristic of the Senegal

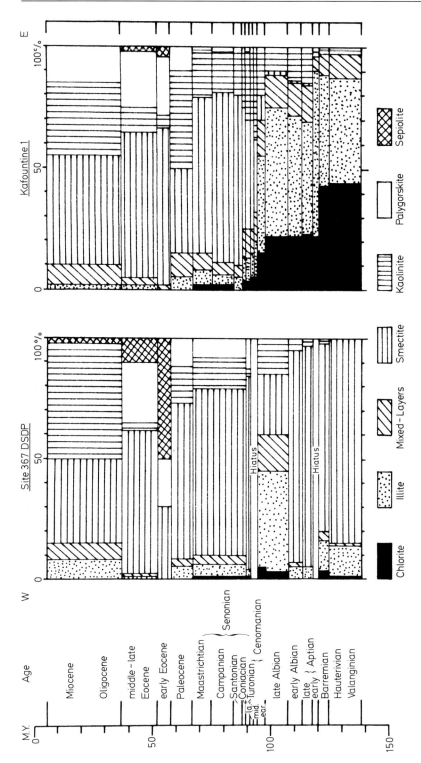

Fig. 20.8. Comparative clay mineralogy at sites DSDP 367 and Kafountine 1. Results are given as average values for successive time intervals. (After Chamley et al. 1988)

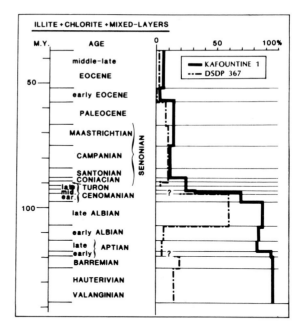

Fig. 20.9. Relative abundance of illite + chlorite + irregular mixed-layers at sites 367 and Kafountine, during the Valanginian to Eocene time interval. (After Chamley et al. 1988)

basin during the lower late Cretaceous (Latil-Brun and Flicoteaux 1986). The smectite abundance rapidly falls from 30 to 10% of the clay minerals from 2000 to 2050 meter depth, while chlorite plus illite abundance suddenly increases from 20 to 50%. More significant, *the strong illite-group increase* observed downwards at site Kafountine during the Cenomanian *is also recorded in the same period at the less buried, deep-sea site 367* (Fig. 20.9). As a consequence, the rapid passage from mineralogical zones I to II results from a *paleoenvironmental cause* rather than from a diagenetic impact.

The major mineralogical break recorded at sites Kafountine and 367, at the boundary between early and middle Cenomanian times at about 2000 meter-depth, correlates to a strong decrease of the subsidence rate and of the crustal thinning-rate on the West African continental margin (Latil-Brun and Flicoteaux 1986; see Chamley et al. 1988). The abundant supply before middle Cenomanian time of minerals directly eroded from crystalline substrates (illite, chlorite, irregular mixed-layers, small-sized quartz, and feldspars) therefore results from the *tectonic instability of the African margin and hinterland.* The instability is contemporaneous with the opening of the Equatorial Atlantic Ocean. Such a widespread tectonic activity is consistent with the increase of similar types of illite during the Albian at both coastal and deep-sea sites. It can also account for the relative abundance below 2000 meter depth of irregular vermiculitic mixed-layers, that may form through weathering in the lower parts of pedogenic profiles and cannot evolve toward smectite under active erosion processes due to tectonics (see Chap. 2). Note that at deep-sea site 367 the tectonic relaxation is followed by a sedimentary hiatus, perhaps brought about by strong bottom currents. *To summarize, the passage from mineralogical zones II to I at middle*

Cenomanian time results from the transition from a tectonically unstable to stable geodynamical context.

Transition between diagenetic and paleoenvironmental influences. The question of any significant diagenetic influence on the clay mineral record does not arise at DSDP site 367 and in the Cenozoic sediments of borehole Kafountine 1 (*zone O*), where clay assemblages typically express various surface conditions (see 19.1). *Zone I* also appears to be protected from burial effects, since the downward chemical changes observed in the smectite composition do not agree with a transformation into illite, and since similar compositions are recorded at both sites Kafountine and 367. The mineralogical *zone II* extends between 2000 and 3500-3900 meter depth, before the subsidence rate decreases. The upper boundary of this zone results from paleoenvironmental rather than from diagenetic causes (see above); but the progressive downward increase of illite and chlorite, as well as other clay mineral changes (Fig. 20.7), suggest that burial effects started within this zone. The paleoenvironmental signal is nevertheless only partly obliterated, especially as far as the climatic and morphologic information is concerned (Fig. 20.10). The comparison between sites 367 and Kafountine suggests that the maximum tectonic activity started in early to late Albian time (see 19.2), and ended at middle Cenomanian time when the subsidence rate decreased. Zone II therefore has probably undergone both synsedimentary and postsedimentary influences, and *represents a transition stage between paleoenvironmental and*

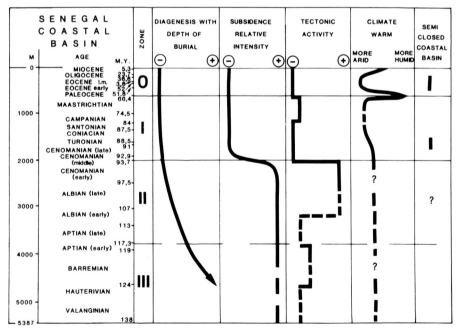

Fig. 20.10. Cretaceous to Paleogene sedimentological expression of diagenesis and paleoenvironment in the Senegal coastal basin (Kafountine borehole) based on comparisons with data from the Cape Verde deep-sea basin. (After Chamley et al. 1988)

diagenetic controls. Such deductions resemble those presented by Rettke (1981) for the mixture of provenance and diagenetic effects on the clay mineral record from shales and bentonites deposited during the Cretaceous in the northern Denver basin, Colorado (USA).

The mineralogical *zone III*, from 3500-3900 meters depth to the bottom of Kafountine 1 borehole (5395 m), *expresses more typically the diagenetic impact due to increasing lithostatic pressure and temperature,* as shown by the low amounts of residual kaolinite, and by the very large abundance of chlorite, illite and associated quartz and Na-feldspars. This conclusion is supported by the comparison with DSDP site 367, at which abundant Al-Fe smectites and less Al-Mg-Na-rich illites occur at the same time (Fig. 20.8). Nevertheless the rapid mineralogical transition recorded between zones II and III, as well as the existence of a similar change at site 367 (Barremian time), suggests that *the paleoenvironmental message is not totally destroyed at 3900-4500 meters depth,* and that a stage of tectonic instability is still slightly registered. A Barremian tectonic event is recognized on both eastern and western sides of the North Atlantic Ocean (Chamley and Debrabant 1984a), as well as in the Senegal basin. The expression of some paleoenvironmental events in Kafountine 1 borehole, at a depth of about 4000 meters below the sediment surface, is consistent with the previously suspected weakness of diagenetic effects.

20.3.2 Generalization to Atlantic and Tethyan Domains

The late Mesozoic clay sedimentation in the North Atlantic Ocean is fairly well known due to numerous studies done on deep-sea boreholes from the Deep Sea Drilling Project (Chap. 19). The paleoenvironmental signal is frequently well expressed, and diagenetic changes seem usually to be restricted to volcanogenic and evaporative sediments, or to isomineral recrystallizations that do not noticeably modify the original composition (Chaps. 9, 13, 14). The late Mesozoic clay sedimentation in the Western Tethyan region is still less investigated than in the Atlantic, and is mainly based on analyses performed on land sections. The diagenetic signal generally increases close to the mountain belts, when folding, metamorphism and burial determine postsedimentary changes.

A comparison of the late Jurassic to late Cretaceous clay assemblages from the Eastern North Atlantic and the French Subalpine ranges permits recognition of some general mechanisms (Fig. 20.11). *A few common characters* occur in both domains. They consist chiefly of an increasing content in smectites from Jurassic to Cretaceous times that could result from a more arid climate, and of a general depletion in chlorite that is probably due to strongly hydrolyzing conditions (see 17.4).

The *differences* between both domains appear to be fairly numerous and significant. By comparison to the Tethyan domain, *the Atlantic region is characterized by the following clay-mineral peculiarities*: (1) *Smectites occur more abundantly,* probably because of the evolved morphological stage of the bordering land-masses favoring the soil development, and of the frequent high distance

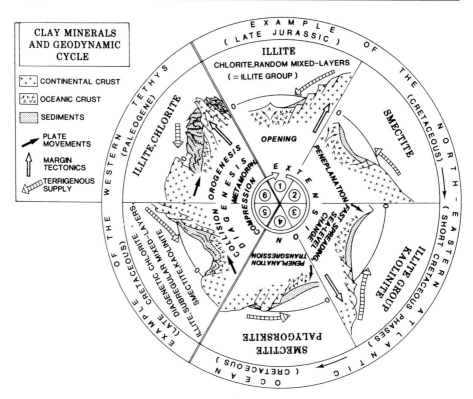

Fig. 20.11. Example of the evolution of clay mineral successions during a tectono-sedimentary cycle controlled by plate tectonics. (After Chamley and Deconinck 1985)

of source areas favoring differential settling processes. (2) *Illite and kaolinite occur less abundantly*, for the same reasons: the lack of sloped relief, and also the presence of abundant carbonate platforms during the early ocean stages, prevent the supply of these minerals to the deep sea, except during tectonically unstable periods. (3) *Palygorskite frequently occurs* as early as the late Cretaceous, while it is uncommon or only locally present in peri-alpine regions. This suggests that the formation of semi-closed basins is favored in the presence of subsiding areas such as those developed at the periphery of the young, distensive Atlantic Ocean. (4). *Clay assemblages are generally simpler and more homogeneous*, suggesting a more stable tectonic context. A mineralogical diversification develops temporarily or locally only, when tectonically unstable stages occur after the basin initiation or during spreading-acceleration periods, or due to local events.

   The differences observed in the clay stratigraphic record of the Atlantic and Tethyan regions appear to chiefly result from the youth of the latter domain and from the oldness of the former (Fig. 20.11. See also Chap. 19). After the rifting stages, the enlargement of the Atlantic Ocean in late Jurassic time is associated with the destabilization and uplift of the young adjacent margins, whose erosion products are rich in illite and chlorite but poor in kaolinite (*step 1*). *Extension stages* determine mainly a peneplanation of adjacent borderlands, on which smectites largely form in soils under the late Jurassic and Cretaceous hydrolyzing climate (*steps 2 and 4*). Palygorskites and/or sepiolites sometimes precipitate in peri-marine basins and are reworked toward the open ocean, especially during transgression periods accompanied by subsiding phases of ocean margins. Only spreading-acceleration stages or local tectonic instability phases can determine the noticeable reworking of minerals from substrates (illite group) and from upstream soils (kaolinite; *step 3*). After a long maturation history, the basin may be submitted to *compression effects*, like those affecting the Western Tethyan Ocean during the late Cretaceous and the Paleogene. Collision events tend to favor the reworking of various mineral species, both rocks and soils being eroded all together through uplift and first diagenetic changes (*step 5*). The further folding and faulting during the paroxystic phases of mountain building favor metamorphism, active erosion and deposition of illite and chlorite (*step 6*).

The clay mineral expression of the successive phases of a geodynamical cycle related to plate tectonics therefore starts and ends with the production and sedimentation of detrital illite and associated mineral species. Intermediate phases are characterized by *various clay assemblages* (Fig. 20.11), *the nature and succession of which help in understanding the development of the geodynamical cycle itself.* These preliminary concepts constitute bases for deciphering internal and external geodynamical events, from the sedimentary record considered on a fairly global scale.

20.4 Late Tertiary Clay Sedimentation in the Tyrrhenian Domain, Western Mediterranean Basin

20.4.1 Western Tyrrhenian Sea

The Tyrrhenian Sea opened within the Western Mediterranean Basin during the late Tertiary period. The history of the opening was particularly investigated during leg 107 of the Ocean Drilling Program (ODP. Kastens, Mascle et al. 1988-1989). Mineralogic, micromorphologic and geochemical studies have been performed on four ODP sites (Chamley et al. 1989), with special reference to the two following boreholes (Fig. 20.12).

– *Hole 652A*, drilled in the westernmost part of the Tyrrhenian Sea, on one of the easternmost tilted blocks at the Sardinia margin – Vavilov basin boundary (40°21.30'N, 12°08.59'E, water depth 3741 m, penetration 721.1 m). Late Miocene sediments occur from 188.2 meters below the sea floor to the bottom of the hole, and comprise more than half a kilometer of barren, sulfate-bearing, calcareous sand mud and mudstone, interbedded with minor evaporitic deposits. Rare paleomagnetic evidence suggests that the whole Pre-Pliocene series is Messinian in age.

– *Hole 654A*, drilled on a tilted block of the upper-eastern margin of Sardinia (40°34.76'N, 10°41.80'E, water depth 2218 m, penetration 483.4 m). Site 654 is located 120 kilometers West-Northwest of site 652, and includes Messinian to Tortonian deposits from 242.7 meters down to the bottom of the hole. Sediments

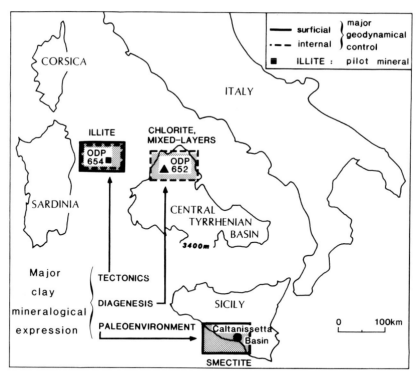

Fig. 20.12. Location of Leg 107 ODP sites 652 and 654, and of Caltanissetta basin. Major geodynamical influences on the uppermost Miocene clay sedimentation in the Central Mediterranean region. (After Chamley et al. 1989)

successively comprise gypsum interbedded with mudstone (down to 312.6 m), dark, laminated organic claystone to siltstone (down to 348.9 m), nannofossil ooze (348.9 – 403.9 m), polymictic sandstone and chalk (403.9 – 415.7 m), reddish mudstone overlying conglomerate, gravel and gravelly mudstone (415.7 – 483.4 m).

Four *parameters and ratios* are especially measured on X-ray diffraction diagrams:

– Illite crystallinity: width of the 10 Å peak at mid-height (mm), glycolated sample.

– Smectite/illite: 17/10 Å peak-height ratio, glycolated sample.

– Kaolinite/chlorite: 3.57/3.54 Å peak-height ratio, air-dried sample.

– Illite-vermiculite mixed-layers/illite: 12/10 Å height ratio above the background, glycolated sample.

The clay mineral results are summarized in Fig. 20.13, in which detailed data are schematically averaged out in seven stratigraphic zones. Late Miocene sediments comprise *diversified clay minerals*: chlorite, illite, various mixed-layers (10-14 Å = illite-smectite to illite-vermiculite types, 14-14 Å = chlorite-smectite to chlorite-vermiculite types), smectite, kaolinite, palygorskite. *At both sites illite, chlorite and smectite represent the most abundant species, smectite occurring*

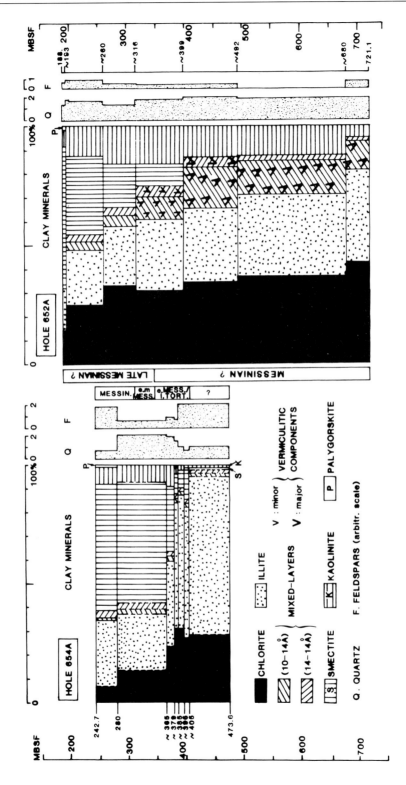

Fig. 20.13. Clay minerals schematic zonation of uppermost Miocene sediments at sites 652 and 654

particularly in the upper part of sedimentary columns. Kaolinite and mixed-layers are subordinate, palygorskite is present only in uppermost Messinian and in Pliocene deposits. Quartz is ubiquitous and often abundant in the clay fraction, feldspars and goethite occur more locally. *Important quantitative variations* are recorded, that appear to be synchronous at both sites and to allow stratigraphic correlations (Chamley et al. 1989). Despite these similarities, sites 652 and 654 reveal systematic mineralogical differences, that are discussed below (20.4.3). Let us first describe the main characters of site 652.

The Pre-Pliocene sediments at site 652 (Fig. 20.13) present some mineralogical characters that vary in a subcontinuous way downhole, or that are very rarely recorded elsewhere in the Mediterranean domain (e.g., Chamley et al. 1978b):

   – *Illite displays a very progressive downward augmentation of abundance, and parallels the increase of chlorite abundance*. Symmetrically *the smectite content tends to progressively decrease* downwards, until the mineral disappears at about 400 meters depth.

   – *Chlorite is much more abundant* throughout hole 652A than at site 654 and generally more than in most Messinian deposits. The chlorite amount equals or exceeds the illite abundance, which is quite unusual.

   – *The kaolinite abundance decreases downwards by two steps*, first close to 400 meters depth where smectite disappears, and then at about 655 meters. Kaolinite constitutes about 5% of the clay fraction in the lower part of hole 652A.

   – *The mixed-layer minerals occur in fairly high amounts* (up to 30% of the clay fraction). They are especially characterized by vermiculitic terms (illite-vermiculite and chlorite-vermiculite types) below 320 meters depth (Fig. 20.18). Regular chlorite-smectite mixed-layers (= corrensite) are identified from 480 to 485 meters, in blackish organic claystones where interstitial waters display the highest contents in the borehole of dissolved Ca, Mg, sulfates and chlorides.

   – *The downward distribution of mixed-layers correlates with the quantitative variations of smectite, illite and chlorite* (Figs. 20.13, 20.17, 20.18). From 188 to 316 meters depth, small amounts of random illite-smectite and chlorite-smectite occur, a very common fact in late Tertiary sediments of the Mediterranean range. Vermiculitic mixed-layers appear at 316 meters at the expense of smectite, but remain in lower amounts than smectitic mixed-layers. Vermiculitic terms predominate over smectitic terms below 400 meters, following the downward disappearance of smectite. Illite-vermiculite types predominate over chlorite-vermiculite terms. Chlorite-vermiculite mixed-layers strongly diminish in abundance below about 500 meters, a depth below which the relative abundance of chlorite increases. A slight decrease in the relative abundance of illite-vermiculite mixed-layers occurs below 680 meters, together with a second diminution of chlorite-vermiculite terms, which correlates with a slight augmentation of both illite and chlorite (from 35 to 38% and from 36 to 42% of the clay fraction, respectively). All these combined mineralogical variations suggest that the behaviors of the different clay species are interconnected at site 652.

Electronmicroscope observations show the downward transition from fleecy smectites to well-outlined illite and chlorite particles (see 20.4.3, Fig. 20.19). *Microprobe analyses* concern samples regularly distributed from the uppermost

Table 20.2. Examples of chemical composition of illites, chlorites and smectites in pre-Pliocene sediments of holes 652 A and 654 A (Western Tyrrhenian Sea), from microprobe analyses (after Chamley et al. 1989) N = number of particles analyzed. Core 20 = slightly below 200 m depth

Core-Section	N	SiO_2	Al_2O_3	MgO	Fe_2O_3	TiO_2	K_2O	Na_2O	CaO
Illites									
Site 652									
20-1	7	46.91	31.48	2.50	6.29	0.31	8.00	2.22	1.71
26-1	8	49.66	31.33	4.00	6.72	0.22	6.99	0.57	0.39
34-1	7	53.00	28.27	3.46	6.97	0.31	6.24	0.77	0.77
42-5	20	50.28	26.54	3.57	6.98	1.41	8.12	1.09	1.47
49-1	14	50.25	25.19	4.75	9.35	0.82	7.38	1.08	0.80
59-1	16	50.01	29.30	3.34	9.14	0.43	6.41	0.75	0.35
68-2	12	53.96	27.70	3.81	5.77	0.49	6.69	1.03	0.56
75-6	18	51.75	30.51	3.71	5.49	0.34	6.78	1.02	0.41
Site 654									
29-1	10	51.42	27.66	3.42	9.38	0.28	6.00	0.62	0.94
42-3	14	51.26	30.90	3.42	4.20	0.52	7.75	1.41	0.56
43-4	20	51.75	31.23	3.13	4.50	0.25	7.67	1.12	0.37
47-3	14	47.89	33.71	2.33	4.40	0.84	9.15	1.53	0.26
52-3	24	46.78	33.31	2.37	7.20	0.75	7.38	1.88	0.21
Chlorites									
Site 652									
20-1	2	27.73	19.92	12.99	32.49	0.13	2.41	1.57	1.79
26-1	2	35.23	21.97	11.93	28.15	0.29	1.05	0.51	0.67
34-1	2	32.37	21.14	8.95	31.94	2.30	1.79	0.54	0.89
42-5	7	39.54	22.94	6.11	23.52	0.84	4.66	0.63	0.79
49-1	5	37.92	21.74	8.00	24.67	1.29	4.18	1.03	0.81
59-1	5	37.93	22.12	11.83	24.32	0.52	2.30	0.44	0.24
68-2	6	32.06	23.56	10.59	31.01	0.39	1.42	0.69	0.30
Site 654									
42-3	7	38.58	23.81	8.82	23.01	0.21	3.26	1.31	0.97
47-3	9	34.01	25.64	9.42	25.79	0.88	2.70	1.07	0.32
52-3	4	27.27	21.73	14.84	33.46	0.53	0.46	1.19	0.28
Smectites									
Site 652									
20-1	11	49.62	19.34	5.06	13.13	0.32	4.98	3.51	3.05
26-1	8	54.37	25.20	4.23	8.85	0.23	4.31	1.18	1.09
34-1	23	57.99	21.49	4.13	7.94	0.72	5.13	1.08	1.16
43-4	8	49.06	28.25	5.39	9.61	0.30	5.46	1.22	0.72
Site 654									
29-1	26	53.13	21.22	4.11	13.15	0.60	4.74	0.86	1.74
42-3	11	52.48	24.19	4.85	9.93	0.78	4.86	1.79	1.10

Messinian (core 20) to the bottom of hole 652A (Table 20.2). *Smectites* are of a common Al-Fe beidellite type, and do not indicate any volcanic or alkaline-evaporitic influence. They tend to disappear below 320 meters depth when vermiculitic mixed-layers appear on X-ray diffraction diagrams, and to be progressively replaced by particles whose chemical formulas are intermediate between those of smectites, vermiculites and illites. *Illites* show an augmentation of Mg and Fe and a decrease of the global Al from 186 to 485 meters depth. The total charge of illite particles diminishes from about 400 to 720 meters depth, while Mg progressively migrates from octahedral to interlayered positions. Some micaceous particles analyzed in the lower part of the hole present fairly abundant Al and K, and correlate with the presence of true muscovite and of subregular illite-smectite mixed-layers. *Chlorites* display a very broad diversity of chemical composition. They comprise both dioctahedral and trioctahedral types, and present various substitutions and layer charges. The presence at 485 meters depth of corrensite correlates with highly magnesian chloritic particles.

The mineralogy and geochemistry of Pre-Pliocene sediments at hole 652A typically suggest the *existence of a diagenetic influence linked to the depth of burial*. The downward disappearance of smectite followed by decreases in kaolinite abundance, the correlative development of mixed-layered-like minerals that become less expandable at depth, the parellel increase of illite and chlorite abundance, the chemical stabilization of some clay minerals and the global grain size increase, altogether suggest the occurrence of thermodynamic effects (see Chap. 15). Notice the existence of some convergence with what happens at depth in some volcano-hydrothermal environments where smectites also tend to be replaced by chlorite in the clay fraction (e.g., Kawahata and Furuta 1985. See 16.5). The steps of the "mixed-layers" evolution seem to express particularly well the successive stages of mineral transformations at site 652. The appearance of vermiculitic terms at about 316 meters depth, coeval with the decrease of smectite abundance, suggests the transition from expandable to nonexpandable minerals. The decrease at 500 meters depth of chlorite-vermiculite types, correlative to the increase of the chlorite abundance, suggests the transition from random mixed-layers to single chlorite particles, the latter mineral being stable at increasing temperature and pressure (e.g., Kisch 1983). The same reasoning applies at about 680 meters depth to the transition from illite-vermiculite to illite.

20.4.2 Comparison with the Senegal Basin

The existence of burial effects at ODP site 652 is quite unexpected, since the total thickness of the borehole is 721 meters and the sediments are young (less than 6 to 7 my). The unforeseen post-sedimentary change at site 652 appears still much extraordinary when compared to the evolution recorded at site Kafountine 1 in the Senegal basin (see 20.3.1). By plotting the results at similar scales, *both sites 652 and Kafountine show very similar patterns* (Fig. 20.14): a downward decrease and disappearance of smectite balanced by a chlorite and illite increase, the appearance close to the middle of the series of vermiculitic mixed-layers, and a downhole decrease of kaolinite abundance. The mineralogical parallelism between the two sites is confirmed by specific measurements on X-ray diagrams, such as the sum of chlorite plus mixed-layers or the kaolinite/chlorite ratio (Fig. 20.15). Surprisingly, *a late Cenozoic sedimentary column deposited in the Tyrrhenian Sea and whose thickness does not exceed 721 meters closely resembles a series deposited in the Senegal basin since the early Cretaceous over a thickness of 5395 meters.*

CENTRAL TYRRHENIAN SEA   WESTERN SENEGAL

SITE 652 ODP   KAFOUNTINE 1 BOREHOLE

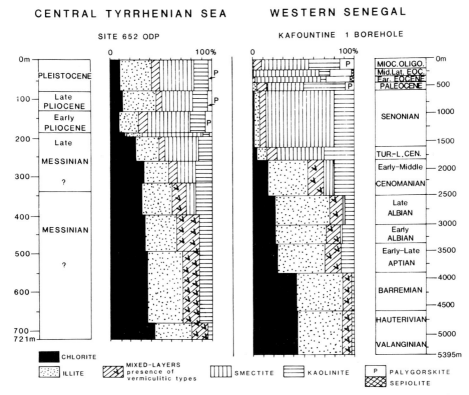

Fig. 20.14. Clay minerals schematic zonation at site 652 A, Tyrrhenian Sea, and site Kafountine 1, Senegal Basin

Both sedimentary columns look like if they had been *influenced by the depth of burial*. Measurements of the illite crystallinity suggest that the two series belong to the late diagenetic zone, and are located significantly above the upper limit or the anchizone: the width at half-height of the 10 Å peak is fairly large (2.5 to 7 mm), and presents larger values in the lower part than in the upper part of the holes. This is compatible with the preservation of some paleoenvironmental messages, such as the tectonic expression of the decreasing subsidence along time at site Kafountine 1 (20.3.1, and Fig. 20.14 and 20.15. See the strong increase of smectite amount during Cenomanian time). Similar evidence may be expected at ODP site 652 (see 20.4.4).

The reality of geothermal effects at site 652 is supported by studies on the organic matter, which stress the fact that the Messinian sequence is thermally mature compared to series from surrounding sites (Emeis et al. 1989). Clay changes controlled by burial effects usually do not occur in sedimentary series thinner than 2 to 3 kilometers (see Chap. 15). The case of site 652, and its resemblance to site Kafountine 1, becomes clarified when considering the geodynamical context and the local geothermal flux. Present-day measurements done aboard the drilling vessel Joides Resolution point to very high heat flow at site

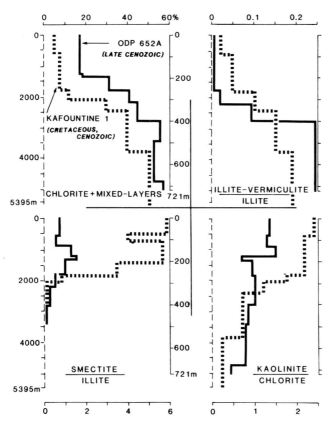

Fig. 20.15. Comparative distribution of some clay minerals and clay mineral ratios at site 652, Tyrrhenian Sea, and site Kafountine 1, Senegal Basin

652 (160 mW/m²), with a *vertical geothermal gradient of about 14°C/100 meters.* This value is five times higher than the geothermal gradient in the region of borehole Kafountine 1 (about 2.8°C/100 m), whose sedimentary thickness is about seven times larger than hole 652A in thickness. Assuming that the heat flow during the early formation of the Western Tyrrhenian Sea in late Miocene times was probably higher than today, it is likely that the difference in thickness between both series is balanced by the influence of different geothermal gradients. Note that the continental crust is particularly thin in the site 652 area, which is located near the oceanic crust of Vavilov basin in the Western Tyrrhenian Sea. Such a situation locally favors strong thermal effects.

To summarize, the Tyrrhenian Sea constitutes a distensive back-arc basin in the general compressive context typical of the Mediterranean domain. The thinning of the continental crust locally favors high heat flows and rapid diagenetic influences linked to burial effects. The final result reached after about six million years resembles what happens in a distensive region like the Senegal basin, where the geothermal gradient is low but compensated by an important depth of burial (about 5 km) and a very large time duration (more than 120 million years).

20.4.3 Regional Comparisons

Clay Sedimentation on the Sardinian Margin
Despite its relative closeness to site 652 (120 km) and some rough similarities (Fig. 20.13) the site 654, that is located more westward on the upper margin of Sardinia (Fig. 20.12), displays noticeable differences in clay mineralogic record. When looking at the average mineral content at both sites, hole 654A sediments comprise more abundant illite, smectite and feldspars, and less abundant chlorite, 10-14 Å and 14-14 Å mixed-layers, kaolinite and quartz (Fig. 20.16).

Detailed measurements on X-ray diagrams indicate the existence, in late Tortonian-early Messinian sediments of *hole 654A*, of a *very rapid upward decrease in the illite content, balanced by a strong increase in smectite abundance* (Fig. 20.17). This major break, associated with rapid but slight changes in chlorite and kaolinite amounts, closely resembles what occurs at site Kafountine in middle Cenomanian time (Figs. 20.7, 20.9). As at Kafountine, abundant, well-outlined but nonpolygonal, fresh and more or less broken particles of illite with some additional chlorite (Fig. 20.19B, C), are quickly replaced by fleecy smectites (Fig. 20.19A). Microprobe data prove the subaerial weathering of illitic minerals in the upper part of the section, above 300 meters depth, in uppermost Miocene sediments marked by abundant smectite (Table 20.2). Micas of a composition close to muscovite become less aluminous and more siliceous and magnesian, and typical Al-micas are nearly lacking in late Messinian Al-Fe beidellite-rich deposits. Chlorite comprises both dioctahedral (Al) and trioctahedral (Fe^{2+}, Mg) types in the lower part of the hole, while the former tends to decrease upwards. All these mineralogic and geochemical changes occur independently of the variations recorded in the lithology or of the transition from continental to marine depositional environment.

As at site Kafountine during late Cretaceous time, the mineralogic, micro-morphologic and microchemical characters at ODP site 654 suggest the *rapid transition* in late Miocene sediments, from 380 to 365 meters depth, *of a mineral assemblage directly derived from the active erosion of fresh crystalline rocks to a*

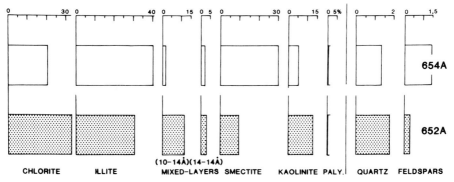

Fig. 20.16. Average contents of clay minerals (%) and of quartz and feldspars (arbitrary scale) in the <2 µm fraction of uppermost Miocene sediments, ODP sites 652 and 654. (After Chamley et al. 1989)

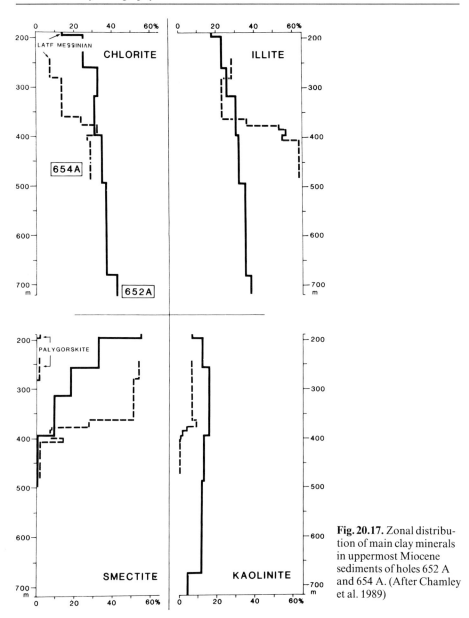

Fig. 20.17. Zonal distribution of main clay minerals in uppermost Miocene sediments of holes 652 A and 654 A. (After Chamley et al. 1989)

mineral suite largely derived from various surficial soils formed at the expense of such rocks. This suggests the *successive occurrence, during late Tortonian-early Messinian times, of tectonic instability, tectonic relaxation and continental weathering stages* in the eastern part of Sardinia island. Such an interpretation agrees with the initial formation steps of the Western Tyrrhenian Sea, whose margins are temporarily destabilized, and correlates with the passage from coarse shallow-water sediments to open-sea hemipelagic deposits (Kastens, Mascle et al. 1988-1989).

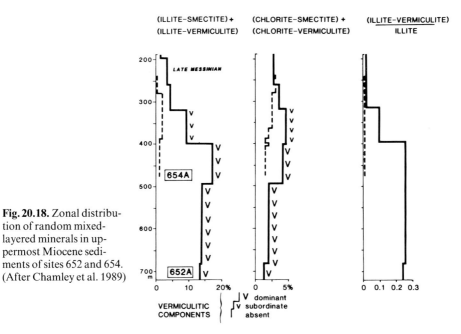

Fig. 20.18. Zonal distribution of random mixed-layered minerals in uppermost Miocene sediments of sites 652 and 654. (After Chamley et al. 1989)

The rapidity of the mineralogical change recorded at site 654 is inconsistent with control by depth of burial. This obviously differs from what is recorded at nearby site 652 (20.4.2). The absence of burial influence at site 654 is confirmed by the lack of nonexpandable mixed-layers of vermiculitic types (Fig. 20.18), by the existence of a typically moderate geothermal gradient on the upper margin of Sardinia ($3.2°C/100$ m), and by the occurrence of a thick underlying continental crust devoid of abnormal heat fluxes (Kastens, Mascle et al. 1988-1989).

To summarize, site 654 area, located on the upper margin of Sardinia, was submitted during late Miocene time to changes in tectonic activity, that are particularly well expressed by rapid variations in the illite content (Fig. 20.20). *By contrast, the nearby site 652 area, located more eastward in the Western Tyrrhenian basin, was submitted at the same time to the influence of thermodynamic diagenesis because of a strong geothermal gradient, which is exactly expressed by the abundance and nature of mixed-layer minerals and chlorite.*

Comparison with Sicily

As in many other areas of the Mediterranean range (9.3.2), the clay sedimentation in Caltanissetta basin, South Sicily (Fig. 20.12), is marked in late Miocene times by large amounts of smectite, whose type is very similar to the flaky particles deposited in the uppermost Messinian period at sites 652 and 654 (Figs. 20.19A and D). After the Tortonian time, the rapid desiccation of the Mediterranean Sea appears to have favored the regional development of subarid conditions and the pedogenic formation of large amounts of Fe-Al smectites (17.3.3). *In Sicily, the permanence of abundant smectite in late Miocene sediments therefore seems to reflect paleoenvironmental conditions predominantly controlled by climate.*

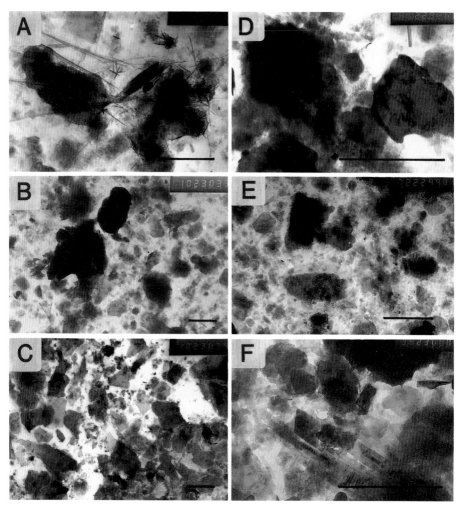

Fig. 20.19. Selected electron micrographs from ODP sites 652 and 654 (bar = 1 μm). **A, B, C** Site 654, foot of Sardinia margin. **A** Late Messinian, core 29-1-12, about 262 m depth. Abundant flaky smectite, associated with well-shaped illite, small hexagonal kaolinite, palygorskite fibers and goethite rosettes. **B** Early Messinian, core 40-4-18, about 352 m depth. Mixture of well-shaped illite and chlorite, and of flaky smectite. Small kaolinite hexagons. **C** Tortonian (?), core 52-1-33, about 464 m depth. Very abundant well-shaped, nonpolygonal, variously sized, and fresh mica-illite and chlorite particles. Few and small fleecy smectites and random mixed-layers. **D, E, F** Site 652, Western Tyrrhenian Basin. **D** Uppermost Messinian, core 22-1-81, about 200 m depth. Fleecy smectite and well-outlined, nonpolygonal, illite and chlorite. **E** Messinian (?), core 44-2-44, about 414 m depth. Mixture of rather large, well-outlined, subhexagonal chlorite and illite particles, and of small, well-shaped particles of subregular vermiculitic mixed-layers. **F** Messinian to Tortonian (?), core 68-2-55, about 646 m depth. Very abundant fresh hexagonal crystals of chlorite (40%) and mica-illite (35%), associated with little kaolinite and illite-smectite mixed-layers. Notice the presence of elongated mica and/or chlorite particles

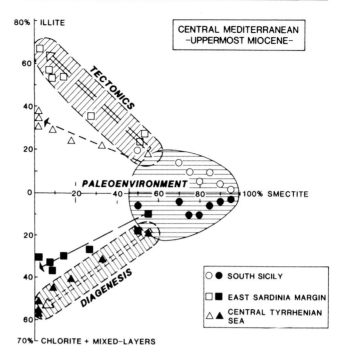

Fig. 20.20. Expression of tectonics, paleoenvironment (mainly climate) and diagenesis by characteristic clay minerals in the Central Mediterranean Region during the uppermost Miocene

In the Central Mediterranean region during late Miocene time, major clay minerals bear very different messages according to location (Figs. 20.20, 20.12). While abundant smectite reflects the surficial geodynamical control of climate in South Sicily, abundant illite occurring to the North expresses the tectonic instability of the Eastern Sardinian margin submitted to the internal geodynamic control of the Tyrrhenian Sea opening. At the same time, another aspect of internal geodynamics, consisting of diagenesis related to high geothermal flux, characterizes the western part of the Tyrrhenian basin above thinned continental crust, and is particularly well expressed by mixed-layers and chlorite (Fig. 20.21).

To summarize, the careful observation of clay mineral changes in an unstable region, like the Western Mediterranean basin during late Miocene time, allows identification of the different types of surficial and internal geodynamical controls at various places, and clarification of the succession of geodynamical events occurring in the course of the time. In the study area , the late Tertiary world-cooling recognized in Sicily during Tortonian time gives place to the regional climatic aridification characteristic of Messinian time, before the refilling of the Mediterranean Sea in early Piocene time brings the region closer again to worldwide conditions. This surficial geodynamical evolution is almost completely preserved in sectors almost devoid of internal geodynamical influence, such as the Caltanissetta basin. On the Sardinia margin the tectonic activity obliterates the climatic and pedogenic signal, except in uppermost Messinian and in Pliocene time after

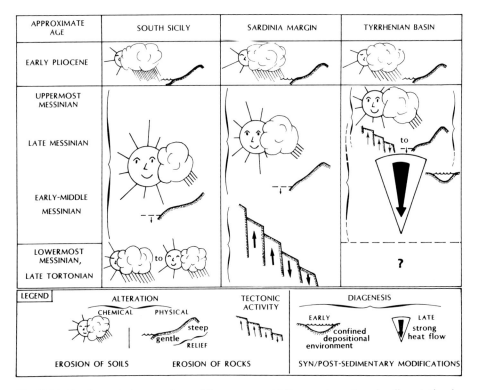

Fig. 20.21. Geodynamical expressions of the uppermost Miocene clay-mineral sedimentation in the Central Mediterranean Region

the tectonic relaxation has taken place. In the nearby Tyrrhenian basin, the first late Miocene stages are not recorded because of lack of sedimentation; most Messinian sediments are dominated by diagenetic effects, the paleoenvironmental expression being recorded only during the last phases of the Tertiary sedimentation.

Significance of Chloritic Minerals in the Tyrrhenian Sea

The abundance of chlorite in the Western Tyrrhenian basin (site 652) *is higher than* at site 654 *on the Sardinia* margin, at site Kafountine in the Senegal basin (20.3.1), as well as in many other regions suject to burial effects (see Chap. 15). Most argillaceous sediments submitted to strong burial effects contain more mica-illite than chlorite (Kisch 1983, Singer and Müller 1983), contrary to what occurs at site 652. Sedimentary series with abundant chlorite in the clay fraction are rather rare (see Millot 1964, 1970), which emphasizes the originality of the site-652 clay fraction. In addition, chlorite is abundant almost throughout hole 652A (25 to 45% of the clay fraction), even in the upper part of the sedimentary column where the mineral is associated with smectite (from 200 to 320 meters depth). Furthermore chloritic minerals at site 652 comprise both *euhedral, hexagonal, sometimes elongated crystals of chlorite in the lower part of the hole* (Fig. 20.19E, F), and regular chlorite-smectite mixed-layers (= *corrensite*) at 480-485 meters depth in a saline-alkaline environment. The presence of hexagonal chlorite is unusual in buried sediments (e.g., site Kafountine, and Chap. 15), except near some evaporative environments such as the early Mesozoic deposits of Mazagan Plateau, Northeast Atlantic margin (Chamley and Debrabant 1984c).

Secondary chlorites and corrensites are sometimes attributed to synsedimentary growth in alkaline-evaporative sediments, such as the Triassic salt-bearing series of Western Europe and North Africa (e.g., Lucas 1962). It is nevertheless intriguing that more recent sediments, such as the common evaporite-bearing deposits of Messinian age that are protected from burial effects, lack abundant chlorite and corrensite (e.g., Chamley et al. 1978b, Chamley and Robert 1980). Less recent evaporite-rich series of early Paleogene age, that did not undergo noticeable burial effects, are generally devoid of abundant amounts of these minerals (e.g., Millot 1964, 1970, Trauth 1977); evaporative sediments from this period frequently favor the authigenic growth of saponite, stevensite, palygorskite, sepiolite and other Mg-rich clay minerals, but not of chlorite or corrensite (see Chap. 9).

These data set up *the question of the real possibility for recently deposited sediments to precipitate chlorite and corrensite in surficial conditions*, as sometimes envisaged (e.g., Lucas 1962, Esquevin and Kulbicki 1963, Kubler 1973). Chlorite and corrensite are often located in central, more buried parts of evaporative basins (in Millot 1964, 1970). Some investigations, such as those conducted by Bodine (1978, 1987) and Bodine and Madsen (1987) on late Silurian, Pennsylvanian and Permian saline sediments of North America, indicate strong relationships between the presence of corrensite, talc or clinochlore, and moderately diagenetic recrystallizations. Bouquillon et al. (1985) also describe the occurrence of regular mixed-layers within buried evaporitic sediments that were submitted to noticeable thermodynamic influences, in Carboniferous sediments from Northern France.

Messinian sediments at site 652 clearly show their dependence on a geothermal control (20.4.2). As Messinian sediments devoid of such a control do not contain especially abundant *chloritic minerals*, it is therefore possible that these minerals *result from the combined influence of an alkaline-evaporative sedimentary environment and of a noticeable diagenetic impact due to depth of burial or to geothermal flux*. This could be taken in account to explain why corrensite and chlorite frequently occur in Triassic evaporative sediments, which have frequently experienced diagenetic effects related to burial or tectonics.

Whatever the origin of Triassic clay minerals, the preferential development of chlorite instead of illite at site 652 appears to result from the conjunction of thermodynamic effects and of an evaporative environment, since magnesium is more easily available at depth in alkaline sediments than potassium. The growth of euhedral, hexagonal chlorite crystals could be favored by the fairly low compaction of young Messinian deposits, as is the case with other euhedral clay minerals in pore spaces of Salton Sea shales, California, during the Plio-Quaternary (Yau et al. 1987b). The local formation of corrensite at 480-485 meters depth in hole 652A possibly results from the combination of suitable thermal and chemical conditions, in a locally ion-rich environment.

To summarize, multidisciplinary investigations on the clay stratigraphy in unstable regions like the Western Mediterranean basin during late Tertiary times reveal various evolution modalities of the surficial and internal geodynamical factors. Modeling such regions through plate tectonics or global chemical exchange concepts (e.g., Siever 1979, Berner et al. 1983, Sheridan 1987) represents a difficult aim. *Combining clay mineral studies to other usual approaches* may constitute a useful tool to progress in such modeling and geodynamical reconstructions.

20.5 Conclusion

1. Recent studies clearly show the *diversity of the possible causes responsible for the presence of a given clay mineral in a given sedimentary environment*. This is obvious for common minerals such as illite, chlorite or smectite, but also applies to less frequent species like palygorskite or sepiolite. Some clay minerals, such as smectites for example, may theoretically present six or more distinct origins in certain sediments, and two or three synchronous complementary provenances are not uncommon in a sedimentary area. Such a potential diversity of origins invites great cautiousness in paleoenvironmental and diagenetic interpretation, and leads to developing multidisciplinary approaches in order to better characterize the mineral assemblages and their significance.

2. *The clay sedimentation in the North Pacific Ocean* during Cenozoic times illustrates the variety of ideas published in the literature on the origin of deep-sea red clay-mineral constituents. Mineralogical, chemical, experimental and theoretical investigations frequently lead to conflicting interpretations. Many approaches, including some presented recently, are based on the idea that smectite systematically derives from an autochthonous origin through volcanism, hydrothermalism, hydrogenous activity or early diagenesis. As a result, frequent interpretations are presented on the terrigenous origin of sediments close to continents opposed to the authigenic formation of open-ocean deposits, or on the successive volcanogenic and then terrigenous provenance of sediments rich first in smectite and then in illite. Some recent investigations demonstrate *the at least partial aeolian origin of deep-sea smectites, and the complex succession of terrigenous, hydrothermal and hydrogenous influences on Cenozoic sedimentation*. Great progress is expected from combined investigations on the deep-sea clay stratigraphy.

3. *The comparison of the late Mesozoic sedimentation in both coastal and deep-sea basins from the Eastern Atlantic Ocean shows how clay stratigraphy may help to understand the successive causes and mechanisms occurring in each area*. For instance, data on Cretaceous sediments in both types of basins demonstrate that surficial (climate, topography, sources) and internal (subsidence, tectonics) geodynamical factors may be expressed in sedimentary series thicker than 3 kilometers, and that various transitional situations may occur where paleoenvironmental messages are superimposed to the diagenesis controlled by depth of burial. Numerous applications may be envisaged from similar studies in other

regions. The generalization of clay-stratigraphy studies in large areas such as the Eastern Atlantic and the Western Tethyan Regions leads to the consideration that the successive phases of the large-scale geodynamical history of a sedimentary basin are punctuated with specific successions of clay-mineral assemblages. *Clay-mineral cycles appear to express the successive steps of the geodynamical cycles in relation to the plate tectonics history*, from the birth of an ocean, marked by the tectonic instability of margins, to its death after diverse processes of collision, mountain-building and metamorphism.

4. *The Western Mediterranean Basin was especially mobile and structurally unstable during the late Tertiary*. Studies on such regions underline how questionable global modeling experiments sometimes are. Detailed clay-mineral studies may constitute a useful tool to better understand geodynamical factors and their chronology. For instance, the comparative clay-mineral study of different sectors from the Tyrrhenian Sea, that opened before the Pliocene in the Mediterranean compressive context, reveals very different situations: crustal thinning and high geothermal gradient in the Western Tyrrhenian basin, determining in young and slightly buried sediments an *accelerated geothermal diagenesis* comparable to what happened in some stable margin basins as thick as 5 kilometers and as old as 120 million years; *successive tectonic destabilization and relaxation* of Eastern Sardinian margins after the initiation of the Tyrrhenian basin; absence of any noticeable internal geodynamical control in basins in South Sicily, allowing the exclusive *expression of successive climates and weathering processes* in exposed sectors. The geothermal effects on clay diagenesis linked to crustal thinning are locally associated with chemical exchanges related to Messinian evaporative conditions, which apparently contributes to the *genesis of euhedral chlorite- and corrensite-rich sediments*. Such mosaic-like geologic regions constitute particularly relevant areas to demonstrate how *the clay filling of sedimentary basins may help to distinguish past allochthonous, autochthonous and diagenetic processes in the sediment formation, and to understand the successive surficial and internal geodynamical controls*.

References

Abed A.M., 1979. – Lower Jurassic lateritic redbeds from central Arabia. Sediment. Geol., 24:149–156.

Académie des Sciences, Paris, Ed., 1984. – Les nodules polymétalliques. Faut-il exploiter les mines océaniques? Gauthier-Villars, Paris:177 pp.

Accarie H., 1987. – Dynamique sédimentaire et structurale au passage plate-forme/bassin. Les faciès crétacés-tertiaires du massif de la Maiella (Abruzzes, Italie). Thèse, Sci. nat., Ecole nat. sup. Mines Paris:162pp. + ann.

Adams C.G., Benson R.H., Kidd R.B., Ryan W.B.F., Wright R.C., 1977. – The Messinian salinity crisis and evidence of late Miocene eustatic changes in the world ocean. Nature (London), 269:383–386.

Adatte T., Rumley G., 1984. – Microfaciès, minéralogie, stratigraphie et évolution des milieux de dépôts de la plate-forme berriaso-valanginienne des régions de Sainte-Croix (VD), Cressier et du Landeron (NE). Bull. Soc. Neuchâteloise Sci. nat., 107:221–239.

Ahn J.H., Peacor D.R., 1985. – Transmission electron microscopic study of diagenetic chlorite in Gulf coast argillaceous sediments. Clays Clay Min., 33:228–236.

Ahn J.H., Peacor D.R., 1986. – Transmission and analytical electron microscopy of the smectite-to-illite transition. Clays Clay Min., 34:165–179.

Ahn J.H., Peacor D.R., 1987a. – Kaolinitization of biotite: TEM data and implications for an alteration mechanism. Am. Miner., 72:353–356.

Ahn J.H., Peacor D.R., 1987b. – Transmission electron microscopic study of the diagenesis of kaolinite in Gulf coast argillaceous sediments. In Schultz L.G., Van Olphen H., Mumptom F.A., Eds., Proc. Int. Clay Conf., Denver, 1985:151–157.

Alimen H., Caillère S., 1964a. – Quelques considérations sur les successions climatiques au Quaternaire déduites de l'étude des argiles des sédiments des Pyrénées de la Bigorre et du Béarn. C.R.Acad. Sci. Paris, 258:5475–5478.

Alimen H., Caillère S., 1964b. – Caractérisation des nappes quaternaires des Pyrénées de la Bigorre et du Béarn par leurs minéraux argileux et essai d'interprétation climatique. Rev. Géogr. pyrén. Sud Ouest, 35:373–396.

Allens V.T., Johns W.D., 1960. – Clays and clay minerals of New England and eastern Canada. Geol. Soc. Am. Bull., 71:75–86.

Aloisi J.-C., Monaco A., 1975. – La sédimentation infralittorale. Les prodeltas nord-méditerranéens. C.R. Acad. Sci., Paris, 280:2833–2836.

Alonso B., 1987. – Late Quaternary depositional patterns of clay minerals in the Ebro margin (Western Mediterranean): paleoclimatic interpretation. In: 6th Meet. Eur. Clay Groups, Sevilla, Proc.:66–68.

Alvinerie J., Latouche C., 1967. – L'influence de la ride de Villagrains-Landiras et de ses prolongements occidentaux sur la répartition des cortèges argileux dans le Miocène inférieur nord-aquitain. C.R. Acad. Sci., Paris, D, 264:1003–1005.

Anderson A.E., Jonas E.C., Odum H.T., 1958. – Alteration of clay minerals by digestive processes of marine organisms. Science, 127:190–191.

Anderson T.F., Donnelly T.W., Drever J.I., Eslinger E., Gieskes J.M., Kastner M., Lawrence J.R., Perry E.A., 1976. – Geochemistry and diagenesis of deep-sea sediments from Leg 35 of the Deep Sea Drilling Project. Nature (London), 261:473–476

André L., Deutsch S., Hertogen J., 1986. – Trace-element and Nd isotopes in shales as indexes of provenance and crustal growth: the early Paleozoic from the Brabant Massif (Belgium). Chem. Geol., 57:101–115.

Andreoli C.Y., Robert M., Pons C.H., 1987. – First steps of smectite-illite transformation with humectation and dessication cycles. In: 6th Meet. Eur. Clay Groups, Sevilla, Proc.:76–78.

Andrews J.E., 1987. – Jurassic clay mineral assemblages and their post-depositional alteration: upper Great Estuarine group, Scotland. Geol. Mag. (Tokyo), 124:261–271.

Aoki S., 1984. – The vertical change of the clay mineral composition in some deep-sea cores raised from the Northern part of the Central Pacific Basin. Geol. Surv. Jpn. Cruise Rep., 20:193–197.

Aoki S., Sudo T., 1973. – Mineralogical study of the core samples from the Indian Ocean, with special reference to the vertical distribution of clay minerals. J. Ocean. Soc. Jpn., 29:87–93.

Aoki S., Oinuma K., Sudo T., 1974a. – The distribution of clay minerals in the recent sediments of the Japan Sea. Deep-Sea Res., 21:299–310.

Aoki S., Kohyama N., Sudo T., 1974b. – An iron-rich montmorillonite in a sediment core from the northeastern Pacific. Deep-Sea Res., 21:865–875.

Aoki S., Kohyama N., Sudo T., 1979. – Mineralogical and chemical properties of smectites in a sediment core from the Southeastern Pacific. Deep-Sea Res., 26:893–902.

Aoyagi K., Chilingarian G.V., 1972. – Clay minerals in carbonate reservoir rocks and their significance in porosity studies. Sediment. Geol., 8:241–249.

Aoyagi K., Kazama T., 1980. – Transformational changes of clay minerals, zeolites and silica minerals during diagenesis. Sedimentology, 27:179–188.

Aprahamian J., Fourneaux J.-C., Lacroix B., Uselle J.-P., 1970. – Observations nouvelles sur les argiles interglaciaires de la vallée de l'Isère. C.R. Acad. Sci., Paris, 271:1071–1074.

April R.H., 1981. – Clay petrology of the upper Triassic/lower Jurassic terrestrial strata of the Newark supergroup, Connecticut valley, U.S.A. Sediment. Geol., 29:283–307.

Arditto P.A., 1983. – Mineral-groundwater interactions and the formation of authigenic kaolinite within the southeastern intake beds of the Great Australian (Artesian) basin, New South Wales, Australia. Sediment. Geol., 35:249–261.

Arnoux A., Chamley H., 1974. – Minéraux des argiles et détergents des eaux interstitielles, dans les sédiments superficiels du golfe du Lion. C.R. Acad. Sci., Paris, 278:999–1002.

Aronson J.L., Hower J., 1976. – Mechanism of burial metamorphism of argillaceous sediment, 2: Radiogenic argon evidence. Geol. Soc. Am. Bull., 87:738–744.

Arrhenius G., 1963. – Pelagic sediments. In: Hill M.N. Ed., The Sea, ideas and observations on progress in the study of the seas. 3. The earth beneath the sea. Interscience, New York:655–727.

Arthur M.A., Dean W.E., Stow D.A.V., 1984a. – Models for the deposition of Mesozoic-Cenozoic fine-grained organic-carbon-rich sediments in the deep sea. In: Stow D.A.V., Piper D.J.W., Eds., Fine-grained sediments: deep-water processes and facies. Geol. Soc. London, Spec. Publ., 15:527–560.

Arthur M.A., Dean W.E., Bottjer D., Scholle P.A., 1984b. – Rhythmic bedding in Mesozoic-Cenozoic pelagic carbonate sequences: the primary and diagenetic origin of Milankovitch-like cycles. In: Berger A.L., Imbrie J., Hays J., Kukla G., Saltzman B., Eds., Milankovitch and climate. Reidel, Dordrecht, 1:191–222.

Arthur M.A., Dean W.E., Schlanger S.O., 1985. – Variations in the global carbon cycle during the Cretaceous related to climate, volcanism, and changes in atmospheric CO_2. In: The carbon cycle and atmospheric CO_2: natural variations Archean to Present. Geophys. Monogr. 32, A.G.U.:504–529.

Aston S.R., 1978. – Estuarine chemistry. In: Riley J.P., Chester R., Eds., Chemical oceanography. Academic Press, New York, London, 7:361–440.

Aston S.R., Chester R., Johnson L.R., Padgham R.C., 1973. – Eolian dust from the lower atmosphere of the Eastern Atlantic and Indian Oceans, China Sea and Sea of Japan. Mar. Geol., 14:15–28.

Ataman G., Gökçen S.L., 1975. – Determination of source and palaeoclimate from the comparison of grain and clay fractions in sandstones: a case study. Sediment. Geol., 13:81–107.

Auffret G.-A., Pastouret L., Chamley H., Lanoix F., 1974. – Influence of the prevailing current regime on sedimentation in the Alboran Sea. Deep-Sea Res., 21:839–849.

Bachman G.O., Machette M.N., 1977. – Calcic soils and calcretes in the southwestern United States. U.S. Geol. Surv., Open-File Rep. 77–794:163 pp.

Badaut D., Risacher F., 1983. – Authigenic smectite on diatom frustules in Bolivian saline lakes. Geochim. Cosmochim. Acta, 47:363–375.

Badaut D., Risacher F., Paquet H., Eberhart J.-P., Weber F., 1979. – Néoformation de minéraux argileux à partir de frustules de Diatomées: le cas des lacs de l'Altiplano bolivien. C.R. Acad. Sci., Paris, 289:1191–1193.

Badaut D., Besson G., Decarreau A., Rautureau R., 1983. – Occurrence of a ferrous, trioctahedral smectite in recent sediments of Atlantis II deep, Red Sea. Clay Min., 20:389–404.

Bailey S.W., 1980 a. – Structures of layer silicates. In: Brindley G.W., Brown G., Eds., Crystal structures of clay minerals and their X-ray identification. Mineral. Soc., London:1–123.

Bailey S.W., 1980 b. – Summary of recommendations of AIPEA nomenclature committee on clay minerals. Am. Mineral., 65:1–7.

Bain D.C., Tait J.M., 1977. – Mineralogy and origin of dust fall on skye. Clay Min., 12:353–355.

Baker E.T., 1973. – Distribution and composition of suspended sediment in the bottom waters of the Washington continental shelf and slope. J. Sediment. Petrol., 43:812–821.

Ballivy G., Poiliot G., Loiselle A., 1971. – Quelques caractéristiques géologiques et minéralogiques des dépôts d'argile du nord-ouest du Québec: Can. J. Earth Sci., 8:1525–1551.

Balsam W.L., Deaton B.C., Barranco F.T. Jr., 1987. – Sediment provenance indicated by visible light spectra. Eos, 68:329.

Baltuck M., 1982. – Provenance and distribution of Tethyan pelagic and hemipelagic siliceous sediments, Pindos mountains, Greece. Sediment. Petrol., 31:63–68.

Baltzer F., 1971. – Relais de la silice en solution par une smectite détritique dans le transport vers la mer des produits d'altération d'un bassin versant tropical sur roches ultra-basiques. C.R. Acad. Sci., Paris, 273:929–932.

Baltzer F., 1975. – Solution of silica and formation of quartz and smectite in mangrove swamps and adjacent hypersaline marsh environments. In: Walsh G.E., Snedaker C.S., Teas H.J., Eds., Proc. Int. Symp. Biology and management of mangroves:482–498.

Barbaroux L., Gallenne B., 1973. – Répartition des minéraux argileux dans les sédiments récents de l'estuaire de la Loire et du plateau continental voisin. C.R. Acad. Sci., Paris, 277:1609–1612.

Barber A.J., Tjokrosapoetro S., Charlton T.R., 1986. – Mud volcanoes, shale diapirs, wrench faults, and melanges in accretionary complexes, Eastern Indonesia. Am. Assoc. Petrol. Geol. Bull., 70:1729–1741.

Barker P.F., Kennett J.P. et al., 1987. – Glacial history of Antarctica. Nature (London), 389:115–116.

Barrett T.J., Taylor P.N., Jarvis I., Lugowski J., 1986. – Pb and Sr isotope and rare earth element composition of selected metalliferous sediments from sites 597 to 601, Deep Sea Drilling Project Leg 92. In: Leinen M., Rea D.K., et al., Eds., Init. Rep. Deep Sea Drill. Proj., 92, Washington (U.S. Gov. Print. Off.):391–407.

Barrie C.Q., 1983. – Late glacial and contemporary deposition of clay-size minerals in Makkovik bay, Labrador. Mar. Geol., 53:199–209.

Barrière J., 1971. – Utilisation des paléosols comme élément de datation des formations quaternaires. C.R. Acad. Sci., Paris, 273:310–313.

Barrios J., Barron V., Peña F., Torrent J., 1987. – Particle size and mineralogical composition of aeolian dusts collected in Spain. In: 6th Meet. Eur. Clay Groups, Sevilla:101–103.

Barron E.J., Arthur M.A., Kauffman E.G., 1985. – Cretaceous rhythmic bedding sequences: a plausible link between orbital variations and climate. Earth Planet. Sci. Lett., 72:327–340.

Bass, M.N., 1976. – Secondary minerals in oceanic basalt, with special reference to Leg 34, DSDP. In: Yeats R.S., Hart S.R., et al., Init. Rep. Deep Sea Drill. Proj., 34, Washington (U.S. Gov. Print. Off.):393–432.

Bassarab D.R., Huff W.D., 1969. – Clay mineralogy of Kope and Fairview formations (Cincinnatian) in the Cincinnati area. J. Sediment. Petrol., 30:1014–1022.

Batten D.J., 1984. – Palynology, climate and the development of late Cretaceous floral provinces in the Northern hemisphere; a review. In: Brenchley P., Ed., Fossils and climate. Wiley & Sons, New York: 127–164.

Bayliss P., Levinson A.A., 1970. – Clay mineralogy and boron determination of the shales from Raindeer Well, Mackenzie River delta, N.W.T., Canada. Can. Petrol. Geol. Bull., 18:80–83.

Bayliss P., Syvitski J.P.M., 1982. – Clay diagenesis in recent marine fecal pellets. Geo-Mar. Lett., 2:83–88.

Beck K.C., Weaver C.E., 1978. – Miocene of the S.E. United States: a model for chemical sedimentation in a peri-marine environment. Reply. Sediment. Geol., 21:154–157.

Behairy A.K., Chester R., Griffiths A.J., Johnson L.R., Stoner J.H., 1975. – The clay mineralogy of particulate material from some surface seawaters of the eastern Atlantic Ocean. Mar Geol., 18:M45–M56.

Bell D.L., Goodell H.G., 1967. – A comparative study of glauconite and the associated clay fraction in modern marine sediments. Sedimentology, 9:169–202.

Bell T.E., 1986. – Microstructure in mixed layer illite/smectite and its relationship to the reaction of smectite to illite. Clays Clay Min., 34:146–154.

Bellaiche G., Chamley H., Rotschy F., 1972. – Quelques particularités de la sédimentation marine quaternaire au sud de l'île du Levant (Var). Tethys, 4:243–250.

Benda L., Meulenkamp J.E., Schmidt R.R., 1982. – Biostratigraphic correlations in the eastern Mediterranean Neogene. 6. Correlation between sporomorph, marine microfossil and mammal associations from some Miocene sections of the Ionian islands and Crete (Greece). Newslett. Stratigr., 11:82–93.

Bennett R.H., Bryant W.R., Keller G.H., 1981. – Clay fabric of selected submarine sediments: fundamental properties and models. J. Sediment. Petrol., 51:217–232.

Benson L.V., Teague L.S., 1982. – Diagenesis of basalts from the Pasco basin, Washington. I. Distribution and composition of secondary mineral phases. J. Sediment. Petrol., 52:595–613.

Berger A., Imbrie J., Hays J., Kukla G., Saltzman B., Eds., 1984. – Milankovitch and climate. Reidel, Dortrecht:895 pp.

Berger B.R., Bethke P., 1985. – Geology and geochemistry of epithermal systems. In: Robertson A.H.F., Ed., Rev. Econ. Geol., 2:298 pp.

Berggren W.A., Hollister C.C., 1977. – Plate tectonics and paleocirculation. Commotion in the Ocean. Tectonophysics, 38:11–48.

Berner R.A., 1971. – Principles of chemical sedimentology. McGraw-Hill, New York:240 pp.

Berner R.A., 1980. – Early diagenesis. Princeton Univ. Press:241 pp.

Berner R.A., Lasaga A.C., Garrels R.M., 1983. – The carbonate-silicate geochemical cycle and its effect on atmospheric carbon dioxide over the past 100 million years. Am. J. Sci., 283:641–683.

Berry R.W., Johns W.D., 1966. – Mineralogy of the clay-sized fractions of some North Atlantic-Arctic ocean bottom sediments. Geol. Soc. Am. Bull., 77:183–196.

Berry R.W., Brophy G.P., Naquasha, 1970. – Mineralogy of the suspended sediment in the Tigris, Euphrates, and Shatt-al-Arab rivers of Iraq, and the recent history of the Mesopotamian plain. J. Sediment. Petrol., 40:131–139.

Berthou P.-Y., Blanc P., Chamley H., 1982: – Sédimentation argileuse comparée au Crétacé moyen et supérieur dans le bassin occidental portugais et sur la marge voisine (site 398 D.S.D.P.): enseignements paléogéographiques et tectoniques. Bull. Soc. géol. Fr., Océans-Paléocéans, Lille 1981, (7) 24:461–472.

Besse D., Desprairies A., Jehanno C., Kolla V., 1981. – Les paragenèses de smectites et de zéolites dans une série pyroclastique d'âge éocène moyen de l'Océan Indien (D.S.D.P., Leg 26, Site 253). Bull. Minéral., 104:56–63.

Bessedik M., 1984. – The Early Aquitanian and Upper Langhian-Lower Serravallian environments in the northwestern Mediterranean region. Paléobiologie continentale, Montpellier, 14:153–179.

Bethke C.M., Altaner S.P., 1986. – Layer-by-layer mechanism of smectite illitization and application to a new rate law. Clays Clay Min., 34:136–145.

Beutelspacher H., van der Marel H.W., 1968. – Atlas of electron microscopy of clay minerals and their admixtures. Elsevier, Amsterdam: 333 pp.

Bhattacharyya D.P., 1983. – Origin of berthierine in ironstones. Clays Clay Min., 31:173–182.

Bhattacharyya D.P., Kakimoto P.K., 1982. – Origin of ferriferous ooids: a SEM study of ironstone ooids and bauxite pisoids. J. Sediment. Petrol., 52:849–857.

Biddle P., Miles J.H., 1972. – The nature of contemporary silts in British estuaries. Sediment. Geol., 7:23–33.

Bignell R.D., 1978. – Genesis of the Red Sea metalliferous sediments. Mar. Min., 1:209–235.

Biscaye P.E., 1976. – Mineralogy and sedimentation of recent deep-sea clay in the Atlantic Ocean and adjacent seas and oceans. Geol. Soc. Am. Bull., 76:803–832.

Biscaye P.E., Eittreim S.L., 1977. – Suspended particulate loads and transports in the nepheloid layer of the abyssal Atlantic ocean. Mar. Geol., 23:155–172.

Biscaye P.E., Chesselet R., Prospero J.M., 1974. – Rb-Sr, $^{87}Sr/^{86}Sr$ isotope system as an index of provenance of continental dusts in the open atlantic ocean. J. Rech. Atmosph., 8:819–829.

Bischoff J.L., 1969. – Red Sea geothermal brine deposits: their mineralogy, chemistry, and genesis. In: Degens E.T., Ross D.A., Eds., Hot brines and recent heavy metal deposits in the Red Sea. Springer, Berlin, Heidelberg, New York:368–401.

Bischoff J.L., 1972. – A ferroan nontronite from the Red Sea geothermal system. Clays Clay Min., 20:217–223.

Bischoff J.L., Rosenbauer R.J., 1977. – Recent metalliferous sediment in the North Pacific manganese nodule area. Earth Planet Sci. Lett., 33:379–388.

Bischoff J.L., Sayles F.L., 1972. – Pore fluid and mineralogical studies of recent marine sediments: Bauer depression region of East Pacific rise. J. Sediment. Petrol., 42:711–724.

Bischoff J.L., Piper D.Z., Heath G.R., Leinen M., 1979. – Geochemistry of deep-sea sediments from the Pacific manganese province: Domes sites A, B and C. In: Bischoff J.L., Piper D.Z., Eds., Marine geology and oceanography of the Pacific manganese nodule province. Plenum, New York:397–436.

Bischoff J.L., Rosenbauer R.J., Aruscavage P.J., Baedecker P.A., Crock J.G., 1983. – Sea-floor massive sulfide deposits from 21°N, East Pacific rise; Juan de Fuca Ridge; and Galapagos rift: bulk chemical composition and economic implications. Econ. Geol., 78:1711–1720.

Bjerkli K., Östmo-Saeter J.S., 1973. – Formation of glauconie in foraminiferal shells on the continental shelf off Norway. Mar. Geol., 14:169–178.

Bjørlykke K., Brensdal A., 1986. – Diagenesis of the Brent sandstone in the Statfjord field, North Sea. Soc. Econ. Mineral. Paleontol., Spec. Publ., 38:157–167.

Bjørlykke K., Dypvik H., 1975. – Geochemical and mineralogical response to the evolution of the Caledonian geosyncline in eugeosynclinal and epicontinental sediments in Norway. In: 9th Congr. Int. Assoc. Sedimentol., Nice, 4:2–6.

Bjørlykke K., Elverhøi A., 1975. – Reworking of Mesozoic clayey material in the North-Western part of the Barents Sea. Mar. Geol., 18:M29–M34.

Blaise B., Mercier E., 1988. – Clay mineralogy of Proterozoic and lower Paleozoic terrains (Ogilvie mountains, Yukon territory, Canada): diagenetic effects and paleoenvironmental implications. Clay Min. (in press)

Blaise B., Bornhold B.D., Maillot H., Chamley H., 1985. – Observations sur les environnements récents au Nord de la dorsale de Juan de Fuca (Pacifique Nord-Est). Rev. Géol. dyn. Géogr. phys., 26:201–213.

Blanc F., Chamley H., Leveau M., 1969. – Les minéraux en suspension témoins du mélange des eaux fluviatiles en milieu marin. Exemple du Rhône. C.R. Acad. Sci. Paris, 269:2509–2512.

Blanc J., Chamley H., 1975. – Remplissages de réseaux karstiques à la grotte de Saint-Marcel d'Ardèche. Bull. Assoc. Fr. Et. Q., 43:71–82.

Blanc-Vernet L., Chamley H., Froget C., 1969. – Analyse paléoclimatique d'une carotte de Méditerranée nord-occidentale. Comparaison entre les résultats de trois études: Foraminifères, Ptéropodes, fraction sédimentaire issue du continent. Palaeogeogr., Climatol., Ecol., 6:215–235

Blanc-Vernet L., Chamley H., Froget C., Le Boulicaut D., Monaco A., Robert C., 1975. – Observations sur la sédimentation marine récente dans la région siculo-tunisienne. Géol. Méditerr., II:31–48.

Blanc-Vernet L., Chamley H., Robert C., 1979. – Minéraux argileux et Foraminifères dans le Pléistocène inférieur de l'île de Zanthe (Grèce occidentale). Ann. Géol. Pays Hellén. (VIth Int. Congr. Medit. Neogene), Athens, t.h.-s., 1:129–138.

Blank M., Leinen M., Prospero J.M., 1985. – Major Asian eolian inputs indicated by the mineralogy of aerosols and sediments in the western North Pacific. Nature (London), 314:84–86.

Bloemendal J., 1987. – The application of rock-magnetism to paleoceanography: an example from the Eastern Equatorial Atlantic. Eos, 68:332.

Bocquier G., Paquet H., Millot G., 1970. – Un nouveau type d'accumulation oblique dans les paysages géochimiques: l'invasion remontante de la montmorillonite. C.R. Acad. Sci., Paris, 270:460–463.

Bodine M.W. Jr., 1978. – Clay-mineral assemblages from drill core of Ochoan evaporites, Eddy County, New Mexico. New Mexico Bur. Min. Mineral Res., Circular 159:21–31.

Bodine M.W. Jr., 1987. – Genesis of clay-mineral assemblages in three Paleozoic marine evaporite deposits in the United States. In: 6th Meet. Eur. Clay Groups, Sevilla, Proc.:118–120.

Bodine M.W. Jr., Madsen B.M., 1987. – Mixed-layers chlorite-smectite from a Pennsylvanian evaporite cycle, Grand County, Utah. In: Proc. Int. Clay Conf., Denver, 1985, Clay Mineral. Soc.:85–93.

Bohor B.F., Triplehorn D.M., 1987. – Flyash: an analog for spherules in K-T boundary clays. In: Lunar and planetary science, Houston, Texas. Lunar Planet. Inst., 18:103–104.

Bohor B.F., Foord E.E., Modreski P.J., Triplehorn D.M., 1984. – Mineralogic evidence for an impact event at the Cretaceous-Tertiary boundary. Science, 224:867–869.

Bohor B.F., Triplehorn D.M., Douglas J.N., Hugh T.M. Jr., 1987. – Dinosaurs, spherules, and the "magic" layer: A new K-T boundary clay site in Wyoming. Geology, 15:896–899.

Boiron M.C., Cathelineau M., 1987. – The crystallization temperature of hydrothermal clay minerals: the data from fluid inclusions and clay geothermometers. In: Proc. 6th Meet. Eur. Clay Groups, Sevilla:124–126.

Boles J.R., Franks S.G., 1979. – Clay diagenesis in Wilcox sandstones of Southwest Texas: implications of smectite diagenesis on sandstone cementation. J. Sediment. Petrol., 49:55–70.

Bolishev N.N., Kapustkina N.A., 1964. – Adsorption capacity of solonetzes. Sov. Soil Sci., 12:1263–1271.

Bonatti E., 1981. – Metal deposits in the oceanic lithosphere. In: Emiliani C., Ed., The sea. Wiley & Sons, New York, 7:639–686.

Bonatti E., Joensuu O., 1968. – Palygorskite from the deep sea: a reply. Am. Mineral., 54:568.

Bonatti E., Zerbi M., Kay R., Rydell H., 1976. – Metalliferous deposits from the Apennine ophiolites: Mesozoic equivalents of modern deposits from oceanic spreading centers. Geol. Soc. Am. Bull., 87:83–94.

Bonatti E., Kolla V., Moore W.S., Stern C., 1979. – Metallogenesis in marginal basins: Fe-rich basal deposits from the Philippine sea. Mar. Geol., 32:21–37.

Bonhomme M., Clauer N., Cotillon P., Lucas J., 1969. – Datation rubidium-strontium de niveaux glauconieux du Crétacé inférieur de Haute-Provence: mise en évidence d'une diagenèse. Bull. Serv. Carte géol. Als.-Lorr., 22:235–247.

Bonneau M., Souchier B., 1979. – Pédologie. 2. Constituants et propriétés du sol. Masson, Paris:459 pp.

Bonnot-Courtois C., 1981. – Géochimie des terres rares dans les principaux milieux de formation et de sédimentation des argiles. Thèse, Sci. nat., Paris-Sud:230 pp.

Bonorino F.G., 1966. – Soil clay mineralogy of the Pampa plains, Argentina. J. Sediment. Petrol., 36:1026–1035.

Bonython C.W., 1956. – The salt of lake Eyre. Its occurrence in Madigan Gulf and its possible origin. Trans. R. Soc. S. Aust., 79:66–92.

Borch C.C. von der, Rex R.W., 1970. – Amorphous iron oxide precipitates in sediments cored during Leg 5: Deep Sea Drilling Project. In: Mac Manus D.A. et al., Eds., Init. Rep. Deep Sea Drill. Proj. 5, Washington (U.S. Gov. Print. Off.):541–544.

Borch C.C. von der, Nesteroff W.D., Galehouse J.S., 1971. – Iron-rich sediments cored during Leg 8 of the Deep Sea Drilling Project. In: Tracey J.I. Jr. et al., Eds., Init. Rep. Deep Sea Drill. Proj., 8, Washington (U.S. Gov. Print. Off.):829–835.

Borchert H., 1965. – Formation of marine sedimentary iron ores. In: Riley J.P., Skirrow G., Eds., Chemical oceanography. Academic Press, New York, London, 2:159–204.

Bornand M., 1978. – Altération des matériaux fluvio-glaciaires. Genèse et évolution des sols sur terrasses quaternaires dans la moyenne vallée du Rhône. Thèse, Sci. Nat., Montpellier:329 pp. dact.

Bornand M., Chamley H., 1974. – Sur les minéraux argileux des terrasses pléistocènes du confluent Rhône-Isère. Bull. Assoc. Fr. Et. Q., 42:61–64.

Bornand M., Chamley H., 1975. – Observations sur la sédimentation argileuse du Miocène supérieur au Pléistocène dans la vallée moyenne du Rhône. Bull. Groupe Fr. Argiles, 27, 2:87–96.

Bornhold B.D., Giresse P., 1985. – Glauconitic sediments on the continental shelf off Vancouver Island, British Columbia, Canada. J. Sediment. Petrol., 55:653–664.

Borst R.L., Keller W.D., 1969. – Scanning electron micrographs of API. Reference clay minerals and other selected samples. Proc. Int. Clay Conf., 1:871–902.

Boström K., 1975. – The origin and fate of ferromanganoan active ridge sediments. Stockholm Contrib. Geol., 27:149–243.

Boström K., Peterson N.M., Joensuu O., Fisher D., 1969. – The origin of the aluminum-poor ferromanganoan sediments in areas of hight heat flow in the East Pacific Rise. Mar. Geol., 7:427–447.

Boström K., Joensuu O., Valdes S., Riera M., 1972. – Geochemical history of South Atlantic Ocean sediments since Late Cretaceous. Mar. Geol., 12:85–121.

Boström K., Joensuu O., Valdes S., Charm W., Glaccum R., 1976. – Geochemistry and origin of East Pacific sediments sampled during DSDP Leg 34. In: Yeats R.S., Hart S.R. et al., Eds., Init. Rep. Deep Sea Drill. Proj., 34, Washington (U.S. Gov. Print. Off.):559–574.

Boström K., Lysen L., Moore C., 1978. – Biological matter as a source of authigenic matter in pelagic sediments. Chem. Geol., 23:11–20.

Bottner P., Paquet H., 1972. – La pédogenèse sur roches-mères calcaires tendres dans les étages bioclimatiques montagnard, subalpin et alpin des Préalpes françaises du Sud. Sci. Sol, 1:63–78.

Bouquillon A., 1987. – Influences continentales et marines dans les sédiments cénozoïques de l'Océan Indien Nord Oriental. Thèse Océanologie, Univ. Lille:270 pp.

Bouquillon A., Chamley H., 1986. – Sédimentation et diagenèse récentes dans l'éventail marin profond du Gange (Océan Indien). C.R. Acad. Sci., Paris, 303:1461–1466.

Bouquillon A., Debrabant P., 1987. – Distribution des minéraux argileux dans l'Océan Indien Nord Oriental. J. Rech. Océanogr., 12:8–11.

Bouquillon A., Labeyrie L., 1988. – Alternance glaciaire/interglaciaire dans les sédiments quaternaires du delta du Gange. 12ème Réun. Sci. Terre, Lille, Soc. géol. Fr. Ed.:22.

Bouquillon A., Chamley H., Debrabant P., Piqué A., 1985. – Etude minéralogique et géochimique des forages de Jeumont et Epinoy (Paléozoïque du Nord de la France). Ann. Soc. géol. Nord, 104:167–179.

Bouquillon A., Chamley H., Fröhlich F., 1989. – Sédimentation argileuse récente dans l'Océan Indien nord-oriental. Oceanol. Acta, 12 (in press)

Bourdier F., 1961. – Le bassin du Rhône au Quaternaire. Géologie et préhistoire. In: C.N.R.S. Ed., 2 vols.:363 pp.

Bouroz A., Spears D.A., Arbey F., 1983. – Essai de synthèse des données acquises sur la genèse et l'évolution des marqueurs pétrographiques dans les bassins houillers. Mém. Soc. géol. Nord, 16:74 pp. + ann.

Bowles F.A., 1975. – Paleoclimatic significance of quartz/illite variations in cores from the eastern equatorial Atlantic. Q. Res., 5:225–235.

Bowles F.A., Fleischer P., 1985. – Orinoco and Amazon river sediment input to the Eastern Caribbean Basin. Mar. Geol., 68:53–72.

Bowles F.A., Angino E.A., Hosterman J.W., Galle O.K., 1971. – Precipitation of deep-sea palygorskite and sepiolite. Earth Planet. Sci. Lett., 11:324–332.

Bown T.M., Kraus M.J.K., 1981. – Lower Eocene alluvial paleosols (Willwood Formation, northwest Wyoming, U.S.A.) and their significance for paleoecology, paleoclimatology, and basin analysis. Palaeogeogr., Climatol., Ecol., 34:1–30.

Boyd R.W., Piper D.J.W., 1976. – Baffin Bay continental shelf clay mineralogy. Mar. Sediment., 12:17–18.

Brauckmann F.J., Fuchtbauer H., 1983. – Alterations of Cretaceous siltstones and sandstones near basalt contacts (Nûgssuaq, Greenland). Sediment. Geol., 35:193–213.

Brell J.M., Doval M., Carames M., 1985. – Clay mineral distribution in the evaporitic Miocene sediments of the Tajo basin, Spain. In: Pozzuoli A., Ed., Miner. Petrol. Acta, 29-A:267–276.

Brewer P.G., Spencer D.W., Biscaye P.E., Hanley A., Sachs P.L., Smith C.L., Kadar S., Fredericks J., 1976. – The distribution of particulate matter in the Atlantic ocean. Earth Planet. Sci. Lett., 32:393–402.

Briat M., Royer A., Petit J.R., Lorius C., 1982. – Late glacial input of eolian continental dust in the Dome C ice core: additional evidence from individual microparticle analysis. Ann. Glaciol., 3:27–31.

Brigatti M.F., Poppi L., 1984. – "Corrensite-like minerals" in the Taro and Ceno Valleys, Italy. Clay Minerals, 19:59–66.

Brindley G.W., 1980. – Order-disorder in clay mineral structures. In: Brindley G.W., Brown G., Ed., Crystal structures of clay minerals and their X-ray identification. Mineral. Soc., London:125–195.

Brindley G.W., Brown G., 1980. – Crystal structures of clay minerals and their X-ray identification. Mineral. Soc., London, 495 pp.

Brongniart A., 1822. – Notice sur la magnésite du Bassin de Paris et sur le gisement de cette roche dans divers lieux. Ann. Mines, 7:291–318.

Brooks R.A., Ferrell R.E. Jr., 1970. – The lateral distribution of clay minerals in lakes Pontchartrain and Maurepas, Louisiana. J. Sediment. Petrol., 40:855–863.

Brown G., 1980. – Associated minerals. In: Brindley G.W., Brown G., Ed., Crystal structures of clay minerals and their X-ray identification. Mineral. Soc., London: 361–410.

Brown G., Catt J.A., Weir A.H., 1969. – Zeolites of the clinoptilolite-heulandite type in sediments of south-east England. Min. Mag. (Tokyo), 37:480–488.

Brown L.F. Jr., Bailey S.W., Cline L.M., Lister S.S., 1977. – Clay mineralogy in relation to deltaic sedimentation patterns of Desmoinesian cyclothems in Iowa-Missouri. Clays Clay Miner., 25:171–186.

Brown L.G., 1953. – Geological aspects of the St Austell granite. Clay Min. Bull., 9:17–21.

Bryant J.P., Dixon J.B., 1964. – Clay mineralogy and weathering of a red-yellow podsolic soil from quartz mica schist in the Alabama piedmont. Proc. Natl. Conf. Clays Clay Minerals, 12:509–522.

Bryant W.R., Bennett R.H., Brukett P.J., Shephard L.E., 1986. – Asian eolian argillite aggregates (shales) in the North Pacific red clays. Eos, 67:900.

Bücher A., Lucas C., 1975. – Poussières africaines sur l'Europe. La Météorologie, Paris, 5:53–69

Bücher A., Dubief J., Lucas C., 1983. – Retombées estivales de poussières sahariennes sur l'Europe. Rev. Géol. dyn. Géogr. phys., 24:153–165.

Burst J.F., 1958. – Mineral heterogeneity of "glauconite" pellets. Am. Miner., 43:481–497.

Burst J.F. Jr., 1959. – Post-diagenetic clay mineral environmental relationships in the Gulf coast Eocene. Clays Clay Min., 6:327–341.

Burst J.F. Jr., 1969. – Diagenesis of Gulf coast clayey sediments and its possible relation to petroleum migration. Bull. Am. Assoc. Petrol. Geol., 53:73–93.

Burton J.H., Krinsley D.H., Pye K., 1987. – Authigenesis of kaolinite and chlorite in Texas Gulf coast sediments. Clays Clay Min., 35:291–296.

Busson G., 1972. – Principes, méthodes et résultats d'une étude stratigraphique du Mésozoïque saharien. Mém. Mus. nat. Hist. nat., C, 26:441 pp.

Busson G., 1984. – Relations entre la sédimentation du Crétacé moyen et supérieur de la plate-forme du nord-ouest africain et les dépôts contemporains de l'Atlantique centre et nord. Eclogae geol. Helv., 77:221–235.

Bustin R.M., Bayliss P., 1979. – Clay mineralogy of the Eureka sound and Beaufort formations, Axel Heiberg and west central Ellesmere islands, eastern Canadian Arctic archipelago. Can. Petrol. Geol. Bull., 27:446–452.

Button A., Tyler N., 1979. – Precambrian paleoweathering and erosion surfaces in Southern Africa: Review of their character and economic significance. U. Wits Ecol. Geol. Res. Unit, Inf. Circ., 135:37 pp.

Butuzova G.Y., Lisitzina N.A., Gradusov B.P., 1977. – Clay minerals in sediments on the profile through the Pacific Ocean. Lithol. Miner. Resourc., 4:3–18.

Butuzova G.Y., Drits V.A., Lisitsina N.A., Tsipursky S.I., Dimitrik A.L., 1979. – Formation dynamics of clay minerals in ore-bearing sediments in the Atlantis II basin, Red Sea. Lithol. Miner. Resourc, 14:23–32.

Buurman P., 1980. – Palaeosols in the Reading Beds (Paleocene) of Alum Bay, Isle of Wight, U.K. Sedimentology, 27:593–606.

Byong-Kwon P., Ingram R.L., 1975. – The clay minerals in recent sediments of North Carolina sounds and estuaries in U.S.A. In: 9ème Congr. Int. Sedimentol., Nice, 5, Abstr.:9.

Cahet G., Giresse P., 1983. – Le rôle des supports organiques marins et en particulier des pelotes fécales dans la minéralisation. Bull. Soc. géol. Fr, (7) 25:523–531.

Caillère S., Hénin S., 1961. – Sépiolite-palygorskite. In: Brindley G.W., Brown G., Eds., Crystal structures of clay minerals and their X-ray identification. Mineral. Soc., London:104–115.

Caillère S., Hénin S., Rautureau M., 1982. – Minéralogie des argiles. 1. Structure et propriétés physico-chimiques. 2. Classification et nomenclature. Masson, Paris:184 pp., 189 pp.

Cairns-Smith A.G., Hartman H., 1986. Clay minerals and the origin of life. Cambridge University Press:192 pp.

Callen R.A., 1984. – Clays of the palygorskite-sepiolite group: depositional environment, age and distribution. In: Singer A., Galan E., Eds., Developments in sedimentology. Elsevier, Amsterdam, 37:1–37.

Calvert S.E., 1983. – Geochemistry of Pleistocene sapropels and associated sediments from the Eastern Mediterranean. Oceanol. Acta, 6:255–267.

Calvert S.E., Price N.B., Heath G.R., Moore T.C. Jr., 1978. – Relationship between ferromanganese nodule compositions and sedimentation in a small survey area of the equatorial Pacific. J. Mar. Res., 36:161–183.

Cann J.R., 1979. – Metamorphism in the ocean crust. In: Talwani M., Harrison C.G., Hayes D.E., Eds., Deep drilling results in the Atlantic Ocean: Ocean crust. A.G.U. M. Ewing Ser., 2:230–238.

Capdecomme L., Kulbicki G., 1954. – Argiles des gîtes phosphatés de la région de Thiès (Sénégal). C.R. Acad. Sci., Paris, 235:187.

Carmouze J.-P., 1976. – La régulation hydrogéochimique du lac Tchad. Thèse Univ. Paris VI:418 p.

Carmouze J.-P., Pédro G., 1977. – Contribution des facteurs géographiques et sédimentologiques à la régulation saline d'un milieu lacustre. Cah. O.R.S.T.O.M., sér. Hydrobiol. 11:231–237.

Carmouze J.-P., Pédro G., Berrier J., 1977 a. – Sur la nature des smectites de néoformation du lac Tchad et leur distribution spatiale en fonction des conditions hydrogéochimiques. C.R. Acad. Sci., Paris, 284:615–618.

Carmouze J.-P., Golterman H.L., Pédro G., 1977 b. – The neoformations of sediments in Lake Chad; their influence on the salinity control. In: Golterman H.L., Ed., Interactions between sediments and fresh water. Proc. Int. Symp., Amsterdam, Sept. 1976:33–39.

Carroll D., 1969. – Chlorite in Central North Pacific Ocean sediments. Proc. Int. Clay Conf., 1:335–338.

Carroll D., 1970. – Clay minerals in Arctic Ocean sea-floor sediments. J. Sediment. Petrol., 40:814–821.

Carroll D., 1979. – Clay minerals, a guide to their X-ray identification. Geol. Soc. Am., Spec. Pap., 126:80 pp.

Carroll D., Starkey H.C., 1958. – Effect of sea-water on clay minerals. Clays Clay Min., 7th Natl. Conf. Pergamon, Oxford:80–101.

Carson B., Arcaro N., 1983. – Control of clay-mineral stratigraphy by selective transport of late Pleistocene-Holocene sediments of Northern Cascadia Basin-Juan de Fuca abyssal plain: implications for studies of clay-mineral provenance. J. Sediment. Petrol., 53:395–406.

Cassan J.P., Lucas J., 1966. – La diagenèse des grès argileux d'Hassi-Messaoud (Sahara): silicification et dickitisation. Bull. Serv. Carte géol. Als.-Lorr., 19:241–253.

Cathelineau M., 1983. – Les minéraux phylliteux dans les gisements hydrothermaux d'uranium. II. Distribution et évolution cristallochimique des illites, interstratifiés, smectites et chlorites. Bull. Minéral., 106:553–569.

Cathelineau M., Nieva D., 1985. – A chlorite solid solution geothermometer. The Los Azufres (Mexico) geothermal system. Contrib. Mineral. Petrol., 91:235–244.

Cathelineau M., Oliver R., Nieva D., Garfias A., 1985. – Mineralogy and distribution of hydrothermal mineral zones in Los Azufres (Mexico) geothermal field. Geothermics, 14:49–57.

Chamley H., 1964. – Remarques sur le minéraux argileux des sédiments fluviatiles et marins de la région du Bas-Rhône. Rec. Trav. St. mar. Endoume, 34:263–270.

Chamley H., 1967. – Possibilités d'utilisation de la cristallinité d'un minéral argileux (illite) comme témoin climatique dans les sédiments récents. C.R. Acad. Sci., Paris, D, 265:184–187.

Chamley H., 1968. – La sédimentation argileuse actuelle en Méditerranée nord-occidentale. Données préliminaires sur la diagenèse superficielle. Bull. Soc. géol. Fr., (7) 10:75–88.

Chamley H., 1969. – Relations entre la nature des minéraux argileux, leur origine pétrographique et leur environnement continental, littoral ou marin. Cas de Nosy Bé (NW de Madagascar). Rec. Trav. Stat. mar. Endoume, fasc. h.s. suppl. 9:193–207.

Chamley H., 1971. – Recherches sur la sédimentation argileuse en Méditerranée. Sci. Géol., Strasbourg, mém. 35:225 pp.

Chamley H., 1974. – Place des argiles marines parmi divers indicateurs paléoclimatiques. In: Coll. Int. C.N.R.S. no 219, "Les méthodes quantitatives d'étude des variations du climat au cours du Pléistocène", Gif-sur-Yvette, 1973:25–37.

Chamley H., 1975 a. – Remarques sur la sédimentation argileuse quaternaire en mer de Norvège. Bull. Union Océanogr. Fr., 7:15–20.

Chamley H., 1975 b. – Sédimentation argileuse en mer Tyrrhénienne au Plio-Pléistocène d'après l'étude du forage JOIDES 132. Bull. Groupe Fr. Arg., 27:97–137.

Chamley H., 1975 c. – Sédimentation argileuse en mer Ionienne au Plio-Pléistocène d'après l'étude des forages 125 DSDP. Bull. Soc. géol. Fr., (7) 17:1131–1143.

Chamley H., 1975 d. – Influence des courants profonds au large du Brésil sur la sédimentation argileuse récente. 9ème Congr. Int. Sédimentol., Nice, Th. 8:13–17.

Chamley H., 1976. – Minéralogie des argiles, lithologie et tectonique, dans le Pliocène du Capo Rossello (Sicile). C.R. somm. Soc. géol. Fr.:39–41.

Chamley H., 1979. – North Atlantic clay sedimentation and paleoenvironment since the late Jurassic. In: Talwani M., Hay W., Ryan W.B.F., Eds., Deep drilling results in the Atlantic Ocean: Continental margins and paleoenvironment. Maurice Ewing ser., Am. Geophys. Union, 3:342–361.

Chamley H., 1980 a. – Clay sedimentation and paleoenvironment in the Shikoku Basin since the middle Miocene (Deep sea drilling project leg 58, North Philippine Sea). In: Klein G. de V., Kobayashi K. et al., Eds., Init. Rep. Deep Sea Drill. Proj., 58, Washington (U.S. Gov. Print. Off.):669–681.

Chamley H., 1980 b. – Clay sedimentation and paleoenvironment in the area of Daito Ridge (Northwest Philippine Sea) since the Early Eocene. In: Klein G. de V., Kobayashi K. et al., Eds., Init. Rep. Deep Sea Drill Proj., 58, Washington (U.S. Gov. Print. Off.):683–693.

Chamley H., 1981. – Long-term trends in clay deposition in the ocean. Oceanologica Acta, Spec.:105–110.

Chamley H., 1986. – Clay mineralogy at the Eocene/Oligocene boundary. In: Pomerol C., Premoli-Silva I., Eds., Terminal eocene events. Elsevier, Amsterdam:381–386.

Chamley H, Bonnot-Courtois C., 1981. – Argiles authigènes et terrigènes de l'Atlantique et du Pacifique NW (Legs 11 et 58 DSDP): apports des terres rares. Oceanol. Acta, 4:229–238.

Chamley H., Colomb E., 1967. – Premières données sur la sédimentation argileuse du Miocène supérieur dans le bassin du Cucuron (Vaucluse). Présence de niveaux lacustres à attapulgite. C.R. Somm. Soc. géol. Fr., (7) 6:230–232.

Chamley H., Debrabant P., 1982. – L'Atlantique Nord à l'Albien: influences américaine et africaine sur la sédimentation. C.R. Acad. Sci., Paris, D, 294:525–528.

Chamley H., Debrabant P., 1984 a. – Paleoenvironmental history of the North Atlantic region from mineralogical and geochemical data. Sediment. Geol., 40:151–167.

Chamley H., Debrabant P., 1984 b. – Paléoenvironnements comparés de l'Atlantique nord-ouest et du golfe du Mexique oriental, depuis le Jurassique supérieur, d'après les données minéralogiques et géochimiques. Bull. Soc. géol. Fr., (7) 36:71–80.

Chamley H., Debrabant P., 1984 c. – Mineralogical and geochemical investigations of sediments on the Mazagan Plateau, Northwestern African margin (Leg 79, Deep sea drilling project). In: Hinz K., Winterer E.L. et al., Eds., Init. Rep. Deep Sea Drill. Proj., 79, Washington (U.S. Gov. Print Off.):497–508.

Chamley H., Debrabant P., 1988. – Continental and marine influences expressed by deep-sea sedimentation off Japan (Kaiko Project). Palaeogeogr., Climatol., Ecol. (in press)

Chamley H., Deconinck J.-F., 1985. – Expression de l'évolution géodynamique des domaines nord-atlantique et subalpin au Mésozoïque supérieur, d'après les successions sédimentaires argileuses. C.R. Acad. Sci., Paris, 300:1007–1012.

Chamley H., Diester-Haass L., 1979. – Upper Miocene to Pleistocene climates in Northwest Africa deduced from terrigenous components of site 397 sediments (DSDP Leg 47 A). In: Rad U. von, Ryan W.B.F. et al., Eds., Init. Rep. Deep Sea Drill. Proj., 47, A, Washington (U.S. Gov. Print. Off.):641–646.

Chamyley H., Giroud d'Argoud G., 1978. – Clay mineralogy in volcanogenic sediments. In: Hsü K.J., Montadert L. et al., Eds., Init. Rep. Deep Sea Drill. Proj., 42, A, Washington (U.S. Gov. Print. Off.):395–397.

Chamley H., Kennett J.P., 1976. – Argiles détritiques, Foraminifères planctoniques et paléoclimats, dans des sédiments quaternaires du golfe du Mexique. C.R. Acad. Sci., Paris, D, 282:1415–1418.

Chamley H., Masse J.-P., 1975. – Sur la signification des minéraux argileux dans les sédiments barrémiens et bédouliens de Provence (SE de la France). 9ème Congr. Int. Sédiment., Nice, 1:25–30.

Chamley H., Picard F., 1970. – L'héritage détritique des fleuves provençaux en milieu marin. Tethys, 2:211–226.

Chamley H., Portier J., 1974. – Minéraux argileux de roches, sols et sédiments fluviatiles dans le bassin versant de l'Ubaye (Alpes de Haute-Provence). Bull. Groupe Fr. Argiles, 26:127–138.

Chamley H., Robert C., 1979. – Late Cretaceous to early Paleogene environmental evolution expressed by the Atlantic clay sedimentation. In: Christensen W.K., Birkelung T., Eds., Cretaceous-Tertiary boundary events Symp., Copenhagen, Proc., 2:71–77.

Chamley H., Robert C., 1980. – Sédimentation argileuse au Tertiaire supérieur dans le domaine méditerranéen. Géol. Méditerr., 7:25–34.

Chamley H., Robert C., 1982. – Paleoenvironmental significance of clay deposits in Atlantic black shales. In: Schlanger S.O., Cita M.B., Eds., Nature and origin of cretaceous carbon-rich facies. Academic Press, New York, London:101–112.

Chamley H., Paquet H., Millot G., 1962. – Minéraux argileux de vases méditerranéennes. Bull. Serv. Carte géol. Als.-Lorr., 15:161–169.

Chamley H., Paquet H., Millot G., 1966. – Minéraux argileux des sédiments marins littoraux et fluviatiles de la région de Tuléar (Madagascar). Bull. Serv. Carte géol. Als.-Lorr., 19:191–204.

Chamley H., Durand J.-P., Roux R.-M., 1971. – Les minéraux argileux du bassin versant de l'Arc. Bull. Mus. Hist. nat. Marseille, 31:105–113.

Chamley H., Froget C.-H., Portier J., 1973. – Minéraux argileux de roches, sols et sédiments fluviatiles dans le bassin versant du Haut-Verdon (Alpes de Haute-Provence). Sci. géol., Bull. Strasbourg, 26:279–303.

Chamley H., Durand J.-P., Trauth N., 1976. – Interstratifiés kaolinite-smectite dans le Valdo-Fuvélien au toit de la bauxite des Alpilles (Provence). C.R. Acad. Sci., Paris, D, 283:439–442.

Chamley H., Diester-Haass L., Lange H., 1977 a. – Terrigenous material in East Atlantic sediment cores as an indicator of NW African climates. "Meteor" Forsch. Ergebn., 28:44–59.

Chamley H., Giroud d'Argoud G., Robert C., 1977 b. – Genèse des smectites messiniennes de Sicile; implications paléoclimatiques. Géol. Méditerr., 4:371–378.

Chamley H., Giroud d'Argoud G., Robert C., 1977 c. – Repercussions of the Plio-Pleistocene tectonic activity on the deep-sea clay sedimentation in the Mediterranean. In: Biju-Duval B., Montadert L., Eds., Int. Symp. Structural history of the Mediterranean Basins, Split 1976, Technip:423–432.

Chamley H., Deynoux M., Giroud d'Argoud G., Trompette R., 1978 a. – Minéraux argileux des formations glaciaires et de leur substratum dans le Précambrien supérieur de l'Adrar de Mauritanie, Afrique Occidentale. Sci. Géol., Strasbourg, 30:207–227.

Chamley H., Dunoyer de Segonzac G., Mélières F., 1978 b. – Clay minerals in Messinian sediments of the Mediterranean area. In: Hsü K.J., Montadert L. et al., Eds., Init. Rep. Deep Sea Drill. Proj., 42, A, Washington (U.S. Gov. Print. Off.):389–395.

Chamley H., Debrabant P., Foulon J., Giroud d'Argoud G., Latouche C., Maillet N., Maillot H., Sommer F., 1979a. – Mineralogy and geochemistry of Cretaceous and Cenozoic Atlantic sediments off the Iberian peninsula (Site 398, DSDP Leg 47 B). In: Sibuet J.-C., Ryan W.B.F. et al., Eds., Init. Rep. Deep Sea Drill. Proj., 47, B, Washington (U.S. Gov. Print. Off.):429–449.

Chamley H., Enu E., Moullade M., Robert C., 1979b. – La sédimentation argileuse du bassin de la Bénoué au Nigéria, reflet de la tectonique du Crétacé supérieur. C.R. Acad. Sci., Paris, 288, D:1143–1146.

Chamley H., Deynoux M., Robert C., Simon B., 1980a. – La sédimentation argileuse du Précambrien terminal au Dévonien dans la région du Hodh (bassin cratonique de Taoudeni, Sud-Est mauritanien). Ann. Soc. géol. Nord, 100:73–82.

Chamley H., Colomb E., Roux M.R., 1980b. – Dépôts lacustres à argiles fibreuses dans le Miocène supérieur de la Basse-Durance (Sud-Est de la France). Ann. Soc. géol. Nord, 100:43–56.

Chamley H., Debrabant P., Foulon J., Leroy P., 1980c. – Contribution de la minéralogie et de la géochimie à l'histoire des marges nord-atlantiques depuis le Jurassique supérieur (sites 105 et 367 DSDP). Bull. Soc. géol. Fr. (7) 22:745–755.

Chamley H., Debrabant P., Candillier A.M., Foulon J., 1983. – Clay mineralogical and inorganic geochemical stratigraphy of Blake-Bahama basin since the Callovian, Site 435 (Deep sea drilling project leg 76). In: Sheridan R.E., Gradstein F.M. et al., Eds., Init. Rep. Deep Sea Drill. Proj., 76, Washington (U.S. Gov. Print. Off.):437–451.

Chamley H., Maillot H., Duée G., Robert C., 1984. – Paleoenvironmental history of the Walvis Ridge at the Cretaceous-Tertiary transition, from mineralogical and geochemical investigations. In: Moore T.C., Rabinowitz P.D. et al., Eds., Init. Rep. Deep Sea Drill. Proj., 74, Washington (U.S. Gov. Print. Off.):685–695.

Chamley H., Coulon H., Debrabant P., Holtzapffel T., 1985a. – Cretaceous interactions between volcanism and sedimentation in the east Mariana Basin, from mineralogical, micromorphological, and geochemical investigations (Site 585, Deep sea drilling project). Init. Rep. Deep Sea Drill. Proj., 89, Washington (U.S. Gov. Print. Off.):414–429.

Chamley H., Cadet J.-P., Charvet J., 1985b. – Nankai trough and Japan trench Late Cenozoic paleoenvironments deduced from clay mineralogic data. In: Kagami H., Karig D.E., Coulbourn W.T. et al., Eds., Init. Rep. Deep Sea Drill. Proj., 87, Washington (U.S. Gov. Print. Off.):633–643.

Chamley H., Meulenkamp J.E., Zachariasse W.J., van der Zwaan G.J., 1986. – Middle to Late Miocene marine ecostratigraphy: clay minerals, planktonic foraminifera and stable isotopes from Sicily. Oceanol. Acta, 9:227–238.

Chamley H., Coudé-Gaussen G., Debrabant P., Rognon P., 1987. – Contribution autochtone et allochtone à la sédimentation quaternaire de l'île de Fuerteventura (Canaries). Bull. Soc. géol. Fr., (8) 3:939–952.

Chamley H., Debrabant P., Flicoteaux R., 1988. – Comparative evolution of the Senegal and eastern central Atlantic Basins, from mineralogical and geochemical investigations. Sedimentology, 35:85–103.

Chamley H., Debrabant P., Robert C., Mascle G., Rehault J.-P., Aprahamian J., 1989. – Mineralogical and geochemical investigations of latest Miocene deposits in the Tyrrhenian Sea (ODP Leg 107). In: Kastens K., Mascle J. et al., Eds., Ocean Drill Progr., Proc. 107 B Proc. (U.S. Gov. Print. Off.) (in press).

Chandler F.W., 1980. – Proterozoic redbed sequences of Canada. Geol. Surv. Can., 311:53 pp.

Chen P.Y., 1978. – Minerals in bottom sediments on the South China Sea. Geol. Soc. Am. Bull., 89:211–222.

Chenet P.Y., Francheteau J., 1980. – Bathymetric reconstitution method: Application to the Central Atlantic basin between 10°N and 40°N. In: Donnelly T., Francheteau J., Bryan W., Robinson P., Flower M., Salisbury M. et al., Eds., Init. Rep. Deep Sea Drill. Proj. 51, 52, 53, Pt. 2, Washington (U.S. Gov. Print. Off.):1501–1513.

Chennaux G., Dunoyer de Segonzac G., Petracco F., 1970. – Genèse de la pyrophyllite dans le Paléozoïque du Sahara occidental. C.R. Acad. Sci., Paris, 270:2405–2408.

Chester R., 1965. – Elemental geochemistry of marine sediments. In: Riley J.P., Skirrow G., Eds., Chemical oceanography. Academic Press, New York, London:23–83.

Chester R., 1982. – Regional trends in the distribution and sources of aluminosilicates and trace metals in recent North Atlantic deep-sea sediments. Bull. Inst. Géol. Bassin d'Aquitaine, Bordeaux, 31:325–335.

Chester R., Elderfield H., Griffin J.J., Johnson L.R., Padgham R.C., 1972. – Eolian dust along the eastern margins of the Atlantic Ocean. Mar. Geol., 13:91–105.

Chester R., Baxter G.G., Behairy A.K.A., Connor K., Cross D., Elderfield H., Padgham R.C., 1977. – Soil-sized dusts from the lower troposphere of the eastern Mediterranean Sea. Mar. Geol., 24:201–217.

Chester R., Sharples E.J., Sanders G.S., 1985. – The concentrations of particulate aluminum and clay minerals in aerosols from the northern Arabian Sea. J. Sediment. Petrol., 58:37–41.

Cheverry C., 1974. – Contribution à l'étude pédologique des polders du lac Tchad. Dynamique des sels en milieu continental sub-aride dans des sédiments argileux et organiques. Thèse, Sci. nat., Strasbourg:275 pp.

Chung Y.S., 1986. – Air pollution detection by satellites: the transport and deposition of air pollutants over oceans. Atmosph. Environ., 20:617–630.

Church T.M., Velde B., 1979. – Geochemistry and origin of a deep-sea Pacific palygorskite deposit. Chem. Geol., 25:31–39.

Churchman G.J., 1980. – Clay minerals formed from micas and chlorites in some New Zealand soils. Clay Minerals, 15, 1:59–76.

Cita M.B., Grignani D., 1982. – Nature and origin of late Neogene Mediterranean sapropels. In: Schlanger S.O., Cita M.B., Eds., Nature and origin of Cretaceous carbon-rich facies. Academic Press, New York, London:165–196.

Cita M.B., Vergnaud-Grazzini C., Robert C., Chamley H., Ciaranfi N., D'Onofrio S., 1977. – Paleoclimatic record of a long deep sea core from the Eastern Mediterranean. Q. Res., 8:205–235.

Claridge G.G.C., 1960. – Clay minerals, accelerated erosion, and sedimentation in the Waipaoa river catchment. N. Z. J. Geol. Geophys., 3:184–191.

Claridge G.G.C., 1965. – The clay mineralogy and chemistry of some soils from the Ross dependency, Antarctica. N. Z.J. Geol. Geophys., 8:186–220.

Clauer N., 1979. – The strontium isotopic composition of recent oceanic smectites and the synsedimentary isotopic homogeneization. 6th Eur. Coll. Geochron., Lillehammer, Norway:16.

Clauer N., Hoffert M., Karpoff A.-M., 1982. – The Rb-Sr isotope system as an index of origin and diagenetic evolution of southern Pacific red clays. Geochim. Cosmochim. Acta, 46:2659–2664.

Clauer N., Giblin P., Lucas J., 1984. – Sr and Ar isotope studies of detrital smectites from the Atlantic Ocean (D.S.D.P., Legs 43, 48 and 50). Isotope Geosci., 2:141–151.

Clément P., Dejou J., de Kimpe C., 1979. – Caractérisation chimico-minéralogique d'une évolution géochimique récente observée dans un site de versant sur les gabbros du Mont Mégantic, Québec (Canada). Présence de gibbsite dans les fractions fines de leurs arènes. C.R. Acad. Sci., Paris, D, 288:1063–1066.

Coakley J.P., Rust B.R., 1968. Sedimentation in an Arctic lake. J. Sediment. Petrol., 38:1290–1300.

Cole T.G., 1983. – Oxygen isotope geothermometry and origin of smectite in the Atlantis II Deep, Red Sea. Earth Planet. Sci. Lett., 66:166–176.

Cole T.G., 1985. – Composition, oxygen isotope geochemistry, and origin of smectite in the metalliferous sediments of the Bauer Deep, southeast Pacific. Geochim. Cosmochim. Acta, 49:221–235.

Cole T.G., Shaw H.F., 1983. – The nature and origin of authigenic smectites in some recent marine sediments. Clay Min., 18:239–252.

Collini B., 1956. – On the origin and formation of Fennoscandian Quaternary clays: Geol. Fören. Stockholm Förh., 78:528–534.

Colman S.M., Dethier D.P., 1986. – Rates of chemical weathering of rocks and minerals. Academic Press, New York, London:603 pp.

Colten-Bradley V.A., 1987. – Role of pressure in smectite dehydration. Effects on geopressure and smectite-to-illite transformation. Am. Assoc. Petrol. Geol., Bull., 71:1414–1427.

Conrad G., Michaud L., 1987. – Contribution à l'histoire climatique, post-hercynienne à villa-franchienne, de l'Afrique tropicale par l'étude des pédogenèses et des diagenèses affectant les argilites du Dévonien du Sahara central (Algérie). C.R. Acad. Sci., Paris, 305:715–719.

Copeland R.A., Frey F.A., Wones D.R., 1971. – Origin of clay minerals in a mid-Atlantic ridge sediment. Earth Planet. Sci. Lett., 10:186–192.

Corbató C.E., Tettenhorst R.T., 1987. – Analysis of illite-smectite interstratification. Clay Min., 22:269–285.

Corliss J.B., Lyle M., Dymond J., Crane K., 1978. – The chemistry of hydrothermal mounds near the Galapagos rift. Earth Planet. Sci. Lett., 40:12–24.

Corliss B.H., Hollister C.D. et al., 1982. – A palaeoenvironmental model for Cenozoic sedimentation in the central North Pacific. In: Scrutton R.A., Talwani M., Eds., The ocean floor. Wiley & Sons, New York:277–304.

Correns C.W., 1939. – Pelagic sediments of the North Atlantic Ocean. In: Recent marine sediments symp. SEPM, Spec. Publ., 4:373–395.

Cossement P., Chamley H., Pastouret L., 1984. – Considérations sur la sédimentation du Quaternaire terminal en mer d'Alboran (Méditerranée occidentale). Ann. Soc. géol. Nord, 104:17–27.

Cotillon P., Ferry S., Gaillard C., Jautée E., Latreille G., Rio M., 1980. – Fluctuation des paramètres du milieu marin dans le domaine vocontien (France Sud-Est) au Crétacé inférieur: mise en évidence par l'étude des formations marno-calcaires alternantes. Bull. Soc. géol. Fr., (7) 22:735–744.

Coudé-Gaussen G., 1982. – Les poussières éoliennes sahariennes. Essai de mise au point. Rev. Géomorphol. dyn., 31:49–69.

Coudé-Gaussen G., Blanc P., 1985. – Présence de grains éolisés de palygorskite dans les poussières actuelles et les sédiments d'origine désertique. Bull. Soc. géol. Fr., (8) 1:571–579.

Coudé-Gaussen G., Rognon P., 1986. – Paléosols et loess du Pléistocène supérieur de Tunisie et d'Israël. Bull. Assoc. Fr. Et. Q., 27–28:223–231.

Goudé-Gaussen G., Hillaire-Marcel C., Rognon P., 1982. – Origine et évolution pédologique des fractions carbonatées dans les loess des Matmata (Sud-Tunisien) d'après leurs teneurs en ^{13}C et ^{18}O. C.R. Acad. Sci., Paris, 295:939–942.

Cloudé-Gaussen G., Le Coustumer M.N., Rognon P., 1984. – Paléosols d'âge Pléistocène supérieur dans les loess des Matmata (Sud-Tunisien). Sci. Géol., Strasbourg, 37:359–386.

Coudé-Gaussen G., Rognon P., Bergametti G., Gomes L., Strauss B., Gros J.M., Le Coustumer M.N., 1987. – Saharan dust of Fuerteventura island (Canaries): chemical and mineralogical characteristics, air mass trajectories, and probable sources. J. Geophys. Res., 92:9753–9771.

Coulon H., Debrabant P., Lefèvre C., 1985. – Données pétrographiques, minéralogiques et géochimiques sur la transition basaltes-sédiments dans l'Atlantique Nord. Ann. Soc. géol. Nord, 104:219–233.

Court J.E., Goldman C.R., Hyne N.J., 1972. – Surface sediments in lake Tahoe, California-Nevada. J. Sediment. Petrol., 42:359–377.

Courtillot V.E., Cisowski S., 1987. – The Cretaceous-Tertiary boundary events: external or internal causes? Eos, 68:193 and 200.

Courtillot V., Besse J., Vandamme D., Montigny R., Jaeger J.-J., Cappetta H., 1986. – Deccan flood basalts at the Cretaceous Tertiary boundary. Earth Planet. Sci. Lett., 80:361–375.

Courtois C., Chamley H., 1978. – Terres rares et minéraux argileux dans le Crétacé et le Cénozoïque de la marge atlantique orientale. C.R. Acad. Sci., Paris, D, 286:671–674.

Courty G., 1981. – Présence de berthiérine dans le minerai de fer llanvirnien d'Halouze (Orne) Ann. Soc. géol. Nord, 100:61–64.

Courty G., 1982. – Caractère primaire du phyllosilicate dans le minerai de fer oolithique du Llanvirnien normand. 107ème Congr. nat. Soc. sav., Brest, Sci:399–410.

Couture R.A., 1977. – Composition and origin of palygorskite-rich and montmorillonite-rich zeolite-containing sediments from the Pacific Ocean. Chem. Geol., 19:113–130.

Couture R.A., 1978. – Miocene of the S.E. United States: a model for chemical sedimentation in a peri-marine environment. Comments. Sediment. Geol., 21:149–154.

Cradwick P.D., Wilson N.J., 1978. – Calculated X-ray diffraction curves for the interpretation of a three-component interstratified system. Clay Minerals, 13:53–63.

Cronan D.S., 1973. – Basal ferruginous sediments cored during Leg 16 (Deep sea drilling project). In: Andel T.H. van, Heath G.R. et al., Eds., Init. Rep. Deep Sea Drill. Proj., 16, Washington (U.S. Gov. Print. Off.):601–604.

Cronan D.S., 1976. – Manganese nodules and other ferro-manganese oxide deposits. In: Riley J.P., Chester R., Eds., Chemical oceanography. Academic Press, New York, London, 5:217–263.

Cronan D.S., Ed., 1986. – Sedimentation and mineral deposits in the Southwestern Pacific Ocean. Academic Press, New York, London:344 pp.

Crook K.A.W., Feary D.A., 1982. – Development of New-Zealand according to the fore-arc model of crustal evolution. Tectonophysics, 87:65–107.

Crossey L.J., Surdam R.C., Lahann R., 1986. – Application of organic/inorganic diagenesis to porosity prediction. Soc. Econ. Miner. Petrol., Spec. Publ., 38:147–155.

Crovisier J.-L., Ehret G., Eberhart J.-P., Juteau T., 1983. – Altération expérimentale de verre basaltique tholéiitique par l'eau de mer entre 3 et 50 °C. Sci. Géol. Bull., 36:187–206.

Culbertson W.C., 1971. – Stratigraphy of the trona deposits in the Green River Formation, southwest Wyoming. Wyoming Univ. Contrib. Geol., 10:15–23.

Cullen D.J., 1967. – The age of glauconite from the Chatham Rise, East of New Zealand. N. Z. J. Mar. Freshwater Res., 1:399–406.

Curtis C., 1987a. – Données récentes sur les réactions entre matières organiques et substances minérales dans les sédiments et sur leurs conséquences minéralogiques. Mém. Soc. géol. Fr., 151:127–141.

Curtis C., 1987b. – Mineralogical consequences of organic matter degradation in sediments: inorganic/organic diagenesis. In: Leggett J.K., Zuffa G.G., Eds., Marine clastic sedimentology. Graham & Trotman, London:108–123.

Curtis C.D., Ireland B.J., Whiteman J.A., Mulvaney R., Whittle C.K., 1984. – Authigenic chlorites: problems with chemical analysis and structural formula calculation. Clay Min., 19:471–481.

Dapples E.C., 1979. – Diagenesis of sandstones. In: Larsen G., Chilingar G.V., Eds., Developments in sedimentology. Elsevier, Amsterdam, 25A:31–97.

Darby D.A., 1975. – Kaolinite and other clay minerals in Arctic Ocean sediments. J. Sediment. Petrol., 45:272–279.

Darsac C., 1983. – La plate-forme berriaso-valanginienne du Jura méridional aux massifs subalpins (Ain-Savoie). Thèse 3ème cycle, Grenoble:316 pp.

Darwin C., 1846. – An account of the fine dust which often falls on vessels in the Atlantic Ocean. Q.J. Geol. Soc. London, 2:26–30.

Datcharry B., 1987. – Histoire du Nord-Est du bassin de Sverdrup (Canada) du Valanginien à l'Albien: précisions lithofaciologiques, palynologiques et minéralogiques. Bull. Soc. géol. Fr., (8) 3:575–581.

Dauphin J.P., 1983. – Eolian quartz granulometry as a paleowind indicator in the Northeast Equatorial Atlantic, North Pacific and Southeast Equatorial Pacific. PhD Thesis, Univ. Rhode Island:335 pp.

Davis E.E., Goodfellow W.D., Bornhold B.D., Adshead J., Blaise B., Villinger H., Cheminant G.M. Le, 1987. – Massive sulfides in a sedimented rift valley, northern Juan de Fuca ridge. Earth Planet. Sci. Lett., 82:49–61.

Dean W.E., Gorham E., 1976. – Major chemical and mineralogical components of profundal surface sediments in Minnesota Lakes. Limnol. Oceanogr., 21:259–284.

Debrabant P., Chamley H., 1982a. – Environnement du fossé hellénique d'après la minéralogie et la géochimie de dépôts superficiels (mission Cyanheat 1979). Géol. Méditerr., 9:11–21.

Debrabant P., Chamley H., 1982b. – Influences océaniques et continentales dans les premiers dépôts de l'Atlantique Nord. Bull. Soc. géol. Fr., Océans-Paléocéans, Lille, (7) 24:473–486.

Debrabant P., Foulon J., 1979. – Expression géochimique des variations du paléoenvironnement depuis le Jurassique supérieur sur les marges nord-atlantiques. Oceanol. Acta, 2:469–476.

Debrabant P., Chamley H., Foulon J., Maillot H., 1979. – Mineralogy and geochemistry of upper Cretaceous and Cenozoic sediments from North Biscay Bay and Rockall Plateau (Eastern North Atlantic), DSDP Leg 48. In: Montadert L., Roberts D.G. et al., Eds., Init. Rep. Deep Sea Drill. Proj., 48, Washington (U.S. Gov. Print. Off.):703–725.

Debrabant P., Chamley H., Foulon J., 1984. – Paleoenvironmental implications of mineralogic and geochemical data in the Western Florida Straits (Leg 77, Deep Sea Drilling Project). In: Buffler R.T., Schlager W. et al., Eds., Init. Rep. Deep Sea Drill. Proj., 77, Washington (U.S. Gov. Print. Off.):377–396.

Debrabant P., Delbart S., Lemaguer D., 1985. – Microanalyses géochimiques de minéraux argileux de sédiments prélevés en Atlantique Nord (forages du DSDP). Clay Min., 20:125–145.

Debrabant P., Krissec L., Bouquillon A., Chamley H., 1989. – Clay mineral supplies in Neogene sediments of the western Arabian Sea: paleoenvironmental implications (O.D.P., Leg 117). Init. Rep. Ocean Drill. Proj., 117, Proc. (in press)

Decarreau A., 1980. – Cristallogenèse expérimentale des smectites magnésiennes hectorite, stévensite. Bull. Minér., 103:579–590.

Decarreau A., 1981. – Cristallogenèse à basse température de smectites trioctaédriques par vieillissement de coprécipités silicométalliques de formule $(Si_{(4-x)}Al_x)M^{2+}O_{14},nH_2O$, où x varie de 0 à 1 et où M^{2+} = Mg-Ni-Co-Zn-Fe-Cu-Mn. C.R. Acad. Sci., Paris, 292:61–64.

Decarreau A., Bonnin D., Badaut-Trauth D., Couty R., Kaiser P., 1987. – Synthesis and crystallogenesis of ferric smectite by evolution of Si-Fe coprecipitates in oxidizing conditions. Clay Min., 22:207–223.

Decommer H., Chamley H., 1981. – Environnements mésozoïques du Nord de la France, d'après les données des argiles et du palynoplancton. C.R. Acad. Sci., Paris, 293:695–698.

Deconinck J.-F., 1987. – Identification de l'origine détritique ou diagénétique des assemblages argileux: le cas des alternances marne-calcaire du Crétacé inférieur subalpin. Bull. Soc. géol. Fr., (8), 3:139–145.

Deconinck J.-F., Chamley H., 1983. – Héritage et diagenèse des minéraux argileux dans les alternances marno-calcaires du Crétacé inférieur du domaine subalpin. C.R. Acad. Sci., Paris, II, 297:589–594.

Deconinck J.-F., Debrabant P., 1985. – Diagenèse des argiles dans le domaine subalpin: rôles respectifs de la lithologie, de l'enfouissement et de la surcharge tectonique. Rev. Géol. dyn. Géogr. phys., 26:321–330.

Deconinck J.-F., Strasser A., 1987. – Sedimentology, clay mineralogy and depositional environment of Purbeckian green marls (Swiss and French Jura). Eclog. geol. Helv., 80:753–772.

Deconinck J.-F., Beaudoin B., Chamley H., Joseph P., Raoult J.-F.,1985. – Contrôles tectonique, eustatique et climatique de la sédimentation argileuse du domaine subalpin français au Malm-Crétacé. Rev. Géol. dyn. Géogr. phys., 26:311–320.

Deconinck J.-F., Strasser A., Debrabant P., 1988. – Formation of illitic minerals at surface temperature in Purbeckian sediments (lower Berriasian, Swiss and French Jura). Clay Min., 23:91–103.

Degens E.T., Ross D.A., Eds., 1969. – Hot brines and recent heavy metal deposits in the Red Sea. Springer, Berlin, Heidelberg, New York:599 pp.

Degens E.T., von Herzen R.P., Wong H.K., 1971. – Lake Tanganyika: water chemistry, sediments, geological structure. Naturwissenschaften, 58:229–241.

Dejou J., Guyot J., Pédro G., Chaumont C., Antoine H., 1968. – Nouvelles données concernant la présence de gibbsite dans les formations d'altération superficielle des massifs granitiques (cas du Cantal et du Limousin). C.R. Acad. Sci., Paris, 266:1825–1827.

De Lange G.J., Rispens F.B., 1986. – Indication of a diagenetically induced precipitate of an Fe-Si mineral in sediment from the Nares abyssal plain, Western North Atlantic. Mar. Geol., 73:85–97.

Delany A.C., Delany A.C., Parkin D.W., Griffin J.J., Goldberg E.D., Reiman B.E.F., 1967. – Airborne dust collected at Barbados. Geochim. Cosmochim. Acta, 31:885–909.

Delmas R.J., Legrand M., Aristarani A.J., Zanolini F., 1985. – Volcanic deposits in Antarctic snow and ice. J. Geophys. Res., 90:12901–12920.

Desperyroux Y., Chamley H., 1986. – Distribution des sédiments récents dans l'estuaire de la Canche (Pas-de-Calais). Ann. Soc. géol. Nord, 105:179–186.

Desprairies A., 1963. – Etude d'un paléo-sol de type ferrugineux tropical, situé dans la région de Brioude (Haute-Loire). 88ème Congr. Soc. Sav.:369–376.

Desprairies A., 1981. – Authigenic minerals in volcanogenic sediments cored during Deep Sea Drilling Project leg 60. In: Hussong D.M., Uyeda S. et al., Eds., Init. Rep. Deep Sea Drill. Proj., 60, Washington (U.S. Gov. Print. Off.):455–466.

Desprairies A., 1983. – Relation entre le paramètre b des smectites et leur contenu en fer et magnésium. Application à l'étude des sédiments. Clay Min., 18:165–175.

Desprairies A., Courtois C., 1980. – Relation entre la composition des smectites d'altération sous-marine et leur cortège de terres rares. Earth Planet. Sci. Lett., 48:124–130.

Desprairies A., Jehanno C., 1983. – Paragenèses minérales liées à des interactions basalte-sédiment-eau de mer (sites 465 et 456 des legs 65 et 60 du D.S.D.P.). Sci. géol. Bull, Strasbourg, 36:93–110.

Desprairies A., Lapierre H., 1973. – Les argiles liées au volcanisme du massif du Troodos (Chypre) et leur remaniement dans sa couverture. Rev. Géogr. phys. Géol. dyn., 15:499–510.

Desprairies A., Bonnot-Courtois C., Jehanno C., Vernhet S., Joron J.L., 1981. – Mineralogy and geochemistry of alteration products in Leg 81 basalts. In: Roberts D.G., Schnitker D. et al., Eds., Init. Rep. Deep Sea Drill. Proj., 81, Washington (U.S. Gov. Print. Off.): 733–742

Deuser W., Ross E., Anderson R., 1981. – Seasonality in the supply of sediment to the deep Sargasso Sea and implications for the rapid transfer of matter to the deep ocean. Deep-Sea Res., 28:495–505.

Deuser W., Brewer P., Jickells T., Commeau R., 1983. – Biological control of the removal of abiogenic particles from the surface ocean. Science, 219:388–391.

Devine S.B., Ferrell R.E.H., Billings G.K., 1973. – Mineral distribution patterns, deep Gulf of Mexico. Am. Assoc. Petrol. Geol. Bull, 57:28–41.

Diester-Haass L., 1975. – Influence of deep oceanic currents on calcareous sands off Brazil. 9th Int. Congr. Sediment., Nice, 8:25–28.

Diester-Haass L., 1976. – Late Quaternary climatic variations in northwest Africa deduced from East Atlantic sediment cores. Q. Res., 6:299–314.

Diester-Haass L., 1979. – DSDP site 397: climatological, sedimentological, and oceanographic changes in the Neogene autochthonous sequence. In: Rad U., Ryan W.B.F. et al., Eds., Init. Rep. Deep Sea Drill. Proj., 47, A, Washington (U.S. Gov. Print. Off.):647–670.

Diester-Haass L., Chamley H., 1978. – Neogene paleoenvironment off NW Africa based on sediments from D.S.D.P. Leg 14. J. Sediment. Petrol., 48:879–896.

Diester-Haass L., Chamley H., 1980. – Oligocene climatic, tectonic and eustatic history off NW Africa (DSDP Leg 41, Site 369). Oceanol. Acta, 3:115–126.

Dietz R.S., 1941. – Clay minerals in recent marine sediments. Am. Min., 27:219–220.

Dilli K., Rao C.N., 1982. – A note on the burial diagenesis of clay minerals in the Bengal fan. Geol. Soc. India, 23:561–566.

Dillon W.P., Sougy J.M.A., 1974. – Geology of West Africa and Canary and Cape Verde Islands. In: Nairn E.M., Stehli F.D., Eds., The Ocean basins and margins, 2: The North Atlantic. Plenum, New York:315–390.

Dixon J.B., Weed S.B., 1977. – Minerals in soil environments. Soil Sci. Soc. Am.:948 pp.

Dogan A.U., Brenner R.L., 1987. – Prediction of reservoir properties using diagenetic analysis of a template unit: example from upper Cretaceous sandstones in Powder river basin, Wyoming. Am. Assoc. Petrol. Geol. Bull., 71:549.

Dominik J., Stoffers P., 1979. – The influence of late Quaternary stagnations on clay sedimentation in the Eastern Mediterranean Sea. Geol. Rundsch., 68:302–317.

Doyle L.J., Sparks T.N., 1980. – Sediments of the Mississippi, Alabama and Florida (Mafla) continental shelf. J. Sediment. Petrol., 50:905–916.

Drever J.I., 1971 a. – Early diagenesis of clay minerals, Rio Ameca basin, Mexico. J. Sediment. Petrol., 41:982–994.

Drever J.I., 1971 b. – Chemical and mineralogical studies, Site 66. In: Winterer E.L. et al., Eds., Init. Rep. Deep Sea Drill. Proj., 7, Washington (U.S. Gov. Print. Off.):965–975.

Drever J.I., 1985. – The chemistry of weathering. NATO ASI Series. Reidel, Dordrecht:324 pp.

Dritz V.A., 1981. – Structural study of minerals by selected area electron diffraction, high resolution electron microscopy. Nauka, Moscow:225 pp.

Droste J.B., 1959. – Clay minerals in Playas of the Mojave desert (California). Science, 130:100.

Droste J.B., 1961. – Clay minerals in sediments of Owens Chinas, Searles, Panamint, Bristol, Cadiz and Danby lake basins, California. Geol. Soc. Am. Bull., 72:1713–1722.

Dubreuilh J., Marchadour P., Thiry M., 1984. – Cadre géologique et minéralogie des argiles des Charentes, France. Clay Min., 19, 1:29–41.

Duce R.A., Unni C.K., Ray B.J., Prospero J.M., Merrill J.T., 1980. – Long-range atmospheric transport of soil dust from Asia to the Tropical North Pacific: temporal variability. Science, 209:1522–1524.

Duchaufour P., Bonneau M., Souchier B., 1977. – Pédologie. 1. Pédogenèse et classification. Masson, Paris:477 pp.

Duncan J.R., Kulm L.D., Griggs G.B., 1970. – Clay mineral composition of late Pleistocene and Holocene sediments of Cascadia Basin, Northeastern Pacific Basin. J. Geol., 78:213–220.

Dunoyer de Segonzac G., 1969. – Les minéraux argileux dans la diagenèse. Passage au métamorphisme. Mém. Serv. Carte géol. Als.-Lorr., 29:320 pp.

Dunoyer de Segonzac G., Abbas M., 1976. – Métamorphisme des argiles dans le Rhétien des Alpes sud-occidentales. Sci. géol., Bull, Strasbourg, 29:3–20.

Dunoyer de Segonzac G., Bernoulli D., 1976. – Diagenèse et métamorphisme des argiles dans le Rhétien sud-alpin et austro-alpin (Lombardie et Grisons). Bull. Soc. géol. Fr., (7), 18:1283–1293.

Dunoyer de Segonzac G., Chamley H., 1968. – Sur le rôle joué par la pyrophyllite comme marqueur dans les cycles sédimentaires. C.R. Acad. Sci., Paris, 267:274–277.

Dunoyer de Segonzac G., Artru P., Ferrero J., 1966. – Sur une transformation des minéraux argileux dans les "terres noires" du bassin de la Durance: influence de l'orogénie alpine. C.R. Acad. Sci., Paris, 262:2401–2404.

Duplay J., 1982. – Populations de monoparticules d'argiles. Thèse 3ème cycle, Poitiers:110 pp.

Duplay J., Desprairies A., Paquet H., Millot G., 1986. – Céladonites et glauconites. Double population de particules dans la céladonite de Chypre. Essai sur les températures de formation. C.R. Acad. Sci., Paris, 302:181–186.

Durand B., Ed., 1984. – Thermal phenomena in sedimentary basins. Technip, Paris, Coll. Colloques et Séminaires, 41:326 pp.

Dyer K.R., 1986. – Coastal and estuarine sediment dynamics. Wiley & Sons, New York:342 pp.

Dymond J., 1981. – Geochemistry of Nazca plate surface sediments: an evaluation of hydrothermal, biogenic, detrital, and hydrogenous sources. Geol. Soc. Am. Mem. 154:133–173.

Dymond J., Eklund W., 1978. – A microprobe study of metalliferous sediment components. Earth Planet. Sci. Lett., 40:243–251.

Dymond J., Biscaye P.E., Rex R.W., 1974. – Eolian origin of mica in Hawaiian soils. Geol. Soc. Am. Bull., 85:37–40.

Dymond J., Lyle M., Finney B., Piper D.Z., Murphy K., Conard R., Pisias N., 1984. – Ferromanganese nodules from MANOP Sites H, S and R. Control of mineralogical and chemical composition by multiple accretionary processes. Geochim. Cosmochim. Acta, 48:931–949.

Dyni J.R., 1976. – Trioctahedral smectite in the Green River Formation, Duchesne Country, Utah. U.S. Geol. Surv. Prof. Pap., 967:14 p.

Eardley A.J., 1938. – Sediments of Great Salt lake. Utah. Am. Assoc. Petrol. Geol. Bull., 22:1305–1411.

Eberhart J.P., 1976. – Méthodes physiques d'étude des minéraux et des matériaux solides. Doin, Paris:507 pp.

Eberl D., Hower J., 1977. – The hydrothermal transformation of sodium and potassium smectite into mixed-layer clay. Clays Clay Min., 26:327–340.

Eberl D.D., Środoń J., Northrop H.R., 1986. – Potassium fixation in smectite by wetting and drying. In: Davis J.A., Hayes K.F., Eds., Geochemical processes at mineral surfaces. Am. Chem. Soc. Symp. ser., 323:296–326.

Ece O.I., 1987. – Petrology of the Desmoinesian Excello black shale of the Midcontinent region of the United States. Clays Clay Min., 35:262–270.

Edman J.D., Surdam R.C., 1986. – Organic-inorganic interactions as a mechanism for porosity enhancement in the upper Cretaceous Ericson sandstone, Green River Basin, Wyoming. Soc. Econ. Paleontol. Min., Spec. Publ., 38:85–109.

Edzwald J.K., O'Melia C.R., 1975. – Clay distributions in recent estuarine sediments. Clays Clay Min., 23:39–44.

Ehlmann A.J., Hulings N.C., Glover E.D., 1963. – Stages of glauconite formation in modern foraminiferal sediments. J. Sediment. Petrol., 33:87–96.

Ehrlich H.L., 1981. – Geomicrobiology. Dekker, New York: 393 pp.

Einsele G., Gieskes J.M., Curray J., Moore D.M., Aguayo E., Aubry M.-P., Fornary D., Guerrero J., Kastner M., Kelts M., Lyle M., Matoba Y., Molina-Cruz A., Niemitz J., Rueda J., Saunders A., Schrader H., Simoneit B., Vacquier V., 1980. – Intrusion of basaltic sills into highly porous sediments, and resulting hydrothermal activity. Nature (London), 283:441–445.

Elderfield H., 1976. – Hydrogenous material in marine sediments, excluding manganese nodules. In: Riley J.P., Chester R., Eds., Chemical oceanography. Academic Press, New York, London, 5:137–215.

El-Heiny I., 1982. – Neogene stratigraphy of Egypt. Newslett. Stratigr., 11:41–54.

Emeis K., 1985. – Particulate suspended matter in major world rivers-II: Results on the rivers Indus, Waikato, Nile, St Lawrence, Yangtse, Parana, Orinoco, Caroni and Mackenzie. Mitt. Geol..-Paläont. Inst. Univ. Hamburg, 58:593–617.

Emeis K.-C., Mycke B., Degens E.T., 1989. – Provenance and maturity of organic carbon in late Tertiary to Quaternary sediments from the Tyrrhenian Sea (ODP Leg 107/Holes 652A and 654A). In: ODP Leg 107, Proc., B (U.S. Gov. Print. Off.). (in press)

Emery K.O., Honjo S., 1979. – Surface suspended matter off western Africa: relations of organic matter, skeletal debris, and detrital minerals. Sedimentology, 26:775–794.

Emery K.O., Lepple F., Toner L., Uchupi E., Rioux R.H., Pople W., Hulburt E.M., 1974. – Suspended matter and other properties of surface waters of the northeastern Atlantic Ocean. J. Sediment. Petrol., 44:1087–1110.

Englund J.O., Jorgensen P., Roaldset E., Aagaard P., 1976. – Composition of water and sediments in lake Mjøsa, South Norway, in relation to weathering processes. In: Golterman H.L., Ed., Interactions between sediments and fresh water. Proc. Int. Symp., Amsterdam, Sept. 6–10, 1976:125–132.

Escalon de Fonton M., 1966. – Du Paléolithique supérieur au Mésolithique dans le Midi méditerranéen. Bull. Soc. préhist. Fr., 63:66–180.

Eslinger E., Sellars B., 1981. – Evidence for the formation of illite from smectite during burial metamorphism in the Belt supergroup, Clark Fork, Idaho. J. Sediment. Petrol., 51:203–216.

Eslinger E.V., Yeh H.-W., 1981. – Mineralogy, O^{18}/O^{16} and D/H ratios of clay-rich sediments from Deep Sea Drilling Project Site 180, Aleutian trench. Clays Clay Min., 29:309–315.

Eslinger E.V., Mayer L.M., Durst T.L., Hower J., Savin S.M., 1973. – A X-ray technique for distinguishing between detrital and secondary quartz in the fine-grained fraction of sedimentary rocks. J. Sediment. Petrol., 43:540–543.

Esquevin J., Kulbicki G., 1963. – Les minéraux argileux de l'Aptien supérieur du Bassin d'Arzacq (Aquitaine). Bull. Serv. Carte géol. Als.-Lorr., 16:197–203.

Estéoule-Choux J., 1967a. – Contribution à l'étude des argiles du Massif Armoricain. Thèse Sci. Nat., Rennes:319 pp.

Estéoule-Choux J., 1967b. – Les minéraux argileux du Tertiaire breton. Bull. Groupe Fr. Argiles, 19:11–24.

Estéoule J., Estéoule-Choux J., Melguen M., Seibold E., 1970. – Sur la présence d'attapulgite dans les sédiments récents du Nord-Est du Golfe Persique. C.R. Acad. Sci., Paris, 271:1153–1156.

Eugster H.P., Hardie L.A., 1978. – Saline lakes.In: Lerman A., Ed., Lakes. Chemistry, geology, physics. Springer, Berlin, Heidelberg, New York:239–293.

Eugster H.P., Kelts K., 1983. – Lacustrine chemical sediments. In: Goudie A.S., Pye K., Eds., Chemical sediments and geomorphology. Academic Press, New York, London:321–368 (439 pp.).

Faas R.W., Crocket D.S., 1983. – Clay fabric development in a deep-sea core: site 515, Deep Sea Drilling Project leg 72. In: Backer P.F., Carlson R.L., Johnson D.A. et al., Eds., Init. Rep. Deep Sea Drill. Proj., 72, Washington (U.S. Gov. Print. Off.):519–535.

Faugères J.-C., Gonthier E., 1981. – Les argiles des sédiments marins du Quaternaire récent dans le golfe d'Aden et la mer d'Oman (mission Orgon IV). Oceanol. Acta, 4:395–399.

Faugères J.-C., Desbruyères D., Gonthier E., Griboulard R., Poutiers J., Resseguier A. de, Vernette G., 1987. – Témoins sédimentologiques et biologiques de l'activité tectonique actuelle du prisme d'accrétion de la Barbade. C.R. Acad. Sci., Paris, 305:115–119.

Faugères L., Robert C., 1976. – Etude sédimentologique et minéralogique de deux forages du golfe thermaïque (mer Egée). Géol. Médit., 3:209–218.

Fedoroff N., 1986. – Un plaidoyer en faveur de la paléopédologie. Bull. Assoc. Fr. Etude Q., 27–28:195–204.

Fedoroff N., Bresson L.M., Courty M.A., Eds., 1987. – Micromorphologie des sols/Soil micromorphology. A.F.E.S., Fr.:686 pp.

Feininger T., 1971. – Chemical weathering and glacial erosion of crystalline rocks and the origin of till. U.S. Geol. Surv. Prof. Pap., 750-C:C65–C81.

Ferguson W.S., Griffin J.J., Goldberg E.D., 1970. – Atmospheric dusts from the North Pacific – a short note on a long-range eolian transport. J. Geophys. Res., 75:1137–1139.

Ferrell R.E. Jr., 1987. – Microtexture of clay-rich sediments from the Oslofjord, Norway. In: Schultz L.G., Olphen H. van, Mumpton F.A., Eds., Proc. Int. Clay Conf., Denver 1985:121–127.

Ferrero J., Kubler B., 1964. – Présence de dickite et de kaolinite dans les grès cambriens d'Hassi-Messaoud. Bull. Serv. Carte géol. Als.-Lorr., 17:247–262.

Ferry S., Rubino J.-L., 1987. – La modulation eustatique du signal orbital dans les sédiments pélagiques. C.R. Acad. Sci., Paris, 305, II:477–482.

Ferry S., Schaaf A., 1981. – The early Cretaceous environment at Deep Sea Drilling Project site 463 (mid-Pacific mountains), with reference to the Vocontian trough (French subalpine ranges). In: Thiede J., Vallier T.L. et al., Eds., Init. Rep. Deep Sea Drill. Proj., 62, Washington (U.S. Gov. Print. Off.):669–682.

Ferry S., Cotillon P., Rio M., 1983. – Diagenèse croissante des argiles dans les niveaux isochrones de l'alternance calcaire-marne valanginienne du bassin vocontien. Zonation géographique. C.R. Acad. Sci., Paris, 297:51–56.

Feuillet J.-P., Fleischer P., 1980. – Estuarine circulation: controlling factor of clay mineral distribution in James river estuary, Virginia. J. Sediment. Petrol., 50:267–279.

Fisher R.V., Schmincke H.U., 1984. – Pyroclastic rocks. Springer, Berlin, Heidelberg, New York:472 pp.

Fishman N.S., Turner-Peterson C.E., Owen D.E., 1987. – Early diagenetic formation of illite: implications for clay geothermometry. Am. Assoc. Petrol. Geol. Bull., 71:557–558.

Folger D.W., 1970. – Wind transport of land-derived mineral, biogenic and industrial matter over the North Atlantic. Deep-Sea Res., 17:337–352.

Folger D.W., 1972. – Characteristics of estuarine sediments of the United States. Geol. Surv. Prof. Pap., 742:94 pp.

Fontes J.C., Fritz P., Letolle R., 1970. – Composition isotopique, minéralogique et genèse des dolomies du Bassin de Paris. Geochim. Cosmochim. Acta, 34:279–294.

Foscolos A.E., Powell T.G., Gunther T.R., 1976. – The use of clay minerals and inorganic and organic geochemical indicators for evaluating the degree of diagenesis and oil generating potential of shales. Geochim. Cosmochim. Acta, 40:953–966.

Fournier-Germain B., 1986. – Les sédiments métallifères océaniques actuels et anciens: caractérisation, comparaisons. Thèse 3ème cycle, Brest:236 pp. + ann.

Frakes L.A., 1979. – Climates throughout geologic time. Elsevier, Amsterdam:310 pp.

Francavilla F., Tomadin L., 1973. – Quelques observations sur la sédimentation des argiles et des pollens d'une carotte de la mer Tyrrhénienne. Rapp. Comm. Int. Mer Médit., C.I.E.S.M., 21:913–915.

Frey M., 1970. – The step from diagenesis to metamorphism in pelitic rocks during Alpine orogenesis. Sedimentology, 15:261–279.

Frey M., Ed., 1987. – Low temperature metamorphism. Blackie, Glasgow:351 pp.

Fritz B., 1985. – Multicomponent solid solutions for clay minerals and computer modeling of weathering processes. In: Drever J.I., Ed., The chemistry of weathering. Reidel, Dordrecht:19–34.

Froget C., 1981. – La sédimentation argileuse depuis l'Eocène sur le plateau Vøring et à son voisinage, d'après le Leg 38 D.S.D.P. (Mer de Norvège). Sedimentology, 28:793–804.

Froget C., Chamley H., 1977. – Présence de sépiolite détritique dans les sédiments récents du golfe d'Arzew (Algérie). C.R. Acad. Sci., Paris, 285:307–310.

Fröhlich F., 1980. – Néoformation de silicates ferrifères amorphes dans la sédimentation pélagique récente. Bull. Minér., 103:596–599.

Fröhlich F., 1982. – Evolution minéralogique dans les dépôts azoïques rouges de l'océan Indien. Relations avec le stratigraphie. Bull. Soc. géol. Fr. (7) 14:563–571.

Frye J.C., Glass H.D., Leonard A.B., Coleman D., 1974. – Caliche and clay mineral zonation of Ogallala formation, Central-Eastern Mexico. New Mexico Inst. Min. Technol., Circul. 144:1–16.

Füchtbauer H., Goldschmidt H., 1959. – Die Tonminerale der Zechsteinformation. Beitr. Miner. Petrogr., 6:320–345.

Gabis V., 1963. – Etude minéralogique et géochimique de la série sédimentaire oligocène du Velay. Bull. Soc. Fr. Minéral. Cristallogr., 86:315–354.

Gac J.-Y., 1979. – Géochimie du bassin du lac Tchad. Bilan de l'altération, de l'érosion et de la sédimentation. Thèse, Sci. nat., Strasbourg:249 pp.

Gac J.-Y., Badaut D., Al-Droubi A., Tardy Y., 1978. – Comportement du calcium, du magnésium et de la silice en solution. Précipitation de calcite magnésienne, de silice amorphe et de silicates magnésiens au cours de l'évaporation des eaux du Chari (Tchad). Sci. Géol. Bull., Strasbourg, 31:185–193.

Galan E., Castillon A., 1984. – Sepiolite-Palygorskite in spanish tertiary basins: genetical patterns in continental environments. In: Singer A., Galan E., Eds., Palygorskite-sepiolite. Occurrences, genesis and uses. Developments in sedimentology. Elsevier, Amsterdam, 37:87–124.

Galan E., Gonzalez Lopez M., Fernandez Nieto C., Gonzalez Diez I., 1985. – Clay minerals of Miocene-Pliocene materials at the Vera Basin, Almeria, Spain. Geological interpretation. Miner. Petrogr. Acta, 29-A:259–266.

Galan E., Perez-Rodriguez J.L., Cornejo J., 1987. – The 6th Meet Eur. Clay Groups, Sevilla. Soc. Espan. Arcillas, Abstr.:632 pp.

Gallenne B., 1974. – Sélection dynamique de la montmorillonite au sein du bouchon vaseux, dans l'estuaire de la Loire. C.R. Acad. Sci., Paris, 278:831–834.

Galliher E.W., 1935. – Geology of glauconite. Bull. Am. Assoc. Petrol. Geol., 19:1569–1601.

Galloway W.E., 1974. – Deposition and diagenetic alteration of sandstone in Northeast Pacific arc-related basins: implications for graywacke genesis. Geol. Soc. Am. Bull., 85:379–390.

Gandais V., 1987. – Clay mineral sources of the Grenada basin, Southeastern Caribbean. Clay Min., 22:395–400.

Gard J.A., 1971. – The electron-optical investigation of clays. Mineral. Soc., London: 383 pp.

Gardner J.V., Dean W.E., Vallier T.L., 1980. – Sedimentology and geochemistry of surface sediments, outer continental shelf, Southern Bering Sea. Mar. Geol., 35:299–329.

Gardner L.R., 1972. – Origin of the Mormon Mesa caliche, Clark Country, Nevada. Geol. Soc. Am. Bull., 83:143–156.

Gastuche M.C., De Kimpe C., 1961. – La genèse des minéraux argileux de la famille du kaolin. II. Aspect cristallin. Coll. Int. C.N.R.S., 105:67–81.

Gaudichet A., Buat-Ménard P., 1982. – Nature minéralogique et origine des particules atmosphériques insolubles du Pacifique Tropical Nord (Atoll d'Enewetak). Etude par microscopie électronique analytique en transmission. C.R. Acad. Sci., Paris, 294:1241–1246.

Gaudichet A., Petit J.-R., Lefèvre R., Lorius C., 1986. – An investigation by analytical transmission electron microscopy of individual insoluble microparticles from Antarctic (Dome C) ice core samples. Tellus, 38B:250–261.

Geisler-Cussey D., Moretto R., 1984. – Evolution paragénétique et géochimique de la phase argileuse dans des dépôts évaporitiques. Exemple du Muschelkalk moyen de Lorraine et du Stampien de Bresse. In: 5th Eur. Reg. Meet. Sedimentology, I.A.S., Marseille, Abstr.:193–194.

Germaneau J., 1969. – Etude de la sédimentation dans l'estuaire de la Seine. II. Origine, déplacement et dépôt des suspensions. Trav. C.R.E.O., 9, 1–4:100 pp.

Gibbs R.J., 1967. – The geochemistry of the Amazon river system. Part I: The factors that control the salinity and the compostion and concentration of the suspended solids. Geol. Soc. Am. Bull., 78:1203–1232.

Gibbs R.J., 1977a. – Transport phases of transition metals in the Amazon and Yukon Rivers. Geol. Soc. Am. Bull., 88:829–843.

Gibbs R.J., 1977b. – Clay mineral segregation in the marine environment. J. Sediment. Petrol., 47:237–243.

Gibbs R.J., 1983. – Coagulation rates of clay minerals and natural sediments. J. Sediment. Petrol., 53:1193–1203.

Gieskes J.M., 1983. – The chemistry of interstitial waters of deep sea sediments: interpretation of Deep Sea Drilling data. In: Riley J.P., Chester R., Eds., Chemical oceanography. Academic Press, New York, London, 8:222–269.

Gile L.H., 1967. – Soils of an ancient basin floor near Las Cruces, New Mexico. Soil Sci., 103:265–276.

Gillot E., 1984. – Signification des glauconies diffuses et granulaires sur la marge celtique au Crétacé (campagne DSDP IPOD 81, sites 549 et 550). Cas de glauconitisation profonde. In: 5th Eur. Reg. Meet Sedimentology, Marseille, Abstr.:197–198.

Gillot E., Magniez-Jannin F., Pascal A., Rat P., 1984. – Peuplements et critères sédimentologiques d'environnement dans l'interprétation d'une séquence transgressive à partir du Barrémien du sondage D.S.D.P. leg 80, site 549 (Atlantique NE). Bull. Soc. géol. Fr., (7) 26:1349–1356.

Girard J.-P., 1985. – Diagenèse hydrothermale tardive des sédiments gréso-argileux du Protérozoïque supérieur du bassin de Taoudeni (Afrique de l'Ouest). Thèse 3ème cycle, Univ. Poitiers:250 pp. + ann.

Giresse P., 1965. – Observations sur la présence de "Glauconie" actuelle dans les sédiments ferrugineux peu profonds du bassin gabonais. C.R. Acad. Sci., Paris, 260:5597–5600.

Giresse P., 1967. – Mécanismes de répartition des minéraux argileux des sédiments marins actuels sur le littoral sud du Cotentin. Mar. Geol., 5:61–69.

Giresse P., 1985. – Le fer et les glauconies au large de l'embouchure du fleuve Congo. Sci. Géol., Bull., Strasbourg, 38:293–322.

Giresse P., 1987. – Les glauconies de la marge atlantique de l'Afrique au Mésozoïque et au Cénozoïque. Rapp. A.T.P. G.G.O., C.N.R.S., Fr., 4782:109 pp.

Giresse P., Odin G.S., 1973. – Nature minéralogique et origine des glauconies du plateau continental du Gabon et du Congo. Sedimentology, 20:457–488.

Giresse P., Lamboy M., Odin G.S. 1980. – Evolution géométrique des supports de glauconitisation. Application à la reconstitution du paléoenvironnement. Oceanol. Acta, 3:251–260.

Giroud d'Argoud G., Chamley H., Masse J.-P., 1976. – Sur la signification des minéraux argileux dans les sédiments de l'Aptien supérieur de Provence. C.R. Acad. Sci., Paris, D, 282:1673–1675.

Glaccum R.A., Prospero J.M., 1980. – Saharan aerosols over the tropical North-Atlantic. Mineralogy. Mar. Geol., 37:295–321.

Glasby G.P., Read A.J., 1976. – Deep-sea manganese nodules. In: Wolf K.H., Ed., Handbook of strata-bound and stratiform ore deposits. Elsevier, Amsterdam, 7:295–340.

Glasby G.P., Stoffers P., Sioulas A., Thijssen T., Friedrich G., 1982. – Manganese nodule formation in the Pacific ocean: a general theory. Geo-Mar. Lett., 2:47–53.

Glenn R.C., Jackson M.L., Hole F.D., Lee G.B., 1960. – Chemical weathering of layer silicate clays in loess-derived Tama silt loam of southern Wisconsin. Clays Clay Min. (8th Natl. Conf., 1959):63–83.

Goldberg E.D., 1961. – Chemical and mineralogical aspects of deep-sea sediments. Phys. Chem. Earth, 4:281–302.

Goldberg E.D., Griffin J.J., 1964. – Sedimentation rates and mineralogy in the South Atlantic. J. Geophys. Res., 69:4293–4309.

Goldberg E.D., Griffin J.J., 1970. – The sediments of the northern Indian Ocean. Deep-Sea Res., 17:513–537.

Goldbery R., 1979. – Sedimentology of the Lower Jurassic flint clay bearing Mishhor formation, Makhtesh Ramon, Israël. Sedimentology, 26:229–251.

Goodchild M.W., Whitaker J.H.McD., 1986. – A petrographic study of the Rotliegendes sandstone reservoir (lower Permian) in the Rough gas field. Clay Min., 21:459–477.

Gorbunova Z.N., 1962. – Clay and associated minerals of the Indian ocean sediments. Trans. Inst. Oceanol. U.R.S.S., 61:93–103.

Gorbunova Z.N., 1963. – Clay minerals in Pacific sediments. Litol. Pol. Iks., 1:28–42. (in Russian)

Gorbunova Z.N., 1966. – Clay mineral distribution in the Indian ocean. Okeanologiia, 6:267–275. (in Russian)

Gorbunova Z.N., 1979. – Clay minerals of the north-western Pacific (materials of Leg 6 of the Glomar Challenger). Okeanologiia, 19:658–665.

Goudie A.S., 1983. – Calcrete. In: Goudie A.S., Pye K., Ed., Chemical sediments and geomorphology: precipitates and residua in the near-surface environement. Academic Press, New York, London:93–131.

Goulart E.P., 1976. – Different smectites types in sediment of the Red Sea. Geol. Jahrb., D17:135–149.

Gouleau D., Kalck Y., Marius C., Lucas J., 1982. – Cristaux d'hydroxyde d'aluminium néoformés dans les sédiments actuels des mangroves du Sénégal (Sine-Saloum et Casamance). Mém. Soc. géol. Fr., 144:147–154.

Graciansky P.C. de, Brosse E., Deroo G., Herbin J.-P., Montadert L., Müller C., Schaaf A., Sigal J., 1982. – Les formations d'âge Crétacé de l'Atlantique Nord et leur matière organique: paléogéographie et milieu de dépôt. Rev. Inst. Fr. Pétrol., 37:275–337.

Graciansky P.C. de, Brosse E., Deroo G., Herbin J.-P., Montadert L., Müller C., Sigal J., Schaaf A., 1987. – Organic-rich sediments and palaeoenvironmental reconstructions of the Cretaceous North Atlantic. In: Brooks J., Fleet A.J., Eds., Marine petroleum source rocks. Geol. Soc. Spec. Publ., 26:317–344.

Gradusov B.P., 1974. – A tentative study of clay mineral distribution in soils of the world. Geoderma, 12:49–55.

Griffault L., Beaufort D., Meunier A., 1984. – Les altérations hydrothermales du socle dans le secteur de Chantejail (Haut-Cézallier). Doc. B.R.G.M., 81–10:31–45.

Griffin G.M., 1962. – Regional clay mineral facies. Products of weathering intensity and current distribution in the Northeastern gulf of Mexico. Geol. Soc. Am. Bull., 73:737–767.

Griffin G.M., 1963. – Occurrence of talc in clay fractions from beach sands of the Gulf of Mexico. J. Sediment. Petrol., 33:231–233.

Griffin J.J., Goldberg E.D., 1963. – Clay-mineral distribution in the Pacific Ocean. In: Hill M.N., Ed., The sea. Interscience, New York, 3:728–741.

Griffin G.M., Ingram R.L., 1955. – Clay minerals of the Neuse river estuary. J. Sediment. Petrol., 25:194–200.

Griffin G.M., Parrot B.S., 1964. – Development of clay mineral zones during deltaic migration. Am. Assoc. Petrol. Geol. Bull., 48:57–69.

Griffin J.J., Koide M., Höhndorf A., Hawkins J.W., Goldberg E.D., 1967. – Sediments of the Lau Basin, rapidly accumulating volcanic deposits. Deep-Sea Res., 19:139–148.

Griffin J.J., Windom H., Goldberg E.D., 1968. – The distribution of clay minerals in the world oceans. Deep-Sea Res., 15:433–459.

Griggs G.B., Kulm, L.D., 1970. – Sedimentation in Cascadia deep-sea Channel. Geol. Soc. Am. Bull., 81:1361–1384.

Grim R.E., 1953. – Clay mineralogy. McGraw-Hill, New York:384 pp.

Grim R.E., 1968. – Clay mineralogy. McGraw-Hill, New York:596 pp.

Grim R.E., Güven N., 1978. – Bentonites. Developments in sedimentology. Elsevier, Amsterdam, 24:256 pp.

Grim R.E., Johns W.D., 1954. – Clay mineral investigations of sediments in the northern Gulf of Mexico. Clays Clay Min., 2nd Natl. Conf. Pergamon, New York:81–103.

Grim R.E., Loughnan F.C., 1962. – Clay minerals in sediments from Sydney Harbour, Australia. J. Sediment. Petrol., 32:240–248.

Grim R.E., Kulbicki G., Carozzi A.V., 1960 a. – Clay mineralogy of the sediments of the Great Salt lake, Utah. Geol. Soc. Am. Bull., 71:515–520.

Grim R.E., Droste J.B., Bradley W.F., 1960 b. – A mixed-layer clay mineral associated with an evaporite. Clays Clay Min., 8th Natl. Conf.:228–236.

Gross D.L., Lineback J.A., Shimp N.F., White W.A., 1972. – Composition of Pleistocene sediments in southern lake Michigan, U.S.A. 25th Int. Geol. Congr., 8:215–222.

Grousset F.E., Chesselet R., 1986. – The Holocene sedimentary regime in the northern Mid-Atlantic Ridge region. Earth Plan. Sci. Lett., 78:271–287.

Grousset F., Donard O., 1984. – Enrichments in Hg, Cd, As, and Sb in recent sediments of Azores-Iceland ridge. Geo-Mar. Lett., 4:117–124.

Grousset F., Duplessy J.-C., 1983. – Early deglaciation of the Greenland Sea during the last glacial to interglacial transition. Mar. Geol., 52:M11–M17.

Grousset F., Duplessy J.-C., 1985. – Early deglaciation of the Greenland Sea during the last glacial to interglacial transition. Reply. Mar. Geol., 62:169–173.

Grousset F., Latouche C., Parra M., 1982. – Late Quaternary sedimentation between the Gibbs fracture and the Greenland basin: mineralogical and geochemical data. Mar. Geol., 47:303–330.

Grousset F., Latouche C., Maillet N., 1983. – Clay minerals as indicators of wind and current contribution to post-glacial sedimentation on the Azores/Iceland Ridge. Clay Min., 18:65–75

Guilbert J.M., Park C.F. Jr., 1986. – The geology of ore deposits. Freeman, New York:985 pp.

Guillemot D., Nesteroff W.D., 1979. –Les dépôts métallifères crétacés de Chypre: comparaison avec leurs homologues actuels du Pacifique. In: Panayiotou A., Ed., Int. Ophiolite Symp., Nicosia:139–146.

Guthrie J.M., Houseknecht D.W., Johns W.D., 1986. – Relationships among vitrinite reflectance, illite crystallinity, and organic geochemistry in Carboniferous strata, Ouachita mountains, Oklahoma and Arkansas. Am. Assoc. Petrol. Geol. Bull., 70:26–33.

Güven N., Hower W.F., Davies D.K., 1980. – Nature of authigenic illites in sandstone reservoirs. J. Sediment. Petrol., 50:761–766.

Gygi R.A., 1981. – Oolitic iron formation: marine or not marine? Ecl. Geol. Helv., 74:233–254.

Habib D., 1982. – Sedimentary supply origin in Cretaceous black shales. In: Schlanger S.O., Cita M.B., Eds., Nature and origin of Cretaceous carbon-rich facies, Academic Press, New York, London:112–127.

Hallam A., 1984. – Continental humid and arid zones during the Jurassic and Cretaceous. Palaeogeogr., Climatol., Ecol., 47:195–223.

Hallam A., 1987. – End-Cretaceous mass extinction event: argument for terrestrial causation. Science, 238:1237–1242.

Haq B.U., 1981. – Paleogene paleoceanography: early Cenozoic oceans revisited. Oceanol. Acta, sp.:71–82.

Harder H., 1976. – Nontronite synthesis at low temperatures. Chem. Geol., 18:169–180.

Harder H., 1978. – Synthesis of iron-layer silicate minerals under natural conditions. Clays Clay Min., 26:65–72.

Harder H., 1980. – Synthesis of glauconite at surface temperatures. Clays Clay Min., 28:217–222.

Hardie L.A., 1968. – The origin of the recent non-marine evaporite deposit of Saline Valley, Inyo County, California. Geochim. Cosmochim. Acta, 32:1279–1301.

Harlan R.W., 1966. – A clay mineral study of recent and Pleistocene sediments from the Sigsbee Deep, Gulf of Mexico. PhD thesis, Texas A. and M. University, College Stn.:140 pp.

Harriss R.C., 1967. – Clay minerals and oceanic evolution. Clays Clay Min., Proc. 15th Natl. Conf. Pergamon, New York:207–213.

Hartmann M., Lange H., Seibold E., Walger E., 1971. – Oberflächen-Sedimente im Persischen Golf von Oman. I – Geologisch-hydrologischer Rahmen und erste sedimentologische Ergebnisse. "Meteor" Forsch. Ergebn., 100:1–76.

Hassouba H., Shaw H.F., 1980. – The occurrence of palygorskite in Quaternary sediments of the coastal plain of north-west Egypt. Clay Min., 15:77–83.

Hathaway J.C., 1972. – Regional clay mineral facies in estuaries and continental margin of the United States east coast. In: Nelson B.W., Ed., Environmental framework of coastal plain estuaries. Geol. Soc. Am. Mem., 133:293–316.

Hathaway J.C., Sachs P.L., 1965. – Sepiolite and clinoptilolite from the mid-Atlantic ridge. Am. Miner., 50:852–867.

ten Haven H.L., Baas M., Kroot M., de Leeuw J.W., Schenck P.A., Ebbing J., 1987. – Late Quaternary Mediterranean sapropels. III: Assessment of source of input and palaeotemperature as derived from biological markers. Geochim. Cosmochim. Acta, 51: 803–810.

Hay R.L., 1966. – Zeolites and zeolitic reactions in sedimentary rocks. Geol. Soc. Am., Spec. Pap. 85:1–130.

Hay R.L., Guldman S.G., 1987. – Diagenetic alteration of silicic ash in Searles lake, California. Clays Clay Min., 35:449–457.

Hay R.L., Moiola R.J., 1963. – Authigenic silicate minerals in Searles lake, California. Sedimentology, 2:312–332.

Hay R.L., Stoessel R.K., 1984. – Sepiolite in the Amboseli basin of Kenya: a new interpretation. In: Singer A., Galan E., Eds., Palygorskite-sepiolite. Occurrences, genesis and uses. Developments in sedimentology. Elsevier, Amsterdam, 37:125–136.

Hay R.L., Wiggins B., 1980. – Pellets, ooids, sepiolite and silica in three calcretes of the south-western United States. Sedimentology, 27:559–576.

Hay R.L., Pexton R.E., Teaguet T., Kiser T.K., 1986. – Spring-related carbonate rocks, Mg clays, and associated minerals in Pliocene deposits of the Amargosa desert, Nevada and California. Geol. Soc. Am. Bull., 97:1488–1503.

Hayes J.B., 1970. – Polytypism of chlorite in sedimentary rocks. Clays Clay Min., 18:285–306.

Haymon R.M., Kastner M., 1981. – Hot spring deposits on the East Pacific rise at 21°N: preliminary description of mineralogy and genesis. Earth Planet. Sci. Lett., 53:363–381.

Haymon R.M., Kastner M., 1986. – The formation of high temperature clay minerals from basalt alteration during hydrothermal discharge on the East Pacific rise axis at 21°N. Geochim. Cosmochim. Acta, 50:1933–1939.

Hays J.D., Peruzza A., 1972. – The significance of calcium carbonate oscillations in eastern equatorial deep-sea sediments for the end of the Holocene warm interval. Q. Res., 2:355–362.

Heath G.R., 1969. – Mineralogy of Cenozoic deep-sea sediments from the equatorial Pacific Ocean. Geol. Soc. Am. Bull., 80:1997–2018.

Heath G.R., Pisias N., 1979. – A method for the quantitative estimation of clay minerals in North Pacific deep-sea sediments. Clays Clay Min., 27:1765–1784.

Heath G.R., Moore T.C. Jr., Opdyke N.D., Dauphin J., 1973. – Distribution of quartz, opal, calcium carbonate and organic carbon in Holocene, 600,000 and Brunhes/Matuyama age sediments of the North Pacific. Geol. Soc. Am., Abstr. Progr., 5:662.

Heath G.R., Moore T.C. Jr., Robert G.L., 1974. – Mineralogy of the surface sediments from the Panama basin, eastern equatorial Pacific. J. Geol., 82:145–160.

Heezen B.C., Nesteroff W.D., Sabatier G., 1960. – Répartition des minéraux argileux dans les sédiments profonds de l'Atlantique Nord et Equatorial. C.R. Acad. Sci., Paris, 251:410–413.

Heezen B.C., Nesteroff W.D., Oberlin A., Sabatier G., 1965. – Découverte d'attapulgite dans les sédiments profonds du golfe d'Aden et de la mer Rouge. C.R. Acad. Sci., Paris, 260:5819–5821.

Hein F.J., Longstaffe F.J., 1985. – Sedimentologic, mineralogic, and geotechnical descriptions of fine-grained slope and basin deposits, Baffin Island Fiords. Geo-Mar. Lett., 5:11–16.

Hein J.R., Scholl D.W., 1978. – Diagenesis and distribution of late Cenozoic volcanic sediment in the southern Bering Sea. Geol. Soc. Am. Bull., 89:197–210.

Hein J.R., Scholl D.W., Gutmacher C.E., 1975. – Neogene clay minerals of the far NW Pacific and Southern Bering Sea: sedimentation and diagenesis. Proc. Int. Clay Conf., 1975:71–80.

Hein J.R., Bouma A.H., Hampton M.A., Ross C.R., 1979a. – Clay mineralogy, fine-grained sediment dispersal, and inferred current patterns, lower look inlet and Kodiak shelf, Alaska. Sediment. Geol., 24:291–306.

Hein J.R., Yeh H.W., Alexander E., 1979b. – Origin of iron-rich montmorillonit from the manganese nodule belt of the equatorial Pacific. Clays Clay Min., 27:185–194.

Hein J.R., Ross C.R., Alexander E., Yeh H.W., 1979c. – Mineralogy and diagenesis of surface sediments from Domes A, B and C areas. Mar. Sci., 9:365–396.

Hekinian R., Rosendahl B.R., Cronan D.S., Dimitriev Y., Fodor R.V., Goll R.M., Hoffert M., Humphris S.E., Mattey D.P., Natland J., Petersen N., Schrarder E.L., Strivastava R.K., Warren N., 1978. – Hydrothermal deposits and associated basement rocks from the Galapagos spreading center. Oceanol. Acta, 1:473–482.

Heling D., 1978. – Diagenesis of illite in argillaceous sediments of the Rhine graben. Clay Miner., 13:211–219.

Heller-Kallai L., Yariv S., Riemer M., 1973. – The formation of hydroxy interlayers in smectites under the influence of organic bases. Clay Min., 10:35–40.

Henmi T., Wada K., 1976. – Morphology and composition of allophane. Am. Mineral., 61:379–390.

Herbillon A.J., Frankart R., Vielvoye L., 1981. – An occurrence of interstratified kaolinite-smectite minerals in a red-black soil topsequence. Clay Min., 16:195–201.

Herron M.M., 1986. – Mineralogy from geochemical well logging. Clays Clay Min., 34:204–213.

Hétier J.M., Yoshinaga N., Tardy Y., 1969. – Présence de la vermiculite-Al, montmorillonite-Al et chlorite-Al, et leur répartition dans quelques sols des Vosges. C.R. Acad. Sci., Paris, 268:259–261.

Hétier J.M., Yoshinaga N., Weber F., 1977. – Formation of clay minerals in andosoils under temperate climate. Clay Min., 12:299–308.

Hiltabrand R.R., Ferrell R.E., Billings G.K., 1973. – Experimental diagenesis of Gulf coast argillaceous sediment. Am. Assoc. Petrol. Geol. Bull., 57:338–348.

Hoffert M., 1980. – Les "argiles rouges des grands fonds" dans le Pacifique centre-est. Sci. géol., Strasbourg, Mém. 61:257 pp.

Hoffert M., Lalou C., Brichet E., Bonte P., Jehanno C., 1975. – Présence en Atlantique Nord de nodules de manganèse à noyaux d'attapulgite et de phillipsite authigènes. C.R. Acad. Sci., Paris, 281:231–233.

Hoffert M., Perseil A., Hékinian R., Choukroune P., Needham H.D., Francheteau J., Le Pichon X., 1978. – Hydrothermal deposits sampled by diving saucer in transform fault "A" near 37°N on the mid-Atlantic ridge, Famous area. Oceanol. Acta, 1:73–86.

Hoffert M., Person A., Courtois C., Karpoff A.-M., Trauth D., 1980. – Sedimentology, mineralogy, and geochemistry of hydrothermal deposits from holes 424, 424A, 424B, and 424C (Galapagos spreading center). In: Rosendahl B.R., Hekinian R. et al., Eds., Init. Rep. Deep Sea Drill. Proj., 54, Washington (U.S. Gov. Print. Off.):339–376.

Hoffert M., Cheminée J.-L., Larqué P., Person A., 1987. – Dépôt hydrothermal associé au volcanisme sous-marin "intraplaque" océanique. Prélèvement effectué avec Cyana, sur le volcan sous-marin actif de Teahitia (Polynésie française). C.R. Acad. Sci., Paris, 304:829–832.

Hoffman J., Hower J., 1979. – Clay mineral assemblages as low-grade metamorphic geothermometers: Application to the thrust faulted disturbed belt of Montana, U.S.A. Soc. Econ. Paleontol. Mineral., Spec. Publ., 26:55–79.

Hoffman J.C., 1979. – An evaluation of potassium uptake by Mississippi river borne clays following deposition in the Gulf of Mexico. Ph. D. Diss., Case Western Reserve Univ., Cleveland, Ohio.

Hollister C.D., Ewing J.T. et a., 1972. – Initial Rep. Deep Sea Drill. Proj., 11, Washington (U.S. Gov. Print. Off.):1977 pp.

Holmes M.A., 1986. – Clay mineralogy of Lower Cretaceous deep-sea fan sediments, Western North Atlantic basin. Am. Assoc. Petrol. Geol. Bull., 70:601–602.

Holmes M.A., 1987. – Clay mineralogy of the Lower Cretaceous deep-sea fan, deep sea drilling project site 603, lower continental rise off North Carolina. Init. Rep. Deep Sea Drill. Proj., 92, Washington (U.S. Gov. Print. Off.):1079–1089.

Holtzapffel T., 1984. – Smectites authigènes et glauconitisation dans les argiles du Gault (Albien du Boulonnais). Ann. Soc. géol. Nord, 104:33–39.

Holtzapffel T., 1985. – Les minéraux argileux. Préparation. Analyse diffractométrique et détermination. Soc. géol. Nord, Publ. 12:136 pp.

Holtzapffel T., 1987. – Diagenèse différentielle des minéraux argileux dans des nodules barytiques et phosphatés et dans leur matrice argileuse. C.R. Acad. Sci., Paris, II, 304:285–288.

Holtzapffel T., Chamley H., 1986. – Les smectites lattées du domaine Atlantique depuis le Jurassique supérieur: gisement et signification. Clay Minerals, 21:133–148.

Holtzapffel T., Ferrière F., 1982. – Minéraux argileux de roches anté-Crétacé supérieur d'Othrys (Grèce continentale): mise en évidence d'une diagenèse. Ann. Soc. géol. Nord, 102:25–32.

Holtzapffel T., Bonnot-Courtois C., Chamley H., Clauer N., 1985. – Héritage et diagenèse des smectites du domaine sédimentaire nord-atlantique (Crétacé, Paléogène). Bull. Soc. géol. Fr., (8) 1:25–33.

Honjo S., 1982. – Seasonality and interaction of biogenic and lithogenic particulate flux of the Panama basin. Science, 218:883–884.

Honjo S., Roman M.R., 1978. – Marine copepod fecal pellets: production, preservation and sedimentation. J. Mar. Res., 36:45–57.

Honjo S., Manganini S.J., Poppe L.J., 1982a. – Sedimentation of lithogenic particles in the deep ocean. Mar. Geol., 50:199–220.

Honjo S., Spencer D.W., Farrington J.W., 1982b. – Deep advective transport of lithogenic particles in Panama basin. Science, 216:516–518.

Honnorez J., 1981. – The aging of the oceanic crust at low temperature. In: Emiliani C., Ed., The oceanic lithosphere. The sea, 7. Wiley & Sons, New York:525–587.

Hower J., 1961. – Some factors concerning the nature and origin of glauconite. Am. Miner., 46:313–334.

Hower J., Eslinger E.V., Hower M.E., Perry E.A., 1976. – Mechanisms of burial metamorphism of argillaceous sediment: 1. Mineralogical and chemical evidence. Geol. Soc. Am. Bull., 87:725–737.

Hsü K.J., Montadert L. et al., 1978. – Initial reports of the deep sea drilling project, 42-A, Washington (U.S. Gov. Print. Off.):1249 pp.

Huang T.W., Chen P.Y., 1975. – The abyssal clay minerals in the West Philippine sea. Acta Oceanogr. Taiw., 5:37–63.

Huang T.C., Stanley D.J., Stuckenrath R., 1972. – Sedimentological evidence for current reversal at the strait of Gibraltar. Mar Techn. J., 6:25–33.

Huang W.H., Keller W.D., 1971. – Dissolution of clay minerals in dilute organic acids at room temperature. Am. Miner., 56:1082–1095.

Huang W.H., Keller W.D., 1972. – Organic acids as agents of chemical weathering of silicate minerals. Nat. Phys. Sci., 239:149–151.

Huang W.L., Bishop A.M., Brown R.W., 1986. – The effect of fluid/rock ratio on feldspar dissolution and illite formation under reservoir conditions. Clay Min., 21:585–601.

Huff W.D., 1974. – Mineralogy and provenance of Pleistocene lake clay in an Alpine region. Geol. Soc. Am. Bull., 85:1455–1460.

Huggett J.M., 1984. – Controls on mineral authigenesis in Coal Measures sandstones of the East Midlands, U.K. Clay Min., 19:343–357.

Huggett J.M., 1986. – An SEM study of phyllosilicate diagenesis in sandstones and mudstones in the Westphalian Coal Measures using back-scattered electron microscopy. Clay Min., 21:603–616.

Hume T.M., 1978. – Clay petrology of Mesozoic to recent sediments of central western North Island, New-Zealand. Ph.D. thesis, Hamilton.

Hume T.M., Nelson C.S., 1986. – Distribution and origin of clay minerals and surficial shelf sediments, Western North Island, New Zealand. Mar. Geol., 69:289–308.

Humphris S.E., Thompson G., 1978. – Trace element mobility during hydrothermal alteration of oceanic basalt. Geochim. Cosmochim. Acta, 42:127–136.

Humphris S.E., Thompson R.N., Marriner G.F., 1980. – The mineralogy and geochemistry of basalt weathering, holes 417A and 418A. Init. Rep. Deep Sea Drill. Proj., 51, 52, 53, Pt. 2, Washington (U.S. Gov. Print. Off.):1201–1218.

Hurst A., Archer J.S., 1986. – Sandstone reservoir description: an overview of the role of geology and mineralogy. Clay Min., 21:791–809.

Icole M., 1973. – Géochimie des altérations dans les nappes d'alluvions du Piémont occidental nord-pyrénéen. Thèse Sci. nat., Paris IV:328 pp.

Icole M., Taieb M., Perinet G., Manega P., Robert C., 1987. – Minéralogie des sédiments du groupe Peninj (lac Natron, Tanzanie). Reconstitution des paléoenvironnements lacustres. Sci. Géol. Bull., Strasbourg, 40:71–82.

Iijima A., 1978. – Geological occurrences of zeolites in marine environments. In: Sand L.B., Mumpton F.A., Eds., Natural zeolites: occurrence, properties, use. Pergamon, Oxford:175–198.

Imbert T., Desprairies A., 1987. – Neoformation of halloysite on volcanic glass in a marine environment. Clay Min., 22:179–185.

Inoue A., 1983. – Potassium fixation by clay minerals during hydrothermal treatment. Clays Clay Min., 31:81–91.

Inoue A., 1987. – Conversion of smectite to chlorite by hydrothermal and diagenetic alterations, Hokuroku Kuroko mineralization area, Northeast Japan. Proc. Int. Clay Conf., Denver 1985:158–164.

Ireland B.J., Curtis C.D., Whiteman J.A., 1983. – Compositional variation within some glauconites and illites and implications for their stability and origins. Sedimentology, 30:769–786.

Irion G., Wunderlich F., Schwedhelm E., 1987. – Transport of clay minerals and anthropogenic compounds into the German Bight and the provenance of fine-grained sediments SE of Helgoland. J. Geol. Soc., 144:153–160.

Isphording W.C., 1973. – Discussion of the occurrence and origin of sedimentary palygorskite-sepiolite deposits. Clays Clay Min., 21:391–401.

Jackson M.L., 1965. – Clay transformations in soil genesis during the Quaternary. Soil. Soc., 99:15–22.

Jacobs M.B., 1974. – Clay mineral changes in Antarctic deep-sea sediments and Cenozoic climatic events. J. Sediment. Petrol., 44:1079–1086.

Jacobs M.B., Ewing M., 1969. – Mineral sources and transport in waters of the Gulf of Mexico and Caribbean Sea. Science, 163:805–809.

Jacobs M.B., Hays J.D., 1972. – Paleo-climatic events indicated by mineralogical changes in deep-sea sediments. J. Sediment. Petrol., 42:889–898.

Jacquin T., 1987. – Les événements anoxiques dans l'Atlantique Sud au Crétacé. Mém. géol. Univ. Dijon, 13:267 pp.

Jakobsson S.P., Moore J.G., 1986. – Hydrothermal minerals and alteration rates at Surtsey volcano, Iceland. Geol. Soc. Am. Bull., 97:648–659.

Janecek T.R., Rea D.K., 1983. – Eolian deposition in the northwest Pacific Ocean: Cenozoic history of atmospheric circulation. Geol. Soc. Am. Bull., 94:730–738.

Janecek T.R., Rea D.K., 1985. – Quaternary fluctuations in the northern hemisphere trade winds and westerlies. Q. Res., 24:150–163.

Jansa L.F., 1981. – Mesozoic carbonate platforms and banks of the eastern North American margin. Mar. Geol., 44:97–117.

Jarvis I., 1985. – Geochemistry and origin of Eocene-Oligocene metalliferous sediments from the central equatorial Pacific: Deep Sea Drilling Project sites 573 and 574. In: Mayer L., Theyer F. et al., Eds., Init. Rep. Deep Sea Drill. Proj., 85, Washington (U.S. Gov. Print. Off.):781–801.

J.C.P.D.S. (Joint Committee on Powder Diffraction Standards), 1978–1979. – Powder diffraction file, search manual and alphabetical index. International Center for Diffraction Data. 1146 and 973 pp.

Jeannette A., Lucas J., 1955. – Sur l'extension au Maroc des niveaux à chlorite dans les argiles du Permo-Trias. Notes Serv. géol. Maroc, 125:129–134.

Jeans C.V., 1971. – The neoformation of clay minerals in brackish and marine environments. Clay Min., 9:209–217.

Jeans C.V., 1978 a. – The origin of the Triassic clay assemblages of Europe with special reference to the Keuper Marl and Rhaetic of parts of England. Proc. Philos. Trans. R. Soc. London Ser. A, 289:549–636.

Jeans C.V., 1978 b. – Silicifications and associated clay assemblages in the Cretaceous marine sediments of Southern England. Clay Min., 13:101–126.

Jeans C.V., Merriman R.J., Mitchell J.G., Bland D.J., 1982. – Volcanic clays in the Cretaceous of Southern England and Northern Ireland. Clay Min., 17:105–156.

Jehanno C., Boclet D., Bonte P., Devineau J., Rocchia R., 1987. – L'iridium dans les minéraux à la limite Crétacé-Tertiaire de plusieurs sites européens et africains. Mém. Soc. géol. Fr., N.S., 150:81–94.

Jennings S., Thompson G.R., 1986. – Diagenesis of Plio-Pleistocene sediments of the Colorado river delta, Southern California. J. Sediment. Petrol., 56:89–98.

Jiranek J., Jirankova J., 1987. – Genesis and geochemistry of the arkose-type kaolins in the Plzen basin, Czechoslovaki. In: 6th Meet. Eur. Clay Groups, Sevilla:310–313.

Johns W.D., Grim R.E., 1958. – Clay mineral composition of recent sediments from the Mississippi River delta. J. Sediment. Petrol., 28:186–199.

Johns W.D., Kurzweil H., 1979. – Quantitative estimation of illite-smectite mixed phases formed during burial diagenesis. Tschermaks Min. Petrol. Mitt., 26:203–215.

Johns W.D., Shimoyama A., 1972. – Clay minerals and petroleum-forming reactions during burial and diagenesis. Am. Assoc. Petrol. Geol. Bull., 56:2160–2167.

Johnson A.G., Kelley J.T., 1984. – Temporal, spatial and textural variation in the mineralogy of Mississippi river suspended sediment. J. Sediment. Petrol., 54:67–72.

Johnson L.R., 1979. – Mineralogical dispersal patterns of North Atlantic deep-sea sediments with particular reference to eolian dusts. Mar. Geol., 29:335–345.

Johnson T.C., Elkins S.R., 1979. – Holocene deposits of the northern North Sea: evidence for dynamic control of their mineral and chemical composition. Geol. Mijnbouwn, 58:353–366.

Johnsson M.J., Reynolds R.C., 1986. – Clay mineralogy of shale-limestone rhythmites in the Scaglia Rossa (Turonian-Eocene), Italian Apennines. J. Sediment. Petrol., 56:501–509.

Jones B.F., Bowser C.J., 1978. – The mineralogy and related chemistry of lake sediments. In: Lerman A., Ed., Lakes. Chemistry, geology, physics. Springer, Berlin, Heidelberg, New York:194–235.

Jones B.F., Weir A.H., 1983. – Clay minerals of lake Abert, an alkaline, saline lake. Clays Clay Min., 31:161–172.

Jones B.F., Doval M., Calvo J.P., Brell J.M., 1986. – Clay mineral authigenesis in lacustrine closed basins: comparison of the Madrid basin with U.S. occurrences. S.E.P.M. Annu. Midyear Meet., III:58.

Jones G.A., 1984. – Advective transport of clay minerals in the region of the Rio Grande Rise. Mar. Geol., 58:187–212.

Jones G.A., Johnson D.A., 1984. – Displaced Antarctic diatoms in Vema Channel sediments: late Pleistocene/Holocene fluctuations in AABW flow. Mar. Geol., 58:165–186.

Jones J.B., Fitzgerald M.J., 1984. – Extensive volcanism associated with the separation of Australia and Antarctica. Science, 226:346–348.

Jones J.B., Fitzgerald M.J., 1987. – An unusual and characteristic sedimentary mineral suite associated with the evolution of passive margins. Sediment. Geol., 52:45–63.

Jouanneau J.-M., 1982. – Matières en suspension et oligo-éléments métalliques dans le système estuarien girondin: Comportement et flux. Thèse, Sci. nat. Bordeaux I:150 pp.

Jouzel J., Merlivat L., Lorius C., 1982. – Deuterium excess in an East Antarctic ice core suggests higher relative humidity at the oceanic surface during the last glacial maximum. Nature (London), 299:688.

Jung J., 1954. – Les illites du bassin oligocène de Salins (Cantal). Bull. Soc. Fr. Min. Cristallogr., 77:1231–1249.

Juteau T., 1984. – De la croûte océanique aux ophiolites. Bull. Soc. géol. Fr., (7) 26:471–488.

Kadko D., 1985. – Late Cenozoic sedimentation and metal deposition in the North Pacific. Geochim. Cosmochim. Acta, 49:651–661.

Kagami H., Ishizuka T., Aoki S., 1983. – Geochemistry and mineralogy of selected carbonaceous claystones in the lower Cretaceous from the Blake-Bahama Basin, North Atlantic. In: Sheridan R.E., Gradstein F.M. et al., Eds., Init. Rep. Deep Sea Drill. Proj., 76, Washington (U.S. Gov. Print. Off.):429–436.

Kantor W., Schwertmann U., 1974. – Mineralogy and genesis of clays in red-black soil toposequences on basic igneous rocks in Kenya. J. Soil Sci., 25:67–78.

Kantorowicz J., 1984. – Nature, origin and distribution of authigenic clay minerals from Middle Jurassic Ravenscar and Brent Group sandstones. Clay Min., 19:359–375.

Kantorowicz J.D., Lievaart L., Eylander J.G.R., Eigner M.R.P., 1986. – The role of diagenetic studies in production operations. Clay Min., 21:769–780.

Kapur S., Cavusgil V.S., Fitzpatrick E.A., 1987. – Soil-calcrete (caliche) relationship on a Quaternary surface of the Cukurova region, Adana (Turkey). In: Fedoroff N. et al., Eds., Soil micromorphology. AFES, Paris:597–603.

Karig D.E., Ingle J.C. Jr. et al., 1975. – Initial reports of the deep sea drilling project, 31, Washington (U.S. Gov. Print. Off.):927 pp.

Karlin R., 1980. – Sediment sources and clay mineral distributions off the Oregon coast. J. Sediment. Petrol., 50:543–560.

Karlin R., Lyle M., Heath G.R., 1987. – Authigenic magnetite formation in suboxic marine sediments. Nature (London), 326:490–493.

Karlsson W., Vollset J., Bjørlykke K., Jorgensen P., 1978. – Changes in mineralogical composition of Tertiary sediments from North-Sea wells. In: Mortland M.M., Farmer V.C., Eds., Int. Clay Conf., 1978. Developments in sedimentology. Elsevier, Amsterdam, 27:281–289.

Karpoff A.-M., 1984. – Miocene red clays of the South Atlantic: dissolution facies of calcareous oozes at Deep Sea Drilling Project sites 519 to 523, Leg 73. In: Hsü K.J., La Brecque J.L. et al., Eds., Init. Rep. Deep Sea Drill. Proj., 73. Washington (U.S. Gov. Print. Off.):515–535.

Karpoff A.-M., Peterschmitt I., Hoffert M., 1980. – Mineralogy and geochemistry of sedimentary deposits on Emperor seamounts, sites 430, 431, and 432: authigenesis of silicates, phosphates and ferromanganese oxides. In: Jackson E.D., Koisumi I. et al., Eds., Init. Rep. Deep Sea Drill. Proj., 55, Washington (U.S. Gov. Print. Off.):463–489.

Karpoff A.-M., Hoffert M., Clauer N., 1981. – Sedimentary sequences at deep sea drilling project site 464: Silicification processes and transition between siliceous biogenic oozes and brown clays. In: Thiede J., Vallier T.L. et al., Eds., Init. Rep. Deep Sea Drill. Proj., 62, Washington (U.S. Gov. Print. Off.):759–771.

Kastens K., Mascle J. et al., 1988–1989. – Ocean Drilling Program, Leg 107, Proc., A and B (U.S. Gov. Print. Off.) (in press).

Kastner M., 1981. – Authigenic silicates in deep-sea sediments: formation and diagenesis. In: Emiliani C., Ed., The sea. Wiley & Sons, New York, 7:915–980.

Kastner M., 1986. – Mineralogy and diagenesis of sediments at site 597: preliminary results. In: Leinen M. Rea D.K. et al., Eds., Init. Rep. Deep Sea Drill. Proj., 92, Washington (U.S. Gov. Print. Off.):345–349.

Kastner M., Asaro F., Michel H.V., Alvarez W., Alvarez L.W., 1984. – The precursor of the Cretaceous-Tertiary boundary clays at Stevns Klint, Denmark, and DSDP hole 465A. Science, 226:137–143.

Kawahata H., Furuta T., 1985. – Sub-sea-floor hydrothermal alteration in the Galapagos spreading center. Chem. Geol., 49:259–274.

Kay M., 1975. – Campbellton sequence, manganiferous beds adjoining the Dunnage melange, Northeastern New-foundland. Geol. Soc. Am. Bull., 86:195–198.

Keller J., Ryan W.B.F., Ninkovich D., Altherr R., 1978. – Explosive volcanic activity in the Mediterranean over the past 200,000 y as recorded in deep-sea sediments. Geol. Soc. Am. Bull., 89:591–604.

Keller W.D., 1970. – Environmental aspects of clay minerals. J. Sediment. Petrol., 40:788–813.

Keller W.D., 1982. – Kaolin – A most diverse rock in genesis, texture, physical properties, and uses. Geol. Soc. Am. Bull., 93:27–36.

Keller W.D., 1988. – Authigenic kaolinite and dickite associated with metal sulfides – probable indicators of a regional thermal event. Clays Clay Min., 36:153–158.

Keller W.D., Reynolds R.C., Inoue A., 1986. – Morphology of clay minerals in the smectite-to-illite conversion series by scanning electron microscopy. Clays Clay Min., 34:187–197.

Kellogg T.B., 1985. – Early deglaciation of the Greenland sea during the last glacial to interglacial transition. Comment. Mar. Geol., 62:167–169.

Kelly W.C., Zumberge J.H., 1961. – Weathering of a quartz diorite at Marble Point, McMurdo Sound, Antarctica. J. Geol., 69:433–446.

Kennett J.P., 1982. – Marine geology. Prentice Hall, Englewood Cliffs, New Jersey:813 pp.

Kennett J.P., Thunell R.C., 1975. – Global increase in quaternary explosive volcanism. Science, 187:497–503.

Kennett J.P., von der Borch C. et al., 1985. – Palaeotectonic implication of increased late Eocene-early Oligocene volcanism from Soth Pacific DSDP sites. Nature, London, 316:507–511.

Khalaf F.I., Al-Ghadban A., Al-Saleh S., Al-Ombran L., 1982. – Sedimentology and mineralogy of Kuwait bay bottom sediments, Kuwait-Arabian gulf. Mar. Geol., 46:71–99.

Kha Nguyen, Paquet H., 1975. – Mécanismes d'évolution et de redistribution des minéraux argileux dans les pélosols. Sci. Géol., Strasbourg, 28:15–28.

Kharaka Y.K., Law L.M., Carothers W.W., Goerlitz D.E., 1986. – Role of organic species dissolved in formation waters from sedimentary basins in mineral diagenesis. Soc. Econ. Pal. Min., Spec. Publ., 38:111–122.

Khoury H.N., Eberl D.D., Jones B.F., 1982. – Origin of magnesium clays from the Amargosa desert, Nevada. Clays Clay Min., 30:327–336.

Kidd R.B., Cita M.B., Ryan W.B.F., 1978. – Stratigraphy of Eastern Mediterranean sapropel sequences recovered during Leg 42A and their paleoenvironmental significance. In: Hsü K.J., Montadert L. et al., Eds., Init. Rep. Deep Sea Drill. Proj., 42, A, Washington (U.S. Gov. Print. Off.):421–443.

Kimpe C.R. de, Miles N., 1987. – Geographic distribution and paragenesis of corrensite in Ontario, Canada. In: 6th Meet. Eur. Clay Groups, Sevilla, Proc.:317–318.

King R.H., 1986. – Weathering of Holocene airfall tephras in the southern Canadian Rockies. In: Colman S.M., Dethier D.P., Eds., Rates of chemical weathering of rocks and minerals. Academic Press, New York, London:239–264.

Kisch H.J., 1983. – Mineralogy and petrology of burial diagenesis (burial metamorphism) and incipient metamorphism in clastic rocks. In: Larsen G., Chilingar G.V., Eds., Diagenesis in sediments and sedimentary rocks, 2. Developments in sedimentology. Elsevier, 25B:289–493.

Klimentidis R.E., Mackinnon I.D.R., 1986. – High-resolution imaging of ordered mixed-layer clays. Clays Clay Min., 34:155–164.

Klinkhammer G., Hudson A., 1986. – Dispersal patterns for hydrothermal plumes in the South Pacific using manganese as a tracer. Earth Planet. Sci. Lett., 79:241–249.

Knebel H.J., Kelly J.C., Whetten J.T., 1968. – Clay minerals of the Columbia river: a qualitative, quantitative and statistical evaluation. J. Sediment. Petrol., 38:600–611.

Knebel H.J., Conomos T.J., Commeau J.A., 1977. – Clay-mineral variability in the suspended sediments of the San Francisco bay system, California. J. Sediment. Petrol., 47:229–236.

Kobayashi K., Oinuma K., Sudo T., 1964. – Clay mineralogy of recent marine sediments and sedimentary rocks from Japan. Sedimentology, 3:233–239.

Kolbe R.W., 1957. – Freshwater diatoms from Atlantic deep-sea sediments. Science, 126:1053–1056.

Kolla V., 1974. – Mineralogical data from sites 211, 212, 213, 214, and 215 of deep-sea equatorial Indian Ocean. In: Borch C.C. von der, Sclater J.G. et al., Eds., Init. Rep. Deep Sea Drill. Proj., Washington (U.S. Gov. Print. Off.), 22:489–501.

Kolla V., Biscaye P.E., 1973. – Clay mineralogy and sedimentation in the eastern Indian Ocean. Deep Sea Res., 20:727–738.

Kolla V., Biscaye P.E., 1977. – Distribution and origin of quartz in the sediments of the Indian Ocean. J. Sediment. Petrol., 47:642–649.

Kolla V., Henderson L., Biscaye P.E., 1976. – Clay mineralogy and sedimentation in the western Indian Ocean. Deep Sea Res., 23:949–961.

Kolla V., Biscaye P.E., Hanley A.F., 1979. – Distribution of quartz in late Quaternary Atlantic sediments in relation to climate. Q. Res., 11:261–277.

Kolla V., Kostecki J.A., Henderson L., Hess L., 1980a. – Morphology and Quaternary sedimentation of the Mozambique Fan and environs, southwestern Indian Ocean. Sedimentology, 27:357–378.

Kolla V., Nadler L., Bonatti E., 1980b. – Clay mineral distributions in surface sediments of the Philippine Sea. Oceanol. Acta, 3:245–250.

Kolla V., Kostecki J.A., Robinson F., Biscaye P.E., Ray P.K., 1981. – Distributions and origins of clay minerals and quartz in surface sediments of the Arabian Sea. J. Sediment. Petrol., 51:563–569.

Konta J., 1985a. – Mineralogy and chemical maturity of suspended matter in major rivers samples under the SCOPE/UNEP project. Mitt. Geol. Paläontol. Inst. Univ. Hamburg, 58:569–592.

Konta J., 1985b. – Crystalline minerals and chemical maturity of suspended solids of some major world rivers. Miner. Petrogr. Acta, 29A:121–133.

Kossovskaya A.G., Drits V.R., 1970. – Micaceous minerals in sedimentary rocks. Sedimentology, 15:83–101.

Kossowskaya A.G., Gushchina E.B., Drits V.A., Dmitrik A.L., Lomova O.S., Serebrennikova N.D., 1975. – Mineralogy and genesis of Meso-Cenozoic deposits of the Atlantic based on materials of deep sea drilling project, leg 2. Lithol. Useful Min., Acad. NAUK S.S.S.R., 6:12–26.

Kounetsron O., Robert M., Bernier J., 1977. – Nouvel aspect de la formation des smectites dans les vertisols. C.R. Acad. Sci., Paris, 284:733–736.

Krinsley D., Biscaye P.E., Turekian K.K., 1973. – Argentine basin sediment sources as indicated by quartz surface texture. J. Sediment. Petrol., 43:251–257.

Krismanndottir H., 1975. – Proceedings of the second United Nations symposium on the development and use of geothermal sources, I, Washington (U.S. Gov. Print. Off.): 441–445

Krone R.B., 1978. – Aggregation of suspended particles in estuaries. In: Kjerfve B., Ed., Estuarine transport processes. Univ. South Carolina Press, Columbia:177–190.

Kubler B., 1968. – Évaluation quantitative du métamorphisme par la cristallinité de l'illite; état des progrès réalisés ces dernières années. Bull. Centre Rech., Pau, 2:385–397.

Kubler B., 1973. – La corrensite, indicateur possible de milieux de sédimentation et du degré de transformation d'un sédiment. Bull. Centre Rech. Pau, S.N.P.A., 7:543–556.

Kulbicki G., Millot G., 1960. – L'évolution de la fraction argileuse des grès pétroliers cambro-ordoviciens du Sahara central. Bull. Serv. Carte géol. Als.-Lorr., 13:147–156.

Kulke H., 1969. – Petrographie und Diagenese des Stubensandsteines (mittlerer Keuper) aus Tiefbohrungen im Raum Memmingen (Bayern). Contrib. Mineral. Petrol., 20:135–163.

Kullenberg B., 1952. – On the salinity of the water contained in marine sediments. Medd. Oceanogr. Inst. Göteborg, 21:1–38.

Lafond R., 1961. – Etude minéralogique des argiles actuelles du bassin de la Vilaine. C.R. Acad. Sci., Paris, 252:3614–3616.

Lafond R., Rivière A., Vernhet S., 1961. – Etude de la composition minéralogique de quelques argiles glaciaires. C.R. Acad. Sci., Paris, 252:3310–3313.

Lambert C.E., Bishop J.K.B., Biscaye P.E., Chesselet R., 1984. – Particulate aluminum, iron and manganese chemistry at the deep Atlantic boundary layer. Earth Planet. Sci. Lett., 70:237–248.

Lamboy M., 1968. – Sur un processus de formation de la glauconite en grains à partir des débris coquilliers. Rôle des organismes perforants. C.R. Acad. Sci., Paris, 266:1937–1940.

Lamboy M., 1976. – Géologie marine du plateau continental au N.O. de l'Espagne. Thèse Sci. nat, Rouen:283 pp.

Lamouroux M., Paquet H., Pinta M., Millot G., 1967. – Notes préliminaires sur les minéraux argileux des altérations et des sols méditerranéens du Liban. Bull. Serv. Carte géol. Als.-Lorr., 20:277–292.

Lancelot Y., Hathaway J.C., Hollister C.D., 1972. – Lithology of sediments from the Western North Atlantic Leg 11 Deep Sea Drilling Project. In: Hollister C.D., Ewing J. et al., Eds., Init. Rep. Deep Sea Drill. Proj., 11, Washington (U.S. Gov. Print. Off.):901–949.

Lancelot Y., Seibold E. et al., 1978. – Initial reports of the deep sea drilling project, 41, Washington (U.S. Gov. Print. Off.):1529 pp.

Lange H., 1982. – Distribution of chlorite and kaolinite in eastern Atlantic sediments off North Africa. Sedimentology, 29:427–431.

Lange H., 1985. – Clay mineralogy of slope sediments off Mauritania. Terra Cognita, 5, Geol. Vereinigung (75th ann. Meet. Kiel), Abstracts:72.

Langford-Smith E.T., 1978. – Silcrete in Australia. Dep. Geography, Univ. N.S.W. Armidale:304 pp.

Larqué P., Weber F., 1984. – Paléosols sur sédiments continentaux paléogènes du Bassin du Puy-en-Velay, Massif Central français. 5ème Eur. Reg. Meet. Int. Assoc. Sediment., Marseille, Abstr.:251–252.

Larsen G., Chilingar G.V., 1979, 1983. – Diagenesis in sediments and sedimentary rocks. Elsevier, Dev. Sediment., 25A:579 pp.; 25B:572 pp.

Latil-Brun M.V., Flicoteaux R., 1986. – Subsidence de la marge sénégalaise, ses relations avec la structure de la croûte. Comparaison avec la marge conjuguée américaine au niveau du Blake-Plateau. Bull. Centre Rech. Pau, 110:64–82.

Latouche C., 1971. – Les argiles des bassins alluvionnaires aquitains et des dépendances océaniques. Thèse, Sci. nat., Bordeaux I:415 pp.

Latouche C., 1972. – La sédimentation argileuse marine au voisinage de l'embouchure de la Gironde. Interprétation et conséquences. C.R. Acad. Sc., Paris, 274:2929–2932.

Latouche C., Maillet N., 1982. – Essai sur l'utilisation des argiles comme témoins des climats néogènes dans la province atlantique nord-orientale. Bull. Soc. géol. Fr., (7) 24:487–496.

Latouche C., Maillet N., 1985. – Le déficit sédimentaire de la période Eocène moyen-Miocène moyen en Atlantique nord-oriental. Discussion sur la base de l'évolution minéralogique des dépôts. Bull. Soc. géol. Fr., (8) 1:21–24.

Laurain M., Meyer R., 1979. – Paléoaltération et paléosol: l'encroûtement calcaire (calcrete) au sommet de la craie, sous les sédiments éocènes de la Montagne de Reims. C.R. Acad. Sci., Paris, 289:1211–1214.

Leggett J.K., 1982. – Geochemistry of Cocos plate pelagic-hemipelagic sediments in hole 487, Deep Sea Drilling Project leg 66. In: Watkins J.S., Moore J.C. et al., Eds., Init. Rep. Deep Sea Drill. Proj., 66, Washington (U.S. Gov. Print. Off.):683–686.

Leguey S., Pozo M., Medina J.A., 1985. – Polygenesis of sepiolite and palygorskite in a fluvio-lacustrine environment in the Neogene basin of Madrid. In: Pozzuoli A., Ed., Min. Petrol. Acta, 29-A:287–302.

Leguey S., Pozo M., Medina J.A., Vigil R., 1987. – Evolution of paleosols with sepiolite development in the Madrid Basin (Spain). 6th Meet. Eur. Clay Groups, Sevilla:335–337.

Leikine M., Velde B., 1974. – Les transformations post-sédimentaires des minéraux argileux du Sénonien, dans le NE algérien. Existence probable d'un métamorphisme anté-Eocène. Bull. Soc. géol. Fr., (7), 16:177–182.

Leinen M., 1981. – Metal-rich basal sediments from northeastern Pacific Deep Sea Drilling Project sites. In: Yeats R.S., Haq B.U. et al., Eds., Init. Rep. Deep Sea Drill. Proj., 63, Washington (U.S. Gov. Print. Off.):667–676.

Leinen M., 1985. – Quartz content of Northwest Pacific Hole 576A and implications for Cenozoic eolian transport. In: Heath G.R., Burckle L.H. et al., Eds., Init. Rep. Deep Sea Drill. Proj., 86, Washington (U.S. Governm. Print. Off.):581–588.

Leinen M., 1987. – The origin of paleochemical signatures in North Pacific pelagic clays: partitioning experiments. Geochim. Cosmochim. Acta, 51:305–319.

Leinen M., Heath G.R., 1981. – Sedimentary indicators of atmospheric activity in the Northern hemisphere during the Cenozoic. Palaeogeogr., Climatol., Ecol., 36:1–21.

Leinen M., Cwienk D., Heath G.R., Biscaye P.E., Kolla V., Thiede J., Dauphin J.P., 1986a. – Distribution of biogenic silica and quartz in recent deep-sea sediments. Geology, 14:199–203.

Leinen M., Rea D.K. et al., 1986b. Init. Rep. Deep Sea Drill. Proj., 92, Washington (U.S. Gov. Print. Off.):617 pp.

Lelong F., Millot G., 1966. – Sur l'origine des minéraux micacés des altérations latéritiques. Diagenèse régressive – Minéraux en transit. Bull. Serv. Carte géol. Als.-Lorr., 19:271–287.

Lemoalle J., Dupont B., 1973. – Iron-bearing oolites and the present conditions of iron sedimentation in lake Chad (Africa). In: Amstutz G., Bernard A.J., Eds., Ores in sediments. Springer, Berlin, Heidelberg, New York.

Lemoine M., Arnaud-Vanneau A., Arnaud H., Létolle R., Mevel C., Thieuloy J.-P., 1982. – Indices possibles de paléo-hydrothermalisme marin dans le Jurassique et le Crétacé des Alpes occidentales (océan téthysien et sa marge continentale européenne): essai d'inventaire. Bull. Soc. géol. Fr., (7) 24:641–647.

Lemoine M., Bourbon M., Graciansky P.-C. de, Létolle R., 1983. – Isotopes du carbone et de l'oxygène de calcaires associés à des ophiolites (Alpes occidentales, Corse, Apennin): indices possibles d'un hydrothermalisme océanique téthysien. Rev. Géol. dyn. Géogr. phys., 24:305–314.

Lenôtre N., Chamley H., Hoffert M., 1985. – Clay stratigraphy at Deep Sea Drilling Project Sites 576 and 578, Leg 86 (Western North Pacific). In: Heath G.R., Burckle L.H. et al., Eds., Init. Rep. Deep Sea Drill. Proj., 86, Washington (U.S. Gov. Print. Off.):571–579.

Lerman A., Mackenzie F.T., Bricker O.P., 1975. – Rates of dissolution of aluminosilicates in seawater. Earth Planet. Sci. Lett., 25:82–88.

Lever A., McCave I.N., 1983. – Eolian components in Cretaceous and Tertiary North Atlantic sediments. J. Sediment. Petrol., 53:811–832.

Levert J., Ferry S., 1988. – Diagenèse argileuse complexe dans le Mésozoïque subalpin révélée par cartographie des proportions relatives d'argiles selon des niveaux isochrones. Bull. Soc. géol. Fr. (8) 4:1029–1038.

Liebling R.S., Scherp H.S., 1976. – Chlorite and mica as indicators of depositional environment and provenance. Geol. Soc. Am. Bull., 87:513–514.

Lineback J.A., Dell C.I., Gross D.L., 1979. – Glacial and postglacial sediments in lakes Superior and Michigan. Geol. Soc. Am. Bull., 90:781–791.

Lisitzin A.P., 1972. – Sedimentation in the world ocean. Soc. Econ. Paleontol. Miner., Spec. Publ., 17:218 pp.

Lisitzina N.A., Butuzova G.Y., 1980. – Authigenous zeolites in sedimentary mantle of World Ocean. 26th Int. Geol. Congr., Paris. Abstr., 2:502.

Logvinenko N.V., Lazurkin V.M., Gerasimov V.N., Shumenko S.I., 1979. – Red deep-water clays of the northern part of the Pacific Ocean. Int. Geol. Rev., 21:1149–1158. (in Russian)

Lombard A., Vernet J.-P., 1969. – Etude des sédiments du lac de Mauvoisin (Valais). C.R. séances Soc. P.H.N. Genève, 4:55–61.

Lomova O.S., 1975. – Abyssal palygorskite clays of the Eastern Atlantic and their genetic relation to alkalic volcanism (from data of legs 2 and 14 of the Glomar Challenger). Litol. Pol. Isk., 4:10–27.

Lonnie T.P., 1982. – Mineralogic and chemical comparison of marine, nonmarine, and transitional clay beds on south shore of Long Island, New York. J. Sediment. Petrol., 52:529–536.

López Galindo A., 1986. – Mineralogia de series cretacicas de la zona subbetica. Algunas consideraciones paleogeograficas derivadas de la composicion quimica de Las Esmectitas. Estud. geol., 42:231–238.

López Galindo A., Comas Minondo M.C., Fenoll Hach-Ali P., Ortega Huertas M., 1985. – Pelagic cretaceous black-greenish mudstones in the Southern Iberian paleomargin, Subbetic zone, Betic Cordillera. Miner. Petrogr. Acta, 29 A:245–257.

Louail J., 1984. – La transgression crétacée au Sud du Massif Armoricain. Mém. Soc. géol. minéral. Bretagne, 29:333 pp.

Louail J., Estéoule J., Estéoule-Choux J., 1979. – The origin of clay minerals in Cenomanian littoral deposits around the Armorican massif. Int. Clay Conf., 1978. Elsevier, Amsterdam:291–300.

Loughnan F.C., 1969. – Chemical weathering of the silicate minerals. Elsevier, Amsterdam:154 pp.

Loughnan F.C., 1971. – Kaolinite claystones associated with the Wongawilli seam in the southern part of the Sydney basin. J. Geol. Soc. Aust., 18:293–302.

Loughnan F.C., Roberts F.I., 1986. – Dickite- and kaolinite-bearing sandstones and conglomerates in Illawara Coal Measures of the Sydney basin, New South Wales. Aust. J. Earth Sci., 33:325–332.

Loveland P.J., 1984. – The soil clays of Great Britain: I. England and Wales. Clay Min., 19:681–707.

Lowe D.J., 1986. – Controls on the rates of weathering and clay mineral genesis in airfall tephras: a review and New Zealand case study. In: Colman S.M., Dethier D.P., Eds., Rates of chemical weathering of rocks and minerals. Academic Press, New York, London:265–330.

Loÿe-Pilot M.D., Martin J.-M., Morelli J., 1986. – Influence of Saharan dust on the rain acidity and atmospheric input to the Mediterranean. Nature (London), 321:427–428.

Lucas J., 1962. – La transformation des minéraux argileux dans la sédimentation. Etude sur les argiles du Trias. Mém. Serv. Carte géol. Als.-Lorr., 23:202 pp.

Lucas J., Ataman G., 1968. – Mineralogical and geochemical study of clay mineral transformations in the sedimentary Triassic Jura Basin (France). Clays Clay Min., 16:365–372.

Lucas J., Prévôt L., 1975. – Les marges continentales, pièges géochimiques; l'exemple de la marge atlantique de l'Afrique à la limite Crétacé-Tertiaire. Bull. Soc. géol. Fr., (7) 17:496–501.

Lucas J., Camez T., Millot G., 1959. – Détermination pratique aux rayons X des minéraux argileux simples et interstratifiés. Bull. Serv. Carte géol. Als.-Lorr., 12:21–33.

Lunkad S.K., 1986. – Clay minerals in late Quaternary anoxic facies sediment from the Eastern Mediterranean Sea: their diagenetic and palaeoclimatic significance. Clay Res., 5:53–63.

Lyle M., Leinen M., Owen R.M., Rea D.K., 1987. – Late Tertiary history of hydrothermal deposition at the East Pacific rise, 19°S: correlation to volcano-tectonic events. Geophys. Res. Lett., 14:595–598.

Macaire J.-J., 1986. – Apport de l'altération superficielle à la stratigraphie. Exemple des formations alluviales et éoliennes plio-quaternaires de Touraine (France). Bull. Assoc. Fr. Et. Q., 23–24:233–245.

Macaire J.-J., Estéoule-Choux J., Estéoule J., 1977. – Sur la présence de zéolites détritiques dans les alluvions quaternaires de la Creuse et de la Claise. C.R. Acad. Sci., Paris, 285:949–952.

Mac Evan D.M.C., 1949. – Some notes on the recording and interpretation of X-ray diagrams of soil clays. J. Soil. Sci., 1:90–103.

Mackenzie F.T., Garrels R.M., 1966a. – Silica-bicarbonate balance in the ocean and early diagenesis. J. Sediment. Petrol., 36:1075–1084.

Mackenzie F.T., Garrels R.M., 1966b. – Chemical mass balance between rivers and oceans. Am. J. Sci., 264:507–525.

Mackenzie R.C., 1970. – Differential thermal analysis of clays. 1. Fundamental aspects. 2. Applications. Academic Press, New York, London.

Mackin J.E., 1986. – Control of dissolved Al distributions in marine sediments by clay reconstitution reactions: experimental evidence leading to a unified theory. Geochim. Cosmochim. Acta, 50:207–214.

Mackin J.E., Aller R.I., 1984. – Dissolved Al in sediments and waters of the East China Sea: implications for authigenic mineral formation. Geochim. Cosmochim. Acta, 48:281–297.

Maglione G., 1974. – Géochimie des évaporites et silicates néoformés en milieu continental confiné. Thèse, Sci. nat., Paris VI:331 pp.

Maillot H., 1982. – Les paléoenvironnements de l'Atlantique Sud: apport de la géochimie sédimentaire. Thèse Sci. nat., Lille I:316pp.

Maldonado A., Stanley D.J., 1981. – Clay mineral distribution patterns as influenced by depositional processes in the Southeastern Levantine Sea. Sedimentology, 28:21–32.

Malley P., Juteau T., Blanco-Sanchez J.-A., 1983. – Hydrothermal alteration of submarine basalts: from zeolitic to spilitic facies in the upper Triassic pillow-lavas of Antalya, Turkey. Sci. Géol. Bull., Strasbourg, 36:139–163.

Mamy J., Gaultier J.P., 1975. – Etude de l'évolution de l'ordre cristallin dans la montmorillonite en relation avec la diminution de l'échangeabilité du potassium. Proc. Int. Clay Conf., Mexico City:149–155.

Manickam S., Barbaroux L., Ottmann F., 1985. – Composition and mineralogy of suspended sediment in the fluvio-estuarine zone of the Loire river, France. Sedimentology, 32:721–741.

Mann U., Fischer K., 1982. – The triangle method. Semiquantitative determination of clay minerals. J. Sediment. Petrol., 52:654–657.

Mann U., Müller G., 1980. – Composition of sediments of the Japan trench transect, Legs 56 and 57, Deep Sea Drilling Project. In: Honza E. et al., Eds., Init. Rep. Deep Sea Drill. Proj., 56–57, Washington (U.S. Gov. Print. Off.):939–977.

Marchig V., Rösch H., 1983. – Formation of clay minerals during early diagenesis of a calcareous ooze. Sediment. Geol., 34:283–299.

Mariner R.H., Surdam R.C., 1970. – Alkalinity and formation of zeolites in saline alkaline lakes. Science, 170:977–980.

Marius C., Lucas J., 1982. – Evolution géochimique et exemple d'aménagement des mangroves au Sénégal (Casamance). Oceanol. Acta, Spec. Symp. Int. lag. côt.:151–160.

Martin A., 1963. – Les dépôts fluvio-glaciaires du Bas-Dauphiné: leurs minéraux argileux, leur radioactivité. Bull. Soc. géol. Fr., (7) 5:538–540.

Martin J.-M., 1971. – Contribution à l'étude des apports terrigènes d'oligoéléments stables et radioactifs à l'océan. Thèse, Sci. nat., Paris:161 pp.

Martin de Vidales J.L., Casas J., Galvan J., Herrero F., Hoyos M.A., 1987. – Weathering products of volcanic rocks from Campo de Calatrava (Ciudad Real, Central Spain). In: 6th Meet. Eur. Clay Groups, Sevilla:378–380.

Mary G., Grenèche J.M., 1986. – Les formations alluviales anciennes de la Sarthe en aval du Mans et leur degré d'altération. Bull. Assoc. Fr. Et. Q., 23–24:247–255.

Mârza I., Ghergariu L., Mînzararu L., 1966. – Considérations sur la genèse et la composition des bentonites de Razoare (R.S. de Roumanie). Bull. Serv. Carte géol. Als.-Lorr., 19:213–220.

Mascle G., 1973. – Etude géologique des Monts Sicani (Sicile). Thèse Sci. nat., Paris VI:691 pp.

Matter A., 1974. – Burial diagenesis of pelitic and carbonate deep-sea sediments from the Arabian Sea. In: Whitmarsh R.B., Weser O.E., Ross D.A. et al., Eds., Init. Rep. Deep Sea Drill. Proj., 23, Washington (U.S. Gov. Print. Off.):421–469.

Matti J.C., Zemmels I., Cook H.E., 1974. – X-ray mineralogy data, northeastern part of the Indian Ocean, Leg 22, Deep Sea Drilling Project. In: Von der Borch C.C., Sclater J.G. et al., Eds., Init. Rep. Deep Sea Drill Proj., 22, Washington (U.S. Gov. Print. Off.):693–710.

Maurel P., 1968. – Sur la présence de gibbsite dans les arènes du massif du Sidobre (Tarn) et de la Montagne Noire. C.R. Acad. Sci., Paris, 266:652–653.

McCave I.N., 1975. – Vertical flux of particles in the ocean. Deep-Sea Res., 22:491–502.

McCave I.N., 1984. – Erosion, transport and deposition of fine-grained marine sediments. In: Stow D.A.V., Piper D.J.W., Eds., Fine-grained sediments: deep-water processes and facies. Blackwell, Oxford:35–69.

McDonald C.C., 1980. – Mineralogy and geochemistry of a Precambrian regolith in the Athabasca basin. M. Sci. Thesis, Saskatchewan Univ.:151 pp.

McDowell D.S., Elders W.A., 1980. – Authigenic layer silicate minerals in borehole Elmore 1, Salton Sea geothermal field, California, U.S.A. Contrib. Mineral. Petrol., 74:293–310.

McHardy W.J., Wilson M.J., Tait J.M., 1982. – Electron microscope and X-ray diffraction studies of filamentous illitic clay from sandstones of the Magnus Field. Clay Min., 17:23–29.

McHargue T.R., Price R.C., 1982. – Dolomite from clay in argillaceous or shale-associated marine carbonates. J. Sediment. Petrol., 52:873–886.

McLean D.M., 1985. – Deccan traps mantle degassing in the Terminal Cretaceous marine extinctions. Cretaceous Res., 6:235–259.

McMaster R.L., Betzer P.R., Carder K.L., Miller L., Eggimann D.W., 1977. – Suspended particle mineralogy and transport in water masses of the West African shelf adjacent to Sierra Leone and Liberia. Deep-Sea Res., 24:651–665.

McMurtry G.M., Fan P.F., 1975. – Clays and clay minerals of the Santa Ana river drainage basin, California. J. Sediment. Petrol., 44:1072–1078.

McMurtry G.M., Yeh H.-W., 1981. – Hydrothermal clay mineral formation of East Pacific rise and Bauer basin sediments. Chem. Geol., 32:189–205.

McMurtry G.M., Wang C.-H., Yeh H.-W., 1983. – Chemical and isotopic investigations into the origin of clay minerals from the Galapagos hydrothermal mounds field. Geochim. Cosmochim. Acta, 47:475–489.

McRae S.G., 1972. – Glauconite. Earth Sci. Rev., 8:397–440.

Meade R.H., 1969. – Landward transport of bottom sediments in estuaries of the Atlantic coastal plain. J. Sediment. Petrol., 39:222–234.

Mégnien C., 1974. – Le passage latéral du gypse au calcaire de Champigny dans le Nord de la Brie et son interprétation paléogéographique. Bull. Inf. Géol. Bass. Paris, 41:47–65.

Meilhac A., Tardy Y., 1970. – Genèse et évolution des séricites, vermiculites et montmorillonites au cours de l'altération des plagioclases en pays tempéré. Bull. Serv. Carte géol. Als.-Lorr, 23:145–161.

Melguen M., Thiede J., 1975. – Influence des courants profonds au large du Brésil sur la distribution des faciès sédimentaires récents. 9th Int. Congr. Sedimentology, Nice, 8:51–56.

Melguen M., Debrabant P., Chamley H., Maillot H., Hoffert M., Courtois C., 1978. – Influence des courants profonds sur les faciès sédimentaires du Vema Channel (Atlantique sud) à la fin du Cénozoïque. Bull. Soc. géol. Fr., (7) 20:121–136.

Mélières F., 1973. – Les minéraux argileux de l'estuaire du Guadalquivir (Espagne). Bull. Gr. Fr. Argiles, 25:161–172.

Mélières F., 1974. – Recherches sur la dynamique sédimentaire du golfe de Cadix (Espagne). Thèse Sci. nat., Paris VI:235 pp. + ann.

Mélières F., 1978. – X-ray mineralogy studies, leg 41, Deep Sea Drilling Project, Eastern North Atlantic ocean. In: Lancelot Y., Seibold A. et al., Eds., Init. Rep. Deep Sea Drill. Proj., 41, Washington (U.S. Gov. Print. Off.):1065–1086.

Mélières F., Martin J.-M., 1969. – Les minéraux argileux dans l'estuaire de la Gironde. Bull. Gr. Fr. Argiles, 21:114–126.

Mélières F., Chamley H., Coumes F., Rouge P., 1979. – X-ray mineralogy studies, Leg 42 A, Deep Sea Drilling Project, Mediterranean Sea. In: Hsü K.J., Montadert L. et al., Eds., Init. Rep. Deep Sea Drill. Proj., 42, A, Washington (U.S. Gov. Print. Off.):361–383.

Mélières F., Deroo G., Herbin J.-P., 1981. – Organic-rich and hypersiliceous Aptian sediments from western Mid-Pacific mountains, Deep Sea Drilling Project Leg 62. In: Thiede J., Vallier T.L. et al., Eds., Init. Rep. Deep Sea Drill. Proj., 62, Washington (U.S. Gov. Print. Off.):903–921.

Mercier-Castiaux M., Chamley H., Dupuis C., 1988. – La sédimentation argileuse tertiaire dans le Bassin Belge et ses approches occidentales. Ann. Soc. géol. Nord, 107:139–154.

Meunier A., 1982. – Superposition de deux altérations hydrothermales dans la syénite monzonitique du Bac de Montmeyre (sondage INAG 1, Massif Central, France). Bull. Min., 105:386–394.

Meunier A., Velde B., 1982. – Phengitization, sericitization and potassium-beidellite in a hydrothermally-altered granite. Clay Min., 17:285–299.

Meunier A., Velde B., Beaufort D., Parneix J.-C., 1987. – Dépôts minéraux et altérations liés aux microfracturations des roches: un moyen pour caractériser les circulations hydrothermales. Bull. Soc. géol. Fr., (8), 3:971–979.

Meyer R., 1976. – Continental sedimentation, soil genesis and marine transgression in the basal beds of the Cretaceous in the East of the Paris basin. Sedimentology, 23:235–253.

Meyer R., 1981. – Rôle de la paléoaltération, de la paléopédogenèse et de la diagenèse précoce au cours de l'élaboration des séries continentales. Présentation d'exemples choisis dans quelques formations sédimentaires françaises. Thèse, Sci. nat., Nancy I:229 pp.

Meyer R., 1987. – Paléoaltérites et paléosols. B.R.G.M., Man. Méth., 13:163 pp.

Michaud L., Flicoteaux R., 1987. – Les minéraux argileux du Crétacé du Bassin sénagalais (entre Sénégal et Gambie) – Facteurs paléogéographiques de genèse et diagenèse. Sediment. Geol., 51:279–295.

Middleton N.J., Goudie A.S., Wells G.L., 1986. – The frequency and source areas of dust storms. In: Nickling W.G., Ed., Aeolian geomorphology. Allen & Unwin, New York:237–259.

Millar C.E., Turk L.M., Foth H.D., 1958. – Fundamentals of soil science, 3rd edn. Wiley & Sons, New York.

Miller A.R., Densmore C.D., Degens E.T., Hathaway J.C., Manheim F.T., McFarlin P.F., Pocklington R., Jokela A., 1966. – Hot brines and recent iron deposits in deeps of the Red Sea. Geochim. Cosmochim. Acta, 30:341–359.

Miller K.G., Fairbanks R.G., Mountain G.S., 1987. – Tertiary oxygen isotope synthesis, sea level history, and continental margin erosion. Paleoceanography, 2:1–19.

Milliman J.D., Müller J., 1973. – Precipitation and lithification of magnesian calcite in the deep-sea sediments of the eastern Mediterranean Sea. Sedimentology, 20:29–45.

Millot G., 1949. – Relations entre la constitution et la genèse des roches sédimentaires argileuses. Thèse Sci. nat., Nancy, et Géol. Appl. Prosp. Min., 2:1–352.

Millot G., 1953. – Minéraux argileux et leurs relations avec la géologie. Rev. Inst. Fr. Pétrole, 8, Spéc.:75–86.

Millot G., 1964. – Géologie des argiles. Masson, Paris:499 pp.

Millot G., 1967. – Les deux grandes voies de l'évolution des silicates à la surface de l'écorce terrestre. Rev. Quest. Sci., 138:337–357.

Millot G., 1970. – Geology of clays. Springer, Berlin, Heidelberg, New York; Masson, Paris:425pp.

Millot G., 1980. – Les grand aplanissements des socles continentaux dans les pays subtropicaux, tropicaux et désertiques. Mém. h.-sér. Soc. géol. Fr., 10:295–305.

Millot G., 1982. – Weathering sequences. "Climatic" planations. Leveled surfaces and paleosurfaces. In: Van Olphen H., Veniale F., Eds., International clay conference 1981. Developments in sedimentology, 35. Elsevier, Amsterdam:585–593.

Millot G., Nahon D., Paquet H., Ruellan A., Tardy Y., 1977. – L'épigénie calcaire des roches silicatées dans les encroûtements carbonatés en pays subaride, Antiatlas (Maroc). Sci. Géol., Bull., Strasbourg, 30, 3:129–152.

Milne I.H., Early J.W., 1958. – Effect of source and environment on clay minerals. Am. Assoc. Petrol. Geol. Bull., 42:328–338.

Milton C., 1971. – Authigenic minerals of the Green River formation. Wyoming Univ. Contrib. Geol., 10:57–63.

Moberly R.J.R., 1963. – Amorphous marine muds from tropically weathered basalt. Am. J. Sci., 261:767–772.

Moberly R.J.R., Kimura H.S., McCoy F.W., 1968. – Authigenic marine phyllosilicates near Hawaii. Geol. Soc. Am. Bull., 79:1449–1460.

Moberly R., Schlanger S.O. et al., 1985. – Init. Rep. Deep Sea Drill. Proj., 89, Washington (U.S. Gov. Print. Off.):678 pp.

Mocek L., Vandorpe B., 1984. – Séquences géochimiques de métaux piégés dans les gels siliceux amorphes déposés sur les grains de quartz littoraux. Précipitations sélectives et successives. C.R. Acad. Sci., Paris, 299:697–700.

Moinereau J., 1977. – Absorption des composés humiques par une montmorillonite H^+, Al^{3+}. Clay Min., 12:75–82.

Molina-Cruz A., 1977. – The relation of the southern tradewinds to upwelling processes during the last 75,000 years. Q. Res., 8:324–339.

Monaco A., 1967. – Etude sédimentologique et minéralogique des dépôts quaternaires du plateau continental et des rechs du Roussillon. Vie Milieu, B-18:33–62.

Monaco A., 1970. – Sur quelques phénomènes d'échanges ioniques dans les suspensions argileuses au contact de l'eau de mer. C.R. Acad. Sci., Paris, 270:1743–1746.

Monaco A., 1971. – Contribution à l'étude géologique et sédimentologique du plateau continental du Roussillon (Golfe du Lion). Thèse, Sci. nat., Montpellier:295 pp.

Monaco A., Valette J.N., Hoffert M., Picot P., 1979. – Héritage et néoformation dans les dépôts volcano-sédimentaires et hydrothermaux avoisinant l'île de Vulcano. Oceanol. Acta, 2:75–90.

Monaco A., Mear Y., Murat A., Fernandez J.-M., 1982. – Critères minéralogiques et géochimiques pour la reconnaissance des turbidites fines. C.R. Acad. Sci., Paris, 295, II:43–46.

Montadert L., De Charpal O., Roberts D., 1979. – Northeast Atlantic passive continental margins: Rifting and subsidence processes. In: Talwani M., Hay W., Ryan W.B.F., Eds., Deep drilling results in the Atlantic Ocean: continental margins and paleoenvironment. Am. Geophys. Union, Maurice Ewing Ser., 3:154–186.

Monty C., 1973. – Les nodules de manganèse sont des stromatolithes océaniques. C.R. Acad. Sci., Paris, 276:3285–3288.

Moore J.E., 1961. – Petrography of northeastern lake Michigan bottom sediments. J. Sediment. Petrol., 3:402–436.

Morad S., AlDahan A.A., 1987a. – A SEM study of diagenetic kaolinization and illitization of detrital feldspars in sandstones. Clay Min., 22:237–243.

Morad S., AlDahan A.A., 1987b. – Diagenetic chloritization of feldspars in sandstones. Sediment. Geol., 51:155–164.

Moriarty K.C., 1977. – Clay minerals in Southeast Indian Ocean sediments, transport mechanisms and depositional environments. Mar. Geol., 25:149–174.

Morley J.J., Pisias N.G., Leinen M., 1987. – Late Pleistocene time series of atmospheric and oceanic variables recorded in sediments from the Subarctic Pacific. Paleoceanography, 2:49–62.

Morrison S.J., Parry W.T., 1986. – Dioctahedral corrensite from Permian red beds, Lisbon valley, Utah. Clays Clay Min., 34:613–624.

Morton J.P., 1985. – Rb-Sr evidence for punctuated illite/smectite diagenesis in the Oligocene Frio formation, Texas gulf coast. Geol. Soc. Am. Bull., 96:11–122.

Morton J.P., Long L.E., 1984. – Rb-Sr ages of glauconite recrystallization: dating times of regional emergence above sea level. J. Sediment. Petrol., 54:495–506.

Morton R.A., 1972. – Clay mineralogy of Holocene and Pleistocene sediments, Guadalupe delta of Texas. J. Sediment. Petrol., 42:85–88.

Mount J.F., Ward P., 1986. – Origin of limestone/marl alternations in the upper Maastrichtian of Zumaya, Spain. J. Sediment. Petrol., 56:228–236.

Muffler L.P.J., White D.E., 1969. – Active metamorphism of Upper Cenozoic sediments in the Salton Sea geothermal field and the Salton Trough, southeastern California. Geol. Soc. Am. Bull., 80:157–182.

Müller C., Blanc-Vernet L., Chamley H., Froget C., 1974. – Les Coccolithophorides d'une carotte méditerranéenne. Comparaison paléoclimatologique avec les Foraminifères, les Ptéropodes et les Argiles. Tethys, 6:805–828.

Müller G., 1961. – Palygorskit und Sepiolith in tertiären und quartären Sedimenten von Hadram (S. Arabien). N. Jahrb. Mineral. Abh., 97:275–288.

Müller G., Förstner U., 1973. – Recent iron ore formation in lake Malawi, Africa. Mineral. Depos., 8:278–290.

Müller G., Förstner U., 1975. – Heavy metals in the Elbe and Rhine estuaries: mobilization or mixing effect? Environ. Geol., 1:33–39.

Müller G., Quakernaat J., 1969. – Diffractometric clay mineral analysis of recent sediments of lake Constance (Central Europe). Contrib. Mineral. Petrol., 22:268–275.

Müller G., Stoffers P., 1974. – Mineralogy and petrology of Black Sea basin sediments. Am. Assoc. Petrol. Geol., Mem. 20:200–248.

Murdmaa J.O., Demidenko Y.L., Kurnosov V.B., Faustov S.S., 1977. – Composition and rates of accumulation of clayey sediments in the Philippine Sea. Oceanology, 17:318–321.

Murray J. Renard A.F., 1891. – Report on the scientific results of the voyage of H.M.S. Challenger during the years 1873–76. Deep-sea deposits. Johnson, London:525 pp.

Nadeau P.H., Reynolds R.C., Jr., 1981. – Volcanic components in pelitic sediments. Nature (London), 294:72–74.

Nadeau P.H., Wilson M.J., McHardy W.J., Tait J.M., 1984a. – Interstratified clays as fundamental particles. Science, 225:923–925.

Nadeau P.H., Tait J.M., McHardy W.J., Wilson M.J., 1984b. – Interstratified XRD characteristics of physical mixtures of elementary clay particles. Clay Min., 19:67–76.

Nadeau P.H., Wilson M.J., McHardy W.J., Tait J.M., 1985. – The conversion of smectite to illite during diagenesis: evidence from some illitic clays from bentonites and sandstones. Mineral. Mag., 49:393–400.

Nahon D., 1986. – Evolution of iron crusts in tropical landscapes. In: Colman S.M., Dethier D.P., Eds., Rates of chemical weathering of rocks and minerals. Academic Press, New York, London:169–191.

Nahon D., Noack Y., 1983. – Pétrologie des altérations et des sols. 1. Pétrologie expérimentale. Sci. Géol., Mém., Strasbourg, 71:160 2. Pétrologie des séquences naturelles. Sci. Géol., Mém., Strasbourg, 72:170 3. Pédologie des altérations et des sols. Sci. Géol., Mém., Strasbourg, 73:208 pp.

Nahon D., Carozzi A.V., Parron C., 1980. – Lateritic weathering as a mechanism for the generation of ferruginous ooids. J. Sediment. Petrol., 50:1287–1298.

Naidu A.S., Mowatt C., 1983. – Sources and dispersal patterns of clay minerals in surface sediments from the continental-shelf areas off Alaska. Geol. Soc. Am. Bull., 94:841–854.

Naidu A.S., Burrell D.C., Hood D.W., 1971. – Clay mineral composition and geological significance of some Beaufort Sea sediments. J. Sediment. Petrol., 41:691–694.

Naidu A.S., Creager J.S., Mowatt T.C., 1982. – Clay mineral dispersal patterns in the North Bering and Chukchi Seas. Mar. Geol., 47:1–15.

Nair R.R., Hashimi N.H., Rao V.P., 1982. – Distribution and dispersal of clay minerals on the western continental shelf of India. Mar. Geol., 50:M1–M9.

Nathan Y., Flexer A., 1977. – Clinoptilolite, paragenesis and stratigraphy. Sedimentology, 24:845–855.

Neiheisel J., Weaver C.E., 1967. – Transport and deposition of clay minerals, Southeastern United States. J. Sediment. Petrol., 37:1084–1116.

Nelson B.W., 1960. – Clay mineralogy of the bottom sediments, Rappahannock river, Virginia. Clays Clay Min., 7th Natl. Conf. Pergamon, Oxford, New York:135–148.

Nelson C.H., 1967. – Sediments of crater lake, Oregon. Geol. Soc. Am. Bull., 78:833–848.

Nelson C.S., Hume T.M., 1977. – Relative intensity of tectonic events revealed by the Tertiary sedimentary record in the North Wanganui basin and adjacent areas, New-Zealand. J. Geol. Geophys. 20:369–392.

Nesteroff W.D., 1973. – Petrology and mineralogy of sapropels. In: Ryan W.B.F., Hsü K.J. et al., Eds., Init. Rep. Deep Sea Drill. Proj., 13, Washington (U.S. Gov. Print. Off.):713–720.

Nesteroff W.D., Sabatier G., 1962. – "Apport" et "néogenèse" dans la formation des argiles des grands fonds marins. In Genèse et synthèse des argiles. Coll. Int. C.N.R.S., 105:149–158.

Nesteroff W.D., Sabatier G., Heezen B.C., 1964. – Les minéraux argileux, le quartz et le calcaire dans quelques sédiments de l'océan Arctique. C.R. Acad. Sci., Paris, 258:991–993.

Newman A.C.D., Ed., 1987. – Chemistry of clays and clay minerals. Min. Soc., Monogr., 6:480 pp.

Nichols M.M., 1972. – Sediments of the James river estuary, Virginia. Geol. Soc. Am. Mem. 133:169–212.

Nicolas J., 1956. – Contribution à l'étude géologique et minéralogique de quelques gisements de kaolins bretons. Thèse Sci. Nat., Paris:254 pp.

Nilsen T.H., Kerr D.R., 1978. – Paleoclimatic and paleogeographic implications of a lower Tertiary laterite (latosol) on the Iceland-Faeroe ridge, North Atlantic region. Geol. Mag., 115:153–236.

Nir Y., Nathan Y., 1972. – Mineral clay assemblages in recent sediments of the Levantine Basin, Mediterranean Sea. Bull. Gr. Fr. Argiles, 24:187–194.

Noack Y., 1981. – La palagonite. Bull. Minéral., 104:36–46.

Noack Y., 1983. – Palagonitization and time. Sci. géol., Bull., Strasbourg, 36:111–116.

Noack Y., 1985. – Différenciation entre altération météorique et altération hydrothermale. Application aux minéraux ubiquistes: talc, chlorites, serpentines. Pétrologie, minéralogie, géochimie. Thèse Sci. Nat., Poitiers:156 pp.

Norris R.J., Carter R.M., Turnbull I.M., 1978. – Cenozoic sedimentation in basins adjacent to a major continental transform boundary in southern New-Zealand. J. Geol. Soc. London, 135:191–205.

O'Brien N.R., 1987. – The effects of bioturbation on the fabric of shale. J. Sediment. Petrol., 57:449–455.

O'Brien N.R., Nakazawa K., Toluhashi S., 1980. – Use of clay fabric to distinguish turbiditic and hemipelagic siltstones and silts. Sedimentology, 27:47–61.

Odigi M.I., 1986. – Mineralogical and geochemical studies of Tertiary sediments from the eastern Niger delta and their relationship to petroleum occurrence. J. Petrol. Geol., 10:101–114.

Odin G.S., 1975. – Les glauconies: constitution, formation, âge. Thèse Sci. nat., Paris VI:277 pp.

Odin G.S., 1982. – The Phanerozoic time scale revisited. Episodes, N3:3–9.

Odin G.S., 1985. – La "verdine", faciès granulaire vert, marin et côtier, distinct de la glauconie: distribution actuelle et composition. C.R. Acad. Sci., Paris, 301:105–108.

Odin G.S., Ed., 1988. – Green marine clays. Developments in sedimentology. Elsevier, Amsterdam, 45:445 pp.

Odin G.S., Matter A., 1981. – De glauconiarum origine. Sedimentology, 28:611–641.

Odin G.S., Stephan J.-F., 1981. – The occurrence of deep water glaucony from the Eastern Pacific: the result of in situ genesis or subsidence? In: Watkins J.S., Moore J.C. et al., Eds., Init. Rep. Deep Sea Drill. Proj., 66, Washington (U.S. Gov. Print. Off.):419–428.

Odin G.S., Bailey S.W., Amouric M., Fröhlich F., Waychunas G., 1988. – Mineralogy of the verdine facies. In: Odin G.S., Ed., Green marine clays. Developments in sedimentology. Elsevier, Amsterdam:445 pp.

Oinuma K., Aoki S., 1977. – The vertical variations in the clay mineral compositions of the sediment core samples from the Japan Sea. J. Tokyo Univ., 20:1–16.

Onoratini G., Chamley H., Escalon de Fonton M., 1973. – Note préliminaire sur la signification paléoclimatique des minéraux argileux dans le remplissage de l'abri Cornille (B. du Rh.). C.R. Soc. géol. Fr., 15:59–61.

Orians K.J., Bruland K.W., 1986. – The biogeochemistry of aluminum in the Pacific Ocean. Earth Planet. Sci. Lett., 78:397–410.

Orsolini P., Chamley H., 1980. – Alluvionnement argileux et dynamique sédimentaire dans la région de Saint-Tropez (Var). Géol. Méditerr., 7:155–159.

Owens J.P., Stefansson K., Sirkin L.A., 1974. – Chemical, mineralogic and palynologic character of the Upper Wisconsinian-Lower Holocene fill in parts of Hudson, Delaware and Chesapeake estuaries. J. Sediment. Petrol., 44:390–408.

Pacey N.R., 1984. – Bentonites in the chalk of central eastern England and their relation to the opening of the northeast Atlantic. Earth Planet. Sci. Lett., 67:48–60.

Packham R.F., Rosaman D., Midgley H.G., 1961. – A mineralogical examination of suspended solids from nine English rivers. Clay Min., 4:239–242.

Pacquet A., 1968. – Analcime et argiles diagénétiques dans les formations sédimentaires de la région d'Agadès (République du Niger). Mém. Serv. Carte géol. Als.-Lorr., 27:221 pp.

Palais J.M., 1985. – Particle morphology, composition and associated ice chemistry of tephra layers in the Byrd ice core: evidence for hydrovolcanic eruptions. Ann. Glaciol., 7:42–48.

Palomo Delgado I., Ortega Huertas M., Fenoll Hach-Ali P., 1985. – The significance of clay minerals in studies of the evolution of the Jurassic deposits of the Betic Cordillera, SE Spain. Clay Min., 20:39–52.

Paquet H., 1970. – Evolution géochimique des minéraux argileux dans les altérations et les sols des climats méditerranéens et tropicaux à saisons contrastées. Mém. Serv. Carte géol. Als.-Lorr., 30:212 pp.

Paquet H., Bocquier G., Millot G., 1966. – Néoformation et dégradation des minéraux argileux dans certains solonetz solodisés et vertisols du Tchad. Bull. Serv. Carte géol. Als.-Lorr., 19:295–322.

Paquet H., Ruellan A., Tardy Y., Millot G., 1969. – Géochimie d'un bassin versant au Maroc oriental. Evolution des argiles dans les sols des montagnes et des plaines de la Basse Moulouya. C.R. Acad. Sci., Paris, 269:1839–1842.

Paquet H., Coudé-Gaussen G., Rognon P., 1984. – Etude minéralogique de poussières sahariennes le long d'un itinéraire entre 19° et 35° de latitude nord. Rev. Géol. dyn. Géogr. phys., 25:257–265.

Parham W.E., 1966. – Lateral variations of clay mineral assemblages in modern and ancient sediments. Proc. Int. Clay Conf., Jerusalem, Isr., 1:135–145.

Parker A., Sellwood B.W., Eds., 1984. – Sediment diagenesis. Reidel, Dordrecht, Nato Asi ser., C, 115:427 pp.

Parker J.I., Edgington D.N., 1976. – Concentration of diatom frustules in lake Michigan sediment cores. Limnol. Oceanogr., 21:887–893.

Parkin D.W., Shackleton N.J., 1973. – Trade wind and temperature correlations down a deep-sea core off the Sahara coast. Nature (London), 245:455–457.

Parmenter C., Folger D.W., 1974. – Eolian biogenic detritus in deep-sea sediments: possible index of equatorial Ice Age aridity. Science, 185:695–698.

Parneix J.C., Beaufort D., Dudoignon P., Meunier A., 1985. – Biotite chloritization process in hydrothermally altered granites. Chem. Geol., 51:89–101.

Parra M., Delmont P., Ferragne A., Latouche C., Pons J.C., Puechmaille C., 1985. – Origin and evolution of smectites in recent marine sediments of the NE Atlantic. Clay Miner., 20:335–346.

Parra M., Pons J.-C., Ferragne A., 1986. – Two potential sources for Holocene clay sedimentation in the Caribbean Basin: the lesser Antilles Arc and the South American continent. Mar. Geol., 72:287–304.

Parron C., Amouric M., 1987. – TEM and microchemical study of the glauconitization process. In: 6th Meet. Eur. Clay Groups, Sevilla, Proc.:426–427.

Parron C., Triat J.-M., 1977. – Nouvelles conceptions sur le Crétacé supérieur du Gard. Répercussions sur la stratigraphie, la paléogéographie et la tectonique de la découverte de trois phases d'altération continentale. Rev. Géogr. phys. Géol. dyn., 19:241–250.

Parron C., Triat J.-M., 1978. – Paléoaltérations continentales et sédimentogenèse marine dans le Crétacé supérieur du massif d'Uchaux (Vaucluse). Bull. B.R.G.M., (1), 1:47–56.

Parry W.T., Reeves C.C., 1968 a. – Clay mineralogy of Pluvial lake sediments, Southern High Plains, Texas. J. Sediment. Petrol., 38:516–529.

Parry W.T., Reeves C.C., 1968 b. – Sepiolite from Pluvial Mound lake, Lynn and Terry counties, Texas. Am. Miner., 53:984–993.

Pastouret L., Auffret G.A., Hoffert M., Melguen M., Needham H.D., Latouche C., 1975. – Sédimentation sur la ride de Terre-Neuve. Can. J. Earth Sci., 12:1019–1035.

Pastouret L., Auffret G.A., Chamley H., 1978 a. – Microfacies of some sediments from the Western North Atlantic: paleoceanographic implications (Leg 44 DSDP). In: Benson W.E., Sheridan R.E. et al., Eds., Init. Rep. Deep Sea Drill. Proj., 44, Washington (U.S. Gov. Print. Off.):477–501.

Pastouret L., Chamley H., Delibrias G., Duplessy J.-C., Thiede J., 1978 b. – Late Quaternary climatic changes in Western tropical Africa deduced from deep-sea sedimentation off the Niger delta. Oceanol. Acta, 1:217–232.

Pearson M.J., Small J.S., 1988. – Illite-smectite diagenesis and paleotemperatures in Northern Sea Quaternary to Mesozoic shale sequences. Clay Min., 23:109–132.

Pearson M.J., Watkins D., Small J.S., 1982. – Clay diagenesis and organic maturation in northern North Sea sediments. In: van Olphen H., Veniale F., Eds., Developments in sedimentology. Elsevier, Amsterdam, 35:665–675.

Pédro G., 1968. – Distribution des principaux types d'altération chimique à la surface du globe. Présentation d'une esquisse géographique. Rev. Géogr. phys. Géol. dyn., (10), 5:457–470.

Pédro G., 1979. – Caractérisation générale des processus de l'altération hydrolytique. Base des méthodes géochimique et thermodynamique. Science du sol. Bull. A.F.E.S., 2–3:93–105.

Pédro G., 1981. – Les grands traits de l'évolution cristallochimique des minéraux au cours de l'altération superficielle des roches. Rendiconti Soc. It. Min. Petrol., 37:633–666.

Pédro G., 1984. – La genèse des argiles pédologiques, ses implications minéralogiques, physico-chimiques et hydriques. Sci. géol. Bull., 37:333–347.

Pédro G., Carmouze J.P., Velde B., 1978. – Peloidal nontronite formation in recent sediments of lake Chad. Chem. Geol., 23:139–149.

Perry E., Hower J., 1970. – Burial diagenesis in Gulf coast pelitic sediments. Clays Clay Min., 18:165–177.

Perry E.A. Jr., 1974. – Diagenesis and the K-Ar dating of shales and clay minerals. Geol. Soc. Am. Bull., 85:827–830.

Perry E.A. Jr., Hower J., 1972. – Late-stage dehydration in deeply buried pelitic sediments. Am. Assoc. Petrol. Geol. Bull., 56:2013–2021.

Perry E.A. Jr., Beckles E.C., Newton R.M., 1976. – Chemical and mineralogical studies, Sites 322 and 325. In: Hollister C.D. et al., Eds., Init. Rep. Deep Sea Drill. Proj., 35, Washington (U.S. Gov. Print. Off.):465–469.

Pertsev N.N., Rusinov V.L., 1980. – Mineral assemblages and processes of alterations in basalts at Deep Sea Drilling Project. Sites 417 and 418. In: Donnelly T., Francheteau J., Bryan W., Robinson P., Flower M., Salisbury M. et al., Eds., Init. Rep. Deep Sea Drill. Proj., 51, 52, 53, Pt 2, Washington (U.S. Gov. Print. Off.):1219–1242.

Petersen L., Rasmussen K., 1980. – Mineralogical composition of the clay fraction of two fluvio-glacial sediments from east Greenland. Clay Min., 15:135–145.

Peterson M.N.A., Griffin J.J., 1964. – Volcanism and clay minerals in the Southeastern Pacific. J. Mar. Res., 22:13–21.

Peterson M.N.A., Edgar N.T., von der Borch C., Rex R.W., 1970. – Cruise leg summary and discussion. In: Peterson M.N.A., Edgar N.T. et al., Eds., Init. Rep. Deep Sea Drill. Proj., 2, Washington (U.S. Gov. Print. Off.):413–430.

Petit J.-R., Briat M., Royer A., 1981. – Ice age aerosol content from East Antarctic ice core samples and past wind strength. Nature (London), 293:391–393.

Pevear D.R., 1972. – Source of recent nearshore marine clays, southeastern United States. In: Nelson B.W., Ed., Environmental framework of coastal plain estuaries. Geol. Soc. Am. Mem., 133:317–335.

Pevear D.R., Dethier D.P., Frank D., 1982. – 1980 Mt. St Helens eruption: clay minerals and early-stage alteration. In: Van Olphen H., Veniale F., Eds., Int. Clay Conf. 1981. Developments in sedimentology, Elsevier, Amsterdam, 35:557–563.

Péwé T.L., 1981. – Desert dust. Geol. Soc. Am. Spec. Pap., 186:303 pp.

Pierce J.W., Siegel F.R., 1969. – Quantification in clay mineral studies of sediments and sedimentary rocks. J. Sediment. Petrol., 39:187–193.

Pierce J.W., Stanley D.J., 1975. – Suspended-sediment concentration and mineralogy in the central and western Mediterranean and mineralogic comparison with bottom sediment. Mar. Geol., 19:M15–M25.

Pierce J.W., Nelson D.D., Colquhoun D.J., 1972. – Mineralogy of suspended sediment off the southeastern United States. In: Swift, Duane, Pilkey, Eds., Shelf sediment transport. Dowden, Hutchinson & Ross, Stroudsburg:281–305.

Pinet P.R., Morgan W.P. Jr., 1979. – Implications of clay-provenance studies in two Georgia estuaries. J. Sediment. Petrol., 49:575–580.

Pinet P.R., Morgan W.P. Jr., 1980. – Implication of clay provenance studies in two Georgia estuaries. Reply. J. Sediment. Petrol., 50:995–996.

Pinsak A.P., Murray H.H., 1960. – Regional clay mineral patterns in the Gulf of Mexico. Clays Clay Min., 7th Natl. Conf. Pergamon, Oxford, New York:178–184.

Pinta M., 1971. – Spectrométrie d'absorption atomique. ORSTOM, 2 vols. Masson, Paris:285 and 293 pp.

Piper D.J.W., Pe G.G., 1977. – Cenozoic clay mineralogy from D.S.D.P. holes on the continental margin of the Australia-New Zealand sector of Antarctica. N.Z. J. Geol. Geophys., 20:905–917.

Piper D.J.W., Slatt R.M., 1977. – Late Quaternary clay-mineral distribution on the eastern continental margin of Canada. Geol. Soc. Am. Bull., 88:267–272.

Piper D.Z., 1974. – Rare earth elements in ferromanganese nodules and other marine phases. Geochim. Cosmochim. Acta, 38:1007–1022.

Piper D.Z., Rude P.D., Monteith S., 1987. – The chemistry and mineralogy of haloed burrows in pelagic sediment at domes site A: the Equatorial North Pacific. Mar. Geol., 74:41–55.

Pokras E.M., Mix A.C., 1985. – Eolian evidence for spatial variability of late Quaternary climates in tropical Africa. Q. Res., 24:137–149.

Pokras E.M., Mix A.C., 1987. – Earth's precession cycle and Quaternary climatic change in tropical Africa. Nature (London), 326:486–487.

Pollastro R.M., 1981. – Authigenic kaolinite and associated pyrite in chalk of the cretaceous Niobrara formation, eastern Colorado. J. Sediment. Petrol., 51:553–562.

Pollastro R.M., 1985. – Mineralogical and morphological evidence for the formation of illite at the expense of illite/smectite. Clays Clay Min., 33:265–274.

Pollastro R.M., Barker C.E., 1986. – Application of clay-mineral, vitrinite reflectance, and fluid inclusion studies to the thermal and burial history of the Pinedale anticline, Green River basin, Wyoming. In: Roles of organic matter in sediment diagenesis. Soc. Econ. Pal. Min., Spec. Publ., 38:73–83.

Pollastro R.M., Martinez C.J., 1985. – Whole-rock, insoluble residue, and clay mineralogies of marl, chalk, and bentonite, Smoky Hill shale member, Niobrara formation near Pueblo, Colorado – depositional and diagenetic implications. Soc. Econ. Pal. Min., 2nd Ann. Midy. Meet.:215–222.

Pollastro R.M., Pillmore C.L., 1987. – Mineralogy and petrology of the Cretaceous-Tertiary boundary clay bed and adjacent clay-rich rocks, Raton basin, New Mexico and Colorado. J. Sediment. Petrol., 57:456–466.

Pomerol C., 1967. – Esquisse paléogéographique du Bassin de Paris, à l'ère tertiaire et aux temps quaternaires. Rev. Géogr. phys. Géol. dyn., (9) 1:55–86.

Poppe L.J., Hathaway J.C., Parmenter C.M., 1983. – Talc in the suspended matter of the North-Western Atlantic. Clays Clay Miner., 31:60–64.

Porrenga D.H., 1966. – Clay minerals in recent sediments of the Niger delta. Clays Clay Min., 14th nat. conf., Pergamon, Oxford, New York:221–233.

Porrenga D.H., 1967. – Glauconite and chamosite as depth indicators in the marine environment. Mar. Geol., 5:495–501.

Post J.L., 1978. – Sepiolite deposits of the Las Vegas, Nevada area. Clays Clay Miner., 26:58–64.

Post J.L., 1984. – Saponite from Near Ballarat, California. Clays Clay Miner., 32:147–153.

Postma H., 1967. – Sediment transport and sedimentation in estuaries environment. In: Lauff G.H., Ed., Estuaries. Am. Assoc. Adv. Sci., publ. 83:158–184.

Potter P.E., Heling D., Shimp N.E., van Wie W., 1975. – Clay mineralogy of modern alluvial muds of the Mississippi River Basin. Bull. Centre Rech., Pau, 9:353–389.

Potter P.E., Maynard J.B., Pryor W.A., 1980. – Sedimentology of shale. Springer, Berlin, Heidelberg, New York:306 pp.

Pouteau C., Beaufort D., Meunier A., 1985. – Les différents types d'altération hydrothermale des formations volcano-sédimentaires de Bossa (Haïti). In: Géodynamique des Caraïbes. Symp. Paris, Technip, Paris:341–352.

Powell T.G., Foscolos A.E., Gunther P.R., Snowdon L.R., 1978. – Diagenesis of organic matter and fine clay minerals: a comparative study. Geochim. Cosmochim. Acta, 42:1121–1197.

Power P.E., 1969. – Clay mineralogy and paleoclimatic significance of some red regoliths and associated rocks in western Colorado. J. Sediment. Petrol., 39:876–890.

Powers M.C., 1954. – Clay diagenesis in the Chesapeake bay area. Clays Clay Miner., 2nd Natl. Conf. Pergamon, Oxford, New York:68–80.

Powers M.C., 1957. – Adjustement of land-derived clays to the marine environment. J. Sediment. Petrol., 27:355–372.

Powers M.C., 1959. – Adjustement of clays to chemical change and the concept of the equivalence level. Clays Clay Miner., 6th Natl. Conf. Pergamon, Oxford, New York:309–326.

Powers M.C., 1967. – Fluid release mechanisms in compacting marine mudrocks and their importance in oil exploration. Am. Assoc. Petrol. Geol. Bull., 51:1240–1254.

Pozzuoli A., Ed., 1985. – Proceedings, clays and clay minerals. Mineral. Petrogr. Acta, Bologna, 29-A:732 pp.

Price N.B., 1976. – Chemical diagenesis in sediments. In: Riley J.P., Chester R., Eds., Chemical oceanography. Academic Press, New York, London, 6, 30:1–58.

Primmer T.J., Shaw H.F., 1987. – Diagenesis in shales: evidence from backscattered electron microscopy and electron microprobe analysis. Proc. Int. Clay Conf., Denver, 1985:135–143.

Prospero J.M., 1981a. – Eolian transport to the world ocean. In: Emiliani C., Ed., The sea, VII: The oceanic lithosphere. Wiley & Sons, New York:801–874.

Prospero J.M., 1981b. – Arid regions as sources of mineral aerosols in the marine atmosphere. Geol. Soc. Am. Spec. Pap., 186:71–86.

Pryor W.A., 1975. – Biogenic sedimentation and alteration of argillaceous sediments in shallow marine environments. Geol. Soc. Am. Bull., 86:1244–1254.

Pye K., 1983a. – Early post-depositional modification of aeolian dune sands. In: Brookfield M.E., Ahlbrandt T.S., Eds., Eolian sediments and processes. Developments in sedimentology. Elsevier, Amsterdam, 38:197–221.

Pye K., 1983b. – Red beds. In: Goudie A.S., Pye K., Eds., Chemical sediments and geomorphology: precipitates and residua in the near-surface environment. Academic Press, New York, London:227–263.

Pye K., 1987. – Aeolian dust and dust deposits. Academic Press, New York, London:334 pp.

Pye K., Krinsley D.H., Burton J.H., 1986. – Diagenesis of U.S. Gulf coast shales. Nature (London), 324:557–559.

Quakernaat J., 1968. – X-ray analysis of clay minerals in some recent fluviatile sediments along the coast of central Italy. Publ. Fys. Geogr. Lab. Univ. Amsterdam, 12:105 pp.

Quantin P., Badaut-Trauth D., Weber F., 1975. – Mise en évidence de minéraux secondaires, argiles et hydroxydes, dans les andosols des Nouvelles-Hébrides, après la déferrification par la méthode de Endredy. Bull. Groupe Fr. Argiles, 27:51–67.

Quantin P., Gautheyrou J., Lorenzoni P., 1987. – Halloysite formation through the in situ weathering of volcanic glass, from trachytic pumices, Vico's vulcano, Italy. In: 6th Meet. Eur. Clay Groups, Sevilla:451–452.

Radczewski O.E., 1939. – Eolian deposits in marine sediments. In: Trask P.D., Ed., Recent marine sediments. Am. Assoc. Petrol. Geol., Tulsa:496–502.

Rampino M.R., Reynolds R.C., 1983. – Clay mineralogy of the Cretaceous Tertiary boundary clay. Science, 219:405–498.

Ramseyer K., Boles J.R., 1986. – Mixed-layer illite/smectite minerals in Tertiary sandstones and shales, San Joaquin basin, California. Clays Clay Min., 34:115–124.

Rangin C., Desprairies A., Fontes J.C., Jehanno C., Vernhet S., 1983. – Metamorphic processes in sediments in contact with young oceanic crust – East Pacific rise, leg 65. In: Lewis B.T.R., Robinson P. et al., Eds., Init. Rep. Deep Sea Drill. Proj., 65, Washington (U.S. Gov. Print. Off.):375–389.

Rao V.P., 1987. – Mineralogy of polymetallic nodules and associated sediments from the Central Indian Ocean basin. Mar. Geol., 74:151–157.

Rao N.V.N.D., Srihari Y., 1980. – Clay mineralogy of the late Pleistocene red sediments of the Visakhapatnam region, east coast of India. Sediment. Geol., 27:213–227.

Rasplus L., Estéoule-Choux J., Estéoule J., 1976. – Les minéraux argileux de l'Eocène continental de la Grande Brenne (Indre). C.R. Acad. Sci., Paris, D, 283:901–904.

Rateev M.A., Emel'Janov E.M., Kheirov M.B., 1966. – Peculiarities of the formation of clay minerals in the recent sediments of the Mediterranean Sea. Lit. Palezn. Isk. Akad. Nauk S.S.S.R., 4:6–23. (in Russian)

Rateev M.A., Gorbunova Z.N., Lisitzin A.P., Nosov G.I., 1968. – Climatic zonality of the argillaceous minerals in the World Ocean sediments. Okeanol., Akad. Nauk S.S.S.R., 18:283–311. (in Russian)

Rateev M.A., Gorbunova Z.N., Lisitzin A.P., Nosov G.I., 1969. – The distribution of clay minerals in the oceans. Sedimentology, 13:21–43.

Rateev M.A., Timofeev P.P., Rengarten N.V., 1980. – Minerals of the clay fraction in Pliocene-Quaternary sediments of the East Equatorial Pacific. In: Rosendahl B.R., Hekinian R. et al., Eds., Init. Rep. Deep Sea Drill. Proj., 51, Washington (U.S. Gov. Print. Off.):307–318.

Rautureau M., Steinberg M., 1985. – Détermination de la composition et de l'homogénéité des phyllosilicates par microscopie électronique analytique à balayage (S.T.E.M.). J. Microsc. Spectrosc. Electron., 10:181–192.

Rawi G.J., Sys C., 1967. – A comparative study between Euphrates and Tigris sediments in the Mesopotamian flood plain. Pedol., Ghent, 17:187–211.

Rea D.K., Bloomstine M.K., 1986. – Neogene history of the South Pacific tradewinds: evidence for hemispherical asymmetry of atmospheric circulation. Palaeogeogr., Climatol., Ecol., 55:55–64.

Rea D.K., Janecek T.R., 1981. – Late Cretaceous history of eolian accumulation in the mid-Pacific Mountains, central North Pacific Ocean. Palaeogeogr., Climatol., Ecol., 36:55–67.

Rea D.K., Leinen M., Janecek T.R., 1985. – Geologic approach to the long-term history of atmospheric circulation. Science, 227:721–725.

Redondo C.P., 1973. – Contribution à l'étude des paléosols des vallées de la Bléone et de la Durance (Alpes de Haute-Provence). C.R. Acad. Sci., Paris, 277:289–292.

Redondo C.P., 1986. – Etude des sédiments détritiques du Crétacé supérieur marin de la Provence occidentale et recherche des zones d'apport. Sédimentologie; pétrographie; minéralogie. Thèse Sci. nat., Univ. Aix-Marseille I:493 pp. + ann.

Retallack G.J., 1983. – A paleopedological approach to the interpretation of terrestrial sedimentary rocks: The mid-Tertiary fossil soils of Badlands National Park, South Dakota. Geol. Soc. Am. Bull., 94:823–840.

Rettke R.C., 1981. – Probable burial diagenetic and provenance effects on Dakota Group clay mineralogy, Denver Basin. J. Sediment. Petrol., 51:541–551.

Revel J.-C., Margulis H., 1972. – Minéraux silicatés et pédogenèse des sols calcaires rouges méditerranéens. C.R. Acad. Sci., Paris, 275:539–541.

Rex R.W., Goldberg E.E., 1958. – Quartz contents of pelagic sediments of the Pacific Ocean: Tellus, 10:153–159.

Rex R.W., Murray B., 1970. – X-ray mineralogy studies. Leg 4. In: Bader R.G. et al., Eds., Init. Rep. Deep Sea Drill. Proj., 4, Washington (U.S. Gov. Print. Off.):325–369.

Rex R.W., Syers J.K., Jackson M.L., Clayton R.N., 1969. – Eolian origin of quartz in soils of Hawaiian Islands and in Pacific sediments. Science, 163:277–279.

Reynolds R.C., 1971. – Clay mineral formation in an alpine environment. Clays Clay Min., 19:361–374.

Reynolds R.C., 1980. – Interstratified clay minerals. In: Brindley G.W., Brown G., Eds., Crystal structures of clay minerals and their X-ray identification. Mineral. Soc., London: 249–303.

Reynolds R.C., 1983. – Calculation of absolute diffraction intensities for mixed-layered clays. Clays Clay Min., 31:233–234.

Reynolds R.C. Jr., Hower J., 1970. – The nature of interlayering in mixed-layer illite-montmorillonites. Clays Clay Min., 18:25–36.

Reynolds S., 1987. – Problems in interpretation of clay fabrics. Am. Assoc. Petrol. Geol. Bull., 71:606–607.

Reyre Y., 1980. – Peut-on estimer l'évolution des climats jurassiques et crétacés d'après la palynologie? Mém. Mus. Nat. Hist. Nat., B, 27:247–260.

Rhoads D.C., Boyer L.F., 1982. – The effects of marine benthos on physical properties of sediments: a successional perspective. In: McCall P.L., Tevesz M.J.S., Eds., Animal-sediment relations, the biogenic alteration of sediments. Plenum, New York:3–52.

Richardson J.L., Richardson A.E., 1972. – History of an African rift lake and its climatic implications. Ecol. Monogr., 42:499–534.

Riech V., Rad U. von, 1979. – Silica diagenesis in the Atlantic Ocean: diagenetic potential and transformations. In: Talwani M., Hay W., Ryan W.B.F., Eds., Deep drilling results in the Atlantic Ocean: continental margins and paleoenvironment. Am. Geophys. Union, Maurice Ewing Ser., 3:315–340.

Rieck R.L., Winters H.A., Mokma D.L., Mortland M.M., 1979. – Differentiation of surficial glacial drift in southeastern Michigan from 7-Å/10-Å X-ray diffraction ratios of clays. Geol. Soc. Am. Bull., 90:216–220.

Rieke H.H., Chilingarian G.V., 1974. – Compaction of argillaceous sediments. Elsevier, Amsterdam:424 pp.

Riemann F., 1983. – Biological aspects of deep-sea manganese nodule formation. Oceanol. Acta, 6:303–311.

Rivière A., Vernhet S., 1951. – Sur la sédimentation des minéraux argileux en milieu marin en présence de matières humique. Conséquences géologiques. C.R. Acad. Sci., Paris, 233:807–808.

Roaldset E., 1987. – Role of the clay in the petroleum generation and exploration. In: 6th Meet. Eur. Clay Groups, Sevilla, Proc.:12–14.

Roberson H.E., 1974. – Early diagenesis: expansible soil clay-sea water reactions. J. Sediment. Petrol., 44:441–449.

Robert C., 1974. – Contribution à l'étude de la sédimentation argileuse en Méditerranée orientale. Thèse 3ème cycle, Aix-Marseille II:82 pp.

Robert C., 1980a. – Observations sur la sédimentation argileuse récente dans la région Nord-Egéenne. Géol. Méd., 7:179–186.

Robert C., 1980b. – Climats et courants cénozoïques dans l'Atlantique Sud d'après l'étude des minéraux argileux (DSDP legs 3, 39 et 40). Oceanol. Acta, 3:369–376.

Robert C., 1981. – Santonian to Eocene paleogeographic evolution of the Rio Grande Rise (South Atlantic) deduced from clay-mineralogical data (DSDP legs 3 and 39). Paleogeogr., Climatol., Ecol., 33:311–325.

Robert C., 1982. – Modalité de la sédimentation argileuse en relation avec l'histoire géologique de l'Atlantique Sud. Thesis. Univ. Aix-Marseille II:141 pp.

Robert C., 1987. – Clay mineral associations and structural evolution of the South Atlantic: Jurassic to Eocene. Paleogeogr., Climatol., Ecol., 58:87–108.

Robert C., Chamley H., 1974. – Gypse et sapropels profonds de Méditerranée orientale. C.R. Acad. Sci., Paris, 278:843–846.

Robert C., Chamley H., 1986. – La kaolinite des sédiments est-atlantiques, témoin des climats et environnements cénozoïques. C.R. Acad. Sci., Paris, II, 303:1563–1568.

Robert C., Chamley H., 1987. – Cenozoic evolution of continental humidity and paleoenvironment, deduced from the kaolinite content of oceanic sediments. Palaeogeogr., Climatol., Ecol., 60:171–187.

Robert C., Maillot H., 1983. – Paleoenvironmental significance of clay mineralogical and geochemical data, Southwest Atlantic, Deep Sea Drilling Project Legs 36 and 71. In: Ludwig W.J., Krasheninnikov V.A., Eds., Init. Rep. Deep Sea Drill. Proj., 71, Washington (U.S. Gov. Print. Off.):317–343.

Robert C., Herbin J.-P., Deroo G., Giroud d'Argoud G., Chamley H., 1979. – L'Atlantique Sud au Crétacé d'après l'étude des minéraux argileux et de la matière organique (Legs 39 et 40 DSDP). Oceanol. Acta, 2:209–218.

Robert C., Gauthier A., Chamley H., 1984. – Origine autochtone et allochtone des argiles récentes de haute altitude en Corse. Géol. Médit., 11:243–253.

Robert C., Stein R., Acquaviva M., 1985. – Cenozoic evolution and significance of clay associations in the New-Zealand region of the South Pacific, Deep Sea Drilling Project, Leg 90. In: Kennett J.P., von der Borch C.C. et al., Eds., Init. Rep. Deep Sea Drill. Proj., 90, Washington (U.S. Gov. Print. Off.):1225–1238.

Robert C., Caulet J.-P., Maillot H., 1988. – Evolution climatique et hydrologique en mer de Ross (Site DSDP 274) au Néogène, d'après les associations de radiolaires, la minéralogie des argiles et la géochimie minérale. C.R. Acad. Sci., Paris, II, 306:437–442.

Robert M., Cabidoche Y.-M., Berrier J., 1980. – Pédogenèse et minéralogie des sols de haute montagne cristalline (étages alpin et subalpin. Alpes-Pyrénées). Sci. Sol, 4:313–337.

Roberts H.H., 1985. – Clay mineralogy of contrasting mudflow and distal shelf deposits on the Mississippi River delta front. Geo-Mar. Lett., 5:185–191.

Robertson A.H.F., 1983. – Latest Cretaceous and Eocene paleoenvironments in the Blake-Bahama Basin, Western North Atlantic. In: Sheridan R.E., Gradstein F.M. et al., Eds., Init. Rep. Deep Sea Drill. Proj., 76, Washington (U.S. Gov. Print. Off.):763–780.

Robertson A.H.F., Bliefnick D.M., 1983. – Sedimentology and origin of lower Cretaceous pelagic carbonates and redeposited clastics, Blake-Bahama formation, Deep Sea Drilling Project, site 534, Western Equatorial Atlantic. In: Sheridan R.E., Gradstein F.M. et al., Eds., Init. Rep. Deep Sea Drill. Proj., 76, Washington (U.S. Gov. Print. Off.):795–828.

Robertson A.H.F., Boyle J.F., 1983. – Tectonic setting and origin of metalliferous sediments in the mesozoic Tethys ocean. In: Rona P.A., Boström K., Laubier L., Smith A.G., Eds., Hydrothermal processes at seafloor spreading centers. Nato Conf. Ser., 12. Plenum, New York:595–663.

Robillard D., Piqué A., 1981. – Epimétamorphisme du Permo-Trias dans le Moyen-Atlas septentrional (Maroc). Mise en évidence et zonation régionale. Rev. Géol. dyn. Géog. phys., 23:301–308.

Robin A.-M., 1979. – Genèse et évolution des sols podzolisés sur affleurements sableux du Bassin Parisien. Thèse, Sci. nat., Nancy I:173 pp.

Robinson D., Ed., 1985. – Diagenesis and low-temperature metamorphism. Miner. Mag., 49:301–498.

Robinson D., Wright V.P., 1987. – Ordered illite-smectite and kaolinite-smectite: pedogenic minerals in a lower Carboniferous paleosol sequence, South Wales. Clay Min., 22:109–118.

Rögl F., Steininger F.F., 1983. – Vom Zerfall der Tetyhs zu Mediterran und Paratethys. Die neogene Palaogeographie und Palinspastik des zirkum-mediterranen Raumes. Ann. Naturhist., Mus. Wien, 85 A:135–163.

Rohrlich V., Price N.B., Calvert S.E., 1969. – Chamosite in the recent sediments of Loch Etive, Scotland. J. Sediment. Petrol., 39:624–631.

Rolfe B.N., Hadley R.F., 1963. – Weathering and transport of sediment in the Cheyenne River Basin, Eastern Wyoming. Clays Clay Min., 11th Natl. Conf.:83.

Rona P.A., 1978. – Criteria of recognition of hydrothermal mineral deposits in oceanic crusts. Econ. Geol., 73:135–160.

Rona P.A., 1980. – TAG hydrothermal field: mid-Atlantic ridge crest at latitude 26°N. J. Geol. Soc. London, 137:385–402.

Rosato V.J., Kulm L.D., 1981. – Clay mineralogy of the Peru continental margin and adjacent Nazca plate: implications for provenance, sea level changes, and continental accretion. Geol. Soc. Am., Mem. 154:545–568.

Rosenqvist, I.T., 1961. – What is the origin of the hydrous micas of Fennoscandia? Uppsala Univ. Geol. Inst. Bull., 40:265–268.

Ross C.S., Kerr P.F., 1931. – The kaolin minerals. U.S. Geol. Surv. Prof. Pap., 165E:151–170.

Ross G.M., Chiarenzelli J.R., 1985. – Paleoclimatic significance of widespread Proterozoic silcretes in the Bear and Churchill provinces of the Northwestern Canadian shield. J. Sediment. Petrol., 55:196–204.

Rossignol-Strick M., 1983. – African monsoons, an immediate climate response to orbital insolation. Nature (London), 304:46–49.

Rotschy F., Chamley H., 1971. – Comparaison des données des Foraminifères planctoniques et des minéraux argileux dans une carotte nord-atlantique. Ecl. Geol. Helv., 64:279–289.

Rotschy F., Vergnaud-Grazzini C., Bellaiche G., Chamley H., 1972. – Etude paléoclimatologique d'une carotte prélevée sur un dôme de la plaine abyssale ligure ("structure Alinat"). Palaeogeogr., Climatol., Ecol., 11:125–145.

Rouchy J.-M., 1981. – La genèse des évaporites messiniennes de Méditerranée. Thèse Sci. nat., Mus. nat. Hist. nat., Paris:295 pp.

Roux R.-M., Vernier E., 1977. – Répartition des minéraux argileux dans les sédiments du golfe de Fos. Géol. Médit., 4:365–370.

Rudert M., Müller G., 1981. – Mineralogy and provenance of suspended solids in estuarine and nearshore areas of the Southern North Sea. Senckenberg. Mar., 13:57–64.

Ruhe R.V., 1983. – Clay minerals in thin loess, Ohio river basin. U.S.A. In: Brookfield M.E., Ahlbrandt T.S., Eds., Eolian sediments and processes, Developments in sedimentology. Elsevier, Amsterdam, 38:91–102.

Rupke N.A., 1975. – Deposition of fine-grained sediments in the abyssal environment of the Algero-Balearic basin, Western Mediterranean Sea. Sedimentology, 22:95–109.

Rusinov V.L., Laputina I.P., Muravitskaya G.N., Zvjagin B.B., Gradusov B.P., 1980. – Clay minerals in basalts from Deep Sea Drilling Project, sites 417 and 418. In: Donnelly T., Francheteau J., Bryan W., Robinson P., Flower M., Salisbury M. et al., Eds., Init. Rep. Deep Sea Drill. Proj., 51, 52, 53, Pt 2, Washington (U.S. Gov. Print. Off.):1265–1272.

Russell K.L., 1970. – Geochemistry and halmyrolysis of clay minerals, Rio Ameca, Mexico. Geochim. Cosmochim. Acta, 34:893–907.

Rutherford G.K., Watanabe Y., 1966. – On the clay mineralogy of two soil profiles on different age formed on volcanic ash in the territory of Papua and New Guinea. Proc. Int. Clay Conf., Jerusalem, Isr., VI:209–219.

Ryan W.B.F., Cita M.B., 1977. – Ignorance concerning episodes of ocean-wide stagnation. Mar. Geol., 23:197–215.

Sage L., Chamley H., 1977. – Minéraux argileux et dynamique sédimentaire au large du Var (S-E France). Ann. Soc. Sci. nat. Arch. Toulon, Var, 29:133–139.

Sakamoto W., 1972. – Study on the process of river suspension from flocculation to accumulation in estuary. Bull. Ocean Res. Inst., Tokyo, 5:46 p.

Salomons W., Hofmann P., Boelens R., Mook W.G., 1975. – The oxygen isotopic composition of the fraction less than 2 microns (clay fraction) in recent sediments from western Europe. Mar. Geol., 18:M23–M28.

Sancetta C., Heusser L., Labeyrie L., Naidu A.S., Robinson S.W., 1985. – Wisconsin-Holocene paleoenvironment of the Bering Sea: evidence from diatoms, pollen, oxygen isotopes and clay minerals. Mar. Geol., 62:55–68.

Sand L.B., Mumpton F.A., Eds., 1978. – Natural zeolites, occurrence, properties, use. Pergamon, Oxford, New York:546 pp.

Sarkisyan S.G., 1972. – Origin of authigenic clay minerals and their significance in petroleum geology. Sediment. Geol., 7:1–22.

Sarnthein M., 1978. – Sand deserts during glacial maximum and climatic optimum. Nature (London), 272:43–46.

Sarnthein M., Diester-Haass L., 1977. – Eolian-sand turbidites. J. Sediment. Petrol., 47:868–890.

Sarnthein M., Koopman B., 1980. – Late Quaternary deep-sea record of northwest African dust supply and wind circulation. Palaeoecol. Afr. Surround. Is., 12:239–253.

Sarnthein M., Tetzlaaf G., Koopman B., Wolter K., Pflaumann U., 1981. – Glacial and interglacial wind regimes over the eastern sub-tropical Atlantic and northwest Africa. Nature (London), 293:193–196.

Sarnthein M., Erlenkeuser H., Zahn R., 1982a. – Termination I: the response of continental climate in the subtropics as recorded in deep-sea sediments. Bull. Inst. Géol. Bassin d'Aquitaine, Bordeaux, 31:393–407.

Sarnthein M., Thiede J., Pflaumann U., Erlenkeuser H., Fütterer D., Koopman B., Lange H., Seibold E., 1982b. – Atmospheric and oceanic circulation patterns off Northwest Africa during the past 25 million years. In: Von Rad V., Hinz K., Sarnthein M., Seibold E., Eds., Geology of the Northwest African continental margin. Springer, Berlin, Heidelberg, New York:545–604.

Sass B.M., Rosenberg P.E., Kittrick J.A., 1987. – The stability of illite-smectite during diagenesis: an experimental study. Geochim. Cosmochim. Acta, 51:2103–2115.

Sassi S., 1974. – La sédimentation phosphatée au Paléocène dans le Sud et le Centre-Ouest de la Tunisie. Thèse Sci. nat., Paris-Sud:292 pp.

Scafe D.W., Kunze G.W., 1971. – A clay mineral investigation of six cores from the Gulf of Mexico. Mar. Geol., 10:69–85.

Scheidegger K.F., Krissek L.A., 1982. – Dispersal and deposition of eolian and fluvial sediments off Peru and northern Chile. Geol. Soc. Am. Bull., 93:150–162.

Scheinfeld R.A., Adams J.K., 1980. – Implications of clay-provenance studies in two Georgia estuaries. Discussion. J. Sediment. Petrol., 50:993–995.

Schlanger S.O., Cita M.B., Eds., 1982. – Nature and origin of Cretaceous carbon-rich facies. Academic Press New York, London:229 pp.

Schlanger S.O., Jenkyns H.C., 1976. – Cretaceous anoxic events: causes and consequences. Geol. Mijnbouw, 55:179–184.

Schmidt V., McDonald D.A., 1979. – The role of secondary porosity in the course of sandstone diagenesis. Soc. Econ. Pal. Miner., Spec. Publ., 26:175–207.

Schneider H.I., Angino E.E., 1980. – Trace element, mineral, and size analysis of suspended flood materials from selected Eastern Kansas rivers. J. Sediment. Petrol., 50:1271–1278.

Schneider W., Schumann D., 1979. – Tonminerale in Normalsedimenten, hydrothermal beeinflußten Sedimenten und Erzschlämmen des Roten Meeres. Geol. Rundsch., 68:631–648.

Scholle P.A., Schluger P.K., Eds., 1979. – Aspects of diagenesis. Soc. Econ. Pal. Miner., Spec Publ., 26:443 pp.

Schoonmaker J., 1986. – Clay mineralogy and diagenesis of sediments from deformation zones in the Barbados accretionary wedge (Deep Sea Drilling Project Leg 78A). Geol. Soc. Am., Mem. 166:105–116.

Schoonmaker J., Mackenzie F.T., Manghnani M., Schneider R.C., Kim D., Weiner A., To J., 1985. – Mineralogy and diagenesis: their effect on acoustic and electrical properties of pelagic clays, deep sea drilling project leg 86. In: Heath G.R., Burckle L.H. et al., Eds., Init. Rep. Deep Sea Drill. Proj., 86, Washington (U.S. Gov. Print. Off.):549–570.

Schoonmaker J., Mackenzie F.T., Speed R.C., 1986. – Tectonic implications of illite/smectite diagenesis, Barbados accretionary prism. Clays Clay Min., 34:465–472.

Schopf J.W., Ed., 1983. – Earth's earliest biosphere, its origin and evolution. Princeton Univ. Press N.J.

Schöttle M., 1969. – The sediments of the Gnadensee. Arch. Hydrobiol., Suppl., 35:255–308.

Schrader E.L., Rosendahl B.R., Furbish W.J., Mattey D.P., 1980. – Mineralogy and geochemistry of hydrothermal and pelagic sediments from the mounds hydrothermal field, Galapagos spreading center: DSDP leg 54. J. Sediment. Petrol., 50:917–928.

Schramm C.T., Leinen M.S., 1987. – Eolian transport to hole 595A from the late Cretaceous through the Cenozoic. In: Menard H.W., Natland J., Jordan T.H., Orcutt J.A. et al., Eds., Init. Rep. Deep Sea Drill. Proj., 91, Washington (U.S. Gov. Print. Off.):469–473.

Schultz L.G., 1978. – Mixed-layer clay in the Pierre Shale and equivalent rocks, northern Great Plains Region. U.S. Geol. Surv. Prof. Pap., 1064-A:28 pp.

Schütz L., Jaenicke R., Pietrek H., 1981. – Saharan dust transport over the North Atlantic Ocean. Geol. Soc. Am. Spec. Pap., 186:87–100.

Sebastian E.M., Sorice F.J., 1987. – The evolution and distribution of the clay minerals in the NE sector of the Basa basin (Betic Cordillera, Granada, S Spain). In: 6th Meet Eur. Clay Groups, Sevilla, Proc.:494–496.

Senkayi A.L., Ming D.W., Dixon J.B., Hossner L.R., 1987. – Kaolinite, opal-CT, and clinoptilolite in altered tuffs interbedded with lignite in the Jackson group, Texas. Clays Clay Min., 35:281–290.

Serdyuchenko D.P., 1968. – Metamorphosed weathering crusts of the Precambrian, their metallogenic and petrographic features. 23rd Int. Geol. Congr., 4:37–42.

Seyfried W.E., Shanks W.C. III, Dibble W.E. Jr., 1978. – Clay mineral formation in Deep Sea Drilling Project leg 34 basalt. Earth Planet. Sci. Lett., 41:265–276.

Shadfan H., Dixon J.B., 1984. – Occurrence of palygorskite in the soils and rocks of the Jordan valley. In: Singer A., Galan E., Eds., Palygorskite-sepiolite. Occurrences, genesis and uses. Developments in sedimentology. Elsevier, Amsterdam, 37:187–198.

Shaw H.F., 1978. – The clay mineralogy of the recent surface sediments from the Cilicia Basin, Northeastern Mediterranean. Mar. Geol., 26:M51–M58.

Shepard F.P., Moore D.G., 1954. – Central Texas coast sedimentation: characteristics of sedimentary environment, recent history and diagenesis. Bull. Assoc. Petrol. Geol., 39:1463–1593.

Sheppard R.A., Gude A.J. III, 1968. – Distribution and genesis of authigenic silicate minerals in tuffs of Pleistocene lake Tecopa, Inyo County, California. U.S. Geol. Surv. Prof. Pap. 597:38 pp.

Sheppard R.A., Gude A.J. III, Hay R.L., 1970. – Makatite, a new hydrous sodium silicate mineral from lake Magadi, Kenya. Am. Mineral., 55:358–366.

Sheridan R.E., 1987. – Pulsation tectonics as the control of long-term stratigraphic cycles. Paleoceanography, 2:97–118.

Sheridan R.E., Gradstein F.M. et al., 1983. – Init. Rep. Deep Sea Drill. Proj., 76, Washington (U.S. Gov. Print. Off.):947 pp.

Sholkovitz E.R., Price N.B., 1980. – The major-element chemistry of suspended matter in the Amazon estuary. Geochim. Cosmochim. Acta, 44:163–171.

Sieffermann G., Jehl G., Millot G., 1968. – Allophanes et minéraux argileux des altérations récentes des basaltes du mont Cameroun. Bull. Groupe Fr. Argiles, 20:109–129.

Siegel F.R., Pierce J.W., Urien C.M., Stone I.C., 1968. – Clay mineralogy in the estuary of the Rio de la Plata, South America. 23rd Int. Geol. Congr., 8:51–59.

Siever R., 1979. – Plate-tectonic controls on diagenesis. J. Geol., 87:127–155.

Siever R., Kastner M., 1967. – Mineralogy and petrology of some mid-Atlantic ridge sediments. J. Mar. Res., 25:263–278.

Siffert B., 1962 a. – Quelques réactions de la silice en solution: la formation des argiles. Mém. Serv. Carte géol. Als. Lorr., 21:86 pp.

Siffert B., 1962 b. – Synthèse d'une sépiolite à température ordinaire. C.R. Acad. Sci., Paris, 254:1460–1462.

Sigl W., Chamley H., Fabricius F., Giroud d'Argoud G., Müller J., 1978. – Sedimentology and environmental conditions of sapropels. In: Hsü K.J., Montadert L. et al., Eds., Init. Rep. Deep Sea Drill. Proj., 42, A, Washington (U.S. Gov. Print. Off.): 445–464.

Simoneit B.R.T., 1986. – Biomarker geochemistry of black shales from Cretaceous oceans. An overview. Mar. Geol., 70:9–41.

Singer A., 1980. – The paleoclimatic interpretation of clay minerals in soils and weathering profiles. Earth Sci. Rev., 15:303–326.

Singer A., 1984. – The paleoclimatic interpretation of clay minerals in sediments. A review. Earth Sci. Rev., 21:251–293.

Singer A., Galan E., 1984. – Palygorskite-sepiolite. Occurrences, genesis and uses. Developments in sedimentology. Elsevier, Amsterdam, 37:352 pp.

Singer A., Müller G., 1983. – Diagenesis in recent sediments. In: Larsen G., Chilingar G.V., Eds., Diagenesis in sediments and sedimentary rocks. Developments in sedimentology. Elsevier, Amsterdam, 25B:115–212.

Singer A., Stoffers P., 1980. – Clay mineral diagenesis in two East African lake sediments. Clay Miner., 15:291–307.

Singer A., Stoffers P., 1987. – Mineralogy of a hydrothermal sequence in a core from the Atlantis II deep, Red Sea. Clay Miner., 22:251–267.

Singer A., Gal M., Banin A., 1972. – Clay minerals in recent sediments of lake Kinneret (Tiberias), Israel. Sediment. Geol., 8:289–308.

Singer A., Stoffers P., Heller-Kallai L., Szafranek D., 1984. – Nontronite in a deep-sea core from the South Pacific. Clays Clay Min., 32:375–383.

Sittler C., 1965a. – Le Paléogène des fossés rhénan et rhodanien. Etudes sédimentologiques et paléoclimatiques. Thèse Sci. nat., Strasbourg: 392 pp.

Sittler C., 1965b. – La sédimentation argileuse fluvio-lacustre à la limite du Crétacé et de l'Eocène en Provence et au Languedoc. Rapport avec le problème de la disparition des Dinosauriens. Bull. Serv. Carte géol. Als.-Lorr., 18:3–14.

Sladen C.P., Batten D.J., 1984. – Source-area environments of late Jurassic and early Cretaceous sediments in Southeast England. Proc. Geol. Assoc., 95:149–163.

Sleep N.H., Morton J.L., Burns L.E., Wolery T.J., 1983. – Geophysical constraints on the volume of hydrothermal flow at ridge axes. In: Rona P.A., Boström K., Laubier L., Smith K.L. Jr., Eds., Hydothermal processes at seafloor spreading centers. Plenum, New York:53–70.

Sliter W.V., 1985. – Cretaceous redeposited benthic foraminifers from Deep Sea Drilling Project Site 585 in the East Mariana Basin, Western Equatorial Pacific, and implications for the geologic history of the region. In: Moberly R., Schlanger S.O. et al., Eds., Init. Rep. Deep Sea Drill. Proj., 89, Washington (U.S. Gov. Print. Off.):327–361.

Smoot T.W., 1960. – Clay mineralogy of pre-Pennsylvanian sandstones and shales of the Illinois Basin, Part III. Clay minerals of various facies of some Chester formations. Ill. State Geol. Surv., 293:1–19.

Smoot T.W., Narain K., 1960. – Clay mineralogy of pre-Pennsylvanian-sandstones and shales of the Illinois Basin, Part II. Relation between clay mineral suites of oil-bearing and non-oil-bearing rocks. Ill. State Geol. Surv., 287:1–14.

Snoussi M., 1986. – Nature, estimation et comparaison des flux de matières issus des bassins versants de l'Adour (France), du Sebou, de l'Oum-Er-Rbia et du Souss (Maroc). Thèse, Sci. nat., Bordeaux I:456 pp. + ann.

Sommer F., 1975. – Histoire diagénétique d'une série gréseuse de Mer du Nord. Datation de l'introduction des hydrocarbures. Rev. Inst. Fr. Pétrol., 30:729–741.

Song W., Yoo D., Dyer K.R., 1983. – Sediment distribution, circulation and provenance in a macrotidal bay: Garolim bay, Korea. Mar. Geol., 52:121–140.

Sorokin V.I., Vlasov V.V., Varfolomeeva E.K., Urasin M.A., 1979. – Effect of the medium on development of glauconite composition. Lithol. Min. Res., 14:690–693.

Spears D.A., Kanaris-Sotiriou R., 1979. – A geochemical and mineralogical investigation of some British and other European tonsteins. Sedimentology, 26:407–425.

Środoń J., 1980. – Precise identification of illite/smectite interstratifications by X-ray powder diffraction. Clays Clay Min., 28:401–411.

Środoń J., 1981. – X-ray identification of randomly interstratified illite/smectite in mixtures with discrete illite. Clay Min., 16:297–304.

Środoń J., 1984a. – X-ray powder diffraction identification of illitic materials. Clays Clay Min., 32:337–349.

Środoń J., 1984b. – Mixed-layer illite-smectite in low-temperature diagenesis: data from the Miocene of the Carpathian foredeep. Clay Min., 19:205–215.

Środoń J., 1987. – Illite/Smectite in the rock cycle. In: 6th Meet. Eur. Clay Groups, Sevilla, Proc.:48–50.

Środoń J., Eberl D.D., 1984. – Illite. In: Bailey S.W., Ed., Micas, Rev. Miner., 13, Miner. Soc. Am.:495–544.

Stackelberg U. von, 1984. – Significance of benthic organisms for the growth and movement of manganese nodules, Equatorial North Pacific. Geo Mar. Lett., 4:37–42.

Stanley D.J., Liyanage A.N., 1986. – Clay-mineral variations in the Northeastern Nile delta, as influenced by depositional processes. Mar. Geol., 73:263–283.

Stanley D.J., Sheng H., Blanpied C., 1981. – Palygorskite as a sediment dispersal tracer in the eastern Mediterranean. Geo Mar. Lett., 1:49–55.

Stakes D.S., O'Neil J.R., 1982. – Mineralogy and stable isotope geochemistry of hydrothermally altered oceanic rocks. Earth Planet. Sci. Lett., 57:285–304.

Starkey H.C., Blackmon P.D., 1984. – Sepiolite in Pleistocene lake Tecopa, Inyo County, California. In: Singer A., Galan E., Eds., Palygorskite-Sepiolite. Occurrences, genesis and uses. Developments in sedimentology. Elsevier, Amsterdam, 37:137–147.

Staudigel H., Hart S.R., Richardson S.M., 1981. – Alteration of the oceanic crust: processes and timing. Earth Planet. Sci. Lett., 52:311–327.

Stein R., 1985. – The post-Eocene sediment record of DSDP Site 366: implications for African climate and plate tectonic drift. Geol. Soc. Am. Mem., 163:305–315.

Stein R., Robert C., 1985. – Siliciclastic sediments at Sites 588, 590 and 591: Neogene and Paleogene evolution in the Southwest Pacific and Australian climate. In: Kennett J.P., von der Borch C. et al., Eds., Init. Rep. Deep Sea Drill. Proj., 90, Washington (U.S. Gov. Print. Off.):1437–1455.

Stein R., Sarnthein M., 1984. – Late Neogene oxygen-isotope stratigraphy and flux rates of terrigenous sediments at hole 544B off Morocco. In: Hinz K., Winterer E.L. et al., Eds., Init. Rep. Deep Sea Drill. Proj., 79, Washington (U.S. Gov. Print. Off.):385–394.

Stein R., Rullkötter J., Welte D.H., 1986. – Accumulation of organic-carbon-rich sediments in the late Jurassic and Cretaceous Atlantic ocean. A synthesis. Chem. Geol., 56:1–32.

Steinberg M., Mpodozis-Marin C., 1978. – Classification géochimique des radiolarites et des sédiments siliceux océaniques, signification paléocéanographique. Oceanol. Acta, 1:359–367.

Steinberg M., Fogelsang J.-F., Courtois C., Mpodozis C., Desprairies A., Martin A., Caron D., Blanchet R., 1977. – Détermination de l'origine des feldspaths et des phyllites présents dans des radiolarites mésogéennes et des sédiments hypersiliceux océaniques par l'analyse des terres rares. Bull. Soc. géol. Fr., (7) 19:735–740.

Steinberg M., Holtzapffel T., Rautureau M., Clauer N., Bonnot-Courtois C., Manoubi T., Badaut D., 1984. – Croissance cristalline et homogénéisation chimique de monoparticules argileuses au cours de la diagenèse. C.R. Acad. Sci., Paris, II, 299:441–446.

Steinberg M., Holtzapffel T., Rautureau M., 1987. – Characterization of overgrowth structures formed around individual clay particles during early diagenesis. Clays Clay Min., 35:189–195.

Stewart M.T., Mickelson D.M., 1976. – Clay mineralogy and relative age of tills in North-Central Wisconsin. J. Sediment. Petrol., 40:200–205.

Stoffers P., Müller G., 1978. – Mineralogy and lithofacies of Black Sea sediments, Leg 42B Deep Sea Drilling Project. In: Ross D.A., Neprochnov Y.P. et al., Eds., Init. Rep. Deep Sea Drill. Proj., 42, B, Washington (U.S. Gov. Print. Off.):373–411.

Stoffers P., Ross D.A., 1979. – Late Pleistocene and Holocene sedimentation in the Persian Gulf – Gulf of Oman. Sediment. Geol., 23:181–208.

Stoffers P., Singer A., 1979. – Clay minerals in lake Mobutu Sese Seko (lake Albert). Their diagenetic changes as an indicator of the paleoclimate. Geol. Rundsch., 68:1009–1024.

Stoffers P., Lallier-Vergès E., Plüger W., Schmitz W., Bonnot-Courtois C., Hoffert M., 1985. – A "fossil" hydrothermal deposit in the South Pacific. Mar. Geol., 62:133–151.

Stoffyn-Egli P., 1982. – Dissolved aluminum in interstitial waters of recent terrigenous marine sediments from the North Atlantic Ocean. Geochim. Cosmochim. Acta, 46:1345–1352.

Stone I.C. Jr., Siegel F.R., 1969. – Distribution and provenance of minerals from continental shelf sediments off the South Carolina coast. J. Sediment. Petrol., 39:276–296.

Stonecipher S.A., 1976. – Origin, distribution and diagenesis of phillipsite and clinoptilolite in deep-sea sediments. Chem. Geol., 17:307–318.

Stonecipher S.A., 1977. – Origin, distribution, and diagenesis of deep-sea clinoptilolite and phillipsite. Ph. D. Thesis, Scripps Inst. Oceanogr., La Jolla, California:223 pp.

Strom R.N., Upchurch S.B., Rosenzweig A., 1981. – Paragenesis of "Box-work geodes", Tampa Bay, Florida. Sediment. Geol., 30:275–289.

Stucki J.W., Banwart W.L., 1980. – Advanced chemical methods for soil and clay minerals research. Reidel, Dordrecht:477 pp.

Studer M., Bertrand J., 1981. – Métamorphisme des sédiments marneux en bordure de filons basiques (Haut-Atlas central, Maroc). Schweiz Min. Petrol. Mitt., 61:51–80.

Styrt M.M., Brackmann A.J., Holland H.D., Clark B.C., Pisutha-Arnond V., Eldridge C.S., Ohmoto H., 1981. – The mineralogy and the isotopic composition of sulfur in hydrothermal sulfide/sulfate deposits on the East Pacific rise, 21°N latitude. Earth Planet. Sci. Lett., 53:382–390.

Suchecki R.K., Perry E.A., Hubert J.F., 1977. – Clay petrology of Cambro-Ordovician continental margin, Cow Head Klippe, Western Newfoundland. Clays Clay Min., 25:163–170.

Sudo T., Shimoda S., 1978. – Clays and clay minerals of Japan. Elsevier, Amsterdam:326 pp.

Sudo T., Shimoda S., Yotsumoto H., Aita S., 1981. – Electron micrographs of clay minerals. Developments in sedimentology. Elsevier, Amsterdam, 31:203 pp.

Summerfield M.A., 1983. – Silcrete as a paleoclimatic indicator: evidence from southern Africa. Palaeogeogr., Climatol., Ecol., 41:65–79.

Surdam R.C., Crossey L.J., 1987. – Integrated diagenetic modeling: A process-oriented approach for clastic systems. Annu. Rev. Earth Planet. Sci., 15:141–170.

Surdam R.C., Eugster H.P., 1976. – Mineral reactions in the sedimentary deposits of the lake Magadi Region, Kenya. Geol. Soc. Am. Bull., 87:1739–1752.

Surdam R.C., Sheppard R.A., 1978. – Zeolites in saline, alkaline lake deposits. In: Sand L.B., Mumpton F.A., Eds., Natural zeolites: occurrence, properties, use. Pergamon, New York, Oxford:145–174.

Sutherland H.E., Calvert S.E., Morris R.J., 1984. – Geochemical studies of the recent sapropel and associated sediment from the Hellenic outer ridge, eastern Mediterranean Sea. I: Mineralogy and chemical composition. Mar. Geol., 56:79–92.

Swain F.M., 1966. – Bottom sediments of lake Nicaragua and lake Managua, Western Nicaragua. J. Sediment. Petrol., 36:522–540.

Swindale L.D., Fan P.-F., 1967. – Transformation of gibbsite to chlorite in ocean bottom sediments. Science, 157:799–800.

Swindale L.D., Jackson M.L., 1960. – A mineralogical study of soil formation in four rhyolite-derived soils from New Zealand. New Zeal. J. Geol. Geophys., 3:141–183.

Swineford A., Frye J.C., 1955. – Petrographic comparison of some loess samples from western Europe with Kansas loess. J. Sediment. Petrol., 25:3–23.

Syvitski J.P.M., Lewis A.G., 1980. – Sediment ingestion by *Tigriopus californicus* and other zooplankton: mineral transformation and sedimentological considerations. J. Sediment. Petrol., 50:869–880.

Taggart M.S. Jr., Kaiser A.D. Jr., 1960. – Clay mineralogy of Mississippi river deltaic sediments. Geol. Soc. Am. Bull., 71:521–530.

Taguchi K., Shimoda S., Itihara Y., Imoto N., Ishiwatari R., Shimoyama A., Akiyama M., Suzuki N., 1986. – Relationship of organic and inorganic diagenesis of Neogene Tertiary rocks, Northeastern Japan. Soc. Econ. Pal. Min., Spec. Publ., 38:47–64.

Tardy Y., 1969. – Géochimie des altérations. Etude des arènes et des eaux de quelques massifs cristallins d'Europe et d'Afrique. Mém. Serv. Carte géol. Als. Lorr., 31:199 pp.

Tardy Y., 1981. – Silice, silicates magnésiens, silicates sodiques et géochimie des paysages arides. Bull. Soc. géol. Fr., (7) 23:325–334.

Tardy Y., Gac J.-Y., 1968. – Minéraux argileux et vermiculite – Al dans quelques sols et arènes des Vosges. Hypothèse sur la néoformation des minéraux à 14 Å. Bull. Serv. Carte géol. Als.-Lorr., 21:285–304.

Tardy Y., Touret O., 1987. – Hydration energies of smectites: a model for glauconite, illite and corrensite formation. Proc. Int. Clay Conf., Denver 1985. Clay Min. Soc., U.S.A.:46–52.

Tardy Y., Paquet H., Millot G., 1970. – Trois modes de genèse des montmorillonites dans les altérations et les sols. Bull. Groupe Fr. Arg., 22:69–78.

Tardy Y., Cheverry C., Fritz B., 1974. – Néoformation d'une argile magnésienne dans les dépressions interdunaires du lac Tchad. Application aux domaines de stabilité des phyllosilicates alumineux, magnésiens et ferrifères. C.R. Acad. Sci., Paris, 278:1999–2002.

Tazaki K., Fyfe W.S., 1987. – Primitive clay precursors formed on feldspar. Can. J. Earth Sci., 24:506–527.

Tazaki K., Fyfe W.S., Heath G.R., 1986. – Palygorskite formed on montmorillonite in North Pacific deep-sea sediments. Clay Sci., 6:197–216.

Teale C.T., Spears D.A., 1986. – The mineralogy and origin of some Silurian bentonites, Welsh Borderland, U.K. Sedimentology, 33:757–765.

Tefry J.H., Trocine R.P., Klinkhammer G.P., Rona P.H., 1985. – Iron and copper enrichment of suspended particles in dispersed hydrothermal plumes along the mid-Atlantic ridge. Geophys. Res. Lett., 12:506–509.

Termier P., 1890. – Note sur la leverriérite. Bull. Soc. Fr. Min., 13:325–330 et Ann. Mines, 17:372.

Tessier F., Triat J.-M., 1973. – Conceptions nouvelles sur l'origine et sur l'âge des ocres d'Apt (Vaucluse). C.R. Acad. Sci., Paris, 276:1135–1138.

Tettenhorst R., Moore G.E. Jr., 1978. – Stevensite oolites from the Green River formation of central Utah. J. Sediment. Petrol., 48:587–594.

Teyssen T.A.L., 1984. – Sedimentology of the Minette oolitic ironstones of Luxembourg and Lorraine: A Jurassic subtidal sandwave complex. Sedimentology, 31:195–211.

Thiede J., 1979. – Wind regimes over the late Quaternary southwest Pacific Ocean. Geology, 7:259–262.

Thiede J., van Andel T.H., 1977. – The paleoenvironment of anaerobic sediments in the Late Mesozoic South Atlantic Ocean. Earth Planet. Sci. Lett., 33:301–309.

Thiede J., Dean W.E., Claypool G.E., 1982. – Oxygen-deficient depositional paleoenvironments in the mid-Cretaceous tropical and subtropical Pacific Ocean. In: Schlanger S.O., Cita M.B., Eds., Nature and origin of Cretaceous carbon-rich facies. Academic Press New York, London:79–100.

Thiry M., 1981. – Sédimentation continentale et altérations associées: calcitisations, ferruginisations et silicifications. Les argiles plastiques du Sparnacien du Bassin de Paris. Sciences géol., Mém., Strasbourg, 84:173 pp.

Thiry M., Trauth N., 1976 a. – Evolution historique de la notion d'argile à silex. Bull. Inf. Géol. Bass. Paris, 13:41–48.

Thiry M., Trauth N., 1976 b. – Les sédiments paléocènes et éocènes inférieurs du Bassin de Paris. Rôle des argiles dans la rétention d'éléments traces. Sci. Géol. Bull., Strasbourg, 29:33–43.

Thiry M., Cavelier C., Trauth N., 1977. – Les sédiments de l'Eocène inférieur du bassin de Paris et leurs relations avec la paléoaltération de la craie. Sci. Géol., Bull., Strasbourg, 30:113–128.

Thisse Y., 1982. – Sédiments métallifères de la fosse Atlantis II (Mer Rouge). Contribution à l'étude de leur contexte morpho-structural et de leurs caractéristiques minéralogiques et géochimiques. Thèse 3ème cycle, Orléans:155 pp.

Thomas M., 1986. – Diagenetic sequences and K/Ar dating in Jurassic sandstones, Central Viking graben: effects on reservoir properties. Clay Min., 21:695–710.

Thomassin J.-H., Baillif P., Touray J.-C., 1983. – Modification of residual glass and precipitation of a neoformation layer: experimental evidence for a double origin of the "palagonite" derived from the alteration of basaltic glass by sea-water. Sci. géol., Bull., 36:165–171.

Thompson G., 1983. – Hydrothermal fluxes in the ocean. In: Riley J.P., Chester R., Eds., Chemical oceanography. Academic Press, New York, London, 8:271–337.

Thompson G., Mottl M.J., Rona P.A., 1985. – Morphology, mineralogy and chemistry of hydrothermal deposits from the TAG area, 26°N mid-Atlantic ridge. Chem. Geol., 49:243–257.

Thompson L.G., 1977. – Microparticles, ice sheets and climate. Inst. Polar Stud., Ohio State Univ., Columbus, Ohio, Rep. 64.

Thorez J., 1975. – Phyllosilicates and clay minerals. Lelotte, Dison:579 pp.

Thorez J., 1983. – Qualitative clay mineral analyses biased by sample treatments. In: 5th Meet. Eur. Clay Groups, Prague:383–389.

Thunell R.C., Williams D.F., Belyea P.R., 1984. – Anoxic events in the Mediterranean sea in relation to the evolution of late Neogene climates. Mar. Geol., 59:105–134.

Tiercelin J.-J., 1977. – Stratigraphie minéralogique du Pléistocène glacio-torrentiel de la Haute-Durance provençale. Corrélations avec le Pléistocène glaciaire du bassin de Laragne-Sisteron. Geobios, 10:489–492.

Tiercelin J.-J., Chamley H., 1975. – Minéraux argileux du Pléistocène glaciaire et interglaciaire de Laragne-Sisteron (Alpes de Haute-Provence). C.R. Acad. Sci., Paris, 280:2293–2296.

Tiercelin J.-J., Vincens A. et al., 1987. – Le demi-graben de Baringo-Bogoria, rift Gregory, Kenya. 30000 ans d'histoire hydrologique et sédimentaire. Bull. Centre Rech. Explor. Prod. Elf Aquitaine, 11:249–540.

Timofeev P.P., Eremeev V.V., Rateev M.A., 1978. – Palygorskite, sepiolite, and other clay minerals in leg 41 oceanic sediments: mineralogy facies and genesis. In: Lancelot Y., Seibold E. et al., Eds., Init. Rep. Deep Sea Drill. Proj., 41, Washington (U.S. Gov. Print. Off.):1087–1101.

Tissot B., Demaison G., Masson P., Delteil J., Combaz A., 1980. – Paleoenvironment and petroleum potential of middle Cretaceous black shales in Atlantic Basins. Am. Assoc. Petrol. Geol. Bull., 64:2051–2063.

Tissot B.P., Welte D.H., 1978. – Petroleum formation and occurrence. Springer, Berlin, Heidelberg, New York:538 pp.

Tlig S., 1982 a. – Géochimie comparée de sédiments de l'Océan Indien et de l'Océan Pacifique. Thèse Sci. nat., Paris Sud:229 pp.

Tlig S., 1982 b. – Distribution des terres rares dans les fractions de sédiments et nodules de Fe et Mn associés dans l'Océan Indien. Mar. Geol., 50:257–274.

Tlig S., Steinberg M., 1982. – Distribution of rare-earth elements (REE) in size fractions of recent sediments of the Indian Ocean. Chem. Geol., 37:317–333

Tobias C., Mégie C., 1980–81. – Les lithométéores au Tchad. Cah. ORSTOM, Pédol., 18:71–81.

Tomadin L., 1969. – Richerche sui sedimenti argillosi fluviali dal Brenta al Reno. Giorn. Geol., (2) 36:159–184.

Tomadin L., 1974. – Les minéraux argileux dans les sédiments actuels de Mer Tyrrhénienne. Bull. Gr. Fr. Arg., 26:219–228.

Tomadin L., Borghini M., 1987. – Source and dispersal of clay minerals from present and late Quaternary sediments of Southern Adriatic sea. In: Proc. 6th Meet. Eur. Clay Groups, Sevilla:537–538.

Tomadin L., Lenaz R., Landuzzi V., Mazzucotelli A., Vannuci R., 1984. – On wind-blown dusts over the central Mediterranean. Oceanol. Acta, 7:13–24.

Tomadin L., Gallignani P., Landuzzi V., Oliveri F., 1985. – Fluvial pelitic supplies from the Apennines to the Adriatic sea. I – The rivers of the Abruzzo region. Miner. Petrogr. Acta, 29A:277–286.

Tomadin L., Cesari G., Fuzzi S., Landuzzi V., Lenaz R., Lobietti A., Mandrioli P., Mariotti M., Mazzucotelli A., Vannucci R., 1987. – Eolian dusts from the seawater-air interface of the Red Sea. In: Paleoclimatology and paleometeorology, NATO Adv. Res. Workshop, Oracle, Arizona, Abstr.

Tompkins R.E., Shephard L.E., 1979. – Orca basin: depositional processes, geotechnical properties and clay mineralogy of Holocene sediments within an anoxic hypersaline basin, Northwest Gulf of Mexico. Mar. Geol., 33:221–238.

Torrent G., Dejou J., Manier M., Larroque P., Gibert J.-P., 1982. – Présence d'imogolite dans la composition minéralogique des fractions fines extraites des croûtes d'altération observées sur le basalte de Roudadou, près d'Aurillac, Cantal. Clay Min., 17:185–194.

Toth J.R., 1980. – Deposition of submarine crusts rich in manganese and iron. Geol. Soc. Am. Bull., I, 91:44–54.

Trauth N., 1977. – Argiles évaporitiques dans la sédimentation carbonatée continentale et épicontinentale tertiaire. Bassins de Paris, de Mormoiron et de Salinelles (France), Jbel Ghassoul (Maroc.) Sci. géol., Strasbourg, Mém. 49:203 pp.

Trauth N., Paquet H., Lucas J., Millot G., 1967. – Les montmorillonites des vertisols lithomorphes sont ferrifères: conséquences géochimiques et sédimentologiques. C.R. Acad. Sci., Paris, 264:1577–1579.

Trauth N., Sommer F., Lucas J., 1969. – Evolution géochimique d'une série sédimentaire paléogène dans les Bassin de Paris. Bull. Serv. Carte géol. Als.-Lorr., 22:272–310.

Trefry J.H., Trocine R.P., Klinkhammer G.P., Rona P.H., 1985. – Iron and copper enrichment of suspended particles in dispersed hydrothermal plumes along the Mid-Atlantic Ridge, Geophys. Res. Lett. 12:506–509.

Triat J.-M., 1982. – Paléoaltérations dans le Crétacé supérieur de Provence rhodanienne. Sci. géol., Mém., Strasbourg, 68:202 pp.

Triat J.-M., Trauth N., 1972. – Evolution des minéraux argileux dans les sédiments paléogènes du bassin de Mormoiron (Vaucluse). Bull. Soc. Fr. Minéral. Cristallogr., 95:482–494.

Tribovillard N.-P., 1988. – Géochimie organique et minérale dans les Terre Noires calloviennes et oxfordiennes du bassin dauphinois (France SE): mise en évidence de cycles climatiques. Bull. Soc. géol. Fr., (8) 4:141–150.

Trichet J., 1970. – Contribution à l'étude de l'altération expérimentale des verres volcaniques. Ecole Normale Sup., Paris, Trav. lab. géol., 4:152 pp.

Triplehorn D.M., 1966. – Morphology, internal structure, and origin of glauconite pellets. Sedimentology, 6:247–266.

Tsirambides A.E., 1986. – Detrital and authigenic minerals in sediments from the western part of the Indian Ocean. Min. Mag., 50:69–74.

Tucholke B.E., Mountain G.S., 1979. – Seismic stratigraphy, lithostratigraphy and paleosedimentation patterns in the North American Basin. In: Talwani M., Hay W., Ryan W.B.F., Eds., Deep drilling results in the Atlantic Ocean: continental margins and paleoenvironment. Am. Geophys. Union, Maurice Ewing Ser., 3:58–86.

Tucholke B.E., Vogt P.R., 1979. – Site 387: Cretaceous to recent sedimentary evolution of the western Bermuda rise. In: Tucholke B.E., Vogt P.R., et al., Eds., Init. Rep. Deep Sea Drill. Proj., 43, Washington (U.S. Gov. Print. Off.):323–391.

Turner P., 1980. – Continental red beds. Developments in sedimentology. Elsevier, Amsterdam, 29:562 pp.

Uematsu M., Duce R.A., Prospero J.M., Chen L., Merrill J.T., McDonald R.L., 1983. – Transport of mineral aerosol from Asia over the North Pacific Ocean. J. Geophys. Res., 88:5343–5352.

Ugolini F.C., 1977. – The Protoranker soils and the evolution of an ecosystem at Kar Plateau, Antarctica. In: Llano G.A., Ed., Adaptation within Antarctic ecosystems, Proc. 3rd S.C.A.R., Symp. Antarctic biology, 7:1091–1110.

Ugolini F.C., Jackson M.L., 1982. – Weathering and mineral synthesis in Antarctic soils. In: Craddock C., Loveless J.K., Vierima T.L., Crawford K.A., Eds., Antarctic geoscience. Symp. Antarctic geology and geophysics, Madison, Wis., U.S.A., 1977. Int. Union Geol. Sci., B, 4:1101–1108.

Ugolini F.C., Deutsch W., Harris H.J.H., 1981. – Chemistry and clay mineralogy of selected cores from the Antarctic dry valley drilling project. In: Mc Ginnis L.D., Ed., Dry Valley Drilling Project, Antarctic Res. Ser., 33:315–329.

Vail P.R., Mitchum M.R., Thompson S., 1977. – Seismic stratigraphy and global changes of sea-level. Part 4: Global cycles of relative changes of sea-level. Am. Assoc. Petrol. Geol., Mem 26:83–97.

Valeton J., 1972. – Bauxites. Elsevier, Amsterdam:226 pp.

Valeton E., Stütze B., Goldbery R., 1983. – Geochemical and mineralogical investigations of the Lower Jurassic flint-clay bearing Mishhor and Ardon formations, Makhtesh Ramon, Israël. Sedimentol. Geol., 35:105-152.

Valette J.N., 1972. – Etude minéralogique et géochimique des sédiments de mer d'Alboran: résultats préliminaires. C.R. Acad. Sci., Paris, 275, D:2287–2290.

van Andel T., Postma H., 1954. – Recent sediments of the gulf of Paria. Rep. Orinoco Shelf Expedition. North Holland, Amsterdam, 1:244 pp.

van den Heuvel R.C., 1966. – The occurrence of sepiolite and attapulgite in the calcareous zone of a soil near Las Cruces, New Mexico. Clays Clay Min., 13:193–207.

van der Gaast S.J., Jansen J.H.F., 1984. – Mineralogy, opal, and manganese of middle and late Quaternary sediments of the Zaire (Congo) deep-sea fan: origin and climatic variation. Neth. J. Sea Res., 17:313–341.

van der Gaast S.J., Mizota C., Jansen J.H.F., 1986. – Curled smectite in soils from volcanic ash in Kenya and Tanzania: a low-angle X-ray powder diffraction study. Clays Clay Min., 34:665–671.

van der Marel H.W., Beutelspacher H., 1976. – Atlas of infrared spectroscopy of clay minerals and their admixtures. Elsevier, Amsterdam:396 pp.

van Houten F.B., 1982. – Ancient soils and ancient climates. In: Climate in earth history, geophysics study committee, geophysics research board. Proc. Natl. Acad. Sci. USA:112–117.

van Houten F.B., 1986. – Search for Milankovitch patterns among oolitic ironstones. Paleoceanography, 1:459–466.

van Houten F.B., Purucker M.E., 1984. – Glauconitic peloids and chamositic ooids. Favorable factors, constraints, and problems. Earth Sci. Rev., 20:211–243.

van Moort J.E., 1971. – A comparative study of the diagenetic alteration of clay minerals in Mesozoic shales from Papua, New Guinea, and in Tertiary shales from Louisiana, U.S.A. Clays Clay Min., 19:1–20.

van Nieuwenhuise D.S., Yarus J.M., Przygocki R.S., Ehrlich R., 1978. – Sources of shoaling in Charleston harbor, Fourier grain shape analysis. J. Sedimentol. Petrol., 48:373–383.

van Olphen H., Fripiat J.J., 1979. – Data handbook for clay materials and other non-metallic minerals. Pergamon, Oxford:346 pp.

Vaslet N., Thouin C., Fouquet Y., Brichard P.J., Kalala T., Mondeguer A., Tiercelin J.-J., 1987. – Découverte de sulfures massifs d'origine hydrothermale dans le rift Est-Africain. Minéralisations sous-lacustres dans le fossé du Tanganyika. C.R. Acad. Sci., Paris, 305:885–891.

Vaudour J., 1979. – La région de Madrid. Altérations, sols et paléosols. Thèse, Géogr., Aix-Marseille:390 pp.

Vaudour J., 1983. – Vieux sols et paléosols des niveaux villafranchiens méditerranéens. Bull. Assoc. Fr. Et. Q., 14–15:95–102.

Velde B., 1977. – Clay and clay minerals in natural and synthetic systems. Elsevier, Amsterdam:218 pp.

Velde B., 1985. – Clays minerals. A physical-chemical explanation of their occurrence. Elsevier, Amsterdam:427 pp.

Veniale F., Soggetti F., Pigorini B., Dal Negro A., Adami A., 1972. – Clay mineralogy of bottom sediments in the Adriatic sea. 4th Int. Clay Conf., Madrid, Spain:301–312.

Venkatarathnam K., Biscaye P.E., 1973. – Clay mineralogy and sedimentation in the eastern Indian Ocean. Deep-Sea Res., 20:727–738.

Venkatarathnam K.V., Ryan W.B.F., 1971. – Dispersal patterns of clay minerals in the sediments of the Eastern Mediterranean. Mar. Geol., 11:261–282.

Vernet J.-P., 1969. – Etude pétrographique des matières en suspension dans le Rhône et ses affluents. Bull. Univ. Lausanne, Suisse, 177:1–7.

Vernier E., Froget C., 1984. – Sédimentation argileuse dans le Sud-Ouest du bassin de Somalie depuis le Crétacé supérieur (sites D.S.D.P. 240 et 241). Rev. Géol. dyn. Géogr. phys., 25:339–348.

Verrechia E., Freytet F., 1987. – Interférence pédogenèse-sédimentation dans les croûtes calcaires. Proposition d'une nouvelle méthode d'étude: l'analyse séquentielle. In: Fedoroff N., Bresson L.M., Courty M.A., Eds., Micromorphologie des sols. A.F.E.S., France:555–561.

Viéban F., 1983. – Installation de la plate-forme urgonienne (Hauterivien-Bédoulien) du Jura méridional aux chaînes subalpines (Ain, Savoie, Haute-Savoie). Thèse 3ème cycle, Grenoble:291 pp.

Visser J.P. de, Chamley H., 1989. – Clay mineralogy of the Pliocene and Pleistocene of site 653 A, Western Tyrrhenian Sea (ODP Leg 107). In: Kastens K., Mascle J. et al., Eds., Ocean Drill. Progr., Proc., 107B, Washington (U.S. Gov. Print. Off.) (in press).

Visser J.P. de, Ebbing J.H.J., Gudjonsson L., Hilgen F.J., Jorissen F.J., Verhallen P.J.J.M., Zevenboom D., 1988. – The origin of rhythmic bedding in the Pliocene Trubi of Sicily, Southern Italy. Palaeogeogr., Climatol., Ecol. 69:45–66

Voisin L., 1981. – Le modèle schisteux en zone froide et tempérée. Thèse Géogr., Paris IV, I:498 pp.

Volkoff B., Melfi A.J., 1978. – Evolution des micas dans la région équatoriale: problème de la présence de smectites dans les sols ferrallitiques fortement désaturés de l'Amazonie brésilienne. C.R. Acad. Sci., Paris, 286:837–840.

Wada K., 1977. – Allophane and imogolite. In: Dixon J.B., Weed S.B., Eds., Minerals in soil environments. Soil Sci. Soc. Am:603–638.

Wada K., Aomine S., 1973. – Soil development on volcanic materials during the Quaternary. Soil Sci., 116:170–177.

Wada H., Okada H., 1983. – Nature and origin of deep-sea carbonate nodules collected from the Japan Trench. Am. Assoc. Petrol. Geol., Mem. 34:661–672.

Walker T.R., Waugh B., Grone A.J., 1978. – Diagenesis in first-cycle desert alluvium of Cenozoic age, southwestern United States and northwestern Mexico. Geol. Soc. Am. Bull., 89:19–32.

Walter P., Stoffers P., 1985. – Chemical characteristics of metalliferous sediments from eight areas on the Galapagos rift and East Pacific rise between 2°N and 42°S. Mar. Geol., 65:271–287.

Watts N.L., 1980. – Quaternary pedogenic calcretes from the Khalahari (Southern Africa): mineralogy, genesis and diagenesis. Sedimentology, 27:661–686.

Weaver C.E., 1956. – The distribution and identification of mixed-layer clays in sedimentary rocks. Am. Miner., 41:202–221.

Weaver C.E., 1959. – The clay petrology of sediments. Clays Clay Min., 6th Natl. Conf. Pergamon, Oxford:154–187.

Weaver C.E., 1960. – Possible uses of clay minerals in search for oil. Bull. Am. Assoc. Petrol. Geol., 44:1505–1518.

Weaver C.E., 1963. – Interpretative value of heavy minerals from bentonites. J. Sedimentol. Petrol., 33:343-349.

Weaver C.E., 1967 a. – Variability of a river clay suite. J. Sedimentol. Petrol., 37:971–974.

Weaver C.E., 1967 b. – Potassium, illite and the ocean. Geochim. Cosmochim. Acta, 31:2181–2196.

Weaver C.E., 1980. – Fine-grained rocks: shales or physilites. Sediment. Geol., 27:301–313.

Weaver C.E., 1984. – Origin and geologic implications of the palygorskite deposits of SE United States. In: Singer A., Galan E., Eds., Developments in sedimentology. Elsevier, Amsterdam, 37:39–73.

Weaver C.E., Beck K.C., 1971. – Clay-water diagenesis during burial: how mud becomes gneiss. Geol Soc. Am. Spec. Pap., 134:96 pp.

Weaver C.E., Beck K.C., 1977. – Miocene of the S.E. United States: a model for chemical sedimentation in a peri-marine environment. Sedimentol. Geol., 17, Spec. Iss.:234 pp.

Weaver C.E., Pollard L.D., 1973. – The chemistry of clay minerals. Developments in sedimentology. Elsevier, Amsterdam, 15:213 pp.

Weir A.H., Ormerod E.C., El Mansey M.I., 1975. – Clay mineralogy of sediments of the Western Nile delta. Clay Min., 10:369–386.

Wezel F.C., Vannucci S., Vannucci R., 1981. – Découverte de divers niveaux riches en iridium dans la "Scaglia rossa" et la "Scaglia bianca" de l'Apennin d'Ombrie-Marches (Italie). C.R. Acad. Sci., Paris, II, 293:837–844.

Whitehouse U.G., McCarter R.S., 1958. – Diagenetic modification of clay mineral types in artificial sea water. Clays Clay Min., 5th Natl. Conf. Pergamon, Oxford:81–119.

Whitehouse U.G., Jeffrey L.M., Debrecht J.D., 1960. – Differential settling tendencies of clay minerals in saline waters. Clays Clay Min., 7th Natl. Conf.:1–80.

Whiteside D.I., Robinson D., 1983. – A glauconitic clay-mineral from a speleological deposit of late Triassic age. Palaeogeogr., Climatol., Ecol., 41:81–85.

Whitney G., Northrop H.R., 1987. – Diagenesis and fluid flow in the San Juan basin. New Mexico – regional zonation in the mineralogy and stable isotope composition of clay minerals in sandstone. Am. J. Sci., 287:353–382.

Wilson M.D., Pittman E.D., 1977. – Authigenic clays in sandstones: recognition and influence on reservoir properties and paleoenvironmental analysis. J. Sedimentol. Petrol., 47:3–31.

Wilson M.J., 1987. – A handbook of determinative methods in clay mineralogy. Blackie, Glasgow:308 pp.

Wilson M.J., Nadeau P.H., 1985. – Interstratified clay minerals and weathering processes. In: Drever J.J., Ed., The chemistry of weathering. Reidel, Dordrecht:97–118.

Wilson M.J., Bain D.C., Mitchell W.A., 1968. – Saponite from the Dalradian meta-limestones of north-east Scotland. Clay Min., 7:343–349.

Wilson M.J., Bain D.C., McHardy W.J., Berrow M.L., 1972. – Clay-mineral studies on some carboniferous sediments in Scotland. Sediment. Geol., 8:137–150.

Wilson M.J., Bain D.C., Duthie D.M.L., 1984. – The soil clays of Great Britain. II. Scotland. Clay Min., 19:709–735.

Windley B.F., 1977. – The evolving continents. Wiley, London, New York:385 pp.

Windom H.L., 1969. – Atmospheric dust records in permanent snowfields: implications to marine sedimentation. Geol. Soc. Am. Bull., 80:761–782.

Windom H.L., 1975. – Eolian contributions to marine sediments. J. Sedimentol. Petrol., 45:520–529.

Windom H.L., 1976. – Lithogenous material in marine sediments. In: Riley J.P., Chester R., Eds., Chemical oceanography. Academic Press, New York, London, 5:103–135.

Windom H.L., Chamberlain F.G., 1978. – Dust storm transport of sediments to the North Atlantic Ocean. J. Sedimentol. Petrol., 48:385–388.

Windom H.L., Griffin J.J., Goldberg E.D., 1967. – Talc in atmospheric dusts. Environ. Sci. Tech., 1:923–926.

Windom H.L., Neal W.J., Beck K.C., 1971. – Mineralogy of sediments in three Georgia estuaries. J. Sedimentol. Petrol., 41:497–504.

Wollast R., Mackenzie F.T., Bricker O.P., 1968. – Experimental precipitation and genesis of sepiolite at earth surface conditions. Am. Min., 53:1645–1662.

Wright P.L., 1974. – The chemistry and mineralogy of the clay fraction of sediments from the Southern Barents Sea. Chem. Geol., 13:197–216.

Wright V.P., 1982. – Calcrete palaeosols from the lower Carboniferous Llanelly formation, South Wales. Sedimentol. Geol., 33:1–33.

Wright V.P., Ed., 1986. – Paleosols. Their recognition and interpretation. Blackwell, Oxford, New York:315 pp.

Wybrecht E., Duplay J., Piqué A., Weber F., 1985. – Mineralogical and chemical evolution of white micas and chlorites, from diagenesis to low-grade metamorphism; data from various size fractions of greywackes (Middle Cambrian, Morocco). Min. Mag., 49:401–411.

Yaalon D.H., Ganor E., 1973. – The influence of dust on soils during the Quaternary. Soil. Sci., 116:146–155.

Yashkin Y.K., 1967. – Upper Proterozoic montmorillonoidal clays in middle portion of pre-Dniestr Region. Sov. Geol., 2:122–128.

Yau Y.-C., Peacor D.R., Essene E.J., Lee J.H., Kuo L.-C., Cosca M.A., 1987 a. – Hydrothermal treatment of smectite, illite and basalt to 460 °C: comparison of natural with hydrothermally formed clay minerals. Clays Clay Min., 35:241–250.

Yau Y.-C., Peacor D.R., McDowell S.D., 1987b. – Smectite-to-illite reactions in Salton Sea shales: a transmission and analytical electron microscopy study. J. Sediment. Petrol., 57:335–342.

Yau Y.-C., Peacor D.R., Beane R.E., Essene E.J., McDowell S.D., 1988. – Microstructures, formation mechanisms, and depth-zoning of phyllosilicates in geothermally altered shales, Salton Sea, California. Clays Clay Min., 36:1–10.

Yeh H.W., Savin S.M., 1977. – Mechanism of burial metamorphism of argillaceous sediments: I.O-isotope evidence. Geol. Soc. Am. Bull., 88:1321–1330.

Yeroshchev-Shak V.A., 1961. – Kaolinite in sediments of the Atlantic Ocean. Dokl. Akad. Nauk S.S.S.R., 137:695–697. (in Russian)

Yin J., Okada H., Labeyrie L., 1987. – Clay mineralogy of slope sediments around the Japanese Islands. Geosci. Rep. Shizuoka Univ., 13:41–65.

Yoshinaga N., Aomine S., 1962. – Imogolite in some andosols. Soils Sci. Plant. Nutrit. Tokyo, 8:6–13.

Yuretich R.F., 1979. – Modern sediments and sedimentary processes in lake Rudolf (lake Turkana), eastern rift valley, Kenya. Sedimentology, 26:313–331.

Zachariasse W.J., Spaak P., 1983. – Middle Miocene to Pliocene paleoenvironmental reconstruction of the Mediterranean and adjacent Atlantic Ocean: planktonic foraminiferal record of southern Sicily. In: Meulenkamp J.E., Ed., Reconstruction of marine paleoenvironments. Utrecht Micropal. Bull., 30:91–110.

Zemmels I., Cook H.E., 1973. – X-ray mineralogy of sediments from the central Pacific Ocean: In: Winterer E.L., Ewing J.I. et al., Eds., Init. Rep. D.S.D.P., 17, Washington (U.S. Gov. Print. Off.):517–560.

Zemmels I., Cook H.E., 1976. – X-ray mineralogy data from the Southeast Pacific basin, Leg 35 D.S.D.P. In: Hollister C.D., Craddock C. et al., Eds., Init. Rep. Deep Sea Drill. Proj., 35, Washington (U.S. Gov. Print. Off.):747–754.

Zierenberg R.A., Shanks W.C., 1983. – Mineralogy and geochemistry of epigenetic features in metalliferous sediments, Atlantis II deep, Red Sea. Econ. Geol., 78:57–72.

Zimmerman H.B., 1972. – Sediments of the New England continental rise. Geol. Soc. Am. Bull., 83:3709–3724.

Zimmerman H.B., 1982. – Fine-grained sediment distribution in the late Pleistocene/Holocene North Atlantic. Bull. Inst. Géol. Bassin Aquitaine, Bordeaux, 31:337–357.

Subject Index

n. b. Cross-references to major clay minerals concern mainly their characterization or synthetic considerations.

A

acidolysis 22, 32
aeolian dust 53, 54, 55, 134, 533
alcalinolysis 22, 35
aliettite 17
aluminum 264, 386, 489
aluminum oxides 18, 114, 509
allevardite 17, 365
allophane 7, 38
alitization 25
amesite 7
amphibolite-facies 297
analcite 19, 83, 264, 297, 366, 409, 507
anchimetamorphism 359
andosol 38
antigorite 7
attapulgite 35
azonal soil 37

B

batavite 10
bauxite 31
beidellite 9, 14, 34, 88, 199, 298, 307, 339, 451, 452, 510, 535
bentonite 411
berthierine 10, 214, 228, 379
biotite 10, 11, 367
birnessite 261, 303, 310
bisialitization 25, 29, 34
black shales 251
boehmite 18
bowlingite 10
bravaisite 17
brittle mica 12
brownstone 291
brown soil 28

C

calcrete 35, 46, 48
celadonite 9, 11, 230, 293, 306, 316, 325, 414, 507
cerium anomaly 265, 268, 280, 281, 283, 300, 309, 320, 340
chamosite 15, 214, 228, 402

chemical weathering 21
chestnut soil 28
chlorite 4, 12, 14, 164, 166, 214, 297, 299, 315, 361, 392, 409, 414, 558
  swelling chlorite 15, 396, 402
chloritization 351, 359, 379, 391, 397, 401, 405, 419, 559
chrysotile 7, 315
cronstedtite 7
clay and décollement zone 483
clay and petroleum 382, 383, 401, 407, 408
clay charge 4, 5
clay dewatering 382
clay dissolution 379, 385
clay fabric 235, 379, 398, 403, 406, 408
clay fraction 5
clay minerals 6
clay particle 5
clay size 64, 65, 126, 217, 361
clay sorting 103, 117, 122, 183, 431
clay structure 3
clay unit 5
clinoptilolite 19, 71, 264, 277, 300, 317, 342, 347, 366, 409, 507
clintonite 13
cookeite 15
corrensite 17, 195, 299, 365, 397, 415, 559

D

damouzite 9
diagenetic 333
dickite 7, 400
dioctahedral 4, 9
dombassite 15

E

extra-terrestrial impact 518, 520

F

ferricrete 31, 32, 35
fibrous clay 15, 276, 317, 352, 415, 494
fireclay 7
flocculation 103, 124, 126

G

geothermal gradient   359, 366, 368, 383, 391, 394, 414, 537, 551
gibbsite   18, 30, 40
glacial/interglacial clay   59, 75, 76, 78, 111, 152, 273, 427, 437, 439, 465, 468
glauconite   9, 11, 79, 197, 213, 227
glauconitization   222
glaucony   214
geothite   18, 261, 286, 303, 316
greenalite   7
greenschist-facies   297

H

halloysite   7, 39, 414
halmyrolysis   244, 259, 291
hectorite   10, 14
hematite   18
heulandite   19, 297, 409
hydrobiotite   17
hydrogenous   259, 260, 269, 271, 276, 285, 305, 311, 322, 333, 532
hydrolysis   22, 23, 24, 429, 443
hydromorphic   38, 443
hydrotalcite   329
hydrothermal   259, 261, 267, 270, 291, 302, 316, 317, 320, 397, 414, 532

I

illite   9, 10, 12, 168, 527
   illite crystallinity   350, 360, 386, 391, 395, 426, 432, 446, 535
illitization   91, 361, 365, 368, 370, 372, 382, 397, 400, 402, 408
imogolite   7, 38
index D   489, 491
index Mn *   489, 491, 497
interlayer   4
interstratification   16
ion exchange   97, 114, 334, 365
iron concretions   261, 310
iron oxides   19, 31, 260, 275, 284, 285, 291, 300, 307, 310, 313, 316, 320, 482
ironstone   48, 227

K

kaolinite   4, 5, 7, 8, 12, 30, 40, 46, 163, 165, 167, 171, 436, 445
kaolinitization   380, 399, 402, 408, 411, 414, 418

L

laterite   30, 43, 45, 48
latosol   30
layer   4, 361
leaching   22
ledikite   10

lepidocrocite   18
lepidolite   10
limestone-marl alternations   453
limonite   18
lithosol   27
loess   56, 57

M

magadiite   83
maghemite   18
manganese nodules   261, 310, 313, 322
margarite   13
mica   8, 9, 10, 12
minnesotaite   10
mixed-layer clay   16, 17, 28, 30, 361, 364, 373, 394, 410, 436, 548
monosialitization   25
montmorillonite   9, 13, 39, 409
muscovite   9, 12

N

nacrite   7
neoformation   24
nontronite   9, 14, 34, 39, 79, 88, 115, 262, 268, 274, 292, 303, 306, 308, 310, 313, 315

O

octahedra   3, 5
organic degradation of clay   111, 127, 236, 242, 413
opal   18, 342, 347, 509
oxygen isotopes   304, 319, 336, 401, 433, 438, 450, 515

P

paleosol   41, 42
palagonitization   328
palygorskite   15, 16, 35, 196, 255, 276, 317, 344, 470, 492, 501, 512, 528, 544
   aeolian palygorskite   56, 78, 139, 141, 146, 150, 169, 174, 188, 278
paragonite   9, 359, 392
pedologic discordance   25, 26
phengite   11, 414
phillipsite   19, 91, 260, 262, 268, 286, 291, 316
phlogopite   10, 11
phyllite C   214
phyllite V   213
physical weathering   21, 27, 33
plate tectonic control   496, 500, 507, 530
podzol   32
precipitation   259, 373
pressure gradient   391
pyrophyllite   8, 9, 12, 316, 359, 394

Q

quartz 18
  aeolian quartz   133, 139, 145, 147, 151, 153, 156

R

red beds   72
rectorite   17

S

salinolysis   22, 37
saponite   10, 14, 200, 231, 292, 297, 396, 414, 510
sauconite   10
sea-level change   439, 465, 514, 522
sepiolite   15, 16, 35, 82, 180, 196, 276, 317, 492
sericite   9
serpentine   7, 8, 12, 293, 298, 316, 414
silcrete   35, 48
silica   19, 33, 275
smectite   4, 5, 9, 10, 12, 13, 28, 39, 46, 170, 171, 196, 336, 436, 447, 451, 459, 527
  buried smectite   395
  smectite morphology   337, 343, 349, 355, 376, 488, 489, 501, 548, 556
  volcanogenic smectite   76, 79, 83, 91, 92, 174, 231, 262, 265, 270, 291, 294, 344, 352, 409, 411, 414, 443, 480, 507, 519
smectitization   28, 32, 39, 79, 87, 196, 199, 200, 265, 292, 295, 304, 310, 313, 343, 412, 509
soil   22, 27
solonetz   37
solontchak   37
stevensite   10, 14, 88, 200, 207, 314
stilpnomelane   476

strontium isotopes   269, 319, 336, 342, 348, 363, 438
substitution   5
sudoite   15
sulfate deposits   295, 304, 314
sulfide deposits   303, 312, 314, 316

T

talc   8, 10, 12, 297, 315, 397, 414
tchernozems   34
tectonic rejuvenation   41, 430, 470, 473, 478, 490, 500, 502, 505, 514, 522, 539, 544, 553
tetrahedra   3, 5
todorokite   19, 261, 303
tonstein   411
tosudite   17
transformation   24, 195, 244, 247, 372
trioctahedral   4, 10

V

vein filling   415
verdine   213, 219
vermiculite   9, 10, 12, 13, 28, 414
vertisol   34
vitrinite reflectance   368, 385, 386

W

wall rock minerals   415
weathering   21, 36, 57, 59, 176, 425, 444, 499, 502
wetting and drying   21, 91, 209

X

xylotile   16

Z

zeolites   19, 83, 295, 297, 299
zonal soil   26